AF689456

Principles of Coordination Polymerisation

Principles of Coordination Polymerisation

Heterogeneous and homogeneous catalysis in polymer chemistry – polymerisation of hydrocarbon, heterocyclic and heterounsaturated monomers

Witold Kuran†

Department of Polymer Chemistry and Technology,
Faculty of Chemistry,
Warsaw University of Technology

JOHN WILEY & SONS, LTD

Chicthester · New York · Weinheim · Brisbane · Singapore · Toronto

Copyright © 2001 by John Wiley & Sons Ltd,
Baffins Lane, Chichester,
West Sussex PO19 1UD, England

National 01243 779777
International (+44) 1243 779777
e-mail (for orders and customer service enquiries): cs-books@wiley.co.uk
Visit our Home Page on http://www.wiley.co.uk
or http://www.wiley.com

All Rights Reserved. No part of this publication may be reproduced, stored in a retrieval system, or transmitted, in any form or by any means, electronic, mechanical, photocopying, recording, scanning or otherwise, except under the terms of the Copyright, Designs and Patents Act 1988 or under the terms of a licence issued by the Copyright Licensing Agency Ltd, 90 Tottenham Court Road, London, UK W1P 0LP, without the permission in writing of the publisher and the copyright holder.

Other Wiley Editorial Offices

John Wiley & Sons, Inc., 605 Third Avenue,
New York, NY 10158-0012, USA

WILEY-VCH GmbH, Pappelallee 3,
D-69469 Weinheim, Germany

John Wiley & Sons Australia Ltd, 33 Park Road, Milton,
Queensland 4064, Australia

John Wiley & Sons (Asia) Pte Ltd, 2 Clementi Loop #02-01,
Jin Xing Distripark, Singapore 129809

John Wiley & Sons (Canada) Ltd, 22 Worcester Road,
Rexdale, Ontario M9W 1L1, Canada

Library of Congress Cataloging-in-Publication Data

British Library Cataloguing in Publication Data

A catalogue record for this book is available from the British Library

ISBN 0-470-84141-9

Typeset in 10/12pt Times from the authors' disks by Kolam Information Services Pvt. Ltd.,
Printed and bound in Great Britain by Biddles Ltd, Guildford and King's Lynn
This book is printed on acid-free paper responsibly manufactured from sustainable forestry, in which at least two trees are planted for each one used for paper production.

Contents

Tribute to Witold Kuran

The author of this book, Witold Kuran, died on 19 November 1999 at the age of 58. For the last 25 years of his life he was a professor at the Faculty of Chemistry Warsaw University of Technology, where he chaired the Division of Polymer Synthesis and Processing. During his early studies he became fascinated with the discoveries of Karl Ziegler and Gulio Natta and the great opportunities provided by organometallic catalysts for the progress of the chemistry and technology of high polymers, and he concentrated on this subject throughout his professional life. Studies of the reactions of organoaluminium and organozinc compounds and the utilisation of these derivatives as catalysts in polymerisation processes of polar vinyl and heterocyclic monomers were the main focus of his work. At the end of the 1960s and beginning of the 1970s, he carried out pioneering studies of the alternating copolymerisation of acrylic monomers with vinyl chloride, olefins and dienes, catalysed by alkyaluminium chlorides, and found that these processes are initiated radically as a result of homolytic dissociation of metal–carbon bonds. In the mid-1970s he began to study the copolymerisation of carbon dioxide with heterocyclic monomers. In a short time, the team led by him developed a number of original and very active catalysts for these processes. During further studies it appeared that these catalysts could be successfully applied in the coordination polymerisation of oxiranes, cyclic carbonates and other heterocyclic monomers. His own broad experimental results and also extremely precise analysis of the results published by other researchers made it possible to form a general concept of organometallic catalysis in the polymerisation of heterocyclic monomers, which he presented in a number of monographic publications. Professor Kuran's output also includes a number of valuable publications dealing with other fields of organometallic chemistry and polymer chemistry. He is the coauthor of fundamental works on palladium(0) complexes, which appeared during his fellowship in the school of Gulio Natta in Milan, and also of many works devoted to the utilisation of cyclic carbonates in the synthesis of condensation polymers. Although he past away at the height of his creative powers at a mere 58, he left a very rich scientific output. He was the author or coauthor of two books and over 110 scientific papers and book chapters, and the holder of 30 patents.

In the Faculty of Chemistry of Warsaw University of Technology he conducted lectures on polymer chemistry and technology. In the mid-1980s, when it became apparent that catalytic polymerisation processes would have a decisive effect on the production of plastics, he made a decision to introduce a new lecture devoted to this field, entitled *Principles of Coordination Polymerisation*. This lecture gained great popularity among graduate and doctoral students, and its basic thesis was also presented in the form of a small monograph and a broad chapter in the Polish textbook 'Polymer Chemistry'. At the beginning of the 1990s, Professor Kuran made the decision to write a monograph, in which he intended comprehensively to present both basic notions and the latest achievements in the field of coordination polymerisation of various classes of monomers. The work on this book initially proceeded very quickly, and the publishing house John Wiley & Sons expressed an intention to publish it. Unfortunately, with time, deadlines for completing the work were postponed several times, since the development of new fields of studies on catalytic polymerisation processes was so rapid that, after nearly every visit to the library, corrections and supplements had to be introduced. The systematic deterioration in the health of Professor Kuran was also a key factor in slowing down the work on the book. The basic version of the manuscript was completed in May 1999 and, after it had been read by close coworkers, the decision was taken, together with Professor Kuran, to finalise the edition of the book. However, these plans could not be fulfilled. After Professor Kuran's death we undertook to complete his work, following as far as possible his instructions, as well as taking into account the comments of the reviewers. We hope that the small changes introduced by us do not disturb the original concept of the author, who was our teacher and friend, and that the great work of his life will prove to be useful both to students and to those professionally involved in polymer chemistry and technology.

Professor Zbigniew Florjańczyk

Preface

The book has been prepared on the basis of lectures dealing with polymer chemistry, coordination polymerisation, catalytic polymerisation processes, polymerisation stereochemistry and organic technology that I have been giving to students of various courses, mostly at the Faculty of Chemistry, Warsaw University of Technology, for more than 25 years. The lack of a single-volume book that would cover the entire area of the coordination polymerisation and collect the relevant literature in the field, with suitable comment, has induced me to do this work. Actually, only a few existing textbooks contain individual chapters connected with coordination polymerisation. However, these chapters omit the coordination polymerisation of some important classes of monomers and do not treat all important problems exhaustively and instructively enough, which makes it difficult to master this important subject. There is a great unfilled gap between the very great achievements and progress in the field of coordination polymerisation, in terms of both basic and industrial research after Ziegler's and Natta's discoveries, and respective textbooks that might deliver suitable data together with explanation and proper comment. The spectacular breakthrough made possible by these discoveries, honoured with the Nobel Prize in 1963, revolutionised polymer science and technology in the second half of the twentieth century. However, although the mentioned discoveries resulted in coordination polymerisation becoming one of the most active and exciting areas of research in academic and industrial laboratories, having a significant impact on progress in the plastics and rubber industries, Ziegler and Natta did not live to see a textbook dealing with coordination polymerisation that might be worthy of the name. Moreover, outstanding scientific achievements and innovations and process improvements in the mid-1980s revolutionised polyolefin technology for the second time. We are now witnessing the next phase of this revolution, which was initiated, in the mid-1980s, by the discoveries of metallocene single-site catalysts which could be tailored to produce polymers of virtually all feasible microstructures. These achievements offer attractive potential for the development of basic research in terms of mechanistic studies as well as for the introduction of new or improved industrial processes in various branches of the plastics and rubber industries. Thus, the lack of a suitable textbook lessens the chance of proper mastery of the important discipline of coordination polymerisation, and the training that most chemists receive seems to be inadequate, which is deplorable. The present textbook represents an effort to overcome these barriers by offering a unified, possibly most complete, view of coordination polymerisation. This book aims to cover this field, addressing almost all the area of coordination polymerisation, i.e. polymerisation with a catalyst involved in the monomer coordination at each polymerisation step. It arose from the need for a text, devoted to coordination polymerisation, to accompany introductory courses as well as advanced courses in polymer science, catalysis and in polymerisation catalysis. It is thought to be useful for research students and research workers in industry, and will also be helpful as a reference book. The main inspiration that has enabled me, after many years, to prepare this book has been my constant interest in coordination polymerisation. This interest was kindled during my stay at the Istituto di Chimica Industriale at the Polytechnics of Milan as a post-doctoral fellow in the early 1970s and my visit to the Max Planck Institut für Kohlenforschung in Mülheim/Ruhr later on. Successively collected literature data, my own studies and the results of work carried out in the Institute of Organic Chemistry and Technology and the Division of Polymer Synthesis and Processing at the Faculty of Chemistry, Warsaw University of Technology, headed by me, were of assistance in the preparation of this work.

The textbook is focused on the polymerisation of monomers belonging to all the important classes of hydrocarbon as well as non-hydrocarbon monomers, in the presence of coordination catalysts of various kinds, and emphasises the growing importance of coordination polymerisation for basic research as well as for industrial processes. A vast amount of material, covering coordination polymerisation almost in its entirety and considering the whole range of monomers susceptible to polymerisation with coordination catalysts and the whole range of respective more important coordination catalysts, has been collected and selected conscientiously to suit readers with various levels and breadths of knowledge of the subject. Considering that several generations of catalysts and processes have been introduced on a commercial scale to produce a large variety of most widespread polymeric materials ranging from commodity thermoplastics to rubbers, to some extent the emphasis is laid on the coordination polymerisation of alkenes, particularly of 1,3-dialkenes, in the textbook. Chapters dealing with polymerisations of these monomers present the methods that have been applied for the implementation of industrial processes and show the evolution of these processes, since there are particularly strong interrelations between the basic and industrial research in this field.

The contents of the book are systematised, taking into account basic features of the mechanisms that operate in coordination polymerisations of various monomers and depend on the kind of monomer and catalyst. This is, essentially, consistent with the alignment of monomers according to their properties resulting from their structure. Thus, after introductory Chapter 1 and Chapter 2, dealing with the general characteristics of coordination polymerisation, subsequent chapters are devoted to the coordination polymerisation of unsaturated hydrocarbon monomers, such as olefins, considering primarily the stereospecific polymerisation of α-olefins (Chapter 3), vinylaromatic monomers, especially styrene (Chapter 4), conjugated dienes (Chapter 5), cycloolefins (Chapter 6) and alkynes (Chapter 7), as well as the coordination polycondensation of divinyl monomers via the acyclic diene metathesis reaction and the coordination coupling polycondensation of functional aromatic compounds by Heck's reaction, including carbonylation coupling polycondensation (Chapter 8). Finally, Chapter 9 is concerned with the coordination polymerisation of non-hydrocarbon monomers, such as heterocyclic and heterounsaturated monomers. Although polymerisations and copolymerisations of these monomers in the presence of coordination catalysts do not represent high-mileage industrial processes (they have been restricted to the production of polyether elastomers on a larger scale), and have not influenced the progress in the plastics and rubber industries to such an extent as those processes involving hydrocarbon monomers, their importance for the development of new ideas in coordination polymerisation and the broad synthetic feasibilities are giving grounds for more detailed treatment of this subject, especially considering its instructive character. Taking the above into consideration, the individual chapters of the book are not equal, but differ in volume according to the importance and breadth of the problems presented as regards the chemistry rather than the technology.

Each chapter presents a unified view of monomers, catalysts and mechanisms, with a special emphasis on stereochemistry, the structures of polymers and the utility of coordination polymerisation, as well as attempts to delineate the current trends of research and actual or potential commercial applications. Problems are listed at the end of each chapter in order to enable students or other readers to absorb its contents more easily.

Taking into account one of the most important features of coordination catalysts, which is their capability of producing stereoregular polymers, when discussing the polymerisation of individual groups of monomers, special attention is paid to the structures of the catalyst active species and the stereocontrol mechanism of these polymerisations. These key questions for an understanding of the nature of coordination polymerisation are treated with special emphasis in the textbook, especially when considering that the utility of coordination polymerisation for producing stereoregular polymers is well established.

The rather difficult material covering the subject of coordination polymerisation has been presented in the book in a way that should be approachable and intelligible to undergraduate students who have mastered general subjects in chemistry. Introductory treatment of all questions, comprehensible for undergraduate students, is followed by more advanced considerations demanding deeper reflection that it is hoped will be rewarding to graduate students and others. The idea is for the text to be eminently teachable for instructors and easily comprehensible for students. By far the majority of colleges and universities have no courses in coordination polymerisation, no staff member conducting research in this area and only cursory mention of polymers in other courses. There are, needless to say, many exceptions, ranging from the isolated effort of a single staff member to such major centres for polymer research as those where a dozen or more staff members constitute a formal or informal polymer research group. However, there are rather rare cases of such academic centres gathering more than a few specialists in the field of coordination polymerisation.

The approach to referencing the literature in the book is such that the most recent bibliography is cited as References, including most of the latest review articles and books; the most important original works are also listed, however. More general reviews and book chapters are also listed as Further Reading. Since this book is not a monograph, the references are by no means exhaustive, but the reader will easily find additional references by looking up those cited, since many citations, as mentioned, are of review articles and books. In order to avoid a long list of References and Further Reading at the end of the textbook, they are listed at the end of each individual chapter. The textbook, by condensing much of the latest knowledge in the field into a more convenient form, may therefore also enable specialists to benefit from the book.

The book is intended for MSc and PhD students in catalysis, and in polymerisation catalysis, for undergraduate and research students in polymer science in general, for college and university instructors and for industrial researchers.

List of Symbols and Abbreviations

A	activator for Ziegler–Natta catalyst precursor
Ac	acyl
Acac	acetylacetonyl
All	allyl
Ar	aryl
Bbn	9-borabicyclo[3.3.1]nonyl
Bu	butyl
Bz	benzyl
Bzo	benzo
CD	α-cyclodextrin
Chx	cyclohexyl
Cod	cycloocta-1,5-diene
Cp	cyclopentadienyl
Cp*	pentamethylcyclopentadienyl
Cp′	substituted or nonsubstituted Cp
Cp″	substituted or nonsubstituted Cp
C*	chiral carbon atom
C^*	total concentration of the active sites
C_p^*	concentration of the propagation active sites
Dmon	dimethanooctahydronaphthalene
Dmpe	1,2-bis(dimethylphosphine)ethane
DOX	1,4-dioxane
E	energy
EB	ethyl benzoate
Et	ethyl
Flu	9-fluorenyl
h	monomer head
Hx	hexyl
Ind	1-indenyl
k_a	rate constant of chain transfer with activator
k_{H_2}	rate constant of chain transfer with hydrogen
k_i	rate constant of chain initiation

k_m	rate constant of chain transfer with monomer
k_p	rate constant of chain propagation
k_s	rate constant of spontaneous chain transfer
k_t	average rate constant of chain transfer
k_{11}	monomer 1 homopropagation rate constant
k_{22}	monomer 2 homopropagation rate constant
k_{21}	monomer 1 heteropropagation rate constant
k_{12}	monomer 2 heteropropagation rate constant
K_p	propagation equilibrium constant
L	ligand
LA	Lewis acid
LB	Lewis base
Ln	lanthanide
M	monomer
M_1	comonomer 1
M_2	comonomer 2
m, *M*	mesogenic
Me	methyl
Mt	metal
$\bar{M}_n$	number-average molecular weight
$\bar{M}_w$	weight-average molecular weight
Nbd	2,5-norbornadiene
Np	neopentyl
Nph	naphthyl
Ph	phenyl
ppm	parts per million
Pr	propyl (or symbol of the element)
Py	pyridine
P_h	monomer head-linked propagating species
P_t	monomer tail-linked propagating species
P_n	polymer chain
P_x	polymer chain
P_m	conditional probability of forming *m* diad
P_r	conditional probability of forming *r* diad
$\bar{P}_n$	number-average degree of polymerisation
r_1	relative reactivity ratio of monomer 1
r_2	relative reactivity ratio of monomer 2
R	alkyl (hydrogen)
r, *R*	racemic
R	right-handed
R_p	overall polymerisation rate
S	left-handed
t	monomer tail
Tbp	2,2′-thiobis(4-methyl-6-*t*-butylphenolate)

THF	tetrahydrofuran
Thind	1-(4,5,6,7-tetrahydroindenyl)
Tmdn	trimethanododecahydronaphthalene
T_g	glass transition temperature
t	time
X	substituent
Z	substituent
Δ	right-handed
Λ	left-handed
Θ_A	fraction of centres complexed with activator
Θ_M	fraction of centres complexed with monomer
□	coordination vacancy

1 Introduction

Although naturally abundant polymers have been exploited by mankind from time immemorial, their structure, even in a broad outline, has been incomprehensible for a long time. The application of physical methods to investigate such materials as rubbers, polysaccharides or proteins as recently as the twentieth century, demonstrating that these materials are characterised by large molecular weights, did not prompt scientists to conclude that they are built from large molecules. Despite the findings, many well-known physicochemists of that time considered the very large molecular weights of the materials studied to have resulted from the association of their low-molecular constituents. This view was still held as late as the end of the 1920s, one decade later, when Staudinger proposed that polymeric materials were actually composed of giant molecules that he called macromolecules [1]. At this juncture, when problems connected with the nature of polymeric molecules and with the explanation of their structure were the most important problems engaging the minds of investigators, the concept of the mechanism in chemistry as a whole was hardly developed. This matters, of course, when considering that natural rubber, which is a relatively simple polymeric substance, has attracted the attention of scientists for nearly one century.

The development of modern polymer chemistry dates from the beginning of the 1930s when Carothers [2] accomplished the synthesis of polymers from polyfunctional monomers by various condensation reactions of organic chemistry, thereby finally confirming the macromolecular hypothesis raised earlier by Staudinger and Tritschi. Since that time, many discoveries have been made or hypotheses developed that have been of fundamental significance not only in the field of polymer chemistry but for the development of chemistry as a whole. It is worth drawing attention to work on the mechanisms of radical chain processes (Staudinger – Nobel Prize winner in 1953, Melville), condensation reactions (Carothers, Flory – Nobel Prize winner in 1974), reactions with carbocations (Whitemore, Evans, Olah – Nobel Prize winner in 1994), reactions with carbanions (Szwarc) and coordination polymerisations of olefins and related unsaturated hydrocarbon monomers (Ziegler and Natta – Nobel Prize winners in 1963 [3,4]). It is also worth noting the significant contribution to the progress of chemistry from biochemists dealing with macromolecular living systems. Among them, quite a large number have been honoured with Nobel

Prize, e.g. Watson and Crick [5] for their discovery of the double helical architecture of DNA.

We should realise, however, that, for reactions concerning the formation of macromolecular systems, the same laws are obeyed as those governing systems consisting of low molecular weight substances with which organic chemistry deals, and there can be no dichotomy of mechanisms and structures in terms of their division into polymer chemistry and organic chemistry.

In polymer chemistry, especially that involving coordination polymerisation, the mechanistic considerations always take into account configurational and conformational aspects as well as thermodynamic and kinetic aspects. However, stereochemical problems are of particular interest in coordination polymerisation, since this polymerisation often leads to stereoregular polymers which is not the common case in other polymerisations. Owing to the tremendous development of spectroscopic techniques, it has become possible to investigate the stereoisomerism in polymers, which concerns both natural and synthetic polymers. It is commonly known now that natural rubber is composed of polyisoprene of predominantly *cis*-1,4 structure [6] but also natural gutta percha – from its *trans*-1,4 isomer [7]. Cellulose, the most abundant naturally occurring polymer (its annual growth on earth is estimated to be in the region of 10^{11}t), appeared to be 1,4-*threo*-disyndiotactic poly(*D*-glucopyranose) or, as we know from organic chemistry, the polymer of *D*-glucopyranose containing *β*-1,4-*D*-glucopyranoside units [8]. On the other hand, amylose, the isomeric polymer of *D*-glycopyranose, containing *α*-1,4-*D*-glycopyranoside units [9], is 1,4-*erythro*-diisotactic poly(*D*-glucopyranose) according to the terminology developed in the field of synthetic polymers since the mid-1950s.

The discovery by Karl Ziegler and his group from the Max Planck Institut für Kohlenforschung in Mülheim/Ruhr, in 1953, that the activation of transition metal compounds with aluminium alkyls led to the formation of unusual organometallic catalysts that were capable of polymerising ethylene [10] was a pioneering advance in the coordination polymerisation of unsaturated hydrocarbon monomers. This achievement was soon followed by the discovery by Giulio Natta and his group from the Istituto Chimica Industriale del Politecnico di Milano, in 1954, that hydrocarbon monomers containing stereogenic carbon atoms such as propylene and higher *α*-olefins or conjugated dienes formed stereoregular polymers when organometallic catalysts of this new type were used for the polymerisation [11]. Ziegler–Natta catalysts, together with alkyl activator-free catalysts, discovered independently during the early 1950s at laboratories of Standard Oil of Indiana (Amoco) [12] and Phillips Petroleum Co. [13,14], initiated the rapid growth of the interest in coordination polymerisation, especially that leading to stereoregular polymers, in both university and industrial laboratories, and caused a breakthrough that had a significant impact on the advancement of polymer science and technology and expansion of the plastics and rubber industries. The production of polyolefin thermoplastic materials exhibiting a broad range of properties as well as ethylene/

propylene and conjugated diene-based rubbers has grown enormously. Synthetic stereoregular polymers such as poly(α-olefin)s exhibit stereoisomerism resulting from their tacticity and, like conjugated diene polymers, exhibit *cis–trans* isomerism. Apart from its practical and industrial aspects, stereospecific polymerisation has introduced a highly sophisticated chemistry into macromolecular science and has made a substantial contribution to the development and success in this field [15,16].

The discovery of stereospecific *ansa*-metallocene-based catalysts by Kaminsky *et al.* [17] and Ewen [18] in the mid-1980s initiated tailored polymerisation processes, involving the cheapest unsaturated hydrocarbon monomers that might afford polymers of desired structure and properties on a large scale. These new metallocene-based catalysts essentially consist of one type of catalytically active site, which may be tailored to polymerise hydrocarbon monomers to polymers that can be characterised by virtually all feasible microstructures. Syndiotactic, hemiisotactic and stereoblock poly(α-olefin)s, apart from those of isotactic microstructure, as well as novel cycloaliphatic polyolefins, are now becoming available in commercial quantities. This also concerns syndiotactic polystyrene which can be easily obtained with metallocene-based catalysts. Moreover, microstructure, molecular weight, end-group composition and comonomer incorporation can be varied independently when polymerising hydrocarbon monomers with metallocene-based catalysts. In contrast to multisite supported catalysts, the novel generation of metallocene single-site catalysts creates greater possibilities of gaining an insight into the basic reaction mechanisms accounting for stereochemical and molecular weight control. The versatility of metallocene and other single-site catalysts found and developed recently will likely result in their utilisation in a wide range of polymerisation processes in industry. It is clear that the potential of metallocene and related polymerisation coordination catalysts has been only partially exploited [19].

Conjugated diene polymerisation processes with coordination catalysts have not been developed to such an extent as those concerning monoalkenes. The progress connected with the introduction of very efficient neodymium-based Ziegler–Natta catalysts into industrial practice in the mid-1980s has changed this [20]. However, significant progress has also been made in the ring-opening metathesis polymerisation of cycloolefins, especially bicyclic olefins [21,22]. Recent advances concerning RIM processes for dicyclopentadiene polymerisation and stereospecific norbornene polymerisations, together with great achievements concerning alkene and dialkene polymerisations in the presence of Ziegler–Natta multisite catalysts and single-site metallocene catalysts, have greatly stimulated research in the fields of organometallic chemistry, catalysis, polymer chemistry and polymer engineering. The recent landmark discoveries offer attractive potential for the development of new and environmentally harmless versatile classes of hydrocarbon-based polymeric materials and expand the frontiers of polymer technology.

It must be emphasised that coordination polymerisation is now considered as one of the most important research frontiers in polymer chemistry and technology. It has become the backbone of several significant branches of the plastics and rubber industries. Although this is of prime concern to numerous homopolymerisation and copolymerisation processes of various hydrocarbon monomers [20–27] (and also carbon monoxide as the comonomer [28–30]), the coordination polymerisation and copolymerisation processes of heterocyclic monomers should also be taken into account. The latter processes, which, among the high-mileage industrial processes, are limited to the coordination copolymerisation of epoxides, yielding polyether elastomers [31,32], are of lesser significance. However, extensive mechanistic studies on these processes as well as exploratory investigations for obtaining better, especially high-performance catalysts in terms of commercial development aspects, are constantly being carried out. The application of coordination catalysts for the polymerisation of heterocyclic monomers has broadened the synthetic feasibility of these polymerisations as regards the possibility of obtaining high molecular weight homopolymers that may be characterised by a high degree of regio- and stereoregularity, and the possibility of obtaining high molecular weight copolymers with heterounsaturated monomers (such as, for example, carbon dioxide [33]) that cannot be obtained in other ways [34].

A very important aspect of the coordination polymerisation of heterocyclic monomers is connected with certain advantageous properties of the polymers obtained from such monomers as potentially useful materials for biomedical and pharmaceutical applications in terms of their low toxicity and favourable biodegradability. Moreover, recently, very efficient catalysts have been applied successfully for the coordination polymerisation of heterocyclic monomers such as β-butyrolactone, which proceeded easily at ambient temperature and thus constituted a convenient method for the preparation of poly[(*R*)-3-hydroxybutyrate] [35], which is the most common member of the group of poly(β-hydroxyalkanoate)s present in a wide range of microorganisms. Note in this connection the friendly environmental behaviour of polymers derived from such monomers. Thus, the coordination polymerisation and copolymerisation of heterocyclic and heterounsaturated monomers has also acquired great importance, although such polymerisation has other purposes [34] than those concerning the coordination polymerisation of hydrocarbon monomers.

It will not be overstating the significance of coordination polymerisation to say that this polymerisation now occupies the most prominent place in polymer science and technology. Although coordination polymerisation is characterised by a relatively short period of development as a branch of polymer chemistry, it is making a major contribution towards the further development of polymer science. One is becoming conscious nowadays of the significant role being played by the stereoisomerism of macromolecules in modern technology. As nature affords us macromolecules, with some characterised by certain useful properties and others, being isomeric to them, characterised by completely different usable features, so modern technology has started to afford stereoisomers of synthetic polymers that exhibit new valuable properties. Coordination polymerisation processes occupy a significant place in such technology.

References

1. Staudinger, H. and Fritschi, J., *Helv. Chim. Acta*, **5**, 785 (1922).
2. Carothers, W. H., *J. Am. Chem. Soc.*, **51**, 2548 (1929).
3. Ziegler, K., *Angew. Chem.*, **76**, 545 (1964).
4. Natta, G., *Science*, **147**, 261 (1965).
5. Watson, J. D. and Crick, F. H. C., *Nature*, **171**, 737, 1964.
6. Schildknecht, C. E., Zoss, A. O. and McKinley, E., *Ind. Eng. Chem.* **39**, 180.
7. Schildknecht, C. E., Gross, S. T., Davidson, H. R., Lambert, J. M. and Zoss, A. O., *Ind. Eng. Chem.*, **40**, 2104 (1948).
8. Freudenberg, K. and Blomqvist, G., *Chem. Ber.*, **68**, 2070 (1935).
9. Freudenberg, K., Friedrick, K., Baumann, I. and Soff, K., *Ann. Chem.*, **494**, 41 (1932).
10. Ziegler, K., Holzkamp, E., Breil, H. and Martin, H., *Angew. Chem.*, **67**, 541 (1955).
11. Natta, G., Pino, P., Corradini, P., Danusso, F., Mantica, E., Mazzanti, G. and Moraglio, G., *J. Am. Chem. Soc.*, **77**, 1708 (1955).
12. Zletz, A., US Pat. 2 692 257 (to Standard Oil of Indiana) (1951); *Chem. Abstr.*, **49**, 2777d (1955).
13. Hogan, J. P. and Banks, R. L., US Pat. 2 825 721 (to Phillips Petroleum Co.) (1958); *Chem. Abstr.*, **52**, 8621i (1958).
14. Clark, A., Hogan, J. P., Banks, R. L. and Lanning, W. C., *Ind. Eng. Chem.*, **48**, 1152 (1956).
15. Farina, M., *Trends Polym. Sci.*, **2**, 80 (1994).
16. Locatelli, P., *Trends Polym. Sci.*, **2**, 87 (1994).
17. Kaminsky, W., Külper, K., Brintzinger, H. H. and Wild, F. R. W. P., *Angew. Chem. Int. Ed. Engl.*, **24**, 507 (1985).
18. Ewen, J. A., *J. Am. Chem. Soc.*, **106**, 6355 (1984).
19. Horton, A. D., *Trends Polym. Sci.*, **2**, 158 (1994).
20. Andreussi, P., Lauretti, E. and Miani, B., in *Abstracts of the IUPAC International Symposium on Stereospecific Polymerisation, STEPOL '94*, Milan, Italy, 1994, p. 125.
21. Fisher, R. A. and Grubbs, R. H., *Makromol. Chem. Macromol. Symp.*, **63**, 271 (1992).
22. Goodall, B. L., McIntosh, L. H. and Rhodes, L. F., *Macromol. Symp.*, **89**, 421 (1995).
23. Galli, P., *Macromol. Symp.*, **89**, 13 (1995).
24. Karol, F. J., *Macromol. Symp.*, **89**, 563 (1995).
25. Covezzi, M., *Macromol. Symp.*, **89**, 577 (1995).
26. Kaminsky, W., *Catalysis Today*, **20**, 257 (1994).
27. Ishihara, N., *Macromol. Symp.*, **89**, 553 (1995).
28. Sen, A. and Jiang, Z., *Macromolecules*, **26**, 911 (1993).
29. Won, P. K., van Doorn, J. A., Drent, E., Sudmeijer, O. and Stil, H. A., *Ind. Eng. Chem. Res.*, **32**, 986 (1993).
30. Amevor, E., Bronco, S., Consiglio, G. and Di Benedetto, S., *Macromol. Symp.*, **89**, 443 (1995).

31. Owens, K. and Kyllingstad, V. L., 'Polyethers,' in *Kirk-Othmer Encyclopedia of Chemical Technology*, Wiley-Interscience, John Wiley & Sons, New York, 1993, Vol. 8, pp. 1079–1093.
32. Kuran, W., 'Poly(propylene Oxide)', in *The Polymeric Materials Encyclopedia*, CRC Press, Boca Raton, 1996, Vol. 9, pp. 6656–6662.
33. Kuran, W., 'Poly(propylene Carbonate)', in *The Polymeric Materials Encyclopedia*, CRC Press, Boca Raton, 1996, Vol. 9, pp. 6623–6630.
34. Kuran, W., *Prog. Polym. Sci.*, **23**, 919 (1997).
35. Le Borgne, A., Pluta, C. and Spassky, N., *Macromol. Rapid Commun.*, **15**, 955 (1994).

Further Reading

Morawetz, H., 'History of Polymer Science', in *Encyclopedia of Polymer Science and Engineering*, Wiley-Interscience, John Wiley & Sons, New York, 1988, Vol. 7, pp. 722–745.

Pasquon, I., Porri, L. and Giannini, U., 'Stereoregular Linear Polymers', in *Encyclopedia of Polymer Science and Engineering*, Wiley-Interscience, John Wiley & Sons, New York, 1989, Vol. 15, pp. 632–733.

Gavens, P. D., Bottrill, M., Kelland, J. W. and McMeeking, J., 'Ziegler–Natta Catalysis', in *Comprehensive Organometallic Chemistry*, Pergamon Press, Oxford, 1982, Vol. 3, pp. 475–547.

McDaniel, M. P., 'Supported Chromium Catalysts for Ethylene Polymerization', *Adv. Catal.*, **33**, 47–98 (1985).

Tait, P. J. T., 'Monoalkene Polymerization: Ziegler–Natta and Transition Metal Catalysts', in *Comprehensive Polymer Science*, Pergamon Press, Oxford, 1989, Vol. 4, pp. 1–25.

Tait, P. J. T. and Watkins, N. D., 'Monoalkene Polymerization: Mechanisms', in *Comprehensive Polymer Science*, Pergamon Press, Oxford, 1989, Vol. 4, pp. 533–573.

Starkweather, H., 'Olefin–Carbon Monoxide Polymers', in *Encyclopedia of Polymer Science and Engineering*, Wiley-Interscience, John Wiley & Sons, New York, 1987, Vol. 10, pp. 369–373.

Porri, L. and Giarrusso, A., 'Conjugated Diene Polymerization', in *Comprehensive Polymer Science*, Pergamon Press, Oxford, 1989, Vol. 4, pp. 53–108.

Eleuterio, H. S., 'Scientific Discovery and Technological Innovation: an Eclectic Odyssey into Olefin Metathesis Chemistry', *J. Macromol. Sci. – Chem. A*, **28**, 907–915 (1991).

Ivin, K. J., 'Metathesis Polymerization', in *Encyclopedia of Polymer Science and Engineering*, Wiley-Interscience, John Wiley & Sons, New York, 1987, Vol. 9, pp. 634–668.

Schrock, R. R., 'Ring-opening Metathesis Polymerization', in *Ring-Opening Polymerization*, Carl Hanser Publishers, Munich, 1993, pp. 129–156.

Bolognesi, A., Catellani, M. and Destri, S., 'Polymerization of Acetylene', in *Comprehensive Polymer Science*, Pergamon Press, Oxford, 1989, Vol. 4, pp. 143–153.

Costa, G., 'Polymerization of Mono- and Di-substituted Acetylenes', in *Comprehensive Polymer Science*, Pergamon Press, Oxford, 1989, Vol. 4, pp. 155–161.

Brintzinger, H. H., Fischer, D., Mülhaupt, R., Rieger, B. and Waymouth, R. M., 'Stereospecific Olefin Polymerisation with Chiral Metallocene Catalysts', *Angew. Chem., Int. Ed. Engl.*, **34**, 1143–1170 (1995).

Wünsch, J. R., 'Syndiotaktisches Polystyrol', in *Polystyrol*, Carl Hanser Publishers, Munich, 1996, pp. 82–104.

Tsuruta, T. and Kawakami, Y., 'Anionic Ring-opening Polymerization: Stereospecificity for Epoxides, Episulfides and Lactones', in *Comprehensive Polymer Science*, Pergamon Press, Oxford, 1989, Vol. 3, pp. 489–500.

Jérôme, R. and Teyssié, P., 'Anionic Ring-opening Polymerization: Lactones,' in *Comprehensive Polymer Science*, Pergamon Press, Oxford, 1989, Vol. 3, pp. 501–510.

Inoue, S. and Aida, T., 'Anionic Ring-opening Polymerization: Copolymerization', in *Comprehensive Polymer Science*, Pergamon Press, Oxford, 1989, Vol. 3, pp. 533–569.

Vogl, O., 'Aldehyde Polymers', in *Encyclopedia of Polymer Science and Engineering*, Wiley-Interscience, John Wiley & Sons, New York, 1985, Vol. 1, pp. 623–643.

Inoue, S. and Aida, T., 'Catalysts for Living and Immortal Polymerization', in *Ring-Opening Polymerization*, Carl Hanser Publishers, Munich, 1993, pp. 197–215.

Problems

1. What are the milestones determining polymer chemistry development?
2. Why is there no dichotomy between the mechanisms governing reactions proceeding in low molecular weight and high molecular weight systems?
3. What is coordination polymerisation and what are its most characteristic features?
4. Which groups of monomers undergo coordination polymerisation?
5. What are the advantages of coordination polymerisation?
6. Explain why coordination polymerisation revolutionised the plastics and rubber industries.

2 General Features of Coordination Polymerisation

2.1 Monomers and Catalysts – Coordination

Coordination polymerisation was first proposed in 1956 for the unusual, at that time, low-pressure polymerisation of ethylene and polymerisation of propylene with the transition metal catalysts discovered by Ziegler in 1953 [1], and for the ferric chloride catalysed ring-opening polymerisation of propylene oxide to crystalline polymer reported by Pruitt *et al.* in a Dow patent [2].

Polymerisation carried out in the presence of a coordination catalyst is referred to as 'coordination polymerisation', when each polymerisation step involves the complexation of the monomer before its enchainment at the active site of the catalyst. The active site in each coordination catalyst comprises the metal atom (Mt), surrounded with ligands, one of which (X) forms a covalent active bond (Mt–X) with this metal atom. This implies that the growing polymer chain is covalently bound to the metal atom. A characteristic feature of coordination polymerisation is the mutual activation of the reacting bonds of both the monomer (M) and the active site (Mt–X) through the complexation of the monomer with the metal atom at this site, which results in the cleavage of these bonds in the concerted reaction.

$$\mathrm{Mt{-}X} + \mathrm{M} \longrightarrow \begin{matrix}\mathrm{Mt{-}X}\\ \uparrow\\ \mathrm{M}\end{matrix} \longrightarrow \mathrm{Mt{-}M{-}X} \xrightarrow{\mathrm{M}} \begin{matrix}\mathrm{Mt{-}M{-}X}\\ \uparrow\\ \mathrm{M}\end{matrix} \longrightarrow \mathrm{Mt{-}M{-}M{-}X} \quad (1)$$

Thus, monomer enchainment into the metal-bound polymer chain is preceded by, and activated by, the coordination of the monomer with the metal atom. It must be noted in this connection that only monomers capable of forming labile complexes with the active site of the catalyst may undergo coordination polymerisation.

The coordination step proposed in many polymerisation systems with coordination catalysts has not been unequivocally established. Thus, the more general term 'insertion polymerisation' has been used for these many-varied polymerisation systems to imply a hindered propagation site and to avoid implying the unproven coordination aspect. However, the term insertion polymerisation primarily relates to numerous polymerisation and copolymerisation processes of various hydrocarbon monomers, especially ethylene and α-olefins, with the use of Ziegler–Natta and other transition metal-based catalysts in which a *cis* ligand insertion of the coordinating monomer occurs, involving a four-centred transition state. Note that coordination polymerisations of heterocyclic and heterounsaturated monomers do not belong to such processes but involve a nucleophilic attack with inversion of the configuration at the attacked carbon atom of the coordinating heterocyclic monomer in a multicentred transition state, or involve, in most cases (when polymers containing heteroatoms in the main chain are formed), a *trans* ligand insertion of the coordinating heterounsaturated monomer via a multicentred transition state.

Monomer coordination at the active site of the catalyst may occur in varied ways, essentially reducing to a twofold mechanism which is dependent on both the kind of monomer and the catalyst. This appears to encompass cases where unsaturated hydrocarbon monomers coordinate at the metal of transition metal-based catalysts, involving π complex formation, as well as cases where heterocyclic and heterounsaturated monomers are subjected to polymerisation with various coordination catalysts in which monomer complexation proceeds via σ bond formation between the heteroatom and the metal atom.

Hydrocarbon monomers containing π bonds form π complexes with transition metals. Figure 2.1 shows the overlap of the molecular orbitals involved in the formation of a π bond between the simplest unsaturated hydrocarbon monomer, such as alkene, and transition metal [3].

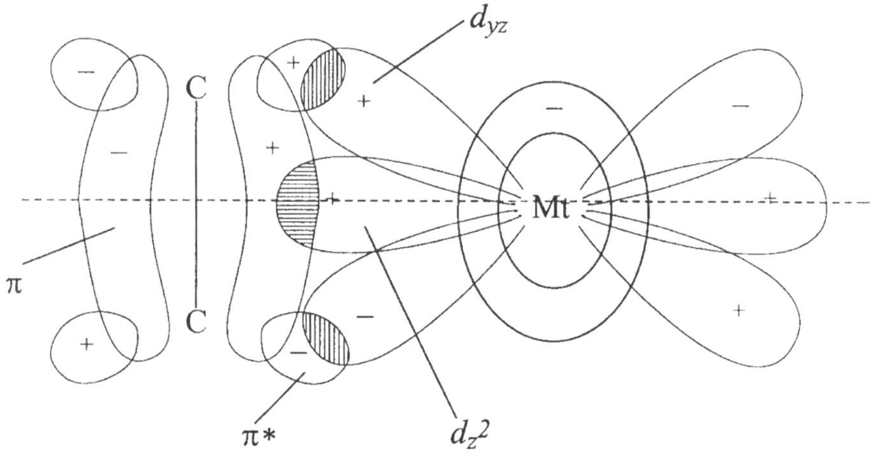

Figure 2.1 Schematic presentation of the olefin coordination at the transition metal

The alkene π-bonding filled orbital donates its electrons to the transition metal by overlapping with ns, np, $(n-1)d_{z^2}$ and $(n-1)d_{x^2-y^2}$ metal orbitals, or with the metal hybridised $(n-1)d_x nsnp_y$ orbital formed from these orbitals, which results in an enhancement of the electron density at the metal atom. As a result of the overlapping of the occupied metal $d_\pi(d_{xy}, d_{xz}, d_{yz})$ orbitals with the π^*-antibonding unfilled alkene orbital, a back-donating π re-donor bond is formed. This causes an enhancement of the electron density at the alkene π^*-antibonding orbital and its decrease at the metal atom. The formation of the above bonds of two types results in a lowering of energy in the original π complex in relation to the respective original metal $(n-1)$d orbitals and monomer π orbitals which become readily available to accept an electron from the excited Mt–X (X = alkyl) active bond and thus to accommodate the *cis* insertion of the coordinating monomer into this bond in a concerted way [4–11].

Coordination polymerisation via π complexes comprises polymerisation and copolymerisation processes with transition metal-based catalysts of unsaturated hydrocarbon monomers such as olefins [11–19], vinylaromatic monomers such as styrene [13, 20, 21], conjugated dienes [22–29], cycloolefins [30–39] and alkynes [39–45]. The coordination polymerisation of olefins concerns mostly ethylene, propylene and higher α-olefins [46], although polymerisation of cumulated diolefins (allenes) [47, 48], isomerisation 2, ω-polymerisation of α-olefins [49], isomerisation 1,2-polymerisation of β-olefins [50, 51] and cyclopolymerisation of non-conjugated α,ω-diolefins [52, 53] are also included among coordination polymerisations involving π complex formation.

The same group of coordination polymerisations in which alkene undergoes π complex formation with the metal atom includes the copolymerisation of ethylene, α-olefins, cycloolefins and styrene with carbon monoxide in the presence of transition metal-based catalysts [54–58]. In this case, however, the carbon monoxide comonomer is complexed with the transition metal via the carbon atom. Coordination bond formation involves the overlapping of the carbon monoxide weakly antibonding and localised mostly at the carbon atom σ orbital (electron pair at the carbon atom) with the unoccupied hybridised metal orbitals and the overlapping of the filled metal d_π orbitals with the carbon monoxide π^*-antibonding orbital (re-donor π bond) [59]. The carbon monoxide coordination with the transition metal is shown in Figure 2.2.

Similar behaviour to carbon monoxide is displayed by other heterounsaturated monomers of carbene-like structure, isocyanides, which homopolymerise in the presence of nickel-based catalysts, yielding polymers with a carbon–carbon main chain, poly(iminomethylene)s [60].

Polymerisations and copolymerisations of heterounsaturated and heterocyclic monomers in the presence of coordination catalysts constitute a distinct group of coordination polymerisation processes. Considering the nature of the coordination bond of the σ type between the monomer heteroatom (beyond carbon monoxide [60]) and the metal atom, the complexes formed differ essentially from the π complexes of unsaturated hydrocarbon monomers with transition metals.

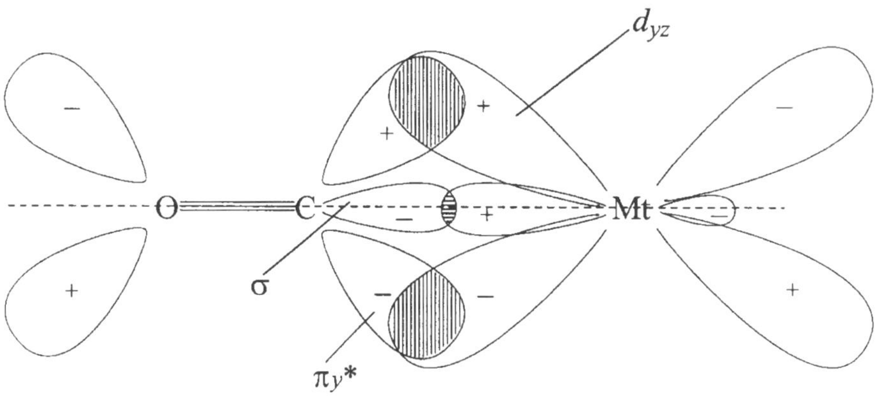

Figure 2.2 Schematic presentation of the carbon monoxide coordination at the transition metal

Coordination polymerisation of heterocyclic monomers comprises polymerisation and copolymerisation processes of such monomers as oxacyclic monomers, especially epoxides [2,61–71], thiacyclic monomers like episulphides [72–76], azacyclic monomers [77,78] and phosphacyclic monomers [79]. Monomers with an exocyclic oxygen atom, such as cyclic esters like lactones [80–90] and lactide [90–92], cyclic acid anhydrides [93–98], cyclic carbonates [99,100] and related monomers, belong to oxacyclic monomers undergoing coordination polymerisation or copolymerisation.

The coordination polymerisation of heterounsaturated monomers, such as aldehydes [101–103] and ketones [104], isocyanates [105] and ketenes [106,107], in homopolymerisation systems has not been widely described in the literature. However, the coordination copolymerisation of heterounsaturated monomers not susceptible to homopropagation, such as carbon dioxide [71,108–113], with heterounsaturated monomers such as cyclic ethers has been successfully carried out and is of increasing interest.

Catalysts for the coordination polymerisation of heterocyclic and heterounsaturated monomers are based on both the main group metal compounds and the transition metal compounds. Coordination catalysts may be heterogeneous as well as homogeneous. In most cases these catalysts are two- or three component systems, and their active species are formed in reactions proceeding only after mixing of the catalyst components, sometimes in the presence of the polymerised monomer. Active sites appear to be present only in some specific spots at the surface of the heterogeneous catalyst. In the case of homogeneous catalysts, the central atom surrounded with respective ligands forms active sites. However, distinct forms of homogeneous catalyst can be supermolecular or sometimes colloidal carriers for these sites. Such catalyst molecules may be large enough to accept and to dissipate the energy that will be released during the coordination and enchainment of the monomer.

2.2 Polymerisation of Coordinated Monomers

The active bond (Mt–X) of the coordination catalyst is stable in the uncomplexed state, but monomer coordination results in enhanced cleavage susceptibility of this bond as well as the respective bond in the coordinating monomer, which results in the occurrence of the polymerisation initiation step, followed by the subsequent chain propagation steps. Thus, the mutual activation of the monomer and the catalyst when they form the complex is to be emphasised as a characteristic feature of coordination polymerisation.

Apart from the activation of the coordinating monomer for its polymerisation, catalysts appear to play the role of polymerisation initiators. The initiating substituent (X) at the metal atom remains in each polymer chain as the end-group [reaction (**1**)]. It is worth mentioning that, to a pedant, coordination polymerisation might not seem to be a 'true' catalytic process, since the 'catalyst', at least in most cases, is not recoverable and unchanged after the polymerisation. To a production chemist, this is more than an academic nicety, since catalyst moieties or residues present in the polymer (largely those originating from the metal and its non-initiating ligands) can exert a serious effect on its properties. Therefore, the term 'catalyst' is used in a broad sense, considering the capacity of catalysts for monomer coordination and for enchainment of the coordinating monomer as well as their function as the initiator, independent of whether they are regenerated in an active form as the 'purists' require.

In this connection, one ought to recall that radical or ionic initiators do not require, as a principle, any coordination (activation) of the monomer prior to its incorporation into a polymer chain. It is to be emphasised that coordination polymerisation appears to encompass cases where the initiating substituent or the polymer chain end is covalently bound to the metal atom. Thus, terms such as 'pseudo-radical' or 'pseudo-ionic' ('pseudo-anionic') should not be used for the determination of the coordination polymerisation, irrespective of the mechanism of the coordinating monomer enchainment. Also, the term 'cationic coordination' is unsuitable and misleading for the determination of the cationically propagated polymerisation that can proceed in some cases involving respective cationic species in coordination catalyst systems. One has to take into consideration that 'cationic coordination' polymerisation ought to require the positively charged metal substituent ($X^{\delta+}$) with respect to the charge on the metal atom which, from the very nature of things, cannot be realised with any coordination polymerisation initiating substituent. Let us consider that the polymer chain, propagating according to the coordination mechanism, possesses a neutral end but holds a partial negative charge with respect to the metal ($Mt^{\delta+}-X^{\delta-}$).

Finally, note that coordinative interactions between monomers and catalysts usually result in a specific spatial orientation of the coordinating monomer. In many cases, owing to the determined mode of configuration of the monomer with respect to the growing polymer chain at the active site, the propagation leads to a stereoregular polymer.

Because of the great diversity of monomers undergoing coordination polymerisation in the presence of various catalysts, the different behaviour of various coordination polymerisation systems is to be noted.

2.2.1 Hydrocarbon Monomers

Hydrocarbon monomers undergoing coordination polymerisation with transition metal-based catalysts involving π bonds for the coordination may be included, from the point of view of the polymerisation initiation and propagation mechanism, into a few groups, such as olefins [5,17–19], conjugated dienes [25–27], cycloolefins [36,37] and alkynes [43–45]. Most of the catalysts used for the polymerisation of these monomers are characterised by the appearance of the initiating substituent bound to the transition metal atom at the active site via the carbon atom. It should be mentioned in this connection, however, that there are also catalysts with a hydrogen atom as well as a heteroatom, such as, for example, chlorine, as the polymerisation initiating substituent. However, as regards catalysts containing a metal–carbon bond that holds the initiating substituent, there is quite a large diversity of this type of bond. The diversity of the metal–carbon bond appearing at the active sites of coordination catalysts gives rise to varied mechanisms of enchainment of hydrocarbon monomers coordinating via a π bond with these sites.

In the coordination polymerisation of olefins, the active site of the catalyst usually contains an alkyl group as the metal substituent forming with the metal an Mt–C active bond of the σ type. The polymerisation consists in the insertion of the coordinated monomer into this bond with the regeneration of a metal–carbon bond of the same character [5]. The initiation and propagation steps in the coordination polymerisation of olefins in the presence of catalysts containing an ethyl initiating group bound to the metal atom are as follows:

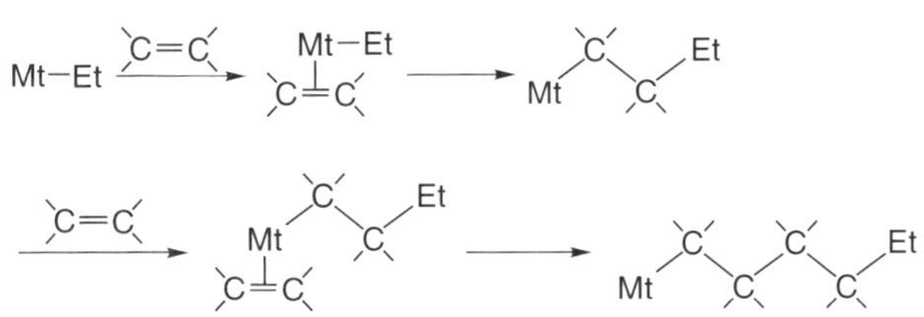

(2)

It is evident that each polymerisation step involves the insertion of the coordinating olefin into the active metal–carbon σ bond.

It is to be added in this connection that, like alkenes, acetylene can also undergo insertion polymerisation in the presence of some coordination catalysts, following a reaction analogous to (**2**) [43,44].

In the case of catalysts for the polymerisation of conjugated dienes, the initiating substituent or the ultimate monomer unit of the growing polymer chain can be bound with the transition metal atom via a metal–carbon bond of the variable type: the σ bond between the metal and monodentate allyl-type ligand [$Mt-(\eta^1-All)$] and the π bond between the metal and tridentate allyl-type ligand [$Mt-(\eta^3-All)$]. The monomer coordination at the metal atom forming the π bond with the initiating substituent or the growing polymer chain causes a change in the nature of this bond which acquires a σ bond character, and the insertion of the coordinated monomer into this bond proceeds with the regeneration of the π allyl-type active bond [25–27]. Scheme (**3**) shows the initiation and propagation steps in 1,3-alkadiene coordination polymerisation (leading to a polymer of *cis*-1,4 structure) with the catalyst containing an allyl initiating substituent at the metal atom:

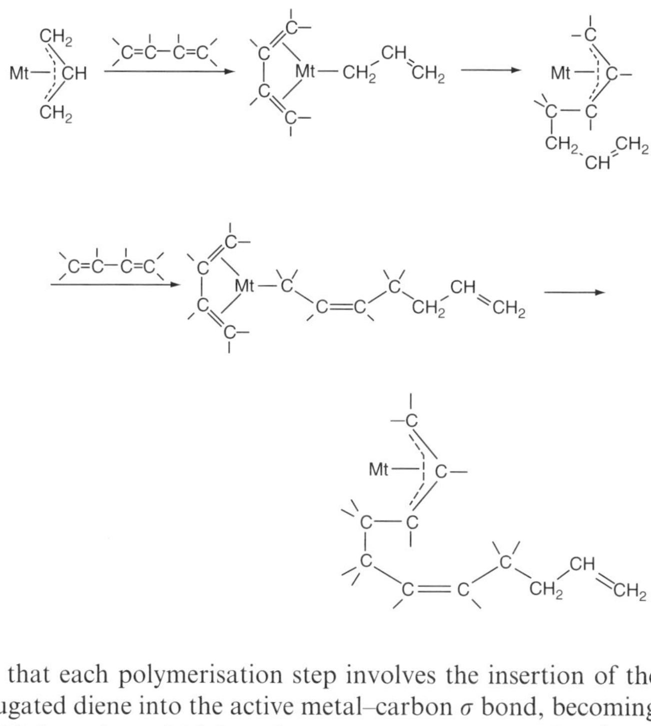

(3)

It is clear that each polymerisation step involves the insertion of the coordinated conjugated diene into the active metal–carbon σ bond, becoming a π bond owing to this insertion, which by subsequent monomer coordination reverts to a σ bond.

Another type of metal–carbon bond, the metal carbene bond (with carbene of an electrophilic or nucleophilic character), appears to be the active bond in transition metal-based catalysts for the ring-opening metathesis polymerisation of cycloolefins. Such a bond, which is co-originated with metal by the sp^2-hybridized carbon atom, possesses a σ, π double bond character ($Mt = C$) [34,35]. The enchainment of the coordinating cycloolefin at the active site with the metal carbene species proceeds by a different mechanism to that operating in the case of alkene or 1,3-dialkene insertion into the σ metal–carbon bond. Namely, the coordinated cycloalkene monomer, being π-complexed with the metal atom bearing the carbene substituent, becomes incorporated into the chain via σ bonds in the metallacycle (metallacyclobutane) formed in the first step. In the second step, the metallacycle σ bonds undergo conversion into the metal carbene species [38]. The course of the initiation and propagation steps in such a two-step transalkylidenation (metathesis) cycloalkene polymerisation (leading to the polymer of *cis* structure) in the presence of a catalyst with a metal diphenylcarbene initiating substituent is as follows:

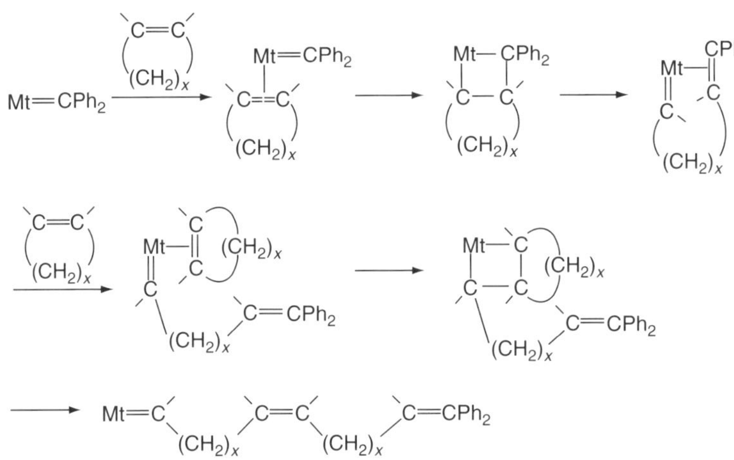

(4)

A characteristic feature of cycloolefin ring-opening metathesis polymerisation is alteration of the metal–carbon active bonds from the metal carbene σ, π bond into metallacycle σ bonds, and vice versa, as polymerisation progresses. It is worth mentioning, in this connection, that metallacyclobutanes can be successfully used as catalysts for this polymerisation [36,37].

An interesting case is the coordination polymerisation of acetylene and higher alkynes. It may proceed by a mechanism quite similar to the metathesis polymerisation of cycloalkenes involving metal carbene and metallacycle (metallacyclobutene) species [45]. The initiation and propagation steps in alkyne polymerisation (leading to a polymer of *cis* structure) in the presence of a catalyst with a diphenylcarbene initiating ligand are as follows:

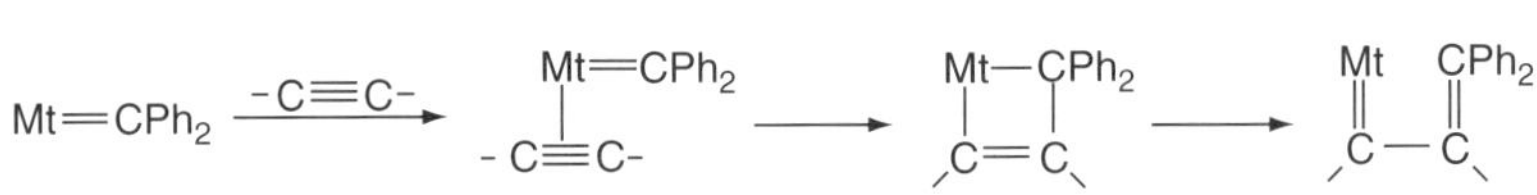

(5)

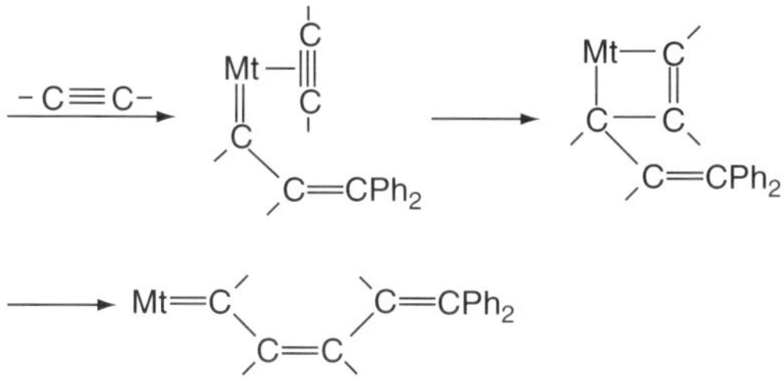

(5)

Note that the transformation of metal carbene σ, π bonds into metallacycle σ bonds, and vice versa, during polymerisation occurs as in the case of cycloalkene ring-opening metathesis polymerisation. Considering the mechanism of metathesis polymerisation of acetylenic monomers, it is worth noting that catalysts containing a transition metal carbyne bond (Mt=C) can induce polymerisation only when this bond is transformed into the respective metal carbene bond (Mt=C) [39].

2.2.2 Non-hydrocarbon (Heterocyclic and Heterounsaturated) Monomers

Coordination catalysts, which are usually applied for the polymerisation and copolymerisation of heterocyclic and heterounsaturated monomers, involve a wide range of metal derivatives characterised by a moderate nucleophilicity and relatively high Lewis acidity. Compounds of group 2 and 3 metals, such as zinc, cadmium and aluminium, as well as transition metals, such as iron, are representative coordination catalysts. The appropriate Lewis acidity of the metal and the appropriate nucleophilicity of the metal substituent in the catalyst make the monomer coordination favourable prior to the nucleophilic attack of the metal substituent on the monomer not yet coordinated.

The differentiation between the coordination polymerisation of heterocyclic and heterounsaturated monomers and their nucleophilically initiated anionic polymerisation lies in the covalent nature of the metal–heteroatom bond (but being polarised as in the $Mt^{\delta+} - X^{\delta-}$ bond) in the coordination catalyst, which activates the monomer by its coordination, enhancing the nucleophilicity of the metal substituent simultaneously, and the ionic character of the metal–heteroatom bond in the nucleophilic initiators.

Catalysts for the polymerisation of heterocyclic and heterounsaturated monomers contain a heteroatom, such as, for example, Cl (i.e. with an Mt–Cl active bond), or the group bound to the metal atom via the heteroatom (Mt–X; for example, X = O, S, N) as the initiating substituent. However, there are also catalysts that possess an alkyl initiating substituent at the metal atom, especially when they are used for the polymerisation of heterounsaturated monomers.

The metal–carbon bond in these catalysts becomes the metal–heteroatom bond after the monomer enchainment.

The coordination polymerisation of heterocyclic and heterounsaturated monomers consists in the nucleophilic attack of the metal initiating substituent (or the growing polymer chain) on the carbon atom of the coordinated monomer. Scheme 2.6 shows initiation and propagation steps in the coordination polymerisation of epoxides, as the most representative heterocyclic monomers with an endocyclic heteroatom, with catalysts containing an Mt–X active bond [68,114,115]:

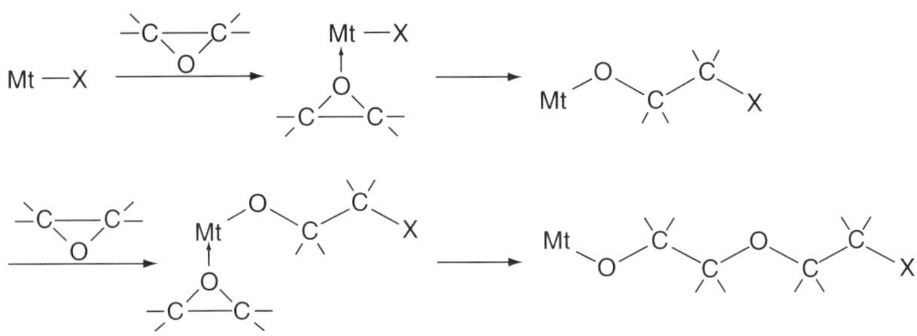

(6)

Note that the mechanism of incorporation of the coordinating epoxide into the Mt–X active bond actually occurs in a multicentred transition state but not in a four-centred transition state. This is implied by the participation of another metal atom of the same or different catalyst molecule in this transition state. As a consequence, an inversion of the configuration at the nucleophilically attacked carbon atom of the coordinated epoxide takes place [63,68,71,116–121].

Other heterocyclic monomers with an endocyclic heteroatom, such as higher cyclic ethers as well as cyclic sulfides like episulfides, undergo, similarly to epoxides, the coordination polymerisation in which the metal–heteroatom σ bond is regenerated in subsequent polymerisation steps [122,123].

Heterocyclic monomers containing both endocyclic and exocyclic heteroatoms such as cyclic esters (lactones, lactide, carbonates) and cyclic anhydrides undergo coordination polymerisation or copolymerisation involving complex formation between the metal atom and the exocyclic heteroatom [100,124]. Polymerisation of β-lactones is representative of such coordination polymerisations with catalysts containing an Mt–X active bond; the initiation and propagation steps are as follows:

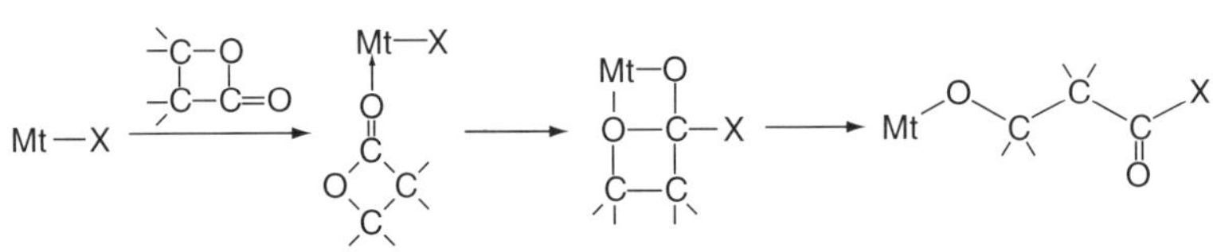

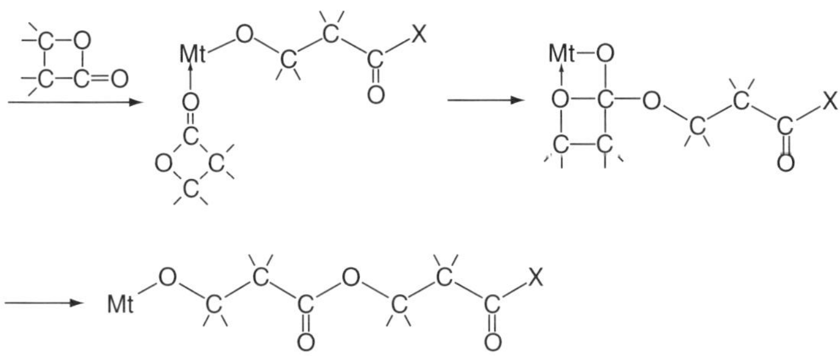

(7)

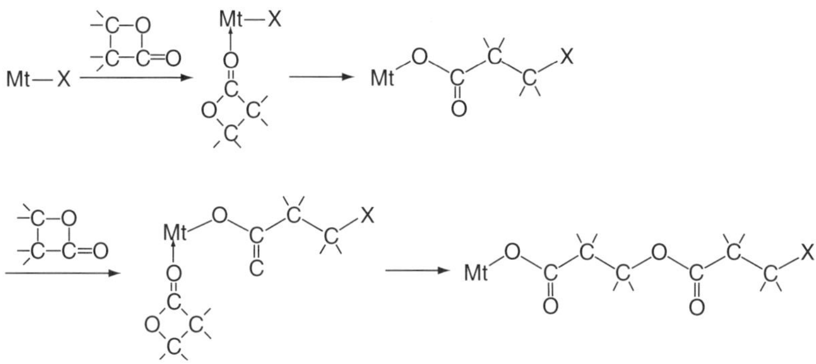

(8)

The mode of lactone ring opening depends on the kind of catalyst. It is characteristic that β-lactone polymerisation with a catalyst containing a metal alkoxide active bond (Mt–X; X = OR) involves C(O)–O bond scission in the coordinating monomer (via the metal orthocarbonate species) with regeneration of the metal alkoxide active bond [scheme (**7**)] [87]. On the other hand, the application of a catalyst with a metal carboxylate active bond [Mt–X; X = OC(O)R] for β-lactone polymerisation results in $C_\beta - O$ bond scission in the coordinating monomer with regeneration of the metal carboxylate active bond [scheme (**8**)] [88–90].

Analogously to the mechanism of epoxide polymerisation, the mechanism of the polymerisation of β-lactone and other heterocycles with both endocyclic and exocyclic heteroatoms involves multicentred transition states with the participation of at least two metal atoms.

Heterounsaturated monomers such as aldehydes or carbon dioxide polymerise and/or copolymerise with the participation of at least two metal atoms in multicentred transition states. Scheme (**9**) shows the initiation and propagation steps in the coordination polymerisation of carbonyl monomers with catalysts containing an Mt–X active bond [125]:

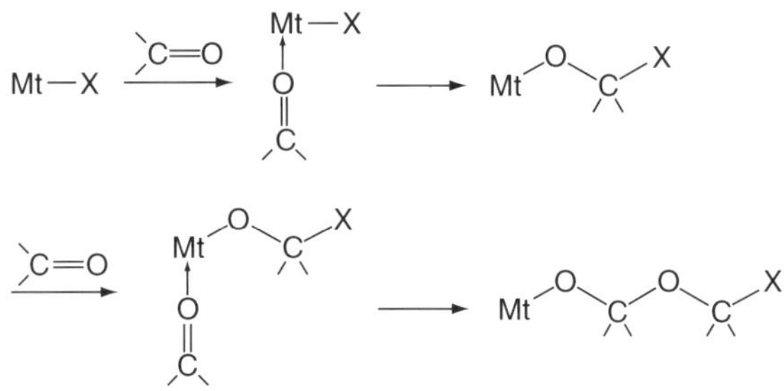

(9)

As mentioned above, such polymerisation involves *trans* ligand insertion which coincides with the polymerisation of heterocyclic monomers as regards the polymerisation mechanism.

2.3 Stereoregularity of Polymers

Stereochemistry deals with the three-dimensional aspects of molecular structure and the effects it has on physical and chemical properties, especially reactivity. As a part of general stereochemistry, the stereochemistry of macromolecules obeys some well-established principles of general validity. However, as it relates to other aspects of the same science, its application to particular cases requires some additional explanations.

Let us recall that organic stereochemistry concerns low molecular weight compounds containing only one or a few sites of stereoisomerism, usually tetrahedral or trigonal carbon atoms, with a special emphasis on the reactivity and reaction mechanism. As regards inorganic stereochemistry, this concerns higher coordination numbers and more complex structures, such as square or trigonal bipyramids, octahedra, etc. In both cases, molecules have discrete dimensions in all directions.

However, in macromolecular stereochemistry, two different conditions also have to be taken into consideration [126]. One is that macromolecules are quasi-one-dimensional molecules, the length of which is a few orders of magnitude greater than the cross-section diameter. If regular repetition of structural segments is detectable in the chain, the best way to examine it is by line symmetry rather than the point symmetry used for discrete molecules. Hence, the most suited model for its study is one with an infinite chain length (or a non-terminal section of the chain). The second point is the presence of intra- or intermolecular disorder of macromolecules. It is to be emphasised that polymers are not pure substances in the traditional chemical sense but mixtures of similar molecules that differ in length and in details of chemical and stereochemical structure. Therefore, studies based on the concept of microtacticity of short chain segments proved to be very useful [127].

It is worth noting in this connection that polymer stereochemistry started as an autonomous discipline in the mid-1950s with Natta's discovery of tactic poly(α-olefin)s [128, 129]. However, several discussions and experimental attempts were reported earlier in the literature. The first statement regarding stereoisomerism in synthetic polymers was most likely made by Staudinger [130]. The use of polymerisation conditions to control stereoregularity was then discussed by Huggins *et al.* [131] and Schildknecht *et al.* [132] who noted that heterogeneous catalysts could lead to the formation of a crystalline product. The most important contribution of this period was, according to Farina [126], made by Frisch *et al.* [133] who noted the non-chirality of vinyl polymers as a result of the pseudo-asymmetric nature of tertiary carbon atoms. Starting from the analysis of asymmetric polymerisation, they proposed two extreme mechanisms upon which all the probabilistic interpretations of stereoregular polymer formation are based: the chain end effect (asymmetric initiation) and the catalyst effect (asymmetric growth).

Let us now take into consideration various types of stereoisomerism in polymers, introducing in this connection some basic stereochemical definitions and notations relating to polymers of various groups [134]. Before that, the differentiation between the conformation and configuration of a polymer molecule should be clearly stated. The conformation of a polymer molecule describes the geometrical arrangement of the atom in the chain that is achieved through rotations about or stretching of its chemical bonds and bending of its valence angles. Examples of different polymer conformations are the fully extended planar zigzag, randomly coiled, helical and folded chain arrangements. Configuration denotes the stereochemical arrangements of the atoms in the polymer chain. The configuration of a polymer molecule cannot be altered without breaking and reforming chemical bonds.

A regular polymer is a linear polymer whose molecules can be substantially described by a unique species of constitutional units in a unique sequential arrangement. A stereoregular polymer is a regular polymer whose molecules can be described in terms of only one species of stereorepeating unit in a single sequential arrangement. A stereorepeating unit is a configurational unit having a defined configuration at all sites of stereoisomerism in the main chain of a macromolecule. It must be emphasised that polymer stereoisomerism can be considered only for regular polymers. Thus, regular polymers of head-to-tail, head-to-head and tail-to-tail structures can exhibit stereoisomerism [134].

However, note that not all of the regular polymers are able to form stereoisomers. In the case of polymers obtained from monomers not containing prochiral or chiral carbon atoms, such as, for example, ethylene, formaldehyde, ethylene oxide or β-propiolactone, no stereochemical considerations are possible, since there are two identical substituents, and not different ones, at the monomer carbon atom. This is exemplified by the following:

(10)

(11)

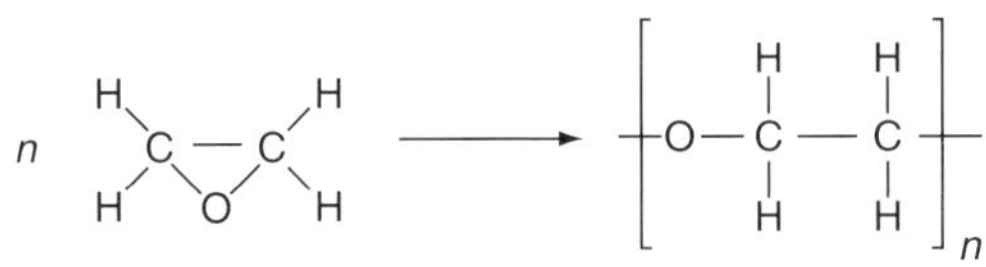

(12)

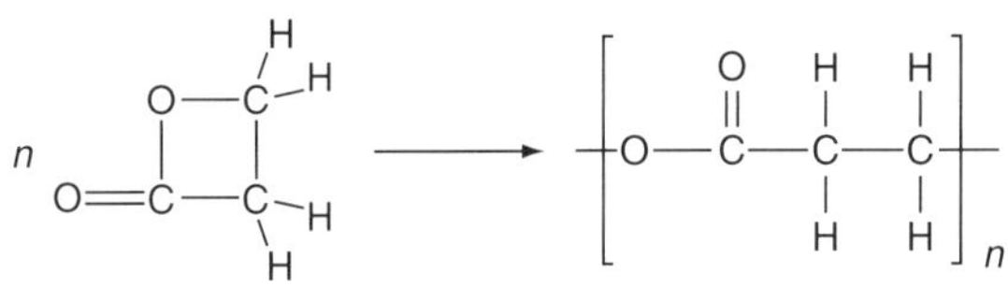

(13)

Stereoisomerism appears in polymers formed in the stereospecific polymerisation of prochiral monomers (e.g. propylene, acetaldehyde) and in the stereoselective polymerisation of chiral monomers (e.g. propylene oxide, β-butyrolactone, 3-methylcyclopentene). Stereospecific polymerisation is a polymerisation in which a tactic polymer is formed. A tactic polymer is a regular polymer whose molecules can be described in terms of only one species of configurational repeating unit in a single sequential arrangement. A configurational repeating unit is the smallest set of one, two or more successive configurational base unit that prescribes configurational repetition at one or more sites of stereoisomerism in the main chain of a polymer molecule. A configurational base unit is a constitutional repeating unit whose configuration is defined at one or more sites of stereoisomerism in the main chain of a polymer molecule. In a regular polymer, a configurational base unit corresponds to the constitutional repeating unit [134].

Taking into account the above basic definitions, let us consider the structure of polymers formed in the polymerisation of prochiral monomers such as propylene or acetaldehyde:

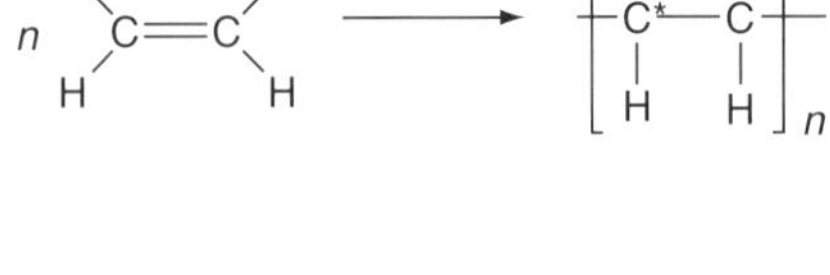

(14)

 (15)

These prochiral monomers have a carbon atom with different substituents which becomes a site of stereoisomerism owing to the formation of new bonds between successive monomer units in the polymer chain as a consequence of the polymerisation. Each site of this type consists of a carbon atom, denoted by C*, being attached to four different substituents: H, R and the two polymer chain segments $+CH(R)CH_2+_x$ and $+CH_2CH(R)+_x$ of different lengths. Carbon atoms such as C* do not give rise to any optical activity, since the optical activity is determined only by the first few atoms (α, β and γ) of each of the substituents attached to C*. In our case, the first few atoms of the two chain segments substituting the C* atom are the same:
$-CH(R)CH_2(R)CH_2-C^{*}H(R)-CH_2CH(R)CH_2CH(R)-$. This is why such polymers exhibit no optical activity. Such sites of stereoisomerism as C* were previously referred to as pseudo-asymmetric centres, and they were then referred to as pseudo-chiral centres.

Remember that the C* atom in the monomer has been referred to as a prochiral atom, since it becomes a pseudo-chiral atom upon polymerisation. As already mentioned, each pseudo-chiral C* atom is a site of stereoisomerism in polymers obtained from prochiral monomers such as α-olefins, higher aldehydes, etc. Each such carbon atom present in the polymer chain can exhibit either of two different configurations. The two configurations are usually referred to as *R*- and *S*-configurations which are arbitrarily assigned. However, these absolute configurations, *R* and *S*, connected with the stereochemical rules of precedence, are practically useless as regards their significance for the polymer properties resulting from the polymer steric isomerism. For the polymer properties resulting from its stereoisomerism, only the relative configurations of successive pseudo-chiral centres are of importance. The regularity in the configuration of successive pseudo-chiral carbon atoms determines the overall order of tacticity of the polymer chain. If the successive pseudo-chiral carbon atoms along the polymer chain have the same configuration, an isotactic polymer structure occurs. On the other hand, a syndiotactic polymer structure occurs when the configuration of the pseudo-chiral carbon atoms alternates from one repeating monomer unit to the next with an opposite configuration. In other words, an isotactic polymer is referred to as a regular polymer, the molecules of which can be described in terms of only one species of configurational base unit in a single arrangement. One should note that, in an isotactic polymer, the configurational repeating unit is identical to the configurational base unit. A syndiotactic polymer is a regular polymer, the molecules of which can be described in terms of alternation of configurational base units that are enantiomeric. Note that in a syndiotactic polymer the configurational repeating unit consists of two configurational base units that are enantiomeric. Thus, according to the definition, tacticity is the orderliness of the succession of configurational repeating units in the main chain of the polymer molecule. An atactic polymer is a regular polymer whose molecules have equal numbers of possible configurational base units in a random sequence distribution [134].

Chiral monomers, such as 3-methylcyclopentene, propylene oxide and β-butyrolactone, undergo ring-opening polymerisation leading to polymers that do not contain any internal plane of symmetry. This is shown by the following:

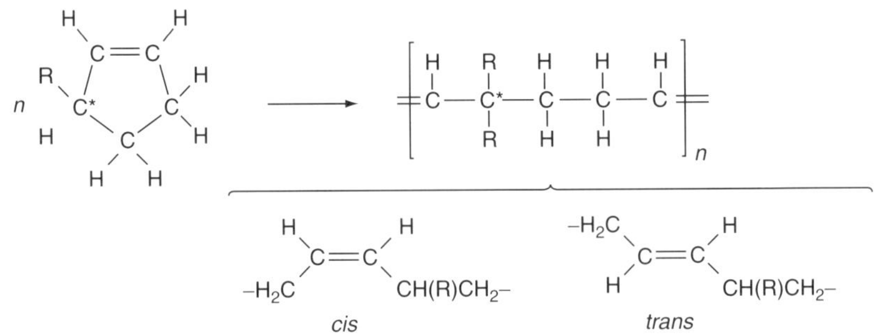

(16)

(17)

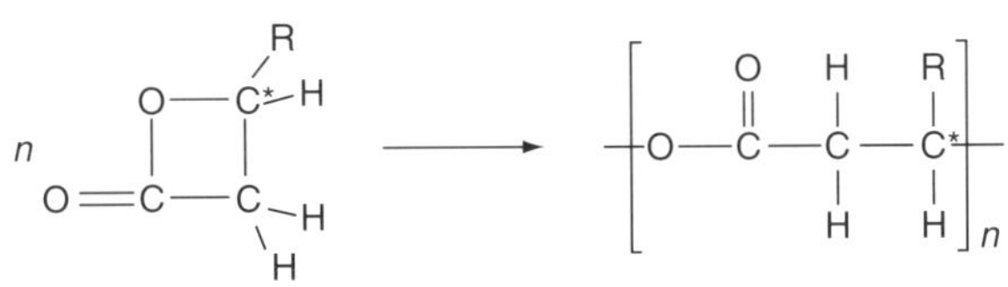

(18)

Polymers obtained from chiral monomers possess a true chiral site in the main chain, which is the carbon atom denoted as C*. The presence of an unsaturated carbon atom or oxygen atom attached to this chiral carbon atom, apart from its other three substituents, H, R and CH_2-terminated chain segment, makes the near environment of the C* atom in the polymer chain grossly dissimilar, unlike the situation in polymer chains formed from prochiral monomers. Because of this asymmetry, the isotactic polymers of the chiral monomers are capable of optical activity. It is worth noting that the stereoregularity of polymers obtained in polymerisations of chiral monomers [schemes (16) to (18)] does not arise from the generation of a chiral site as a consequence of bond formation in the polymerisation, as is the case in polymerisations of prochiral monomers [schemes (14) and (15)]. The stereoregularity of polymers formed by ring-opening polymerisation of chiral cyclic monomers according to schemes (16) to (18) is due to the initial presence of the chiral site, the C* atom, in the polymerised monomer. In this connection, stereoselective polymerisation should be defined here. Stereoselective polymerisation is a polymerisation in which a polymer molecule is formed from a mixture of stereoisomeric monomer molecules by preferential incorporation of one species into a growing polymer chain. However, polymerisation in which stereoisomerism present in the monomer is merely retained in the polymer is not to be regarded as stereospecific. For example, the polymerisation of a chiral monomer, e.g. *D*- or *L*-propylene oxide, with retention of the configuration is not referred to as stereospecific polymerisation; but the selective polymerisation, with retention of the configuration, of one of the enantiomers present in a mixture of *D*- and *L*-propylene oxide molecules is so classified [134]. In stereoselective polymerisation, the stereoisomeric sites to be selected can also be located as pendant groups, e.g. as in asymmetric α-olefin, $CH_2 = CHR^*$.

The stereoregularity of polymers relates not only to the configuration of four substituents attached to saturated carbon atoms in the polymer chains but also to the geometric isomerism, resulting from the presence of unsaturated carbon atoms in the polymer chains. Such isomerism appears in chains of polymers formed in the 1,4 polymerisation of conjugated dienes [scheme (19)] and the polymerisation of acetylenes [scheme (20)] as well as the ring-opening polymerisation of cycloolefins [scheme (16)]:

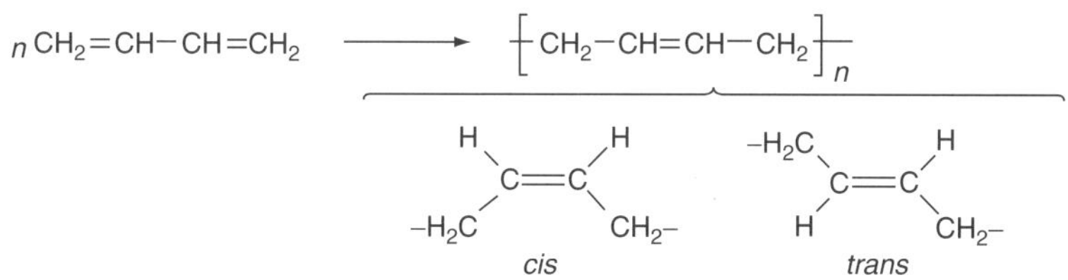

(19)

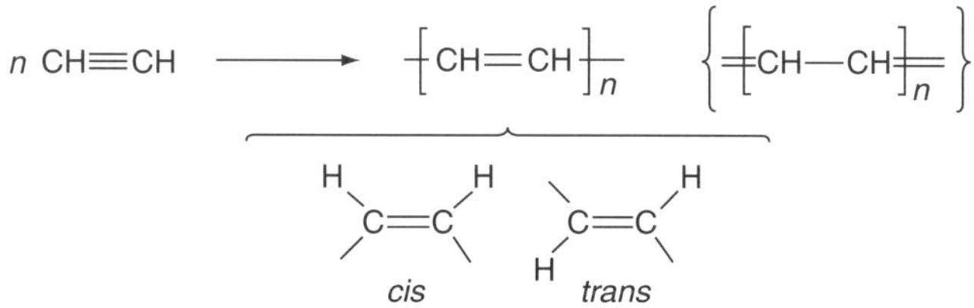

(20)

A cistactic polymer is a tactic polymer in which the main chain double bonds of the configurational base units are entirely in the *cis* arrangement. A transtactic polymer is a tactic polymer in which the main chain double bonds of the configurational base units are entirely in the *trans* arrangement [134]. Geometrical *cis–trans* isomerism also arises from different configurations of substituents on a cyclic structure.

It should be mentioned, in connection with definitions of stereoregular and tactic polymers, that a stereoregular polymer is always a tactic polymer, but a tactic polymer is not always a stereoregular polymer, because a tactic polymer does not need to have all sites of stereoisomerism defined.

When considering synthetic stereoregular polymers, one should realise that their chains contain monomer units incorporated in error with respect to predominant chain stereoregularity. The 'errors' can be both of a chemical nature (e.g. head-to-head linking instead of head-to-tail linking, isomerised monomer units) and of a steric nature at the sites of stereoisomerism. Therefore, each synthetic stereoregular polymer is characterised by some degree of stereoregularity (usually very high). The nomenclature recommended applies to the predominant structural features of the real polymer molecule. It must be noted in this connection that the presence of stereochemical defects in the structure of polymer chains is a valuable source of information about the stereocontrol mechanism in polymerisations yielding stereoregular polymers.

It has already been mentioned that macromolecular stereochemistry has some additional conditions to be taken into consideration, which is connected with the specific nature of polymer chains compared with low molecular weight organic molecules. Therefore, macromolecular stereochemistry requires a distinction to be made between stereoisomerism and chirality. This distinction was strongly stressed not long ago [135]. The term stereogenic has been proposed for a multivalent atom able to form stereoisomers, regardless of any symmetry consideration, and the term chirotopic has been proposed for any atom that contains no element of mirror symmetry (such an atom resides in a local chiral environment and the macromolecule as a whole must not be chiral). Thus, the asymmetric or chiral carbon atom is preferred now to be referred to as stereogenic chirotopic, and the pseudo-asymmetric or pseudo-chiral carbon atom as stereogenic achirotopic or non-chirotopic [126].

2.4 Coordination Polymerisation in Polymer Chemistry and Technology

During the early 1950s, research groups from Standard Oil of Indiana [14], Phillips Petroleum Co. [15] and Karl Ziegler with co-workers from the Max Planck Institut für Kohlenforschung in Mülheim/Ruhr [12] discovered independently three different catalysts that allowed high molecular weight polyethylenes to be produced at low pressure and temperature. Such polymers became known as high-density polyethylenes, in contrast to low-density polyethylenes produced by the extensively commercialised high-pressure, free radical initiated processes. These discoveries laid the basis for the coordination polymerisation of ethylene, which continued to diversify. Of the three mentioned discoveries concerning Standard Oil catalysts (molybdenum oxide supported on alumina), Phillips catalysts (chromium oxide supported on silica) and Ziegler catalysts (derived from reactions of transition metal halides, especially titanium chlorides, with organometallic compounds or metal hydrides of 1–3 main groups, especially aluminium alkyls), the latter two have been extensively commercialised.

Although the phenomenon of stereoisomerism is not applicable to polyethylene, which has no stereogenic carbon atoms, the use of coordination catalysts for ethylene polymerisation produced a polymer that is different from low-density polyethylene. High-density polyethylene is essentially a linear polymer composed of long chains that crystallise in a lamellar form. Such a polymer contains very few of the branches typical of low-density polyethylene obtained in a free radical high-pressure polymerisation process. High-density polyethylene crystallises to a higher degree and has a higher melting point because the polymer chains are far less branched than in the case of low-density high-pressure polyethylene. Moreover, high-density polyethylene has higher strength, greater solvent and chemical resistance and is less brittle at low temperature.

Coordination catalysts allowed for the first time the copolymerisation of ethylene with other olefins such as 1-butene, 1-hexene or 1-octene, which, by introducing side branches, reduces the crystallinity and allows a linear low-density polyethylene to be produced at comparatively low pressures [136]. Figure 2.3 shows schematic structures for the three polyethylenes, with the main features exaggerated for emphasis [46].

The use of coordination catalysts, especially homogeneous vanadium-based catalysts, for the copolymerisation of ethylene and propylene, with an ethylene content of 15–75 mol.-% in the feed, made it possible to produce amorphous

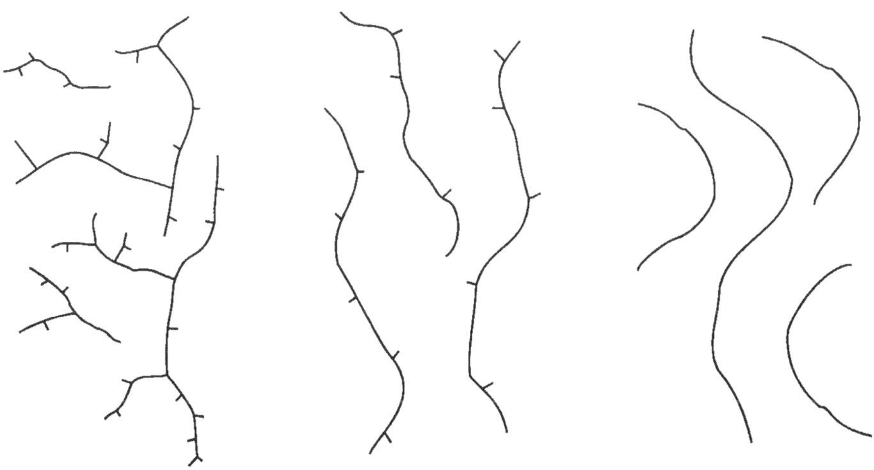

Figure 2.3 Schematic molecular structures of low-density polyethylene (LDPE), linear low-density polyethylene (LLDPE) and high-density polyethylene (HDPE)

ethylene–propylene random copolymers with desirable elastomeric properties [137]. When small amounts of a non-conjugated diene, e.g. 1,4-hexadiene, are added to the polymerisation system, ethylene–propylene–diene terpolymers are obtained. Such terpolymers, vulcanisable with sulphur, are commercially important synthetic rubbers [138].

One year after Ziegler's discovery in 1953, it was Giulio Natta and his group at the Istituto di Chimica Industriale del Politecnico, Milan, who first introduced the concept of stereospecific polymerisation of α-olefins. They prepared and characterised isotactic polypropylene and other poly (α-olefin)s in 1954 [13]. In contrast to the polymerisation of ethylene, only coordination catalysts are successful in the polymerisation of propylene to a crystalline polymer. Natta *et al.* [13] managed to synthesise isotactic crystalline polypropylene with a combination of catalysts that had previously been discovered by Ziegler. He thus achieved a breakthrough for the technical application of polypropylene. Isotactic polypropylene, which is characterised by a high degree of crystallinity owing to the steric order in its chains, exhibits a relatively high melting point and does not dissolve in low-boiling aliphatic hydrocarbons. Ethylene–propylene block copolymers, which are mostly mixtures of the copolymer and both homopolymers, are obtained using catalysts that are active in the isospecific polymerisation of propylene [139]. These ethylene–propylene block copolymers (polyallomers), because of their attractive physical properties, are produced commercially on a large scale.

Ziegler–Natta catalysts constitute a remarkable group of catalysts. They are the only catalysts that produce stereoregular polymers by stereospecific polymerisation of α-olefins, i.e. propylene, 1-butene, 4-methyl-1-pentene and higher 1-alkenes. Poly(1-butene) is similar to polyethylene and polypropylene in many properties. Poly(4-methyl-1-pentene) has a much lower density than the other polymers (ca $0.83\,\mathrm{g/cm^3}$ [140]). Its melting point is even higher than that of polypropylene, making it suitable for high-temperature applications. It should be emphasised that α-olefins are not polymerised by either free radical or ionic initiators.

Many cycloolefins have also been polymerised successfully in the presence of coordination catalysts. The polymerisation of cyclic olefins by double bond opening with ring retention was discovered by Natta and his co-workers shortly after the invention of Ziegler–Natta catalysts. They were able to obtain cycloolefin homopolymers of various microstructures [32] as well as alternating copolymers of non-homopropagating cycloolefins with ethylene [141]. However, these polymers did not attract much interest, which was connected with the difficulties of selecting catalysts to give exclusively double bond opening of the polymerised cycloolefin. Much more interest has been shown in the ring-opening metathesis polymerisation of cyclic olefins. Although this polymerisation emerged from the studies of coordination catalysts for alkene polymerisation [32, 142], the mechanistic details of the transalkylidenation appear to be more firmly developed. In fact, the studies of metathesis polymerisation have produced new possible models for polymerisation mechanisms [38, 143]. Ring-opening metathesis polymerisation has attracted a considerable research effort flourishing not only in developing the mechanistic aspects but in application in the plastics and rubber industries [144].

In 1954, 1,4-*cis*-polyisoprene, the synthetic equivalent of natural rubber, was obtained in the laboratories of Goodrich-Gulf [22] by isoprene polymerisation with new catalysts developed by Natta, and later on 1,4-*trans*-polyisoprene, a synthetic analogue of gutta percha, was obtained by Natta *et al.* [23].

At the same time, syndiotactic polypropylene was also isolated by Natta *et al.* and characterised [145]. Syndiotactic polypropylene which was obtained by low-temperature polymerisation using soluble vanadium-based catalysts [146] could not be, however, commercialised, although it had a blend of interesting usable properties.

It may be of interest that isotactic polystyrene formed by styrene polymerisation with Ziegler–Natta catalysts [13] did not appear to be a polymer that could exhibit significantly better usable properties compared with atactic polystyrene produced in free radical styrene polymerisation processes.

The Ziegler–Natta catalysts have acquired practical importance particularly as heterogeneous systems, mostly owing to the commercial production of linear high- and low-density polyethylenes and isotactic polypropylene. Elastomers based on ethylene–propylene copolymers (with the use of vanadium-based catalysts) as well as 1,4-*cis*-and 1,4-*trans*-poly(1,3-butadiene) and polyisoprene are also produced. These catalysts are extremely versatile and can be used in many other polymerisations of various hydrocarbon monomers, leading very often to polymers of different stereoregularity. In 1963, both Ziegler and Natta were awarded the Nobel Prize in chemistry.

Following the early pioneering advances in the transition metal-catalysed polymerisation of hydrocarbon monomers, especially ethylene and propylene, most of the catalyst research was aimed at improving catalyst productivities and selectivities. A major breakthrough in catalyst research occurred during the 1970s when super highly active catalysts were developed for isospecific propylene polymerisation and introduced into industrial processes. The coordination polymerisation and copolymerisation of ethylene, especially the production of linear low-density polyethylene, also started to attract marketing attention and to compete with high-pressure polyethylene. Modern catalysts, used in a fluidised bed in gas-phase polymerisation processes, also allow the control of polymer particle size and porosity, and thus the production of of pellet-sized particles [147–150]. Polyethylene worldwide total production capacities in 1990 were $33\,782 \times 10^3$ t/year (low-density polyethylene: $15/598 \times 10^3$ t/year, linear low-density polyethylene: 6754×10^3 t/year, high-density polyethylene: $11/430 \times 10^3$ t/year) [46]. As regards polypropylene (isotactic), the worldwide total manufacturing capacity (in 1989) was $13\,300 \times 10^3$ t/year [46].

During the 1980s, catalyst and process developments focused mostly on the formation of polymers with a controlled molecular weight distribution and comonomer incorporation to modify the mechanical properties and rheological behaviour. However, the discovery of novel metallocene single-site catalysts, which contain stereospecific active sites of essentially one type, represented another important milestone crucial to the development of new hydrocarbon polymer materials. This relates to linear low-density polyethylene obtained by copolymerisation of ethylene with 1-butene, 1-hexene and/or 1-octene [151–154], ethylene–propylene copolymers [155–157], polyolefins of varied tacticity, especially isotactic [11, 158], syndiotactic [159, 160] and stereoblock isotactic–atactic polypropylene [161–163], ethylene–cycloolefin copolymers [164–168], novel cycloaliphatic polyolefins [168–172] including α,ω-diene cyclopolymers [173–176], syndiotactic polystyrene [177–181], ethylene–styrene copolymers [182], 1,4-*cis*-, 1,4-*trans*-or 1,2-polymers of 1,3-dialkenes of varied tacticity [29, 183] and even functionalised polyolefins [184–190].

Studies on metallocene-catalysed polymerisations of hydrocarbon monomers, in particular alkenes, have derived much of their impetus from the desire to model the reaction mechanism of heterogeneous polymerisation catalysts. Alkylaluminium-activated metallocene complexes of the group 4 transition metals (dicyclopentadienyltitanium dichloride, diethylaluminium chloride), which have been known since 1955 to catalyse the polymerisation of ethylene, appeared to be very attractive as model systems [191, 192]. The limitation of homogeneous metallocene-based catalysts to the polymerisation of ethylene was a crucial obstacle to progress in this field for many years. Around 1980, Kaminsky and Sinn [193–196] found that metallocenes might be activated for the polymerisation of propylene and higher α-olefins by methylaluminoxanes. Further important discoveries by Brintzinger *et al.* [197,198], Ewen *et al.* [11,159,199] and Kaminsky *et al.* [158,169] as well as mechanistic studies by Pino [200–202] and Zambelli [203,204], which certainly are the major achievements after the early discoveries by Ziegler and Natta, have opened a new era in coordination polymerisation which now affords a unique possibility of tailoring the structure of polyolefins and other hydrocarbon polymers through manipulation of the catalyst structure almost at will.

The homogeneous metallocene catalysts had been referred to as single-site or uniform-site catalysts. They allowed a much better insight into the alkene polymerisation mechanism, including stereochemistry. However, at ordinary polymerisation conditions the homogeneous catalyst can exist in several conformational states which may result in side reactions causing a loss of stereochemical control. Many side reactions can be suppressed by attaching the catalyst to an inert support. The evolution of advanced catalysts and coordination polymerisation processes based on rational models is now beginning to overtake heterogeneous Ziegler–Natta catalysis. Such an approach also has the attraction of rendering the catalyst commercially useful in gas-phase polymerisation processes. Site isolation has been achieved with metallocene attached on silica or alumina gel [205–209]. Instead of the polymer dust obtained in the presence of a dissolved metallocene catalyst, the particles of the solid metallocene-supported catalyst generate coherent polymer grains [210]. Heterogenised metallocene catalysts, for example those supported on silica, can thus be readily used in existing Ziegler–Natta production facilities, for instance in solvent-free slurry or gas phase polymerisation systems [211–214]. The single-site catalyst technology based upon metallocenes activated with methylaluminoxanes or with other activators provides unprecedented control of the polymerisation process, in particular of regio-and stereospecificities, molecular weights, molecular weight distributions and comonomer incorporation. This enables new polyolefin materials to become available on a commercial scale. Despite the practical advantages of supported catalysts, interactions between support materials and catalyst complexes are only partly understood on a molecular level.

The first studies on the coordination polymerisation of non-hydrocarbon monomers such as heterocyclic monomers like epoxides, e.g. propylene oxide, date back to the early 1950s [1,2,215]. The first coordination catalyst used in the polymerisation of higher epoxides, known as the Pruitt–Baggett adduct, consisted of a complex of ferric chloride and propylene oxide [117]. A variety of other catalysts for epoxide polymerisation have subsequently been described, and in the 1960s quite a large number of catalysts were used for epoxide polymerisation. Most of them included a combination of alkylmetal compound with mono-and/or multiprotic compound, e.g. alcohol, water, diol, etc., in binary or ternary systems. Some of the epoxide polymerisation catalysts, such as metal alkoxides and bimetallic μ-oxoalkoxides, have been prepared in other ways, without using metal alkyls. Heterogeneous or homogeneous catalysts formed in systems with or without the application of a metal alkyl compound contain multinuclear associated species, mostly involving aluminium or zinc atoms, and are capable of producing high molecular weight epoxide homopolymers or copolymers of relatively broad molecular weight distribution [115,124]. In the 1980s, new homogeneous epoxide polymerisation catalysts, such as metalloporphyrins of aluminium and zinc, were described by Inoue and Aida. These catalysts, containing mononuclear non-associated species, were capable of promoting living polymerisation which led to low molecular weight epoxide polymers and copolymers characterised by a narrow molecular weight distribution [69,216]. Later on, other homogeneous mononuclear catalysts such as Schiff's base aluminium derivatives [70,217] and calix[4]arene aluminium derivatives [71] were successfully applied for epoxide stereospecific polymerisation.

The progress in the polymerisation of epoxides, which are considered to be very important heterocyclic monomers, concerns both mechanistic studies and technological studies aimed at commercialising certain epoxide polymerisation processes. Although wide studies have dealt with the catalyst structure, only a limited number of catalysts have been structurally determined in the solid state as well as in solution. The epoxide polymerisation mechanism, including stereochemical aspects, has been satisfactorily explained on the molecular level by Tsuruta *et al.* [218–222], who used well-defined catalysts containing multinuclear species. Furthermore, the mechanism of epoxide polymerisation with catalysts containing mononuclear species has also been explained [69–71,115,124,216,217].

So far, high-mileage industrial processes, including epoxide coordination polymerisation, have consisted of the production of elastomers using epichlorohydrin, ethylene oxide and/or propylene oxide, and eventually allyl glycidyl ether as the vulcanisable monomer unit [223–225]. Worldwide production of epichlorohydrin polymers and copolymers (hydrin rubbers) [223] and propylene oxide copolymers (polyether rubbers) [225] exceeded 12 000 t/year in 1990. Furthermore, the structure of Vandenberg catalysts, trialkylaluminium–water and trialkylaluminium–water–acetylacetone or other similar chelating agent, discovered in the late 1950s and used for industrial epoxide coordination polymerisation processes, is still uncertain, in spite of the great effort that has been made to resolve this problem.

Studies on the coordination polymerisation and copolymerisation of epoxides have been widely extended to other heterocyclic monomers and also heterounsaturated monomers since the early 1960s. Most epoxide polymerisation catalysts appeared also to promote the coordination polymerisation and copolymerisation of other oxacyclic and thiacyclic monomers and heterounsaturated monomers. Extensive investigations of these coordination polymerisations in terms of mechanistic studies and exploratory studies for obtaining improved catalysts and processes as well as for commercialising new processes are constantly being carried out. The application of coordination catalysts for the polymerisation of heterocyclic monomers has allowed the production of high molecular weight as well as low molecular weight polymers characterised by a high degree of regio-and stereoregularity. Apart from potential uses as plastics and rubbers, polymers derived from heterocyclic and heterounsaturated monomers may be of importance to the production of new, potentially useful materials in terms of their biomedical and/or pharmaceutical applications. Another important goal that might be reached by the application of coordination catalysts for the polymerisation of heterocyclic monomers seems to be the unique possibility of enchaining heterounsaturated monomers, which are non-susceptible to homopropagation (e.g. carbon dioxide [226]), via their copolymerisation with heterocyclic monomers. It should be emphasised that polymers obtained from non-hydrocarbon monomers such as heterocyclic and heterounsaturated monomers cannot compete with high-performance hydrocarbon polymers. However, the coordination polymerisation of cyclic and acyclic heteroatom-containing monomers is guided by different factors to the polymerisation of hydrocarbon monomers, and therefore both types of coordination polymerisation are claimed to complement each other in terms of potential industrial applications as well as basic academic research.

References

1. Price, C. C. and Osgan, M., *J. Am. Chem. Soc.*, **78**, 4787 (1956).
2. Pruitt, M. E., Jackson, L. and Baggett, J. M., US Pat. 2 706 181 (to Dow Chemical Co.) (1956).
3. Chatt, J. and Duncanson, L. A., *J. Chem. Soc.*, **1953**, 2939 (1956).
4. Cossee, P., *Tetrahedron Lett.*, **17**, 12 (1960).
5. Cossee, P., *J. Catal.*, **3**, 80 (1964).
6. Natta, G., Farina, M. and Peraldo, M., *Chim. Ind. (Milan)*, **42**, 255 (1960).
7. Miyazawa, T. and Ideguchi, T., *J. Polym. Sci., B*, **1**, 389 (1963).
8. Zambelli, A., Giongo, M. G. and Natta, G., *Makromol. Chem.*, **112**, 183 (1968).
9. Zambelli, A. and Tosi, C., *Fortschr. Hochpolym. Forsch.*, **15**, 31 (1974).
10. Böhm, L. L., *Polymer*, **19**, 545 (1978).
11. Ewen, J. A., *J. Am. Chem. Soc.*, **106**, 6355 (1984).
12. Ziegler, K., Holzkamp, E., Breil, H. and Martin, H., *Angew. Chem.*, **67**, 541 (1955).
13. Natta, G., Pino, P., Corradini, P., Danusso, F., Mantica, E., Mazzanti, E. and Moraglio, G., *J. Am. Chem. Soc.*, **77**, 1708 (1955).
14. Zletz, A., US Pat. 2 692 257 (to Standard Oil of Indiana) (1954); *Chem. Abstr.*, **49**, 2777d (1955).
15. Hogan, J. P. and Banks, R. L., US Pat. 2 825 721 (to Phillips Petroleum Co.) (1958); *Chem. Abstr.*, **52**, 86211 (1958).
16. Fischer, M., Ger. Pat. 874 215 (to BASF A.G.) (1943); *Chem. Abstr.*, **51**, 10124g (1957).
17. Rodriguez, L. A. M. and van Looy, H. M., *J. Polym. Sci., A-1*, **4**, 1951, 1971 (1966).
18. Henrici-Olivé, G. and Olivé, S., *Adv. Polym. Sci.*, **6**, 421 (1969).
19. Pino, P. and Rotzinger, B., *Makromol. Chem., Suppl.*, **7**, 41 (1984).
20. Natta, G. and Corradini, P., *Makromol. Chem.*, **16**, 17 (1955).
21. Ishihara, N., Kuramoto, M. and Uoi, M., *Macromolecules*, **21**, 2464 (1986).
22. Horne Jr, S. E., Kichl, J. P., Shipman, J. J., Volt, V. L. and Gibbs, C. T., *Ind. Eng. Chem.*, **48**, 784 (1956).
23. Natta, G., Corradini, P. and Porri, L., *Rend. Accad. Naz. Lincei*, **8**, 728 (1956).
24. Porri, L. and Aglietto, M., *Makromol. Chem.*, **177**, 1465 (1976).
25. Destri, S., Gatti, G. and Porri, L., *Makromol. Chem., Rapid Commun.*, **2**, 605 (1981).
26. Destri, S., Bolognesi, A., Porri, L. and Wang. F., *Makromol. Chem., Rapid Commun.*, **3**, 187 (1982).
27. Porri, L., Giarrusso, A. and Ricci, G., *Makromol. Chem., Macromol. Symp.*, **48/49**, 239 (1991).
28. Yasuda, H. and Nakamura, A., *Angew. Chem., Int. Ed. Engl.*, **26**, 723 (1987).
29. Ricci, G., Porri, L. and Giarrusso, A., *Macromol. Symp.*, **89**, 383 (1995).
30. Anderson, A. W. and Merckling, N. C., US Pat. 2 721 189 (to Du Pont de Nemours) (1955); *Chem. Abstr.*, **50**, 3008 (1956).
31. Eleuterio, H. S., US Pat. 3 074 918 (to Du Pont de Nemours) (1957); *Chem. Abstr.*, **55**, 16005 (1957).
32. Natta, G., Dall'Asta, G., Mazzanti, G. and Motroni, G., *Makromol. Chem.*, **69**, 163 (1963).
33. Dall'Asta, G. and Motroni, G., *Eur. Polym. J.*, **7**, 707 (1971).
34. Fischer, E. O. and Dotz, K. H., *Chem. Ber.*, **105**, 3966 (1972).
35. Casey, C. P. and Burkhardt, T. J., *J. Am. Chem. Soc.*, **96**, 7808 (1974).

36. Gilliom, L. R. and Grubbs, R. H., *J. Am. Chem. Soc.*, **108**, 733 (1986).
37. Wallace, K. C. and Schrock, R. R., *Macromolecules*, **20**, 450 (1987).
38. Herisson, J. L. and Chauvin, Y., *Makromol. Chem.*, **141**, 161 (1970).
39. Katz, T. J., Ho, H.-T., Shih, N.-Y., Ying, Y.-C. and Van Stuart, I. W., *J. Am. Chem. Soc.*, **106**, 2659 (1984).
40. Natta, G., Mazzanti, G. and Corradini, P., *Atti Acad. Naz. Lincei, Cl. Sci. Fis. Mat. Nat.*, **25**, 3 (1958).
41. Luttinger, L. B., *Chem. Ind. (Lond.)*, **1960**, 1135 (1960).
42. Ito, T., Shirakawa, H. and Ikeda, S., *J. Polym. Sci., Polym. Chem. Ed.*, **12**, 11 (1974).
43. Schen, M. A., Karasz, F. E. and Chien, J. C. W., *J. Polym. Sci., Polym. Chem. Ed.*, **21**, 2787 (1983).
44. Clarke, T. C., Yannoni, C. S. and Katz, T. J., *J. Am. Chem. Soc.*, **105**, 7787 (1983).
45. Masuda, T., Sasaki, N. and Higashimura, T., *Macromolecules*, **8**, 717 (1975).
46. Whiteley, K. S., Heggs, T. G., Koch, H., Mawer, R. L., Goldbach, G. and Immel, W., 'Polyolefins', in *Ullmann's Encyclopedia of Industrial Chemistry*, VCH Publishers, Inc., Weinheim, 1992, Vol. A21, pp. 487–577.
47. Baker Jr, W. P., *J. Polym. Sci., A*, **1**, 655 (1963).
48. Havinga, R. and Schors, A., *J. Macromol. Sci. – Chem. A*, **2**, 1, 31 (1968).
49. Möhring, V. M. and Fink, G., *Angew. Chem., Int. Ed. Engl.*, **24**, 1001 (1985).
50. Endo, K., Ueda, R. and Otsu, T., *J. Polym. Sci., Polym. Chem. Ed.*, **29**, 807, 843 (1991).
51. Endo, K., Ueda, R. and Otsu, T., *Makromol. Chem.*, **193**, 539 (1992).
52. Marvel, C. S. and Stille, J. K., *J. Am. Chem. Soc.*, **80**, 1740 (1958).
53. Makowski, H. S., Shim, B. K. C. and Wilchinsky, Z. W., *J. Polym. Sci., A*, **2**, 1549, 4973 (1964).
54. Sen, A. and Lai, T.-W., *J. Am. Chem. Soc.*, **104**, 3520 (1982).
55. Lai, T.-W. and Sen, A., *Organometallics*, **3**, 866 (1984).
56. Drent, E., van Broekhoven, J. A. M. and Doyle, M. J., *J. Organomet. Chem.*, **417**, 235 (1991).
57. Batistini, A., Consiglio, G. and Suter, U. W., *Angew. Chem. Int. Ed. Engl.*, **31**, 303 (1992).
58. Bartolini, S., Carfagna, C. and Musco, A., *Macromol. Rapid Commun.*, **16**, 9 (1995).
59. Fenske, R. F., *Pure Appl. Chem.*, **27**, 61 (1971).
60. Nolte, R. J. M. and Drenth, W., *Recl. Trav. Chim. Pays-Bas*, **92**, 788 (1973).
61. Osgan, H. and Price, C. C., *J. Polym. Sci.*, **34**, 153 (1959).
62. Furukawa, J., Tsuruta, T., Sakata, T., Saegusa, T. and Kawasaki, A., *Makromol. Chem.*, **32**, 90 (1959).
63. Vandenberg, E. J., *J. Polym. Sci.*, **47**, 486 (1960).
64. Colclough, R. O., Gee, G. and Jagger, A. H., *J. Polym. Sci.*, **48**, 273 (1960).
65. Osgan, M. and Teyssié, P., *J. Polym. Sci., B*, **5**, 789 (1967).
66. Inoue, S., Tsuruta, T. and Furukawa, J., *Makromol. Chem.*, **53**, 215 (1962).
67. Kuran, W., Rokicki, A. and Pieńkowski, J., *J. Polym. Sci., Polym. Chem. Ed.*, **17**, 1235 (1979).
68. Kuran, W. and Listoś, T., *Macromol. Chem. Phys.*, **195**, 401 (1994).
69. Aida, T. and Inoue, S., *Macromolecules*, **14**, 1166 (1981).
70. Vincens, V., Le Borgne, A. and Spassky, N., *Makromol. Chem., Rapid Commun.*, **10**, 623 (1989).
71. Kuran, W., Listoś, T., Abramczyk, M. and Dawidek, A., *J. Macromol. Sci. – Chem. A*, **35**, 427 (1998).
72. Machon, J. P. and Sigwalt, P., *Compt. Rend. Acad. Sci. (Paris)*, **260**, 549 (1965).

73. Furukawa, J., Kawabata, N. and Kato, A., *J. Polym. Sci., Polym. Lett. Ed.*, **5**, 1073 (1967).
74. Coulon, C., Spassky, N. and Sigwalt, P., *Polymer*, **17**, 821 (1976).
75. Dumas, P., Sigwalt, P. and Guérin, P., *Makromol. Chem.*, **193**, 1709 (1992).
76. Aida, T., Kawaguchi, K. and Inoue, S., *Macromolecules*, **23**, 3887 (1990).
77. Tsuruta, T., Inoue, S. and Matsuura, K., *Makromol. Chem.*, **63**, 219 (1963).
78. Matsuura, K., Inoue, S. and Tsuruta, T., *Makromol. Chem.*, **80**, 149 (1964).
79. Katużyński, K., Libiszowski, J. and Penczek, S., *Makromol. Chem.*, **178**, 943 (1977).
80. Cherdron, H., Ohse, H. and Corte, F., *Makromol. Chem.*, **56**, 187 (1962).
81. Agostini, D. E., Lando, J. B. and Shelton, J. R., *J. Polym. Sci., A-1*, **9**, 2775 (1971).
82. Tani, H., Yamashita, S. and Teranishi, K., *Polym. J.*, **3**, 417 (1972).
83. Ouhadi, T., Stevens, C. and Teyssié, P., *Makromol. Chem. Suppl.*, **1**, 191 (1975).
84. Benvenuti, M. and Lenz, R. W., *J. Polym. Sci., A, Polym. Chem.*, **29**, 793 (1991).
85. Pajersky, A. D. and Lenz, R. W., *Makromol. Chem. Macromol. Symp.*, **73**, 7 (1993).
86. Le Borgne, A., Pluta, C. and Spassky, N., *Macromol. Rapid Commun.*, **15**, 955 (1994).
87. Asano, S., Aida, T. and Inoue, S., *J. Chem. Soc., Chem. Commun.*, **1985**, 1148 (1985).
88. Yasuda, T., Aida, T. and Inoue, S., *Macromolecules*, **16**, 1792 (1983).
89. Inoue, S. and Aida, T., *Makromol. Chem., Macromol. Symp.*, **73**, 1792 (1983).
90. Le Borgne, A., Vincens, V., Jouglard, M. and Spassky, N., *Makromol. Chem., Macromol. Symp.*, **73**, 37 (1993).
91. Feng, X. D., Song, C. X. and Chen, W. Y., *J. Polym. Sci., Polym. Lett. Ed.*, **21**, 593 (1983).
92. Trofimoff, L. R., Aida, T. and Inoue, S., *Chem. Lett.*, **1987**, 991 (1987).
93. Inoue, S., Kitamura, K. and Tsuruta, T., *Makromol. Chem.*, **126**, 250 (1969).
94. Hsieh, L. H., *J. Macromol. Sci. – Chem. A*, **7**, 1526 (1973).
95. Kuran, W. and Nietsochowski, A., *Polym. Bull.*, **2**, 411 (1980).
96. Kuran, W. and Nietsochowski, A., *J. Macromol. Sci. – Chem. A*, **15**, 1567 (1981).
97. Aida, T. and Inoue, S., *J. Am. Chem. Soc.*, **107**, 1358 (1985).
98. Aida, T., Sanuki, S. and Inoue, S., *Macromolecules*, **18**, 1049 (1985).
99. Kuran, W. and Listoś, T., *Makromol. Chem.*, **193**, 945 (1992).
100. Kricheldorf, H. R., Jenssen, J. and Kreiser-Saunders, I., *Makromol. Chem.*, **192**, 2391 (1991).
101. Furukawa, J., Saegusa, T., Fujii, H., Kawasaki, A., Imai, H. and Fujii, Y., *Makromol. Chem.*, **37**, 149 (1960).
102. Natta, G., Mazzanti, G., Corradini, P. and Bassi, I. W., *Makromol. Chem.*, **37**, 156 (1960).
103. Natta, G., Corradini, P. and Bassi, I. W., *J. Polym. Sci.*, **51**, 505 (1960).
104. Kawai, W., *Bull. Chem. Soc. Jpn*, **35**, 516 (1962).
105. Yilmaz, O., Usanmaz, A. and Alyürük, K., *Eur. Polym. J.*, **28**, 1351 (1992).
106. Natta, G., Mazzanti, G., Pregaglia, G. F., Binaghia, M. and Peraldo, M., *J. Am. Chem. Soc.*, **82**, 4742 (1960).
107. Yamashita, Y. and Nanumoto, S., *Makromol. Chem.*, **58**, 244 (1962).
108. Inoue, S., Koinuma, H. and Tsuruta, T., *Makromol. Chem.*, **130**, 210 (1969).
109. Inoue, S., Koinuma, H. and Tsuruta, T., *J. Polym. Sci. B*, **7**, 287 (1969).
110. Kuran, W., Pasynkiewicz, S., Skupińska, J. and Rokicki, A., *Makromol. Chem.*, **177**, 11 (1976).
111. Kuran, W. and Listoś, T., *Macromol. Chem. Phys.*, **195**, 977 (1994).

112. Koinuma, H. and Hirai, H., *Makromol. Chem.*, **178**, 241 (1977).
113. Aida, T., Ishikawa, M. and Inoue, S., *Macromolecules*, **19**, 8 (1986).
114. Tsuruta, T., *Pure Appl. Chem.*, **53**, 1745 (1986).
115. Kuran, W. and Listoś, T., *Polish J. Chem.*, **68**, 643 (1994).
116. Price, C. C. and Spector, R., *J. Am. Chem. Soc.*, **88**, 4171 (1966).
117. Vandenberg, E. J., *J. Polym. Sci. A-1*, **7**, 525 (1969).
118. Tani, H., Oguni, N. and Watanabe, S., *J. Polym. Sci., B*, **6**, 577 (1968).
119. Hirano, T., Khanh, P. H. and Tsuruta, T., *Makromol. Chem.*, **153**, 331 (1972).
120. Hasebe, S. and Tsuruta, T., *Makromol. Chem.*, **188**, 1403 (1987).
121. Watanabe, Y., Yasuda, T., Aida, T. and Inoue, S., *Macromolecules*, **25**, 1396 (1992).
122. Oguni, N. and Hyoda, J., *Macromolecules*, **13**, 1687 (1980).
123. Coulon, C., Spassky, N. and Sigwalt, P., *Polymer*, **17**, 821 (1976).
124. Kuran, W. and Listoś, T., *Polish J. Chem.*, **68**, 1071 (1994).
125. Van der Kerk, G. J. M., *Pure Appl. Chem.*, **30**, 389 (1972).
126. Farina, M., *Macromol. Symp.*, **89**, 489 (1995).
127. Bovey, F. A. and Tiers, G. V. D., *J. Polym. Sci.*, **44**, 173 (1960).
128. Natta, G., *Atti Accad. Naz. Lincei, Cl. Sci. Fis. Mat. Nat.*, **4**, 61 (1955).
129. Natta, G. and Corradini, P., *Atti Accad. Naz. Lincei, Cl. Sci. Fis. Mat. Nat.*, **4**, 73 (1955).
130. Staudinger, H., Ashdown, A., Brunner, M. Bruson, H. A. and Wehrli, S., *Helv. Chim. Acta*, **12**, 934 (1929).
131. Huggins, M. L., *J. Am. Chem. Soc.*, **66**, 1991 (1944).
132. Schildknecht, C. E., Gross, S. T., Davidson, H. R., Lambert, J. M. and Zoss, A. O., *Ind. Eng. Chem.*, **40**, 2104 (1948).
133. Frisch, H. L., Schuerch, C. and Szwarc, M. J., *J. Polym. Sci.*, **11**, 559 (1953).
134. Brandrup, J. 'Stereochemical Definitions and Notations Relating to Polymers, in *Polymer Handbook*, Wiley-Interscience, John Wiley & Sons, New York, 1989, pp. I/43–61.
135. Mislow, K. and Siegel, J., *J. Am. Chem. Soc.*, **106**, 3319 (1984).
136. Glenz, W., *Kunststoffe*, **76**, 834 (1986).
137. Natta, G., Crespi, G., Valvassori, A. and Sartori, G., *Rubber Chem. Technol.*, **36**, 1583 (1963).
138. Baldwin, F. P. and Ver Strate, G., *Rubber Chem. Technol.*, **45**, 709 (1972).
139. Heggs, T. G., in *Block Copolymers*, Wiley-Interscience, John Wiley & Sons, New York, 1973, pp. 513–523.
140. Griffith, J. H. and Ranby, B. G., *J. Polym. Sci.*, **44**, 369 (1960).
141. Natta, G., Allegra, G., Bassi, I. W., Corradini, P. and Ganis, P., *Makromol. Chem.*, **58**, 242 (1962).
142. Eleuterio, H. S., *J. Macromol. Sci. – Chem. A*, **28**, 907 (1991).
143. Ivin, K. J., Rooney, J. J., Steweart, C. D., Green, M. L. H. and Mahtab, R., *J. Chem. Soc., Chem. Commun.*, **1978**, 604 (1978).
144. Feast, W. J., 'Metathesis Polymerization: Applications', in *Comprehensive Polymer Science*, Pergamon Press, Oxford, 1989, Vol. 4, pp. 135–142.
145. Natta, G., Pasquon, I., Corradini, P., Peraldo, M., Perogaro, M. and Zambelli, A., *Atti Accad. Naz. Lincei, Cl. Sci. Fis. Mat. Nat.*, **28**, 539 (1960).
146. Natta, G., Pasquon, I. and Zambelli, A., *J. Am. Chem. Soc.*, **84**, 1488 (1962).
147. Galli, P. and Haylock, J. C., *Makromol. Chem., Macromol. Symp.*, **63**, 19 (1992).
148. Galli, P., *Macromol. Symp.*, **89**, 13 (1995).
149. Covezzi, M., *Macromol. Symp.*, **89**, 577 (1995).
150. Karol, J. F., *Macromol. Symp.*, **89**, 563 (1995).
151. Anon, *Modern Plastics Int.*, **1995**, 12 (1995) May.

152. Anon, *Eur. Plastics News*, **1995**, 22 (1995) May.
153. Anon, *Eur. Plastics News*, **1995**, 23–27 (1995) June.
154. Anon, *Modern Plastics Int.*, **1996**, 16 (1996) April.
155. Guidetti, G. P., Busi, P., Giulianelli, I. and Zannetti, R., *Eur. Polym. J.*, **19**, 757 (1983).
156. Avella, M., Martuscelli, E., Volpe, G. D., Segre, A., Rossi, E. and Simonazzi, T., *Makromol. Chem.*, **187**, 1927 (1986).
157. Tsutsui, T., Mizuno, A. and Kashiwa, N., *Polymer*, **30**, 428 (1989).
158. Kaminsky, W., Külper, K., Brintzinger, H. H. and Wild, F. R. W. P., *Angew. Chem., Int. Ed. Engl.*, **24**, 507 (1985).
159. Ewen, J. A., Jones, R. L., Razavi, A. and Ferrara, J. D., *J. Am. Chem. Soc.*, **110**, 6255 (1988).
160. Cheng, H. N. and Ewen, J. A., *Makromol. Chem.*, **190**, 19341 (1989).
161. Mallin, D. T., Rausch, M. D., Lin, Y. -G., Dong, S. and Chien, J. C. W., *J. Am. Chem. Soc.*, **112**, 2030 (1990).
162. Chien, J. C. W., *Makromol. Chem., Macromol. Symp.*, **63**, 209 (1992).
163. Coates, G. W. and Waymouth, R. M., *Science*, **267**, 217 (1995).
164. Kaminsky, W., Bark, A. and Arndt, M., *Makromol. Chem., Macromol. Symp.*, **47**, 83 (1991).
165. Kaminsky, W. and Noll, A., *Polym. Bull.*, **31**, 175 (1993).
166. Kaminsky, W. and Bark, A., *Polym. Int.*, **28**, 251 (1992).
167. Cherdron, H. and Brekner, M., *Macromol. Symp.*, **89**, 543 (1995).
168. Anon, *Modern Plastics Int.*, **1993**, 8 (1993) May.
169. Kaminsky, W. and Spiehl, R., *Makromol. Chem.*, **190**, 515 (1989).
170. Kaminsky, W. and Möller-Lindenhof, N., *Bull. Soc. Chim. Belg.*, **99**, 103 (1990).
171. Collins, S. and Kelly, W. M., *Macromolecules*, **25**, 233 (1992).
172. Kelly, W. M., Taylor, N. J. and Collins, S., *Macromolecules*, **27**, 4477 (1994).
173. Resconi, L. and Waymouth, R. M., *J. Am. Chem. Soc.*, **112**, 4953 (1990).
174. Mogstad, A. L. and Waymouth, R. M., *Macromolecules*, **25**, 2282 (1992).
175. Kesti, M. R. and Waymouth, R. M. *J. Am. Chem. Soc.*, **114**, 3565 (1992).
176. Coates, G. W. and Waymouth, R. M., *J. Am. Chem. Soc.*, **115**, 91 (1993).
177. Ishihara, N., Seimiya, T., Kuramoto, M. and Uoi, M., *Macromolecules*, **19**, 2464 (1986).
178. Pellecchia, C., Longo, P., Proto, A. and Zambelli, A., *Makromol. Chem., Rapid Commun.*, **13**, 265 (1992).
179. Ready, T. E., Day, R. O., Chien, J. C. W. and Rausch, M. D., *Macromolecules*, **26**, 5822 (1993).
180. Zambelli, A., Pellecchia, C. and Proto, A., *Macromol. Symp.*, **89**, 373 (1995).
181. Ishihara, N., *Macromol. Symp.*, **89**, 553 (1995).
182. Lee, D.-H., Yoon, K. -B., Lee, E. -H., Noh, S. -H., Byun, G. -G. and Lee, C. -S., *Macromol. Rapid Commun.*, **16**, 265 (1995).
183. Oliva, L., Longo, P., Grassi, A. and Zambelli, A., *Makromol. Chem., Rapid Commun.*, **11**, 519 (1990).
184. Yasuda, H., Yamamoto, H., Yokota, K., Miyake, S. and Nakamura, A., *J. Am. Chem. Soc.*, **114**, 4908 (1992).
185. Yasuda, H., Furo, N., Yamamoto, H., Nakamura, A., Miyake, S. and Kibino, N., *Macromolecules*, **25**, 5115 (1992).
186. Yasuda, H., Yamamoto, H., Yamashita, M., Yokota, K., Nakamura, A., Miyake, S., Kai, Y. and Tanehisa, N., *Macromolecules*, **26**, 7134 (1993).
187. Collins, S. and Ward, D. A., *J. Am. Chem. Soc.*, **114**, 5460 (1992).
188. Giardello, M. A., Yamamoto, Y., Brard, L. and Marks, T. J., *J. Am. Chem. Soc.*, **117**, 3276 (1995).

189. Kesti, M. R., Coates, G. W. and Waymouth, R. M., *J. Am. Chem. Soc.*, **114**, 9679 (1992).
190. Chung, T. C., *Macromol. Symp.*, **89**, 151 (1995).
191. Natta, G., Pino, P., Mazzanti, G. and Giannini, U., *J. Am. Chem. Soc.*, **79**, 2975 (1957).
192. Breslow, D. S. and Newburg, N. R., *J. Am. Chem. Soc.*, **79**, 5072 (1957).
193. Andresen, A., Cordes, H. G., Herwig, J., Kaminsky, W., Merck, A., Mottweiler, R., Pein, J., Sinn, H. and Vollmer, H. J., *Angew. Chem., Int. Ed. Engl.*, **15**, 630 (1976).
194. Sinn, H. and Kaminsky, W., *Adv. Organomet. Chem.*, **18**, 99 (1980).
195. Sinn, H., Kaminsky, W., Vollmer, H. J. and Woldt, R., *Angew. Chem., Int. Ed. Engl.*, **19**, 396 (1980).
196. Kaminsky, W., Miri, M., Sinn, H. and Woldt, R., *Makromol. Chem., Rapid Commun.*, **4**, 417 (1983).
197. Wild, F. R. W. P., Zsolnai, R., Huttner, G. and Brintzinger, H. H., *J. Organomet. Chem.*, **232**, 233 (1985).
198. Wild, F. R. W. P., Zsolnai, R., Huttner, G. and Brintzinger, H. H., *J. Organomet. Chem.*, **288**, 63 (1985).
199. Ewen, J. A., Haspeslagh, L., Atwood, J. L. and Zhang, H. J., *J. Am. Chem. Soc.*, **109**, 6544 (1987).
200. Pino, P., Cioni, P. and Wei, J., *J. Am. Chem. Soc.*, **109**, 6189 (1987).
201. Pino, P. and Galimberti, M., *J. Organomet. Chem.*, **370**, 1 (1989).
202. Waymouth, R. M. and Pino, P., *J. Am. Chem. Soc.*, **112**, 4911 (1990).
203. Longo, P., Grassi, A., Pellecchia, C. and Zambelli, A., *Macromolecules*, **20**, 1015 (1987).
204. Grassi, A., Zambelli, A., Resconi, L., Albizzati, E. and Mazzocchi, R., *Macromolecules*, **21**, 617 (1988).
205. Kaminaka, M. and Soga, K., *Makromol. Chem., Rapid. Commun.*, **12**, 367 (1991).
206. Soga, K. and Kaminaka, M., *Makromol. Chem.*, **194**, 1745 (1993).
207. Chien, J. C. W. and He, D., *J. Polym. Sci., A, Polym. Chem.*, **29**, 1603 (1991).
208. Kaminsky, W. and Renner, F., *Makromol. Chem., Rapid Commun.*, **14**, 239 (1993).
209. Soka, K., Kim, H. J. and Shiono, T., *Macromol. Rapid Commun.*, **15**, 139 (1994).
210. Antberg, M., Dolle, V., Haftka, S., Rohrmann, J., Spaleck, W., Winter, A. and Zimmermann, H. J., *Makromol. Chem., Macromol. Symp.*, **48/49**, 333 (1991).
211. Richardson, K., *Chem. in Britain*, **30**, 87 (1994).
212. Anon, *Modern Plastics Int.*, **1994**, 29 (1994) December.
213. Anon, *Modern Plastics Int.*, **1993**, 12 (1993) November.
214. Anon, *Eur. Plastics News*, **1995**, 11 (1995) October.
215. Price, C. C., Osgan, M., Hughes, R. E. and Shambelan, C., *J. Am. Chem. Soc.*, **78**, 690 (1956).
216. Inoue, S., 'Metalloporphyrin Catalysts for Control Polymerisation', in *Catalysis in Polymer Synthesis*, ACS Symp. Ser. 496, Washington, DC, 1992, pp. 194–204.
217. Vincens, V., Le Borgne, A. and Spassky, N., 'Oligomerization of Oxiranes with Aluminium Complexes as Initiators', in *Catalysis in Polymer Synthesis*, ACS Symp. Ser. 496, Washington, DC, 1992, pp. 205–214.
218. Ishimori, M., Higawara, T., Tsuruta, T., Kai, Y., Yasuoka, N. and Kasai, N., *Bull. Chem. Soc. Jpn*, **49**, 1165 (1976).
219. Ishimori, M., Higawara, T. and Tsuruta, T., *Makromol. Chem.*, **179**, 2337 (1978).
220. Higawara, T., Ishimori, M. and Tsuruta, T., *Makromol. Chem.*, **182**, 501 (1978).
221. Hasegawa, H., Miki, K., Tanaka, N., Kasai, N., Ishimori, M., Heki, T. and Tsuruta, T., *Makromol. Chem., Rapid Commun.*, **3**, 947 (1982).

222. Kageyama, H., Kai, Y., Kasai, N., Suzuki, C., Yoshino, N. and Tsuruta, T., *Makromol. Chem. Rapid Commun.*, **5**, 89 (1984).
223. Vandenberg, E. J., 'Catalysis: a Key to Advances in Applied Polymer Science', in *Catalysis in Polymer Synthesis*, ACS Symp. Ser. 496, Washington, DC, 1992, pp. 2–23.
224. Owens, K. and Kyllingstad, V. L., 'Polyethers', in *Kirk-Othmer Encyclopedia of Chemical Technology*, Wiley-Interscience, John Wiley & Sons, New York, 1993, Vol. 8, pp. 1079–1093.
225. Meissner, B., Schätz, M. and Brajko, V., 'Synthetic Rubbers: Epichlorohydrin Rubber', in *Studies in Polymer Science*. Vol. 1. *Elastomers and Rubber Compounding Materials*, Elsevier, Amsterdam, 1989, pp. 274–278.
226. Kuran, W., 'Poly(propylene Carbonate)', in *The Polymeric Materials Encyclopedia*, CRC Press, Inc., Boca Raton, 1996, Vol. 9, pp. 6623–6630.

Further Reading

Vandenberg, E. J., 'Coordinate Polymerization', in *Encyclopedia of Polymer Science and Engineering*, Wiley-Interscience, John Wiley & Sons, New York, 1986, Vol. 4, pp. 174–175.

Pino, P., Giannini, U. and Porri, L., 'Insertion Polymerization', in *Encyclopedia of Polymer Science and Engineering*, Wiley-Interscience, John Wiley & Sons, New York, 1987, Vol. 8, pp. 147–220.

Pasquon, I., Porri, L. and Giannini, U., 'Stereoregular Linear Polymers', in *Encyclopedia of Polymer Science and Engineering*, Wiley-Interscience, John Wiley & Sons, New York, 1989, Vol. 15, pp. 632–733.

Gavens, P. D., Bottrill, M., Kelland, J. W. and McMeeking, J., 'Ziegler–Natta Catalysis', in *Comprehensive Organometallic Chemistry*, Pergamon Press, Oxford, 1982, Vol. 3, pp. 475–547.

McDaniel, M. P., 'Supported Chromium Catalysts for Ethylene Polymerization', *Adv. Catal.*, **33**, 47–98 (1985).

Tait, P. J. T., 'Monoalkene Polymerization: Ziegler–Natta and Transition Metal Catalysts', in *Comprehensive Polymer Science*, Pergamon Press, Oxford, 1989, Vol. 4, pp. 1–25.

Tait, P. J. T. and Watkins, N. D., 'Monoalkene Polymerization: Mechanisms', in *Comprehensive Polymer Science*, Pergamon Press, Oxford, 1989, Vol. 4, pp. 533–573.

Tait, P. J. T. and Berry, I. G., 'Monoalkene Polymerization: Copolymerisation', in *Comprehensive Polymer Science*, Pergamon Press, Oxford, 1989, Vol. 4, pp. 575–584.

Corradini, P., Busico, V. and Guerra, G., 'Monoalkene Polymerization: Stereospecificity', in *Comprehensive Polymer Science*, Pergamon Press, Oxford, 1989, Vol. 4, pp. 29–50.

Kaminsky, W., 'Polyolefins', in *Handbook of Polymer Synthesis*, M. Dekker, New York, 1992, Part A, pp. 1–76.

Starkweather, H., 'Olefin–Carbon Monoxide Polymers', in *Encyclopedia of Polymer Science and Engineering*, Wiley-Interscience, John Wiley & Sons, New York, 1987, Vol. 10, pp. 369–373.

Porri, L. and Giarrusso, A., 'Conjugated Diene Polymerization', in *Comprehensive Polymer Science*, Pergamon Press, Oxford, 1989, Vol. 4, pp. 53–108.

Natta, G. and Porri, L., 'Diene Elastomers', in *Polymer Chemistry of Synthetic Elastomers*, Wiley-Interscience, John Wiley & Sons, New York, 1969, Part 2, pp. 597–678.

Natta, G., Valvassori, A. and Sartori, G., 'Ethylene–Propylene Rubbers', in *Polymer Chemistry of Synthetic Elastomers*, Wiley-Interscience, John Wiley & Sons, New York, 1969, Part 2, pp. 679–702.
Kaminsky, W., 'Polymeric Dienes', in *Handbook of Polymer Synthesis*, M. Dekker, New York, 1992, Part A, pp. 385–431.
Natta, G. and Dall'Asta, G., 'Elastomers from Cyclic Olefins', in *Polymer Chemistry of Synthetic Elastomers*, Wiley-Interscience, John Wiley & Sons, New York, 1969, Part 2, pp. 703–726.
Ivin, K. J., 'Metathesis Polymerization', in *Encyclopedia of Polymer Science and Engineering*, Wiley-Interscience, John Wiley & Sons, New York, 1987, Vol. 9, pp. 634–668.
Amass, A. J., 'Metathesis Polymerization: Chemistry', in *Comprehensive Polymer Science*, Pergamon Press, Oxford, 1989, Vol. 4, pp. 109–134.
Kricheldorf, H. R., 'Metathesis Polymerization of Cycloolefins', in *Handbook of Polymer Synthesis*, M. Dekker, New York, 1992, Part A, pp. 433–479.
Schrock, R. R., 'Ring-opening Metathesis Polymerization', in *Ring-opening Polymerization*, Carl Hanser Publishers, Munich, 1993, pp. 129–156.
Bolognesi, A., Catellani, M. and Destri, S., 'Polymerization of Acetylene', in *Comprehensive Polymer Science*, Pergamon Press, Oxford, 1989, Vol. 4, pp. 143–153.
Costa, G., 'Polymerization of Mono- and Di-substituted Acetylenes', in *Comprehensive Polymer Science*, Pergamon Press, Oxford, 1989, Vol. 4, pp. 155–161.
Brintzinger, H. H., Fischer, D., Mülhaupt, R., Rieger, B. and Waymouth, R. M., 'Stereospecific Olefin Polymerisation with Chiral Metallocene Catalysts', *Angew. Chem., Int. Ed. Engl.*, **34**, 1143–1170 (1995).
Kaminsky, W., 'Zirconocene Catalysts for Olefin Polymerization', *Catal. Today*, **20**, 257–271 (1994).
Arndt, M. and Kaminsky, W., 'Microstructure of Poly(cycloolefin)s Produced by Metallocene/Methylaluminoxane (mao) Catalysts'. *Macromol. Symp.*, **97**, 225–246 (1995).
Wünsch, J. R., 'Syndiotaktisches Polystyrol,' in *Polystyrol*, Carl Hanser Publishers, Munich, 1995, pp. 82–104.
Tsuruta, T., 'Stereoselective and Asymmetric-selective (or Stereoelective) Polymerisations', *J. Polym. Sci., D*, **6**, 179–250 (1972).
Tsuruta, T. and Kawakami, Y., 'Anionic Ring-opening Polymerization: Stereospecificity for Epoxides, Episulfides and Lactones', in *Comprehensive Polymer Science*, Pergamon Press, Oxford, 1989, Vol. 3, pp. 489–500.
Jérôme, R. and Teyssié, P., 'Anionic Ring-opening Polymerization: Lactones', in *Comprehensive Polymer Science*, Pergamon Press, Oxford, 1989, Vol. 3, pp. 501–510.
Inoue, S. and Aida, T., 'Anionic Ring-opening Polymerization: Copolymerization', in *Comprehensive Polymer Science*, Pergamon Press, Oxford, 1989, Vol. 3, pp. 533–569.
Vogl, O., 'Aldehyde Polymers', in *Encyclopedia of Polymer Science and Engineering*, Wiley-Interscience, John Wiley & Sons, New York, Vol. 1, pp. 623–643.
Inoue, S. and Aida, T., 'Catalysts for Living and Immortal Polymerization', in *Ring-opening Polymerization*, Carl Hanser Publishers, Munich, 1993, pp. 197–215.
Kuran, W., 'Coordination Polymerization of Heterocyclic and Heterounsaturated Monomers', *Prog. Polym. Sci.*, **23**, 919–992 (1996).
Randall, J. C., 'Microstructure', in *Encyclopedia of Polymer Science and Engineering*, Wiley-Interscience, John Wiley & Sons, New York, 1987, Vol. 9, pp. 795–824.
Jenkins, A. D. and Loening K. L., 'Nomenclature', in *Comprehensive Polymer Science*, Pergamon Press, Oxford, 1989, Vol. 1, pp. 13–54.
Ciardelli, F., 'Optically Active Polymers,' in *Encyclopedia of Polymer Science and Engineering*, Wiley-Interscience, John Wiley & Sons, New York, 1987, Vol. 10, pp. 463–493.

Ciardelli, F., 'Optical Activity', in *Comprehensive Polymer Science*, Pergamon Press, Oxford, 1989, Vol. 1, pp. 561–571.

Bovey, F. A., 'Isomerism in Polymer Chains', in *High Resolution NMR of Macromolecules*, Academic Press, New York, 1972, pp. 53–64.

Bovey, F. A., 'The Study of the Conformations of Vinyl Polymer Chains and Model Compounds by NMR', in *High Resolution NMR of Macromolecules*, Academic Press, New York, 1972, pp. 182–204.

Bovey, F. A. and Kwei, T. K., 'Microstructure and Chain Conformation of Macromolecules', in *Macromolecules – an Introduction to Polymer Science*, Academic Press, New York, 1979, pp. 207–271.

Corradini, P., 'The Impact of the Discovery of Stereoregular Polymers in Macromolecular Science', *Macromol. Symp.*, **89**, 1–11 (1995).

Wilke, G., 'Karl Ziegler – the Last Alchemist', in *Ziegler Catalysts*, Springer-Verlag, Berlin, 1995, pp. 1–14.

Mülhaupt, R., 'Novel Polyolefin Materials and Processes: Overview and Prospects', in *Ziegler Catalysts*, Springer-Verlag, Berlin, 1995, pp. 35–55.

Problems

1. What are the differences in the coordination of hydrocarbon unsaturated monomers and heteroatom-containing cyclic and acyclic monomers? Give examples of catalysts and monomers.
2. What are the differences in the coordination of carbon monoxide and carbon dioxide? Give examples of catalysts.
3. Give an example of coordination insertion polymerisation. Write down the scheme.
4. Give an example of coordination polymerisation via transalkylidenation. Write down the scheme.
5. Give an example of coordination polymerisation involving nucleophilic attack on the growing polymer chain terminated by the coordinating monomer. Write down the scheme.
6. Indicate monomers that might be polymerised to stereoregular polymers:

$CH_2{=}CHR$ $\qquad$ $H_2C{=}O$

$CH_2{=}CR_2$ $\qquad$ $RCH{=}O$

$CH_2{=}CH{-}CH{=}CH_2$ $\qquad$ 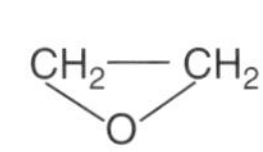

$CH_2{=}CH{-}CH_2{-}CH{=}CH_2$ $\qquad$ 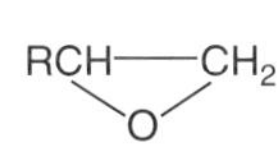

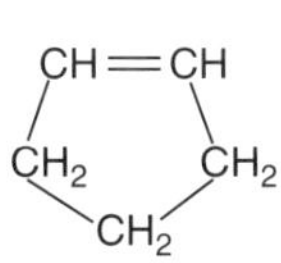 $\qquad$ 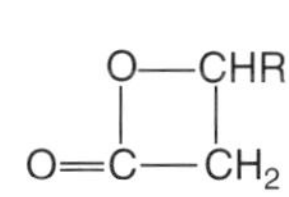

7. Distinguish between stereoregular and tactic polymers. Give examples of each.
8. Distinguish between stereospecific and stereoselective polymerisations. Give examples of each.
9. Why do stereoregular polymers differ from non-stereoregular polymers in their physicochemical and mechanical properties?
10. Name important polymers produced in coordination polymerisation processes and characterise their structure.

3 Coordination Polymerisation of Olefins

3.1 Stereoisomerism of α-Olefin Polymers

Stereoisomerism of regioregular head-to-tail α-olefin polymers, $\lbrack CH(R)-CH_2 \rbrack_n$, which lies in the appearance of isotactic and syndiotactic poly(α-olefin)s [1–3], can be illustrated by two different pictorial representations (Figure 3.1). The upper representations are projections adapted from Fischer notations by rotation through 90°. One should recall here that Fischer projections are displayed only vertically. The horizontal line in the rotated projections corresponds to the polymer main chain. This line represents carbon–carbon bonds directed behind the plane of the page by imagining the rotation of each carbon–carbon bond in the polymer chain into the eclipsed conformation (the upper Newman representation) as opposed to the staggered conformation (the lower Newman representation) that actually exists (as zigzag–helix). The vertical lines in the projections adapted from Fischer notations represent bonds coming behind or in front of the plane, which are the bonds between the stereogenic main chain tertiary carbon atom and the α-olefin alkyl substituent. Vertical lines can appear on both sides of the horizontal line, above and/or below it, depending on the relative configuration of the tertiary carbon atom. The use of rotated Fischer projection corresponds to the common practice of using horizontal lines to denote polymer backbone bonds, but this does not give any immediately visual impression of the zigzag chain.

The lower projections on the left in Figure 3.1 are planar zigzag projections of the backbone carbon atoms of hypothetical fully extended poly(α-olefin) chain segments in a staggered conformation (the lower Newman representation on the right). The alkyl substituents at the tertiary carbon atoms can appear in front of and/or behind the plane in which the polymer backbone lies, depending on the relative configuration of these carbon atoms. Planar projections of hypothetical extended zigzag chains occasionally give a clearer indication of the three-dimensional arrangement.

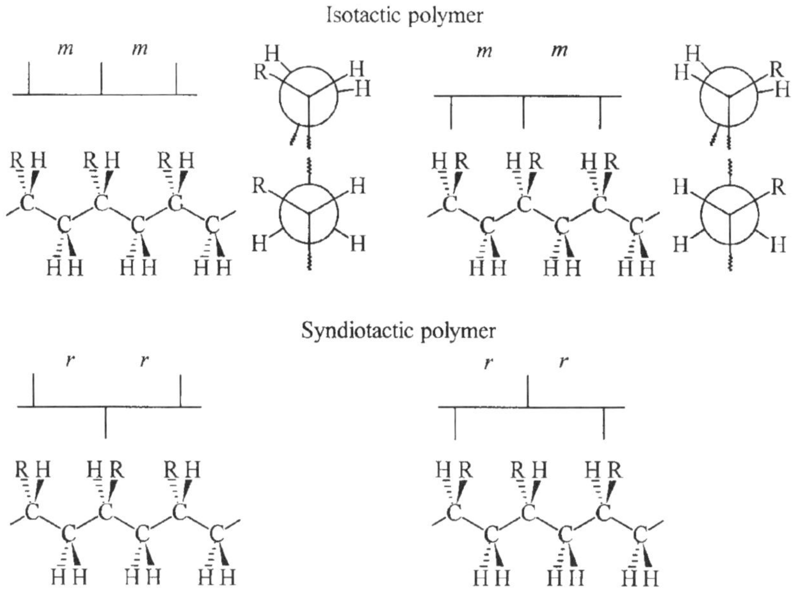

Figure 3.1 Stereoisomerism of poly(α-olefin)s {-[-CH(R)–CH_2-]$_n$-}. Isotactic and syndiotactic polymers. Reproduced by permission from Elsevier Science, from Ref. 1

If the successive vertical lines are located on the same side of the horizontal line, above or below it, in the adapted Fischer projection, or if the successive alkyl substituents appear in front of or behind the plane of the extended polymer backbone in the flat zigzag projection (Figure 3.1), then an isotactic poly(α-olefin) structure occurs. These two representations show that the isotactic polymer contains tertiary carbon atoms of the same relative configuration. One can see that an isotactic order in the polymer chain is characterised by the appearance of stereodiads of the same relative configuration, which are referred to as *meso* (*m*) diads. Therefore, an isotactic poly (α-olefin) contains only *m* diads (or isotactic *mm* triads) along its chains, irrespective of the absolute configuration of their stereogenic tertiary carbon atoms. Note in this connection that an isotactic poly (α-olefin) consists of isotactic macromolecules with tertiary carbon atoms of the opposed configuration.

If the successive vertical lines are located alternately on the opposite sides of the horizontal line, above and below it, in the adapted Fischer projection, or if the successive alkyl substituents appear alternately in front of and behind the plane of the extended polymer backbone in the flat zigzag projection (Figure 3.1), then a syndiotactic poly(α-olefin) structure occurs. These two representations show that the syndiotactic polymer contains neighbouring tertiary carbon atoms of opposite relative configurations. The syndiotactic poly(α-olefin) chain is characterised by the appearance of stereodiads of the opposed relative

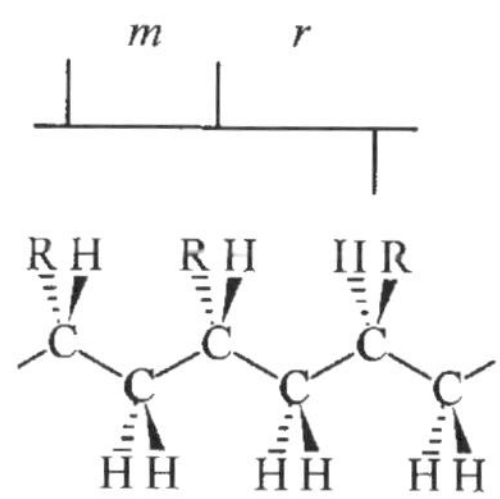

Figure 3.2 Atactic poly(α-olefin)

configuration, which are referred to as *racemo* (*r*) diads. Thus, a syndiotactic poly(α-olefin) contains only *r* diads (or syndiotactic *rr* triads) along its chains.

If there is no regular arrangement of vertical lines in the adapted Fischer projection or alkyl substituents in the planar zigzag projection (Figure 3.2), then an atactic structure occurs. Atactic poly(α-olefin)s are characterized by the appearance in their chains of *m* diads and *r* diads in equal amounts. These diads constitute heterotactic *mr* triads and *rm* triads as well as isotactic *mm* triads and syndiotactic *rr* triads which also appear in equal amounts.

The term 'tacticity of poly(α-olefin)s' was introduced by Natta, and the polymerisation leading to tactic polymers was referred to as 'stereospecific polymerisation'. An atactic polymer is formed if the monomers fall into place with complete stereochemical randomness. In other words, the expected statistical distribution of stereogenic tertiary carbon atoms of opposed relative configurations is formed and the polymer as a whole is optically inactive. If, upon polymerisation, all the repeating units take up the same relative configuration, then the polymer is isotactic. In principle it is optically active, but among synthetic polymers optical activity is not usually observed because, in most cases, for every macromolecule produced with a given relative configuration a macromolecule of the opposed relative configuration is also formed. Nevertheless, an isotactic polymer possesses regularity of structure, and therefore crystallisability, which confers upon it quite different properties from those of the atactic form.

Isotactic poly(α-olefin)s crystallise in a helical conformation, and, in the case of polypropylene, with three units per turn [4,5]. Isotactic polypropylene has a melting point of 175°C and does not dissolve in boiling *n*-heptane [6,7]. Note that, depending upon the configuration of the tertiary carbon atom of the polymer main chains, the poly(α-olefin) helices will be characterised by right-handedness or left-handedness. It should be mentioned that the helical structure of the poly(α-olefin) chain per se is sufficient for the appearance of chirality of such a macromolecule [8]. Figure 3.3 presents the helical conformation of chains of isotactic poly(α-olefin)s in the crystalline state (with three units per turn – the case of polypropylene) [5].

In syndiotactic polypropylene the relative configuration of stereogenic tertiary carbon atoms alternates and thus, as in the isotactic polymer, provides regularity of structure although in a different form. Syndiotactic polypropylene

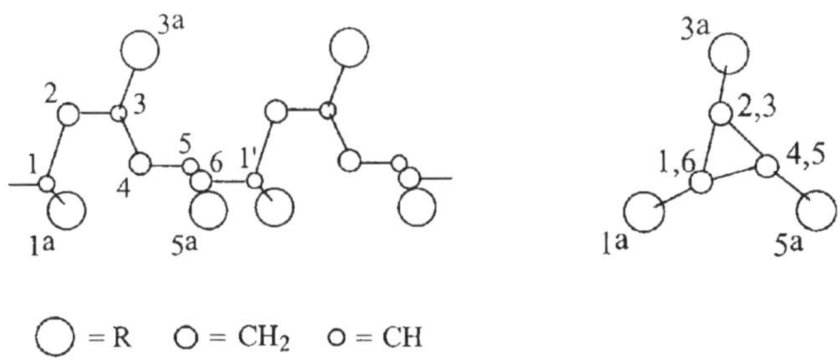

Figure 3.3 Schematic presentation of the conformation of chains of isotactic poly(α-olefin)s, such as polypropylene, in the crystalline state

crystallises in a binary helix conformation with four units per turn (Figure 3.4) [9,10]. The polymer chain conformation changes as a consequence of the changing relative configuration at successive tertiary carbon atoms along this chain. One should emphasise the appearance of right-handed and left-handed binary helices of syndiotactic polypropylene. A crystalline modification of syndiotactic polypropylene in which the macromolecules have a planar zigzag conformation is also known [11]. Syndiotactic polypropylene has a melting point of 135°C and dissolves in boiling *n*-hexane.

Considering the correlation between the polymer properties and its tacticity, which is connected with relative configurations of stereogenic tertiary carbon atoms of poly(α-olefin)s, one can see the usefulness of operating with relative configurations of stereogenic carbon atoms in order to characterise the polymer microstructure. It is to be added that only in the early 1960s, after Bovey's pioneering work [12–14], did nuclear magnetic resonance spectroscopy allow a stereochemical analysis of short segments (diads, triads, etc.), i.e. determination of the polymer microtacticity, and *m–r* nomenclature become popular among polymer chemists. Note that this nomenclature concerns the designation of the relative configurations of consecutive, but not necessarily contiguous, constitu-

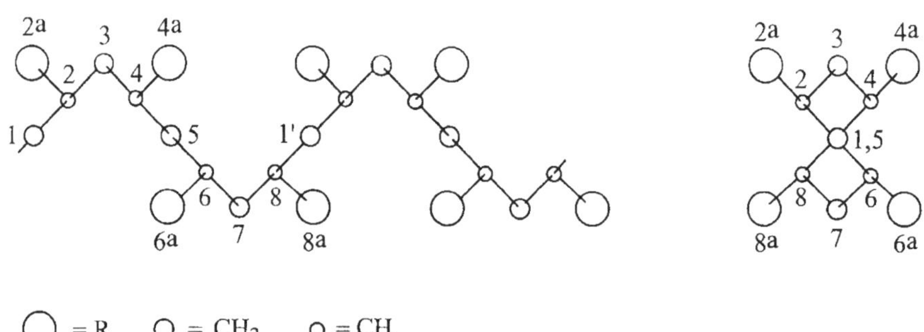

Figure 3.4 Schematic presentation of the conformation of chains of syndiotactic polypropylene in the crystalline state

tionally equivalent carbon atoms that have a symmetrically constituted group (if any). The term '*racemo*' is introduced as a logical prefix for the designation of an arrangement that is analogous to racemic in the sense defined above. It is unfortunate that the meaning of the term 'racemic' current in organic chemistry is not directly applicable to polymers, but the use of the prefix *racemo* should not cause confusion because of the special context [15].

It is worth noting that the absolute configuration of the tertiary carbon atoms in isotactic or syndiotactic polypropylene molecules has, in principle, no practical significance for the polymer properties, which might result from the presence of these carbon atoms in the main chain of the macromolecule.

Whether or not the stereogenic tertiary carbon atoms in isotactic and syndiotactic polypropylene are chirotopic depends on the model chosen for their representation [16]. Stereochemical analysis, which followed Natta's discovery, involved two models of the polymer chain. Stereoregular polymers were considered, on the one hand, as the extrapolation towards high molecular weights of well-studied organic molecules, e.g. such as trihydroxyglutaric acid:

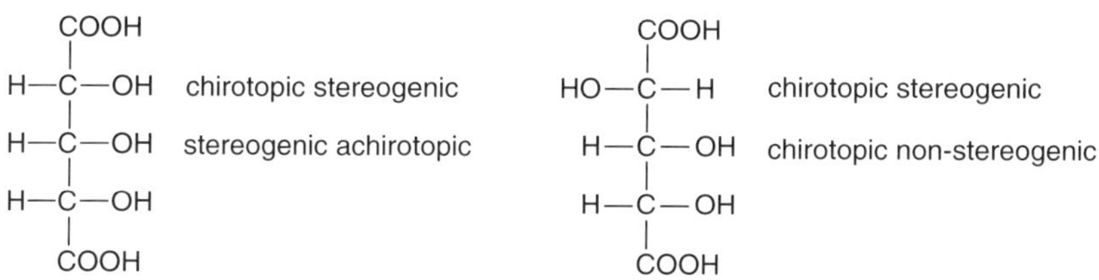

two stereoisomers of trihydroxyglutaric acid

Therefore, a stereochemical model consisting of a finite chain with equal end-groups was used [17]. Isotactic polypropylene is described as a *meso*-compound with the two half-molecules being a mirror image of each other.

On the other hand, the concept of the infinite chain used by crystallographers does not take into account the effect of the end-groups but is only concerned with the symmetry elements of the chain itself. An attempt to reconcile the two concepts was made [18]; small cyclic molecules are suitable models for linear macromolecules and have the advantage that they can be studied by means of point symmetry, much better known than the line symmetry required for infinite chains. From this point of view, 1,3,5-*cis*-trimethylcyclohexane is a better configurational model of isotactic polypropylene than 3,5,7-trimethylnonane and its homologues:

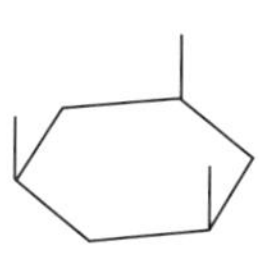

1,3,5-*cis*-trimethylcyclohexane

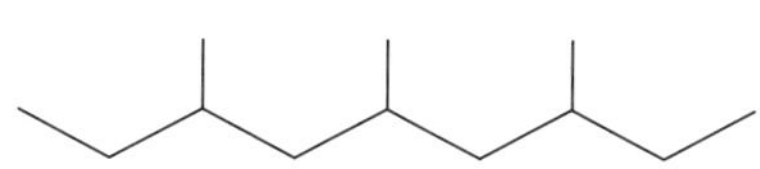

3,5.7-trimethylnonane

The cyclic model is unsuitable for local structures but should be applied only to long chains in such a way that translational symmetry is not lost.

For the infinite chain model (and the equivalent cyclic model), both polypropylene stereoisomers, isotactic and syndiotactic polymers, are achiral and the tertiary carbon atoms are achirotopic. For the model with equal end-groups, the isotactic polymer is non-chiral (*meso*), but the tertiary carbon atoms are chirotopic (except the central one, if the number of stereogenic carbon atoms is odd). However, in such a case, optical activity, or another indicator of chirality, decreases with molecular weight and vanishes at very high molecular weights, behaviour that was referred to as cryptochiral [19]. The infinite chain model (or equivalently, the cyclic model) is the most effective test for evaluating the chirality of a given structure: its response is very simple (yes/no), without any

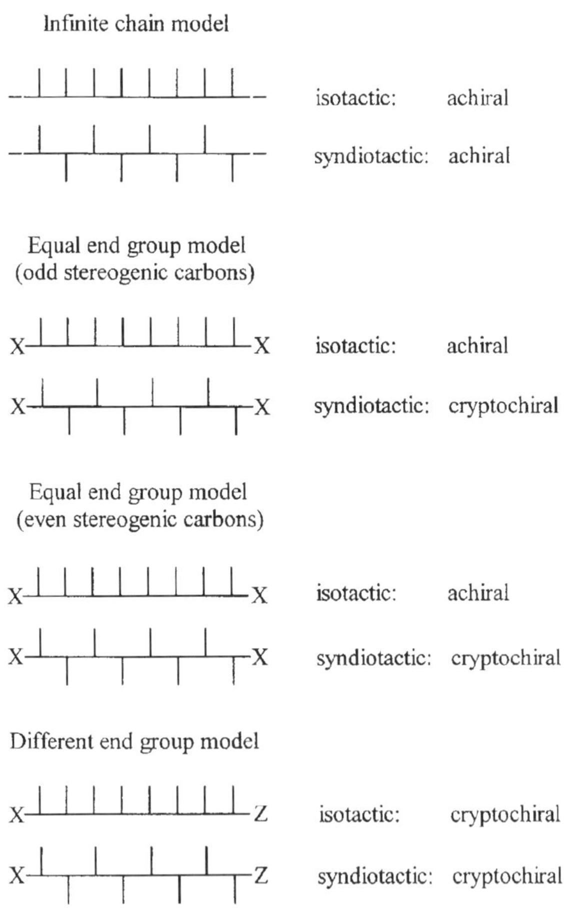

Figure 3.5 Schematic models for the stereoisomerism of polypropylene

implication of cryptochirality. The *meso* character of isotactic polypropylene is not an intrinsic feature of the polymer but is related to the polymer model system [16]. The analysis of syndiotactic polypropylene with the finite chain model is rather questionable, since with an even number of stereogenic carbon atoms in the chain of this polymer the structure is chiral (or better still cryptochiral), while with an odd number it is achiral [16].

The model with different end-groups is not realistic: in all cases isotactic polypropylene and syndiotactic polypropylene are chiral, or more precisely, their structure is cryptochiral. This model is to be chosen when examining oligomers, and especially when studying the polymerisation mechanism where the structure of the reactive chain end is of extreme importance [16].

Important conclusions that can be drawn from considerations of polypropylene stereoisomerisms using various models are shown in Figure 3.5 [16].

Stereoregular poly(α-olefin)s can also be characterised by the appearance of longer or shorter isotactic sequences with the same relative configuration of tertiary carbon atoms in the polymer chain (isoblock polymer) as well as with configurations alternating with each other (stereoblock polymer). Also isosyndioblock polymers are known. Such stereoregular polymers are presented schematically in Figure 3.6.

Isoblock polypropylene is characterised by the appearance of stereosequences of *m* diads bridged by pairs of *r* diads [20,21]. Such polypropylene, containing a small number of long blocks, is usually referred to as isotactic polypropylene. Note that the polymer chains in this case contain helix sequences of the same handedness (right or left) in the chain, which is due to the same relative configuration of tertiary carbon atoms in these sequences. Stereoblock polypropylene, which is also called block-isotactic polypropylene,

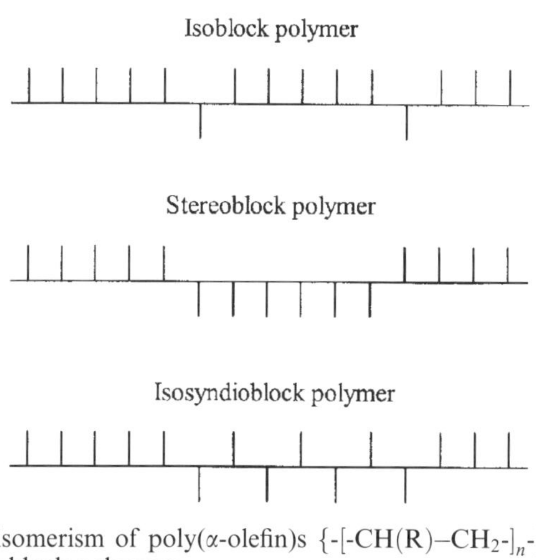

Figure 3.6 Stereoisomerism of poly(α-olefin)s {-[-CH(R)–CH$_2$-]$_n$-}. Isoblock, stereoblock and isosyndioblock polymers

consists of short stereosequences of *m* diads bridged by isolated *r* diads [22,23]. Chains of this polymer contain helix sequences alternating their handedness (from right to left, and vice versa) along the chain, which is due to alterations of the relative configuration of tertiary carbon atoms in these sequences.

Polymers containing two distinct series of tertiary carbon atoms alternating with each other are called hemitactic: in one series the arrangement of the substituents is exactly defined, whereas in the other series their arrangement is random [24]. In terms of microtacticity, a hemiisotactic poly(α-olefin) consists of a statistical succession of *mm* and *rr* triads, whereas a hemisyndiotactic polymer consists of statistical successions of non-overlapping *mr* and *rm* triads (Figure 3.7) [2]. Note in this connection that the isosyndioblock polymer (Figure 3.7) is a particular case of the hemi-isotactic polymer (Figure 3.7).

Being acquainted with the structure of poly(α-olefin)s, one may reasonably explain some of the differences in their physicochemical properties. For example, isotactic polypropylene, the chains of which in the helical conformation can be closely packed, has rather a high density (0.92–0.94 g/cm^3) and melting point (175°C) and is insoluble in low-boiling aliphatic hydrocarbons at boiling point. Syndiotactic polypropylene, consisting of chains in the form of binary helices, which cannot be packed so closely as in the previous case, has a density of 0.89–0.91 g/cm^3 and a melting point of 135°C, which is 40 k lower from that of isotactic polypropylene; syndiotactic polypropylene is also moderately soluble in

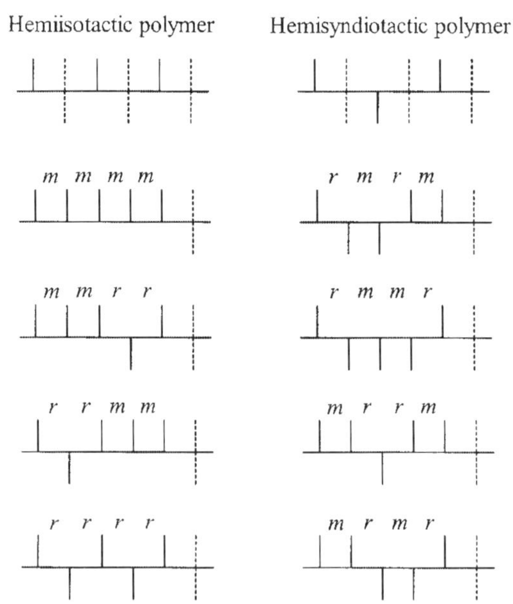

Figure 3.7 Stereoisomerism of poly(α-olefin)s {-[-CH(R)–CH_2-]$_n$-}. Hemiisotactic and hemisyndiotactic polymers

hydrocarbons at room temperature. However, atactic polypropylene is a waxy amorphous substance characterised by a density of 0.85–0.90 g/cm^3, easily soluble in hydrocarbons at room temperature. In turn, stereoblock polypropylene, the chains of which are characterised by the appearance of sequences with alternating handedness of their helical structure, cannot be closely packed. Because of this, the melting point of stereoblock polypropylene, characterised by a small number of long blocks, does not exceed 155 °C, and its solubility in hydrocarbon solvents is moderate. However, stereoblock polypropylene containing a large number of short blocks has a softening point not far above room temperature and is easily soluble in hydrocarbon solvents. Owing to such physical properties, which may be considered as opposite to the respective properties of isotactic polypropylene, stereoblock (or block-isotactic) polypropylene is also referred to as anisotactic polypropylene. Note in this connection that the microstructure of stereoblock polypropylenes consisting of very short blocks resembles that of hemitactic propylene polymers [25].

Ditactic polymers belong to the same group of stereoregular poly (α-olefin)s comprising tactic polymers. A ditactic polymer is a tactic polymer that contains two sites of defined stereoisomerism in the main chain of the configurational base unit. The most suitable example of ditactic polymers are diisotactic and disyndiotactic poly(deutero-α-olefin)s formed in the stereospecific polymerisation of some deutero-α-olefins, such as *cis*-and *trans*-1-deuteropropene [26,27] as well as *cis*-and *trans*-1,2,3,3,3-pentadeuteropropene [22]. Opening of the double bond of a monomer, e.g. CH(R)=CH(R′), where R = CH_3 and R′ = ^{2}H, can yield monomeric units comprising two pairs of enantiomers originated from two isomers, *cis* and *trans*, i.e. monomeric units of four types:

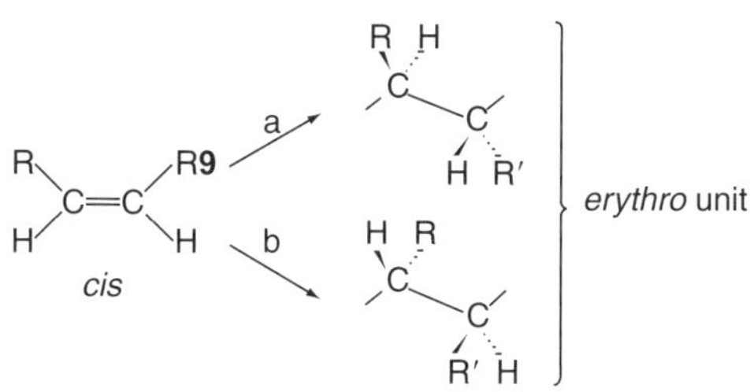

(1)

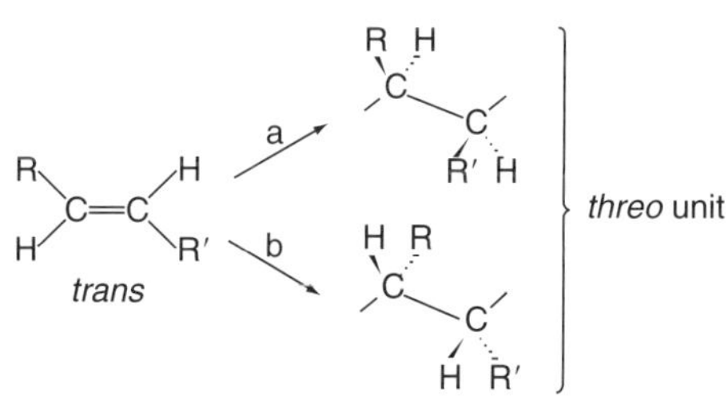

(2)

The relative configurations at two contiguous carbon atoms in these monomer units are designated by the prefix *erythro* or *threo*.

The head-to-tail arrangement of units of the same type yields diisotactic polymers. A diisotactic polymer is an isotactic polymer that contains two stereogenic, chirotopic or achirotopic, carbon atoms with defined stereochemistry in the main chain of the configurational base unit. A diisotactic polymer may be *erythro*-diisotactic, characterised by a succession of units with a configuration like that formed according to scheme (1a) or (1b), or *threo*-diisotactic, characterised by a succession of units with a configuration like that formed according to scheme (2a) or (2b). The relative configurations of the two stereogenic tertiary carbon atoms in the *erythro*-diisotactic polymers are the same, whereas they are opposed to each other in the *threo*-diisotactic polymer (Figure 3.8) [2,15]. Chains in diisotactic polymers in the crystalline state assume a helical conformation [2].

The regularly alternating head-to-tail arrangement of a pair of enantiomeric units formed according to schemes (1a) and (1b) or schemes (2a) and (2b) gives rise to disyndiotactic polymers. A disyndiotactic polymer is a syndiotactic polymer that contains two stereogenic, chirotopic or achirotopic, carbon atoms with defined stereochemistry in the main chain of the configurational base unit. Disyndiotactic polymers may be *erythro*-disyndiotactic and *threo*-disyndiotactic polymers in which the configurations at the two stereogenic tertiary carbon atoms (of the enchained monomer unit) are the opposite and the same respectively (Figure 3.8). However, a close insight into these two disyndiotactic structures shows them to be identical except for the difference in the end-groups. Therefore, from the practical point of view, considerations

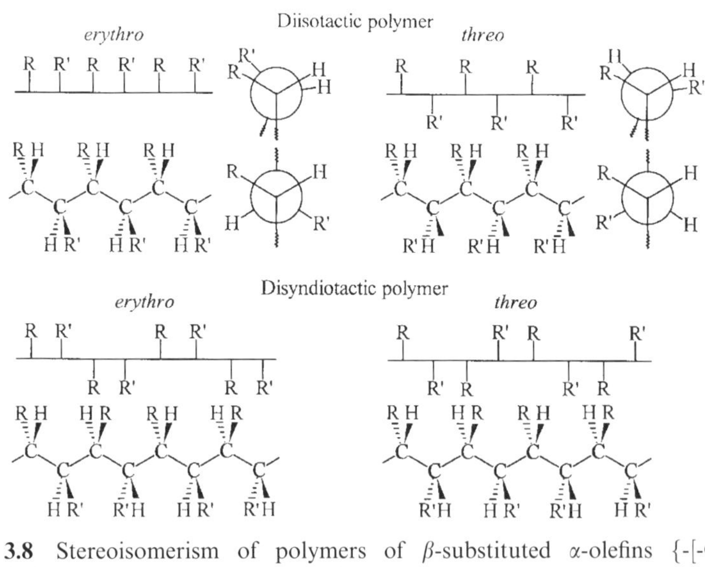

Figure 3.8 Stereoisomerism of polymers of β-substituted α-olefins {-[-CH(R)–CH(R′)-]$_n$-}. Diisotactic and disyndiotactic polymers

of the stereoregularity of disyndiotactic polymers may concern only one disyndiotactic structure [2,15].

It should also be mentioned that atactic polymers may be formed from monodeuteropropylene and pentadeuteropropylene, similarly to the case of the polymerisation of non-deuterated propylene. However, atactic structures with respect to one stereogenic tertiary carbon atom, with the other possibly arranged to form tactic polymer sequences, are not known. Also, ditactic polymers, which might be characterised by the appearance of two distinct series of tacticity as regards both stereogenic tertiary carbon atoms, i.e. the isotactic with respect to one stereogenic carbon atom and syndiotactic with respect to the other one, and vice versa, are not known.

3.2 Polymerisation Catalysts

Although it is common practice now to refer to olefin polymerisation catalysts as Ziegler–Natta catalysts, it is considered to be of more scientific value to divide coordination catalysts for ethylene and α-olefin polymerisations into four general types: Ziegler–Natta catalysts (heterogeneous and homogeneous); homogeneous catalysts without an organometallic or metal hydride activator; supported metallocene and related catalysts; and supported catalysts involving transition metal oxides such as Phillips catalysts. Such an arrangement of catalysts results from the classic definition of Ziegler–Natta catalysts. As will be appreciated, homogeneous Ziegler–Natta catalysts and homogeneous metallocene catalysts without an organometallic activator as well as heterogeneous supported Ziegler–Natta catalysts and supported metallocene catalysts overlap with each other to some extent.

3.2.1 Ziegler–Natta Catalysts

Ziegler–Natta catalysts are defined as the products formed in reactions of transition metal compounds of groups 4 to 8 (procatalysts, catalyst precursors) with organometallic compounds or metal hydrides of groups 1 to 4 (activators). These reactions are carried out in an inert medium and under inert (anaerobic) conditions:

Group 4–8 transition metal compound	+	Group 1–4 metal alkyl, aryl or hydride	⟶	Group 4–8 transition metal alkyl,aryl or hydride	(3)
(Procatalyst or catalyst precursor)		(Activator)		(Ziegler–Natta catalyst)	

The above definition is very broad and not all possible combinations of the catalyst precursor with the activator will result in the formation of active catalysts for the polymerisation of alkenes. However, a number of Ziegler–Natta catalysts, soluble or insoluble in an inert reaction medium, as well as not requiring any reaction medium as in a bulk liquid propylene medium or gas-phase processes, are reported in the literature [28].

In reporting a Ziegler–Natta catalyst, the kind of transition metal compound should not be omitted. Group 4–8 transition metal compounds, such as halides, oxyhalides, alkoxides, acetylacetonates, etc., have been used as catalyst precursors with activators such as alkyl derivatives or hydrides of group 1–4 metals. Titanium chlorides and triethylaluminium are most commonly applied for the preparation of heterogeneous catalysts in an aliphatic hydrocarbon medium. Also, vanadium oxychloride or acetylacetonate and dialkyaluminium chloride are often used for the preparation of homogeneous catalysts in an aliphatic hydrocarbon or an aromatic hydrocarbon medium.

With the sole exception of the random ethylene–propylene copolymers, for industrial applications heterogeneous catalysts have been used for alkene polymerisations. Ethylene–propylene statistical copolymerisation has been carried out using homogeneous vanadium-based catalysts [28].

On the other hand, the most investigated homogeneous Ziegler–Natta catalysts for alkene polymerisations have been obtained using group 4 metallocenes (especially zirconocenes) and alkylaluminoxanes (most often methylaluminoxanes) in an aromatic hydrocarbon such as toluene as the reaction medium [29,30]. These new catalysts are commonly referred to as single-site (or uniform-site) catalysts as well as metallocene catalysts. It is worth mentioning that some of the metallocenes can promote alkene polymerisations without needing to be used in a combination with an organometallic (or metal hydride) compound as a catalyst activator. Such metallocenes, which are used alone as single-component catalysts [31], or in combination with other compounds (but not organometallics or metal hydrides) [32–35], for alkene polymerisation cannot, by definition, be classified as Ziegler–Natta catalysts. Metallocene catalysts, whether or not they are referred to as Ziegler–Natta catalysts, promise a breakthrough that may in the future change the profile of important branches of industry providing polymers. The first alkene polymerisation processes involving single-site metallocene catalysts (on a silica or alumina support) have now been commercialised [36].

The type of solvent or diluent should be specified in reporting a Ziegler–Natta catalyst system. Alkene polymerisations are usually carried out in inert solvents, such as aliphatic or aromatic hydrocarbons (e.g. some gasoline fractions or toluene). The use of protic or aprotic polar solvents or diluents instead of the hydrocarbon polymerisation medium can drastically alter the reaction mechanism. This usually results in catalyst deactivation for alkene coordination polymerisation. Modern alkene polymerisation processes are carried out in a gas phase, using fluidised-bed catalysts, and in a liquid monomer as in the case of propylene polymerisation [28,37].

The classification of Ziegler–Natta catalysts requires consideration of criteria connected with their properties as catalysts, which might be applied for industrial processes of ethylene and α-olefin polymerisations, as well as criteria concerning the nature of their active sites and polymerisation mechanism.

Industrial and commercial and economic criteria have greatly influenced the development of more efficient and selective catalysts for ethylene and α-olefin polymerisation and copolymerisation. The most important problem was achieving a sufficiently high catalyst activity to eliminate from the production process all procedures of catalyst deactivation and operations of removal of catalyst residues from the polymer. Also of concern was the possibility of eliminating the hydrocarbon polymerisation medium from the process, the application of which was associated with the need for its regeneration, drying and storage as well as with the need for drying of the polymer. Increased yields of high molecular weight stereoregular polymer fraction and thus the elimination of waxy atactic and low molecular weight polymer deposits within the reactor are very important factors for α-olefin polymerisation. As far as propylene polymerisation processes are concerned, it was necessary to enhance the percentage of polypropylene fraction not extracted into boiling *n*-heptane, which is often referred to as the isotactic index [17]. Consequently, not all of the Ziegler–Natta catalysts, which are effective in ethylene polymerisation, can be used for propylene polymerisation, since in the latter case the polymerisation occurs with greater difficulty and there is an additional demand for polymerisation stereospecificity. Instead, practically all of the propylene polymerisation catalysts can usually be applied successfully for ethylene polymerisation. The catalyst effectiveness refers also to the possibility of better control of the molecular weight and molecular weight distribution of the polymer as well as its increased ease of processability. An important factor is the facility of control of the particle shape, size and porosity and bulk density of the polymer, which is crucial for the elimination of the pelletisation step from the last phase of the production process. Ultimately, catalyst effectiveness concerns the development of industrial processes of greater flexibility, thus also allowing the application of different types of catalyst and comonomer in the same reactor. The polymerisation process has to allow the catalyst to generate the broadest range of properties with the minimum constraints. An ideal catalyst should provide the following features: very high activity (low cost) and selectivity (specific product), control of polymer microstructure with respect to molecular weight, molecular weight distribution, chain shape and randomless microtacticity, control of polymer microstructure with respect to morphology (particle size, shape and porosity) and phase distribution and control of polymer properties (broad range of applications) [38].

These catalyst features, important from the industrial-economic viewpoint, are considered as criteria allowing Ziegler–Natta catalysts to be divided into five groups, commonly referred to as the first, second, third, fourth and metallocene (single-site) generations. Thus, the division of Ziegler–Natta catalysts into successive generations of enhancing effectiveness, which have been formed within various, even considerably interrelated, periods, deals only with heterogeneous catalysts that are used for industrial purposes. Catalysts ranked among the first four generations are active in the formation of an isotactic polymer in the case of propylene polymerisation [28,38].

It should be mentioned that many of the requirements necessary for the economic production of polyethylene and polypropylene have been achieved. However, catalysts of greater activity and of greater selectivity in the production of polymers and copolymers can be anticipated. This is of prime concern to alkene polymerisation processes in the presence of single-site metallocene catalysts. Such catalysts, undoubtedly of great scientific and commercial importance, have been developed on a large scale within recent years [29,30].

The classification of Ziegler–Natta catalysts is not only limited to factors connected with their industrial applications but also includes catalysts that have not been used in industrial polymerisation processes. Other criteria are adopted that concern the catalyst solubility or insolubility in the polymerisation system and the nature of the active species influencing the formation of polymers of various tacticity or atactic polymers in the case of α-olefin polymerisation. Therefore, Ziegler–Natta catalysts can be divided into heterogeneous (non-supported and supported) and homogeneous catalysts. Besides this, we distinguish among them catalysts based on transition metal compounds of the halide type (and related compounds) and activators such as metal alkyl, alkylmetal halide or metal hydride, catalysts based on group 4 metallocene and methylaluminoxane or another compound and catalysts consisting of group 3 metallocene, taking into account the nature of the catalyst active species of various stereospecificity in α-olefin polymerisations [28,29].

3.2.1.1 Heterogeneous Non-supported Catalysts

The discoveries by Ziegler [39] were concerned with ethylene polymerisation in the presence of an *in situ* prepared catalyst, which was obtained as a precipitate from the reaction of a soluble transition metal halide, especially $TiCl_4$, with a metal alkyl or alkyl metal halide, primarily $AlEt_3$ or $AlEt_2Cl$, the reaction being carried out at the activator to a procatalyst molar ratio not much greater than unity and in an inert hydrocarbon medium. The resulting insoluble catalyst, forming a suspension in a hydrocarbon medium, is based mostly on the amorphous brown β-$TiCl_3$ modification with a chain structure, which is formed in the $TiCl_4$–AlR_3 system by the following alkylation and reduction reactions:

$$TiCl_4 + AlR_3 \longrightarrow TiRCl_3 + AlR_2Cl \quad (4)$$

$$TiCl_4 + AlR_2Cl \longrightarrow TiRCl_3 + AlRCl_2 \quad (5)$$

$$TiCl_4 + AlRCl_2 \longrightarrow TiRCl_3 + AlCl_3 \quad (6)$$

$$\mathrm{TiRCl_3} \longrightarrow \frac{1}{x}\,(\mathrm{TiCl_3})_x + \mathrm{R}^{\cdot} \begin{cases} \text{recombination,} \\ \text{disproportionation,} \\ \text{etc.} \end{cases} \tag{7}$$

The precipitated brown β-$TiCl_3$ undergoes, in certain spots of its surface, an alkylation reaction with the organoaluminium compound present in the reaction system, which results in the formation of active species with a Ti–C bond (or with a Ti–H bond in the case of a metal hydride activator (in these spots. scheme (8) shows the activation of a catalyst precursor such as $TiCl_3$ with AlR_3 as the catalyst activator:

$$(\mathrm{TiCl_3})_{y+z} + z\mathrm{AlR_3} \longrightarrow (\mathrm{TiCl_3})_y \cdot (\mathrm{TiRCl_2})_z + z\mathrm{AlR_2Cl} \tag{8}$$

Note that only ca 1% of the titanium atoms introduced with $TiCl_3$ into the system give rise to the formation of surface active sites (since most of the titanium atoms remain inside the solid $TiCl_3$ particles) [40]. However, such active sites exhibit rather low stereospecificity in propylene polymerisation. The activity and stereospecificity of catalysts based on the β-$TiCl_3$ modification also depend on the type of alkylaluminium compound used as the activator. The application of triethylaluminium leads to a catalyst of much higher activity but of much lower stereospecificity, and on account of this diethylaluminium chloride is used for the polymerisation of propylene and other α-olefins, while triethylaluminium (and also triisobutylaluminium) is used for ethylene polymerisation [28].

It was Natta, however, who realised that polypropylene isotacticity was directly connected to the uniformity of the catalyst surface, and used preformed solid crystalline transition metal halides in lower valence states for catalyst preparation. Catalysts obtained from such hydrocarbon insoluble precursors, especially crystalline violet forms of $TiCl_3$ with a layered structure (α-, γ-, and δ-modifications), by their activation with alkylaluminium compounds, such as $AlEt_3$ and $AlEt_2Cl$, appeared to polymerise propylene to an isotactic high molecular weight polymer [41,42]. The procatalyst activation in this case [according to scheme (8)] also exploits only ca 1% of the amount of $TiCl_3$ introduced to form active sites [40], but such sites are characterised by relatively high stereospecificity. The isotactic index of polypropylene obtained with catalysts based on crystalline violet modifications of $TiCl_3$ as procatalysts and alkylaluminium compounds such as triethylaluminium as activators is in the range 85–90% [28].

The α–$TiCl_3$ form can be prepared by the reduction of $TiCl_4$ with hydrogen at above 400 °C, usually at 800 °C. Reduction with metallic aluminium in a hydrocarbon medium at 250 °C produces α-$TiCl_3 . \frac{1}{3}AlCl_3$ in which $AlCl_3$ is cocrystallised with α-$TiCl_3$ [28,43]:

$$3\mathrm{TiCl_4} + \mathrm{Al} \longrightarrow 3\mathrm{TiCl_3} \cdot \tfrac{1}{3}\mathrm{AlCl_3} \tag{9}$$

The β-form of $TiCl_3$ is obtained by reducing $TiCl_4$ with aluminium metal or alkylaluminium compounds at lower temperature (usually in the range 0–100 °C) or by reduction with hydrogen. The γ-$TiCl_3$ modification is obtained by heating the β-form at a temperature higher than 150 °C or reducing $TiCl_4$ at temperatures in the range 150–200 °C; reduction of $TiCl_4$ with aluminium metal yields γ-$TiCl_3 \cdot \frac{1}{3}AlCl_3$. The δ-$TiCl_3$ form is obtained by prolonged grinding of α- or β-$TiCl_3$. The use of α-and β-forms of $TiCl_3$ as procatalysts results in the formation of catalysts with lower activities than those obtained with the use of δ-$TiCl_3$, while the β-$TiCl_3$-based catalysts produce polypropylene, which is essentially amorphous [7,28].

Most non-supported catalysts used for alkene slurry polymerisation processes are obtained using the δ-$TiCl_3 \cdot \frac{1}{3}AlCl_3$ solid solution. It was found [44] that ball milling of the product obtained from the reduction of $TiCl_4$ by aluminium metal or aluminium alkyls led to a more active catalyst than pure α-$TiCl_3$. Furthermore, it was revealed [45] that during the milling process both the α-and the γ-forms of $TiCl_3$ were converted to the δ-form, which displayed double-layer stacking. Catalysts of the above type are referred to as first-generation Ziegler–Natta catalysts. Most first-generation commercial catalysts are based on a precursor of the δ-$TiCl_3 \cdot xAlCl_3$ solid solution type and have been dry milled and heat treated. These catalysts have specific surface areas in the range 10–40 m^2/g. Their productivity is a few kilograms of polyethylene or polypropylene (with an isotactic index of 88–93%) per gram Ti/h atm [28,43]. The production processes, involving slurry-type polymerisation, were, however, complicated and expensive and, because of their complexity, not very versatile. Although the catalyst productivities could be much higher for ethylene than for propylene, it was necessary to extract the catalyst residues in order to reduce the quantities of Ti and Cl in the product to acceptable levels.

As regards higher α-olefins (with the exclusion of α,α-disubstituted olefins such as isobutene and its homologues), they undergo isospecific polymerisation in the presence of first generation Ziegler–Natta catalysts, giving rise to high molecular weight crystalline polymers [46,47].

Improvements in both the polypropylene isotacticity and yield were the driving forces for continued research in the new Ziegler–Natta catalyst system. The catalyst effectiveness, in particular with respect to isospecificity in α-olefin polymerisations, can be improved by the addition of certain Lewis bases to the polymerisation system, usually in an amount in the range 0.01–0.1 mol^{-}% with respect to the amount of $TiCl_3$ used as the catalyst precursor. A very wide range of Lewis bases such as ethers, esters, ketones, amines, amides, phosphines and other organophosphorus compounds, polymeric derivatives, etc., can be used and are often ball milled with the δ-$TiCl_3 \cdot xAlCl_3$ procatalyst. The mode of addition of the Lewis base and the conditions of concentration and temperature applied during the preparation stages as well as the intrinsic nature of the Lewis base are all important in determining the catalyst activity. While the use of Lewis bases often gives rise to catalysts characterised by higher stereospecificities for α-olefin polymerisation, they often adversely affect catalyst activity in terms of productivity. However, this is not always the case, and catalysts exhibiting both higher stereospecificities and higher activities can be obtained by the judicious addition of an appropriate Lewis base. Such catalysts, which contain a Lewis base as the third component, are classified as second-generation Ziegler–Natta catalysts. Their productivity usually reaches a dozen or so kilograms of polyethylene or polypropylene (with an isotactic index of 92–97%) per gram Ti/h atm [28,43].

The reasons for an enhancement in catalyst stereospecificity and activity for α-olefin polymerisations in the presence of Lewis bases added to the polymerisation system are complex but may be formulated, in a simplified way, as follows. The increased stereospecificity (isospecificity) may be due to the easier complexation of the more accessible non-stereospecific catalyst sites by the Lewis base by comparison with the less readily accessible more hindered isospecific sites; the latter sites remain unblocked by the Lewis base. Thus, the fraction of isospecific catalyst sites present on the surface of the $TiCl_3$–AlR_3–Lewis base catalyst increases with respect to the overall amount of active sites (aspecific sites are blocked), which gives rise to an enhancement of catalyst isospecificity for α-olefin polymerisation. One can ascertain the lowering of the overall catalyst activity owing to the partial blocking of some surface catalyst active sites by the Lewis base. However, the overall catalyst surface area increases (up to ca 150 m^2/g or more) owing to the reagglomeration of particles of the catalyst precursor such as $TiCl_3 \cdot xAlCl_3$ under the influence of the added Lewis base, which results in an increase in the concentration of catalyst active sites and thus in an enhancement of catalyst activity. The reagglomeration of procatalyst particles occurs by the preferential complexation and extraction of $AlCl_3$ by the Lewis base, leaving behind a porous and weakly bonded $TiCl_3$ matrix. The above simplified considerations of the role of Lewis bases used as the third component of Ziegler–Natta catalysts may become more complete if other factors, such as the transformation of aspecific sites by partial complexation with a Lewis base into isospecific sites, the lowering of the reduction ability of aluminium alkyl activators by their complexation with a Lewis base, etc., are also taken into account [28].

The improvements in polypropylene isotacticity and yield, however, were not sufficient to reduce catalyst residues to a level that would enable the polymer de-ashing process (catalyst removal – reducing the quantities of Ti and Cl) to be eliminated. Industrial processes of slurry polymerisations of ethylene and propylene with non-supported Ziegler–Natta catalysts of the first and second generations incorporate some operations needed for obtaining products suitable for sale. These operations comprise catalyst removal from the polymer, amorphous (atactic) polypropylene separation, unreacted monomer purification, polymer pelletisation and effluent treatment [37,38]. It must be emphasised, however, that the development of second-generation Ziegler–Natta catalysts represented an important intermediary stage for obtaining better catalysts of higher effectiveness.

Considerations of first- and second-generation Ziegler–Natta catalysts should be concluded by stating that catalysts for propylene polymerisation are characterised by the presence of their titanium component almost exclusively as halides – titanium chlorides. However, in the case of ethylene, highly active polymerisation catalysts can also be obtained using titanium alkoxides or alkoxychlorides as the catalyst precursors, but alkylaluminium chlorides, and not aluminium trialkyls, must be used as the activators. Heterogeneous catalysts of this type, e.g. $Ti(OBu)_4$–$AlEt_2Cl$, produce polyethylene, which is characterised by an ultrahigh molecular weight (in the range $3–6 \times 10^6$) [48].

3.2.1.2 Heterogeneous Supported Catalysts

The large-scale production of polyolefins has been enhanced by the development of very high-activity supported Ziegler–Natta catalysts. Higher catalyst activities mean lower production costs owing to possibilities of eliminating some operations from the production technology, or introducing new cheap production technologies (e.g. gas-phase processes).

Since Natta *et al.* [40] demonstrated that only a small percentage of the Ti atoms in the non-supported catalysts (typically less than 1% in the first-generation catalysts) was active for alkene polymerisation, it was realised that much of the procatalyst mass acted simply as a support for the active sites formed by activation [scheme (8)], and that significant improvement could be made by depositing the active titanium species on a support whose residues, unlike those to $TiCl_3$, would be inert and not detrimental to the properties of the polymer [49]. Thus, it was evident that higher catalyst activities were likely to be achieved through the use of transition metal compounds supported on appropriate matrices. Early attempts to support $TiCl_4$ directly on silica, alumina or magnesia did not lead to a sufficient increase in catalyst productivity [50]. The first useful high-yield catalyst involved Mg(OH)Cl as a support [43]. Over the years, a wide range of supported catalysts have been successfully developed and used for the industrial polymerisation of ethylene, and more recently propylene. In the case of ethylene polymerisation, supported catalysts have been implemented into practice relatively easily, but propylene, for which catalyst activities are generally lower and which has the additional criterion of stereoregulation, has proved to be a much more difficult problem. Various metal oxides, chlorides, oxychlorides and alkoxides such as Al_2O_3, SiO_2, $Al_2O_3{\cdot}SiO_2$, MgO, ZnO, ZrO_2, TiO_2, ThO, Mg(OH)Cl, $Mg(OEt)_2$ and $MgCl_2$ have been applied as catalyst supports. The function of the catalyst support may also be played by organic polymers [28,51]. However, a variety of magnesium compounds have been used most successfully as supports [28,50]; pre-eminent among these is $MgCl_2$ or reaction mixtures that can produce this compound, at least on the support surface. The massive increase in activity of $MgCl_2$-supported catalysts has been claimed to be due to an increase in the percentage of Ti atoms forming active sites (approaching nearly 100%) and not to a significant increase in reaction rate at the active site (intrinsic catalyst activity) [52]. This was the third-generation catalyst system, and the discovery of magnesium chloride in the active form, as an ideal support for the fixation of $TiCl_4$ and its derivatives, opened a new era in the field of Ziegler–Natta catalysed polymerisation, from both the industrial and scientific points of view. During the 1980s, active $MgCl_2$-supported catalysts brought about revolutionary developments in the production of polyolefins compared with the first-and second-generation Ziegler–Natta catalysts. The introduction of the catalyst support concept significantly increased the complexity of the catalyst construction but also significantly reduced the complexity of industrial olefin polymerisation processes. The active polymerisation centres could now be dispersed throughout the support surface, making them all essentially accessible to the polymerised monomer. Since all the titanium chloride molecules were now available to take part in the polymerisation, fewer were needed. Consequently, the yield of polymer per gram of titanium was so high that no extraction procedures were necessary in order to produce a viable commercial resin. Elimination of the de-ashing part of the industrial process had obvious economic and environmental advantages and resulted in the explosive growth of polyolefins in the global plastics marketplace [49].

Supports used for obtaining Ziegler–Natta catalysts can differ essentially from one another. Some of the supports may contain reactive surface groups (such as hydroxyl groups present in specially prepared metal oxides) while others do not contain such reactive functional groups (such as pure anhydrous metal chlorides). Therefore, the term 'supported catalyst' is used in a very wide sense. Supported catalysts comprise not only systems in which the transition metal compound is linked to the support by means of a chemical covalent bond but also systems in which the transition metal atom may occupy a position in a lattice structure, or where complexation, absorption or even occlusion may take place [28]. The transition metal may also be anchored to the support via a Lewis base; in such a case the metal complexes the base, which is coordinatively fixed on the support surface [53,54].

Supported precursors for Ziegler–Natta catalysts may be obtained, depending on the kind of support, in two ways: by treatment of the support containing surface hydroxyl groups with a transition metal compound with chemical covalent bond formation, and by the treatment of a magnesium alkoxide or magnesium chloride support with a Lewis base and transition metal compound with coordination bond formation.

The use of supports containing hydroxyl groups such as alumina, silica, Mg(OH)Cl, etc., for chemical fixing of the transition metal compound has been widespread since the early 1960s. Heat treatment (calcination) of such supports can control the number and type of surface hydroxyl groups and indirectly the amount and distribution of transition metal atoms anchored to the surface. The most commonly used Ziegler–Natta catalyst of this type is obtained by application of the precursor yielded from the reaction of Mg(OH)Cl support and $TiCl_4$ [28,51]:

$$—Mg—OH + TiCl_4 \longrightarrow —Mg—OTiCl_3 + HCl \qquad (10)$$

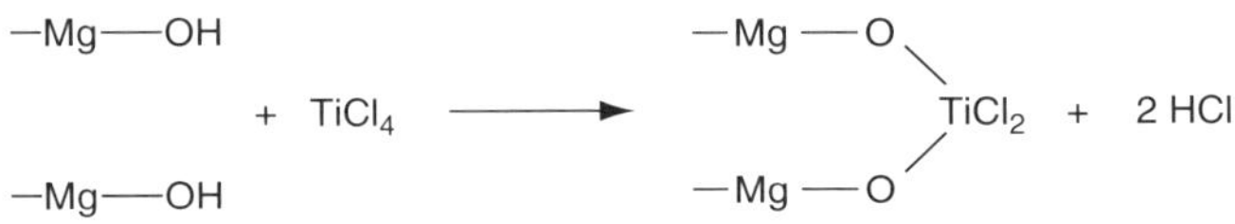

(11)

The obtained supported precursors are then subjected to activation by trialkylaluminium, which can be shown schematically for the $MgOTiCl_3$ species as follows:

$$—Mg—O—TiCl_3 \xrightarrow[-AlR_2Cl]{+AlR_3} —Mg—O—TiRCl_2$$

$$\xrightarrow[-R^{\cdot}]{} —Mg—O—TiCl_2 \xrightarrow[-AlR_2Cl]{+AlR_3} —Mg—O—TiRCl \qquad (12)$$

Supporting the titanium catalyst component has been claimed to lessen the tendency of titanium active site to be reduced by the activator [55].

Many of the catalysts obtained from such precursors and trialkylaluminium as the activator exhibit high activity in the polymerisation of ethylene, but are not, however, useful for the polymerisation of propylene and higher α-olefins.

Catalysts can also be produced by using organic polymers as supports. An interesting series of catalysts has been obtained using a polymer that is insoluble in the polymerisation solvent but that undergoes swelling by it and will allow diffusion of the polymerised material from the catalyst into the polymerisation medium [56,57]. A suitable support is a terpolymer made from ethylene, propylene and non-conjugated diene, on to which poly(methacrylic acid) has been grafted. The high-temperature solution polymerisation of olefins with Ziegler–Natta catalysts on such polymeric supports produced polymers containing practically no catalyst residue (the catalyst is separated from the obtained polymer solution by filtration) [51].

Catalysts that exhibit very high activity in the polymerisation of ethylene (producing ca 100 kg of polyethylene per gram Ti/h atm) have been obtained using magnesium alkoxides, especially magnesium ethoxide, for the reaction with $TiCl_4$. The original structure of the alkoxide is usually destroyed during the reaction, and new species of increased surface area are produced (ca $60\ m^2/g$ for the $Mg(OEt)_2/TiCl_4$ catalyst precursor). These hollow-type structures would account for the complete disintegration of the catalyst particles (catalyst obtained with the $AlEt_3$ activator) observed during polymerisation.

Final catalyst residues were reported to be smaller than 0.0005 μm in diameter [28].

Relatively high Al/Ti ratios are required for maximum activities of catalysts of the $Mg(OR)_2/TiCl_4$–$AlEt_3$ type; the use of Al/Ti molar ratios in the range 80–100:1 was reported [28]. However, it is worth noting in this connection that this is not surprising when one considers an almost equally high Al/Ti ratio but with respect only to the surface Ti atoms in Ziegler–Natta catalysts of the first and second generation.

The most commonly used supported catalysts, especially for propylene polymerisation, have resulted from the use of magnesium chloride as a support. Magnesium chloride has three distinct crystalline forms: α-, β-, and δ-$MgCl_2$. The most common one is the cubic close-packed α-$MgCl_2$. The hexagonal close-packed β-form is thermodynamically less stable. The δ-form is a rotationally disordered modification of β-$MgCl_2$ [58,59]. Magnesium chloride seems to appear suitable as a support for propylene polymerisation catalysts. The reason for this unique behaviour is believed to be due to the similarity in crystal structure between $MgCl_2$ and $TiCl_3$ [58,60]. In particular, δ-$MgCl_2$ resembles δ-$TiCl_3$, and the ionic radii of Ti^{4+} and Mg^{2+} are of 0.68 and 0.65 Å respectively [60]. Together, these facts suggest that $MgCl_2$ should have the ability to mimic the structure of the active titanium trichloride [61].

For a successful catalyst, $MgCl_2$ must first be converted to a form that can efficiently incorporate $TiCl_4$. This is done primarily by ball milling. Ball milling of $MgCl_2$ can be carried out with $TiCl_4$ alone (in the case of ethylene polymerisation catalysts) or with a Lewis base, usually an aromatic ester, e.g. ethyl benzoate, dibutyl phthalate, etc. Such a Lewis base is often referred to as an internal Lewis base (inside or inward Lewis base). During the ball-milling procedure in the presence of a Lewis base, the agglomerates (secondary crystallites) initially break up to yield primary crystallites, when the freshly cleaved surfaces are coated with the Lewis base, preventing reaggregation. The active titanium is incorporated on to the support by refluxing the $MgCl_2$/Lewis base coground mixture with $TiCl_4$, typically at 80–130 °C for a few hours. During treatment with $TiCl_4$, some of the Lewis base is replaced by $TiCl_4$. The above procedure requires a final washing with hydrocarbons in order to remove the excess $TiCl_4$ remaining unbound to the support (partially in the form of a complex with the Lewis base) [28,61]. Calculations indicated that the (100) and (110) faces of $MgCl_2$ are characterised by different Lewis acidities (the latter being more acidic). This has an effect on the ease of removal of the Lewis base from different faces of $MgCl_2$. A Lewis base would be more easily removed from the (100) face and $TiCl_4$ should be anchored there more readily but be inhibited somewhat from fixing to the (110) face. This potential site-blocking pathway is a proposed explanation for the improved stereospecificity of $MgCl_2$/Lewis base-supported catalysts [61–63].

Finally, most of the $MgCl_2$-supported Ziegler–Natta catalysts for propylene polymerisation also require treatment of the supported procatalyst with another Lewis base [referred to as the external (outside or outward) Lewis base] together with trialkylaluminium during the polymerisation phase in order to generate catalyst active sites [activation by AlR_3 (scheme (3)] capable of improved stereocontrol of the polymerisation. One should add that the Al/Ti molar ratio is usually in the range 300–400:1. Besides reducing the Ti valence and forming the active Ti–C bond, the activator, which is usually triethylaluminium, can react with the Lewis base incorporated in the catalyst. In turn, the external Lewis base added with the activator can interact both with the support and with $TiCl_4$ as well as with any byproducts arising in the system. Although the role of the external Lewis bases is complex, the following essential reactions seem to proceed in $MgCl_2$/internal Lewis base/$TiCl_4$–$AlEt_3$/external Lewis base systems: complex formation with trialkylaluminium, leading to a decrease in its reducing capacity with respect to the titanium entities, and deactivation of some specific centres by selective complex formation or even conversion of aspecific sites into isospecific sites. Benzoates, tolyloates, anisates and various amines have been successfully used as an external Lewis base in $MgCl_2$/ethyl benzoate/$TiCl_4$–$AlEt_3$/Lewis base catalysts, whereas various di-and trialkoxysilanes have been used in $MgCl_2$/dibutyl phthalate/$TiCl_4$–$AlEt_3$/Lewis base catalysts. The exact role of external Lewis bases is still unresolved but their presence is crucial, in most cases, for obtaining highly isospecific catalysts. However, an inverse correlation between stereospecificity and productivity is a characteristic feature of these catalysts. The productivity of commercial catalysts of this type is claimed to reach 500–1000 kg of polypropylene (with an isotactic index of ca 95 %) per gram Ti/h [28]. These catalysts are also efficient in the isospecific polymerisation of higher α-olefins (with the exception of α, α-disubstituted olefins).

A new class of super highly active catalysts based on an $MgCl_2$ support have been developed since the late 1970s. These catalysts are characterised by the appearance of surface-active species of practically one type. Their productivities are up to 2000 kg of polypropylene (with an isotactic index of 95–98 %) per gram Ti/h. In addition, lower Al/Ti ratios, i.e. ratios of the trialkylaluminium activator to titanium species, are required, and the usual reciprocal correlation between polypropylene isotactic index and yield is virtually absent [28,38].

Polymer particles obtained in propylene polymerisation often replicate the shape of the catalyst particles [64,65]. Polymer chains begin to grow not only at the active sites being placed on the surface but also at those inside the crystal granule. Therefore, polymer chains grow outwards from the surface as well as inwards, causing the granule to expand progressively. The catalyst is dispersed throughout the polymer particle through breakdown into primary particles. This fragmentation process is a necessary condition for high catalyst activity. A feature of such catalysts is that the catalyst particles grow by a process of replication. This means that the overall shape of each primary particle is maintained as it grows by polymerisation, and thus the distribution of polymer particle sizes is related to the distribution of catalyst particle sizes [43]. In order to provide true replication of the catalyst particle, the mechanical strength of the granule must be in balance with the catalyst polymerisation activity [58,66]. If reactivity is too high, an uncontrolled 'explosion' occurs, the mechanical forces generated by the growing polymer chain fracturing the granule into fine polymer powder. If the mechanical strength of the catalyst particle is too great, a low level of reactivity is observed because the inward active sites cannot generate the polymer owing to the lack of space. Good replication and high activity only occur when the mechanical strength and polymerisation activity of a supported catalyst are well balanced. In the case of the super highly active third-generation Ziegler–Natta catalysts, the replication factor (the ratio between the size of the polymer granule and that of the catalyst granule) can reach values of up to 40–50, compared with 7–10 for conventional catalysts, without forming fine particles [49].

The fourth generation of Ziegler–Natta catalysts has disclosed a 'new dimension' in catalysis: the domain of polymer shape and morphology. It is this control of the architecture or three-dimensional structure of the catalyst particle that has enabled the polymerisation not only to reproduce the shape of the catalyst but also to generate a solid particle with a controlled reproducible porosity. The net result is that after polymerisation the polymer granule can be reproducibly varied from a relatively uniform density to a layered, open structure and even a hollow particle with a thick polymer skin. Polypropylene obtained commercially with fourth-generation Ziegler–Natta catalysts does not require any de-ashing procedure, atactic fraction removal and pelletisation [38,49]. In industrial processes of ethylene polymerisation with these catalysts, the polyethylene produced contains only a few ppm of the transition metal with a catalyst ash level of 50 ppm (excluding the contribution from the activator). These levels are sufficiently low for catalyst removal from the polymer to be avoided. Even levels of chlorine residues (40 ppm) are small enough for polyethylene to be essentially non-corrosive [51]. However, new procedures for obtaining better catalysts are being introduced. For example, spherical silica may serve as a support for $TiCl_4$. Such a catalyst precursor obtained with the use of a Lewis base is then activated by applying a multistep procedure [28]. The application of a SiO_2 support for catalyst preparation allows a decrease in the level of chlorine residues in the polymer as well as very effective catalyst replication. Moreover, the presence of ideally dispersed silica in the polymer results in an improvement in some of its rheological properties, such as its friction properties, for processing. The productivity of superhigh-activity catalysts of the fourth generation is similar to that of third-generation catalysts. The isotactic index of polypropylene obtained with the fourth-generation catalysts is, however, very high and reaches 98 % [49].

The role of Lewis bases in $MgCl_2$-supported Ziegler–Natta catalysts for α-olefin polymerisation is still a subject of continuous interest. Recently [67], the use of 1,3-diethers [of the general formula $R^1OCH_2C(R^2)(R^3)CH_2OR^4$], e.g. 2,2-diisobutyl-1,3-dimethoxypropane, as the internal Lewis base during procatalyst preparation has made it possible to produce catalysts that are both highly active and stereospecific even in the absence of external donors. Commercially developed industrial catalysts of this type, which are obtained without using an external Lewis base, exhibit all the features of the fourth-generation Ziegler–Natta catalysts. Their productivity is ca 2500 kg of polypropylene (with an isotactic index of not less than 97 %) per gram Ti/h [38]. In order to emphasise that these new catalysts need treatment with an internal Lewis base only during their preparation, they have been proposed by their inventors to be included in the fifth generation of Ziegler–Natta catalysts [68].

It is worth noting that, to obtain efficient supported Ziegler–Natta catalysts of the third and fourth generation, it is $TiCl_4$ rather than preformed $TiCl_3$ that is supported. When $TiCl_3$ was supported on $MgCl_2$ (e.g. by deposition from the solution of a $TiCl_3.3Py$ complex) and then activated with an alkylaluminium compound, the obtained catalyst appeared to be aspecific for propylene polymerisation [69]. However, it could be converted into an isospecific catalyst by adding a Lewis base such as ethyl benzoate [70].

Industrial $MgCl_2$-supported super highly active catalysts for propylene polymerisation are characterized by the presence of their titanium component as titanium chlorides. However, other titanium compounds, especially those such as titanium alkoxides or alkoxychlorides, may be fixed on the $MgCl_2$ support and then activated with trialkylaluminium in order to prepare catalysts for ethylene polymerisation. Catalysts obtained in this way, such as $MgCl_2/Ti(OBu)_4$–$Al(i\text{-}Bu)_3$ and $MgCl_2/Ti(OBu)_2Cl_2$–$Al(i\text{-}Bu)_3$, produce ultrahigh molecular weight polyethylene. The productivity of such catalysts is less than that of the corresponding $MgCl_2/TiCl_4$–$Al(i\text{-}Bu)_3$ catalyst; it increases with an increasing number of Cl substituents at the Ti atom, i.e. in reverse order to the ligand electron releasing ability. Therefore, a more productive catalyst for ethylene polymerisation appeared to be that obtained using the respective phenoxide instead of the alkoxide, i.e. $MgCl_2/Ti(OPh)_4$–$Al(i\text{-}Bu)_3$ [71].

At the end of considerations of heterogeneous supported catalysts, mention should be made of Ziegler–Natta catalysts for ethylene and propylene polymerisations supported on carbon-containing carriers, although there is rather scant information in the literature concerning this matter.

3.2.1.3 Soluble Vanadium-based Catalysts

Homogeneous Ziegler–Natta catalysts of the classic type, capable of olefin polymerisation, are usually derived from reactions of a hydrocarbon soluble transition metal compound with an alkylmetal compound in a hydrocarbon solution (heptane, hexane, toluene). However, the polymer formed in the presence of such soluble Ziegler–Natta catalysts is very often precipitated during polymerisation.

The only important industrial applications of such soluble catalysts are of those prepared from VCl_4, $VOCl_3$, $V(Acac)_3$, $VO(OEt)Cl_2$, $VO(OEt)_2Cl$, $VO(OEt)_3$ or $VO(OBu)_3$ as precursors and $AlEt_3$, $AlEt_2Cl$ or $Al(i\text{-}Bu)_2Cl$ as activators, in heptane solution, by which ethylene/propylene copolymers and ethylene/propylene/non-conjugated diene terpolymers are produced [72]. The Al/V molar ratio in these catalysts does not usually exceed a value of 3:1.

Nevertheless, many vanadium-based catalysts and polymerisation systems comprising them have received much academic attention in the hope that they might provide models for heterogeneous catalysts and polymerisation systems, since the problems connected with surface properties and particle size were believed to have been overcome. It must be noted, however, that homogeneous vanadium-based catalysts appeared to be more complex than was thought. There is no decisive evidence on the structure of catalytic sites formed by reaction between the procatalyst and activator.

It may be interesting, in connection with the ethylene/propylene copolymers mentioned above, to present here some homogeneous Ziegler–Natta catalysts formed by soluble complexes of titanium and magnesium chlorides with alkyl phosphates as catalyst precursors and alkylaluminium compounds as activators: $(TiCl_4)_x \cdot (MgCl_2)_y \cdot [O{=}P(OBu)_3]_3$–$Al(i\text{-}Bu)_3$ and $Cl_3TiOMgCl \cdot [O{=}P(OBu)_3]_3$– $Al(i\text{-}Bu)_3$ (Al/Ti molar ratio of ca 10:1). These catalysts have been used for random ethylene/propylene copolymerisation [73].

Attention has also been paid to soluble vanadium-based catalysts for their activity in low-temperature propylene polymerisation leading to a syndiotactic polymer. It was a toluene-soluble VCl_4–$AlEt_2Cl$ catalyst that was first used for syndiospecific propylene polymerisation [10]. Catalysts for the synthesis of syndiotactic polypropylene are based on VCl_4, $VOCl_3$ or $V(Acac)_3$ and $AlEt_2Cl$ or $Al(i\text{-}Bu)_2Cl$. Particularly effective syndiospecific catalysts are obtained when the Al/V molar ratio is in the range 2–10:1 (generally 5:1) and when a weak base, such as anisole, is added in a molar ratio of 1:1 with the vanadium compound [10,74]. Living syndiotactic polypropylene formation with V $(Acac)_3$–AlR_2Cl catalysts at below $-65\,^{\circ}C$ has been claimed [75,76]. Syndiotactic polypropylene is the only crystalline syndiotactic poly(α-olefin) that has been obtained with vanadium-based homogeneous catalysts [2,10]. It must be emphasised that higher α-olefins hardly undergo syndiospecific polymerisation in the presence of vanadium-based soluble Ziegler–Natta catalysts; only the formation of oligomeric syndiotactic poly(1-butene) in rather low yield has been found [2].

Soluble vanadium-based catalysts have also been used for ring-retaining polymerisation of cyclobutene VCl_4–$AlEt_3$ [77,78], $V(Acac)Cl_2$–$AlEt_2Cl$, $VO(OBu)_3$–$AlEt_2Cl$ [78], butadiene 1,4-polymerisation $VCl_3 \cdot 3THF$–$AlEt_2Cl$, $V(Acac)_3$–$AlEt_2Cl$ [79,80] and α-olefin, β-olefin or cycloolefin/conjugated diene alternating copolymerisation VCl_4–$Al(i\text{-}Bu)_3$–$Al(i\text{-}Bu)_2Cl$–anisole [81], $V(Acac)_3$–$AlEt_2Cl$ or $AlChx_2Cl$ [82,83], $VOCl_3$, VCl_4 or $VO(Acac)_2$–AlR_3 or AlR_2Cl [84–86].

3.2.1.4 Homogeneous Single-site Metallocene Catalysts

In the 1980s, soluble zirconocenes activated by methylaluminoxanes have proved to be highly active and versatile homogeneous Ziegler–Natta catalysts for the polymerisation of ethylene and propylene [28–30]. Metallocene-based catalysts have been known, however, since the mid-1950s. The most representative of them are those based on biscyclopentadienyltitanium dichloride, Cp_2TiCl_2, which is bent metallocene with two η^5-cyclopentadienyl ligands bound to the transition metal atom. It was discovered at that time that Cp_2TiCl_2, activated by trialkylaluminium or alkylaluminium chlorides, could polymerise ethylene but not propylene [87–89]. Subsequent investigations on this and other soluble catalysts have been conducted, and many catalyst systems have been developed and studied. A feature of these catalysts has been their relatively low productivity (a few kilograms of polyethylene per gram Ti/h atm) [28]. The first efficient metallocene catalyst for polymerising ethylene was discovered in 1976 [90,91]. It was reported earlier, in 1973, that contamination of the polymerisation system with traces of water, usually known as severe catalyst poison, gave increased catalyst productivities [92], and this unusual finding was attributed to the formation of aluminoxanes resulting from the partial hydrolysis of alkyaluminium compounds [93]. Following these reports, Kaminsky and Sinn found that methylaluminoxane used as an activator with Cp_2TiCl_2 as the procatalyst afforded highly effective ethylene polymerisation catalysts [90,91,94,95].

Alkylaluminoxanes, which are prepared by careful treatment of trialkylaluminium with water, are usually defined as compounds containing at least one bridging oxo group between two or more aluminium centres. They constitute mixtures in which appear species such as $R_2AlO–[Al(R)–O]_x–AlR_2$ and $–[Al(R)–O]_x–$ $(x = 2–20)$ appear, with or without branching of the –[Al(O–O]– type; the mentioned species are three-dimensional cage compounds with four-coordinate aluminium centres [37,96–99]. In spite of the importance of alkylaluminoxanes, in particular the methyl derivative, as activators for metallocene procatalysts, their structure is not yet sufficiently clarified. Studies on the reaction of trimethylaluminium with water led to the isolation of oligomeric methylaluminoxanes, varying in their degree of condensation and structure element. These oligomeric methylaluminoxanes form associates that show in benzene a cryoscopically measured molecular weight of 1000–1550 g/mol. Two or three molecules with 6–10 Al–O units form the complex. One or two trimethylaluminium molecules are also complex bonded, and could not be removed by vacuum [29,37,96]. Therefore, alkylaluminoxanes and among them methylaluminoxanes are denoted as $[Al(R)O]_x$ and $[Al(Me)O]_x$ respectively.

Methylaluminoxane-activated zirconocenes were first reported to exhibit extraordinarily high activities in ethylene polymerisation, much higher than those of heterogeneous catalysts, and retained over prolonged periods of time. The productivity of the zirconocene–methylaluminoxane catalysts, which were prepared in a toluene solution, reached ca 25 000–40 000 kg of polyethylene per gram Zr/h atm. Under the conditions applied, every zirconium atom forms, on average, about 20 000 polyethylene macromolecules, which have a molecular weight in the range 100 000–200 000. Especially above 50 °C, zirconocene-based catalysts are more active for ethylene polymerisation than hafnium- or titanium-based catalysts [37].

The zirconocene–methylaluminoxane catalysts also appeared capable of polymerising propylene and copolymerising ethylene with propylene and higher α-olefins (with the exception of α,α-disubstituted olefins). However, only an atactic polymer was formed in propylene polymerisation when using catalysts based on achiral zirconocenes such as Cp_2TiX_2 [100] under the temperature conditions (in the range 30–100 °C) usually applied for propylene polymerisation with heterogeneous Ziegler–Natta catalysts. Note that polymers obtained with achiral zirconocenes appeared to be very pure atactic polypropylenes, in contrast to atactic polymers containing isotactic and syndiotactic sequences, usually formed in the presence of heterogeneous Ziegler–Natta catalysts [37].

Attempts to attain the necessary steric control with homogeneous group 4 (Ti, Zr, Hf) metallocene–$[Al(Me)O]_x$ catalysts have increased appreciably since 1980, with encouraging results. Ewen [22] found that stereoblock isotactic polypropylene could be obtained with catalysts of the Cp_2TiX_2–$[Al(Me)O]_x$ type by low-temperature propylene polymerisation (in the range from –78 to –30 °C). The discovery, about the mid-1980s, by Ewen [22] and by Kaminsky and Brintzinger [101] that methylaluminoxane-activated chiral *ansa*-metallocenes of C_2-symmetry, such as *rac.*-ethylenebis[1-(4,5,6,7-tetrahydroindenyl)]-zirconium dichloride [*rac.*-$(ThindCH_2)_2ZrCl_2$] [101] and *rac.*-ethylenebis(1-indenyl)zirconium dichloride [*rac.*\hskip -3pt-\hskip -3pt$(IndCH_2)_2ZrCl_2$] [22], lead to isotactic polypropylene, followed by the development by Ewen *et al.* [23] of a C_s-symmetry *ansa*-metallocene as a catalyst precursor, isopropylidene (cyclopentadienyl)(9-fluorenyl)zirconium dichloride [$Me_2C(Cp)(Flu)ZrCl_2$], which yielded, with methylaluminoxane, an excellent catalyst for syndiospecific propylene polymerisation, led to a phenomenal growth of interest in stereospecific polymerisation with homogeneous catalysts.

Isospecific metallocene catalysts *rac.*-$(IndCH_2)_2ZrCl_2$–$[Al(Me)O]_x$ and *rac.*-$(ThindCH_2)_2[ZrCl_2$–$[Al(Me)O]_x$, usually polymerise propylene at rates 4–5 times slower than those for ethylene [30,102]. However, this preference for ethylene is less pronounced than that observed with a methylaluminoxane-activated catalyst based on an unsubstituted metallocene precursor, i.e. Cp_2TiCl_2 [95,103]. On the other hand, the syndiospecific $Me_2C(Cp)(Flu)ZrCl_2$–$[Al(Me)O]_x$ catalyst is unique in that it exhibits higher activity when polymerising propylene than when polymerising ethylene [102]. Higher α-olefins such as 1-butene [101] and 1-hexene, and even α-olefins branched in their 4-position such as 4-methyl-1-hexene [104], undergo polymerisation with *ansa*-metallocene–methylaluminoxane catalysts at rates only moderately lower than those for propylene polymerisation [30]. Efficient homogeneous alkene polymerisation catalysts are obtained by mixing bent metallocenes Cp'_2MtX_2 (Cp' =cyclopentadienyl or substituted cyclopentadienyl; Mt = Zr, Hf, Ti; X = Cl, Me) with excess $[Al(Me)O]_x$ in an inert solvent such as toluene. A number of characteristic features may be mentioned about these catalysts, and they are as follows [105].

Catalyst activity increases with an increasing Al/Mt ratio [95,101,106]. A large excess of methylaluminoxane (Al/Mt ratio > 500) is generally required for acceptable activity; a particularly high catalyst productivity, reaching 2260 kg of polyethylene per mmol Zr/h atm (24 775 kg per gram Zr/h atm), was obtained at 90 °C using an Al/Zr ratio of ca 67 000 [22].

A decrease in the catalyst Al/Mt ratio and an increase in polymerisation temperature both result in a lower average polymer molecular weight and, in the case of stereospecific α-olefin polymerisation, decreased stereoregularity [22,107].

The catalyst activity generally decreases in the following order: Zr > Hf > Ti [29,30,108]. Titanium is rapidly deactivated above −50 °C owing to the reduction of Ti^{4+} to Ti^{3+} [93].

In contrast to aliphatic hydrocarbons, toluene dissolves methylaluminoxane in a sufficient amount to ascertain its proper concentration in the system. The extremely high activity of Cp'_2MtX_2–$[Al(Me)O]_x$ catalysts and therefore the very low concentration of metallocene needed in order to obtain an effective alkene polymerisation system (in the range of hundredths of ppm with respect to the monomer) are the reasons for the very high susceptibility of these catalysts to deactivation by tracer amounts of oxygen, water, etc. Remember that heterogeneous Ziegler–Natta catalysts are usually prepared by using a titanium compound in an amount in the region from a dozen or so to a few hundred ppm with respect to the monomer.

In connection with the above characteristic features of metallocene–methylaluminoxane catalysts, it must be emphasised that alternative, potentially cheaper alkylaluminoxanes, such as ethyl or *i*-butyl derivatives, which are more soluble in aliphatic hydrocarbons than methylaluminoxane, or other alkylaluminium compounds used as activators for metallocene procatalysts, show inferior activity.

The need to use a large excess of methylaluminoxane with respect to group 4 metallocene in order to obtain a highly active alkene polymerisation catalyst is understandable, considering the catalytically active nature of discrete d^0 metallocenes. After reaction with $[Al(Me)O]_x$, group 4 metallocene is assumed to be converted into a formally d^0 14-electron cationic species, $[Cp'_2MtMe]^+$, forming an active site with the active Mt–C bond, being loosely associated with methylaluminoxane-derived anion $[Al_x(Me)_{x-1}O_xX_2]^-$ [109–114]:

$$Cp'_2MtX_2 + [Al(Me)O]_x \rightleftharpoons [Cp'_2MtMe]^+ [Al_x(Me)_{x-1}O_xX_2]^- \tag{13}$$

When toluene solutions of Cp_2ZrCl_2 are treated with $[Al(Me)O]_x$, a fast initial ligand exchange reaction primarily generates the monomethyl derivative $Cp_2Zr(Me)Cl$ [115], and then the excess methylaluminoxane leads to

Cp_2ZrMe_2 [95,100,101]. It is generally assumed that some of the Al centres in $[Al(Me)O]_x$ have an exceptionally high propensity to abstract the methyl anion from Cp_2ZrMe_2 and to sequester it in a weakly coordinating ion $[Al_x(Me)_{x-1}O_xX_2]^-$ [30]. Therefore, it is clear that the equilibrium constant of the reaction proceeding according to scheme (13), and thus the concentration of active sites, will be higher if the $[Al(Me)O]_x/Cp'_2MtX_2$ ratio is increased.

During the last decade, a variety of new catalysts have been presented for the stereospecific polymerisation of α-olefins, based on non-bridged metallocene or stereorigid *ansa*-metallocene as the procatalyst and a methylaluminoxane activator [29,30,37,105–107,112–114,116–135]. Apart from isotactic [118,119,124, 131,132] and syndiotactic [23,118,124,133] polypropylenes and other poly(α-olefin)s [121], hemiisotactic [112,121,124], isoblock [131,132,134], syndioisoblock (stereocopolymer) [127], stereoblock isotactic [135] and stereoblock isotactic–atactic [116,128,129] polypropylenes have been obtained using these new catalysts.

The symmetry of the metallocene and also the kind of procatalyst metal atom, the nature of the catalyst activator and the polymerisation temperature determine the polypropylene tacticity. The general stereoregulation behaviour of metallocene catalysts may be explained in terms of the local chirality, or chirotopicity, of the catalytic sites bonded to the same metal atom. For this analysis, the structure of metallocenes as catalysts should be considered.

A catalyst active site such as $[Cp'_2MtMe]^+[Al_x(Me)_{x-1}O_xX_2]^-$ reacts with an alkene monomer, giving rise to a tetrahedral growing centre with a cationic metallocene moiety of the structure $[Cp'_2Mt(P_n)M]^+$ and a methylaluminoxane-derived counterion, where P_n denotes the growing polymer chain and M denotes the coordinating monomer molecule [136]. In general terms, the structures of the parent metallocene, Cp'_2MtX_2, and the cationic catalytic site, $[Cp'_2Mt(P_n)M]^+$, may be represented as $Cp'(Cp'')MtL_2$, where Cp′ and Cp″ denote two equal or different cyclopentadienyl-type ligands or two moieties of bidentate ligand, and L denotes X or P_n and M. Stereochemical analysis of $Cp'(Cp'')MtL_2$ metallocenes leads to the conclusion that they can be divided into five classes [16,122]. Schematic Fischer projections of these metallocenes belonging to classes I to V are shown in Figure 3.9 [122].

The number of classes, equal to 5, is derived considering all the possible conditions for chirotopicity of the catalytic sites corresponding to L. If they are not chirotopic, i.e. if they are achirotopic (e.g. they are bisected by a horizontal mirror plane), there are two possibilities only: the two sites are equal (class I catalysts) or different from each other (class II catalysts). If, on the contrary, they are chirotopic, three possibilities exist: the two catalytic sites are homotopic (equal) – related by a twofold symmetry axis (class III catalysts), enantiotopic – related by a vertical mirror plane (class IV catalysts) or diastereotopic (different from each other) – no symmetry element is present (class V catalysts). As a consequence, only five classes of metallocene catalysts may exist if interconversion among stereoisomers is not taken into account [122].

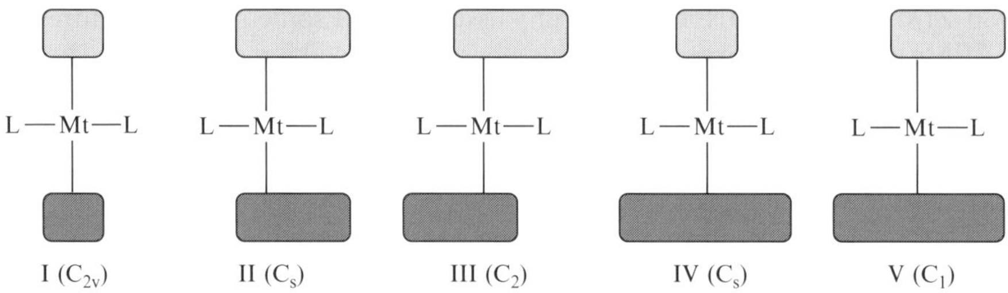

Figure 3.9 Schematic Fischer projections for the bent metallocene catalysts representing five classes (I–V). The grey areas correspond to the cyclopentadienyl-like ligands

Propylene polymerisation with class I and class II catalysts gives rise, in principle, to atactic polypropylene, with class III catalysts to isotactic polypropylene [22] and with class IV catalysts to syndiotactic polypropylene [23], while for the less symmetric class V catalysts no general prediction is possible. In specific cases, however, isotactic, hemiisotactic, stereoblock isotactic–attactic as well as syndiotactic polypropylenes can be obtained with class V catalysts, depending on their kind [107,112,116,124,127,137].

The range of polypropylene microstructures available by procatalyst ligand (and/or metal) variation in the metallocene–methylaluminoxane system is illustrated in Table 3.1 [22,23,101,105,107,112,113,124,127,132,137]. The ligands shown in Table 3.1, representative of the particular symmetry and class of the catalyst, are given as examples only. There are a variety of other metallocenes that have been successfully used to obtain polypropylenes of various stereostructures.

Class I catalysts obtained from achirotopic metallocene precursors of the Cp_2MtX_2 type, which produce atactic polypropylene at elevated temperature in the range of ca 50–70 °C (characteristic of propylene polymerisation in the presence of heterogeneous catalysts), can yield stereoblock isotactic polypropylene at lowered temperature, e.g. −45 °C (Table 3.1) [22].

Class II catalysts, producing atactic polypropylene at elevated temperature but capable of producing stereoblock isotactic polypropylene at lowered temperature, are obtained from achirotopic metallocene precursors of the *meso*-$(IndCH_2)_2MtX_2$ type (Table 3.1) [22].

Class III catalysts are obtained from chirotopic metallocene precursors of the *rac.*-$(IndCH_2)_2MtX_2$ type (Table 3.1) [22]. Catalysts belonging to class III and capable of isospecific propylene polymerisation have also been found and developed, obtained using other precursors such as *rac.*-diethylsilylenebis[1-(3-methylcyclopentadienyl)]zirconium dichloride [*rac.*-$Et_2Si(MeCp)_2ZrCl_2$] [112], *rac.*-dimethylsilylenebis(1-indenyl)zirconium dichloride [*rac.*-$Me_2Si(Ind)_2ZrCl_2$] [102,118], *rac.*-dimethylsilylenebis[1-(2,4-dimethylcyclopentadienyl)] zirconium dichloride [*rac.*-$Me_2Si(Me_2Cp)_2ZrCl_2$], *rac.*-dimethylsilylenebis[1-(2-methylindenyl)]zirconium dichloride [*rac.*-$Me_2Si(MeInd)_2ZrCl_2$],

Table 3.1 Relationship between the kind of bent metallocene procatalyst and the polypropylenex microstructure

Metallocene procatalyst[a]	Class (symmetry)	Polypropylene
 Cp_2ZrX_2	I (C_{2v})	 atactic
 Cp_2TiPh_2	I (C_{2v})	 stereoblock isotactic[b]
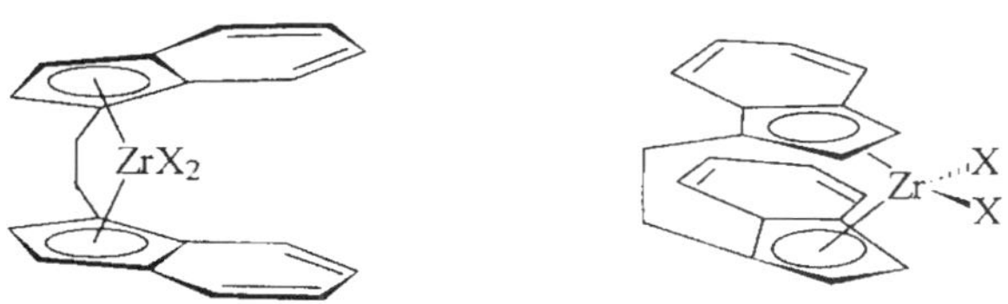 *meso*-$(IndCH_2)_2ZrX_2$	II (C_s)	 atactic

continues overleaf

Table 3.1 (*continued*)

Metallocene procatalyst[a]	Class (symmetry)	Polypropylene
 *rac.*-$(IndCH_2)_2ZrX_2$	III (C_2)	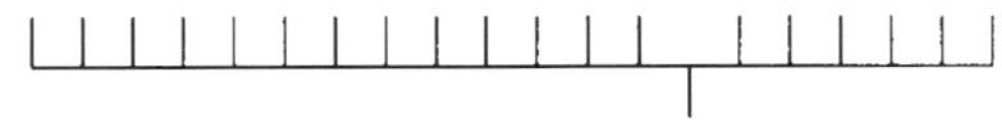 isotactic
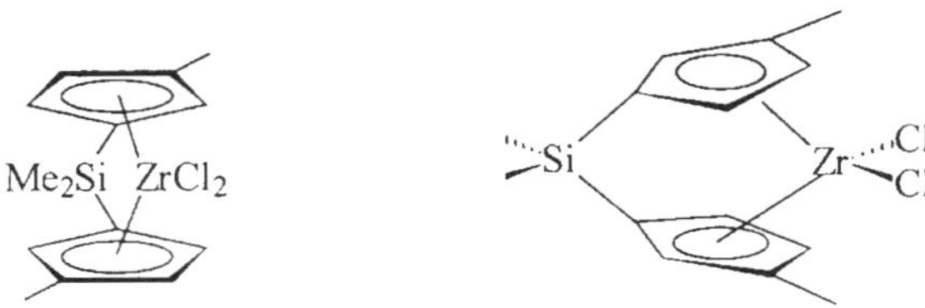 *rac.*-$Me_2Si(MeCp)_2ZrCl_2$	III (C_2)	 isoblock
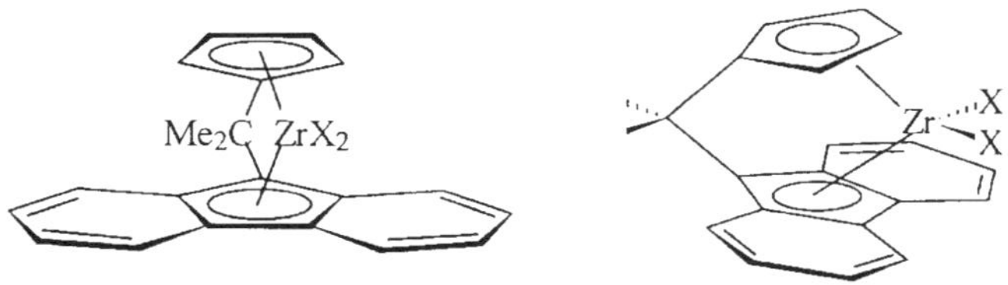 $Me_2C(Cp)(Flu)ZrX_2$	IV (C_s)	 syndiotactic

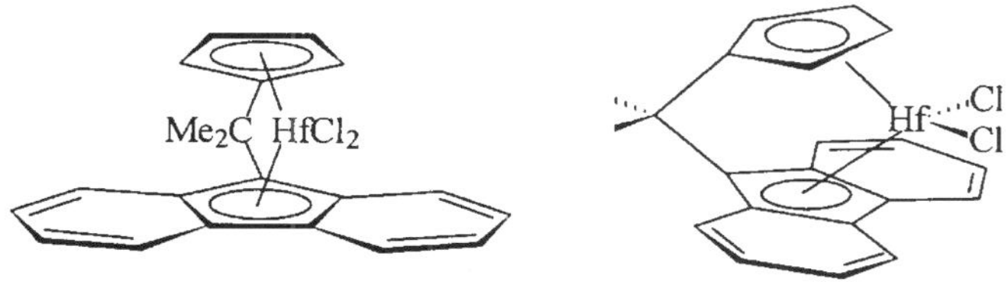

Me$_2$C(Cp)(Flu)HfCl$_2$

IV (C$_s$)

syndioisoblock

Me$_2$C(MeCp)(Flu)ZrX$_2$

V (C$_1$)

hemiisotactic

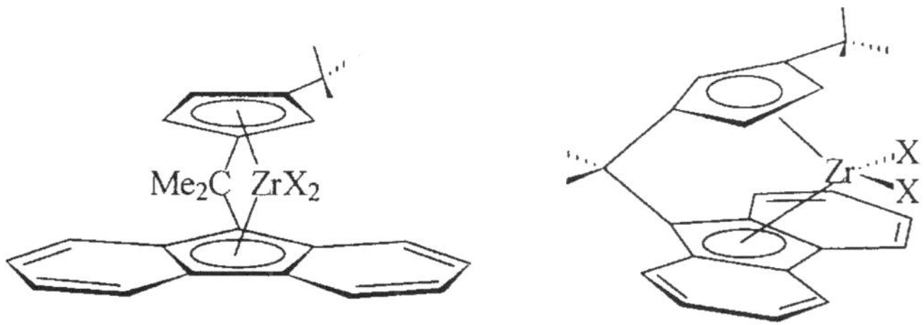

Me$_2$C(*t*-BuCp)(Flu)ZrX$_2$

V (C$_1$)

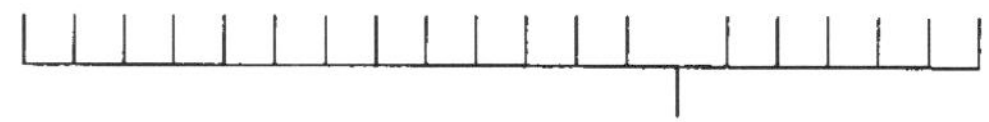

isotactic

continues overleaf

Table 3.1 *(continued)*

Metallocene procatalyst[a]	Class (symmetry)	Polypropylene
 *rac.*-MeCH(Ind)(Me_4Cp)$TiCl_2$	V (C_1)	 stereoblock isotactic–atactic
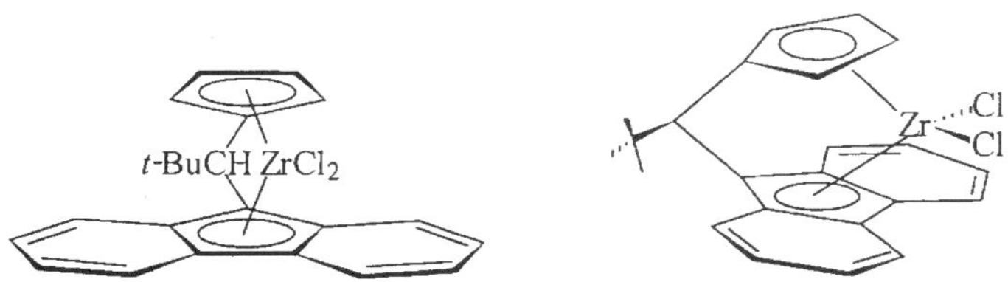 *t*-BuCH(Cp)(Flu)$ZrCl_2$	V (C_1)	 syndiotactic

[a] Front side view (corresponding to Fischer projections) and side view respectively, X = Cl, Me.
[b] Obtained by low-temperature polymerisation.
Reproduced by permission from Ref. 22. Copyright 1994 American Chemical Society.

rac.-dimethylsilylenebis[1-(2-methyl-4-naphthylindenyl)]zirconium dichloride [*rac.*-Me_2Si $(Me, NphInd)_2ZrCl_2$],*rac.*-tetramethyldisilylenebis[1-(2-methyl-4,5-benzoindenyl)]zirconium dichloride [*rac.*-$(MeBzoIndMe_2Si)_2ZrCl_2$] [118, 125,132,138,139] and related *ansa*-metallocenes with cyclopentadienyl-like ligands connected by an ethylene, a isopropylidene, a dialkylsilylene or a double dialkylsililene (tetramethyldisilylene) bridge. It is worth noting that an excessive increase in the polymerisation temperature or the use of $[Al(Me)O]_x$ in an unsufficient amount with respect to $Cp'(Cp'')MtX_2$ in the polymerisation system may result in the production of stereoblock isotactic polypropylene [22,107, 140,141].

In the case of propylene polymerisation with class III catalysts based on *rac.*-dimethylsilylenebis[1-(3-methylcyclopentadienyl)]zirconium dichloride [*rac.*-$Me_2Si(MeCp)_2ZrCl_2$] (Table 3.1) or *rac.*-dimethylsilylenebis[1-(3-methyl-cyclopentadienyl)]hafnium dichloride [*rac.*-$Me_2Si(MeCp)_2HfCl_2$] as a procatalyst, isoblock polypropylene is obtained [132].

Class IV catalysts, which are obtained from precursors with enantiotopic sites of the $Me_2C(Cp)(Flu)ZrX_2$ type (Table 3.1), produce syndiotactic polypropylene at a temperature in the region of ca 50–70 °C [23,102,142]. It is interesting that a catalyst based on a precursor of the same basic structural characteristics but with a two-atom ethylene bridge instead of a one-atom isopropylidene bridge, i.e. ethylene(cyclopentadienyl)(9-fluorenyl)zirconium dichloride [$(CpCH)_2(FluCH)_2ZrCl_2$], efficiently promotes the polymerisation of propylene to syndiotactic polypropylene [143]. Higher α-olefins, both linear and β-branched (but not α-branched), e.g. 1-butene or 4-methyl-1-pentene, also afford highly syndiotactic polymers with $Me_2C(Cp)(Flu)ZrX_2$–$[Al(Me)O]_x$ catalysts [117].

It is noteworthy that the class IV catalyst based on isopropylidene (cyclo pentadienyl)(9-fluorenyl)hafnium dichloride [$Me_2C(Cp)(Flu)HfCl_2$] (Table 3.1), activated by methylaluminoxane, gives rise to syndioisoblock polypropylene [127].

Class V catalysts are obtained from precursors with diastereotopic sites such as isopropylidene[1-(3-alkylcyclopentadienyl)](9-fluorenyl)zirconium dichloride [$Me_2C(Cp)(Flu)ZrX_2$] and dimethylsilylene[1-(3-alkylcyclopentadienyl)](9-fluorenyl) zirconium dichloride [$Me_2Si(RCp)(Flu)ZrX_2$], where the alkyl substituent R = Me, *t*-Bu. These metallocenes are referred to as pseudohelicenes [144,145]. Note that the pseudohelicene ligands produce only the desired *racemic* enantiomers. They cannot form *meso* isomers, having a mirror plane, like class II catalysts. Class V catalysts based on the mentioned pseudohelicenes with the methyl R substituent, i.e. $Me_2C(MeCp)(Flu)ZrCl_2$ (Table 3.1) and $Me_2Si(MeCp)(Flu)ZrCl_2$, yield hemiisotactic polypropylene [112,124,146]. On the other hand, class V catalysts based on pseudohelicenes containing the *t*-butyl R substituent, i.e. $Me_2C(t-BuCp)(Flu)ZrCl_2$ (Table 3.1) and $Me_2Si(i\text{-}BuCp)(Flu)ZrCl_2$, produce isotactic polypropylene [124,127,146]. Since the pseudohelicenes do not form *meso* isomers, the latter catalysts therefore have an advantage over the isospecific catalysts of class III, which are based on helicenes requiring separation of the non-stereospecific *meso* stereoisomers from the *racemic* forms [146].

Class V catalysts based on precursors such as *rac*.-ethylidene(1-tetramethylcyclopentadienyl)(1-indenyl)titanium dichloride [*rac*.-MeCH(Me_4Cp)(Ind)TiCl_2] (Table 3.1) or *rac*.-ethylidene(1-tetramethylcyclopentadienyl)(1-indenyl)dimethyltitanium [*rac*.-MeCH(Me_4Cp)(Ind)TiMe_2] yield, at temperatures in the range 25–50 °C, partially stereoregular polymers with alternating isotactic and atactic stereosequences, which exhibit thermoplastic elastomeric properties [116,120]. The formation of such stereoblock isotactic–atactic propylene polymers is possible by two-state propagation (isospecific, non-stereospecific), which involves a chiral steric environment to propagating chains provided by these catalysts [107].

Another class V catalyst, derived from 2,2-dimethylpropylidene(cyclopentadienyl) (9-fluorenyl)zirconium dichloride [*t*-BuCH(Cp)(Flu)ZrCl_2] (Table 3.1), appeared to be an excellent catalyst for syndiospecific propylene polymerisation [107,137].

Although the most efficient catalysts to date appear to contain a ligand bridge in a metallocene precursor, e.g. class III *rac*.-Me_2Si(Me,NphInd)$_2$ ZrCl_2, the metallocene does not have to possess this attribute for isospecific catalysis [139]. The suitable substituents in unbridged metallocenes may create and maintain a high degree of stereorigidity of the metallocene. Apparently, the non-bonded repulsive interaction exerted by these substituents acts as an invisible and fictitious bridge in the metallocene [143]. Table 3.2 shows a few of the representative non-bridged metallocene procatalysts, which, combined with methylaluminoxane, can yield stereoregular polypropylene of high isotactic index [119,128,129,143,147].

A catalyst based on a non-bridged metallocene, such as bis(1-methylfluorenyl)zirconium dichloride [$(MeFlu)_2ZrCl_2$] (Table 3.2), is capable of producing polypropylene of high isotacticity index (90%) at a relatively high polymerisation temperature of 60 °C. The high isospecificity of this catalyst is connected with its relatively high Zr–(1-methylfluorenyl) ligand rotational barrier, enabling its existence in a relatively stable conformation, which corresponds to the structure of bridged metallocenes forming class III catalysts [143]. The other unbridged zirconocene, with an enhanced rotational barrier of the Zr–cyclopentadienyl ligand, temporarily exhibiting chirotopicity as in class III bridged metallocenes and a lack of chirotopicity at an isomeric class II catalytic site, is

---

[a] For the temporal conformation corresponding to the configuration of the respective *ansa*-metallocene.
[b] Chirotopic conformation (*racemic*) responsible for the isospecific polymerisation.
[c] Chirotopic conformation (*racemic*) partially responsible for the isospecific polymerisation at lowered temperature.
[d] Achirotopic conformation (*mesogenic*) partially responsible for the isospecific polymerisation at lowered temperature.
[e] Obtained by low-temperature polymerisation.
[f] Achirotopic conformation (*mesogenic*) responsible for the non-stereospecific polymerisation.

Table 3.2 Non-bridged bent metallocene procatalysts and the polypropylene microstructures

<table>
<tr><th>Metallocene procatalyst</th><th>Class[a]</th><th>Polypropylene</th></tr>
<tr><td>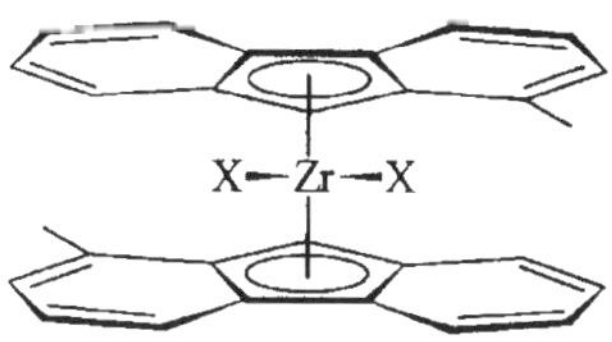

[b]
(MeFlu)$_2$ZrCl$_2$</td><td>III</td><td>isotactic</td></tr>
<tr><td rowspan="2">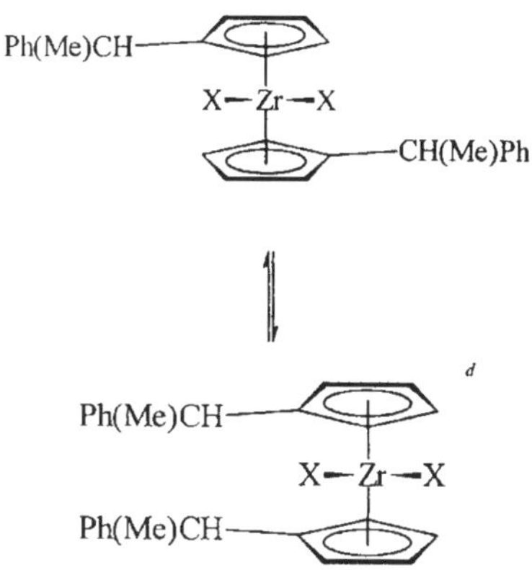

[d]
rac.-[Ph(Me)CHCp]$_2$ZrCl$_2$</td><td>III</td><td rowspan="2">isotactic[e]</td></tr>
<tr><td>II</td></tr>
<tr><td rowspan="2">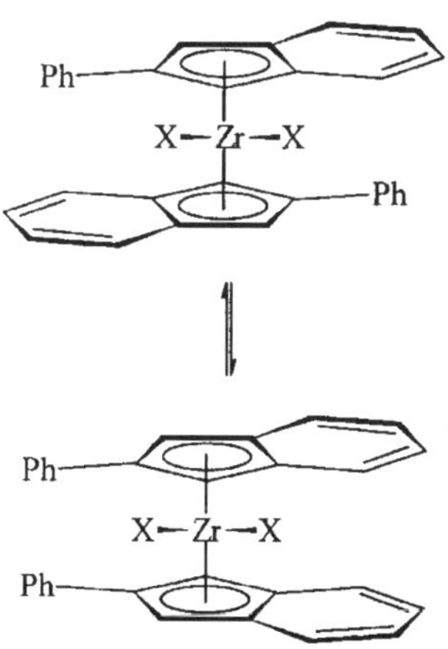

[b] [f]
(PhInd)$_2$ZrCl$_2$</td><td>III</td><td rowspan="2">stereoblock isotactic–atactic</td></tr>
<tr><td>II</td></tr>
</table>

rac.-bis[(1-phenylethyl)cyclopentadienyl]zirconium dichloride {*rac.*-[Ph(Me)CHCp]$_2$ZrCl$_2$ (Table 3.2). It has been used with a methylaluminoxane activator for propylene polymerisation, leading at −50 °C to isotactic polypropylene; with increasing temperature, polymerisation isospecificity practically disappeared as in the case of class II catalysts [119].

It is interesting that the use of a catalyst based on an unbridged metallocene precursor such as bis(2-phenylindenyl)zirconium dichloride [(PhInd)$_2$ZrCl$_2$] (Table 3.2) for propylene polymerisation at 20 °C gives rise to stereoblock isotactic–atactic polypropylene with thermoplastic elastomeric properties [128,129]. The (PhInd)$_2$ZrCl$_2$–[Al(Me)O]$_x$ catalyst appears to isomerise by restricted rotation of its 2-phenylindenyl ligands between coordination geometries during chain growth; the temporal chirotopic site, corresponding to the class III catalyst site of bridged metallocenes, leads to isotactic blocks, whereas the temporal achirotopic site, corresponding to the class II catalyst site of bridged metallocenes, leads to atactic blocks of the polypropylene formed. Similarly, the catalyst based on pentamethylcyclopentadienyl(2-phenylindenyl)zirconium dichloride [(Cp*)(PhInd)ZrCl$_2$] produces stereoblock isotactic–atactic polypropylene, but in lower yield than that obtained in the presence of the catalyst based on (PhInd)$_2$ZrCl$_2$ [147].

Extensive efforts have also been made to develop olefin polymerisation catalysts based on metallocenes with only one ligand of the cyclopentadienyl type. Ethylene-,dimethylsilylene- or tetramethyldisilylene-bridged mono(1-tetra -methylcyclopentadienyl), mono(1-indenyl) or mono(9-fluorenyl)-amidotitanium complexes, such as dimethylsilylene(1-tetramethylcyclopentadienyl)(*t*-butyl)amidotitanium dichloride [Me$_2$Si(Me$_4$Cp)N(*t*-Bu)TiCl$_2$] (Figure 3.10), have recently attracted both industrial and scientific interest as precursors for methylaluminoxane-activated catalysts, which polymerise ethylene and copolymerise ethylene with 1-butene, 1-hexene and 1-octene [30,105,148–152].

These monocyclopentadienyl–amidotitanium complexes, which are classified as constrained-geometry catalysts, are capable of producing low-density polyethylene (ethylene copolymers with C$_4$, C$_6$ or C$_8$ 1-alkenes) that also contain long-chain branches, in contrast to strictly linear low-density polyethylene (ethylene copolymers with C$_4$, C$_6$ or C$_8$ 1-alkenes) produced by bent metallocene-based catalysts [30,105,148,149].

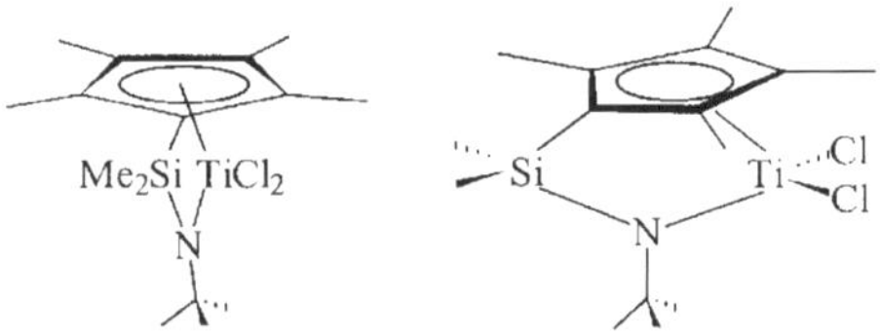

Figure 3.10 Monocyclopentadienyl amidotitanium procatalyst, dimethylsilylene(1-tetra -methylcyclopentadienyl)(*t*-butyl)amidotitanium dichloride {Me$_2$Si(Me$_4$Cp)N(*t*-Bu) TiCl$_2$}, for obtaining low-density polyethylene. Front side view and side view respectively

As regards propylene polymerisation in the presence of methylaluminoxane-activated catalysts based on the monocyclopentadienyl (Me_4Cp, Ind, Flu) amido complexes of group 4 metals, it is known that catalysts of this type can in some instances give isotactic [153], syndiotactic [154] and atactic [155] polypropylenes.

3.2.1.5 Homogeneous Single-site Non-metallocene Catalysts

Although extensive efforts to develop catalysts without any ligands of the cyclopentadienyl type as alternatives to ubiquitous bent metallocenes have been made, they have met with limited success. However, a few homogeneous Ziegler–Natta catalysts based on new group 4 metal complexes of the non-metallocene type as procatalysts and methylaluminoxane or another alkylaluminium compound as the activator have recently been found and utilised for olefin polymerisation. These new alternative complexes can contain a diversity of ligands such as an amide ligand [156], an aminopyridinato ligand [157], a Schiff's base–bisenolate ligand [158], a bisphenolate ligand [159] or a calixarenediolate ligand [160]. Also, benzyl derivatives of group 4 transition metals, such as tetrabenzylziroconium, can be applied, in a combination with methylaluminoxane or aluminium trialkyl, as homogeneous catalysts for alkene polymerisation [161,162].

Single-site catalysts based on non-cyclopentadienyl ancillary ligands, which are analogous to the metallocene catalyst systems and contain cationic metal species, are moderately active in ethylene polymerisation and rather poorly active in propylene polymerisation. However, some of them, e.g. those formed in the bis{[di(trimethylsily)]amido}zirconium dichloride $[(Me_3Si)_2N]_2ZrCl_2$–$[Al(Me)O]_x$ [156] and tetrabenzylzirconium $\{Zr(CH_2Ph)_4\}$–$[Al(Me)O]_x$ [161] systems, have been claimed to yield highly isotactic polypropylene and higher poly (α-olefin)s.

Such examples have shown that the role of the cationic group 4 metal complexes in the coordination polymerisation of ethylene and α-olefins with homogeneous single-site Ziegler–Natta catalysts must not be limited to those containing cyclopentadienyl-like ligands.

3.2.2 Homogeneous Catalysts without a Metal Alkyl (or Hydride) Activator

Homogeneous alkylaluminium (alkylaluminoxane)-free olefin polymerisation catalysts are commonly referred to as catalysts containing group 4 metallocene cationic species. These catalysts resemble the structure and properties of the respective metallocene-based Ziegler–Natta catalysts but, by definition, are not included among the latter catalysts. The development of alkylaluminium-free metallocene-based olefin polymerisation catalysts is connected with discoveries of simpler two-component systems based on group 4 metallocene dialkyls and a stoichiometric (or near-stoichiometric) amount of triphenylborate, ammonium tetraphenylborate salts (eventually the respective pentafluorophenyl derivative) [32,33,110–113,163–171] or triphenylmethyl tetrakis(perfluorophenyl)borate [35], generating cationic metallocene alkyls such as $[Cp'_2ZrR]^+$, capable of olefin polymerisation. Even earlier, at the beginning of the 1980s, single-component catalysts consisting of Lewis base complexes of cationic group 4 metallocenes, such as $[Cp'_2Zr(THF)R]^+$, with a tetraphenylborate counterion [109, 172–174] or isoelectronic neutral group 3 and lanthanide (Ln) metallocene alkyls, e.g. $Cp^*{}_2LnR$, or hydrides, e.g. $(Cp^*{}_2LnH)_2$ [31,175–180], have been found and proved for their catalytic activity in olefin polymerisation.

Homogeneous alkylaluminium-free olefin polymerisation catalysts also comprise non-metallocene cationic group 4 metal complexes such as those with benzyl ligands [162]. A distinct group of alkylaluminium-free homogeneous olefin polymerisation catalysts consists of nickel complexes [181–183].

3.2.2.1 Single-site Group 4 Metallocene Catalysts

The essential factor determining the synthesis of cationic group 4 metallocene alkyls, $[Cp'_2ZrR]^+$, capable of olefin polymerisation is the utilisation of robust and poorly coordinating anions showing good solubility in inert aromatic solvents. The unusually low coordinating ability of such anions in ion pairs with cationic metallocenes is crucial for the catalytic activity [184].

Early attempts to use traditional 'non-coordinating' anions, such as $[BF_4]^-$ or $[PF_6]^-$, appeared to be unsuccessful as fluorine transfer to the cation led to deactivation [109]. Even for large, weakly coordinating anions such as tetraphenylborate, $[B(C_6H_5)_4]^-$, and a carborane anion such as $[C_2B_9H_{12}]^-$, fairly strong interactions have been observed with cationic zirconocene alkyls [33,110,169,185]. Polymerisation systems containing $[Cp_2ZrMe]^+$ cations combined with $[B(C_6H_5)_4]^-$ or $[C_2B_9H_{12}]^-$ anions thus polymerise propylene only at low rates, if at all [33,109,163–169]. A breakthrough in this regard was the application of perfluorinated tetraphenylborate as a counterion [110,186]. An ion pair $[Cp'_2ZrMe]^+[B(C_6F_5)_4]^-$ is formed by reaction of Cp'_2ZrMe_2 with dimethylanilinium (or tributylammonium) tetrakis(perfluorophenyl)borate [scheme (14)] or triphenylmethyl tetrakis(perfluorophenyl)borate [scheme (15)]:

$$Cp'_2MtMe_2 + [PhNHMe_2]^+[B(C_6F_5)_4]^- \longrightarrow [Cp'_2MtMe]^+[B(C_6F_5)_4]^- + PhNMe_2 + MeH \qquad (14)$$

$$Cp'_2MtMe_2 + [Ph_3C]^+[B(C_6F_5)_4]^- \longrightarrow [Cp'_2MtMe]^+[B(C_6F_5)_4]^- + Ph_3CMe \qquad (15)$$

These were the first well-defined zirconocene catalysts capable of polymerising ethylene, propylene and higher α-olefins at high rates without the addition of any further activator. Metallocene catalysts obtained by abstraction of the Me^- anion from Cp'_2ZrMe_2 by a strong Lewis acid, $B(C_6F_5)_3$, were likewise found to exhibit high activity in ethylene and α-olefin polymerisation [30,105]:

$$Cp'_2MtMe_2 + B(C_6F_5)_3 \longrightarrow [Cp'_2MtMe]^+[MeB(C_6F_5)_3]^- \quad (16)$$

The behaviour in olefin polymerisation of the organometallic activator-free metallocene catalysts is in principle similar to that displayed by the catalysts based on methylaluminoxane-activated metallocenes. This is valid in particular for systems with the crowded ethylenebisindenyl and related ligands {e.g. the $[rac.\text{-}(IndCH_2)_2ZrMe]^+$ cation}; similar activity, polymer microstructure (mainly isotactic) and melting point (polymer high molecular weight and low molecular weight distribution) indicate that the $[B(C_6H_5)_4]^-$ anion exhibits similar (poor) coordinating ability to the counter-anion generated using $[Al(Me)O]_x$. Therefore, in both cases the conditions for the α-olefin coordination and its incorporation into the growing chain at the cation $[rac.\text{-}(IndCH_2)_2 ZrMe]^+$ are similarly favourable [105].

However, sterically less crowded metallocenes, such as those generating cations like $[Me_2C(Cp)(Flu)ZrMe]^+$ (syndiospecific), appear to be more sensitive to the effects of anion coordination, the strength of which increases in the following order: $[Al_x(Me)_{x+1}O_x]^- < [B(C_6H_5)_4]^- < [MeB(C_6H_5)_3]^-$. For this and other more 'open' cations, stronger coordination of the anion leads to decreased activity, decreased stereospecificity (and polymer melting point) and decreased polymer average molecular weight [112,113]. Therefore, methylaluminoxane-free catalysts based on sterically less crowded metallocenes do not equal respective methylaluminoxane-activated Ziegler–Natta metallocene catalysts regarding their activity and stereospecificity [105].

3.2.2.2 Single-site Group 3 and Lanthanide Series Metallocene Catalysts

Catalysts that consist of group 3 and lanthanide series metallocene alkyls or hydrides are, at least conceptionally, of greater simplicity than those olefin polymerisation catalysts that are formed in two-component group 4 metallocene systems. Neutral metallocenes such as Cp_2MtR and $Cp_2{}^*MtR$ (Mt = Sc, Y, Lu; R = Me, Et) contain formally d^0 14-electron species, isoelectronic to species present in group 4 metallocene cations. Thus, group 3 metallocenes may act as single component catalysts for olefin polymerisation. Indeed, in most cases they produce high molecular weight polyethylene, but low molecular weight polypropylene [175–180]. Such homogeneous metallocene catalysts are uniquely suited to a study of the rates and mechanisms of individual olefin insertion steps into Mt–C active bonds.

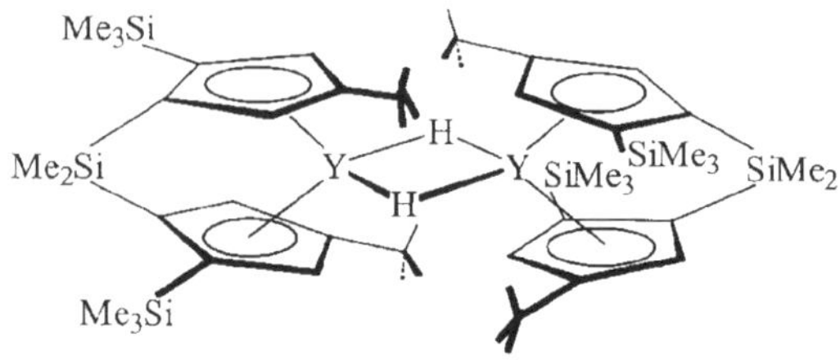

Figure 3.11 Single-component catalyst, dimeric homochirotopic *rac.*-(*S*, *S*)-dimethylsilylenebis[1-(2-trimethylsilyl-4-*t*-butylcyclopentadienyl)]yttrium hydride $\{[rac.\text{-}Me_2Si(Me_3Si,t\text{-}BuCp)_2YH]_2\}$, for obtaining highly isotactic poly(α-olefins)s. Side view. Reproduced by permission from Ref. 31. Copyright 1992 American Chemical Society

Even more interesting are group 3 *ansa*-metallocenes. Dimeric homochirotopic *rac.*-dimethylsilylenebis[1-(2-trimethylsilyl-4-*t*-butylcyclopentadienyl)]yttrium hydride $\{[rac.\text{-}Me_2Si(Me_3Si, t\text{-}BuCp)_2YH]_2\}$ (Figure 3.11) appeared to polymerise propylene, 1-butene, 1-pentene and 1-hexene to highly isotactic, moderately high molecular weight polymers [31,187].

The polymerisation proceeds quite slowly, presumably owing to the inactivity of the formally d^0 16-electron yttrium species of the dimeric catalyst. A formally d^0 14-electron monomeric hydride or alkyl derivative is probably required for olefin polymerisation [187].

Dimeric homochirotopic $[rac.\text{-}Me_2Si(Me_3Si, t\text{-}BuCp)_2YH]_2$ is the first single-component isospecific catalyst; it is suitable to study the subtle steric factors that govern the remarkably high stereospecificities exhibited in the polymerisation of α-olefins by this and related two-component class III catalysts of C_2 symmetry based on group 4 metallocenes.

3.2.2.3 Group 4 Metal Benzyl Catalysts

Although σ-homoleptic hydrocarbyls such as $TiBz_4$ and $ZrBz_4$ were the first recognised homogeneous catalysts for the isospecific polymerisation of α-olefins [188], their low catalytic activities, even when used as a two-component catalyst with an alkylaluminium activator, discouraged further research in this field. Recently, however, following the results concerning the synthesis of cationic group 4 metallocene complexes, cations such as $[Mt(CH_2Ph)_3]^+$ (Mt = Ti, Zr) have been synthesised and proved to be catalytically active in olefin polymerisation [162,189,190]. Group 4 metal benzyl catalysts containing cationic species, e.g. $[ZrBz_3]^+[BzB(C_6F_5)_3]^-$, afford (only at elevated temperature) high molecular weight polyethylene, which is characterised by an extremely broad molecular weight distribution ($M_w/M_n = 65$) [162]. Note in this connection that polyethylene afforded by metallocene-based Ziegler–Natta and methylaluminoxane-free catalysts is characterised by a narrow molecular weight distribution.

Cationic $[ZrBz_3]^+$ species are capable of polymerising α-olefins such as propylene, 1-butene and 1-pentene at elevated temperature. It is worth emphasising that the non-metallocene $[Zr(CH_2Ph)_3]^+[PhCH_2B(C_6F_5)_3]^-$ catalyst, which is a cationic arene zirconium complex [189], is capable, at least partially, of isospecific α-olcfin polymerisation at relatively high temperature [162,189,190].

3.2.2.4 Nickel-complex Catalysts

Complexes of nickel constitute a distinct group of homogeneous alkylaluminium-free catalysts for olefin polymerisation. An efficient catalyst for ethylene polymerisation is formed in the reaction of bis(cycloocta-1,5-diene)nickel(0) $[Ni(Cod)_2]$ with phosphorus-ylid and triphenylphosphine in toluene solvent [181]:

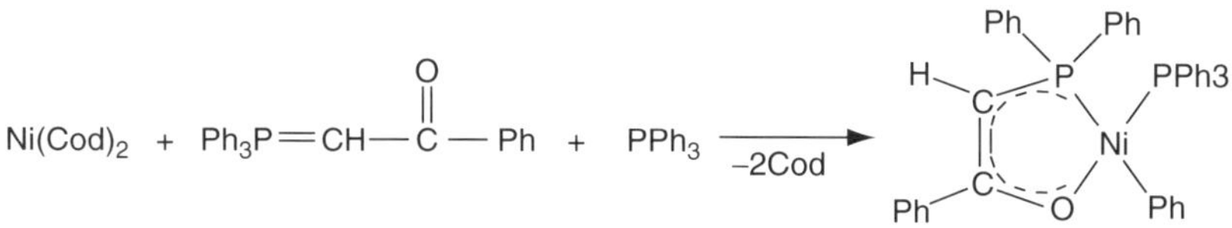

(17)

The obtained Ni(II) complex, containing an Ni–C (Ni–Ph) active bond, polymerises ethylene to high molecular weight high-density polyethylene, especially when precipitated with aliphatic hydrocarbon as the polymerisation medium.

However, homogeneous catalysts obtained from nickel (II) complexes, e.g. bisallylnickel $\{Ni(All)_2\}$, and bis(trimethylsilyl)aminobis (trimethylsilylimino) phosphorane $\{(Me_3Si)_2NP(=NSiMe_3)_2\}$ in toluene solution deserve special attention [182]:

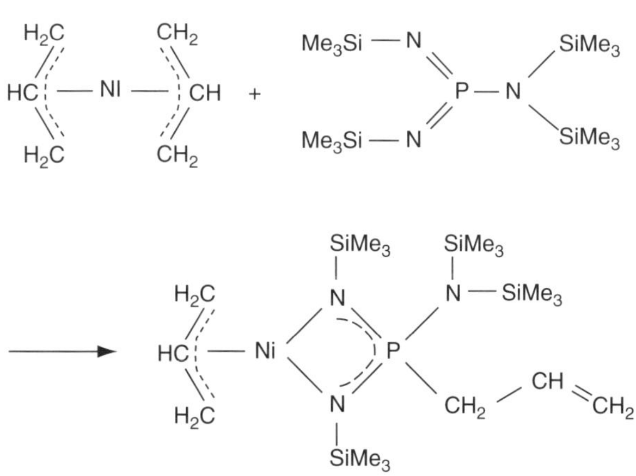

(18)

Similar catalysts can be obtained from nickel(0) complexes, e.g. $Ni(Cod)_2$, and $(Me_3Si)_2NP(=NSiMe_3)_2$.

The polymerisation of ethylene in the presence of an Ni(II) complex containing a ligand originating from aminobis(imino)phosphorane leads to short-chain branched polyethylene [182]. This is due to the copolymerisation of ethylene with short-chain 1-alkenes formed in such a system.

The catalyst derived from the reaction according to scheme (18) and related catalysts appeared capable of polymerising α-olefins; surprisingly, the structure of the poly(α-olefin)s formed is consistent with a 2, ω-coupling [191] but not with the usual 1,2-coupling of the monomers:

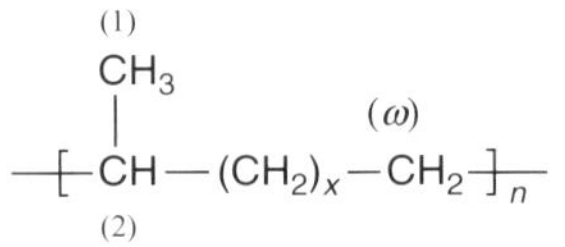

2,ω-coupled poly(α-olefin)

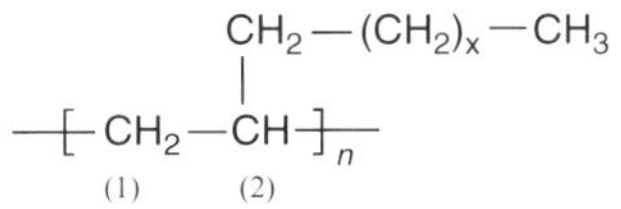

1,2-coupled poly(α-olefin)

Nickel(II) complexes obtained from nickel(0) or nickel(II) compounds and the $(Me_3Si)_2NP(=NSiMe_3)_2$ phosphorane component, which promote the isomerisation 2, ω-polymerisation of α-olefins, are referred to as nickel migratory catalysts. It must be emphasised that this is a peculiarity of the $(Me_3Si)_2NP(=NSiMe_3)_2$ phosphorane; other phosphoranes of this type in which only one silicon atom is substituted by a carbon atom give an inactive catalyst [183].

It may be interesting to note that nickel(II) compounds also display isomerisation properties towards β-olefins; they form, in a combination with Ziegler–Natta catalysts, systems, e.g. $TiCl_3$–$AlEt_3$ (or $AlEt_2Cl$–$NiCl_2$ (Ti/Al/Ni molar ratio 1:3:1), in which a β-olefin such as 2-butene isomerises to the respective α-olefin (1-butene), then undergoing polymerisation to poly(α-olefin) [poly(1-butene)] with predominant usual 1,2-coupling of the monomers [192–195].

3.2.3 Supported Metallocene Catalysts

Although homogeneous metallocene single-site catalysts make it possible to produce a variety of polyolefins of any structure in extremely high yields, their direct utilisation for industrial olefin polymerisation processes has encountered various difficulties. Major disadvantages of the homogeneous catalysts are the lack of polymer morphology control and reactor fouling. Supported catalysts allow control of morphology, may be utilised in a cheaper gas-phase or bulk process (propylene) and may permit the use of less methylaluminoxane or even the use of trialkylaluminium as the activator while still offering simple control of polymer properties by ligand variation. Therefore, much attention has been paid to modifying metallocene catalyst systems for industrial applications. The evolution of advanced catalysts and catalytic processes based on rational model hypotheses has begun to be carried over to heterogeneous Ziegler–Natta catalysts.

The solid supported metallocene catalyst particles generate polymer coherent grains, instead of the polymer dust produced by a dissolved homogeneous metallocene catalyst. As with a heterogeneous non-metallocene Ziegler–Natta catalyst, these grains appear to be enlarged replicas of the catalyst particles [134]. Metallocene catalysts that are heterogenised, e.g. on a silica gel support, can thus be readily used in existing Ziegler–Natta production facilities, for instance in solvent-free slurry or gas-phase reaction systems [30].

In order to obtain supported catalysts, the metallocene is preadsorbed on alumina or silica gels [196–201] as well as on magnesium dichloride [200, 201] using an impregnation method. Supporting metallocenes such as *rac.*-$(ThindCH_2)_2ZrCl_2$, $Me_2C(Cp)(Flu)ZrCl_2$ and Cp_2ZrCl_2 on Al_2O_3 or $MgCl_2$, followed by activation with ordinary aluminium trialkyls (AlR_3; R = Me, Et, *i*-Bu), leads to catalysts affording isotactic, syndiotactic and atactic polypropylene in fairly good yields [200, 201]. It should be emphasised that methylaluminoxane-free metallocene-based supported catalysts, activated with common trialkylaluminium, can be used. The use of chlorine-containing alkylaluminium compounds always makes the catalysts inactive [201]. However, supported metallocene catalysts exhibiting similar high activity and stereospecificity to soluble catalysts have been obtained by pretreatment of silica with methylaluminoxane, followed by reaction of the formed $SiO_2{-}[Al(Me)O]_x$ with metallocenes of various symmetry, giving solids that might be activated with common aluminium trialkyls [198–209].

Although some important aspects concerning the nature of metallocene supported catalysts are not clear at the present stage, the profile of the formation of catalysts (catalyst precursors) with $MgCl_2$, Al_2O_3 and SiO_2 supports and with a methylaluminoxane-pretreated SiO_2 support can be represented as follows [201, 202]:

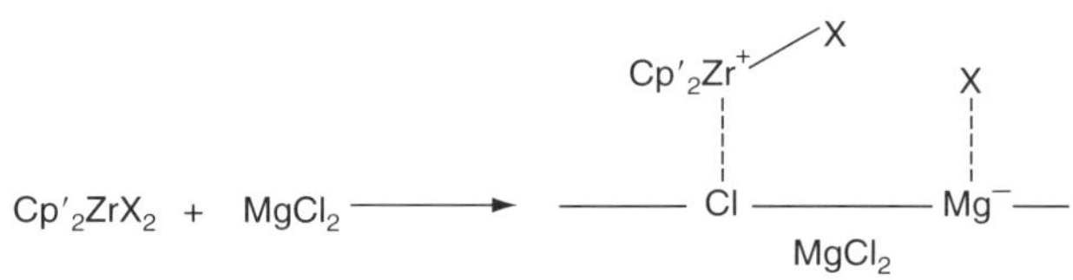

(19)

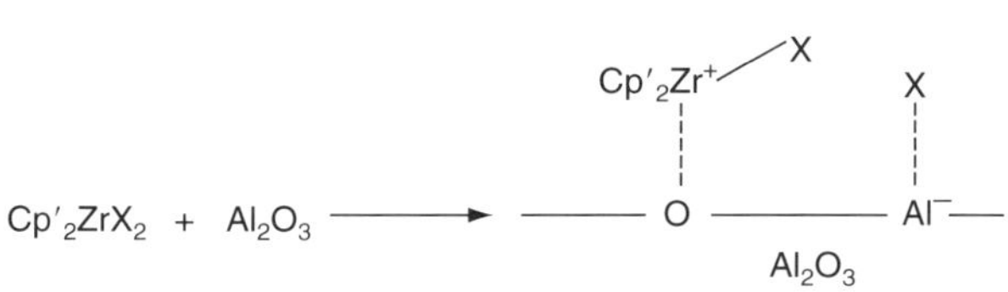

(20)

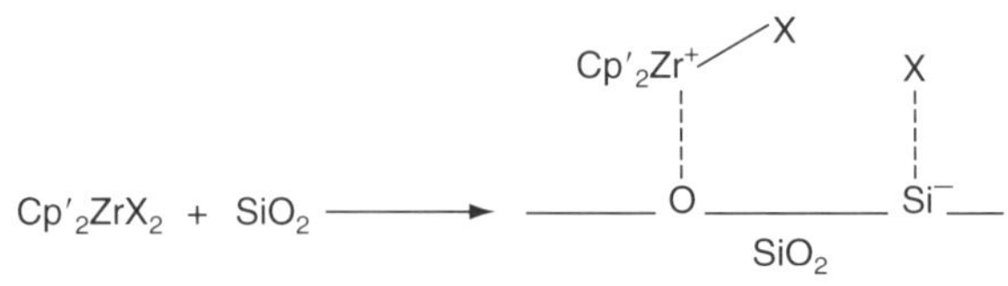
(21)

$$\equiv\text{Si—OH} + [\text{Al(Me)O}]_x \xrightarrow[-\text{MeH}]{} \equiv\text{Si—O}\!+\!\text{Al}_x(\text{Me})_{x-1}\text{O}_x]$$
$$(\text{SiO}_2)$$

$$\xrightarrow{\text{Cp}'_2\text{ZrX}_2} \equiv\text{Si—O}\!+\!\text{Al}_x(\text{Me})_{x-1}\text{O}_x\text{X}]^- \; [\text{Cp}'_2\text{ZrX}]^+ \qquad (22)$$

Most of the supported metallocene catalysts reported so far were devised to immobilise the metallocene on the surface of inorganic carriers, utilising the ionic interaction between the Cl ligands of the metallocene and the surface active site [schemes (19) to (21)]. Similarly, in the methylaluminoxane-pretreated catalyst, the metallocene is immobilised by an analogous ionic interaction [scheme (22)]. Therefore, it is obvious that catalyst precursors formed according to schemes (19) to (22) can be easily activated with common aluminium trialkyls.

The most advanced of such catalysts require only limited excess ratios of trialkylaluminium (or methylaluminoxane) activators per zirconocene unit; Al/Zr ratios of roughly 100–400:1, much below those used for homogeneous catalyst systems, appear to give entirely satisfactory activities for these surface-supported metallocene catalysts. The overall productivities of these heterogenised metallocene catalysts, per gram of support and catalyst, are about 50% of those of recent heterogeneous, $MgCl_2$-supported Ziegler–Natta catalysts [30].

A special interest has been devoted to SiO_2-supported catalysts. The metallocene can be immobilised by reaction with SiO_2 surface hydroxyl groups when present [205]:

$$\underset{(\text{SiO}_2)}{\equiv\text{Si—OH}} + \text{Cp}'_2\text{ZrX}_2 \xrightarrow[-\text{HX}]{} \equiv\text{Si—O}^-[\text{Cp}'_2\text{ZrX}]^+ \qquad (23)$$

It may be interesting that even a homogeneous zirconocene catalyst can be activated by common trialkylaluminium when trimethylsilanol (Me_3SiOH) is used as a modifier [210].

Metallocene catalysts supported on SiO_2 are practically free of chlorine. This could be an advantage from an ecological standpoint when the chemical or thermal recycling of polymeric products reaches a large scale [211].

The silica gel-supported catalysts can be further improved by SiO_2 modification in order to prepare more tightly immobilised metallocene catalysts. A series of catalysts whose ligands are chemically bound to the solid surface was obtained by the following schemes [207, 208]:

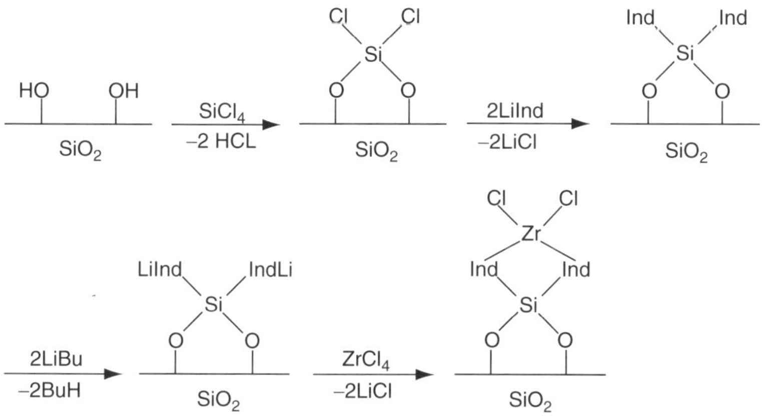

(24)

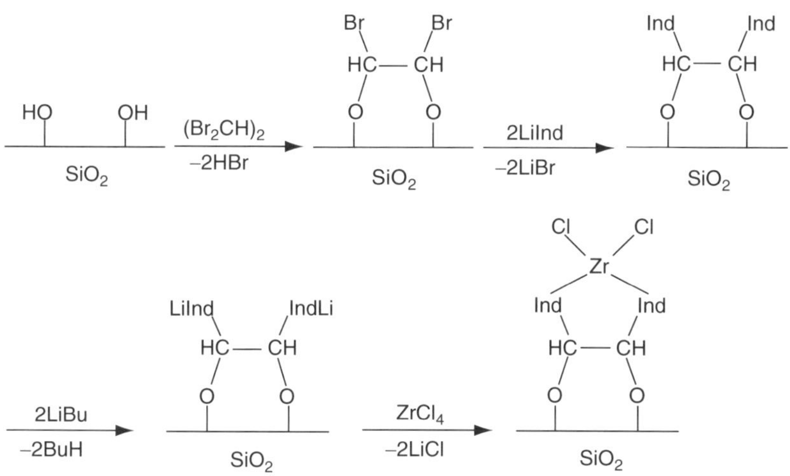

(25)

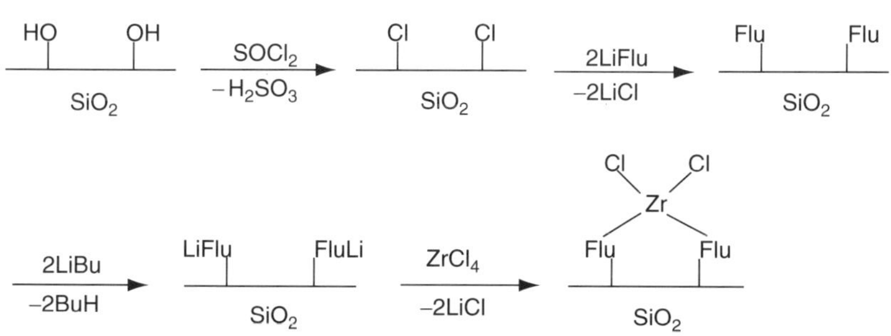

(26)

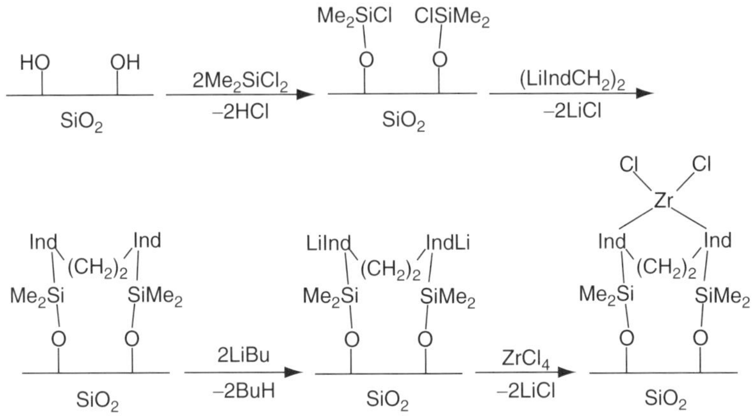

(27)

Precursors formed according to schemes (24) to (27) are activated with $[Al(Me)O]_x$ or AlR_3, which leads to highly active stereospecific (isospecific) olefin polymerisation catalysts. In some cases, however, the isotactic polypropylene formed displayed two melting points, which might result from the occurrence of catalyst species of two types: single-bound species, in addition to the double-bound species formed according to schemes (24) to (27), owing to insufficient surface hydroxyl groups [208].

Since the discovery of Ziegler–Natta catalysts, polyolefin industries have been developed mainly by using titanium-based catalysts. However, after the appearance of metallocene catalysts, much emphasis has been placed on those based on zirconium owing to the superiority of their performance over titanocene catalysts.

More recently, titanocene-based SiO_2-supported catalysts, analogous to that formed according to scheme (24), have seen prepared. Their precursors appear as follows:

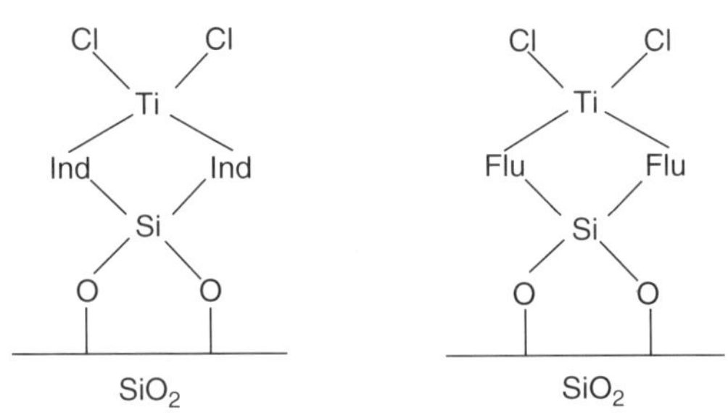

Although the activity of these catalysts is not so high, they give highly isotactic polypropylene of considerably high molecular weight [208].

On the other hand, as regards the bridge connecting the ligands of metallocene compounds, a one-membered silicon bridge and two-membered ethylene bridge have been preferentially used so far. This was considered, however, with a variety of metallocene catalysts with different types of bridge, to be synthesised and tested for olefin polymerisation. For example, the use of an Sn-bridged procatalyst analogous to that formed according to scheme (24), which was activated with $Al(i\text{-}Bu)_3$, made it possible to obtain isotactic polypropylene displaying only one melting point (at around 160–162 °C). A higher reactivity of $SnCl_4$ towards the silica hydroxyl groups might lead to more uniform active sites on the surface [208].

Polyolefins possessing properties distinctly different from those produced in homogeneous solution, characterised by an unusually high molecular weight, have recently been obtained by using catalysts with *ansa*-metallocene precursors supported directly on an unpretreated silica gel carrier [204,206]. This may be due to the active site isolation effect, i.e. to the strict suppression of participation of other sites in the polymerisation [30].

Metallocene supported catalysts are not limited only to those involving inorganic catalyst carriers. Starch [212] and α-cyclodextrin [213] have also been successfully used as supports for zirconocene–$AlMe_3$ [212] and zirconocene–$[Al(Me)O]_x$ [213] catalysts capable of producing high molecular weight polyethylene.

At the end of considerations of metallocene supported catalysts, it should be noted that the catalyst derived from chromocene, biscyclopentadienylchromium (Cp_2Cr), deposited on silica supports, also exhibits very high activity for the polymerisation of ethylene [214]. The formation of the catalyst has been formulated as follows [215]:

$$\underset{(SiO_2)}{\equiv Si-OH} \xrightarrow[-CpH]{Cp_2Cr} \equiv Si-O-CrCp \tag{28}$$

Moreover, transition metal benzyl and allyl compounds can react with hydroxyl groups on the surface of silica and alumina gels to give active olefin polymerisation catalysts [216–218]. For example, alumina-supported tetrabenzyltitanium ($TiBz_4$), tetraallylzirconium $[Zr(All)_4]$ and triallylchronium $[Cr(All)_3]$ as well as silica supported tetraallylhafnium $[Hf(All)_4]$ exhibit high activity for ethylene polymerisation [219], whereas alumina-supported $TiBz_4$ and triallylzirconium bromide $[Zr(All)_3Br]$ polymerise propylene to a partially isotactic polymer [216].

3.2.4 Supported Catalysts for Ethylene Polymerisation – Phillips Catalysts

The use of supports in heterogeneous catalysis was well understood by 1950 and flourished with the discovery in 1954 by Hogan and Banks [220,221] of highly active chromium trioxide catalysts supported on silica, which could polymerise ethylene under mild conditions of temperature and pressure to produce predominantly linear, high-density polyethylene.

Catalyst preparation initially involves the impregnation of a high surface support (generally SiO_2 or SiO_2/Al_2O_3 of 0.87:0.13 composition) with an aqueous CrO_3 solution, which is followed by heating (calcination) in a current of air with increasing temperature to a final value of ca 800 °C. At this temperature, all of the physically adsorbed water and most of the surface hydroxyl groups are driven off [222] and chromium becomes present at the support, involving direct linking of each chromium atom to the surface. By the reaction of CrO_3 with surface hydroxyl groups, both silyl chromate $[(\equiv Si-)_2CrO_4]$ and silyl dichromate $[(\equiv Si-)_2Cr_2O_7]$ species are formed [223,224]:

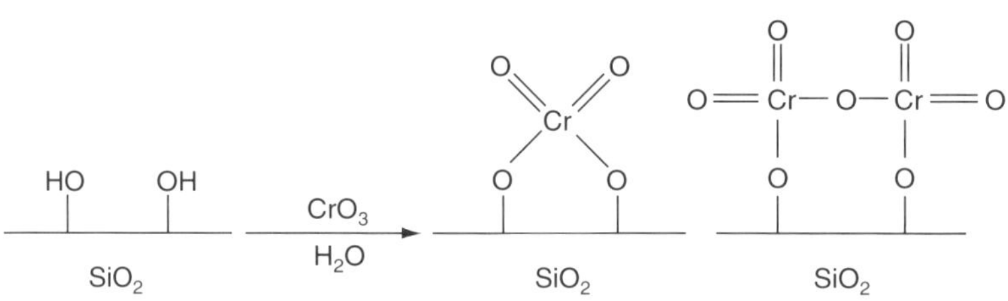

(29)

It should be noted in this connection that hydroxyl groups are not necessarily required to be present on the SiO_2 surface, since completely dehydroxylated silica gel appears still to be an excellent support for active CrO_3-based ethylene polymerisation catalysts [225].

Supported CrO_3 catalysts, referred to as Phillips catalysts, are important industrial catalysts and are employed in high-density polyethylene production. Phillips catalysts polymerise ethylene with an induction period, which has been ascribed to the slow reduction of Cr(VI) by the monomer and to the displacement of oxidation products (mainly formaldehyde) from the catalytic species [226]. The prereduction of the catalyst with the use of H_2 or CO enables the induction period to be eliminated. Active sites thus formed involve surface low-valence Cr(II) and Cr(III) centres, which can appear as mononuclear (formed from chromate species) and binuclear (formed from dichromate species) [227–232].

The selection and treatment of the support is fundamental to the process, and a plant may use catalysts made from a variety of supports to produce a whole range of products. Catalyst productivities are of the order of 5 kg of polyethylene per gram of catalyst or higher, with a corresponding chromium content of 2 ppm or less. The percentage of Cr atoms that form active polymerisation centres has been estimated as 12 % [43]. Typically, commercial Phillips catalysts contain ca 1 % total Cr and have particle sizes of 30–150 μm [224].

Ethylene polymerisation in the presence of Phillips catalysts is carried out in a slurry using a hydrocarbon medium at pressures up to 40 atm and temperatures from 80 to above 100 °C. The temperature of catalyst preparation in its particular stages and the polymerisation temperature are the primary control of molecular weight of the polyethylene formed.

Although the active species containing a Cr–C bond is difficult to determine, an oxidative addition of ethylene to low-valence coordinatively unsaturated chromium ions is proposed to produce such species [219]:

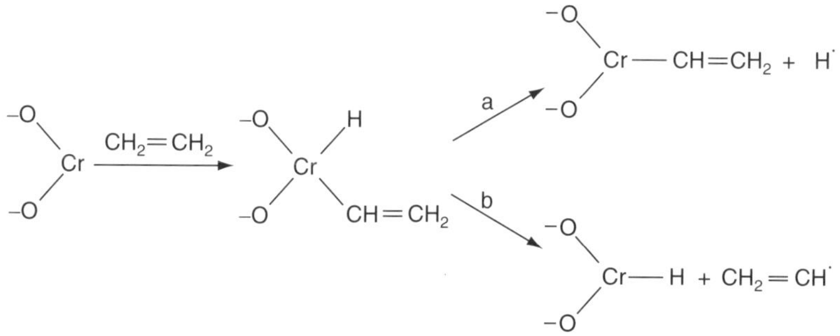
(30)

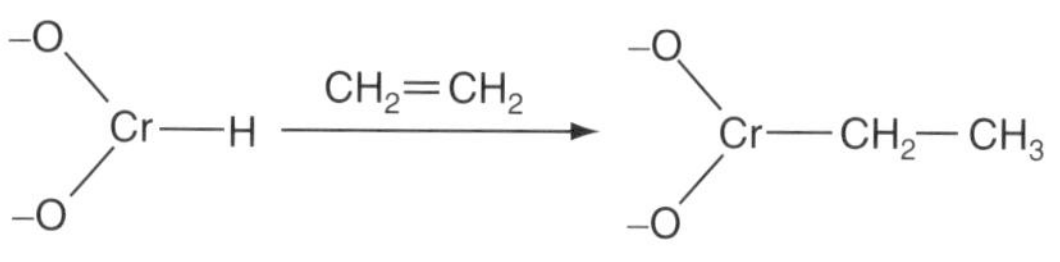
(31)

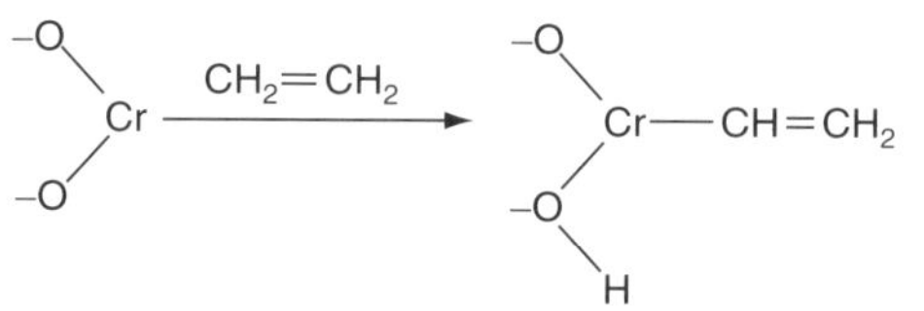
(32)

Another possible pathway for the formation of Cr–C bonds involves allyl species [233–235]:

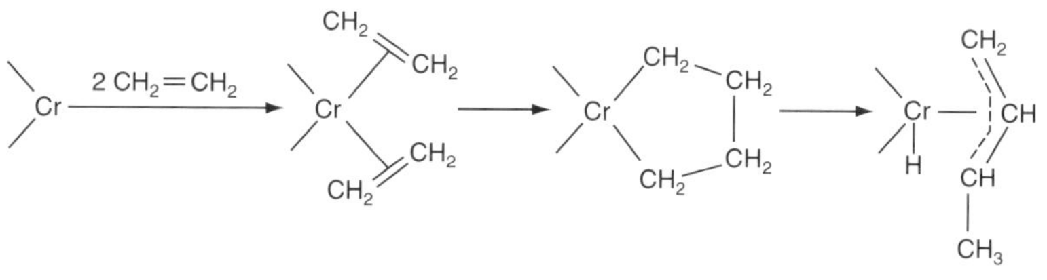
(33)

The chromium active species during polymerisation are believed to involve Cr(II) or Cr(III), or perhaps a combination of both [224].

Phillips catalysts are active in the polymerisation of propylene and higher α-olefins, yielding tacky polymers with irregular structure and small amounts of crystalline polymers; in the case of polypropylene, a small amount of crystalline fraction appeared to constitute the isotactic polymer [236].

Also, the copolymerisation of ethylene and α-olefins can be readily performed using Phillips catalysts. The copolymerisation of ethylene with 1-butene or 1-hexene is the basis of the large expanding linear low-density polyethylene market [28,37,43,237].

Although both Phillips and Ziegler–Natta catalysts copolymerise ethylene and propylene, the latter catalysts are more widely used because they can be more readily tailored to produce a narrow distribution of compositions and molecular weight [43].

There have been further developments of chromium catalysts that fall conceptually in the field of Phillips catalysts. One of them is the use of aluminium phosphate, $AlPO_4$, as a support because of its close similarity to silica, being isoelectronic and isostructural [224].

It was reported also that bis(triphenylsilyl) chromate $[(Ph_3Si)_2CrO_4]$, which is closely related to the proposed active species of the Phillips catalyst, polymerises ethylene at high pressure [238]. When supported on silica, it forms a very active catalyst for low-pressure ethylene polymerisation [226].

Somewhat similar to the Phillips catalyst is the Standard Oil of Indiana catalyst, which was the first olefin polymerisation coordination catalyst to be discovered [239]. It typically consists of MoO_3 supported on alumina or silica and calcined in air at high temperature. Unlike the Phillips catalyst, it is necessary to reduce the catalyst precursor thus obtained with hydrogen at elevated temperature before using it for polymerisation. Despite extensive studies, Standard Oil of Indiana catalysts have not been widely commercialised [43].

3.3 Kinetic–Mechanistic Aspects of Polymerisation with Ziegler–Natta Catalysts

Chemical kinetics deals with the reaction rate, including all factors influencing this rate, and with the explanation of the rate value in terms of the reaction mechanism. Kinetic studies of olefin polymerisation with Ziegler–Natta catalysts have been especially useful in the industrial development of olefin polymerisation processes but have contributed relatively little to the understanding of the polymerisation mechanism [240,241]. Such a situation arises from the very complex nature of Ziegler–Natta catalysts, especially those heterogeneous catalysts in which active species display widely different activities and geometric location. Changes in the number and structure of active sites can take place with changing polymerisation temperature and time. Therefore, the time dependences of the catalyst intrinsic activity, of the overall polymerisation and of the operating diffusion processes complicate the kinetic behaviour of such polymerisation systems. Moreover, monomer adsorption (coordination) as well as adsorption of the activator (complexation), byproducts formed during procatalyst alkylation and the Lewis base (internal and external) in the case of heterogeneous Ziegler–Natta catalysts often make the polymerisation kinetics too complicated.

No precise information about the olefin polymerisation mechanism has been obtained from kinetic measurements in systems with heterogeneous catalysts; analysis of kinetic data has not yet afforded consistent indications either concerning monomer adsorption on the catalyst surface or concerning the existence of two steps, i.e. monomer coordination and insertion of the coordinated monomer, in the polymerisation [scheme (2) in chapter 2]. Note that, under suitable conditions, each step can be, in principle, the polymerisation rate determining step [241]. Furthermore, no π complexes have been directly identified in the polymerisation process. Indirect indications, however, may favour particular steps [242]. Actually, no general olefin polymerisation mechanism that may be operating in the presence of Ziegler–Natta catalysts exists, but rather the reaction pathway depends on the type of catalyst, the kind of monomer and the polymerisation conditions.

In the case of olefin polymerisation with homogeneous Ziegler–Natta catalysts, especially with single-site (metallocene) catalysts, the kinetic analysis may become simpler than in the case of polymerisation with heterogeneous catalysts, and in some instances can serve as a very useful tool for uncovering the true polymerisation mechanism [30,243].

3.3.1 Phenomenological Features

A comparison of kinetic data concerning olefin polymerisation with Ziegler–Natta catalysts of various kinds permits the classification of typical kinetic curves in coordinates of rate and time. Such kinetic rate–time profiles are significant in Ziegler–Natta olefin polymerisations in that their particular shape may be characteristic of a particular catalyst or catalyst–monomer system. A number of different types of kinetic rate–time profiles are encountered in studies of Ziegler–Natta catalysts, and some typical examples are presented in Figure 3.12 (heterogeneous non-supported catalysts) and Figure 3.13 (heterogeneous supported catalysts) [240,241].

It can be seen from Figure 3.12 that the use of the AlR_3 activator produces a more active but less stable catalyst than that with the AlR_2Cl activator. Polymerisations in the presence of these catalysts show an acceleration (or settling) period, the rate of which increases to reach maximum ($TiCl_3$–AlR_3 catalyst) or to reach a more or less steady value ($TiCl_3$–AlR_2Cl catalyst). The nature of this cceleration period is connected with an increase in the catalyst surface area. Spaces between loosely linked primary crystallites of $TiCl_3$ (of 0.03–0.70 μm size in the case of δ-$TiCl_3$), which form agglomerates (of 20–40 μm size in the case of δ-$TiCl_3$), are quickly filled with the polymer formed during the initial polymerisation period. Under the mechanical forces exerted by the growing polymer chains, the agglomerates undergo disintegration into the

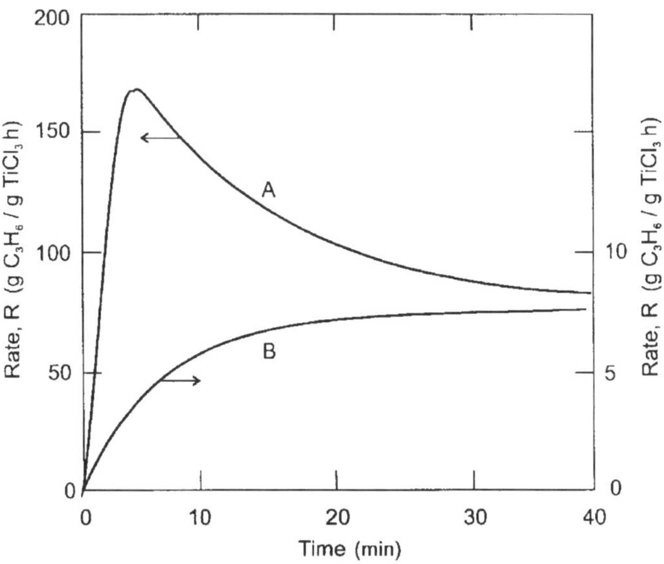

Figure 3.12 Variation in the rate of propylene polymerisation with time: A, $TiCl_3$–$AlEt_3$ catalyst; B, $TiCl_3$–$AlEt_2Cl$ catalyst. Reproduced by permission of John Wiley & Sons, Inc. from Ref. 241. Copyright 1989 Wiley New York

primary crystallites, which results in an increasing number of active sites and thus in a higher polymerisation rate [28,240].

Curves presented in Figure 3.13 testify to the large specificity of supported Ziegler–Natta catalysts regarding the kind of monomer; some centres that polymerise ethylene do not polymerise propylene (or higher α-olefins, which may also be differentiated by particular catalyst centres, depending on the structure of the α-olefin, e.g. branched as in 3-methyl-1-butene or not branched). Therefore, no hints about the monomer reactivity can be obtained by simple comparison of polymerisation rates without simultaneous estimation of the concentration of active sites [241].

Many supported highly active catalysts show behaviour similar to case B in Figure 3.13; the polymerisation rate may also start at a maximum value and then decrease more or less rapidly with time. Such kinetic behaviour is also characteristic of some homogeneous catalysts. Other polymerisation systems show no acceleration period but have a polymerisation rate that remains almost constant with time; this is a rare case and relates, for instance, to 4-methyl-1-pentene polymerisation with $MgCl_2$-supported catalysts containing phthalate esters as well as to ethylene polymerisation with the Cp_2TiCl_2–$[Al(Me)O]_x$ catalyst (apart from a short settling period in the latter case) [240].

Ethylene and propylene polymers yielded by heterogeneous Ziegler–Natta catalysts are characterised by a rather large distribution of molecular weights.

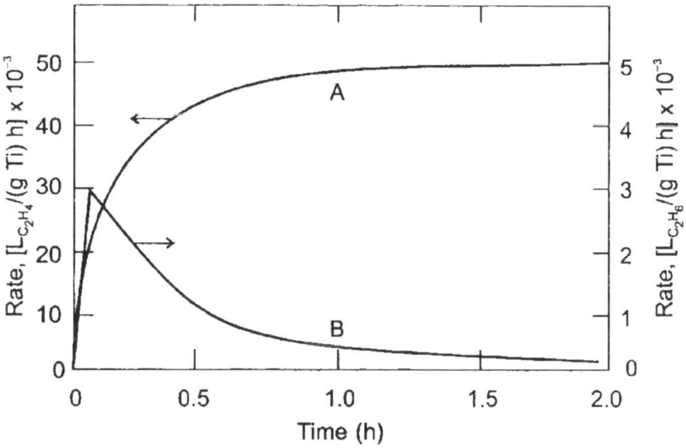

Figure 3.13 Variation in the rate of olefin polymerisation in the presence of the $MgCl_2/TiCl_4$–$Al(i\text{-}Bu)_3$ catalyst with time: A, ethylene; B, propylene. Reproduced by permission of John Wiley & Sons, Inc. from Ref. 241. Copyright 1989 Wiley New York

The $\bar{M}_w/\bar{M}_n$ ratio is usually equal to 5–10 for polyethylene [49,64,66,67, 123,244–247]. A much lower polydispersity is displayed by polymers obtained in polymerisation with homogeneous metallocene catalysts; the $\bar{M}_w/\bar{M}_n$ ratio usually does not significantly exceed a value of 2 [22,95,101,112,138,140]. By polymerising propylene with soluble vanadium-based Ziegler–Natta catalysts at low temperature, a very narrow molecular weight distribution of the polypropylene has been found (the $\bar{M}_w/\bar{M}_n$ ratio usually reaches values of 1.15–1.25) and a linear increase in its $\bar{M}_n$ with time has been observed, indicating a noticeable living character of the polymerisation [75,76,241].

The change in polyolefin molecular weight with time in polymerisations run with heterogeneous Ziegler–Natta catalysts may be, in principle, of two types: the molecular weight increases for an initial period of time and then reaches a constant value (e.g. the α-$TiCl_3$–$AlEt_3$ catalyst) or the molecular weight continues to increase throughout the duration of polymerisation although the rate of this increase diminishes with time (e.g. the δ-$TiCl_3$–$AlEt_2Cl$ catalyst). The latter behaviour can be explained in terms of the failure to reach a steady state in the polymerisation system [240].

Although the polymerisation rate increases with increasing temperature, ethylene and α-olefin polymerisations in the presence of most Ziegler–Natta catalysts arc carried out at moderately elevated temperature, usually not exceeding 100 °C. This is due to destabilisation of the system, which occurs when temperature is raised beyond a certain critical value. There are, however, few catalysts that operate in industrial polymerisation processes at temperatures above 200 °C [51,240].

3.3.2 Reactions Undergone during Polymerisation

3.3.2.1 Polymerisation with Heterogeneous Catalysts

The polymerisation of olefins in the presence of Ziegler–Natta catalysts (with an organometallic activator, e.g. AlR_3) involves individual reaction steps, which may be presented as follows: [240]:

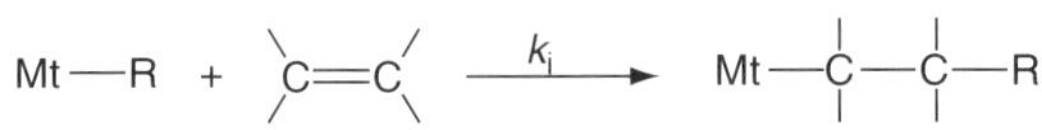

(34)

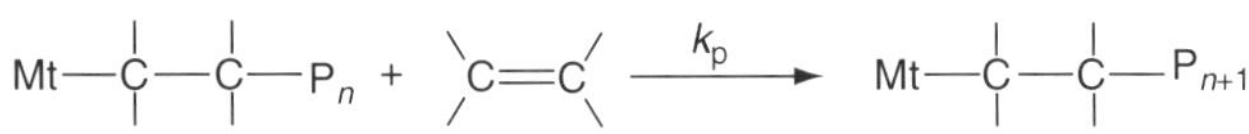

(35)

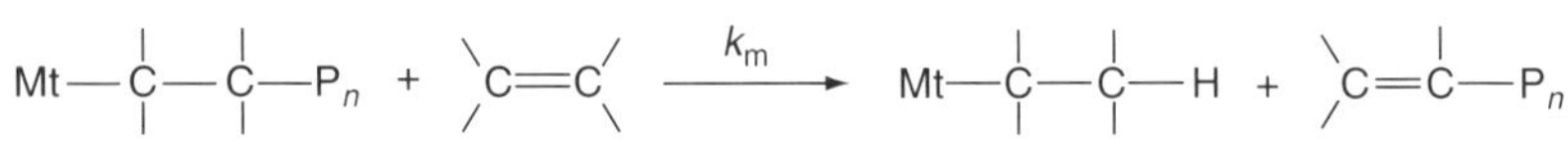

(36)

(37)

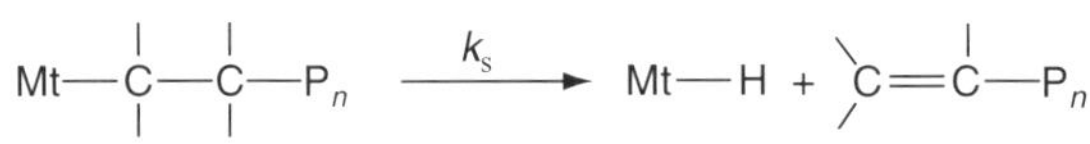

(38)

The chain initiation step [scheme (34)] is assumed to be the insertion of the first coordinating monomer molecule into a transition metal–carbon (or metal–hydrogen) bond at the catalyst active site. The chain propagation step [scheme (35)] is regarded as the insertion of a coordinating monomer molecule into a transition metal–carbon bond at a propagation centre. All propagation centres are regarded as equally active with the same propagation rate constant, k_p, independent of their geometric location and of their degree of polymerisation. However, since it has been established that not all centres have equal activities, the values of k_p must be regarded as average values [240].

In the case of α-olefin polymerisation, the monomer insertion into the Mt–C bond in the propagation step can proceed by two pathways – via monomer 1,2-enchainment (primary insertion) or 2,1-enchainment (secondary insertion):

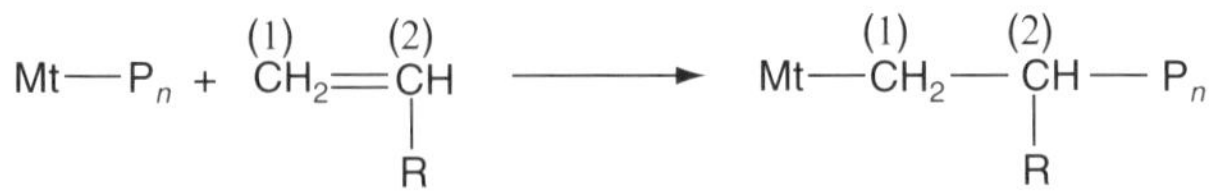

(39)

(40)

In general, Ziegler–Natta catalysts are highly regiospecific in the polymerisation of α-olefins, which means that each insertion mode (primary or secondary) leads to repeating monomer units with equal constitution in the poly(α-olefin) chain formed. In these instances, heterogeneous Ziegler–Natta catalysts polymerise α-olefins by primary insertion of the α-olefin [scheme (39)], leading to Mt–CH_2–CH(R)–[CH_2–CH(R)]$_n$– chains (of isotactic structure), but soluble vanadium-based Ziegler–Natta catalysts polymerise α-olefins (propylene) by secondary insertion of the α-olefin [scheme (40)], leading to Mt–CH(R)–CH_2–[CH(R)–CH_2]$_n$– chains (of syndiotactic structure, if formed at low temperature) [1].

Where no specific chain transfer agent has been added to the polymerisation system, three chain transfer reactions are usually considered: transfer with the monomer [scheme (36)], transfer with the trialkylaluminium activator [scheme (37)] and spontaneous transfer [scheme (38)]. The transfer with the monomer and the spontaneous transfer involve a reaction of *β*-hydrogen elimination from the growing polymer chain, whereas the transfer with the activator is the exchange reaction of substituents at two metals [240,241].

The relative importance of the chain transfer process depends on the polymerisation conditions and on the kind of catalyst used. For the $MgCl_2/TiCl_4$–AlR_3 catalyst, over a large range of activator and monomer concentrations, chain transfer with the monomer [scheme (36)] is by far the most important chain transfer reaction. Other chain transfer reactions, particularly the spontaneous transfer, become detectable only under extreme conditions [241,248]. With $TiCl_3$–ZnR_2 catalysts, chain transfer with the activator appeared to be the most significant chain transfer reaction [249].

Chain transfer reactions have also been started by the addition of chain transfer agents to polymerisation systems with Ziegler–Natta catalysts. In such cases, hydrogen is commonly used as an important regulator to lower the molecular weights of polyethylene and polypropylene produced with heterogeneous catalysts, although it is known slightly to reduce the activity of Ziegler–Natta catalysts [37]. Average molecular weights of typical medium-flow polyolefins produced with the use of modern catalysts are of the following values: for high-density and linear low-density polyethylenes, $\bar{M}_w \approx 96\,000$ and $\bar{M}_n \approx 18\,000$–$23\,000$; for typical medium-flow polypropylene (isotactic), $\bar{M}_w = 460\,000$ and $\bar{M}_n = 54\,000$ [43].

The chain transfer reaction with hydrogen is as follows:

$$\mathrm{Mt{-}\overset{|}{\underset{|}{C}}{-}\overset{|}{\underset{|}{C}}{-}P}_n + \mathrm{H_2} \xrightarrow{k_{\mathrm{H_2}}} \mathrm{Mt{-}H} + \mathrm{H{-}\overset{|}{\underset{|}{C}}{-}\overset{|}{\underset{|}{C}}{-}P}_n \qquad (41)$$

Termination of the olefin polymerisation with heterogeneous Ziegler–Natta catalysts by the addition of carbon monoxide to the system is often used in the laboratory to determine the active centres of the catalyst.

Note that chain transfer reactions proceeding according to schemes (36) to (38) and (41) terminate individual polymer chain growth but do not terminate the polymerisation kinetic chain, since the new centres formed in these reactions initiate further polymerisation.

It may be interesting that chain termination with hydrogen, which is utilised in Ziegler–Natta polymerisations, does not operate in polymerisation systems with Phillips catalysts; no influence of hydrogen to control the molecular weight of polyethylene in the latter case was achieved [37].

It is apparent from schemes (36) and (38) that, in some instances, polymer chains with terminal unsaturation are formed as a result of chain transfer reactions. Alternatively, such vinyl terminated chains can undergo, as α-olefins, an insertion into the Mt–C bond at the active site, resulting in the formation of branched polymer chains. Actually, unsaturated chain insertion into a metal–ethyl bond may generate ethyl (short) branching, whereas insertion of this unsaturated chain into a metal–polymer chain bond may generate long branching. The formation of short and long branches in polyethylene is as follows [51]:

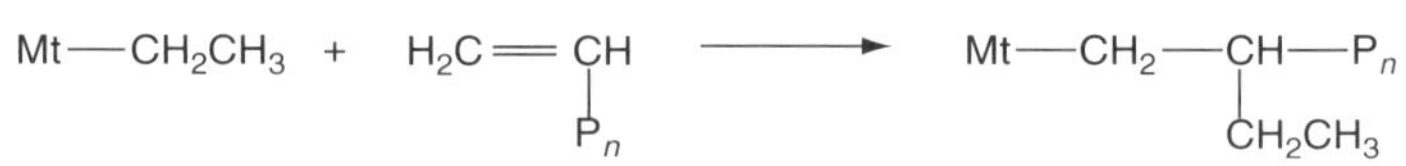

(42)

(43)

However, considering the much lower reactivity of α-olefins than that of ethylene in copolymerisation systems [250], high-density polyethylene is essentially free of both long and short branchings, especially of the former, although very small amounts can be deliberately incorporated to achieve specific product targets. Typically, high-density polyethylenes contain 1–3 long branches per 500 monomer units, which is far less than the 10–20 long branches in low-density polyethylenes produced by high-pressure radical polymerisation processes.

Linear low-density polyethylene, which contains small amounts of 1-butene, 1-hexene or 1-octene units incorporated by copolymerisation with ethylene, has branching of uniform length, which is randomly distributed along a given chain, but there is a spread of average concentrations between chains, the highest concentration of branches being generally in shorter chains [43].

For the majority of olefin polymerisations with heterogeneous Ziegler–Natta catalysts, the polymerisation rates, R_p, are proportional to the concentrations of procatalyst (MtX_n) and monomer (M), but do not depend on the concentration of alkylaluminium activator (A) as long as a threshold concentration is maintained [37]:

$$R_p = k_p \times [MtX_n]^1 \times [M]^1 \times [A]^0$$

This means that there is practically no dependence of the olefin polymerisation rate on the activator/procatalyst molar ratio over a wide range. In some instances, however, such a dependence may be observed, when the concentrations of the procatalyst and the activator are a few orders higher than those normally applied for olefin polymerisations [251,252].

The equation describing the polymerisation rate is simply expressed in terms of the dependence on the concentration of propagation active sites, $C_p{}^*$, and the monomer concentration, [M], where $C_p{}^*$ is assumed to be of constant value ($dC_p{}^*/dt = 0$), i.e. the polymerisation system is under steady state conditions:

$$R_p = k_p \times C_p{}^* \times [M]$$

Such an expression is based on the simple kinetic model of olefin polymerisation and is valid for systems with non-supported Ziegler–Natta catalysts (of moderate activity). Some important limiting expressions can be deduced when [M] is high and when $k_i \cong k_p >> k_m$, and $C_p{}^* \cong C^*$, where C^* denotes the total concentration of active sites, and $\bar{P}_n$ denotes the number-average degree of polymerisation:

$$R_p = k_p \times C^* \times [M]$$

and

$$\bar{P}_n = k_p/k_m$$

i.e. $\bar{P}_n$ at high [M] becomes independent of all polymerisation variables [240].

Under more normal experimental conditions:

$$1/\bar{P}_n = k_m/k_p + k_a \times [A]/(k_p \times [M]) + k_s/(k_p \times [M])$$

Therefore, a plot of $1/\bar{P}_n$ versus 1/[M] for polymers obtained at constant [A] should be a straight line with (k_m/k_p) as the intercept and $(k_a[A]/k_p + k_s/k_p)$ as the slope. Furthermore, when $1/\bar{P}_n$ is plotted versus [A] for polymers obtained at constant [M], a straight line should be derived with $(k_m/k_p + k_s/k_p\ [M])$ as the intercept and $(k_a/k_p\ [M])$ as the slope [240].

Kinetic models referred to as adsorption models have been proposed, especially for olefin polymerisation with highly active supported Ziegler–Natta catalysts, e.g. $MgCl_2$/ethyl benzoate/$TiCl_4$–AlR_3. These models include reversible processes of adsorption of the monomer (olefin coordination at the transition metal) and adsorption of the activator (complexation via briding bonds formation). There are a variety of kinetic models of this type, most of them considering the actual monomer and activator concentrations at the catalyst surface, Θ_M and Θ_A respectively, described by Langmuir–Hinshelwood isotherms. It is to be emphasised that Θ_M and Θ_A must not be the same as the respective bulk concentrations [M] and [A] in solution. Therefore, fractions of surface centres complexed by the monomer and the activator, but not bulk concentrations in solution, are assumed to represent the actual monomer and activator concentrations respectively. This means that the polymerisation rate equation based on the simple polymerisation model should take into account the actual concentration of monomer at the propagation centres, expressed by Θ_M, giving a relationship that might be more strictly applicable [240]:

$$R_p = k_p \times C_p^* \times \Theta_M$$

The importance of active centre concentration determinations, which are crucial to any fundamental characterisation of Ziegler–Natta catalysts, should be emphasised. Knowledge of this concentration allows evaluation of the appropriate values of the polymerisation rate (propagation rate) and the growing time (average lifetime) of the polymer chains. When more than one type of active centre exists, the overall polymerisation rate is expressed as the sum of the particular rates of polymerisation proceeding at individual active centres of different types. The simplest case of this is the polymerisation of propylene by a catalyst producing an atactic and isotactic polymer. It should also be realised, however, that it is not only C_p^* that may change with time; the intrinsic activity of the active centre may also vary with time, and this will be reflected in a change in the value of the propagation rate constant k_p [240].

The activation energy of the overall olefin polymerisation process has been determined, though the C_p^* and other values may also change with temperature in Ziegler–Natta systems; the activation energy ranges between 43 and 59 kJ/mol. Dividing k_p by the average chain transfer rate constant, k_t, gives the number-average molecular weight of the polymer chain. The k_m/k_p values were found to be between 4×10^{-4} and 1.2×10^{-3} for catalysts with an α-$TiCl_3$ precursor and between 10^{-5} and 2×10^{-4} for the δ-$TiCl_3$–$AlEt_2Cl$ catalyst in propylene polymerisations at 70 °C [37].

Most commonly used methods for determination of the active centre concentration are kinetic methods (variation in the number-average degree of polymerisation and variation in the polymer yield with time), quenching methods (labelled quenching agent such as titrated alcohol or water, deuterated alcohol or water, $^{131}I_2$), radio-tagging methods (use of ^{14}C-labelled AlR_3 as the catalyst activator, treating the system with ^{14}CO or $^{14}CO_2$ and quenching with alcohol) and simultaneous adsorption and rate measurements (poisons such as allene and CO). Note in this connection that values obtained for C_p^* using different determination methods differ from one another [37, 240].

3.3.2.2 Polymerisation with Homogeneous Metallocene Catalysts

In the case of olefin polymerisation in the presence of homogeneous metallocene-based catalysts, the individual polymerisation stages have not been very thoroughly investigated. However, kinetic studies have helped, among others, to define the nature of the active centres and to establish the occurrence of some polymerisation elementary steps in a quantitative way.

Polymerisation in homogeneous systems has been found to occur very rapidly; the time of chain growth is approximately 10^{-3}–10^{-2} s. With a degree of polymerisation of 10 000, a single act of insertion lasts 10^{-6}–10^{-5} s, which corresponds to the duration of fast biological processes [95]. With the Cp_2ZrCl_2–$[Al(Me)O]_x$ catalyst, a few dozen tons of polyethylene per gram Zr/h can be obtained; one Zr atom is capable of producing dozens of thousands of macromolecules per hour [37].

In polymerisations run with metallocene-based catalysts, the chain propagation step involving primary α-olefin insertion into the transition metal–carbon bond [scheme (39) has been found to operate in all the studied systems, regardless of the tacticity of the polymers formed (isotactic, syndiotactic, etc.) [29,30]. However, metallocene–methylaluminoxane systems have appeared not to be very highly regiospecific catalysts, and misinsertions [secondary insertions according to scheme (40)] occur, the consequence of which is the appearance of Mt–CH(R)–CH_2–P_n species in the polymer chains propagating predominantly via the Mt–CH_2–CH(R)–P_n species. Insertion of propylene into the secondary Zr–CH(Me)–CH_2 unit has been revealed to be ca 100 times slower than that into a primary Zr–CH_2–CH(Me) unit [114, 253]. As a consequence, secondary propylene insertions (2,1-enchainment) according to scheme (40) with an incidence frequency of only 1 % are sufficient to tie up about 90 % of the catalyst in secondary Zr–CH(Me)–CH_2 units when the polymerisation is conducted in the absence of hydrogen. Considering this, substantially increased activities can be expected for a catalyst that will be unimpeded by occasional 2,1-misinsertions [30].

Occasional regioerrors appear significantly to inhibit the polymerisation of α-olefins by methylaluminoxane-activated metallocene catalysts [114, 138, 253–261]. In order to reduce the number of secondary Zr–CH(R)–CH_2 species, and therefore to accelerate the polymerisation, advantage has been taken of the chain transfer reaction with hydrogen:

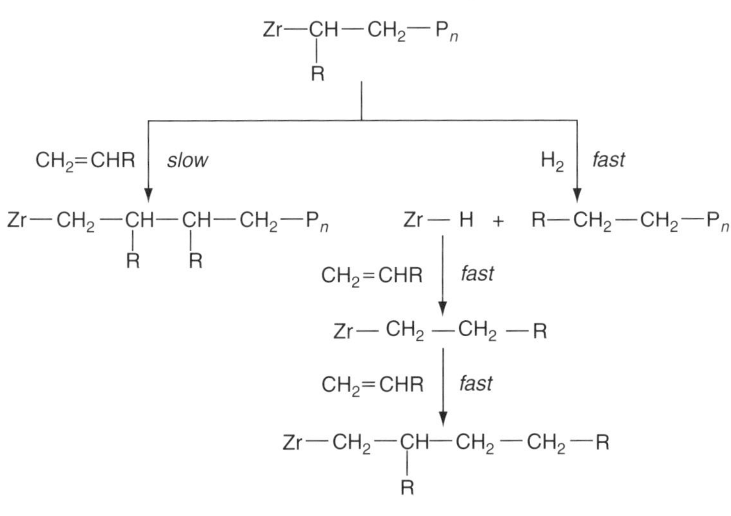

(44)

For example, the rates of propylene and 1-butene polymerisation by the *rac.*-$(IndCH_2)_2ZrCl_2$–$[Al(Me)O]_x$ catalyst increase in the presence of hydrogen roughly by a factor of 10–60 respectively [260]. These polymers were found to be free of misinserted units; hydrogenolysis apparently eliminates the slowly inserted 2,1-enchained monomer units and allows the start of a new, fast-growing polymer chain [scheme (44)] [30].

Kinetic studies concerning olefin polymerisation in the presence of homogeneous *ansa*-metallocene–methylaluminoxane catalysts have revealed that β-hydride transfer from the growing polymer chain directly to a coordinating olefin, i.e. chain transfer with the monomer [scheme (36)], has been suggested as the predominating chain transfer reaction while the metallocene precursor is free of α-substituent (in the 2-position of the cyclopentadienyl-like ligand) [253, 256, 262]. However, for *ansa*-metallocene-based catalysts with an α-substituted cyclopentadienyl-like ligand, direct β-hydride transfer from the growing polymer chain to the metal, i.e. spontaneous chain transfer [e.g. scheme (38)], occurs almost exclusively [256, 263].

The chain transfer reaction with the activator, i.e. $[Al(Me)O]_x$, during olefin polymerisation in homogeneous systems with metallocene catalysts proceeds (analogously to scheme (37)] to a greater extent than that involving the AlR_3 activator in heterogeneous Ziegler–Natta systems [264].

Chain transfer reactions in homogeneous olefin polymerisation systems with metallocene catalysts, which terminate individual polymer chains, in some instances can also terminate the polymerisation kinetic chain. For example, chain transfer with the monomer in propylene oligomerisation or polymerisation, which involves σ bond metathesis reactions between the Mt–C species of the growing polymer chain and the C–H species of methyl [scheme (45)] or vinyl [scheme (46)] groups in the monomer, gives rise to temporally inactive metal–allyl or metal–vinyl species respectively [177, 241, 264]:

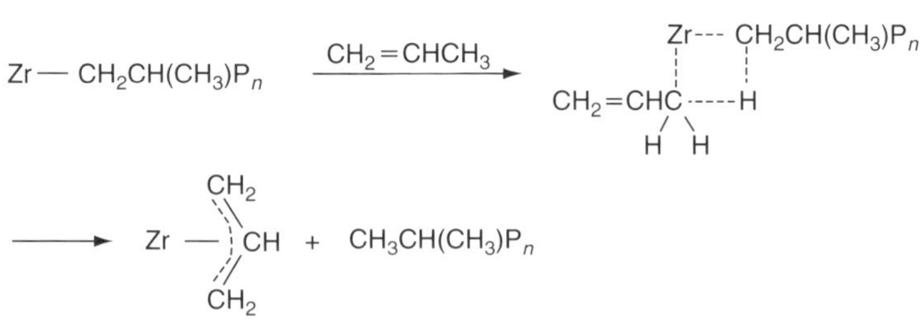

(45)

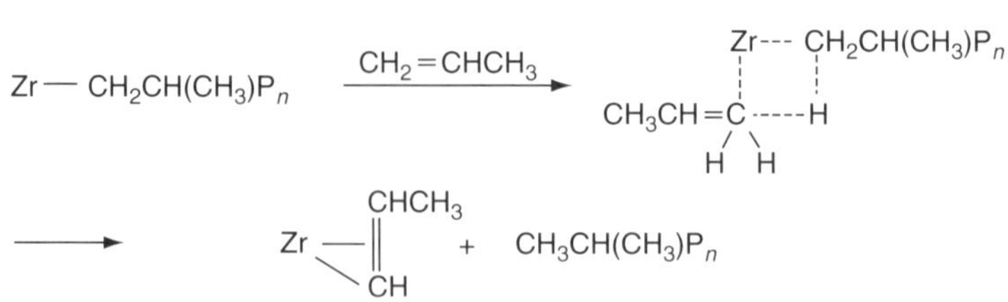

(46)

It has already been mentioned that metallocene-based catalysts differ from heterogeneous Ziegler–Natta catalysts in that they produce polyolefins of narrow molecular weight distribution. The broad molecular weight distribution of polyolefins obtained with heterogeneous catalysts is usually associated with the non-uniformity of active centres [123, 265]. However, in the case of olefin polymerisation with metallocene catalysts, a polydispersity of 2 is predicted by Schulz–Flory statistics, since polymers arise from identical catalyst centres with fixed rates of chain propagation and chain termination. Therefore, such a value of polydispersity is regarded as an indication that only a single catalyst site contributes to the propagation reaction in a homogeneous polymerisation system. Even though the deactivation and reactivation reactions are likely to generate a diversity of complex species in these polymerisation systems under steady-state conditions, only one of these species, presumably the $[(Cp'_2ZrR).(olefin)]^+$ cation, appears to contribute to chain propagation [30].

While ethylene is polymerised with non-bridged metallocene catalysts of the Cp_2ZrX_2–$[Al(Me)O]_x$ type to give polyethylenes of high molecular weights in the range 100 000–1 000 000 [90, 91, 95, 103], propylene polymers obtained at room temperature with these catalysts are of rather low molecular weight in the range 200–1000 [103]. However, the molecular weight of poly (α-olefin)s is strongly influenced by the kind of metallocene catalyst employed. The application of *ansa*-metallocene-based catalysts for propylene polymerisation results in an increased molecular weight of the polymers obtained. For example, *rac.*-$(IndCH_2)_2ZrCl_2$–$[Al(Me)O]_x$ and *rac.*-$(ThindCH_2)_2ZrCl_2$–$[Al(Me)O]_x$ catalysts polymerise propylene at room temperature to polymers with $\bar{M}_w \approx 50\,000$ [101, 254, 255]. Higher molecular weight polypropylenes were obtained with analogous *ansa*-hafnocene-based catalysts, although at the expense of substantially decreased catalytic activities [140, 266]. However, the additional α-methyl substituents at each cyclopentadienyl-like ligand in a zirconocene procatalyst result in the polymers having an increased molecular weight while still displaying high catalytic activity [132, 138]; catalysts obtained using precursors such as *rac.*-$Me_2Si(MeInd)_2ZrCl_2$ or *rac.*-$Me_2Si(Me,NphInd)_2ZrCl_2$ (Figure 3.14) produce, at a polymerisation temperature of 50 °C, polypropylene with a molecular weight $\bar{M}_w$ of 100 000–400 000 [256, 263] and 1000 000 [263] respectively.

The molecular weight of a polyolefin can be easily lowered by increasing the temperature, increasing the metallocene concentration or decreasing the monomer concentration in the polymerisation system. The molecular weight is also lowered by the addition of small amounts (0.1–2 mol. %) of hydrogen; for instance, in the case of ethylene polymerisation with the Cp_2ZrCl_2–$[Al(Me)O]_x$ catalyst, the polyethylene molecular weight $\bar{M}_w$ is 170 000 without H_2 in the polymerisation system, but in a system with H_2 (0.5 mol.-%) it diminishes to 42 000 [267].

At the end of considerations of olefin polymerisation with homogeneous metallocene catalysts, it should be added that it is necessary to differentiate between the soluble catalyst system itself and the polymerisation system. For

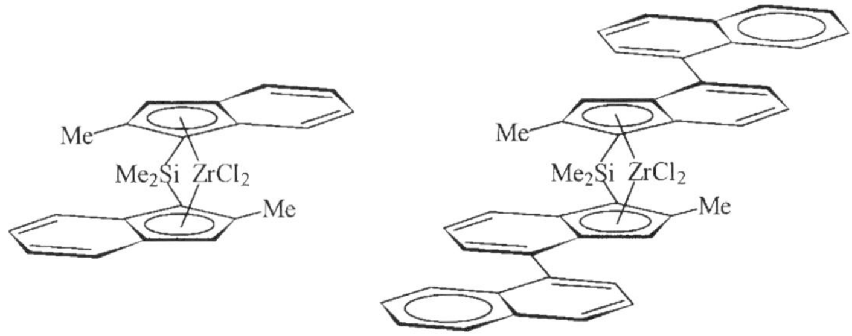

Figure 3.14 Additionally α-substituted *ansa*-metallocene precursors, *rac.*-dimethylsilylenebis[1-(2-methylindenyl)]zirconium dichloride [*rac.*-$Me_2Si(MeInd)_2ZrCl_2$] and *rac.*-dimethylsilylenebis[1-(2-methyl-4-naphthylindenyl)]zirconium dichloride [*rac.*-Me_2 Si(Me,NphInd)$_2$ $ZrCl_2$], for obtaining high-molecular weight polypropylene. Front side view. Reproduced by permission from Ref. 263. Copyright 1994 American Chemical Society

instance, the well-defined bis(cyclopentadienyl)titanium-based system is soluble, but unfortunately it becomes heterogeneous when polyethylene is formed [37].

3.4 Models of Active Sites and the Polymerisation Mechanism

Although a two-step mechanism, involving monomer coordination and enchainment by the insertion of the coordinating monomer [scheme (2) in chapter 2], is commonly accepted for olefin polymerisation with coordination catalysts, there is no general uniform mechanism that might operate in all polymerisation systems and under all polymerisation conditions applied. As already mentioned, even questions concerning the rate controlling steps have yet to be resolved [58, 242, 268–271]. However, for most common polymerisations under common conditions, the insertion of the coordinated monomer is assumed to be the rate determining step [268–270].

Although conflicting conclusions have also been drawn from investigations using *ab initio* molecular orbital methods in combination with molecular mechanics methods [272–274], most studies agree that the insertion step is the rate controlling one. The insertion of coordinated olefin proceeds very fast; an energetically favourable spatial arrangement of ligands, metal–carbon bond and coordinated monomer determines high rates of insertion. The insertion involves the cleavage of the metal–carbon bond and the double bond in the monomer molecule, and the formation of a new metal–carbon bond. It is characteristic that olefin polymerisations with both heterogeneous and homogeneous catalysts involve *cis*-insertion of the C=C bond in the coordinated olefin into the Mt–C bond in all cases, irrespective of the monomer enchainment mode (1,2 or 2,1) and the tacticity of the poly(α-olefin) formed. A *cis*-opening of the olefin double bond has been proved conclusively by using deuterated propylene for the polymerisation and determining the chain microstructure of the polymers formed. For example, poly(1-deuteropropylene) of *erythro*-diisotactic structure (Figure 3.8) is formed in the polymerisation of *cis*-1-(^{2}H)-propene with heterogeneous Ziegler–Natta catalysts [26, 275]. Such a polymer structure results from the monomer *cis*-insertion:

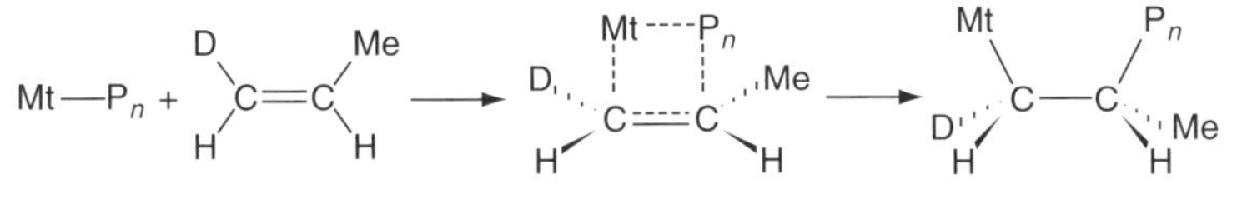

(47)

If the polymerisation of *cis*-1-(^{2}H)-propene occurred with monomer enchainment via another route (namely via the addition of the *trans* type), poly(1-deuteropropylene) of *threo*-diisotactic structure [scheme (48)] would be formed, but this is not the case:

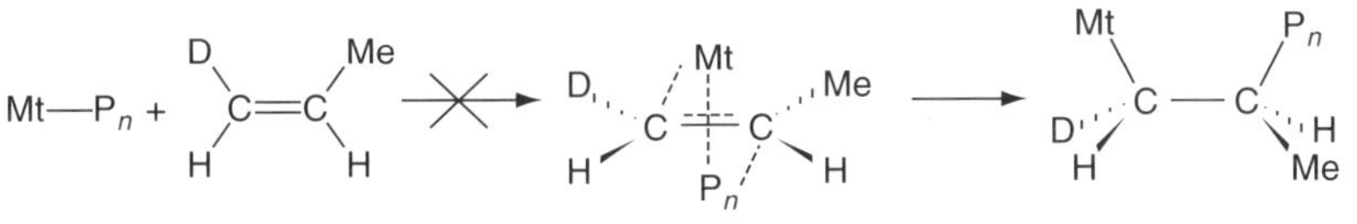

(48)

Similarly, in the polymerisation of *cis*-1,2,3,3,3-penta-1-(^{2}H)-propene run with the homogeneous $(IndCH_2)_2ZrCl_2$–$[Al(Me)O]_x$ catalyst, the respective poly(-pentadeuteropropylene) of *erythro*-diisotactic structure is formed, which confirms the insertion to be of the *cis* type [22].

On the other hand, the polymerisation of *trans*-1-(^{2}H)-propene with heterogeneous Ziegler–Natta catalysts leads to the formation of the respective polymer of *threo*-diisotactic structure [scheme (49)], which proves the *cis*-insertion of the monomer [26,275]:

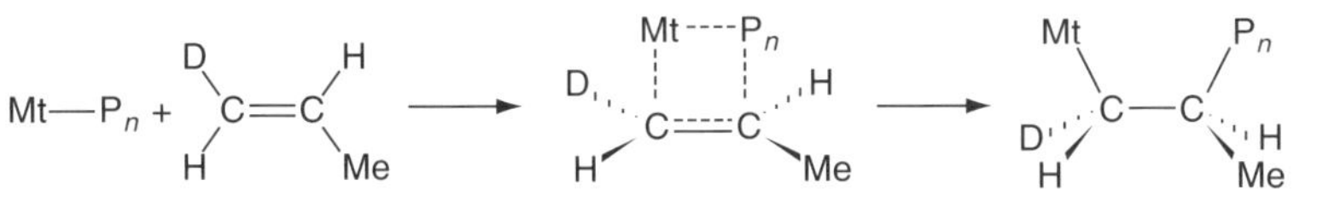

(49)

Furthermore, studies of the microstructure of copolymers formed by the low-temperature copolymerisation of *cis*-1-(^{2}H)-propene (or *trans* isomer) and perdeuteropropene in the presence of soluble vanadium-based Ziegler–Natta catalysts showed syndiospecific propagation to involve a monomer insertion of the *cis* type [27].

Since olefin insertion into the metal–carbon bond has been established to be of the *cis* type, it has been considered to proceed by a concerted mechanism involving the formation of a four-membered transition state. However, various models of active centres and of the insertion mechanism have been proposed for olefin polymerisation systems with coordination catalysts.

3.4.1 Polymerisation with Heterogeneous Ziegler–Natta Catalysts

A number of models of the active centres in Ziegler–Natta catalysts have been postulated. The diversity of these models arises from the multitude of products found to be formed or believed to be formed in the reaction of the catalyst precursor with the activator [e.g. schemes (4) to (8) and (12)]. The proposed active centres fall into either of two general categories: those containing monometallic species with the central transition metal atom (e.g. Ti), and those containing bimetallic species with the central transition metal atom linked via bridges with the metal atom originating from the activator (e.g. Al).

The most widely accepted olefin polymerisation mechanism that has been postulated to operate in systems with monometallic active centres is that proposed by Cossee [268,276–278]. According to this proposal, concerning polymerisation with catalysts based on layered violet titanium trichloride (α-$TiCl_3$), the active centres are located on lateral surfaces and formed by the replacement of the singly bonded non-bridging Cl atom protruding from the titanium trichloride surface by an alkyl group from the activator [scheme (8)]. Hence, the active centre contains the Ti atom surrounded by four Cl atoms from the crystal lattice (bridged to two other Ti atoms), an alkyl group, which forms an active Ti–C bond, with the metal, and a coordination vacancy (□). The titanium atom in such a site, Cl_4 Ti (□)R, attains an octahedral configuration when the coordination vacancy becomes occupied by the coordinating olefin molecule, namely Cl_4 Ti(olefin)R. The olefin is postulated to be coordinated at the vacant site with the double bond parallel to an octahedral axis. The polymerisation of olefin according to the mechanism devised by Cossee is as follows:

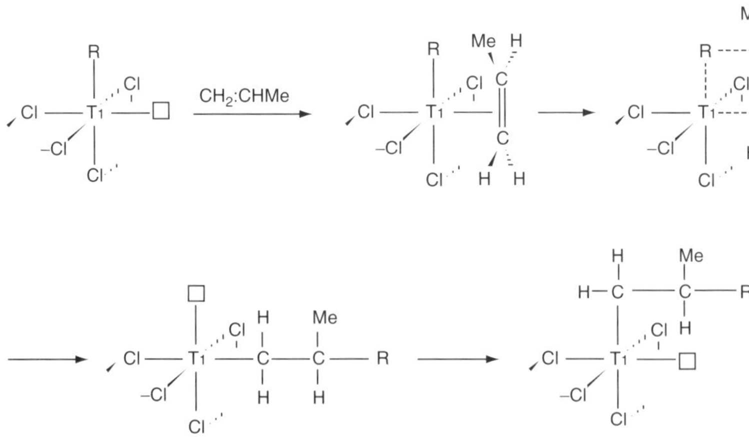

(50)

The metal–carbon (Ti–R) bond at the uncomplexed Cl_4 Ti (□)R active centre is relatively stable. The coordination of the olefin molecule at this centre, due to π bond formation (Figure 2.1), leads to lowering of the energy in the resultant π complex, Cl_4 Ti(olefin)R; the molecular orbital originating from the mixing of $3d_\pi$ orbitals from titanium with π^* orbitals from the olefin is sufficiently lower in energy with respect to the original $3d_\pi$ titanium orbitals for an electron easily to be transferred into it from the excited Ti–C active bond. Hence, the resultant alkyl radical attaches itself to the nearest carbon atom of the coordinated olefin in a concerted way, involving a four-centred transition state. Therefore, Cossee's mechanism assumes that, in the insertion step, the growing polymer chain migrates to the position previously occupied by the coordinating olefin molecule. Such a mechanism is referred to as the chain migratory insertion mechanism. The *cis*-migration of the chain to the π-alkene ligand, with π–σ rearrangement, gives a new coordinatively unsaturated species that is able to coordinate another olefin molecule. However, in the last stage of the insertion, a back skip of the chain has been postulated by Cossee to take place in order to fulfil stereochemical requirements of α-olefin isospecific polymerisation that the growing chain always occupies the same position when a new monomer molecule is coordinated [scheme (50)].

It should be emphasised that strong support has been provided for this mechanism of α-olefin polymerisation, involving sites located on lateral α-$TiCl_3$ faces, from molecular orbital calculations [268]. Analogous model sites, but located on layers in relief with respect to lateral faces of layered $TiCl_3$ structures [279–282], as well as for catalysts with the $MgCl_2/TiCl_4$ precursor [279], have also been considered in molecular mechanics studies.

Although it is apparent that for isospecific polymerisation the polymer chain must skip back to the original position after each chain migratory *cis*-insertion in order to maintain sterically identical propagation steps, such a stage has been considered as a weakness of Cossee's mechanism. In connection with this, it has been hypothesised that the insertion of a π-complexed olefin molecule is initiated by another incoming monomer molecule; according to this hypothesis, a coordinated olefin is unable to undergo insertion without the aid of another incoming monomer molecule [283].

Monometallic centres as in the model proposed by Cossee have been widely accepted, though this model does not take into consideration the presence of an activator or products formed from it in Ziegler–Natta polymerisation systems [e.g. schemes (4) to (6)]. Bimetallic models give plausible answers concerning the role of activators and are in closer agreement with many experimental data [37]. The most commonly known olefin polymerisation mechanism involving bimetallic active centres is that postulated by Rodriguez and Van Looy [285, 286]. According to this mechanism, which is based on Cossee's mechanism, a molecule of the activator or the product originating from it (AlR_2 Cl) is complexed to the transition metal (Ti) via the Cl bridge; the chain propagation, however, is monometallic:

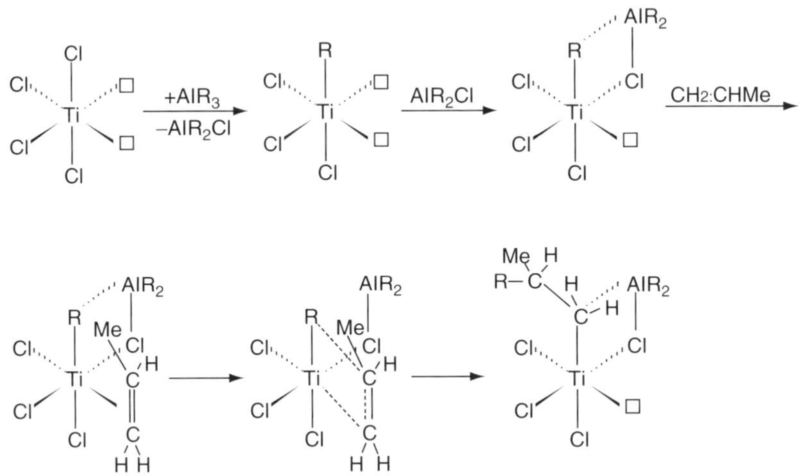

(51)

Variations in monometallic and bimetallic mechanisms have been proposed. The monometallic mechanisms seem to be inherently simpler than bimetallic mechanisms except for requiring a migration step. On the other hand, bimetallic mechanisms that consider the presence of the activator in the polymerisation system can be more convincing in some instances.

A plausible concept of active sites and the mechanism of propylene polymerisation in the presence of heterogeneous non-supported Ziegler–Natta catalysts, e.g. δ-$TiCl_3$–$AlEt_3$, as well as supported Ziegler–Natta catalysts, e.g. $MgCl_2$/ethyl benzoate/$TiCl_4$–$AlEt_3$ or $MnCl_2$/ethyl benzoate/$TiCl_4$–$AlEt_3$, has been given by Doi *et al.* [287]. Two types of active site were assumed: a strictly stereospecific site (Figure 3.15) yielding highly isotactic polypropylene and a non-stereospecific site, which can be interconverted between isospecific and syndiospecific centres (Figure 3.16), producing atactic polypropylene consisting of isotactic and syndiotactic stereoblocks of short sequence length [287].

In the stereospecific site (Figure 3.15) the surface metal (Mg, Mn or Ti) is coordinatively saturated, whereas in the non-stereospecific site (Figure 3.16) it has a chlorine vacancy; therefore, a reversible migration of dialkylaluminium moiety in the latter site is possible. The frequency of such migration with respect to the rates of subsequent propagation steps determines the sequence length of stereoblocks in the atactic polypropylene formed. When a temporary syndiospecific stereocontrol by the syndiospecific centre at the non-stereospecific site lasts long enough with respect to the isospecific steric control performed by the isospecific centre at this site (Figure 3.16), i.e. when interconversion from the syndiospecific to the isospecific centre occurs slowly, the formation of a predominant syndiotactic fraction in atactic polypropylene fraction can be expected. Actually, syndiotactic polypropylene is almost always formed, in variable amounts, as a byproduct of isotactic polypropylene, when the polymerisation

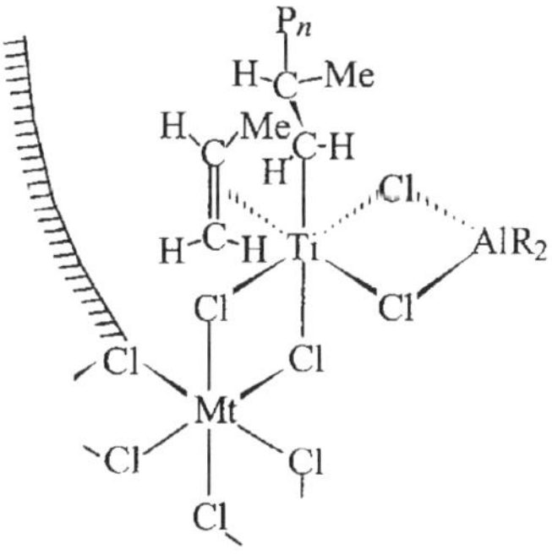

isospecific propagation

Figure 3.15 Possible structure for active sites in heterogeneous Ziegler–Natta catalysts producing highly isotactic polypropylene. Mt = Mg or Mn ($MgCl_2$- or $MnCl_2$-supported catalysts), or Ti (non-supported $TiCl_3$-based catalysts); ┬┬┬┬┬-surface of $MgCl_2$, $MnCl_2$ or $TiCl_3$

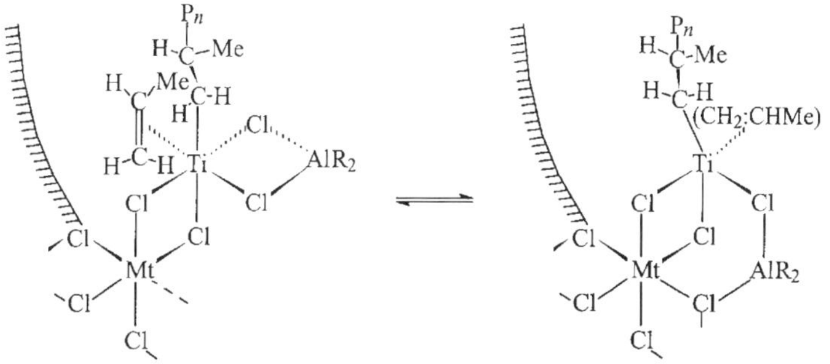

isospecific propagation

syndiospecific propagation

Figure 3.16 Possible structures for active sites in heterogeneous Ziegler–Natta catalysts producing atactic polypropylene. Mt = Mg or Mn ($MgCl_2$- or $MnCl_2$-supported catalysts), or Ti (non-supported $TiCl_3$-based catalysts); ┬┬┬┬┬-surface of $MgCl_2$, $MnCl_2$ or $TiCl_3$

is carried out in the presence of heterogeneous Ziegler–Natta catalysts based on either $TiCl_3$ [9,287] or $TiCl_4$ supported on $MgCl_2$ [287,288].

Note, in this connection, the stereoblock structure of atactic polypropylene that is yielded as a byproduct in addition to the isotactic polymer in polymerisations run with heterogeneous Ziegler–Natta catalysts of the first, second and third generations [289]; this structure differs from the structure of completely atactic polypropylene obtained in the presence of aspecific heterogeneous catalysts with precursors derived by supporting $TiCl_3$ on $MgCl_2$ [68,69] as well as homogeneous metallocene-based catalysts, such as *rac.*-ethylenebis[1-(3-

methylindenyl)]zirconium dichloride–methylaluminoxane *rac.*-$(MeIndCH_2)_2ZrCl_2-[Al(Me)O]_x$ [112].

As already mentioned, commercially important $MgCl_2$-supported Ziegler–Natta catalysts for the polymerisation of propylene require ball-milling of $MgCl_2$ in the presence of an internal Lewis base and then reaction with $TiCl_4$ during the preparation of the procatalyst, which is followed by activation with $AlEt_3$ in the presence of an external Lewis base. Although many studies have been carried out on the structure of such catalysts, interactions of the Lewis base with Mg and Ti atoms at active centres have not been unambiguously characterised. A few models have been proposed for active sites present in $MgCl_2$/Lewis base/$TiCl_4$–$AlEt_3$ catalysts. These models may fall into two basic categories: models in which active sites with the Ti(III) species are directly bound to the $MgCl_2$ surface, and those in which active sites with the Ti(III) species are bonded with the surface, involving the Lewis base as a complexing agent [1,37,51–54,61,66–68,70,240,290].

Corradini *et al.* [279–281,291], who examined, by calculating the non-bonding interactions, Cossee's model of active centres in non-supported (or 'self-supported') $TiCl_3$-based Ziegler–Natta catalysts, applied this model to $MgCl_2$-supported catalysts [283]. This implies the possibility of an epitactic coordination of $TiCl_4$ and, after its reduction, $TiCl_3$ species on lateral coordinatively unsaturated faces of $MgCl_2$ crystals, giving rise to reliefs crystallographically coherent with the matrix [1]. However, no proof has been found for the existence of such species [54,62,292]. Moreover, it could be expected that ionic species are also present in heterogeneous $MgCl_2$-supported Ziegler–Natta catalysts [293]. The concept of the ionic nature of the active centres in Ziegler–Natta catalysts was first formulated for the metallocene-based Cp_2TiCl_2–AlR_2Cl catalyst in the early 1960s [294–297]. Undoubtedly, the positive charge on the central metal atom affects the interaction of olefin with the active centre. It has been suggested that, in the case of $MgCl_2$-supported catalysts, $MgCl_2$ is not an 'inert' support but modifies the electronic state of Ti, increasing its electrophilicity and availability for olefin coordination [298]. However, a more essential factor is the charge on the metal atom, which changes the energy profile of the Ti–C bond, thus lowering the activation barrier of olefin insertion into this bond (which has been assumed also for metallocene-based catalysts [299]).

As regards the role of Lewis bases in the formation of active centres in $MgCl_2$-supported catalysts, this may be manifold; Lewis bases complete with $TiCl_4$ in adsorption on the different crystalline faces of $MgCl_2$, poison or modify the aspecific catalytic sites and occur in establishing complex equilibria in solution and in the adsorbed phase with titanium chlorides ($TiCl_4$, $TiCl_3$, $TiCl_2$) and $AlEt_3$. As a final result, the adsorbed $TiCl_4$ molecules are reduced to $TiCl_3$, with a geometry very close to that of the catalytic sites already proposed for self-supported $TiCl_3$-based Ziegler–Natta catalysts [122].

One has to realise the extreme chemical complexity of catalysts such as $MgCl_2$/internal Lewis base/$TiCl_4$–$AlEt_3$/external Lewis base. Moreover, the picture is further complicated by the well-known inhomogeneity of the active sites, i.e. the presence of a variety of active sites, differing in kinetic and steric features, on the catalyst surface [300]. As a consequence, many fundamental aspects of the polymerisation mechanism and the intimate structure of active sites are still poorly understood, although some models of the active sites have been proposed on the basis of polymerisation results [301,302] and on the basis of non-bonded interaction calculations [303].

It should be noted in this connection that, for ethylene polymerisation in the presence of a non-supported catalyst such as $TiCl_4$–AlR_3 (R = Et, *i*-Bu), polymerisation proceeds at the maximum rate when the average valence state of titanium is 2 [304]. However, while catalysts with divalent titanium compounds as precursors are known to be highly active for the polymerisation of ethylene, their activities for the polymerisation of α-olefins are generally low [51,240].

A consensus on or explanation for the influence of the oxidation state of titanium on olefin polymerisation activity has not been reached. The absence of any insertion of the coordinating ethylene into the Ti–C bond in Ti(II) species is noteworthy; instead, two ethylene molecules, which coordinate at two coordination sites at Ti(II) species, undergo an oxidative addition, and thus the respective metallacycle, titanacyclopentane, is formed [305]. Such a reaction for dimethyltitanium complexed by 1,2-bis(dimethylphosphione)ethane [Dmpe] is as follows [305]:

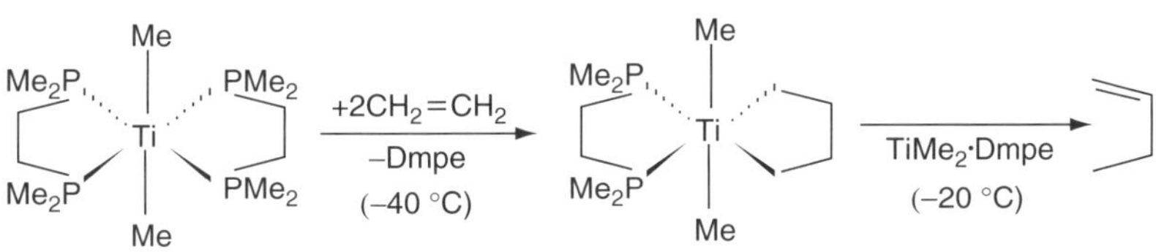

(52)

Decomposition of the titanacyclopentane complex via β-hydride shift, yielding 1-butene, is to be noted.

In light of the finding that Ti(II) compounds react with ethylene not by its insertion into Ti–C bonds but by forming titanacyclopentanes, mechanistic steps for generating the active polymerisation catalyst in $TiCl_2.2THF$–$AlMe_2Cl$ and $TiCl_2 \cdot 1.8MgCl_2 \cdot 6THF$–$AlMe_2Cl$ systems have been proposed by Eisch *et al.* [306]. These steps involve desolvation and phase separation, resulting in the formation of solid $TiCl_2$ [scheme (53)] (or $TiCl_2/MgCl_2$), oxidative ethylene addition at $TiCl_2$ surface species, resulting in the appearance of Ti(IV) species [scheme (54)], β-hydride elimination and hydrotitanation of the succeeding ethylene molecule [scheme (55)] and formation of ionic species [scheme (56)] capable of polymerising ethylene via its coordination and insertion into the Ti–C σ bond.

$$TiCl_2 \cdot 2\,THF + 2\,AlMe_2Cl \longrightarrow TiCl_2 + 2\,THF \cdot AlMe_2Cl \tag{53}$$

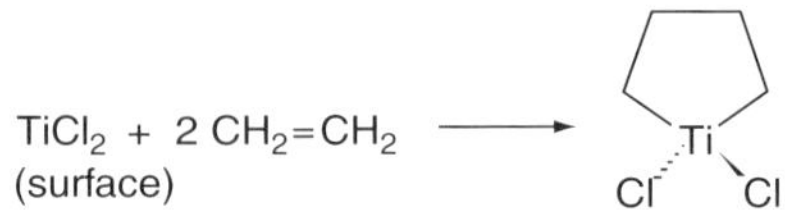

(54)

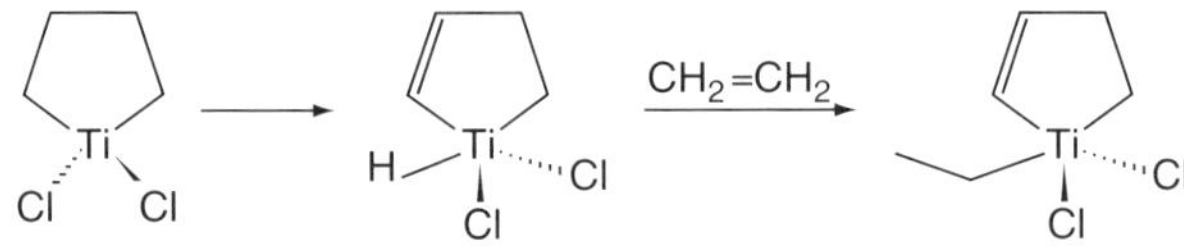

(55)

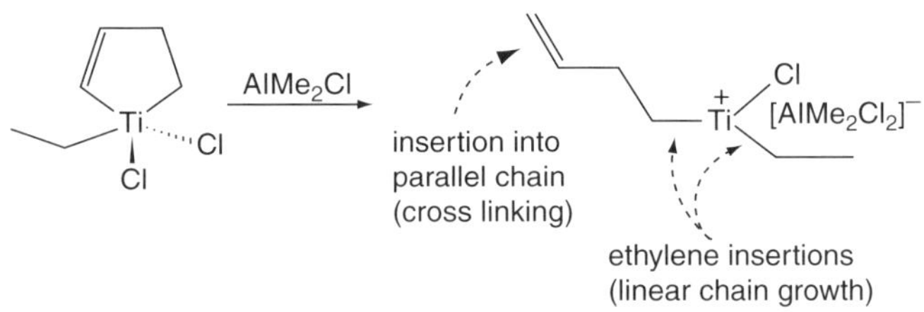

(56)

It must be emphasised that the scheme presented by (53) to (56) culminates in the formation of cationic Ti(IV) intermediate species, paired with an $[AlMe_2Cl_2]^-$ counterion, which is attached to $TiCl_2$ particles or dispersed on $MgCl_2$ particles when present [cationic Ti(III) species may then be formed by the reduction of Ti(IV) species via alkyl radical abstraction] [306]. Not only does the active species have one or two Ti–C σ bonds [in Ti(III) or Ti(IV) cationic species respectively] that can undergo ethylene insertion to permit linear chain growth but it also contains a terminal C=C bond to permit insertion into an extramolecular Ti–C bond and thus result in chain crosslinking. It has been found that ethylene polymerisation with $TiCl_2$-based catalysts can yield highly insoluble, presumably highly crosslinked polyethylene melting above 200 °C [306].

It may be worth mentioning that an earlier proposal concerning the nature of the active site and the propagation mechanism in ethylene polymerisation with Ziegler–Natta catalysts, involving the formation of a titanacyclopentane intermediate species [dispersed on support (or 'self-support') particles via two Ti–Cl bonds], assumed alkyl group migration in such species, resulting in chain growth during reduction of this species to Ti(II) and thus releasing two coordination vacancies [307,308]:

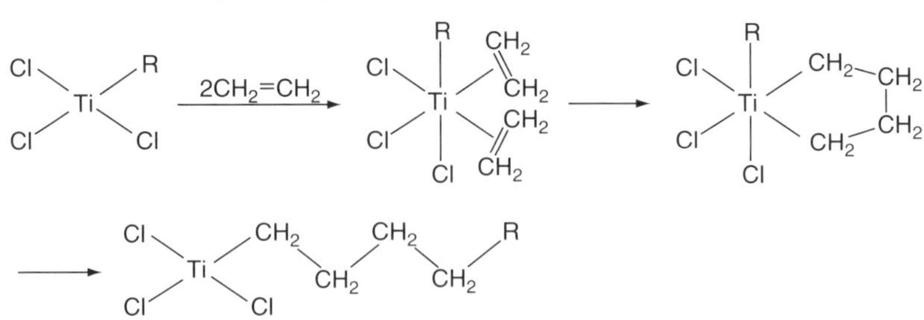

(57)

Semiempirical molecular orbital calculations on this model [309] suggest that, in the case of propylene polymerisation, equatorial 2,4-substitution of the metallacyclopentane ring is the most stable form; this would lead to regioselective head-to-tail propagation during the polymerisation of propylene and, moreover, to the formation of isotactic polypropylene [51]. Such calculations concern a case, however, that has not been confirmed by experiments; a coordination of propylene at Ti(II) species and subsequent reaction according to the above scheme is not as obvious as that of ethylene.

Considering the vast number of various heterogeneous Ziegler–Natta catalysts and olefin polymerisation systems, there is probably no general mechanism that operates in all olefin polymerisations, but rather the polymerisation pathway depends on the kind of catalyst, the kind of monomer and the polymerisation conditions. In fact, other mechanisms have been devised for olefin polymerisation with heterogeneous Ziegler–Natta catalysts. Of these mechanisms, one deserves particular attention. It postulates the propagation to occur via metal carbene and hydride species and metallacyclobutane intermediates [310–312]. According to this mechanism, often referred to as the 'Green–Rooney' mechanism, the metal–alkyl bond first undergoes an α-hydride migration to yield a metal hydride and carbene species. Olefin coordinates at a vacant site and is then involved with metallocarbene to form a metallacyclobutane intermediate. The hydride ligand then migrates back to the carbon atom, giving rise to polymer chain growth. Such a polymerisation mechanism may be illustrated by as follows:

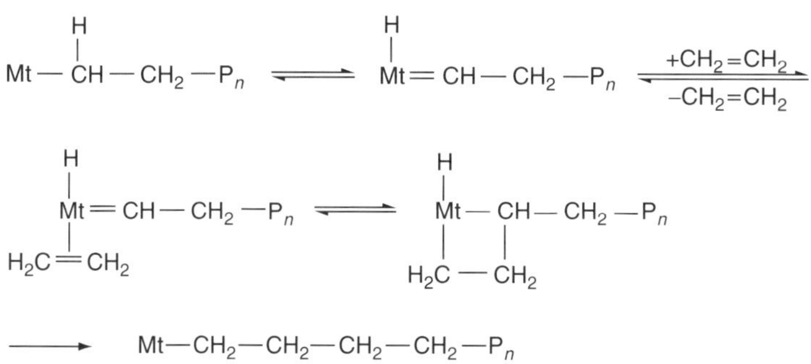

(58)

If the latter step, namely the back migration of the hydride ligand, is slow, the catalyst becomes a metathesis catalyst and chain growth hardly proceeds. Although the carbenic mechanism may be relevant in particular cases, it is to be noted that the insertion mechanism rather than this mechanism is favoured in olefin polymerisation with Ziegler–Natta catalysts [51]. However, it is worth emphassing that the hypothesis of metallocarbene and metallacyclobutane intermediates may also be applied to the isospecific polymerisation of propylene with heterogeneous Ziegler–Natta catalysts [1].

3.4.2 Polymerisation with Phillips Catalysts

The typical Phillips catalyst comprises chemically anchored chromium species on a silica support. The formation of a surface silyl chromate, and eventually silyl dichromate [scheme (29)], is significant during the catalyst preparation, because at the calcination temperature chromium trioxide would decompose to lower-valent oxides. Chromium trioxide probably binds to the silica as the chromate initially, at least for the ordinary 1% loading. However, some re-arrangement to the dichromate at high temperature may occur. It is incorrect to regard only one particular valence state of chromium as the only one capable of catalysing ethylene polymerisation. On the commercial CrO_3/silica catalyst the predominant active species after reduction by ethylene or carbon monoxide [scheme (59)] is probably Cr(II), but other species, particularly Cr(III), may also polymerise ethylene under certain conditions:

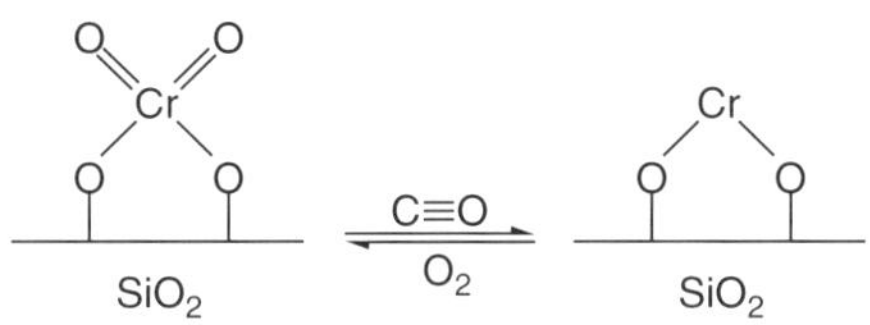

(59)

The catalyst reduced by carbon monoxide or ethylene becomes air and moisture sensitive; it undergoes an easy oxidation to Cr(VI) species when exposed to air.

The reduction brought about by treatment with carbon monoxide leads to the catalyst being capable of polymerising ethylene immediately without an induction period; ethylene polymerisation in the presence of a catalyst prepared by reduction with ethylene is characterised by an induction period, which is connected with the reduction [28,37,43,224,240].

One should realise, however, that the environment of the chromium species, i.e. the type and order of arrangement of ligands, and not only the formal chromium valence, is important for the catalytic activity; probably, there are different families of the chromium species. The existence of systems such as Phillips catalysts, which will polymerise olefins in the absence of organometallic or metal hydride activators, provides evidence of propagation occurring at the transition metal–carbon bond. Although there is no direct proof of the ethylene polymerisation mechanism with Phillips catalysts, it may be considered to involve predominantly the formation of metallocarbene and metallacyclobutane intermediates [chain growth at the metal hydride and carbene species according to the 'Green–Rooney pathway' as shown in scheme (58)] [313]; however, the insertion mechanism [chain growth at the σ metal–carbon species as shown in scheme (2) in Chapter 2] cannot be excluded.

3.4.3 Polymerisation with Soluble Vanadium-based Ziegler–Natta Catalysts

In 1962, Natta *et al.* [10] disclosed that syndiotactic polypropylene could be made at low temperature using a soluble VCl_4–$AlEt_2Cl$ catalyst. Later on, other catalysts based on VCl_4, $VOCl_3$ or $V(Acac)_3$ as precursors and on $AlEt_2Cl$ or $Al(i\text{-}Bu)_2Cl$ as activators appeared to be effective catalysts for the low-temperature syndiospecific polymerisation of propylene. To produce highly syndiotactic polypropylene, the Al/V molar ratio needs to be in-between 2 and 10; when a weak Lewis base such as anisole is added, a ratio of 1:1 is sufficient [10,74,314–318]. Remember that syndiotactic polypropylene was first isolated by Natta *et al.* [9] at the end of the 1950s by the fractionation of a polymer obtained with heterogeneous catalysts and containing a predominant isotactic fraction.

Let us recall also that vanadium-based soluble Ziegler–Natta catalysts have found widespread industrial application for the manufacture of elastomeric ethylene/propylene copolymers and ethylene/propylene/diene terpolymers [319–322]. The most commonly used vanadium-based catalysts for random ethylene/propylene copolymerisation are those prepared from VCl_4, $VOCl_3$, $V(Acac)_3$, $VO(OEt)Cl_2$, $VO(OEt)_2Cl$ or $VO(OEt)_3$ as precursors and $AlEt_3$, $AlEt_2Cl$ or $Al(i\text{-}Bu)_2Cl$ as activators, with an Al/V molar ratio not exceeding 3:1 [37, 72].

According to most models proposed for polymerisation with vanadium-based homogeneous Ziegler–Natta catalysts such as VCl_4–AlR_2Cl, the active site may involve pentacoordinated three-valent vanadium species with three Cl atoms, the secondary carbon atom of the last monomer unit inserted (predominantly by 2,1 enchainment of the α-olefin) and the coordinating olefin [323–327]. However, a model of the active site that assumes a hexacoordinated V(III) species with four Cl atoms, the carbon atom of the last chain unit and the coordinating olefin has also been proposed [328].

However, irrespective of the structure of the active site proposed in these models, the polymer chain propagation is believed to involve the insertion of the coordinating olefin into the V–C bond (predominantly secondary insertion as shown in scheme (40)]; the growing polymer chain migrates to the position previously occupied by the coordinating olefin molecule. In other words, the chain migratory insertion mechanism operates in olefin polymerisation with homogeneous vanadium-based Ziegler–Natta catalysts. One should emphasise that there is no back skip of the chain prior to the coordination of the next olefin molecule in this case, which is in contrast to olefin polymerisation with heterogeneous Ziegler–Natta catalysts [scheme (50)].

Note, in this connection, that it is incorrect to ascrise the formation of small amounts of syndiotactic polypropylene fraction in the presence of heterogeneous Ziegler–Natta catalysts to the operation of a mechanism assuming a lack of back skip of the chain in the last stage of insertion in the presence of these catalysts. Remember that a reasonable explanation of syndiotactic polypropylene formation in the presence of heterogeneous Ziegler–Natta catalysts has been devised by Doi *et al.* (Figure 3.16) [287].

3.4.4 Polymerisation with Homogeneous Metallocene Single-site Catalysts

In accordance with Tolman's electronic rules of stable transition metal complexes [329], titanocenes, zirconocenes and hafnocenes of both non-bridged and bridged structures $[Cp'(Cp'')MtX_2]$ are characterised by a formally uniform d^0 16-electron pseudotetrahedral configuration; therefore such metallocenes are coordinatively unsaturated. Cationic group 4 metallocene species, being loosely associated with a methylaluminoxane-derived anion [scheme (13)], are more coordinatively unsaturated, since they have formally a d^0 14-electron configuration [109–112]. Such metallocene cations, paired with the respective anion, are crucial intermediates in olefin polymerisation with homogeneous single-site catalysts [163–174].

Ethylene polymerisation with alkylaluminium-activated metallocene catalysts involves a propagation step characterised by the appearance of species alternating between an inactive ('dormant') state and an active state in which the chain growth actually occurs [330–333]:

$$\begin{array}{ccccc} \mathrm{Ti\,P}_{n\ 1} & \xrightarrow{CH_2=CH_2} & \mathrm{Ti\,P}_{n} & \xrightarrow{CH_2=CH_2} & \mathrm{Ti\,P}_{n+1} \\ \updownarrow & & \updownarrow & & \updownarrow \\ \mathrm{Ti\,P}_{n\ 1} & & \mathrm{Ti\,P}_{n} & & \mathrm{Ti\,P}_{n+1} \end{array} \quad (60)$$

$\underbrace{\hspace{10em}}_{\text{non-propagating species}}$

This 'intermittent growth' model was further elaborated [334–341], and ethylene polymerisation systems were devised to involve consecutive equilibria appearing to convert alkylaluminium halide and metallocene first into donor–acceptor complexes [scheme (61)] equivalent to contact (inner) ion pairs and then into separated (dissociated) ion pairs [scheme (62)] [297,338]:

$$Cp_2TiRCl + AlRCl_2 \rightleftharpoons Cp_2TiRCl{\cdot}AlRCl_2 \quad (61)$$

$$Cp_2TiRCl \cdot AlRCl_2 \rightleftharpoons [Cp_2TiR]^+ [AlRCl_3]^- \quad (62)$$

In these highly dynamic equilibria, only the cation of a separated ion pair exhibits the capability of interacting with a monomer and, hence, of contributing to polymer growth. Contact ion pairs, which appear to dominate in these equilibria, can thus be termed 'dormant' in this regard [30]. It is worth noting in this connection, that the first direct evidence for the generation of ionic particles in such systems and their responsibility for catalytic activity in ethylene polymerisation has been obtained by electrodialysis studies [342].

This model would explain the inability of metallocene–alkylaluminium halide systems to promote the polymerisation of propylene and higher α-olefins [94]; it is obvious that there is insufficient capability of the more weakly coordinating α-olefins to form reactive, olefin-separated ion pairs by displacement of an aluminate anion from the metal centre. At any rate, the limitation of homogeneous catalysts to the polymerisation of only ethylene was a crucial obstacle to progress in this field for many years. This impediment was fortunately overcome, however, by a series of serendipitous observations [90–95, 100, 101,103] that led, around the 1980s, to the discovery by Kaminsky, Sinn *et al.* [90, 91,94,95,100,101] that metallocenes are activated for catalysing the polymerisation of propylene and other α-olefins (without α,α-disubstituted olefins) by methylaluminoxane [30].

One should assume here, in accordance with general opinion, that the productive reaction complex in methylaluminoxane-activated group 4 metallocene catalysts is the $[Cp'_2Mt(R).olefin]^+[Al_x(R)_{x-1}O_xX_2]^-$ complex, which is generated by displacement of an $[Al_x(R)_{x-1}O_xX_2]^-$ ion from its $[Cp'_2Mt(R)]^+$ counterion by the coordinating olefin molecule [30].

The mechanism that is commonly considered to operate in the polymerisation of ethylene and α-olefins in the presence of group 4 metallocene-based catalysts is that devised by Cossee [268, 276, 277] for propylene polymerisation with heterogeneous Ziegler–Natta catalysts, though modifications invoking effects such as α-agostic hydrogen interactions with the metal centre have been proposed [343,344].

After olefin insertion into the Mt–C bond in cationic metallocene, the polymer chain may be located in a position previously occupied by the coordinating olefin molecule (chain migratory insertion mechanism) or it may skip back to its starting position (chain stationary insertion mechanism) prior to the subsequent coordination of the next olefin molecule; it depends on the class of metallocene, in principle, whether or not the chain migrates without the participation of the succeeding monomer molecule, i.e. whether or not the catalytic site undergoes back-skipped isomerisation. A general mechanism has been proposed that assumes free, restricted and prohibited chain migration as a result of the sterically imposed environment of the metallocene central atom for the formation of highly regioregular (but atactic) polypropylene [345] as well as for the formation of syndiotactic, hemiisotactic and isotactic α-olefin polymers [143,146]. We can conclude that, irrespective of the kind of metallocene catalyst, which may act in olefin polymerisation involving a chain migratory insertion and/or a chain stationary insertion (a back-skipped insertion) of the olefin, the mechanism of the insertion follows, in principle, that devised by Cossee for olefin polymerisation with heterogeneous Ziegler–Natta catalysts.

Possible pathways for monomer insertion in olefin polymerisation systems with homogeneous single-site metallocene-based catalysts and heterogeneous Ziegler–Natta catalysts are shown in Figure 3.17 [122].

Pathways (a)–(b)–(c)–(d) and (a′)–(b′)–(c′)–(d′) correspond to the original mechanism proposed by Cossee [268,276,277] and are still valid, apart from some minor modifications [1], for heterogeneous catalysts. For metallocene-based catalysts of classes II and partially V, this mechanism gives rise to successive additions at the same site (from a configurational point of view) and is known as the 'chain stationary insertion' mechanism ('chain skipped insertion' or 'site isomerisation without insertion' mechanism) [143, 146, 345]. The (a)–(b)–(c)–(a′)–(b′)–(c′) pathway corresponds to the 'chain migratory' insertion mechanism found in the case of metallocene catalysts of classes I, III, IV and partially V [143, 146].

One important note concerning the mechanistic aspects of olefin polymerisation should be added, however, in order to enable a proper understanding of the meaning of concepts such as 'chain migratory' insertion, 'chain stationary' or 'chain skipped' insertion and 'site isomerisation without insertion'. These concepts are based on an assumption of the presence of a coordination vacancy at the central metal atom in a catalytic site, as Cossee has proposed. The concept of the existence of the coordination vacancy serves as a useful tool for mechanistic considerations of olefin polymerisation with coordination catalysts (Figure 3.17), but in fact the coordination vacancy cannot be treated as a 'ligand' and a catalytic site with coordination vacancy ought to be seen as a coordinatively unsaturated site. In the case of heterogeneous Ziegler–Natta catalysts, a coordinatively unsaturated site, presented as a metal atom surrounded by five monodentate (η^1) ligands and one coordination vacancy, is actually pentacoordinated and must not adopt a geometry attributable to an octahedral configuration; the saturated octahedral configuration and the geometry attributable to it is attained by such a site when the olefin molecule is coordinated on it. Also, in the case of homogeneous metallocene-based catalysts, the geometry of the site, presented as a cationic metallocene species

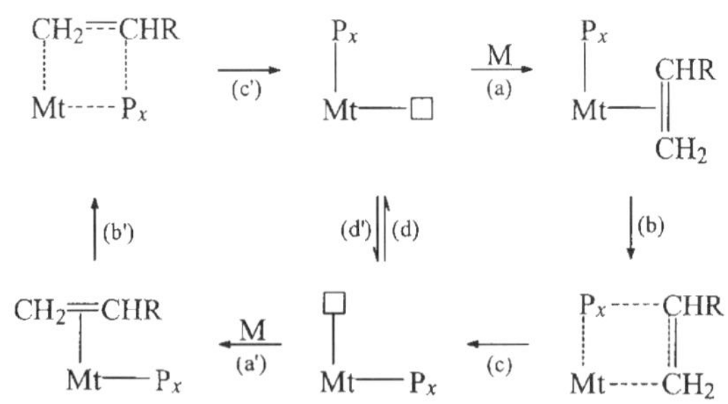

Figure 3.17 Mechanism for monomer insertion in α-olefin polymerisation with coordination catalysts. $P_x = P_n$ and/or P_{n+1}; $M = CH_2 = CHR$; □ = coordination vacancy

having formally a d^0 14-electron configuration, must not be attributable to that of a formally d^0 16-electron species with a coordinated olefin molecule or paired with the counterion. In view of the above, it is obvious that concepts of mechanisms involving 'chain migration' with the participation of the monomer (insertion) and without its participation ('chain back skip' or 'site isomerisation') are based on approximate and conventional concepts. 'Site isomerisation' may be expressed in terms of two possible modes by which the coordinating monomer molecule approaches the metal atom (regarding the spatial configuration of ligands at the metal atom); this may be visualised by pathway (a) or (a′), starting from the actually existing species 'giving' both fictitious species with a coordination vacancy in equilibrium [pathway (d), (d′)] (Figure 3.17).

Similarly to the case of heterogeneous Ziegler–Natta olefin polymerisation catalysts, the coordination of the olefin molecule at the cationic metallocene species, due to π bond formation (Figure 2.1), leads to lowering of the energy of the resultant π complex, e.g. the $[Cp'_2Mt(R)\cdot olefin]^+[Al_x(R)_{x-1}O_xX_2]^-$ complex, which results in activation of the catalyst Mt–C bond and olefin C=C bond for the insertion reaction [136].

It must be emphasised that the relatively low activation energy of the insertion reaction in this complex is characteristic of complexes having formally a d^0 16-electron configuration, which is just adopted by the cationic group 4 metallocene species complexed with a coordinating olefin as the two-electron donor [136].

Also, neutral group 3 and lanthanide series metallocenes form with olefins π complexes having formally a d^0 16-electron configuration, which is favourable for the insertion reaction during olefin polymerisation (or oligomerisation) in the presence of these single-component catalysts [187].

Note that more coordinatively unsaturated metallocene complexes with olefins are characterised by a relatively high activation energy of insertion of the coordinated olefin into the Mt–C bond. Quantum-mechanical calculations showed the activation energy for ethylene insertion into the Ti–Me bond in the $[Cl_2Ti(Me)\cdot(CH_2{=}CH_2)]^+$ complex of formally d^0 8-electron structure to be a few times higher than that for insertion in the $[Cp_2Ti(Me)\cdot(CH_2{=}CH_2)]^+$ complex [136]. This illustrates, among other things, why group 4 metallocene-based catalysts exhibit unique properties for promoting olefin polymerisation.

However, another factor that determines the properties of metallocene-based catalysts for catalysing olefin polymerisation has to be taken into consideration – the already mentioned agostic interaction of one of the α-hydrogen atoms of the metal-bound alkyl chain with the metal centre of the metallocene-based catalyst [343,344].

The α-agostic interaction is such that the electron pair in the C_α–Hσ bond of the alkyl substituent at the metal is partly donated to the electron-poor metal; this may play an important role in making more simple the insertion of a coordinated olefin into the metal–alkyl bond. This hypothesis has been tested in some instances [346–350], showing metal–hydrogen interaction via an α-agostic bond.

Density functional molecular orbital calculations suggest that α-agostic hydrogen interaction with the metal atom in a Zr–CH_3 unit helps to stabilise it and align the methyl group of this unit for interaction with the ethylene π^* orbital [351,352]. The α-agostic transition state has one less rotational degree of freedom, which stabilises it and reinforces for the C–C bond formation.

An extended Hückel molecular orbital study [274] provides evidence of the occurrence of agostic stabilisation of the insertion transition state and indicates that α-agostic hydrogen interaction with the metal in the Zr–CH_2R unit is linked with the increasing electron deficiency of the reaction complex as it progresses along the olefin insertion coordinate from a formally d^0 16-electron to d^0 14-electron species [30]. Ethylene insertion into the Zr–C bond in the silylenebiscyclopentadienyl(methyl)zirconium cation, $[H_2Si(Cp)_2Zr{-}Me]^+$, is shown in Figure 3.18 [272].

It is worth noting the α-agostic interaction in the transition state, but also the β-agostic interaction (of a C_β bond with Zr) in the product formed. Stabilisation of the alkyl product of insertion by β-agostic interaction is suggested [272], although in the case of the n-propyl-substituted species, Zr–CH_2–CH_2–CH_3, at the end of insertion it may be stabilised by γ-agostic interaction (of the C_γH bond with Zr) [136,272–274]. However, a γ-agostic species initially arising

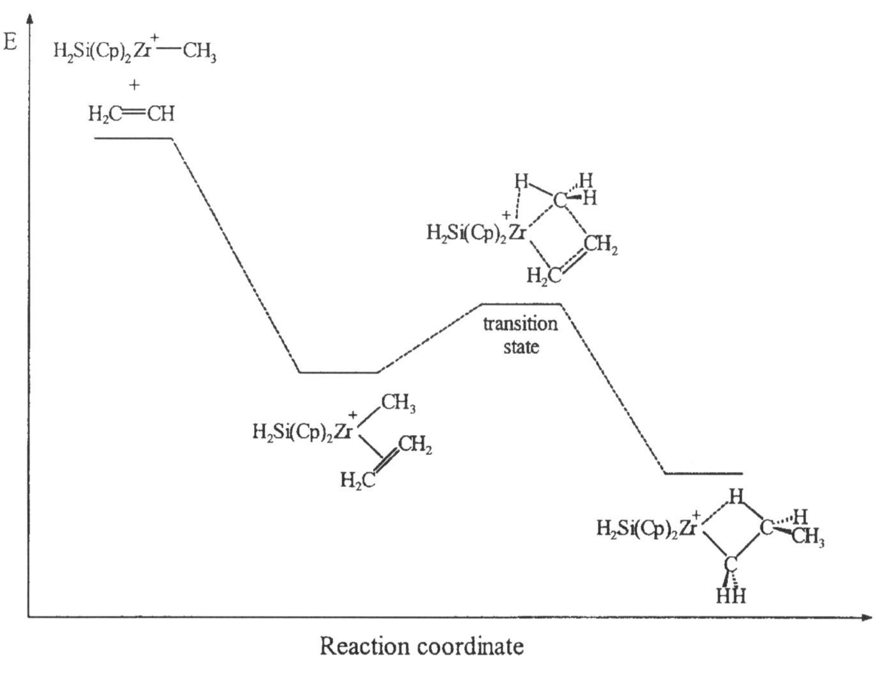

Figure 3.18 Theoretical study of ethylene insertion into the Zr–C bond in the silylenebiscyclopentadienyl(methyl)zirconium cation $[H_2Si(Cp)_2Zr{-}Me]^+$. E = energy

from the insertion reaction is likely to be arranged into a structure with a β-agostic bond, which is an even more stable structure [173,353–356]. The β-agostic species probably represents something like a resting state for a $[Cp_2ZrR]^+$ ion [30,105].

The importance of agostic metal–hydrogen interaction in olefin polymerisation with metallocene-based catalysts may be seen from a molecular modelling study of systems in which the insertion of propylene into a zirconium–secondary alkyl bond, Zr–CH(Me), occurs [357]. It is characteristic that propylene polymerisation with metallocene-based catalysts proceeds via primary insertion involving Mt–CH_2–CH(Me)–P_n species [scheme (39)], irrespective of the tacticity of the polypropylene formed; it must be emphasised at the same time that the formation of an α-agostic bond between hydrogen and zirconium atoms relates only to species that have a metal–primary alkyl bond. However, in the case of propylene insertion involving the Zr–CH(Me)–CH_2–P_n species [scheme (44)], it farely proceeds [114,253], since an α-agostic interaction is impossible in the transition state for the insertion of propylene into a secondary Zr–CH(Me) bond [357]. Instead, the relatively stable β-agostic species, with a blocked coordination site for olefin coordination at the metal atom, is formed:

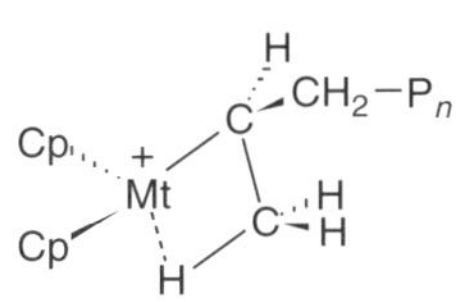

It has also been suggested that the experimental barrier of activation for propagation in olefin polymerisation is due to rearrangement of γ-agostic species into those with α-agostic interactions in order to free the coordination site for the insertion of the next monomer molecule, rather than the actual insertion (Figure 3.19) [358].

Figure 3.19 shows the propagation mechanism that is assumed for olefin polymerisation with group 4 metallocene-based catalysts. The active species, which is the alkylated cationic metallocene and in which an α-agostic metal–hydrogen interaction is possible (interaction with the counterion is excluded), coordinates the olefin at the vacant coordination site to form the respective π complex. In the subsequent step, the olefin is inserted into the Mt–C bond through a four-membered cyclic transition state. The newly formed species, in which γ-agostic metal–hydrogen interaction may initially arise, then undergoes, in the final step, a rotation of the polymer chain to allow for incorporation of the next monomer molecule. The vacant coordination site in the species thus formed, in which an α-agostic metal–hydrogen interaction is possible, is no longer blocked by the polymer chain. One should realise that this species is chemically and configurationally equivalent to the initial species, and it coordinates the succeeding olefin molecule, allowing for the propagation of polymerisation [358].

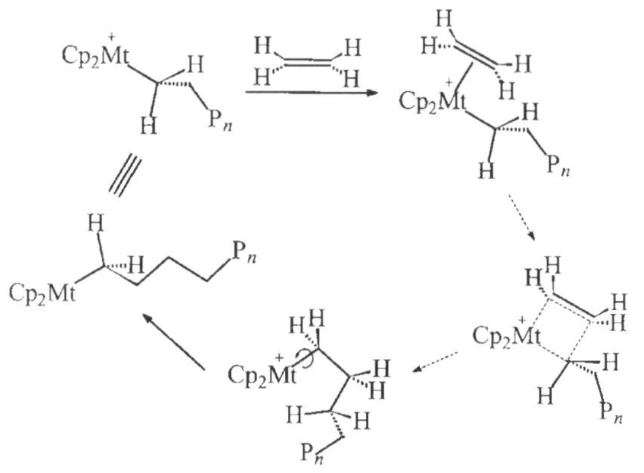

Figure 3.19 Chain migratory insertion mechanism for olefin polymerisation with metallocene-based catalysts

As already mentioned, species in which α-agostic metal–hydrogen interaction occurs are propagating species; by contrast, species in which β-agostic or γ-agostic metal–hydrogen interaction takes place are probably non-propagating species owing to the blocking of the coordination site by the chain:

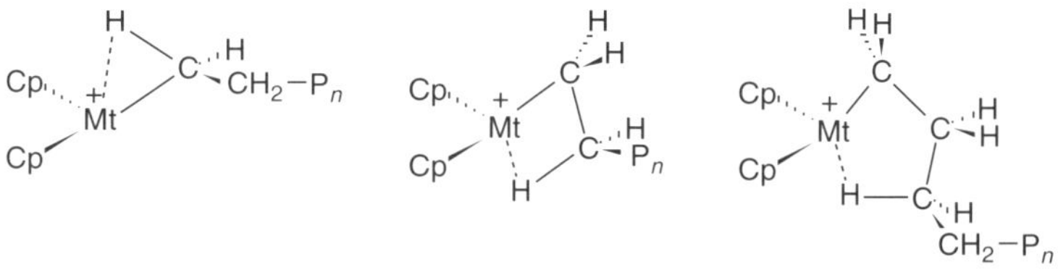

The described chain migratory insertion mechanism, which operates in olefin polymerisation with metallocene-based single-site catalysts, follows that proposed by Cossee [268,277,278] for olefin polymerisation with heterogeneous catalysts; there is, however, no back skip of the polymer chain to the previously occupied position prior to the coordination of the next monomer molecule, but rotation of the chain around the axis of the Mt–CH_2 bond takes place (Figure 3.19) [358].

This mechanism, although understandable in a conceptual sense, is not fully understood in a mechanistic sense. For instance, the exact nature of active species and the role of the activator and/or counterion is a subject of debate; this concerns methylaluminoxane-activated group 4 metallocene systems in particular. Methylaluminoxane may act to generate the active species and remove impurities from the polymerisation system as well as playing a more fundamental role such as assisting in the insertion of each monomer unit or reactivating 'dormant' sites [358].

However, recent studies on alkylaluminoxanes as activators for group 4 metallocenes as catalysts for olefin polymerisation have allowed a more precise definition of their role. For instance, the isolation of non-fluxional aluminoxane compounds, $[(t\text{-Bu})_2AlOAl(t\text{-Bu})_2]_2$ and $[Al(t\text{-Bu})O]_x (x = 6, 7, 8, 9)$, has provided the basis for an investigation of the mode of activity observed for alkylaluminoxanes as activators for the zirconocene-promoted polymerisation of ethylene. The former compound, namely $[(t\text{-Bu})_2AlOAl(t\text{-Bu})_2]_2$, which contains two three-coordinate aluminium atoms, exhibits no reaction with Cp_2ZrMe_2, and no catalytic activity of such a mixture towards ethylene polymerisation is observed. In contrast, the closed-cage compound $[Al(t\text{-Bu})O]_6$, containing all four-coordinated aluminium atoms, reacts reversibly to give the ion pair complex $[Cp_2ZrMe]^+[Al_6(t\text{-Bu})_6O_6Me]^-$, which is active as a catalyst for the polymerisation of ethylene. Polymerisation is also observed for Cp_2ZrMe_2–$[Al(t\text{-Bu})O]_x$ ($x =$ 7, 9). The structure of $[Al(t\text{-Bu})O]_6$ and $[Cp_2ZrMe]^+$ $[Al_6(t\text{-Bu})_6O_6Me]^-$ is presented in Figure 3.20 [98, 359].

It should be noted that the cage *t*-butylaluminoxane, $[Al(t\text{-Bu})O]_6$, is not acidic in a Lewis sense per se, but it possesses a 'latent' Lewis acidity; this means that it possesses the ability to undergo cage opening, via heteroleptic bond cleavage, to generate a Lewis acidic site [98,359].

The Cp_2ZrMe_2–$[Al(t\text{-Bu})O]_6$ catalyst exhibits lower activity in ethylene polymerisation than the Cp_2ZrMe_2–$[Al(Me)O]_x$ catalyst. Reasons for decreased activity of the former catalyst compared with the latter are difficult to determine, since the exact speciation of methylaluminoxane is unknown; however, commercial methylaluminoxane contains a large amount of trimethylaluminium [359, 360].

At the end of considerations of the role of alkylaluminoxanes as activators for metallocenes as Ziegler–Natta olefin polymerisation catalysts, it should be noted that the analogy between methylaluminoxane and simple acidic moieties does not appear to hold. For example, highly Lewis acidic perfluorinated boranes, such as $B(C_6F_5)_3$, show enhanced activity as activators when compared with $[Al(Me)O]_x$, but the stability of the resulting zirconocene-based catalyst is drastically lower than that of the catalyst formed with $[Al(Me)O]_x$ [98].

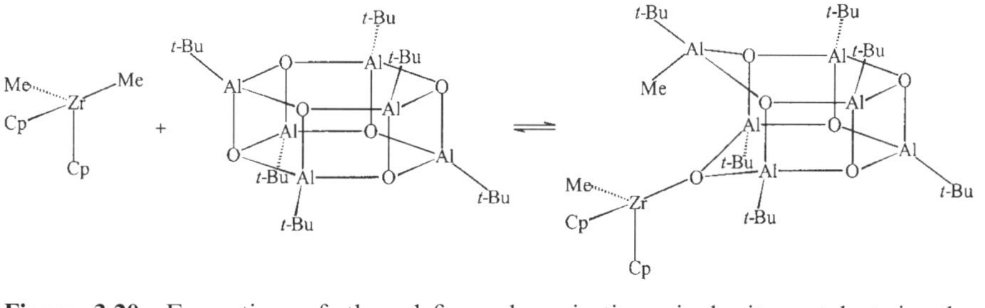

Figure 3.20 Formation of the olefin polymerisation single-site catalyst in the reaction of zirconocene, Cp_2ZrMe_2, with cage *t*-butylaluminoxane, $[Al(t\text{-Bu})O]_6$

Olefin insertion into the metal–alkyl bond in single-component catalysts, scandocenes and yttrocenes, for polymerisation of olefins may also be assisted by an α-agostic metal–hydrogen interaction. Recent findings implicate an α-agostic structure for the transition state for chain propagation [187]. The possible role of an α-agostic interaction is such that it may lower the transition energy required for olefin insertion, as suggested by calculations. Therefore, on the basis of evidence for α-agostic stabilisation of thetransition state for C–C bond formation, a modified Green–Rooney pathway for chain propagation, involving metal hydride and carbene species [343,344,347], has been devised [187] for some Ziegler–Natta olefin polymerisation catalysts [272,274] but not for others [136,273]. Active biscyclopentadienylmetal catalysts are d^0 14-electron derivatives with two vacant orbitals: one to accommodate the incoming olefin, the other for the α-agostic interaction. Thus, the formation of a transition state with the metal coordinating more than four ligands is suggested [187]:

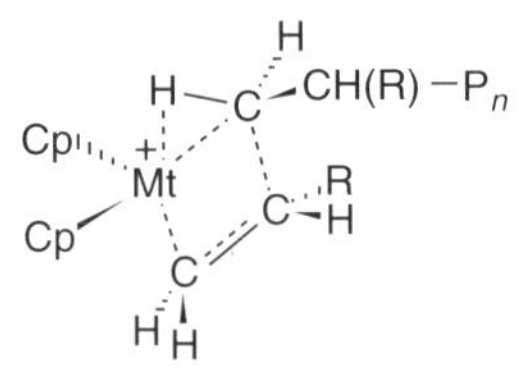

Relatively stable β-agostic interactions have been identified as the preferred ground-state structures for the neutral scandocene, $Cp^*{}_2Sc–CH_2–CH_3$, as well as for cations such as cobaltocene, $[Cp^*(L)Co–CH_2–CH_2R]^+$ [$L = PMe_3$, $P(OMe)_3$] [361], and zirconocene, $[Cp_2(PMe_3)Zr–CH_2–CH_2R]^+$ [173]. Calculations implicate the (less stable owing to ring strain) α-agostic arrangement in the transition state for C–C bond formation. Hence, it may be tentatively concluded that β-agostic structures characterise ground states and α-agostic structures characterise transition states for chain propagation [187]. On the other hand, some tests devised by researchers, e.g. hydrocyclisation of *trans*, *trans*-1,6-dideutero-1,5-hexadiene with the $Cp_2ZrCl_2–[Al(Me)O]_x$ catalyst [350], have not always supported an α-agostic transition state; such findings could be taken as evidence that α-agostic assistance is not always required for C–C bond formation [187].

3.5 Stereoregulation Mechanism

3.5.1 Factors Influencing Polymerisation Stereospecificity

Polymers resulting from successions of (prevailingly) equal additions, either 1,2 [scheme (39)] or 2,1 [scheme (40)], irrespective of the choice of configuration, are said to be regioregular (from a regiospecific or regioselective polymerisation process). In general, Ziegler–Natta and related coordination catalysts are highly regiospecific in the polymerisation of α-olefins but differ widely in stereospecificity. Remember that, if there is regularity in the succession of configurations of the stereogenic tertiary carbon atoms, the macromolecules are said to be stereoregular (from a stereospecific or stereoselective polymerisation process). Moreover, a poly(α-olefin) chain ideally resulting from the regular succession of *m* diads (stereodiads of the same relative configuration) is defined as isotactic, and one from the regular succession of *r* diads (stereodiads of the opposed relative configuration) is defined as syndiotactic (Figure 3.1), but in the absence of stereoregularity the chain is said to be atactic (Figure 3.2).

Until the mid-1980s, it was customary to classify stereospecific Ziegler–Natta catalysts into heterogeneous isospecific and homogeneous syndiospecific catalysts. Considering the progress that has been accomplished owing to the development of metallocene-based catalysts, this class of homogeneous, either isospecific or syndiospecific, Ziegler–Natta and related coordination catalysts was added to the traditional classes [1]. It should also be noted that stereospecific metallocene-based catalysts, when supported on a carrier such as SiO_2, Al_2O_3 and $MgCl_2$, can serve as heterogeneous catalysts for α-olefin polymerisation, maintaining either isospecificity or syndiospecificity (especially when using the silica support) [201, 206].

Regiospecificity of Ziegler–Natta and related coordination catalysts for α-olefin polymerisation implies non-equivalence of the two insertion modes, primary and secondary [schemes (39) and (40) respectively], owing to electronic and/or steric factors.

Stereospecificity, which for a given mode of insertion entails the ability of the catalyst to discriminate between the two faces of the prochiral α-olefin molecule, requires at least one chiral centre to be involved in the polymerisation reacting species. The steric control may be dictated by the chirality of the catalyst active site [Δ and Λ at an octahedral configuration (Figure 3.21) or a tetrahedral configuration (Figure 3.22)] (enantiomorphic site stereocontrol) as well as by the configuration (*R*, *S*) of the stereogenic carbon atom of the last inserted monomer unit (Figure 3.23) (chain end stereocontrol).

Both possibilities, i.e. enantiomorphic site stereocontrol (in the case of an optically inactive catalyst it consists of a racemic mixture of enantiomorphic sites) and chain end stereocontrol, have been verified, depending on the kind of catalyst. These two essential types of stereocontrol mechanism operating in propylene polymerisation with various stereospecific Ziegler–Natta catalysts are presented in Table 3.3.

The main elements of chirality possibly present in the intermediates and transition states that can be hypothesised within this framework are as follows [1]. Firstly, a prochiral α-olefin molecule, e.g. propylene, coordinating via its two faces at the catalyst active site gives rise to non-superposable *re* and *si* diastereoisomeric complexes (Figure 3.24) [362, 363]. According to the considered mechanisms, an isotactic polymer is generated by a long series of

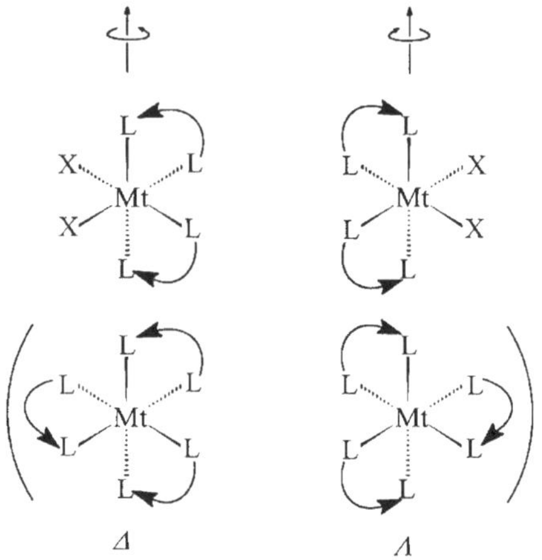

Figure 3.21 Octahedral enantiomorphic sites: *Δ* (*dextrorotatory*) – right-handed; *Λ* (*laevorotatory*) – left-handed

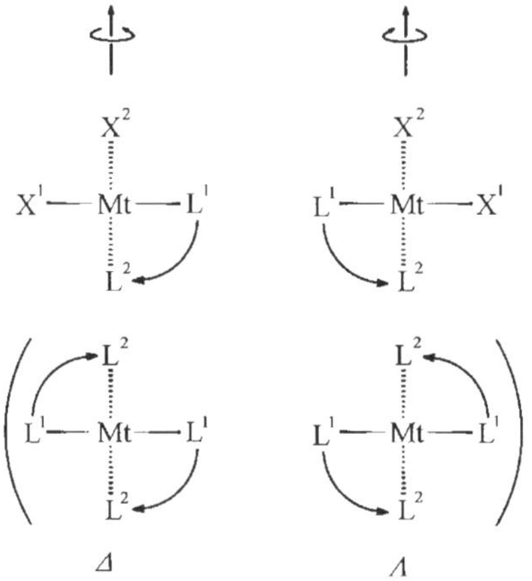

Figure 3.22 Tetrahedral enantiomorphic sites: *Δ* (*dextrorotatory*) – right-handed; *Λ* (*laevorotatory*) – left-handed

insertions of all *re* or all *si* coordinated α-olefins, and a syndiotactic polymer is generated by alternate insertions of *re* and *si* coordinated α-olefins [1].

A second element of chirality is the configuration of the stereogenic tertiary carbon atoms in the growing chain, and in particular of that in the last inserted monomer unit.

Last but not least, the catalytic site may be chiral by itself. An example for the heterogeneous catalysts, first proposed by Arlman [278] and common to other models devised by Allegra [364] and Corradini *et al.* [279,283,345,382], is.

Table 3.3 Stereochemistry of propylene polymerisation with stereospecific Ziegler–Natta catalysts

Catalyst type	Polymerisation	
	Monomer insertion	Type of stereocontrol
heterogeneous isospecific (e.g. $MgCl_2/TiCl_4$–$AlEt_3$)	*cis*-1, 2[a]	Enantiomorphic site
homogeneous syndiospecific (e.g. VCl_4–$AlEt_2Cl$)	*cis*-2, 1[b]	Chain end
homogeneous isospecific (achiral–achirotopic) {e.g.Cp_2MtX_2–$[Al(Me)O]_x$}	*ci*-1, 2[b]	chain end
homogeneous isospecific (chiral–homotopic) {e.g.*rac*-$(IndCH_2)_2MtX_2$–$[Al(Me)O]_x$}	cis-1, 2[b]	Enantiomorphic site
homogeneous syndiospecific (chiral–enantiotopic) {e.g.$Me_2C(Cp)(Flu)MtX_2$–$[Al(Me)O]_x$}	cis-1, 2[b]	Enantiomorphic site
homogeneous isospecific (chiral–diastereotopic) {e.g.$Me_2C(t$-$BuCp)(Flu)MtX_2$–$[Al(Me)O]_x$}	*cis*-1, 2[a]	Enantiomorphic site
homogeneous syndiospecific (chiral–diastereotopic) {e.g.*t*-$BuCH(Cp)(Flu)ZrCl_2$–$[Al(Me)O]_x$}	*cis*-1, 2	Enantiomorphic site/chain end

[a] Chain stationary (skipped) insertion propagation mechanism.
[b] Chain migratory insertion propagation mechanism.

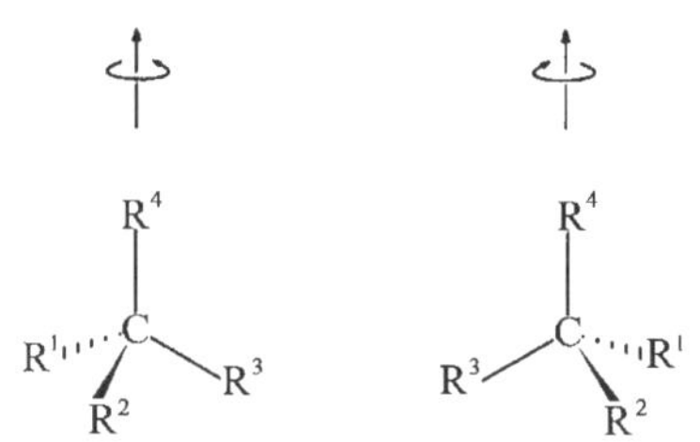

R *S*

Figure 3.23 Configuration of tetrahedral enantiomers: *R* (*rectus*) – right-handed; *S* (*sinister*) – left-handed; carbon R substituents are arranged in decreasing order of priority according to the rules of precedence: $R^1 > R^2 > R^3 > R^4$

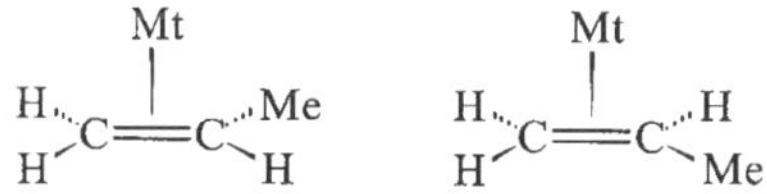

re *si*

Figure 3.24 Two possible chiral coordinations of a propylene molecule to a metal atom: *re* (*rectus*) – right-handed; *si* (*sinister*) – left-handed

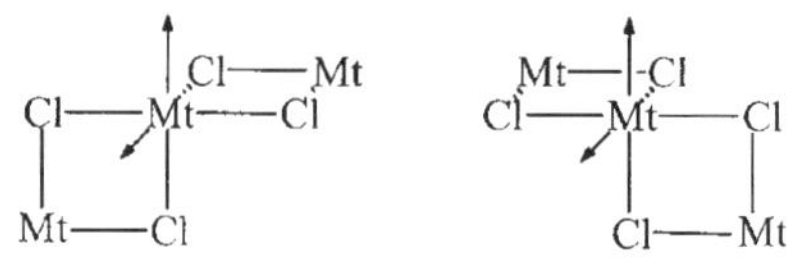

Figure 3.25 Chiral model octahedral sites (Δ and Λ) for the heterogeneous catalysts constituted by the metal atom bridge bonded to two other metal atoms; the two positions occupied by the coordinating monomer and the growing polymer chain are indicated by arrows

shown in Figure 3.25 [1]. A metal atom is bridge-bonded through halogen atoms to two more metal atoms and two coordination positions are occupied by the monomer and the growing polymer chain (Figure 3.25). An example for the homogeneous catalysts with a chiral site may be the model site referring to the stereorigid chiral complexes containing an ethylenebis (1-indenyl) ligand [22,101,112,345,365].

Within this framework, steric control in the polymerisation of α-olefins is to be connected with the energy differences between the diastereoisomeric situations that originate from the combination of two or more of the above elements of chirality [1].

3.5.2 Isospecific Propagation Reaction Stereocontrol in the Presence of Heterogeneous Ziegler–Natta Catalysts

A number of models for catalytic active sites have been proposed to explain the isospecific polymerisation of propylene in the presence of $TiCl_3$-based catalysts [68]. It was Natta who proposed [366], for the first time, that the steric control in isospecific propylene polymerisation is caused by the structure of the active species located on the borders of crystal layers of violet titanium trichloride. Arlman and Cossee [277] suggested that the isospecific active sites are located on lateral crystal surfaces, which in violet titanium trichloride such as α-$TiCl_3$ correspond to (110) planes. Titanium atoms present on the above-defined fracture surface have a vacant octahedral site bonded to five chlorine atoms (Figure 3.26) [1,68].

One chlorine ligand, which protrudes from the surface (in an inward, more hindered, position), is replaced by the alkyl group during the procatalyst activation process [scheme (8)], whereas the other four, bridged to further titanium atoms, are more strongly bound; an octahedral configuration of the titanium atom is attained by the monomer coordination (in an outward, less hindered, position). It is necessary to emphasise the non-equivalency of both inward and outward coordination positions; according to calculations for model sites in heterogeneous catalysts, the minimum energy situation corresponds to the outward propylene coordination, while the energy increase for the

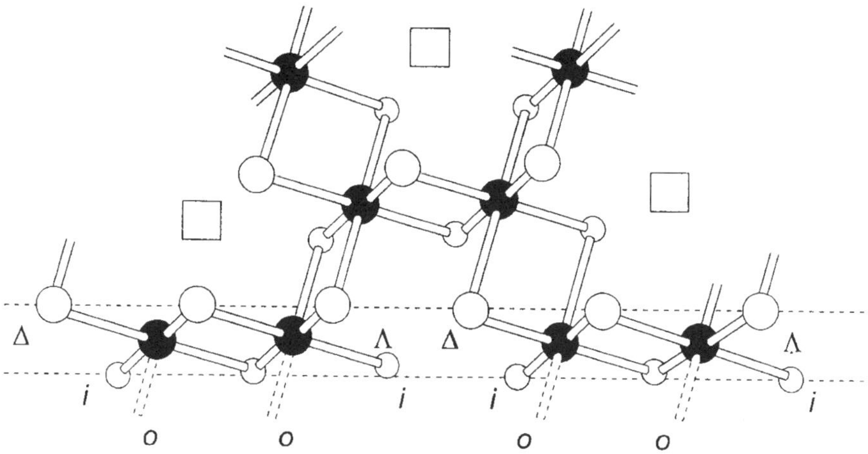

Figure 3.26 Schematic drawing of a lateral cut of a violet $TiCl_3$ layer. In the more hindered inward (*i*) position, the chlorine atom is replaced by the alkyl group from the activator, and within the polymerisation by the growing polymer chain. In the less hindered outward (*o*) position, the α-olefin molecule is coordinated. The chirality, *Δ* and *Λ*, of two titanium atoms in indicated. ● – *Ti;* ❍ and ○ Cl; ❑ – vacant positions. Reproduced (by permission from Elsevier Science) from Ref. 1. Copyright 1989 Pergamon Press

inward coordination is in the range 3–7 kcal/mol depending on the particular site (on isolated structural layers [280,283], on an edge [279] or on the bulk [279] of (110) lateral cleavages) [345]. It should also be noted that neighbouring titanium atoms have opposite chirality (Figure 3.26) [1,68,303].

Remember that the polymerisation mechanism devised by Cossee [268] implies two main steps: coordination of the monomer at the titanium vacant site with the double bond parallel to the Ti–C bond, and chain migratory insertion of the coordinating monomer molecule (with migration of the growing polymer chain to the position previously occupied by the coordinating monomer molecule); isospecificity of the active site is assumed only if the polymer chain skips back to the original position before further insertion [scheme (50)].

One should note here the model proposed by Allegra [364], which assumes the possibility that active sites are located on violet titanium trichloride layers protruding above the two adjacent ones. This model obviates the necessity of the isomerisation step to ensure catalyst site isospecificity; a C_2 symmetry axis locally relates the atoms relevant to the non-bonded interactions with the corresponding monomer and the growing polymer chain. Hence, the two situations resulting from exchanging, in the coordination step, the relative positions of the growing chain and the incoming monomer are identical (Figure 3.27) [68]. It may be interesting to note that similar symmetry properties are exhibited by homogeneous isospecific metallocene catalysts, such as *rac.*-$(IndCH_2)_2ZrCl_2$–$[Al(Me)O]_x$ (Table 3.1).

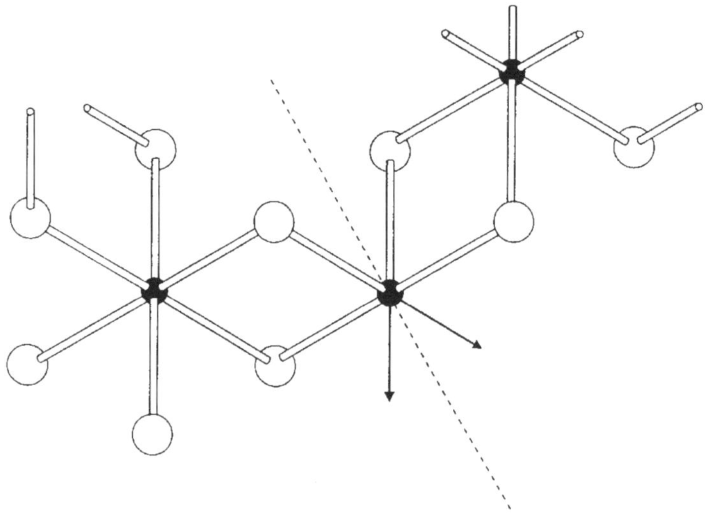

Figure 3.27 $TiCl_3$ layer termination proposed by Allegra (Ref. 364); the arrows indicate the two equivalent coordination positions related by a two fold axis. ● – *Ti;* ○ Cl

It has been suggested by Corradini *et al.* [1], on the basis of some evaluations of non-bonded interactions for the Cossee and Allegra models, that the chiral environment of the metal atom imposes a chiral orientation of the first C–C bond of the polymer chain occupying the inward (more hindered) position; this orientation has been identified as a crucial factor in determining the polymerisation isospecificity. The non-bonding interactions between the methyl group of the coordinating propylene molecule and the carbon atom of the growing polymer chain in a β-position to the metal actually induce a preferential coordination of the enantioface of the monomer, which enables the propylene methyl group and the C_β atom of the chain to be located on opposite sides with respect to the plane defined by the Ti–C and C=C bonds (Figure 3.28) [1,68].

This would determine lower activation energy and hence favour the insertion of the *si* coordinated (Figure 3.28) as opposed to the *re* coordinated (Figure 3.29) propylene molecule on a Λ site [1]. The opposite situation, namely the favoured insertion of the *re* coordinated propylene molecule as opposed to the *si* coordinated molecule, holds for a Δ site [1].

For the discussed model sites, the two situations presenting the outward and inward propylene coordination are enantioselective and non-enantioselective respectively. Hence, these model sites are isospecific only under the assumption that propylene always coordinates at a given coordination position (i.e. that the chain skips back to the starting position after each monomer insertion and prior or simultaneously to the coordination of the new propylene molecule) [345].

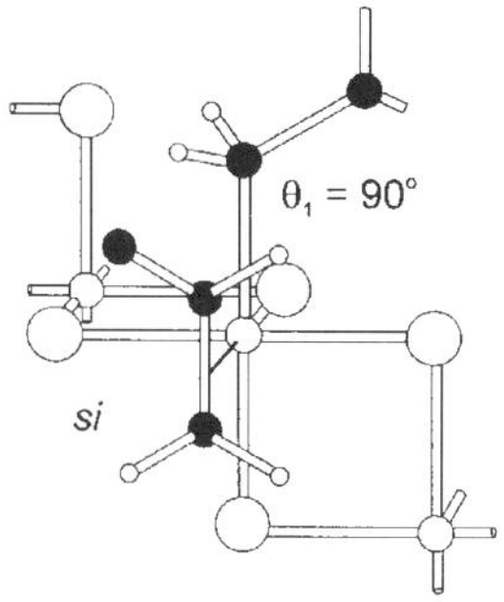

Figure 3.28 Perspective drawing of the coordination of propylene (*si*), suitable for a primary insertion, on a Λ site of Figure 3.26. The growing alkyl chain occupies a more hindered (inward) position; propylene enters the reaction site in a less hindered (outward) position with its methyl substituent away from the C_β atom of the last monomeric unit in the chain. ○ – *Ti*; ◯ – Cl; • – C; ○ – H; ● – CH_3. Reproduced (by permission from Elsevier Science) from Ref. 1. Copyright (1989) Pergamon Press

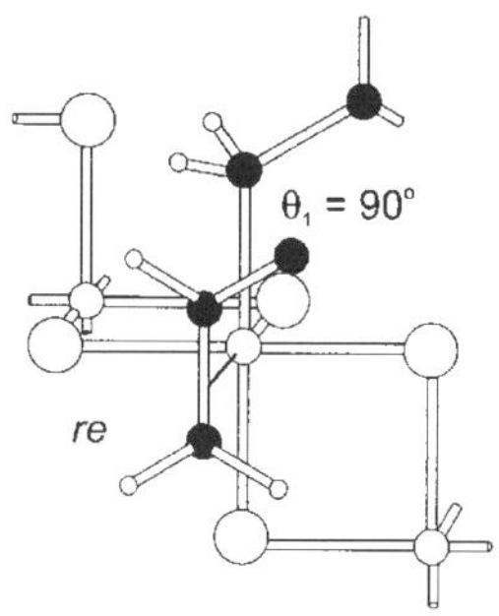

Figure 3.29 Perspective drawing of the coordination of propylene (*re*) on a Λ site of Figure 3.26. ○ – *Ti*; ◯ – Cl; • – C; ○ – H; ● – CH_3. Reproduced (by permission from Elsevier Science) from Ref. 1. Copyright (1989) Pergamon Press

The model devised by Corradini *et al.* [1] is in agreement with the experimental findings concerning the stereochemical structure of the end groups of poly (α-olefin) obtained with the catalyst based on δ-$TiCl_3$ and different ^{13}C-enriched aluminium alkyls [367–369]. The enantioselectivity of the first insertion step of an olefinic C=C bond into the Ti–R bond of the active site in heterogeneous isospecific catalysts (unsupported or supported on $MgCl_2$) depends to a notable extent also on the size of the R group at the Ti atom and increases according in the following order: $R = CH_3 \ll CH_3CH_2 < (CH_3)_2CHCH_2 < P_n$. Propylene insertion into a Ti–CH_3 bond, in the case of the catalyst obtained with the use of $AlMe_3$ as an activator, is not stereospecific; although the methyl group of the Ti–CH_3 bond is placed in chiral sites of the catalyst (Δ or Λ), it is not capable of 'recognising' the proper enantioface of the incoming monomer (*re* or *si* coordination). However, when two carbon atoms constitute the alkyl substituent at the Ti atom, i.e. when it is the ethyl group, propylene insertion into a Ti–CH_2CH_3 bond (in the case of the catalyst with the $AlEt_3$ activator) is partially stereospecific; propylene insertion into a Ti–$CH_2CH(CH_3)_2$ bond (in the case of the catalyst with the Al(*i*-Bu)$_3$ activator) is isospecific, as are successive insertions. Note, in this connection, that the second monomer insertion into a Ti–C bond (which formally starts the propagation step) is always isospecific, irrespective of the size of the initiating alkyl group [281,368]. The model under discussion also justifies the retention of configuration, in the growing polypropylene chain, after insertion of an ethylene unit [370] because this model maintains its behaviour also when a secondary carbon atom is in a β-position in relation to the metal. Remember the proposal by Brookhart and Green [344] that monomer insertion is assisted by strong α-agostic interactions between titanium and the C–H bond of the growing chain [1,2,68]. The enantioselectivity of the first insertion step of an olefinic C=C bond into the Ti–R bond further depends on the bulkiness of the substituent at the α-olefin double bond [2].

To summarise these considerations of the stereochemical control of propylene insertion in isospecific polymerisation with heterogeneous violet $TiCl_3$-based catalysts, the stereocontrol mechanism comes from non-bonded interactions between the methyl substituent of the incoming monomer and the C_β atom of the R ligand at the Ti atom. The chirality of the catalytic sites imposes a fixed chiral orientation on the first C–C bond of the reactive end of the growing chain. This in turn determines the configuration of a new inserted propylene unit. Exposed Ti_2Cl_6 reliefs on the (110) lateral face of the violet α-, γ-or δ-$TiCl_3$ crystals [281] provide the steric requirements for the isospecific mode according to the monometallic mechanism [2].

Considering the above model of the isospecific active site in violet $TiCl_3$-based Ziegler–Natta catalysts, Corradini *et al.* [283], by evaluating non-bonded interactions for the above-proposed active sites, suggested that the isospecific centre in $MgCl_2$-supported Ziegler–Natta catalysts should be derived only from Ti_2Cl_6 dimers epitactically placed on the (100) lateral borders of the layers of $MgCl_2$ crystals. The two adjacent titanium atoms have been considered to be enantiomerically related, and their chirality was assumed to cause control of the stereoregulation during polymerisation [68]. The proposed stereospecific (isospecific) sites exhibited a notable similarity to the Ti_2Cl_6 dimer present on the (110) $TiCl_3$ face (Figure 3.30) [303]. Non-stereospecific models were assumed to result from $TiCl_3$ isolated units of the $MgCl_2$ (100) and (110) faces [279].

It is commonly known that the presence of Lewis bases is essential to improve the stereospecificity of $MgCl_2$-supported catalysts. Many hypotheses have been formulated to explain the stereocontrol of the polymerisation by $MgCl_2$-supported Ziegler–Natta catalysts containing Lewis bases (internal and external) [1,53,61,68,240,290]. These hypotheses involve model sites falling into two basic categories: those with the Ti(III) species directly bound to the $MgCl_2$ surface and those with Ti(III) species bonded with the $MgCl_2$ surface via the Lewis base. Although it is still under investigation how all of the components

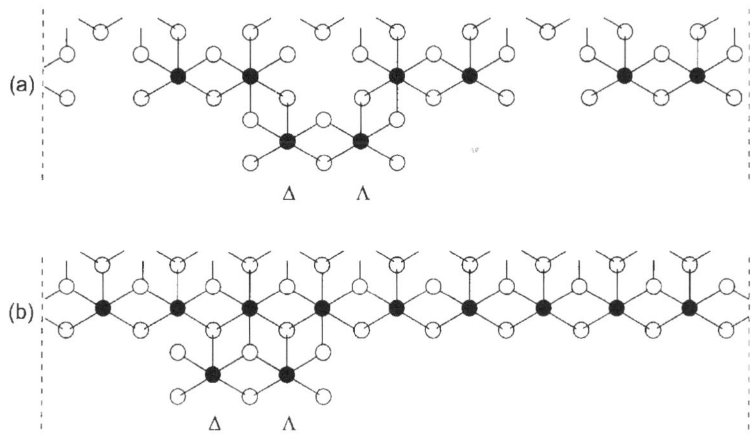

Figure 3.30 Detailed view of (a) a Ti_2Cl_6 group in relief with respect to the (110) cut of $TiCl_3$ and (b) a Ti_2Cl_6 group epitactically placed on a (100) cut of $MgCl_2$. Chiral environments of metal atoms are explicity labelled. • – (a) Ti, (b) Mg or Ti; ○ – Cl

are incorporated in the catalyst to form its isospecific active site, it has been suggested that ionic species are present in $MgCl_2$-supported catalysts [293]. Hence, the model suggested by Corradini *et al.* [1] appears to be oversimplified.

The discussed mechanism of α-olefin isospecific polymerisation in the presence of heterogeneous Ziegler–Natta catalysts, which considers, according to Corradini *et al.* [1], a chiral orientation of the growing polymer chain in the nantiomorphic catalytic site, can also be used to explain the stereoselective behaviour of these catalysts in the isospecific polymerisation of racemic α-olefins [282]. The polymerisation of racemic α-olefins with non-supported [371–374] and supported [375] heterogeneous isospecific Ziegler–Natta catalysts proceeds stereoselectively, i.e. poly(α-olefin)s obtained can be separated into fractions consisting of units derived largely from the same enantiomer (forming isotactic sequences) and having optical activity of opposite signs. The stereoselectivity is high when the chiral carbon atom is in the α-or β-position to the C = C bond [376], as for instance in 3-methyl-1-pentene [CH_2 = CH–C*H(CH_3)–CH_2CH_3] or 3,7-dimethyl-1-octene [CH_2 =CH–C*H (CH_3)–$(CH_2)_3$–CH$(CH_3)_2$] and 4-methyl-1-hexene [CH_2=CH–CH_2–C*H (CH_3)–CH_2CH_3] respectively, and is completely lost when he chiral centre is in the γ-position, as in 5-methyl-1-heptene [CH_2=CH–$(CH_2)_2$–C*H(CH_3)–CH_2CH_3] [1,2].

The stereoselective behaviour of heterogeneous Ziegler–Natta catalysts is caused by the chirality of the catalytic active site but not by chiral atoms in the growing chain. The model devised by Corradini *et al.* [282] can explain the experimental results relative to the first insertion of a chiral α-olefin into an initial Ti–CH_3 bond [377,378], i.e. the absence of enantioselectivity (discrimination between *re* and *si* insertions) but the presence of diastereoselectivity (preference for an $R(S)$ enantiomer upon *re*(*si*) insertion). The results show that the diastereotopic faces of the monomers have a different reactivity [379].

The data concerning the diastereoselectivity of the first step of insertion of the monomer into the Ti–CH_3 bond in the isospecific polymerisation of an α-olefin such as 3-methyl-1-pentene (with the δ-$TiCl_3$–$Al(^{13}CH_3)_3$ catalyst) show the more reactive face to be that corresponding to the front attack in Figure 3.31, i.e. the one having an absolute like configuration of the substituent and the face [diastereomeric monomer units arising from attack of the front face (R,R) and the back face (S,R) of (R)-3-methyl-1-pentene on the growing polymer chain are shown] [377,379].

The R,R and S,S diastereotopic faces of the monomer are two times more reactive than the R,S and S, R faces. Even in the chain propagation steps, an analogous diastereoselectivity is observed [2].

Thus, for both chiral and prochiral α-olefins, the isotactic sequence of the stereogenic tertiary carbon atom of the backbone is due to the enantioselectivity of the chiral active sites to the prochiral carbon atom of the monomer. The stereoselectivity (namely the selection, among the enantiomers, of a racemic

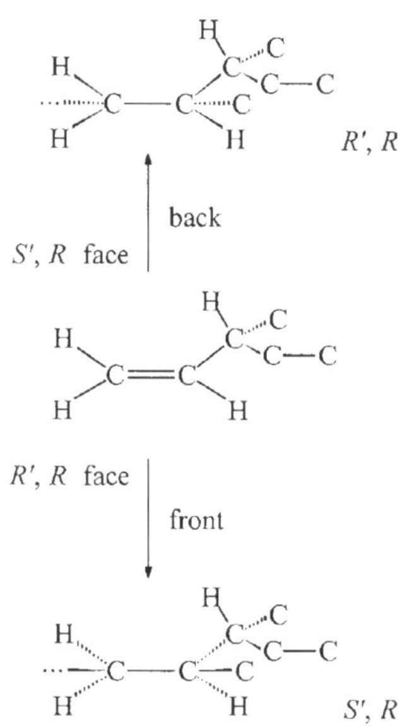

Figure 3.31 Diastereomeric 3-methyl-1-pentene monomer units arising from incorporation of the monomer on the growing polymer chain, depending on the reacting diastereotopic monomer face: attack of the front face (*down*) and of the back face (*up*) of (–)-(R)-3-methyl-1-pentene. For the assignment of the configuration of the backbone-substituted carbon of the monomer unit, it is considered that, soon after the insertion, the C atom bonded to the metal atom of the active site has the highest rank. The configuration of the faces and of the tertiary carbon atoms of the backbone are primed. Reproduced by permission from Ref. 379. Copyright 1995 Springer-Verlag Berlin Heidelberg

monomer by the chiral active species – according to the definition by Pino) is a combined consequence of the enantioselectivity towards the stereogenic tertiary carbon atom and of the intrinsically different reactivity of the diastereotopic monomer faces [379]. Steric interaction of the side group of the α-olefin molecule and the ligands of the active site is the origin of the stereoselection phenomena. This has been stringently shown by the results of copolymerisation of racemic α-olefins with ethylene; the random insertion of ethylene units (22–44%) into the polymer chains formed from 4-methyl-1-hexene and 3,7-dimethyl-1-octene does not cause any appreciable decrease in the degree of stereoselection, although the last chiral carbon atom of the growing polymer chain is placed at least in the δ-position to the titanium atom of the catalyst active site [376,380,381].

In stereoselective (enantiosymmetric) polymerisation, an optically inactive polymer is produced since the catalyst contains equal quantities of active sites of opposite chirality. Although at a given site the rate of polymerisation for one enantiomer can be up to a few times faster than for the other, the polymerisation rate of an *R* monomer at *R* preferred sites is the same as that for an *S* monomer at *S* preferred sites. This situation can be changed by introducing a second centre of chirality into the active site. Diastereomeric centres involving a chiral metal atom and a chiral ligand are likely to give different polymerisation rates. Therefore, the polymerisation of one enantiomer would be faster than that of the other, and an optically active polymer and residual monomer of opposite chirality will result [51]. Hence, if the polymerisation of racemic α-olefins is carried out in the presence of Ziegler–Natta catalysts obtained by the use of organometallic compounds containing optically active groups [382,383] as activators {e.g. bis[(*S*)-2-methylbutyl]zinc} or by the use of optically active Lewis bases [288,375,384,385] as the third component [e.g. (–)-menthyl anisate], it is a stereoelective (enantioasymmetric) polymerisation. In this process, indeed, an optically active polymer is obtained, and the optically active unreacted monomer is recovered [2,51].

As in stereoselectivity, the degree of stereoelection is reduced as the chiral centre is positioned further away from the C = C bond; only racemic α-olefins with a chiral carbon atom in the α- or β-position to the double bond have been stereoselectively polymerised. Note that, in general, stereoelectivity is lower than stereoselectivity.

3.5.3 Syndiospecific Propagation Reaction Stereocontrol in the Presence of Soluble Vanadium-based Ziegler–Natta Catalysts

Although low-temperature syndiospecific polymerisation of propylene with soluble Ziegler–Natta catalysts, based on soluble vanadium compounds and dialkylaluminium chlorides as activators, was first carried out successfully as far back as the early 1960s [10] (syndiotactic polypropylene was isolated earlier by separation from mostly isotactic polypropylenes obtained with heterogeneous catalysts [91]), the problem of the origin of stereoregulation in low-temperature syndiospecific propylene polymerisation has been relatively little investigated. However, the main features of the polymerisation mechanism have been recognised [1,2].

Polymerisation experiments run with catalysts prepared with ^{14}C-labelled $AlEt_2Cl$ [74] and NMR analysis of syndiotactic polymers obtained from $1-(^{13}C)$-enriched propylene [386] show the secondary insertion of monomeric units to occur into a V–C bond [317], i.e. according to the 2,1 mode [scheme (40)] [387,388].

The syndiospecific polymerisation of propylene with soluble vanadium-based Ziegler–Natta catalysts is not completely regiospecific [389–392], i.e. the monomer unit enchainment is not entirely head-to-tail. In addition to syndiotactic stereoblocks, the polymer also contains sterically irregular stereoblocks. The whole polymerisation can be thus described as a copolymerisation with four head-to-tail and tail-to-tail stages [2,379].

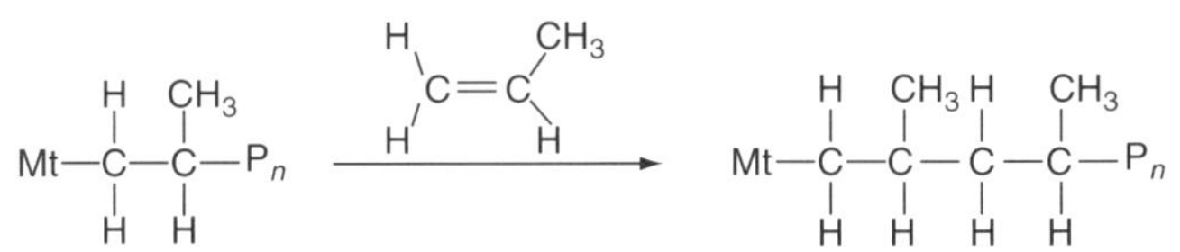
(63)

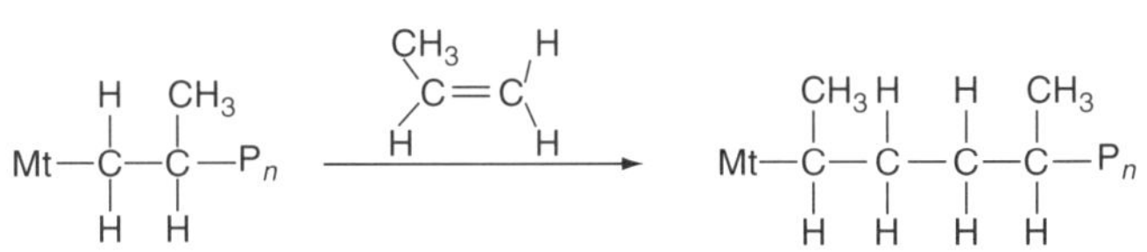
(64)

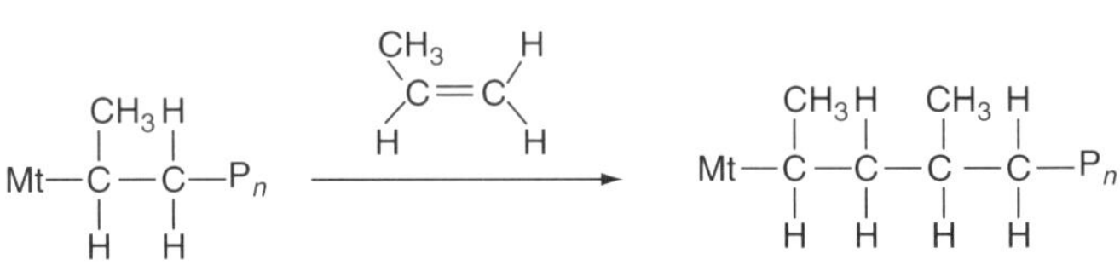
(65)

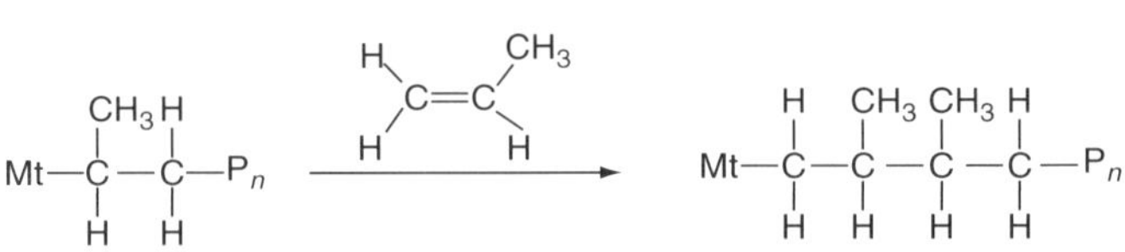
(66)

It must be emphasised that the propagation step according to scheme (65), which involves the secondary insertion of propylene in a V–CH(CH_3) bond, is syndiospecific and accounts for the incorporation into the chain of most monomer units [379].

Regioirregularity arising from the occasional primary insertion of propylene, i.e. according to the 1,2 mode [scheme (66)], is strongly affected by the bulkiness of the alkyl group bonded to the aluminium atom in the alkylaluminium compound used as the activator; the regioirregularity decreases and the syndiospecificity increases as the steric hindrance of the alkyl group of AlR_2Cl increases [74,393]. It should also be added that the syndispecificity of the propylene polymerisation with vanadium-based catalysts increases as the polymerisation temperature is lowered. Moreover, a higher degree of syndiospecificity in the polymerisation is attainable when using *n*-heptane instead of toluene as the reaction medium [2].

As regards the polymerisation of higher linear α-olefins, e.g. 1-butene, with homogeneous vanadium-based syndiospecific catalysts, it rarely occurs and affords only trace amounts of a low molecular weight syndiotactic polymer [394].

In view of the data concerning propylene polymerisation in the presence of homogeneous vanadium-based Ziegler–Natta catalysts, the syndiospecificity of the polymerisation is believed [387,395] to arise from steric repulsions between the last inserted monomer unit of the growing chain and the methyl group of coordinated propylene molecule, i.e. chain end stereocontrol is postulated to play the essential role in the stereoregulation.

The same conclusion as in the case of propylene homopolymerisation has been drawn considering IR [396] and NMR [389,395] spectra of ethylene/propylene copolymers obtained with vanadium-based syndiospecific catalysts. The type of propylene insertion depends on the kind of last inserted monomer unit: secondary insertion [scheme (40)] occurs more frequently when the last monomeric unit of the growing chain is propylene, while primary propylene insertion [scheme (39)] is more frequent when the last monomeric unit of the growing chain is ethylene [2]. The above explains the microstructure of ethylene/propylene copolymers obtained with vanadium-based Ziegler–Natta catalysts. These copolymers contain both *m* and *r* diads when the sequence of propylene units is interrupted by isolated ethylene units; i.e. a propylene insertion after an ethylene insertion is substantially non-stereospecific [327,390,397]. The existence of a steric interaction between the incoming monomer molecule and the last added monomer unit is also confirmed by the fact that the propagation rate for the secondary insertion of propylene in syndiospecific polymerisation is lower than for primary insertion in non-stereospecific polymerisation [398].

A few models of the active site in homogeneous vanadium-based Ziegler–Natta catalysts were proposed in order to explain the syndiospecific polymerisation of propylene. According to the model favoured by most authors, syndiospecific propagation in propylene polymerisation is due to the greater stability of complex (a) in Figure 3.32, in which the methyl group of the coordinated monomer and the methyl group of the last inserted monomeric unit are on opposite sides with respect to the plane defined by the V–C bonds,

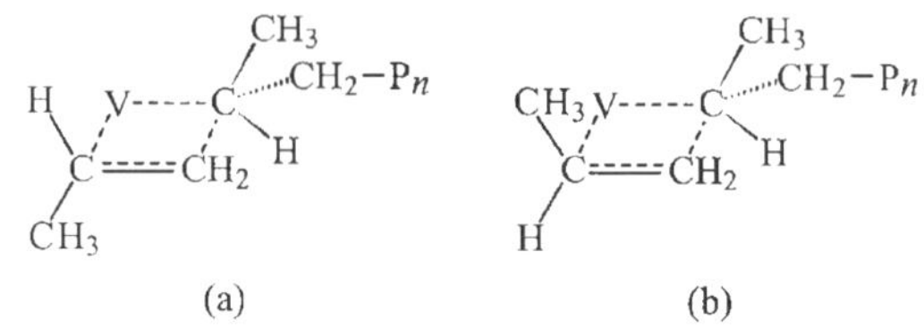

Figure 3.32 Supposed four-centred activated complexes for secondary insertion of a propylene molecule

as compared with the stability of complex (b) in Figure 3.32 in which these methyl groups are on the same side [395].

In the case of the former complex, secondary insertion is preferred provided that the steric hindrance of substituents bound to the central metal atom is not too different, and the *trans* arrangement of methyl groups of the coordinated monomer and of the last monomeric unit gives rise to a syndiotactic polymer. In this case the syndiospecificity is controlled by the chirality of the last carbon atom of the growing chain. The necessity for low temperature in the production of highly syndiotactic polypropylene, usually below $-50\,^{\circ}$C, suggests a small energy difference between different transition states (Figures 3.32 a and b) [1,2].

In the model proposed for the active centre of the propylene syndiospecific polymerisation, the V(III) atom is pentacoordinated [327]. Its ligands include three chlorine atoms (two of which are bridge-bonded to the aluminium atom), the chiral carbon atom of the last monomer unit of the growing chain and the coordinated propylene molecule. Prior to its coordination and after its insertion, the vanadium atom is tetracoordinated. In the alternative similar model, two chlorine atoms are substituted by a bidentate dionate, and the chlorine atom is bridge-bonded to the aluminium atom in dimeric AlR_2Cl [2].

It has also been assumed that the configuration of the last carbon atom of the growing chain influences directly the chirality of the catalytic intermediate complex, which in turn determines the configuration of the incoming monomeric unit [328]. In the model proposed for the active centre in such a case, the V(III) atom is hexacoordinated. Its ligands include four chlorine atoms (assumed to be bridge-bonded to aluminium atoms), apart from the growing chain and the coordinated monomer (Figure 3.33).

Such a catalytic centre is chiral, but interconversion between enantiomeric complexes is assumed to occur, after each insertion step, when the V atom is pentacoordinated. The analysis of the non-bonded interactions at the catalytic site suggests an *si* insertion of the last monomer unit (which generates what is called an *si* chain) to favour the formation of a Λ complex. This complex should in turn favour the *re* coordination and insertion of the successive monomer unit, thus ensuring syndiospecific polypropylene chain propagation. In other words, the chirality of the growing chain (expressed by the configuration of the last inserted monomeric unit) imposes the chirality of the coordinating monomer

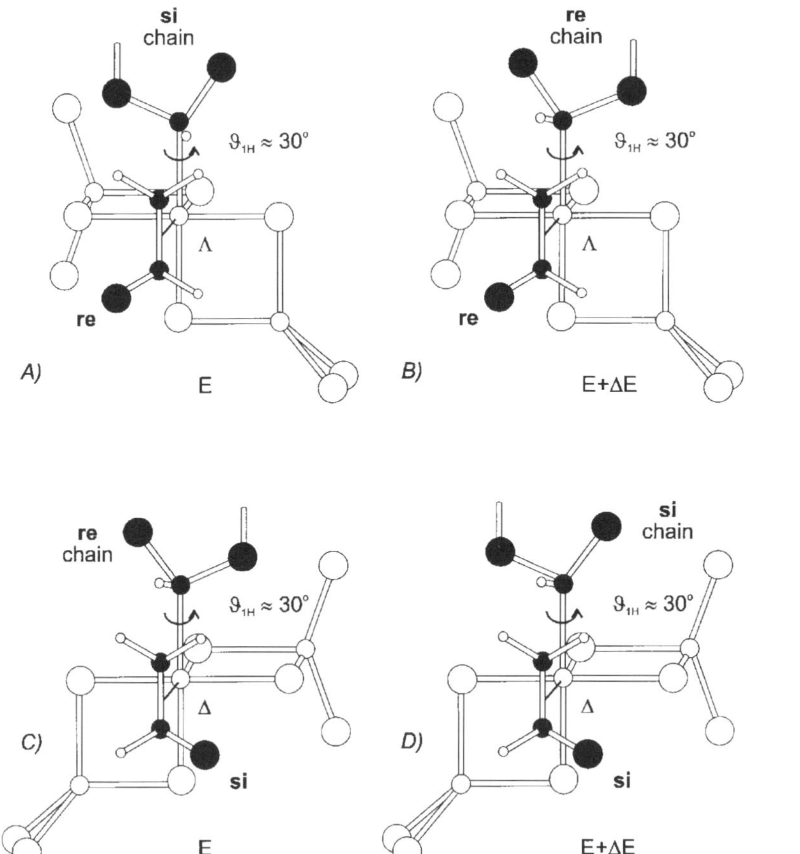

Figure 3.33 Possible diastereoisomeric catalytic complexes for the soluble vanadium-based syndiospecific catalysts, with a *si* chain, presenting suitable orientation of the chain and of the propylene molecule. The corresponding minimum energies are indicated below the models. ○ – V or Al; ◯ – Cl; ● – C; ○ – H; – ● – CH_3 or CH_2. Reproduced by permission from Ref. 328. Copyright 1985 American Chemical Society

(hence the configuration of the successive inserted monomer unit) by influencing directly the chirality of the metal atom [1].

On the basis of the available experimental data, none of the discussed models can be discarded.

3.5.4 Propagation Reaction Stereocontrol in the Presence of Single-site Metallocene Catalysts

The appearance of only one or a few kinds of active site in metallocene-based catalysts and the possibility of studying these catalysts using NMR and X-ray techniques have made it possible to formulate the relationship between the structure of active sites and the stereoregulation mechanism in the polymerisation of α-olefins in the presence of these catalysts.

Remember that the catalytic site is a pseudotetrahedral group 4 metal cation bearing η^5 ligands, the growing polymer chain and the coordination vacancy for the incoming monomer; the synthesis and structural characterisation of cationic metal complexes of the $[Cp'_2MtR]^+[X]^-$ type, containing cyclopentadienyl-like ligands and a non-coordinating anion as a counterion, has led to the conclusion concerning the cationic nature of active sites [33,109,111,164, 166,169]. The cationic nature of active sites is also characteristic of Ziegler–Natta catalysts such as metallocene–methylaluminoxane, which has been concluded on the basis of solid-state CPMAS ^{13}C NMR study of the reaction in the $Cp^*_2Zr(^{13}CH_3)_2$–$[Al(Me)O]_x$ system [399].

As regards the insertion mechanism in α-olefin polymerisation with metallocene-based catalysts, one should recall that a chain migratory mechanism is operating, but occasional skipped insertion or a constant skipped insertion mechanism may also be operating (Figure 3.17), depending on the kind of catalyst.

The correlation between the symmetry of the metallocene active site and polymerisation stereochemistry (Table 3.3) as well as polymer tacticity (Tables 3.1 and 3.2) is now fully understood [1,2,16,29,30,37,68,105,112,117,122–127,143,146,241]; two general mechanisms of stereoregulation can be assumed: chain end stereocontrol for achiral catalysts and enantiomorphic site control for chiral catalysts. In the former case, the stereocontrol, if it takes place, has been attributed to the chirality of the last inserted monomer unit in the growing chain. In the latter case, however, the stereocontrol has been attributed to the chirality of the catalytic complex. The general mechanism for enantioface selectivity in chain migratory insertion during α-olefin polymerisation in the presence of metallocene catalysts, based on chiral or prochiral precursors (Figure 3.17), can be visualised using key-in-the-lock formalism: every active centre with a chiral metal atom has two available coordination sites (the two locks) that can both insert the α-olefin, and that can be different in both shape and chirality. Because of site switching, the monomer (the key) has to be inserted alternately on each site. As a consequence, isotactic, syndiotactic, hemiisotactic and stereoblock polymers are obtained [68].

3.5.4.1 Stereocontrol of Isospecific Propagation with Achirotopic Catalysts

In catalysts obtained from achiral non-bridged metallocenes of class I with C_{2v} molecular symmetry (double helical), such as Cp_2MtX_2, the positions of the coordinated monomer and of the alkyl ligand are not chirotopic and, therefore, the catalyst control is completely lacking (Table 3.1) [68].

After the discovery that the Cp_2TiPh_2–$[Al(Me)O]_x$ catalyst becomes increasingly isospecific in propylene polymerisation with decreasing temperature [22], the chain end stereocontrol mechanism (Table 3.3) has been postulated to operate during low-temperature polymerisation. Moreover, as regards the polymerisation mechanism in this case, the chain migratory insertion mechanism has been postulated to operate. Remarkably, this was the first known case of isotactic chain propagation dictated by the chirality of the last inserted monomeric unit. In this case, stereospecific non-bonded interactions occur between the coordinating monomer and the chiral carbon atom of the last monomeric unit of the growing chain; the chirality of the centre formed on primary (1,2) propylene insertion determines the energetically most favourable enantioface of the coordinated propylene molecule to be inserted (Figure 3.34a); the insertion in complex (b) in Figure 3.34 hardly proceeds at low temperature [23]. In the former complex, the *trans* arrangement of methyl groups of the coordinated monomer and of the last monomeric unit gives rise to isotactic polypropylene. The need to carry out the polymerisation at low temperature (e.g. at $-45\,^{\circ}C$) indicates a small energy difference between the different transition states (Figure 3.34a and b).

A drastic decrease in stereospecificity with increasing size of the alkyl substituent at the α-olefinic double bond is characteristic of polymerisation with achiral metallocene-based catalysts; polypropylene obtained with these catalysts contains more than 80% of *m* diads, while poly(1-butene) contains ca 60% of such diads, but poly(4-methyl-1-pentene) contains only 50% of *m* diads (which testifies to the complete atacticity of the latter polymer). It is obvious that the coordinating larger α-olefin molecule, especially in the case of 4-methyl-1-pentene, is not able to 'recognise' the spatial configuration of the alkyl substituent at the tertiary carbon atom of the last primarily inserted monomeric unit, since this substituent is too similar to the other substituent comprising the growing chain. In fact, the fragment of this alkyl substituent with the two carbon atoms, C_α and C_β($-CH_2$–CH <), is identical to the fragment of the penultimate monomeric unit in the polymer chain, and thus the tertiary carbon atom of the last inserted monomeric unit practically loses its chirality [400]:

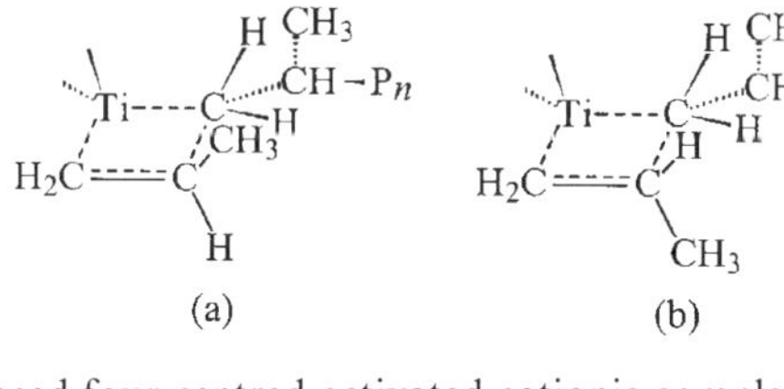

Figure 3.34 Supposed four-centred activated cationic complexes for primary insertion of a propylene molecule

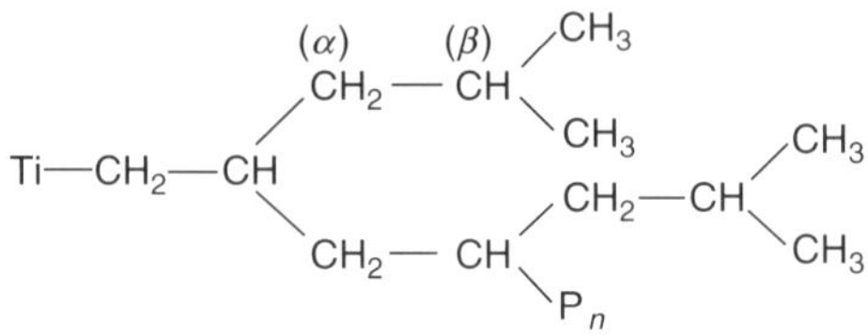

Therefore, either *re* or *si* coordinated 4-methyl-1-pentene molecules are inserted, which results in the formation of atactic poly(4-methyl-1-pentene).

Catalysts obtained from achiral stereorigid metallocenes of class II with C_s molecular symmetry (non-helical), such as *mesogenic* isomers of bridged metallocenes (namely *ansa*-metallocenes) such as *meso*-$(IndCH_2)_2MtX_2$ (Table 3.1) and *meso*-$(ThindCH_2)_2MtX_2$, exhibit similar behaviour to that of achiral non-bridged metallocene-based catalysts in α-olefin polymerisation, and their stereocontrol in the propylene polymerisation is completely lacking; they give rise under normal polymerisation conditions to atactic polypropylene [22,401]. Similarly to achiral non-bridged metallocene-based catalysts, catalysts based on achiral bridged *mesogenic* metallocenes become increasingly isospecific in propylene polymerisation with decreasing temperature [22]; it is obvious that the chain end stereocontrol mechanism (Table 3.3) operates in this case.

It may be of interest that the atactic polypropylene produced with the catalyst based on the *meso*-$(ThindCH_2)_2ZrCl_2$ precursor contains hardly any regioirregularity due to 2,1 or 1,3 insertions, compared with the polymer obtained with the catalyst based on the respective *racemic* metallocene [namely *rac*.-$(ThindCH_2)_2ZrCl_2$] [401]. A molecular mechanics analysis of a model of this *meso*-$(ThindCH_2)_2ZrCl_2$-based catalytic complex with propylene has revealed that the polymer chain should skip back to the starting position after each monomer insertion and prior to the coordination of a new propylene molecule, i.e. the chain stationary insertion mechanism should operate in the propylene polymerisation with this catalyst [345]. It must be emphasised that this is in contrast to propylene polymerisation with the mentioned Cp_2MtX_2-based catalysts in which the chain migratory mechanism, and not the chain stationary mechanism, operates (Figure 3.17).

For the catalytic site based on the *meso*-$(ThindCH_2)_2ZrCl_2$ procatalyst, the two diastereomeric situations resulting from exchanging the relative positions of the monomer and of the growing alkyl chain, which present opposite configurations of the central metal atom, are shown in Figure 3.35 [(A) and (B) respectively] [345].

The growing alkyl chain occupies an open sector of the *meso*-$(ThindCH_2)_2$ ligand framework (less hindered outward position), while the monomer coordinates in the region on the side of the two six-membered rings (more hindered inward position). The lowest energy situation has been found in the case of the inward propylene coordination [complex (A) in Figure 3.35]; situations corresponding the outward propylene coordination [complex (B) in Figure 3.35] are

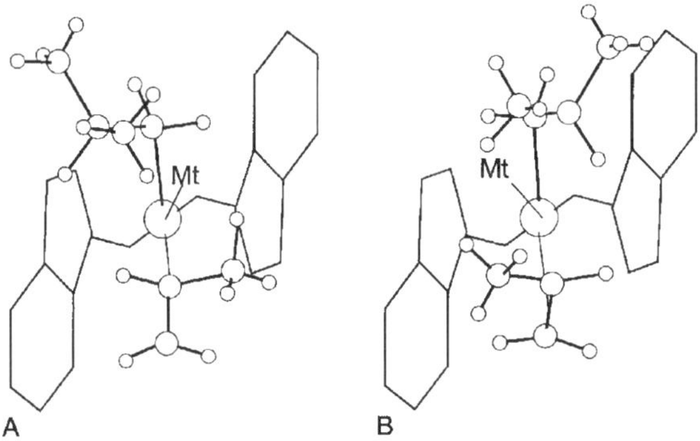

Figure 3.35 Models for the primary insertion of propylene into a polypropylene growing chain, when the catalyst precursor is *meso*-$(ThindCH_2)_2ZrCl_2$. Complex A corresponds to the minimum energy situation and is suitable for primary insertion; complex B (of higher energy) is disfavoured for the insertion and is isomerised to complex A. For the sake of clarity, only the C–C bonds are sketched for the π ligands. ❍ – Zr; ❍ – C; ○ – H. Reproduced by permission from Ref. 345. Copyright 1995 Springer-Verlag Berlin Heidelberg

of higher energy. Furthermore, within the framework of molecular mechanics analysis, the diastereoisomer with inward propylene not only exhibits a lower energy at the coordination step but also exhibits a lower energy for the monomer orientation suitable for primary insertion; a monomer orientation that might be suitable for secondary insertion is strongly disfavoured in this case. The discussed model indicates that a chain back skip after each insertion reaction is favoured prior to the coordination of the next monomer molecule; therefore, the growing chain skips back from the inward towards the outward coordination position, as indicated in the polymerisation scheme of Figure 3.36 [345,402].

Thus, if back skip of the chain occurs at each propagation step in propylene polymerisation with the *meso*-$(ThindCH_2)_2ZrCl_2$-based catalyst, it ought to be highly regioselective (much higher than for the catalyst based on the corresponding *racemic* ligand) according to the discussed model, which does, in fact, take place [345]. Let us repeat that this model is obviously non-enantioselective, because of the lack of chirality of the *meso*-$(ThindCH_2)_2ZrCl_2$-based catalyst site, and atactic poly(propylene) is actually formed (at ambient temperature).

3.5.4.2 Stereocontrol of Isospecific Propagation with Chirotopic (Homotopic) Catalysts

In catalysts obtained from chiral stereorigid metallocenes of class III with C_2 molecular symmetry (helical), such as *racemic* isomers of *ansa*-metallocenes, e.g. *rac*.-$(IndCH_2)_2MtX_2$ (Table 3.1) and *rac*.-$(ThindCH_2)_2MtX_2$, the two coordination positions available for the incoming monomer and the growing

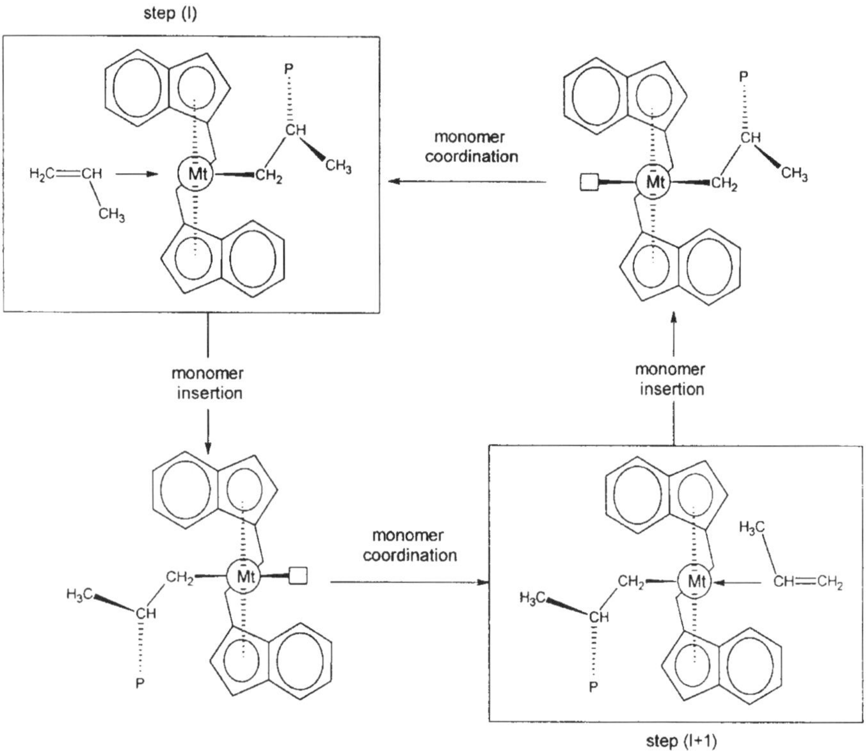

Figure 3.36 Schematic representation of the chain stationary insertion mechanism of propylene polymerisation with the bridged *mesogenic* metallocene-based catalyst. Reproduced by permission from Ref. 402. Copyright 1995 Wiley-VCH Weinheim

chain are homotopic and, therefore, the configuration of the transition metal atom does not change after the chain migratory insertion, i.e. after the shift of the growing chain to the position previously occupied by the coordinated monomer [22,23]. The chain migratory insertion mechanism operates in the polymerisation with the stereochemistry controlled by an enantiomorphic site of this type (Table 3.3); as a consequence, if the constant approach of the same enantioface of the α-olefin molecule is favoured, the resulting polymer is isotactic [68].

It should be emphasised that chiral metallocenes, suitable for isospecific polymerisation of propylene, became available in around 1980 almost by accident. Using ethylene-bridged ligands, Wild *et al.* [403,404] obtained ethylenebis(indenyl)-and ethylenebis(tetrahydroindenyl)titanium complexes, *rac.*-$(IndCH_2)_2TiCl_2$ and *rac.*-$(ThindCH_2)_2TiCl_2$, and their zirconium analogues. The conformationally constrained π ligands in these complexes give them chiral structures, which could be expected to be retained in polymerisation systems.

The obtained chiral *ansa*-metallocenes, after activation with methylaluminoxane, were indeed found, in independent studies by Ewen [22] and by Kaminsky and Külper [101], to polymerise propylene and higher α-olefins to produce highly isotactic polymers [30].

Figure 3.37 shows models corresponding to the two successive polymerisation steps, with exchanged positions of the coordinated propylene molecule and of the polymer chain (in the case of the (R,R)-*rac*.-$(IndCH_2)_2$ ligand) [30]. The growing alkyl (polymer) chain occupies an open sector of the *rac*.-$(IndCH_2)_2$ ligand framework; the C_α–C_β segment of the polymer chain appears to orient itself so as to avoid interaction with the cyclopentadienyl ring substituent in the β-position adjacent to its coordination site (in other words, in the β-position to the bridgehead C atom). Such an orientation of the C_α–C_β chain segment, according to which the C_β atom of the last inserted monomeric unit is placed in the most open sector of the chiral bridged metallocene ligand framework, is energetically favourable (Figure 3.37) [1,30]. As previously reported for heterogeneous Ziegler–Natta catalysts [1,280], repulsive interactions were shown by Guerra *et al.* [402] to force an α-olefin into that enantiofacial approach to the Mt–alkyl unit that holds the olefin substitutent (methyl group in the case of propylene) *trans* to (i.e. away from) the C_β atom of the metal-bound alkyl chain. In other words, the calculations indicated [345,402] that the chiral environment of the metal atom at the catalytic site forces the growing chain to a chiral orientation (for instance, in Figure 3.38 the orientation of the chain indicated with solid lines is allowed, but that indicated with dashed lines is forbidden). The chiral orientation of the chain, in turn, makes favourable the coordination of the propylene molecule with one of its two faces (for the sketched complex with the (R,R)-*rac*.-$(ThindCH_2)_2$ ligand, the *re* coordination would be favoured) [1].

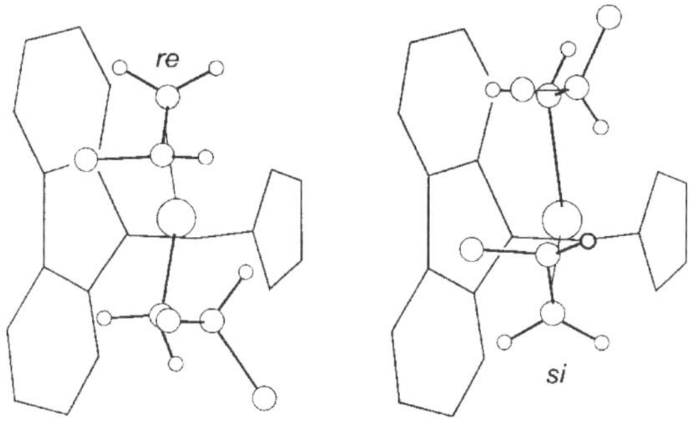

Figure 3.37 Models for the primary insertion of propylene into a polypropylene growing chain in isospecific polymerisation with the (R, R)-*rac*.-$(IndCH_2)_2ZrX_2$-based catalyst. The growing alkyl chain occupies an open sector of the ligand framework; propylene enters the reaction complex with its methyl substituent away from the C_β atom of the last monomeric unit in the chain (the monomer methyl group is directed towards another open sector of the ligand framework). For the sake of clarity, only the C–C bonds are sketched for the π ligands. ◯ – Zr; ○ – C or CH_3; ∘ – H. Reproduced by permission from Ref. 30. Copyright 1995 Wiley-VCH Weinheim

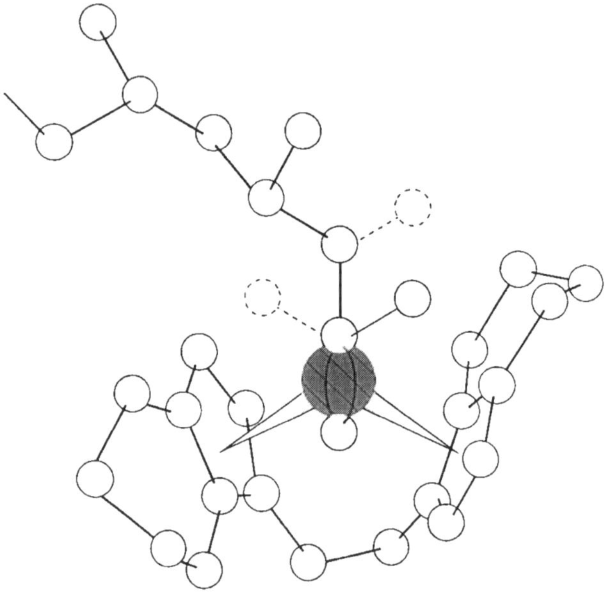

Figure 3.38 Possible model site for isospecific propylene polymerisation with the (*R*, *R*)-*rac*.-$(ThindCH_2)_2$ ligand-containing catalyst. The broken lines indicate the forbidden conformation of the growing chain and the favoured coordination of propylene corresponding to this forbidden chain conformation. For the sake of clarity, hydrogen atoms are omitted. ❍ – Zr ○ – C, CH, CH_2 or CH_3. Reproduced (by permission from Elsevier Science) from Ref. 1. Copyright 1989 Pergamon Press

The behaviour of the discussed model site is in agreement with the available experimental findings concerning the stereochemical structure of end groups of poly(α-olefin) a such as polypropylene and poly(1-butene), obtained with a catalyst based on C_2-symmetric, chiral metallocene catalysts, such as *rac*.-$(IndCH_2)_2TiMe_2$ activated with (^{13}C)-enriched $[Al(CH_3)O]_x$ or $[Al(CH_2CH_3)O]_x$ [117,405,406]. Analogously to the results obtained previously for heterogeneous catalysts, the discussed model appeared to be non-enantioselective for propylene insertion into an initial Mt–CH_3 bond, and partially enantioselective for 1-butene insertion into an initial Mt–CH_3 bond, but completely enantioselective for either propylene or 1-butene insertions into an initial Mt–CH_2CH_3 bond. Moreover, the model is in accordance with the results obtained by Pino *et al.* [407], which concern optical activity measurements on saturated oligomers derived from propylene hydrooligomerisation with the enantiomerically pure catalyst *rac*.-$(IndCH_2)_2ZrMe_2$–$[Al(Me)O]_x$ in the presence of H_2. These results showed that chiral propylene hydrotrimers and hydrotetramers with the predicted absolute configurations were obtained; *re* insertion of the monomer was favoured in the case of *R*,*R* chirality of coordination of the *rac*.-ethylenebisindenyl ligand. The formation of asymmetric oligomers with unsaturated end groups in similar reaction systems without H_2 was found by Kaminsky *et al.* [408] to follow the same rules.

The data obtained for the polymerisation, especially those concerning the enantioselectivity of the first monomer insertion into a metal–alkyl bond of the active site, show that efficient control of the chiral catalyst over the enantiofacial orientation of the incoming α-olefin is evidently contingent upon the presence of an Mt–alkyl segment with at least two C atoms. This segment appears to act as a sort of lever transmitting the effects of the β-substituents at the C_5 ring ligands on the orientation of the prochiral α-olefin [30].

Both equivalent coordination sites of the C_2-symmetric complex are framed by the β-substituents in such a manner that α-olefin insertions at the Zr centre occur with equal enantiofacial preference at both sites. It must be emphasised that unilateral coverage of each coordination site is essential for this stereoselectivity; this is borne out by the finding that atactic polypropylene is formed if both coordination sites are flanked by two β-substituents, as in catalysts based on *rac.*-ethylenebis[1-(3-methylindenyl)]zirconium dichloride [*rac.*-$(MeIndCH_2)_2ZrCl_2$] (Figure 3.39) [112].

The bilateral coverage of each coordination site in such a catalytic active site by a CH and a CH_3 group (both in a β-position to the bridgehead C atom) appears to render them indifferent to the enantiofacial orientation of an incoming α-olefin molecule, and thus atactic poly(α-olefin) is formed [30].

Two-membered carbon bridges (ethylene bridges), as in *rac.*-$(IndCH_2)_2ZrX_2$ or *rac.*-$(ThindCH_2)_2ZrX_2$ metallocene precursors, have the right size for at least medium stereospecificity of catalysts obtained from these precursors. Also, the one-membered silicon bridge in a metallocene precursor, e.g. in *rac.*-$Me_2Si(MeCp)_2ZrCl_2$ (Table 3.1), is suitable for obtaining a highly stereospecific catalyst [132]. It should be noted that other catalysts based on dimethylsilylene-bridged *racemic* zirconocenes maintain high isospecificity even at an elevated temperature of ca 70 °C [118,125,139]; this is connected with higher stereorigidity of these zirconocenes in comparison with those containing two-membered carbon bridges.

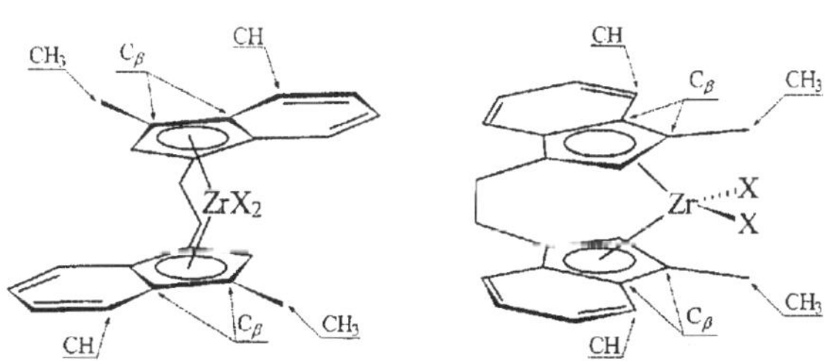

Figure 3.39 Chiral procatalyst of C_2 symmetry containing two β-substituents at the cyclopentadienyl rings, *rac.*-ethylenebis[1-(3-methylindenyl)]zirconium dichloride [*rac.*-$(MeIndCH_2)_2ZrCl_2$], which, in combination with $[Al(Me)O]_x$, produces atactic polypropylene. Front side view and side view respectively

Dimethylsilylene-bridged zirconocene precursors, such as *rac.*-$Me_2Si(Ind)_2ZrCl_2$ and its 4,5-benzo-or 2-methyl-4,5-benzo-substituted analogues, were employed with an $[Al(Me)O]_x/Al(^{13}CH_3)_3$ (5:1 molar ratio) activator to obtain catalysts for the polymerisation of a racemic α-olefin, (*R,S*)-3-methyl-1-pentene, in order to study, in more detail, subtle steric monomer–ligand interactions during the first monomer insertion into a (^{13}C)-enriched Zr–CH_3 bond [409]. The ^{13}C NMR data regarding the stereostructure of the first monomeric units inserted into this bond during polymerisation with a *rac.*-$Me_2Si(Ind)_2ZrCl_2$-based catalyst show the appearance of moderate enantioselectivity in this case. This is in contrast to a completely aspecific first monomer insertion in the polymerisation of α-olefins with heterogeneous $TiCl_3$-based catalysts. The observed moderate enantioselectivity of the first monomer insertion in the polymerisation with the *rac.*-$Me_2Si(Ind)_2ZrCl_2$-based catalyst was ascribed to repulsive direct interactions between the monomer and the ligand framework. As regards diastereoselectivity of the first monomer insertion in the polymerisation of (*R, S*)-3-methyl-1-pentene in the presence of the *rac.*-$Me_2Si(Ind)_2ZrCl_2$-based catalyst, it was found to be similar to that in the case of heterogeneous $TiCl_3$-based catalysts; the *R, R* and *S, S* diastereotopic faces of the monomer are about twice as reactive as the *R, S* and *S, R* faces. Therefore, the first steps of enantioselectivity (i.e. the choice between the two enantiotopic monomer faces) and diastereoselectivity (i.e. the choice among the four diastereotopic monomer faces) have been suggested to be governed by different repulsive interactions with the ligand framework; in other words, the enantioselectivity and diastereoselectivity appear to be independent effects [409].

As already mentioned, the regioselectivity of propylene polymerisation with isospecific *rac.*-$(IndCH_2)_2ZrX_2$-based and *rac.*-$(ThindCH_2)_2ZrX_2$-based catalysts is not as high as that of polymerisation with non-stereospecific catalysts based on corresponding *mesogenic* zirconocenes. It has been established that, in the discussed polymerisation with the *racemic* metallocene-based catalysts, the primary (largely prevailing) and secondary insertions of propylene on a primary growing chain occur preferably with opposite enantiofaces, primary insertion of propylene is also favoured over secondary insertion on a secondary growing chain, and primary insertion of propylene occurs preferentially for a given enantioface, irrespective of the primary or secondary nature of the growing chain (a primary or secondary growing chain is a chain in which the last inserted monomer unit was obtained by a primary or secondary insertion of propylene respectively) [345, 402].

A detailed molecular mechanics analysis of model catalytic sites has rationalised these observed behaviours for homogeneous propylene isospecific polymerisation catalysts based on bridged *rac.*-metallocene [357]. It is interesting that, in contrast to the discussed case of enantioselectivity of the models for the primary insertion of propylene, the enantioselectivity of the models for the secondary insertion of propylene is due to direct interactions of the alkylsubstituent (the methyl group in the case of propylene) of the coordinating monomer with the π ligand in a *rac.-ansa*-metallocene-based catalyst. The model sites with an (*R*, *R*)-*rac.*-$(IndCH_2)_2$ or (*R*, *R*)-*rac.*-$(ThindCH_2)_2$ ligand favour the *re* monomer coordination for the orientation suitable for a primary insertion, while they favour the *si* monomer coordination for the orientation suitable for a fortuitous secondary insertion. The preference for the *si* monomer coordination in the case of the secondary insertion is due to less repulsive interactions of the methyl group of coordinating propylene with the cyclopentadienyl-like ligand (Figure 3.40) [402].

In connection with the occurrence of secondary insertions during propylene polymerisation with homogeneous *rac.-ansa*-metallocene-based catalysts to a higher extent than in the case of polymerisation with heterogeneous Ziegler–Natta catalysts, the 1,3 'misinsertions' of propylene, characteristic of some homogeneous polymerisation systems, which involve reactions of secondary growing chains, should be discussed here. Let us consider that, depending on the polymerisation temperature and on the kind of metallocene-based catalyst applied, one finds either head-to-head enchainment resulting from primary α-olefin insertion on a secondary growing chain [scheme (67a)], or tetramethylene units formed by 1,3 insertions [schemes (67b) and (67c)]. These 1,3 insertions arise from the isomerisation of a secondary Zr–alkyl unit (involving a β-hydride transfer, rotation and then reinsertion) to one with a terminal Zr–alkyl bond [scheme (67b)] prior to insertion of the next propylene molecule [scheme (67c)] [254, 258, 259, 261]:

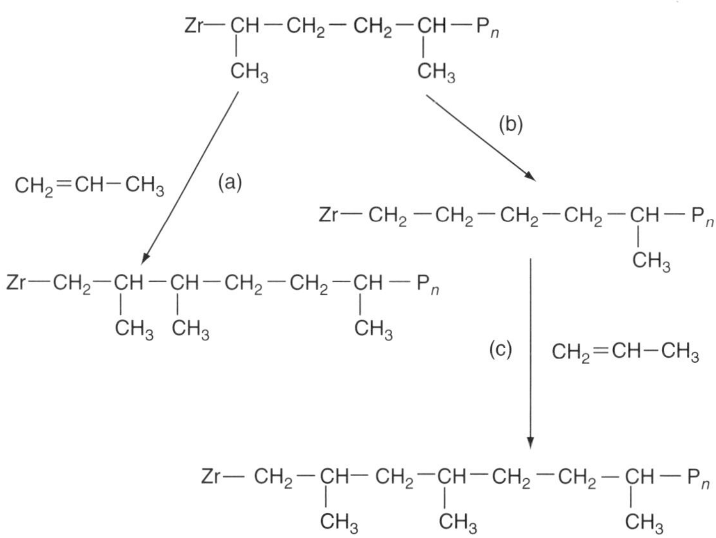

(67)

Different zirconocene-based catalysts cause the occurrence of regioirregularities in different ratios; only 1,3 misinsertions (ca 1%) are found in polypropylene

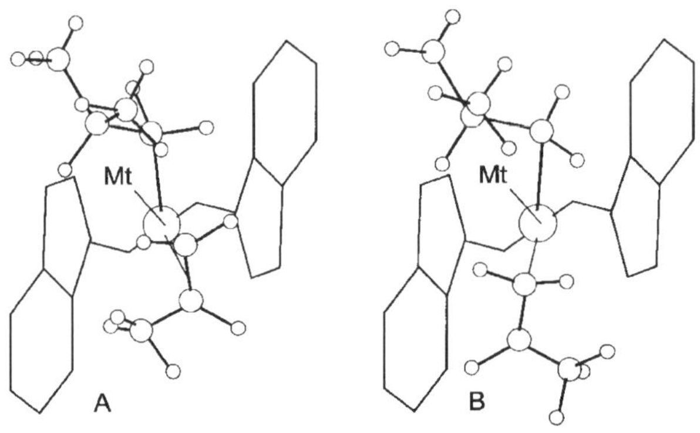

Figure 3.40 Models for the secondary insertion of propylene into a polypropylene growing chain, when the catalyst precursor is (*R*, *R*)-*rac*.-$(ThindCH_2)_2ZrCl_2$. Complex A with an *re* coordinated monomer corresponds to higher energy, while complex B with an *si* coordinated monomer corresponds to lower energy and is suitable for secondary insertion. For the sake of clarity, only the C–C bonds are sketched for the π ligands. ❍ – Zr; ○ – C; ○ – H

obtained with the *rac*.-$(ThindCH_2)_2ZrCl_2$–$[Al(Me)O]_x$ catalyst, and only 2,1 misinsertions (ca 1.6%) are found in polypropylene obtained with the *rac*.-$Me_2Si(Ind)_2ZrCl_2$–$[Al(Me)O]_x$ catalyst, while both types of misinsertion are present in the polymer produced by the *rac*.-$(IndCH_2)_2ZrCl_2$–$[Al(Me)O]_x$ catalyst at 40 °C [30]. A general rule is that with 'fast' catalysts the misinsertions are generally of the 2,1 type, but with 'slow' catalysts they are mainly of the 1,3 type [131, 256].

At the end of considerations of the stereoregulation mechanism in the isospecific polymerisation of α-olefins with homogeneous catalysts obtained from chiral stereorigid metallocenes of class III with C_2 molecular symmetry, let us address isospecific propylene polymerisation catalysts, which are methylaluminoxane-activated non-bridged metallocenes (Table 3.2).

For a highly isospecific catalyst based on the $(MeFlu)_2ZrX_2$ precursor, the absence of the regioirregular 2,1 or 1,3 misinsertions (ubiquitous in the backbone of isotactic polypropylene produced with catalysts based on bridged *rac*.-$(IndCH_2)_2ZrX_2$-type precursors with a flexible structural framework [257]) is considered as evidence that the methyl substituents at fluorenyl rings (in the 1-position) in $(MeFlu)_2ZrX_2$ create and maintain a high degree of stereorigidity, apparently owing to the non-bonding repulsive interaction exerted by the methyl groups [143]. Therefore, methyl-substituted fluorenyl ligands are chirally disposed and apparently so strongly hindered in their mutual rotation that enantiomers of the $(MeFlu)_2ZrX_2$-based catalytic complex are not interconverted during the growth of a polymer chain; the *racemic* isomers of C_2 symmetry constitute the catalyst site, and isotactic polypropylene is thus formed in polymerisation with the discussed catalyst (Table 3.2).

Another unbridged metallocene, such as *rac.*-$[Ph(Me)CHCp]_2ZrCl_2$, which contains substituents with a chiral C atom at the C_5 ring but is unable to restrict sufficiently rotations of the C_5 ring ligands, when activated with $[Al(Me)O]_x$, gives rise to isotactic polypropylene at a lowered temperature of ca $-50°C$ (Table 3.2). A partial control by the chirality of the active site (as in class III bridged metallocene catalysts) and by the configuration of the last inserted monomeric unit in the chain (as in class II bridged metallocene catalysts) over the stereochemistry of the next propylene insertion diminishes with increasing polymerisation temperature [119]. Apparently, insufficiently restricted rotations of the substituted cyclopentadienyl ligands and their substituents and the increasing conformational mobility of the growing polymer chain eliminate this stereocontrol mechanism at ambient temperature [30].

An unbridged metallocene catalyst, $(PhInd)_2ZrCl_2$–$[Al(Me)O]_x$, reported by Waymouth *et al.* [128, 129, 147], which appears to isomerise, by restricted mutual rotation of its phenyl-substituted indenyl ligands between chiral and achiral coordination geometries (corresponding to class III and class II catalysts respectively) during chain growth, yields isotactic–atactic stereoblock polypropylene during polymerisation at ambient temperature (Table 3.2). This catalyst, designed to isomerise between *racemic* (chiral) and *mesogenic* (achiral) isomers (sites of C_2 and C_s symmetry respectively), provides a means of controlling the distribution of isotactic and atactic stereosequences in the polypropylene formed.

3.5.4.3 Stereocontrol of Syndiospecific Propagation with Chirotopic (Enantiotopic) Catalysts

In catalysts obtained from prochiral stereorigid metallocenes of class IV with C_s molecular symmetry (non-helical), such as $Me_2C(Cp)(Flu)MtX_2$ (Table 3.1), a local symmetry plane relates the two enantiotopic coordination positions available for the growing polymer chain and the incoming monomer [23]. If the chain migratory insertion mechanism is operating, the configuration of the complex cation is inverted at each propagation step and, therefore, if enantioselectivity exists, the approach of the coordinating monomer will occur, presenting alternatively the two enantiofaces [68]. Figure 3.41 shows models corresponding to the two successive polymerisation steps with exchanged positions of the coordinated propylene molecule and of the polymer chain (in the case of polymerisation in the presence of the $Me_2C(Cp)(Flu)MtCl_2$-based catalyst) [30].

The two positions of the growing polypropylene chain and of the coordinated monomer molecule are enantiotopic; therefore, if enantioselectivity exists, the model is syndiospecific [23].

The prochiral face selectivity of the catalytic sites is due to the unique spatial arrangement of the stereorigid $Me_2C(Cp)(Flu)$ ligand encompassing the Zr atom and its active coordination positions [Figure 3.42 ($R = H$)].

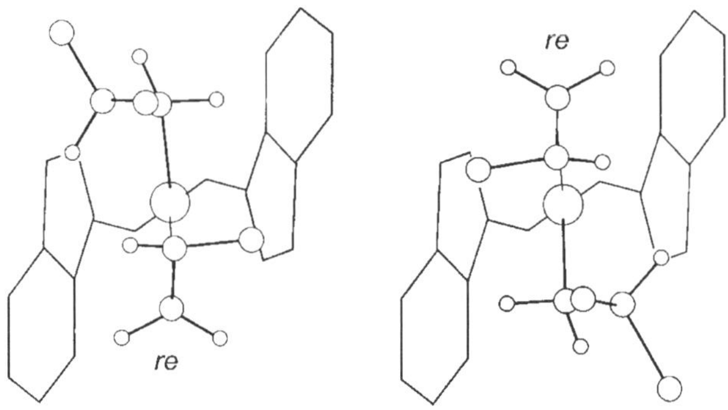

Figure 3.41 Models for the primary insertion of propylene into a polypropylene growing chain in a syndiospecific polymerisation with the $Me_2C(Cp)(Flu)ZrX_2$-based catalyst. The growing alkyl chain occupies an open sector of the ligand framework; propylene enters the reaction complex with its methyl substituent away from the C_β atom of the last monomeric unit in the chain (the monomer methyl group is directed towards the 'mouth' of the two C_6 rings). For the sake of clarity, only the C–C bonds are sketched for the π ligands. ◯ – Zr; ○ – C or CH_3; ∘ – H. Reproduced by permission from Ref. 30. Copyright 1995 Wiley-VCH Weinheim

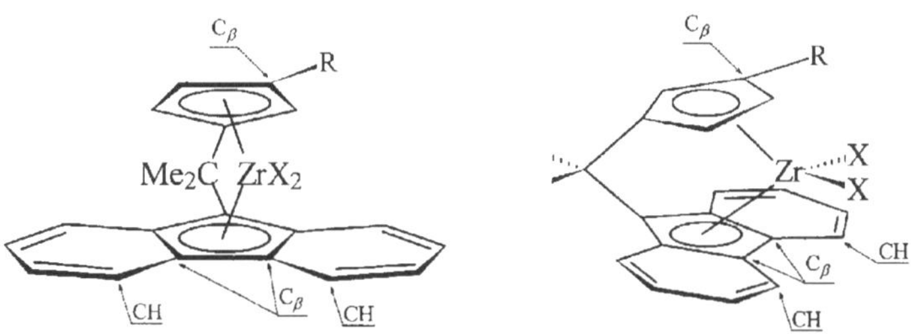

Figure 3.42 Isopropylidene[1-(3-R-cyclopentadienyl)](9-fluorenyl)zirconium dichloride [$Me_2C(RCp)(Flu)ZrX_2$]; R = H, Me, *t*-Bu. Front side view and side view respectively

According to a model proposed by Guerra *et al.* [402], the non-bonded repulsive interactions between the dorsal atom groups of the fluorenyl six-membered rings [CH groups in a β-position to the bridgehead C atom as shown in Figure 3.42 (R = H)] and the atoms of the last inserted monomeric unit (via primary insertion) force the C_β atom of this unit to move away from the sterically congested fluorenyl ligand, enter the open sectors next to the C_5 ring and adopt the chiral orientation, which fits best the chiral environment of the catalytic site with least steric interactions. In this configuration each alkylic enantiomer, with its chirally oriented growing chain, coordinates preferentially one of the prochiral faces (*re* or *si*) of the propylene molecule by repulsive interaction of the C_β atom of this growing chain with the methyl group of propylene (Figure 3.41). The enantiofacial orientation alternates between coordination sites, which, considering that the chain migratory insertion mechanism operates, leads to a syndiotactic polymer [23, 143].

It should be emphasised that the chain migratory insertion mechanism is primordial for syndiospecificity of α-olefin polymerisation with catalysts of class IV with C_s symmetry such as those based on $Me_2C(Cp)(Flu)ZrCl_2$. If any back skip of the chain occurs prior to the coordination of the next monomer molecule, i.e. if the chain stationary insertion mechanism operates in the polymerisation apart from the chain migratory insertion mechanism, short isotactic sequences in the syndiotactic chains are formed (yielding syndio isoblock polypropylene). This is, in fact, the case in propylene polymerisation with the $Me_2C(Cp)(Flu)HfCl_2$-based catalyst (Table 3.1) [127,143].

As regards higher α-olefins, their polymerisation with metallocene-based catalysts of class IV with C_s symmetry affords highly syndiotactic polymers as in the case of propylene [117]. This is a consequence of enantiomorphic site control over the polymerisation stereochemistry.

3.5.4.4 Stereocontrol of Isospecific, Hemiisospecific, Isospecific–Aspecific and Syndiospecific Propagation with Chirotopic (Diastereotopic) Catalysts

In catalysts obtained from chiral stereorigid metallocenes of class V with C_1 molecular symmetry (pseudohelical) (lacking any symmetry element), such as $Me_2C(t\text{-}BuCp)(Flu)MtX_2$ [Table 3.1, Figure 3.42 ($R = t$-Bu)], the diastereotopicity of the coordination positions excludes the formation of isotactic polymer via a chain migratory insertion mechanism. The bulky tertiary butyl group, residing on one of the cyclopentadienyl distal positions (β-position to the bridgehead C atom), interacts strongly with the growing polymer chain in a non-bonded, repulsive manner and hinders its migration to the coordination position underneath the β-substituent, which is available only for propylene coordination [124, 127, 143, 146]. The formation of an isotactic polymer can thus be explained only by assuming back skip of the growing chain to the initial, less hindered position after every monomer insertion [68]. The exclusive availability of only one position for propylene coordination at each active centre and its preference for only one kind of propylene enantioface are the reasons for the formation of isotactic polypropylene with the discussed catalyst [143].

Total blocking of one of the coordination sites in the $Me_2C(t\text{-}BuCp)(Flu)ZrCl_2$-based catalyst with a bulky t-Bu group and a fluorenyl C_6 ring, while keeping the other site unilaterally framed by the other fluorenyl C_6 ring, makes the polymerisation possible only via the chain stationary insertion mechanism; the polymer chain, which occupies an open sector of the ligand framework, is chirally oriented by a β-positioned CH group of the fluorenyl ligand and imposes the respective enantiofacial orientation of the coordinating monomer (suitable for primary insertion) by repulsive interaction forcing an

α-olefin into that enantiofacial approach to the Mt–alkyl unit, which holds the olefin substituent (methyl group in the case of propylene) away from the C_β atom of the metal-bound alkyl chain. In such a situation, a series of insertions of all *re* and all *si* coordinated propylene molecules at respective enantiomorphic sites generates an isotactic polymer. Recently, however, the enantiomorphic site model has been considered to be applicable only for isospecific polymerisation systems with class III catalysts (with C_2 molecular symmetry); a rational explanation for the observed stereoselectivity in the case of isospecific propylene polymerisation with class V catalysts (C_1 molecular symmetry) has been provided by molecular mechanics calculations within the framework of the 'growing chain orientation mechanism' of stereocontrol [1,30,410].

It has been suggested recently [410] that, also in the case of heterogeneous isospecific Ziegler–Natta catalysts, the stereoregulation mechanism assuming a 'growing chain orientation' is more reasonable than that predicted by the enantiomorphic site model [411,412].

The $Me_2C(MeCp)(Flu)ZrCl_2$-based catalyst [Table 3.1, Figure 3.42 (R = Me)], which possesses, analogously to the $Me_2C(t\text{-}BuCp(Flu)ZrCl_2$-based catalyst, one coordination site framed by one β-substituent (CH group in the fluorenyl C_6 ring) and the other one framed by two β-substituents (CH group in the fluorenyl C_6 ring and the CH_3 substituent at the cyclopentadienyl C_5 ring), produces hemiisotactic polypropylene (Figure 3.7). Note that the β-positioned methyl substituent at the cyclopentadienyl ring in the $Me_2C(MeCp)(Flu)$ $ZrCl_2$-based catalyst is of an intermediate size between the hydrogen atom as a β-positioned substituent at the cyclopentadienyl ring in the $Me_2C(Cp)(Flu)$ $ZrCl_2$-based catalyst (syndiospecific catalyst) and the tertiary butyl group as the β-positioned substituent in the $Me_2C(t\text{-}BuCp)(Flu)ZrCl_2$-based isospecific catalyst. The rationalisation of the formation of hemiisotactic polypropylene with the $Me_2C(MeCp)(Flu)ZrCl_2$–$[Al(Me)O]_x$ catalyst necessitates an in-depth look at the microstructure of this polymer. Detailed ^{13}C NMR studies suggest that it consists of non-overlapping stereoblocks of syndiotactic and isotactic sequences of differing lengths (Figure 3.6) [127].

Therefore, according to Razavi *et al.* [127,143], chain migratory insertions and chain stationary insertions would follow each other alternately, probably involving a pen-penultimate mechanism dictated by the periodicity of the helices of the growing chain [413]. If such a coordination site switching mechanism intervenes periodically, the resulting polypropylene chain would expose a syndioisoblock microstructure.

The formation of polypropylene characterised by such a microstructure can be envisaged if the chain migratory insertion is intermittently inhibited as a result of the chain repulsive, non-bonded interaction with the β-positioned methyl substituent at the cyclopentadienyl ring. During the periods where chain migratory insertions occur at the sites of the configuration varied alternately in each propagation step, a syndiotactic block is formed, and, whenever interaction with the β-positioned methyl group inhibits chain migratory insertion, an isotactic block is formed as a consequence of multiple chain stationary insertions at the same site [143].

The formation of hemiisotactic polypropylene with the $Me_2C(MeCp)(Flu)ZrCl_2$–$[Al(Me)O]_x$ catalyst can also be justified according to chain migratory insertion at diasterotopic coordination positions. While preserving its enantioselectivity, the site, which is unilaterally framed by the β-positioned CH group in the fluorenyl C_6 ring (stereospecific site), continues to coordinate and insert preferentially the propylene molecule with only one of its enantiofaces (*re* or *si*). However, the opposite site, framed with the two β-positioned substituents, namely a CH group in the fluorenyl C_6 ring and a CH_3 group substituting the cyclopentadienyl C_5 ring (Figure 3.42), having lost its diastereoface selectivity (non-stereospecific site), coordinates both propylene enantiofaces (*re* and *si*) indiscriminately. As a consequence, the catalyst becomes 'half-stereoselective' and produces, according to the assumed chain migratory insertion mechanism, a hemiisotactic polymer. However, the stereospecific site with its open sector of the ligand framework occupied by the growing polymer chain is more favourable energetically than the non-stereospecific site; therefore, chain stationary insertions have been postulated to compete with chain migratory insertions during the propagation [121].

If 1-butene or 1-hexene is chosen instead of propylene as the monomer polymerising with the $Me_2C(MeCp)(Flu)ZrCl_2$-based catalyst, the polymers obtained become enriched in *m* diads. This has been suggested to testify to the preference of site isomerisation prior to the coordination of the next monomer molecule with increasing size of the polymerising α-olefin [121].

The general mechanism of free, restricted and prohibited chain migration in propylene polymerisation with catalysts such as $Me_2C(RCp)(Flu)ZrCl_2$–$[Al(Me)O]_x$ (R = H, Me, *t*-Bu), which produce syndiotactic, hemiisotactic and isotactic polypropylenes respectively, can be considered in terms of the spatial relations between the distal substituents and the active coordination positions, which are very similar in all three cases [127,143]. In catalysts of class IV with C_s symmetry, based on $Me_2C(Cp)(Flu)ZrCl_2$, because of the smaller size of the β-positioned substituent ($R = H$) at the cyclopentadienyl ring, the chain migratory insertion mechanism governs the polymerisation process with rare stationary insertions (Table 3.1). In catalysts of class V with C_1 symmetry, based on $Me_2C(MeCp)(Flu)ZrCl_2$, the intermediate size of the β-positioned substituent (R = Me) at the cyclopentadienyl ring causes disruption in the operation of the chain migratory insertion mechanism and alternation of chain migratory and chain stationary insertion. This results in the formation of hemiisotactic, syndioisoblock polypropylene. In other catalysts of class V with C_1 symmetry, based on $Me_2C(t\text{-}BuCp)(Flu)ZrCl_2$, the huge β-positioned substituent (R = *t-Bu*) at the cyclopentadienyl ring precludes the chain migratory insertion mechanism, and only the chain stationary insertion mechanism governs the polymerisation, which results in the formation of isotactic polypropylene.

The *rac.*-MeCH(Me_4Cp)(Ind)$TiCl_2$-based catalyst, which produces polypropylene containing isotactic blocks alternating with atactic blocks (Table 3.1), possesses a chiral bridging carbon atom and can have the growing polymer chain either *syn* or *anti* to the methyl group in the bridging ethylidene moiety, and placed in differing surroundings of the tetramethylcyclopentadienyl and indenyl ligands during polymerisation. Therefore, these two positions are non{\hbox-}equivalent and different in non-bonded interactions. Considering this, the catalytic species has been postulated to exist in two isomeric states in equilibrium: one of lower local symmetry (isospecific site) and the other of higher local symmetry (a much less stereospecific or even non-stereospecific site) [116]. The former site has been assumed to coordinate the monomer via the same enantioface (all *re* or all *si*) in each propagation step, leading to isotactic chain sequences, and the latter site to coordinate the monomer non-stereoselectivity (i.e. without differentiation between monomer *re* and *si* enantiofaces) in consecutive propagation steps, leading to atactic chain sequences during polymerisation [107]. In this mechanism, a catalyst site is conceptualised to switch back and forth between two states as propagation proceeds; when such a coordination site switching mechanism intervenes periodically, the resulting polypropylene chain exposes a microstructure of *iso* atactic block poly (α-olefin) [414]. Note that, if the lengths of particular blocks were to decrease (in the case where rapid interconversion of both sites occurred), especially if atactic chain segments were to diminish to those consisting only of *r* diads, the polymer obtained would be hemiisotactic (iso syndioblock) polypropylene; this is, in fact, the case for the Me_2C(RCp)(Flu)$ZrCl_2$-based catalyst [107].

The *t*-BuCH(Cp)(Flu)$ZrCl_2$-based catalyst yields syndiotactic polypropylene (Table 3.1). Note that the catalyst precursor contains a chiral bridging carbon atom in this case. However, the chirality of this atom is not an essential factor for the syndiospecificity of the polymerisation with the discussed catalyst. The theoretical study indicates that the syndiospecific polymerisation of propylene with a chiral *t*-BuCH(Cp)(Flu)$ZrCl_2$-based catalyst cannot be interpreted as solely determined by the enantiomorphism of the catalytic site. By changing the stereochemistry of the C_β atom of the last inserted monomeric unit (primary insertion) from the *R* to the *S* configuration, the preferred prochiral face of the propylene molecule changes from *re* to *si* in the case of syndiospecific propagation. In fact, the syndiotactic monomer enchainment is not changed by the chirality of the bridging carbon atom in *t*-BuCH (Cp)(Flu)$ZrCl_2$; such an enchainment occurs in any combination of the metal centre (*R* or *S*) and the C_β atom (*R* or *S*) of the last monomeric unit in the growing chain. Molecular mechanics calculation confirms that the chirality of the metal centre is not the sole enantioselection determinant. The chiral C_β atom of the growing chain also contributes to select the proper enantioface of the coordinating propylene molecule. In summary, the stereochemistry of the monomer insertion is entirely dictated by the steric energy of the transition state involved, which is determined by the combination of the chirality of the enantiomorphic site, the configuration of the C_β atom of the last inserted monomeric unit and a possible agostic interaction at the site [107,137].

3.5.4.5 General Features of Polymerisation Stereocontrol – Influence of Steric and Electronic Factors

In summary, the current homogeneous single-site catalysts used in the stereospecific polymerisation of propylene are based on metallocenes containing stereoselectively substituted cyclopentadienyl ligands; it is generally acknowledged that the steric environment around the metal contributes to the orientation of the insertion. One may note, in this connection, that the concerted, repulsive and non-bonded interactions between substituents of the ligand, monomer and growing polymer chain ultimately dictate, in the metallacyclo transition state [136], the stereochemistry of the insertion and the microstructure of the resulting polymer chain. It is interesting that calculations concerning the geometries and energy during the monomer insertion process into the Mt–CH_3 bond of metallocene methyl cations such as $[H_2Si(Cp)_2ZrMe]^+$ have suggested the activation energy at the level of the transition state to be higher for the insertion reaction (ca 6 kcal/mol) than that for π complex formation [272]. On the other hand, neither a π complex nor an insertion transition state participation was found by computational methods in the reaction of ethylene with $[Cp_2TiMe]^+$, but a spontaneous process, rationalised by a $2 + 2$ addition, without any relation to monomer coordination or insertion steps [415,416], has been postulated; moreover, the insertion reaction of ethylene has been estimated to occur in an extremely short time (ca 10^{-13} s). Irrespective of the presence or absence of an energy barrier for the insertion reaction of the coordinated monomer, it seems that, the tighter the coordination sphere around the transition metal atom in metallocene-based catalysts, the more active they are for the polymerisation of ethylene rather than propylene.

It may be noted, by analysis of the polymerisation behaviour and structure of chiral catalysts of classes III, IV and V, that the substituents at the β-position to the bridgehead atom of an *ansa*-zirconocene are the most important ones with respect to the effects exerted by metallocene cyclopentadienyl ligands on polypropylene stereoregularities. These β-positioned substituents are in close proximity to the coordination sites at which the growing polymer chain and the coordinating α-olefin molecule are bound. Steric relationships in such metallocene complexes can be described by their 'coordination gap aperture angle', the largest possible angle spanned by two planes through the metal centre, which touch the van der Waals surfaces of the β-substituents at the C_5 ring ligands, and by their 'lateral extension angle', which is limited by the two α-substituents protruding into the coordination gap [417] (Figure 3.43) [30].

It should be emphasised that all open-structure isopropylidene-bridged metallocenes have much lower activities towards ethylene polymerisation

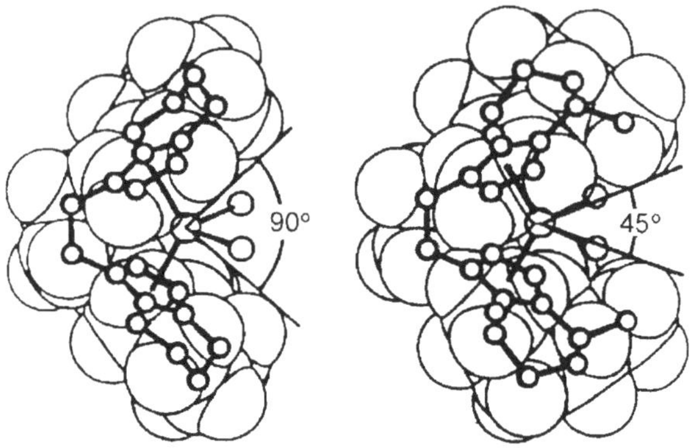

Figure 3.43 Coordination gap aperture angle in *rac.*-ethylenebis[1-(4,5,6,7-tetrahydroindenyl)]zirconium dichloride [*rac.*-$(ThindCH_2)_2ZrCl_2$] (left) and in *rac.*-ethylenebis[1-(4,7-dimethyl-4,5,6,7-tetrahydroindenyl)]zirconium dichloride [*rac.*-$(Me_2$ $ThindCH_2)_2$ $ZrCl_2$] (right). For the sake of clarity, only the C–C bonds are sketched for the π ligands. ❍ – Zr; ❍ – CL; ❍ – C, CH, CH_2 or CH_3. Side view. Reproduced by permission from Ref. 30. Copyright 1995 Wiley-VCH Weinheim

than those obtained from ethylene-bridged metallocenes with smaller coordination gap aperture angles. Characterising the relative stereospecificities of chiral metallocene-based catalysts, one may see that they are in accord with α-olefin π-facial selectivities being governed by the chain end orientations, and with the chain end orientations in turn being enforced by the β-positioned substituents at the cyclopentadienyl rings (enantiomorphic site sterecontrol). Direct contacts between the β-positioned substituents at the cyclopentadienyl rings and the propylene methyl group then additionally reinforce (isospecific catalysts) or diminish (syndiospecific catalysts) stereoregulation [22,112,138,143]. The degree of influence of the direct catalyst–monomer contacts on stereospecificity appears to be critically related to the coordination gap aperture angle (and lateral extension angle) at the catalytic site.

The simple molecular models of catalytic sites, which consider the specific site enantioselectivity via non-bonded contacts of cation-like species with the monomer, are insufficient for explaining all the experimental data obtained. These data could not be interpreted in terms of steric effects only and suggest that the electronic factors are also involved. Apart from the discussed steric factors, such as the size of the substituent of the ancillary ligand, the ionic radii of the transition metal (Ti, Zr, Hf) and the degree of coverage of the active coordination positions by the substituents ought to be taken into consideration [30,112,127,418].

The influence of electronic factors on the stereospecific behaviour of metallocene-based catalysts in α-olefin polymerisation is an important aspect of the stereoregulation mechanism. Two phenomena, lanthanide contraction and relativistic effects, that appear in group 4 transition metal series result in some differences between hafnium compared with zirconium: Hf has slightly smaller atomic and ionic radii, forms stronger σ bonds and is more resistant to reduction [418].

The smaller size of the hafnium cation, combined with a more effective nuclear charge, favours the formation of contact ion pairing. Furthermore, the active coordination positions of the smaller Hf cation receive more efficient ligand coverage and are sterically more crowded. Thus, for both electrostatic and steric reasons, lateral displacements of the growing polymer chain are more hindered on hafnium than in the case of zirconium. For syndiospecific catalysts, such as those obtained from $Me_2C(Cp)(Flu)MtCl_2$, this would mean of course a decrease in stereoregularity by comparison with syndiotactic polypropylene prepared with $Me_2C(Cp)(Flu)HfCl_2$-based catalysts.

On the other hand, for isospecific catalysts such as those obtained from *rac.*-$Me_2C(Ind)_2MtCl_2$ or from *rac.*-$(IndCH_2)_2MtCl_2$, the occasional restriction of chain migration either has no bearing on the stereoregularity of the polymer or will slightly improve it in the case of Hf-containing catalysts; in this case, an 'unnecessary' chain migration can only be detrimental to isospecificity [418].

The replacement of the isopropylidene interannular bridge with an ethylene bridge in syndiospecific catalysts of class IV with non-helical symmetry is accompanied with a decrease in stereospecificity of the catalyst. With the two carbon atoms in the bridge of $(CpCH_2)(FluCH_2)ZrCl_2$, the cyclopentadienyl and fluorenyl moieties are pushed further away and adopt a more parallel position. Under these conditions, more ligand coverage is provided for coordination sites. Note that, the smaller the coordination gap aperture angle, the larger the Cp (centroid)–Mt–Flu (centroid) angle becomes; therefore, since the coordination gap aperture angle is smaller in $(CpCH_2)(FluCH_2)ZrCl_2$ than in $Me_2C(Cp)(Flu)ZrCl_2$, the Cp (centroid)–Mt–Flu (centroid) angle in the former complex is larger than in the latter one. The increase in the centroid–Mt–centroid angle at the same time is accompanied with a change in the frontier orbital energies and hybridisations. Thus, both from a steric and from a valence orbital energetic point of view, the approach and subsequent orbital overlap between chain end hydride σ orbitals and transition metal orbitals for a proper agostic interaction will be less favoured for the $(CpCH_2)(FluCH_2)ZrCl_2$-based catalyst. The lower probability of α-agostic interaction combined with increased structural flexibility of the catalyst framework explains the lower stereospecificity of the catalyst based on the ethylene bridged zirconocene [418].

The replacement of the ethylene bridge with an isopropylidene bridge in isospecific catalysts of class III with helical symmetry generally results in considerable lowering of their activity as well as a decrease in their stereospecificity. A comparison of the structure of the *rac.*-$(IndCH_2)_2ZrCl_2$ and *rac.*-$Me_2C(Ind)_2ZrCl_2$ catalyst precursors shows the relatively large difference between their centroid–Mt–centroid angles, the value of 125.39° for the former precursor being about 7° greater than that found for the latter one. This is believed [418] to be the parameter that should be held responsible for the observed changes in catalytic performance. The closing of the centroid–Mt–centroid angle in the dorsal part of the molecule necessitates an opening of the coordination gap aperture angle.

With change from an ethylene to an isopropylidene bridge, the active sites are relieved partly from the steric encumbrance exerted by the distal substituents and can undergo more easily agostic interactions owing to favourable steric but also to orbital energetic factors. The chances of an adequate overlap between the corresponding metal frontier orbitals and hydride σ orbitals for α-agostic interaction increase in *rac.*-$Me_2C(Ind)_2ZrCl_2$ by comparison with *rac.*-$(IndCH_2)_2ZrCl_2$. This should theoretically increase the stereospecificity of the catalyst based on the former precursor. However, the stereospecificity of this catalyst has been apparently cancelled out by the decreased steric influence of the ligand and its lower directing influence on the conformation of the growing chain and monomer prochiral face selectivity [418]. The α-agostic interactions can thus be considered to be the origin of the stereoselectivity of chiral *ansa*-metallocene-based catalysts. Therefore, the α-agostic model would describe the *re* or *si* orientation of an inserting α-olefin as being controlled by the required placement of the C_β atom of the growing chain in the more open of the two possible positions at the rigid three-membered $Zr–H_\alpha–C_\alpha$ ring [30]. Of the two alternative $Zr–C_\alpha H_2–C_\beta H(CH_3)$ orientations resulting from binding one of the H_α atoms to the metal centre, one orientation is sterically inaccessible since it would cause the growing chain to collide with a β-positioned substituent of the chiral π ligand framework; the other one is sterically unencumbered and practically indistinguishable from that proposed [345,402] on the basis of non-bonding van der Waals repulsions alone (Figures 3.37 and 3.41) [30].

Let us recall that the occurrence of agostic metal–hydrogen bonds has been widely accepted in group 4 metallocene catalysis in α-olefin polymerisation; the electronically and coordinatively unsaturated, highly reactive cationic centres are stabilised by agostic interactions with the hydrogens at the α-(γ^-) or β-position on the reactive end of the polymer chain. These C–H bonds act as ligands with a 'pair' of electrons; three-centred, two-electron Mt–H–C bonds are formed in three-membered ($Mt–H_\alpha–C_\alpha$), four-membered ($Mt–H_\beta–C_\beta$) and five-membered ($Mt–H_\gamma–C_\gamma$ rings. An α- (γ-) agostic interaction that would maintain the active centre intact during the time lapse between two coordinations and yet would give way to the incoming monomer for successive insertion is crucial for the formation of polymers with high molecular weight [418]. One may add in this connection that, as regards β-agostic interactions, their occurrence is detrimental to the formation of longer polymer chains, since these interactions could lead to β-hydride eliminations analogous to those represented by scheme (38).

Considerations of the polymerisation stereochemical behaviour of catalysts belonging to class V with pseudohelical symmetry (based on lopsided metallocene precursors) still warrant more in-depth investigations. However, an assertive statement can be made that, the more unsymmetrical the catalysts are, the more *m* diads there are in syndiotactic polypropylene. On the other hand, a syndiospecific catalyst of class V based on the t-BuCH(Cp)(Flu)$ZrCl_2$ precursor containing a chiral bridging carbon atom involves two arrangements: the first with a dominant effect of the enantiomorphic site, and the second with a dominant effect of non-bonded interactions with methyl groups of the propylene monomer and the C_β atom of the last inserted monomeric unit [137].

It is to be added in this connection that some authors have recently broken away from the symmetry-based models for catalyst stereospecificity, since the simple rules of thumb resulting from metallocene symmetry appeared to be too narrow. A universal model has been developed that accurately describes the experimental microstructures of the poly(α-olefin)s by considering the four lowest-energy conformers of the metallocene species coordinating the propylene molecule (R_{re}, S_{re}, S_{si} and R_{si}) and the positional changes that the polymer chain undergoes during insertion. The relative energy levels of the four diastereomers can be determined by molecular modelling calculations; these energy differences are decisive in determining the microstructure of the resulting polymers [419].

At the end of considerations of the influence of stereoelectronic factors on polymerisation stereocontrol, different contact ion pairing between metallocenium cations and methylaluminoxane-based, perfluorotetraphenylborate or other anions should be addressed. Some previous statements suggest that, for the conversion of group 4 metallocene dichloride or dialkyl complexes into active olefin polymerisation catalysts, the generation of metalloceniumalkyl ions by some strongly Lewis acid reagent appears to be essential [109,341,399]. It is important to consider in this regard the equilibria between a dialkyl complex Cp'_2MtR_2 and its Lewis acid (LA) adduct Cp'_2MtR_2. LA, which is equivalent to the contact ion pair $[Cp'_2Mt\text{–}R]^{\delta+}\cdots[R\text{–}LA]^{\delta-}$, and between the latter and an ion pair separated by a solvent or an olefin molecule, $[Cp'_2Mt\text{-}R(\text{solvent})]^+$ $[R\text{–}LA]^-$ and $[Cp'_2Mt\text{–}R(\text{olefin})]^+$ $[R\text{–}LA]^-$ [341]. The ion pair separated by an olefin molecule is usually considered as an intermediate for the insertion of the monomer into the Mt–R bond. On the other hand, binuclear cations of the type $[(Cp'_2MtMe)\text{–}X\text{–}(Cp'_2MtMe)]^+$ (X = F, Me) might also be present in polymerisation systems containing metallocenium alkyl ions [165,420]. They have been postulated, however, to be involved as possible intermediates contributing to polymer chain termination and catalyst deactivation.

Contact ion pair steric effects on stereoregulation have not been predicted by simple molecular modelling based on assumed geometries of the isolated cations. However, some general remarks are obvious. For instance, as regards the occurrence and the frequency of back-skipped insertions on the same enantiomorphic site, leading to m diads in polypropylene formed in polymerisation systems with syndiospecific catalysts, they depend largely on the extent of the steric interaction of the C_β-positioned substituent in the cyclopentadienyl-like ligand with the growing polymer chain and the effective nuclear charge of the metallocenium site (the preference of the cation to coordinate the anion rather than the monomer or solvent molecules). Consequently, with an increase in the steric bulk of the β-positioned substitutents and/or an increase in the effective nuclear charge of the metal centre, the probability of chain blockage would be promoted and the occurrence of chain stationary insertion could be favoured. This has been realised by the replacement of zirconium in the $Me_2C(Cp)(Flu)ZrCl_2$ catalyst precursor with its hafnium homologue. The smaller size of the Hf cation with a greater charge in $Me_2C(Cp)(Flu)HfCl_2$ requires a shorter chain to β-positioned substituent distance and a more effective ion pairing. Inhibition of the chain migratory insertions will occur more often and also more effectively, resulting in the formation of isotactic sequences in the backbone of the otherwise syndiotactic polypropylene chain (Table 3.1) [127].

Taking the above into consideration, it may be concluded that sterically less crowded metallocenium cations, such as that obtained from the $Me_2C(Cp)(Flu)MtX_2$ precursor, appear more sensitive to the effects of anion coordination, the strength of which (of the coordination) increases in the order $[Al_x(Me)_{x-1}O_xX_2]^- < [B(C_6F_5)_4]^- < [B(Me)(C_6F_5)_3]^-$. For this and other more open cations, stronger coordination leads to decreased activity, decreased stereospecificity (and thus decreased polymer melting point) as well as decreased polymer molecular weight [113].

As regards isospecific *rac.*-$(IndCH_2)_2ZrCl_2$-based catalysts, irrespective of their kind of anionic moiety, they were found [112] to have activities of the same order of magnitude.

Summarising the influence of steric and electronic factors on stereoregulation in α-olefin polymerisation with chiral metallocene-based catalysts, one should note that catalysts obtained from *ansa*-metallocenes are not as stereorigid as the heterogeneous catalysts. There may be fluxional conformers, for instance, owing to torsional twisting of the η^5-cyclopentadienyl-like ligands. There may also be dynamic dissociation/association of methylaluminoxane and changes in the state of coordination, i.e. monodentate, bidentate, etc. These fluctuations all can alter the stereospecific and/or regiospecific centre during the propagation step in α-olefin polymerisation [264].

The effect of a counterion on the inversion barrier of the metallocenium cation is of considerable importance; the inversion barrier is increased when a counterion is present. To probe this effect, the inversion barrier was calculated without the presence of a counterion and was found [273] for [*rac.*-$(IndCH_2)_2ZrR]^+$ to be 4 kcal/mol, while the observed inversion barrier (with $[B(Me)(C_6F_5)_3]^-$) was found [111] to be 18.3 kcal/mol.

The influence of an anionic counterpartner of the metallocenium cation on its stereospecificity can also be seen when considering polymerisation conditions such as the kind of solvent and temperature. For example, the loss of stereospecificity when propylene polymerisation with $Me_2C(Cp)(Flu)ZrCl_2$-based catalysts is performed in polar solvents [142] seems to be caused by inversion of the configuration of the solvent-separated zirconium ion via migration of the growing polymer chain to the initial position before a new monomer molecule insertion. A tight contact between the zirconium complex cation and the counterion in apolar solvents should prevent the above migration. The temperature dependence of the stereospecificity has been attributed to the temperature influence on the equilibrium between the enantiomers constituting the catalytic active site, which have a different stereoregulating ability in propylene polymerisation. In fact, the effect of temperature on polymer stereoregularity is much less pronounced when the rigidity of the ligand framework is enhanced by using single-atom bridges instead of two-atom bridges and by introducing α-positioned substituents (with respect to bridgehead atoms) into cyclopentadienyl-like ligands [68].

3.5.5 Steric Defects in Stereospecific Chain growth Reactions and Analysis of Polymer Stereoregularity

One should realise that none of the Ziegler–Natta and related coordination catalysts is totally regiospecific and stereospecific; there is always a temporary lack of capability of the catalyst for the performance of chemical and steric control during the polymerisation. In stereoregular α-olefin polymers, some 'errors' are thus normally present in the macromolecule. The errors can be both of a chemical nature (head-to-head linking, isomerised monomeric units) and of a steric nature at the sites of stereoisomerism. Each stereoregular α-olefin polymer is characterised by some degree of stereoregularity. The nomenclature recommended applies to the predominant structural features of the real polymer molecule. Terms such as 'isotactic' or 'syndiotactic', when applied to real poly(α-olefin) chains, ought to be intended as 'mainly isotactic' or 'mainly syndiotactic', to account for the occurrence of failures in the catalyst stereocontrol. The distribution of steric defects appearing along polymer chains is indicative of which kind of stereocontrol is operating during the propagation. Bernoullian statistics have been shown [12] to be consistent with chain end stereocontrol, while non-Bernoullian distributions originate from enantiomorphic site stereocontrol [411,412] or growing chain orientation stereocontrol [1,30,410].

The extent of chain regularity in poly(α-olefin)s is determined primarily by high-resolution nuclear magnetic resonance, ^{1}H NMR and especially ^{13}C NMR, spectroscopy. Both the ^{1}H and ^{13}C NMR signals of the macromolecule are most conveniently related to its microstructure [421].

The *r* diad has a twofold axis of symmetry, and consequently the two methylene protons are in equivalent environments in the poly(α-olefin) chain:

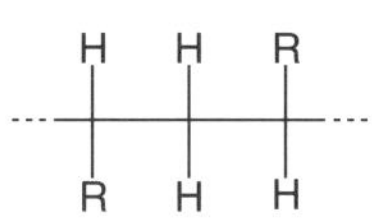

r diad

They thus exhibit the same chemical shift in the ^1H NMR spectrum and appear as a singlet resonance.

The *m* diad has no symmetry axis and so the two methylene protons, H_a and H_b, are non-equivalent, and should in general give different chemical shifts in the ^1H NMR spectrum:

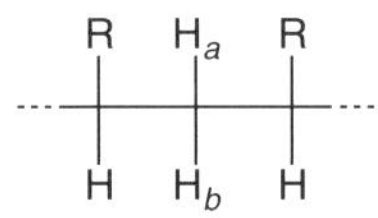

m diad

Thus, ^1H NMR spectroscopy can provide stereochemical information concerning α-olefin polymer chains [421].

Let us recall that triad tacticity describes isotactic (*mm*), syndiotactic (*rr*) and heterotactic stereotriads (*mr*) in the poly (α-olefin) chain:

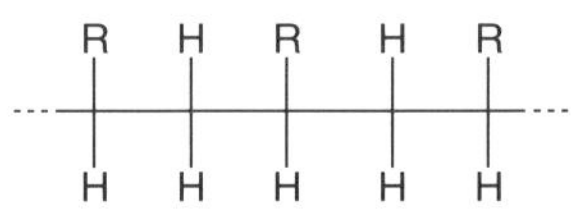

mm triad

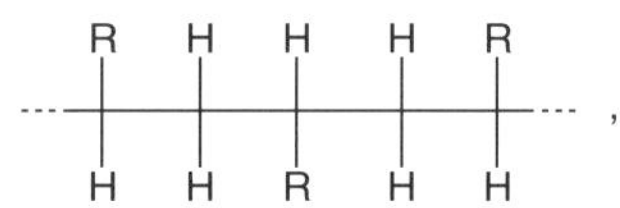

rr triad

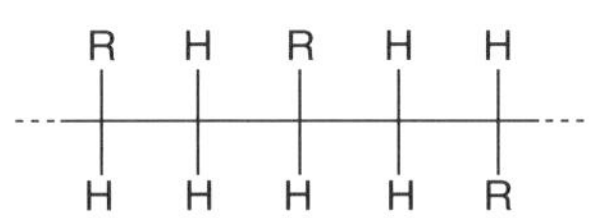

mr triad

Designating by P_m the conditional probability of forming an *m* diad, one may see that the conditional probability of forming an *r* diad, P_r, will be equal to $1–P_m$. Since two monomer insertions are required to form a triad sequence, it can be readily seen that the probability of forming a particular triad is the product of the probabilities of forming the two diads comprising the triad. Thus, the probabilities of forming isotactic (*mm*), syndiotactic (*rr*) and heterotactic triads (*mr*) in a poly (α-olefin) chain are given by the following dependences (the coefficient of 2 for the *mr* triad is because this triad is formed in two ways, namely *mr* and *rm*) [421]:

$$[mm]=(P_m)^2,$$
$$[rr]=(1-P_m)^2,$$
$$[mr]=2 \times P_m \times (1-P_m)$$

For simple poly(α-olefins) of practical importance, the chain stereostructure can be quantitatively determined at least at the stereotriad level [1]. Longer configurational sequences can be observed by nuclear magnetic resonance, particularly using larger magnetic fields or ^{13}C NMR spectroscopy or both. The chemical shifts due to $\underline{C}H_3$, $\underline{C}H_2$ and $\underline{C}H$ carbons are widely spaced. It is mainly the chemical shift of the pendant CH_3 group carbon that is utilised for sequential analysis.

The starting point for ^{13}C NMR analysis is the position of the central CH_3 group of the segment consisting of five propylene units. In polypropylene, the ^{13}C NMR chemical shift of each CH_3 group is determined by the configurations of neighbouring repeating units on either side; each $\underline{C}H_3$ signal is thus assignable to a particular sequence pattern, represented by a respective number (four for pentads) of consequtive *m* or *r* designators framing the CH_3 group under consideration. The pentad distribution is the fingerprint of polypropylene stereoregularity and contains a lot of information about the polymerisation mechanism; the polymer chain microstructure constitutes the best 'record' of the history of the formation of this chain and provides a profound insight into the kind of stereocontrol during polymerisation. One may assume that the larger-sized stereosequence (hexad, heptad) distributions will be analysed with further advancements in nuclear magnetic resonance instrumentation [422, 423].

Let us consider typical steric defects in (mainly) isotactic poly(α-olefin) chains. One can imagine two ways in which a minor configurational error may be introduced into an isotactic polymer chain: a 'steric' propagation error, denoted as

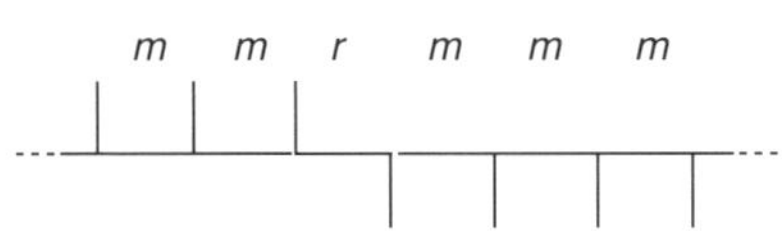

for which one will observe an error tetrad ratio *mmr*:*mrm*=2:1, and a 'template' propagation error, denoted as

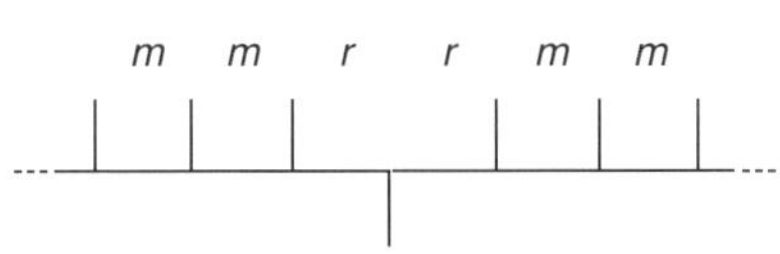

for which *rr* resonance appears and the error tetrad ratio *mmr*:mrr – 1:1 [424]. The first structure implies that the principal force is the steric nature of the chain end itself; an occasional change in the configuration of the last inserted monomeric unit tends to be perpetuated. In other words, once a new *m* unit is inserted beyond the *r* defect, propagation proceeds as before. The second structure implies, on the other hand, that the propagation is under the stereocontrol of the enantiomorphism of the catalyst site, which corrects an *r* propagation error by restoring the absolute relationship of the chain to the site; the occasional change in the configuration of the inserted monomeric unit, having no effect on the site chirality, tends to remain isolated [424].

All 10 possible pentad signals (isotactic: *mmmm*; syndiotactic: *rrrr*; heterotactic: *mmmr*, *rmmr*, *mmrr*, *mmrm*, *rmrr*, *rmrm*, *rrrm* and *mrrm*) (Figure 3.44) are observed in ^{13}C NMR spectra for the randomly configured repeating units of atactic polypropylene (in the region of ca δ = 19.8–21.7 ppm); eight out of the 10 possible pentads give rise to well-separated signals, and only two *mr*-centred pentads, *mmrm* and *rmrr*, merge into a single peak [425–427].

Isotactic polypropylene, on the other hand, is ideally characterised by a single ^{13}C NMR signal for isotactic *mmmm* pentads (at ca δ=21.7 ppm), since its repeating units have identical configuration over long segments of polymer chains [30]. Isotactic polypropylene produced by the heterogeneous third generation Ziegler–Natta catalysts is highly stereoregular with [*mmmm*] > 0.95 (the ratio of the *mmmm* pentad integral to the integral sum of all pentad signals observed). At ambient temperature, typical *ansa*-metallocene–$[Al(Me)O]_x$ catalysts yield polypropylene with stereoregularities of [*mmmm*] $\approx$ 0.8–0.9. At increased polymerisation temperature (such as that applied in the case of propylene polymerisation with heterogeneous Ziegler–Natta catalysts), however, most of these homogeneous metallocene-based catalysts are distinctly less stereospecific than typical heterogeneous Ziegler–Natta catalysts used for propylene polymerisation [30].

Steric defects in isotactic polypropylene, which involve the appearance of isolated *r* diads or pairs of *r* diads, may be considered on a pentad level (Figures 3.45a and b respectively). The ^{13}C NMR signals associated with occasional stereoerrors in the propylene isotactic polymers produced by chiral metallocene-based catalysts (pairs of *r* diads) indicate that the polymerisation stereochemistry is governed by the enantiomorphism of catalytic sites; an error pentad distribution close to *mmmr*:*mmrr*:*mmrm*:*mrrm* = 2:2:0:1 is observed

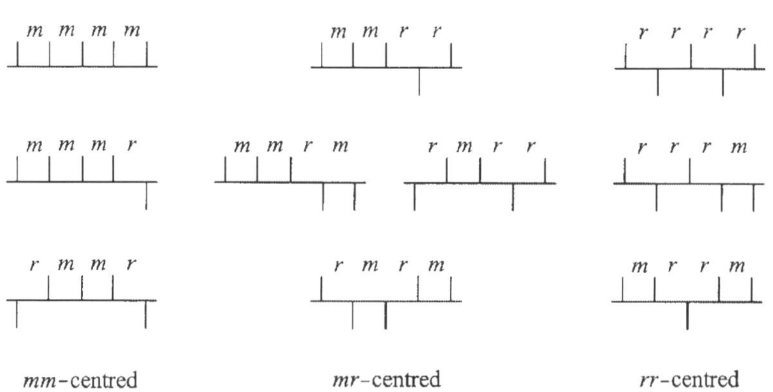

Figure 3.44 Ten possible stereochemical pentads of a poly(α-olefin)

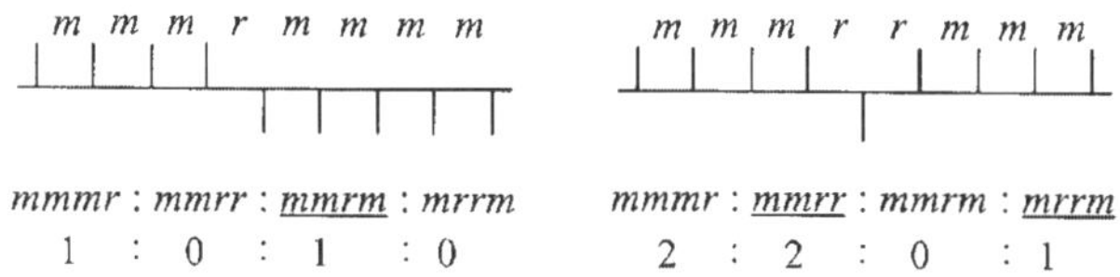

(a) (b)

Figure 3.45 Typical steric defects (pentad distribution of stereoerrors) in a (mainly) isotactic poly(α-olefin) chain: (a) isolated *r* diad, characteristic of chain end stereocontrol; (b) pair of *r* diads, characteristic of enantiomorphic site stereocontrol during the propagation

for these polymers (Figure 3.45b). The same pentad pattern, with *r* stereodiads occurring in pairs, is characteristic of isotactic polypropylene produced by heterogeneous Ziegler–Natta catalysts; the catalytic site of fixed chirality forces the enantiofacial orientation of subsequent α-olefin insertions to return to the previous preference immediately after an occasional steric defect (Figure 3.45b) [1,30]. Such a conclusion is confirmed by ^{13}C NMR analysis of isotactic polypropylene chains containing copolymerised ethylene units in low amounts (Figure 3.46) [428].

If the steric control was due to the chirality of the last inserted monomeric unit, the insertion of a propylene unit following that of an achiral ethylene unit would be non-stereospecific (both cases (a) and (b) in Figure 3.46 would be possible), whereas virtually complete stereospecificity has been observed experimentally (Figure 3.46b) [1,370,395].

Chain end stereocontrol, which gives rise to single-inversion pentads, *mmrm* and *mmmr*, as main error signals in ^{13}C NMR spectra, is characteristic of propylene polymerisation with the Cp_2TiPh_2–$[Al(Me)O]_x$ catalyst at lowered temperature (leading to stereoblock polypropylene); in this case an error pentad distribution is observed close to *mmmr*:*mmrr*:*mmrm*:*mrrm*=1:0:1:0 (Figure 3.35a) [1,30].

Syndiotactic polypropylene produced by typical *ansa*-metallocene–$[Al(Me)O]_x$ catalysts is highly stereoregular with $[rrrr] > 0.9$ (the ratio of the *rrrr* pentad integral to the integral sum of all pentad signals observed); a single ^{13}C NMR signal for syndiotactic *rrrr* pentads appears at ca $\delta = 20.3$ ppm [30,418].

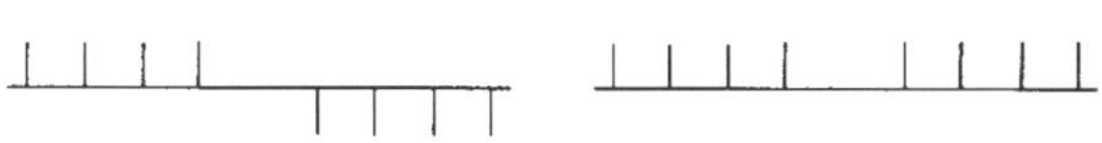

(a) (b)

Figure 3.46 Possible isotactic poly(α-olefin) chain propagation after the insertion of an achiral ethylene unit: (a) chain end stereocontrol; (b) enantiomorphic site stereocontrol during the propagation

Figure 3.47 shows steric defects in (mainly) syndiotactic poly (α-olefin) chains. It was established, by analysis of the chain microstructure using ^{1}H and ^{13}C NMR techniques, that the stereochemical sequences of configurations of the monomeric units in polypropylene macromolecules obtained with soluble vanadium-based Ziegler–Natta catalysts at lowered temperature contained long blocks of *r* diads bridged by isolated *m* diads (Figure 3.47a) [390, 394], which was in accord with the Bernoullian statistical model [12]. It is worth noting that an occasional change in the configuration of the last inserted monomeric unit tends to be perpetuated in such a case.

In agreement with this finding, it has been shown that, in ethylene/propylene copolymerisation with vanadium-based catalysts, propylene insertion after an ethylene insertion is substantially non-stereospecific (both cases (a) and (b) in Figure 3.46 are possible) [1,390].

The ^{13}C NMR analysis of polypropylene obtained with the $Me_2C(Cp)(Flu)ZrCl_2–[Al(Me)O]_x$ catalyst showed that the polymer chains consist of long blocks of *r* diads separated by pairs of *m* diads (Figure 3.47b), which is indicative of polymerisation stereocontrol by the enantiomorphic catalyst site [23]. In this case, it is obvious that an occasional change in the configuration of the last inserted monomeric unit has no influence on the chirality and tends to remain isolated.

The above-discussed problems concerning the appearance of steric defects during stereospecific chain growth reactions have been presented assuming some simplifications; first of all, account has been taken of stereoerrors resulting from 'wrong' selection of the monomer enantioface (formation of diastereomers with the *si* configuration instead of those with the *re* configuration, and vice versa). One should realise, however, that factors connected with change in the insertion mechanism, the inversion of the chirality of the catalyst site, the isomerisation processes, etc., may also influence the appearance of steric defects and leave their trace on the microstructure of formed poly (α-olefin)s, detectable by nuclear magnetic resonance techniques.

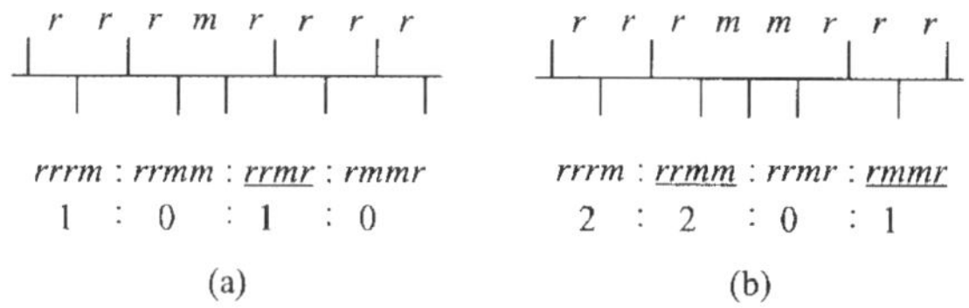

Figure 3.47 Typical steric defects (pentad distribution of stereoerrors) in a (mainly) syndiotactic poly(α-olefin) chain: (a) isolated *m* diad, characteristic of chain end stereocontrol; (b) pair of *m* diads, characteristic of enantiomorphic site stereocontrol during the propagation

3.6 Polymerisation of Higher α-Olefins

As a general rule, all α-olefins, either linear or branched, may be polymerised with coordination catalysts, involving an insertion mechanism, to stereoregular polymers. Although a sufficiently large number of different higher α-olefins (C_4-C_{18}) have been investigated in polymerisation with Ziegler–Natta and related catalysts, the industrial interest in polymers of higher α-olefins is much lower than that for ethylene and propylene polymers. Only isotactic poly(α-olefin)s such as poly(1-butene) and poly(4-methyl-1-pentene) have reached commercial production status [43].

From the scientific point of view, however, polymerisation of higher α-olefins is of considerable interest, since the relationship between the structure of the catalyst and that of the resulting polymers may be explained by considering the influence of the size of the alkyl group in the monomer. Moreover, polymerisation of chiral α-olefins allows study of stereoselection or stereoelection phenomena during polymerisation.

It is worth noting in this respect that NMR spectroscopy, so useful in propylene homopolymer and ethylene/propylene copolymer characterisation, is of limited use in the case of higher poly(α-olefin)s; overlapping resonances of the main chain atoms and pendant groups severely impair resolution and interpretation. Therefore, deductions concerning polymer stereoregularity largely rely on crystallinity-related phenomena, such as X-ray and IR spectra, melting behaviour, solvent fractionation and various physical properties [43].

3.6.1 Activity of Ziegler–Natta Catalysts

Most Ziegler–Natta and related coordination catalysts suitable for propylene polymerisation are effective for the polymerisation of higher α-olefins. The polymerisability of higher α-olefins, however, is appreciably lower than that of propylene under comparable conditions. Therefore, as a general rule, it is necessary to use highly stereospecific catalysts to achieve good yields of high molecular weight, crystalline (tactic) polymers without much atactic by product [43].

Both isotactic and syndiotactic polymers of higher α-olefins are obtained in the presence of Ziegler–Natta catalysts. For obtaining highly isotactic polymers, preferred systems use combinations of $TiCl_3$ with $AlEt_2Cl$, or possibly the newer supported catalysts such as $MgCl_2/LB/TiCl_4–AlEt_3/LB$; catalysts made directly from $TiCl_4$ as the precursor are less stereospecific, similarly to the case of propylene polymerisation [43]. However, homogeneous metallocene-based catalysts also yield isotactic polymers of higher α-olefins. The main difference among the isotactic polymers obtained in the presence of heterogeneous catalysts and those obtained with homogeneous catalysts is that the latter ones are somewhat less regioregular. However, because of the longer alkyl group, higher α-olefins undergo polymerisation more regioselectively than propylene [37].

Syndiotactic polymers of higher α-olefins such as 1-butene and 4-methyl-1-pentene are produced by homogeneous metallocene-based catalysts [117, 429, 430]. In contrast to polymerisation with metallocene-based catalysts, higher α-olefins are much less reactive in polymerisation with soluble vanadium-based catalysts, and already in the case of 1-butene polymerisation only yield trace amounts of low molecular weight syndiotactic polymer [394].

3.6.2 Monomer Polymerisability

Activities of α-olefins in insertion polymerisation with coordination catalysts are generally related to the monomer structure; both electronic and steric factors influence the activity of α-olefins in stereospecific polymerisation, but the steric factor primarily influences the rate of polymerisation, which has been especially shown for isospecific polymerisation with heterogeneous catalysts [46, 250].

The reactivity of the olefinic double bond decreases drastically as the size of the pendant alkyl group increases and as the branching of this group moves close to the double bond; reactivity is completely inhibited when the carbon atom in the 3-position is fully substituted, as, for example, in 3,3-dimethyl-1-butene [2]. Hence, for linear α-olefins the polymerisation rate decreases with increasing chain length; for example, the value of the 1-butene propagation rate is about 3 times smaller than that of propylene [431]. For branched α-olefins, the polymerisation rate depends on the position, number and type of branching, as shown by the following comparison of polymerisation rates of different monomers [47, 241, 432, 433]: 5-methyl-1-heptene $>$ 4-methyl-1-hexene $\gg$ 3-methyl-1-pentene; 1-butene $\gg$ 3-methyl-1-butene $>$ vinylcyclohexane $>$ 3,5,5-trimethyl-1-hexene; 4-methyl-1-hexene $\gg$ 4-phenyl-1-hexene $\gg$ 4,4,4-triphenyl-1-butene.

The α-olefin activity is significantly decreased when both carbons in positions 3 and 4 are substituted; for instance, the activity of 3,4-dimethyl-1-pentene is 5 times lower than that of 3-methyl-1-pentene, which, in turn, is 3 times lower than that of 4-methyl-1-pentene [37].

The polymerisation of racemic α-olefins such as 4-methyl-1-hexene with both heterogeneous [282,372,384] and homogeneous [434] isospecific catalysts tends to incorporate preferentially one of the α-olefin enantiomers into the growing polymer chain; however, some differences in the stereoselective polymerisation of 4-methyl-1-hexene by heterogeneous and homogeneous Ziegler–Natta catalysts occur [435]. As regards the polymerisation of 3-methyl-1-pentene, similar effects have been observed for polymerisation with heterogeneous catalysts [379]; but syndiospecific *ansa*-zirconocene-based catalysts generally exhibit negligible activities in this polymerisation [256].

Polymerisation of racemic 4-methyl-1-hexene with the $Me_2C(Cp)(Flu)ZrCl_2$–$[Al(Me)O]_x$ catalyst was found obviously to produce syndiotactic polymer, which appeared to be a random copolymer of two enantiomers, poly[(*R*,*S*)-4-methyl-1-hexene] [436,437].

3.7 Polymerisation of Allenes

The polymerisation of allene (1,2-propadiene), the simplest cumulene, is of interest not least because polyallene can be considered to be a versatile intermediate polymer 'blank', if formed with the chemical structure resulting from polymerisation by the scheme

(68)

Such polyallene could in theory be converted to either isotactic or syndiotactic polypropylene using asymmetric hydrogenation catalysts [438].

Allene has been polymerised to high molecular weight linear polymers by various Ziegler–Natta catalysts based on compounds of such transition metals as Ti, V, Cr, Mn, Fe, Co and Ni as precursors and alkylaluminium compounds as activators [439–441]. Crystalline and amorphous polymers have been produced, in various proportions, with each of the catalysts used. The crystalline polymers consist predominantly of 1,2-linked (head-to-tail) monomeric units formed by insertion polymerisation as in scheme (68), but some regioirregularities resulting from the 2,1-insertion, leading to head-to-head and tail-to-tail arrangements, may be present in the polymer chain:

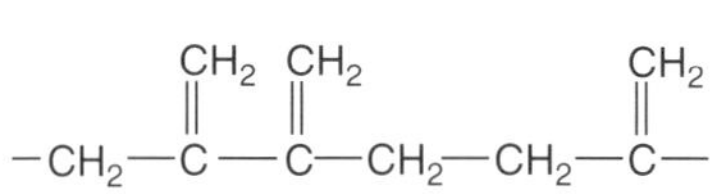

Amorphous allene polymers also contain 1,1-linked monomeric units and *cis*-1,3-linked monomeric units [241]:

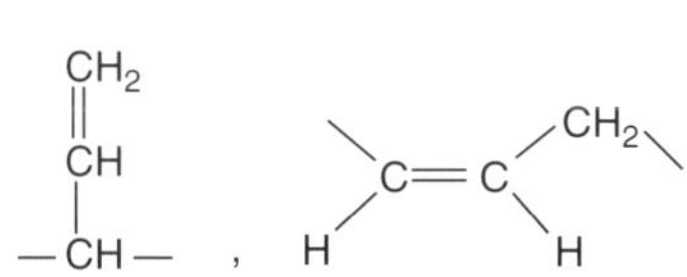

A nickel-based Ziegler–Natta catalyst, $Ni(Acac)_2$–$Al(i\text{-}Bu)_3$, yields a regular allene polymer consisting almost exclusively of head-to-tail linked (1,2-inserted) monomeric units [441].

Various monocomponent catalysts based on nickel compounds, such as $Ni(All)_2$, Ni(All)Cl, $Ni(All)OCOCF_3$ and $Ni(Cod)_2$, have been used for the polymerisation of allene; crystalline polymers of rather high molecular weight were yielded, but they were characterised by lower structural regularity as a result of diads made up of 1,2- and 2,1-linked monomeric units [438,441, 442]. Complexes of other transition metals, especially *cis*-$Rh(CO)_2Cl{\cdot}PPh_3$, $Co(All)_3$, $Co_2(CO)_8$ (in hydrocarbon solutions) and Pd complexes (in acetic acid solutions) are also active in allene polymerisation [241].

Substituted allenes undergo polymerisation to linear polymers in the presence of both Ziegler–Natta catalysts, e.g. $VOCl_3$–Al(*i*-Bu)$_3$ [440], and monometallic complexes such as Ni complexes [442]. Amorphous polymers consisting of 1,2-linked monomeric units (predominant) and 2,3-linked monomeric units have been obtained from 1,2-butadiene ($H_2C{=}C{=}CHCH_3$):

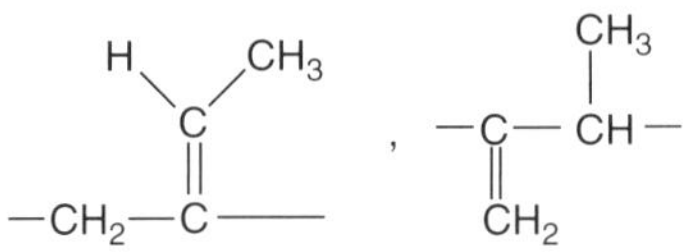

whereas crystalline polymers consisting almost exclusively of 2,3-linked monomeric units have been obtained from 3-methyl-1,2-butadiene [$H_2C{=}C{=}C(CH_3)_2$]:

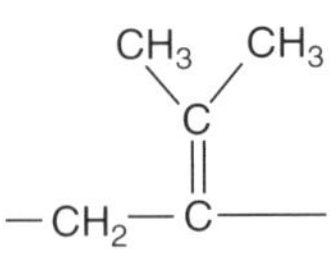

Racemic 2,3-pentadiene [(*R*,*S*)-$CH_3CH{=}C{=}CHCH_3$] yields amorphous polymers consisting of 2,3-linked monomeric units [443]:

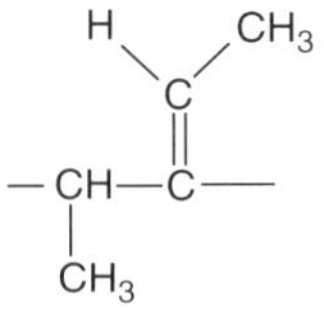

However, when one of the enantiomers is polymerised, the polymer formed contains predominantly stereoregular sequences characterised by an isotactic structure of the main chain and by the same configuration of the CH_3 group at the double bonds [443]:

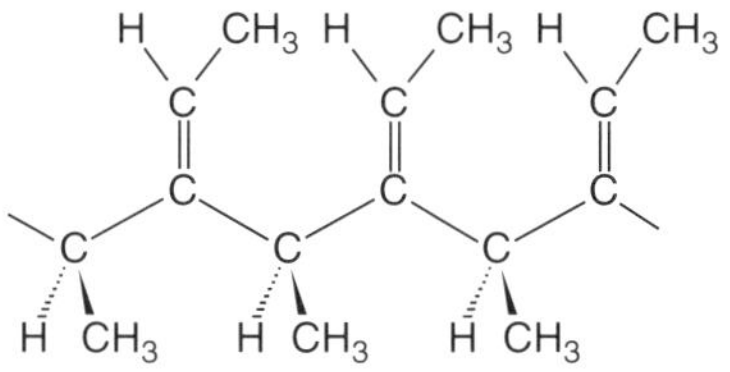

3.8 Isomerisation–Polymerisation of Olefins

As already discussed, most Ziegler–Natta and related coordination catalysts polymerise α-olefins by a 1,2-insertion mechanism, yielding highly regio- and stereoregular polymers. When some nickel-based coordination catalysts are used, however, isomerisation of the active species may accompany the propagation, and polymers containing 2, ω-coupled monomeric units are formed [183,191].

As regards β-olefins, they are usually claimed to be inactive in coordination polymerisation (insertion polymerisation). However, a number of β-olefins undergo a monomer isomerisation–polymerisation with Ziegler–Natta catalysts to give, in moderate yields, polymers containing 1,2-coupled monomeric units; these polymers are characterised, however, by moderate molecular weight and moderate stereoregularity. It is interesting that the yields of the polymer formed can be enhanced and its molecular weight and stereoregularity (isotacticity) increased when group 8 transition metal compounds, especially nickel compounds (as β-olefin isomerisation catalyst precursors), are added to the polymerisation system [192–195, 444–446].

3.8.1 Polymerisation of α-olefins via 2, ω-Coupling

Catalysts capable of 2, ω-linkage of α-olefins are obtained from the reaction of Ni(0) or Ni(II) compounds, such as $Ni(Cod)_2$ or $Ni(All)_2$ respectively, with aminobis(imino)phosphorane, such as $(Me_3Si)_2NP(=NSiMe_3)_2$, preferably used in equimolar ratio; the catalysts are characterised by the appearance of Ni(II) species as has been shown by scheme (18) [182].

When linear α-olefins are polymerised, the polymer contains only methyl branches, regularly distributed along the chain with a separation corresponding to the size of the alkyl substituent in the monomer. It may be interesting that the structure of the polymer formed from 1-pentene, poly[2,5-(1-

pentene)], corresponds to that of an alternating copolymer of ethylene and propylene:

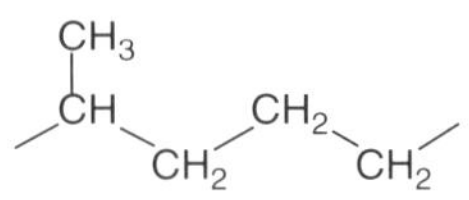

Fink *et al.* [183] developed a scheme explaining the 2,ω-polymerisation pathway, which is represented in the case of 1-butene 2,4-propagation by

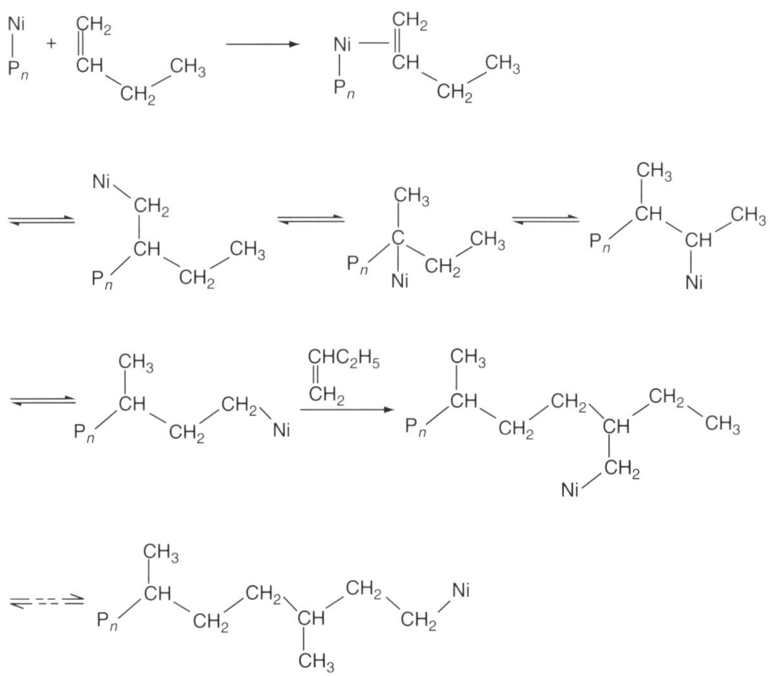

(69)

It is to be noted that the monomer can only be inserted into a primary nickel–alkyl bond at the end of the growing chain, the insertion is regioselective and only C_ω–C_2 coupling of the growing chain with the next monomer molecule occurs, and the nickel species 'migrates' along the polymer main chain (isomerisation) between two insertions. It should be added that, during the isomerisation, transfer reactions can take place, but not insertions.

Considering the necessity of the presence of at least one hydrogen atom at every carbon atom in the α-olefin molecule for the 2,ω-polymerisation, the isomerisation can be explained in terms of a β-elimination/addition mechanism, which is presented in the case of 1-pentene by [183].

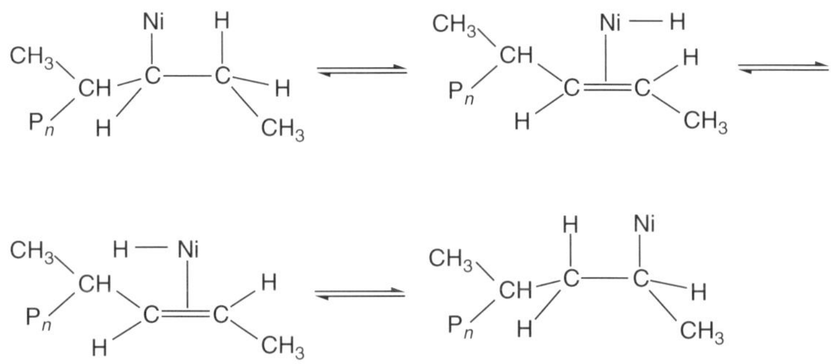

(70)

The β-elimination, according to this mechanism, leads to a nickel hydride species coordinated to the double bond formed in the chain; the rotation of such a coordination complex, followed by subsequent addition, results in the formation of an isomerised chain [scheme (70)]. Once a free nickel hydride species is generated, it can only react with a new α-olefin and thus start a new chain (the transfer reaction to the monomer).

It may be interesting that the activation energy for 2,ω-polymerisation decreases with the length of the α-olefin used. This is interpreted in terms of an increase in the number of exergonic migration (isomerisation) steps with the length of the α-olefin (the insertion step is endergonic) [183].

The use of optically active α-olefins, e.g. 4-methyl-1-hexene [CH_2= $CHCH_2C^*H(CH)_3CH_2CH_3$], indicates that no reaction of pure *R* and *S* enantiomers takes place, whereas their racemic mixture always very easily undergoes 2,6-polymerisation. The resulting optically inactive polymer appeared to consist of alternating *R* and *S* enantiomers in its chains. Hence, the chiral carbon atom in 4-methyl-1-hexene is not racemised during the isomerisation, i.e. the migration of nickel species along the chain proceeds either by retention or inversion of the configuration of the chiral centre. This is in agreement with the previous statement that a nickel hydride species formed during β-elimination is coordinated to the double bond of the polymer chain but is not a free hydride [scheme (70)]; this ensures the preservation of the stereochemical configuration [183].

One problem of 2,ω-polymerisation is that the molecular weight of the product, poly[2,ω-(α-olefin)], is always very low ($\bar{M}_w \approx 6 \times 10^3$), whereas the molecular weight distribution is small ($\bar{M}_w/\bar{M}_n \approx 1.6$). It is possible, however, to increase the molecular weight of the polymer, e.g. the molecular weight of poly[2,6-(1-hexene)], to $\bar{M}_w \approx 90 \times 10^3$ by increasing the reaction pressure to 1400 MPa. The reason for this is the possible kinetic pressure effect in the case of 1-hexene in which the insertion but not the isomerisation is the rate determining step for 2,ω-polymerisation [183].

3.8.2 Polymerisation of β-Olefins via 1,2-Coupling

The polymers of β-olefins with 1,2-linked monomeric units have been obtained by polymerisation in the presence of Ziegler–Natta catalysts, such as $TiCl_3$–AlR_3 (R=Et, *i*-Bu), used preferably in a combination with Ni(II) compounds, especially $NiCl_2$. For instance, monomer isomerisation–polymerisation of *cis*-2-butene with the $TiCl_3/NiCl_2$–$AlEt_3$ (1:1:3) catalyst, carried out in *n*-heptane at 80 °C, produced poly(1-butene) in a yield of ca 72% within 24 h. Polymers obtained under such conditions are characterised by relatively high molecular weight (40×10^3–85×10^3) and contain a significant amount (up to ca 72%) of the isotactic fraction [444]. They also contain head-to-head (tail-to-tail) as well as 2,3-linked monomeric units to some extent [193].

The β-olefin isomerisation–polymerisations are composed of two distinct reactions, isomerisation to the respective α-olefin and α-olefin polymerisation via a 1,2-insertion mechanism, which involve different active sites:

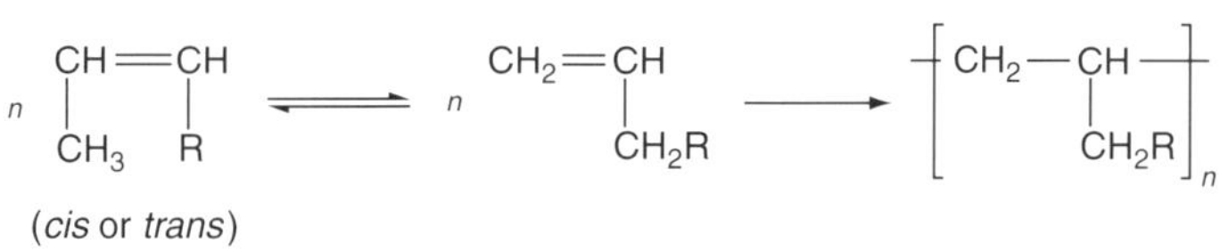

(71)

Readily polymerisable α-olefins consumed by the polymerisation are replenished by the isomerisation of β-olefins [447]. Kinetic studies of β-olefin isomerisation–polymerisation indicate that the polymerisation, and not the isomerisation, is the rate-determining step under the conditions mentioned. The advantage of the β-olefin polymerisation procedure, apart from the properties of the polymers obtained, is the 'one-pot' synthesis of α-olefin polymers using β-olefin monomers [444].

The results obtained from studies of reactions of *cis*- and *trans*-2-butene in the presence of $TiCl_3$–$AlEt_3$, $NiCl_2$–$AlEt_3$ and $TiCl_3/NiCl_2$–$AlEt_3$ catalysts suggest that a certain bimetallic compound of titanium and nickel may be formed in the latter case. Ni(II) and Ti(III) species are postulated to promote β-olefin isomerisation and α-olefin propagation respectively. Moreover, the isomerisation is proposed to involve a β-elimination/addition mechanism. The results obtained with the $NiCl_2$–$AlEt_3$ catalyst indicate that the isomerisation proceeds through σ-alkylnickel compounds, which undergo β-elimination to form Ni–H species. The addition of this species to β-olefin, followed by subsequent β-elimination involving the H_α atom, results in the formation of the isomerisation product, which is α-olefin. The insertion of β-olefins to the Ni–H bond occurs very easily owing to the high d orbital population of nickel compounds. Thus, with nickel-based catalysts the isomerisation takes place readily as compared with titanium-based catalysts [444].

3.9 Copolymerisation

Ever since their original discoveries, Ziegler–Natta catalysts and Phillips catalysts have been used for both the homopolymerisation and the copolymerisation of olefins. Moreover, Ziegler–Natta catalysts also allowed the copolymerisation of olefins with vinylaromatic monomers, conjugated dienes and cycloolefins. Other coordination catalysts such as group 8 metal compounds, especially cationic Pd(II) complexes, enabled the alternating copolymerisation of olefins and carbon monoxide [2,29,30,37,43,46,241,448–450].

Compared with homopolymerisations, copolymerisations offer a deeper insight into the nature of the catalyst and the polymerisation as well as wider synthetic possibilities resulting in polymers of alternative and useful properties. Indeed, copolymers are arguably of far greater commercial and industrial significance than homopolymers. Note that coordination copolymerisation enables an enchainment of some non-homopropagating monomers, such as carbon monoxide and, in particular cases, some cycloolefins, which do not homopropagate via 1,2-insertion owing to steric hindrances [448–450].

In the copolymerisation, each monomer competes for the available catalytic active species, and the composition, structure and molecular weight of the copolymer produced reflect this competition. The mechanistic features of the copolymerisation are in principle similar to those of the homopolymerisation. In the copolymerisation, however, the effect of the kind of last inserted monomeric unit and the kind of comonomer undergoing insertion (M_1 or M_2) should be taken into consideration; the propagation step can proceed in at least four ways [448]:

$$\mathrm{Mt{-}M_1{-}P}_n + \mathrm{M_1} \xrightarrow{k_{11}} \mathrm{Mt{-}M_1{-}M_1{-}P}_n \tag{72}$$

$$\mathrm{Mt{-}M_1{-}P}_n + \mathrm{M_2} \xrightarrow{k_{12}} \mathrm{Mt{-}M_2{-}M_1{-}P}_n \tag{73}$$

$$\mathrm{Mt{-}M_2{-}P}_n + \mathrm{M_1} \xrightarrow{k_{21}} \mathrm{Mt{-}M_1{-}M_2{-}P}_n \tag{74}$$

$$\mathrm{Mt{-}M_2{-}P}_n + \mathrm{M_2} \xrightarrow{k_{22}} \mathrm{Mt{-}M_2{-}M_2{-}P}_n \tag{75}$$

In the above schemes, k_{11} and k_{22} are the respective monomer homopropagation rate constants, and k_{12} and k_{21} are the monomer heteropropagation rate constants. The expression for the copolymer composition in terms of comonomer feed concentrations is as follows:

$$\frac{d[M_1]}{d[M_2]} = \frac{[M_1]}{[M_2]} \times \frac{r_1 \times [M_1] + [M_2]}{r_2 \times [M_2] + [M_1]}$$

where $d[M_1]/d[M_2]$ denotes the ratio of M_1 to M_2 monomeric units in the copolymer, $[M_1]$ and $[M_2]$ denote the concentration of the respective comonomer in the feed and $r_1 = k_{11}/k_{12}$ and $r_2 = k_{22}/k_{21}$ denote the respective comonomer relative reactivity ratio.

When the product of monomer relative reactivity ratios is approximately one ($r_1 \times r_2 = 1$), the last inserted monomeric unit in the chain does not influence the next monomer incorporation and Bernoullian statistics govern the formation of a random copolymer. When this product tends to zero ($r_1 \times r_2 = 0$), there is some influence from the last inserted monomeric unit (when first-order Markovian statistics operate), or from the penultimate inserted monomeric units (when second-order Markovian statistics operate), and an alternating copolymer formation is favoured in this case. Finally, when the product of the reactivity ratios is greater than one ($r_1 \times r_2 > 1$), there is a tendency for the comonomers to form long segments and block copolymer formation predominates (or even homopolymer formation can take place) [448].

3.9.1 Copolymerisation of Ethylene and α-Olefins

Olefin copolymerisations, especially those of ethylene with propylene and/or another α-olefin, lead to copolymers that are of great practical importance; the total production volume of such olefin copolymers is comparable with that of olefin homopolymers [30].

A variety of catalysts have been used for olefin copolymerisations. Ethylene/propylene copolymerisations run in the presence of heterogeneous Ziegler–Natta catalysts such as $TiCl_3$–AlR_3 tend to form copolymers with monomeric units distributed in separate blocks, and thus the copolymers, even with an equimolar ethylene and propylene content, exhibit crystallinity due to blocks of both comonomers; values of the product of monomer relative reactivity ratios are greater than one ($r_1 \times r_2 > 1$) in such cases [451]. In addition, the non-uniform active sites of heterogeneous non-supported catalysts such as $TiCl_3$–AlR_3 [451] as well as $MgCl_2$-supported catalysts [452] have in general different copolymerisation characteristics; different catalytic sites display different comonomer selectivity. Therefore, different monomer reactivity ratios are expected for different catalytic sites and no meaningful extrapolation concerning the relative comonomer reactivity can be attempted on the basis of the average r_1 and r_2 values measured with heterogeneous Ziegler–Natta catalyst systems [241].

Of particular interest are random copolymers of ethylene and propylene, which are obtainable with soluble vanadium-based Ziegler–Natta catalysts such as $V(Acac)_3$–$AlEt_2Cl$ [453], $VOCl_3$–$AlEt_2Cl$ [72], $VOCl_3$–$Al(i\text{-}Bu)_2Cl$

[27], VCl_3–$Al(n\text{-}Hx)_3$ [454] and $VO(OEt)_3$–$Al(i\text{-}Bu)_2Cl$ [455]; values of the product of monomer reactivity ratios are smaller than one ($r_1 \times r_2 < 1$) in most cases [27,72,453,454] or greater than one ($r_1 \times r_2 > 1$) in the case of vanadyl alcoholate-based catalysts [455]. This could indicate that ethylene/propylene copolymers obtained with soluble vanadium-based catalysts may differ slightly from a random distribution of comonomers, for which $r_1 \times r_2 = 1$; depending on the kind of catalyst, a tendency to form some alternating or block sequences of comonomers may occur. Note in this connection, however, that soluble vanadium-based Ziegler–Natta catalysts are also characterised by the appearance of non-uniform active sites (usually $\bar{M}_w/\bar{M}_n = 3$–10 for ethylene/propylene copolymers obtained with vanadium-based catalysts). In all cases of ethylene/propylene copolymerisation, the reactivity of ethylene is greater than that of propylene. Vanadium-based catalysts display the lowest r_1 value (for ethylene). Characteristic of soluble vanadium-based catalysts is their lack of stereospecificity and rather poor regioselectivity; there are 2,1-inserted propylene units present in ethylene/propylene copolymer chains, apart from the 1,2-inserted propylene units present in ethylene/propylene copolymer chains, apart from the 1,2-inserted propylene units. It must be emphasised, however, that, in the field of polyolefin elastomers, stereoregular placements in the polymer chain have never been desired [37].

Random ethylene/propylene copolymers are amorphous and represent an interesting class of synthetic elastomers. The introduction of double bonds, useful for sulphur vulcanisation in the copolymer, can be achieved by copolymerisation of ethylene and propylene with non-conjugated dienes containing only one double bond capable of insertion; for instance, 1,4-hexadiene, dicyclopentadiene and 5-ethylidene-2-norbornene (endocyclic double bond)

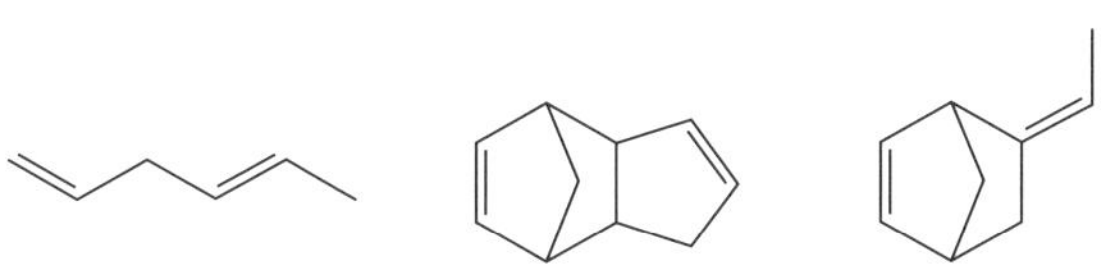

are most applicable [37,271]. For the above-mentioned reasons, soluble vanadium-based catalysts are used for the industrial production of ethylene/propylene and ethylene/propylene/diene random copolymers [37].

It is interesting that some heterogeneous superhigh-activity Ziegler–Natta catalysts such as $MgCl_2/TiCl_4/LB$–$AlEt_3$ also yield random ethylene/propylene copolymers. These copolymers, however, exhibit a blocky nature and highly isotactic propylene sequences (with no 2,1-inserted propylene units) that contribute to undesired crystallinity [68,456].

The extent of comonomer selectivity of the heterogeneous Ziegler–Natta catalysts is particularly evident in the copolymerisation of the two enantiomers of a racemic α-olefin. In fact, with $TiCl_3$–AlR_2Cl [374] or $MgCl_2/TiCl_4$–$AlEt_3$ [375], macromolecules containing a prevalence of monomeric units arising from one or other enantiomer are formed when the chiral carbon atom is present in an α- or a β-position with respect to the double bond. Furthermore, when mixtures of chiral monomers of different structure are used, only enantiomers of the same chirality substantially copolymerise [453]. Also, when a racemic α-olefin is copolymerised with one enantiomer of another α-olefin, only enantiomers of the same chirality copolymerise; the enantiomer possessing the opposite chirality yields a homopolymer [241].

Random ethylene copolymers with small amounts (4–10 wt-%) of α-olefins, e.g. 1-butene, 1-hexene, 1-octene and 4-methyl-1-pentene, are referred to as linear low-density polyethylene, which is a commercially relevant class of polyolefins. Such copolymers are prepared by essentially the same catalysts used for the synthesis of high-density polyethylene [241]. Small amounts of α-olefin units incorporated in an ethylene copolymer have the effect of producing side chains at points where the α-olefin is inserted into the linear polyethylene backbone. Thus, the copolymerisation produces short alkyl branches, which disrupt the crystallinity of high-density polyethylene and lower the density of the polymer so that it simulates many of the properties of low-density polyethylene manufactured by high-pressure radical polymerisation of ethylene [448] (Figure 2.3).

In copolymerisations carried out with heterogeneous catalysts, the higher α-olefin is preferentially incorporated at sites producing shorter chains; higher α-olefins are thus enriched in the low-molecular fractions of the copolymers obtained. This increases the content of extractables and tends to make these copolymers sticky, two generally undesirable properties [30].

Many attempts have been made to synthesise ethylene/propylene block copolymers, referred to as polyallomers, with isospecific Ziegler–Natta catalysts. However, true block copolymers can hardly be synthesised. This is due to the short life of the growing polymer chains [68,241]. Therefore, only in a few cases, when the copolymers are synthesised by adding two comonomers sequentially and under very specific conditions in order to reduce chain transfer reactions, does unambiguous evidence for true block copolymer formation with isospecific catalysts exist [457].

Copolymerisations of propylene [68,458–460] and 1-butene [47] with higher α-olefins have also been carried out in order to obtain copolymers offering a range of properties not available from the homopolymers alone. As regards the copolymerisability of α-olefins, which decreases with increasing size of the alkyl substituent at the double bond, it results mainly from steric effects [458].

The different reactivities of the olefins are important for the copolymerisation. The comonomer reactivity ratio, r_2, in copolymerisation with ethylene appears to decrease with increasing steric hindrance around the double bond in the α-olefin in to the following order [250]: ethylene > propylene > 1-butene > linear α-olefin > branched α-olefin.

Normally, propylene reacts 5–100 times more slowly than ethylene, and 1-butene 3–10 times more slowly than propylene [37]. Relative copolymerisation

Table 3.4 Relative reactivity ratios ($r_2 = k_{22}/k_{21}$) of ethylene and higher α-olefins (M_2) with respect to propylene (M_1)

Monomer	Monomer relative reactivity ratio, r_2
Ethylene	8–20
Propylene	1
1-Butene	0.2–0.6
1-Pentene	0.2–0.5
1-Hexene	0.2–0.4
1-Decene	0.1–0.3
1-Tetradecene	0.1–0.2
1-Octadecene	0.1–0.15
4-Methyl-1-pentene	0.15
3-Methyl-1-butene	0.02–0.06
3-Methyl-1-pentene	0.05

activities, expressed in the value of r_2, for higher α-olefins with respect to propylene are presented in Table 3.4 [37,43].

Ethylene/propylene and higher α-olefin copolymers can be obtained using metallocene catalysts for the copolymerisation. Both homogeneous catalysts such as *rac.*-$(IndCH_2)_2ZrCl_2$–$[Al(Me)O]_x$ and *rac.*-$(IndH_4CH_2)_2ZrCl_2$–$[Al(Me)O]_x$ [141,264,461,462] and supported catalysts such as SiO_2/*rac.*-$(IndCH_2)_2ZrCl_2$–$[Al(Me)O]_x$ and SiO_2/*rac.*-$(IndH_4CH_2)_2ZrCl_2$–$Al(i\text{-}Bu)_3$ [200,202,462] have been successfully used to obtain the ethylene/propylene and ethylene/propylene/diene copolymers. It is characteristic that metallocene catalysts show the highest polymerisation activities. Furthermore, they have much higher reactivity towards propylene or higher α-olefin in comparison with soluble vanadium-based catalysts. Ethylene/propylene and higher α-olefin copolymers obtained with metallocene-based catalysts are characterised by a narrow molecular weight distribution and do not substantially contain 2,1-inserted propylene units in their chains. Finally, metallocene catalysts, both chiral and achiral, irrespective of whether they are used as homogeneous solutions or supported on SiO_2 or $MgCl_2$, appeared to be the best catalysts for achieving copolymers with a random distribution of comonomers [126].

It must be emphasised that copolymers produced by metallocene-based catalysts display comonomer distributions essentially independent of the chain lengths; the copolymers obtained with these catalysts contain larger fractions of higher α-olefins than those obtained with heterogeneous Ziegler–Natta catalysts under comparable conditions [30].

The copolymerisation characteristics of metallocene-based catalysts vary with the kind of catalyst used. Higher α-olefins are taken up more readily by isospecific, chiral *ansa*-metallocene-based catalysts than by the non-bridged ones. An even greater tendency for the enchainment of higher α-olefins is evident for the syndiospecific $Me_2C(Cp)(Flu)ZrCl_2$–$[Al(Me)O]_x$ catalyst [30].

Table 3.5 Monomer relative reactivity ratios, r_1 (ethylene) and r_2 (propylene), for ethylene/propylene copolymerisations with various Ziegler–Natta catalysts[a]

Catalyst	r_1	r_2	$r_1 \times r_2$
δ-$TiCl_3$–$AlEt_3$	7.3	0.76	5.5
$MgCl_2/TiCl_4$–$AlEt_3$	4		
$SiO_2/MgCl_2/TiCl_4$–$AlEt_3$	5–10	0.2–0.34	1.9
$TiCl_4$–$Al(n\text{-}Hx)_3$	33.4	0.032	1.07
VCl_3–$Al(n\text{-}Hx)_3$	5.6	0.145	0.81
$VOCl_3$–$Al(i\text{-}Bu)_2Cl$	20.3	0.022	0.45
$VOCl_3$–$AlEt_2Cl$	12.1	0.018	0.22
Cp_2ZrCl_2–$[Al(Me)O]_x$	48	0.015	0.72
$(IndCH_2)_2ZrCl_2$–$[Al(Me)O]_x$	2.57	0.39	1.0
$(ThindCH_2)_2ZrCl_2$–$[Al(Me)O]_x$	2.90	0.28	0.81
$Me_2C(Cp)(Flu)ZrCl_2$–$[Al(Me)O]_x$	1.3	0.20	0.26
$ThindCH_2)_2ZrCl_2$–$[Al(Me)O]_x$	59	0.012	0.71
$Me_2Si(Ind)_2ZrCl_2$–$[Al(Me)O]_x$	25	0.016	0.4

[a] For heterogeneous catalysts, values of $r_1 \times r_2 > 1$ indicate that the comonomers are incorporated in blocks; for soluble vanadium-based catalysts and homogeneous metallocene-based catalysts, random or alternating sequences are indicated by values $r_1 \times r_2 \leq 1$.

Reproduced by permission of John Wiley & Sons, Inc. from Ref. 30. Copyright 1995 Wiley-VCH Weinheim

Typical values of comonomer relative reactivity ratios in ethylene/propylene copolymerisations run with various heterogeneous and homogeneous Ziegler–Natta catalysts are listed in Table 3.5 [30, 72, 454].

One should realise that metallocenes offer a unique opportunity to expand the traditional application field of soluble vanadium-based and supported titanium-based Ziegler–Natta catalysts. As a consequence of the new copolymer microstructures, new physicomechanical behaviours are observed. A number of copolymers were found to have elastomeric properties, without the need for vulcanisation (for instance, a copolymer containing 60 wt-% ethylene units and 40 wt-% propylene units) [126]. Metallocene-based catalysts also offer 'one-pot' synthesis of linear low-density polyethylene with a bimodal composition distribution, which is of practical importance. Studies of ethylene/1-hexene copolymerisation with the Cp_2ZrCl_2–$[Al(Me)O]_x$ catalyst suggest that its active sites exist in two forms, one being of tetrahedral configuration (less reactive towards 1-hexene) and the other of octahedral configuration (more reactive towards 1-hexene). Therefore, a bimodal composition distribution can be achieved in the copolymers formed [463].

3.9.2 Copolymerisation of Ethylene and β-Olefins

The ability of a β-olefin to copolymerise with ethylene in the presence of Ziegler–Natta catalysts arises from minimisation of steric interactions at the catalytic active site by ethylene units; the steric hindrance, which prevents homopropagation of the β-olefin, is overcome when the β-olefin monomer is inserted into the active Mt–C bond in a propagating chain that has ethylene as the terminal unit. On the other hand, the ethylene monomer is incorporated readily in a propagating chain that has a β-olefin terminal unit. Thus, the ethylene–β-olefin system undergoes copolymerisation on account of a tendency towards alternation caused by steric factors [448]. Typically, ethylene/2-butene copolymers are obtained with titanium- and vanadium-based Ziegler–Natta catalysts (activated with AlR_2Cl) [2]. Because of the high relative reactivity ratio of ethylene, compared with that of 2-butene, a considerable excess of the latter monomer is needed to avoid the formation of sequences of ethylene units in the copolymer chains. An alternating propagation in the copolymerisation of ethylene and 2-butene (*cis* or *trans* isomers) is presented as follows:

$$\mathrm{Mt-CH_2CH_2-P}_n \xrightarrow{\mathrm{CH_3CH{=}CHCH_3}} \mathrm{Mt-CH(CH_3)CH(CH_3)-CH_2CH_2-P}_n$$

$$\xrightarrow{\mathrm{CH_2{=}CH_2}} \mathrm{Mt-CH_2CH_2-CH(CH_3)CH(CH_3)-CH_2CH_2-P}_n \qquad (76)$$

It should be added that alternating ethylene/2-butene copolymers can exhibit stereoregularity; namely the ethylene/*cis*-2-butene copolymer, which possesses an *erythro*-diisotactic structure and is a crystalline polymer. It may be interesting to note that from the formal point of view the alternating *erythro*-diisotactic ethylene/*cis*-2-butene copolymer, i.e. *erythro*-diisotactic poly[ethylene-*alt*-(*cis*-2-butene)], can be treated as isotactic head-to-head and tail-to-tail polypropylene. Isomeric *trans*-2-butene gives atactic amorphous copolymers with ethylene [2,82].

It should be mentioned that some Ziegler–Natta catalysts containing Ni(II) compounds, which are capable of inducing β-olefin isomerisation to the corresponding α-olefin, e.g. 2-butene to 1-butene, readily promote the copolymerisation of the α-olefin formed with ethylene [464–466].

3.9.3 Copolymerisation of Ethylene and Cycloolefins

Cycloolefins having rings with more than four carbon atoms do not homopolymerise in the presence of Ziegler–Natta catalysts based on titanium or vanadium compounds as precursors and alkylaluminium activators. However, these cycloolefins may copolymerise with ethylene via the double bonds while preserving the cycloolefin ring; ethylene is able to compensate the steric hindrance at the C_α atom of the growing chain after and before the 1,2-insertion of the cycloolefin [2].

Alternating copolymers have been obtained by copolymerisation of ethylene and cycloolefins (using a large excess of cycloolefin) in the presence of vanadium-based Ziegler–Natta catalysts such as $V(Acac)_3$–$AlEt_2Cl$ and VCl_4–$AlEt_2Cl$:

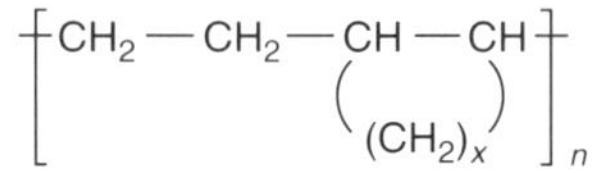

Only cycloolefins with rings containing an odd number of carbon atoms, such as cyclopentene ($x = 3$) and cycloheptene ($x = 5$), yield crystalline copolymers with an *erythro*-diisotactic configuration. Cycloolefins with an even number of carbon atoms in the ring, such as cyclobutene ($x = 2$) and cyclooctene ($x = 6$), give amorphous copolymers [241].

The rate of copolymerisation of ethylene and odd-membered ring cycloolefins is higher than the rate of copolymerisation involving even-membered cycloolefins [467]. This indicates that both the polymerisation kinetics and the spatial configuration of the copolymer are influenced by steric factors [2].

Real progress in the copolymerisation of ethylene (or α-olefin) and cycloolefin has been made utilising metallocene catalysts. These catalysts appeared to be about 10 times more active than soluble vanadium-based catalysts in ethylene/cycloolefin copolymerisation involving double bond opening with preservation of the cycloolefin ring. Moreover, metallocene catalysts appeared to produce high molecular weight copolymers of ethylene and monocyclic olefins such as cyclobutene, cyclopentene, cycloheptene and cyclooctene as well as copolymers of ethylene and polycyclic olefins such as 2-norbornene, 1,4,5,8-dimethano-1,2,3,4,4a,5,8,8a-octahydronaphthalene (Dmon) and 1,4,5,6,9,10-trime thano-1,2,3,4,4a,5,5a,6,9,9a,10,10a-dodecahydronaphthalene (Tmdn) [468]:

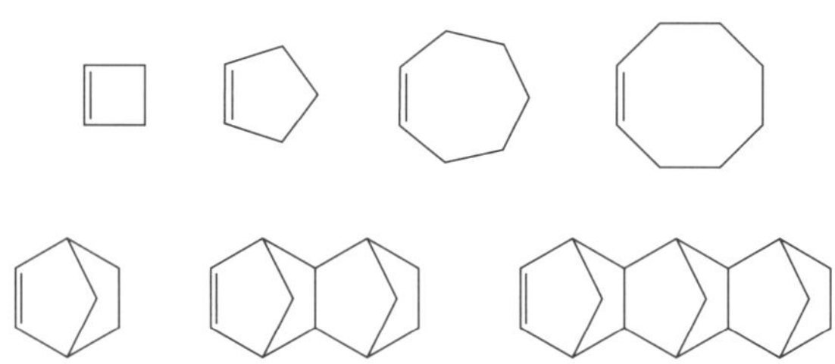

Copolymers of ethylene and cycloolefins represent a new class of transparent and thermoplastic polymers. By using metallocene catalysts it is possible to design the copolymerisation process as well as to vary the properties of copolymers over a wide range. The ethylene/cycloolefin copolymers are amorphous when more than 10–15 mol.-% of the cyclic monomer is incorporated. These copolymers are transparent and highly resistant to chemicals and heat, as well as exhibiting very high glass transition temperatures and desirable elastic moduli. Such properties make them suitable as potential polymeric materials for optical applications [29, 30, 449, 468].

While ethylene is vastly more reactive than cyclopentene in copolymerisation with the $(IndCH_2)_2ZrCl_2$–$[Al(Me)O]_x$ catalyst (relative reactivity ratio of ethylene $r_1 = 80$–300), norbornene is quite readily incorporated, with

$r_1 = 1.5$–3.2 (an r_1 value of 2 means that ethylene is enchained only twice as fast as norbornene) [468–470]. Norbornene/ethylene copolymers are most interesting for technical uses because of the readily available monomers. It is interesting that, as in some other cases, the rate of homopolymerisation of ethylene in the presence of the diphenylmethylene(cyclopentadienyl)(9-fluorenyl)zirconium dichloride $[Ph_2C(Cp)(Flu)ZrCl_2]$–$[Al(Me)O]_x$ catalyst is lower than the rate of the corresponding copolymerisation (ethylene/norbornene) [471].

The relative monomer reactivity ratios, r_1 and r_2, for ethylene/cycloolefin copolymerisation with zirconocene–methylaluminoxane catalysts are presented in Table 3.6 [468].

Metallocene catalysts show low r_1 values, which allows easy incorporation of bulky cycloolefins into the growing copolymer chain. Surprisingly, the ethylene reactivity ratio in copolymerisation with cyclopentene in the presence of a $(ThindCH_2)_2ZrCl_2$-based catalyst ($r_1 = 2.2$) and in copolymerisation with norbornene in the presence of catalysts characterised by C_s and C_1 symmetry ($r_1 \approx 3.4$ and 3.1 respectively) is considerably lower than that for the copolymerisation of ethylene with propylene ($r_1 = 6.6$ at 37 °C). Various catalysts produce copolymers of structures that are between statistical and alternating [468].

Note that ethylene/norbornene copolymerisation with the $Me_2C(t\text{-}BuCp)(Flu)ZrCl_2$–$[Al(Me)O]_x$ catalyst produces a strongly alternating copolymer (when an excess of norbornene in the comonomer feed is used). The tendency towards cross-propagation of the comonomers is connected with the presence of the bulky *t*-butyl substituent at the C_β atom of the cyclopentadienyl ring in the catalyst molecule. The alternating ethylene/norbornene copolymer is crystalline and exhibits a melting point of 295 °C, glass transition temperature of 145 °C, good heat resistance and resistance to non-polar solvents [468].

Ethylene has also been copolymerised successfully with substituted polycyclic olefins, such as 5-phenyl-2-norbornene (*endo*/*exo* = 2.3:1) and 2-phenyldimethanooctahydronaphthalene (*endo*/*exo* = 2.3:1), in the presence of C_s-symmetric

Table 3.6 Monomer relative reactivity ratios, r_1 (ethylene) and r_2 (cycloolefin), for ethylene/cycloolefin copolymerisations with zirconocene–$[Al(Me)O]_x$ catalysts

Cycoolefin	Catalyst precursor	Temperature (°C)	r_1	r_2	$r_1 \times r_2$
Cyclopentene	$(ThindCH_2)_2ZrCl_2$	25	2.2	< 1	~ 1
Norbornene	$Me_2Si(Ind)_2ZrCl_2$	30	2.6	< 2	~ 1
Norbornene	$Me_2C(Cp)(Flu)ZrCl_2$	30	3.4	0.06	0.2
Norbornene	$Ph_2C(Cp)(Flu)ZrCl_2$	30	3.0	0.05	0.15
Norbornene	$Me_2C(t\text{-}BuCp)(Flu)ZrCl_2$	30	3.1	0	0
DMON[a]	$Ph_2C(Cp)(Flu)ZrCl_2$	50	7.0	0.02	0.14
DMON[a]	$Ph_2C(Cp)(Ind)ZrCl_2$	50	6.4	0.10	0.64
TMDN[b]	$Ph_2C(Cp)(Flu)ZrCl_2$	50	15.6	0.06	0.94

[a] Dimethanooctahydronaphthalene.
[b] Trimethanododecahydronaphthalene.

and C_2-symmetric zirconocene catalysts. The C_s-symmetric catalysts, such as $Me_2C(Cp)(Flu)ZrCl_2$–$[Al(Me)O]_x$ and $Ph_2C(Cp)(Flu)ZrCl_2$–$[Al(Me)O]_x$, appeared to be very suitable for yielding amorphous copolymers (transparent) with very high glass transition temperatures (ca 230 °C) for an ethylene/phenyldimethanooctahydronaphthalene copolymer containing more than 30 mol.-% of the cycloolefin (and with a molecular weight greater than 100×10^3). This is the highest glass transition temperature observed for ethylene/cycloolefin copolymers and is superior to other materials used today for polymer fibres and discs [472].

3.9.4 Copolymerisation of Ethylene and α-Olefins with Carbon Monoxide

Carbon monoxide, which is characterised by a carbene-like structure

does not homopolymerise but readily undergoes copolymerisation with olefins in the presence of coordination catalysts based on Ni(II) [473,474] and Pd(II) [475,476] complexes [241].

In the case of ethylene/carbon monoxide copolymerisation with nickel- and palladium-based catalysts, a strictly alternating high molecular weight copolymer is formed (average molecular weight in the range 10×10^3–100×10^3). When more developed catalysts are used, the copolymerisation conditions can be mild: a temperature of 25 °C combined with a pressure of ca 20 atm. The obtained copolymer, poly(ethylene-*alt*-carbon monoxide), poly(1-oxytrimethylene)

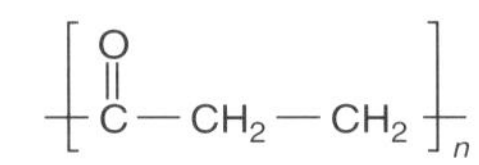

is known as polyketone. It is characterised by a high melting point (about 260 °C), very high crystallinity and is insoluble in all the usual organic solvents. Owing to the economically attractive feedstock, especially carbon monoxide, which contributes half to the copolymer weight, the copolymer offers interesting engineering resin applications; it can be potentially used as a starting material for the preparation of other functionalised polymers [241,250,477].

Catalysts for ethylene/carbon monoxide copolymerisation were initially obtained from Ni(II) derivatives, such as $K_2Ni(CN)_4$ and $(n\text{-}Bu_4N)_2Ni(CN)_4$, and Pd(II) derivatives, such as $[(n\text{-}Bu_3P)PdCl_2]_2$, $Pd(CN)_2$ and $HPd(CN)_3$, often combined with alcohol or protonic acid as a cocatalyst [241]. It must be emphasised that, in contrast to titanium-, zirconium- or vanadium-based catalysts, nickel- and palladium-based catalysts tolerate polar functional groups (including hydroxyl, carboxylic and sulfonic groups) and do not require aluminium alkyl activators. Lately, much effort has been made to improve catalysts for the production of perfectly alternating copolymers of ethylene with carbon monoxide; developed Pd(II)-based catalysts are formed by the combination of a suitable Pd(II) complex (with a phosphorous and/or nitrogen ligand) with tetrafluoroboric or other acid and by the equimolar combination of a bidentate phosphine ligand, preferably 1,3-bis(diphenylphosphino)propane, with a Pd(II) species, in which the counter-anions are weakly coordinating [107,478–481]. Examples of such catalysts are as follows: $[(R_3P)_nPd(MeCN)_{4-n}]^{+2}2[BF_4]^-$–MeOH [475], *trans*-$[(Chx_3P)_2Pd(H)(H_2O)]^+$ $[BF_4]^-$–MeOH (in tolune, $ClCH_2CH_2Cl$ or MeOH) [482], $Pd(Acac)_2/PPh_3$–*p*- $MeC_6H_4SO_2OH$ (in MeOH) [483], $(p\text{-}MeC_6H_4SO_2O)_2Pd(MeCN)_2/Ph_2P(CH_2)_3PPh_2$ (in MeOH) and $Pd(OAc)_2/Ph_2P(CH_2)_3PPh_2$–$CF_3SO_2OH$ (in MeOH) [481].

Active sites present in palladium-based catalysts, which promote the alternating insertion of coordinating comonomers, ethylene to the acyl Pd–C(O) bond and carbon monoxide to the alkyl Pd–CH_2 bond, appear to be cationic Pd(II) species with a square planar, formally d^0 8-electron structure, $[L_2(M)Pd(II)–P_n]^+$, accompanied with weakly coordinating counter-anions [478–480,484].

The formation of the initiating $[Pd(II)–OMe]^+$ and $[Pd(II)–H]^+$ sites is shown respectively by the following schemes [107]:

$$Pd^{2+} + CH_3OH \longrightarrow [Pd-OCH_3]^+ + H^+ \tag{77}$$

$$[Pd-OCH_3]^+ \longrightarrow [Pd-H]^+ + O{=}CH_2 \tag{78}$$

In the absence of methanol, the Pd^{+2} species is activated by the water–gas reaction (traces of water are usually present in the reaction medium) according to the scheme [107]

$$2Pd^{2+} + CO + H_2O \longrightarrow 2[Pd-H]^+ + CO_2 \tag{79}$$

The initiating $[Pd(II)–H]^+$ species can also be already present in the catalyst precursor, e.g. as in $HPd(CN)_3$ [241].

Initiation reactions in ethylene/carbon monoxide copolymerisation systems with palladium-based catalysts are presented by the schemes [107]

$$[Pd-OCH_3]^+ + CO \longrightarrow [Pd-\overset{\displaystyle O}{\overset{\|}{C}}-OCH_3]^+ \tag{80}$$

$$[Pd-H]^+ + CH_2{=}CH_2 \longrightarrow [Pd-CH_2CH_3]^+ \tag{81}$$

Propagation reaction (cross-propagation) proceeds as follows [107]:

(82)

(83)

It should be added, for the sake of correctness, that ethylene and carbon monoxide insertions in initiation and propagation steps, shown in schemes (80) to (83), are preceded by the coordination of ethylene (Figure 2.1) and carbon monoxide (Figure 2.2) respectively:

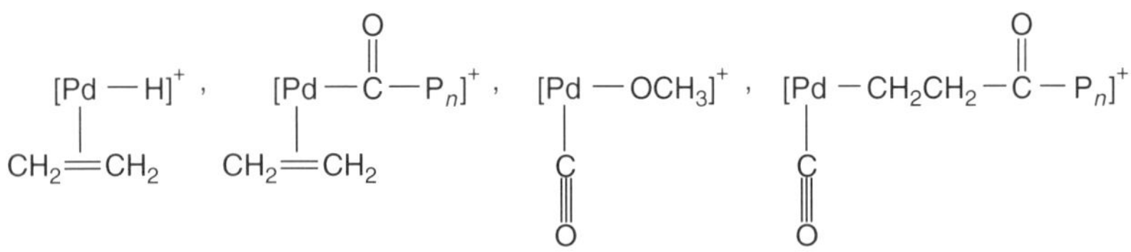

The ethylene/carbon monoxide copolymerisation, involving alternating insertion reactions in palladium-acyl–olefin complexes [scheme (82)] and palladium-alkyl–carbon monoxide complexes [scheme (83)], was found to proceed according to the chain migratory insertion mechanism [478–480]. Previously unobserved acyl migration to coordinated ethylene [scheme (82)] as well as alkyl migration to coordinated carbon monoxide [scheme (83)] (and to ethylene, analogously to scheme (2) in chapter 2, when ethylene is used in an excess with respect to carbon monoxide) has recently been pointed out [484]. A relatively high barrier for alkyl olefin migratory insertion (analogous to scheme (2) in chapter 2) with respect to alkyl carbonyl migratory insertion [scheme (83)] plays a role in enforcing the tendency of the ethylene/carbon monoxide copolymerisation to cross-propagate during chain growth [484] in addition to the preferred coordination of carbon monoxide (relative to ethylene) at the Pd–CH_2 species [478].

Chain transfer/termination steps involve reactions with methanol [schemes (84) to (86)], spontaneous transfer [scheme (87)] or reaction with acids [scheme (88)] [107]:

(84)

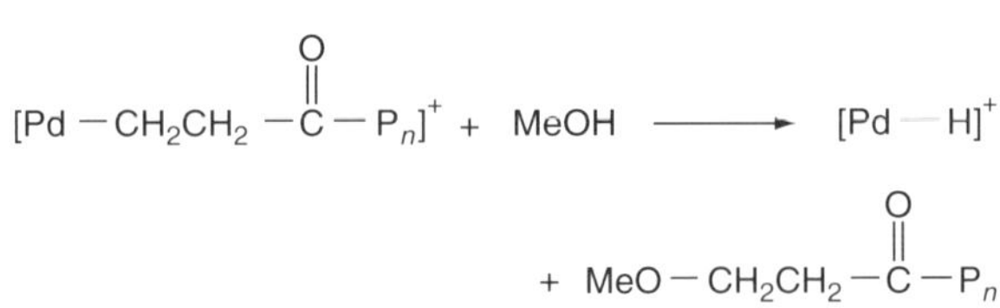
(85)

$$[\mathrm{Pd-CH_2CH_2-\overset{\overset{\displaystyle O}{\|}}{C}-P}_n]^+ + \mathrm{MeOH} \longrightarrow [\mathrm{Pd-OMe}]^+ + \mathrm{CH_3CH_2-\overset{\overset{\displaystyle O}{\|}}{C}-P}_n \tag{86}$$

$$[\mathrm{Pd-CH_2CH_2-\overset{\overset{\displaystyle O}{\|}}{C}-P}_n]^+ \longrightarrow [\mathrm{Pd-H}]^+ + \mathrm{CH_2{=}CH-\overset{\overset{\displaystyle O}{\|}}{C}-P}_n \tag{87}$$

$$[\mathrm{Pd-CH_2CH_2-\overset{\overset{\displaystyle O}{\|}}{C}-P}_n]^+ + \mathrm{H}^+ \longrightarrow \mathrm{Pd}^{+2} + \mathrm{CH_2CH_2-\overset{\overset{\displaystyle C}{\|}}{C}-P}_n \tag{88}$$

Palladium-based catalysts for ethylene/carbon monoxide copolymerisation are very effective; high rates with conversions of more than 1×10^6 mol of ethylene and carbon monoxide per Pd active site were obtained [481].

The terpolymerisation of ethylene, propylene and carbon monoxide leads to corresponding alternating olefin/carbon monoxide terpolymers, in which propylene units randomly replace ethylene units along the chain. These terpolymers are also characterised by a high melting point; however, the melting point is lower than that of the ethylene/carbon monoxide copolymer (257 °C). The lowering of the melting point is directly proportional to the number of propylene units built into the terpolymer chain. For instance, at 6 and 17 wt-% propylene enchainment, the melting points of the terpolymers are around 220 and 170 °C respectively [481].

When propylene or other α-olefin is subjected to the copolymerisation with carbon monoxide, three different regiochemical arrangements are possible within an olefin diad [485]. The carbonyl group can be embedded between two primary carbon atoms (tail-to-tail), between a secondary and a primary carbon atom (head-to-tail) or between two secondary carbon atoms (head-to-head). These three different regiochemical situations are easily determinable by ^{13}C NMR spectroscopy in the region of resonances assignable to the carbonyl group [485–488]. In the case of propylene, additionally a 1,3-enchainment of monomeric units is possible in copolymers with carbon monoxide [481]. Catalysts derived from the L_2PdX_2 or $(L–L)PdX_2$ precursors, which are effective for the copolymerisation of ethylene and carbon monoxide, display almost no regiochemical control when used for propylene/carbon monoxide copolymerisation. In fact, the ratio among the different diads observed for the systems containing 1,3-propanediylbis(diphenylphosphine), $Ph_2P(CH)_{23}PPh_2$, is about 1:2.6:1 [485]. However, the use of more basic 1,3-propanediylbis(dialkylphosphine) ligands makes it possible to obtain an alternating regioregular (head-to-tail) propylene/carbon monoxide copolymer, poly(1-oxo-2-methyltrimethylene) [489]:

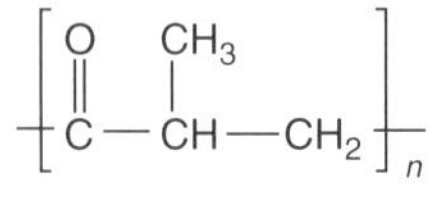

Steric control of the propylene/carbon monoxide copolymerisation, catalysed by palladium-based complexes, towards the production of overwhelmingly isotactic copolymers is possible, particularly when catalysts with a palladium central atom bearing optically active atropisomeric ligands are used. The stereoregularity of the formed propylene/carbon monoxide copolymer is determined by the enantioface selection of propylene during the propagation [489], very much like Ziegler–Natta catalysis of propylene homopolymerisation [385]. A concentration of isotactic tetrads in optically active poly(1-oxo-2-methyltrimethylene) obtained with a Pd-based catalyst, containing (*R*)-(6,6′-dimethoxybiphenyl-2,2′-diyl)bis(dicyclohexylphosphine) as the modifying ligand, was evaluated to be higher than 90%; this corresponds to an enantioface selection by the catalyst of ca 98% [489].

According to the literature data available, no catalyst has yet been found that might make it possible to obtain a prevailingly syndiotactic propylene/carbon monoxide copolymer.

Under certain conditions of propylene/carbon monoxide copolymerisation with certain catalysts, poly(spiroketal) structural units

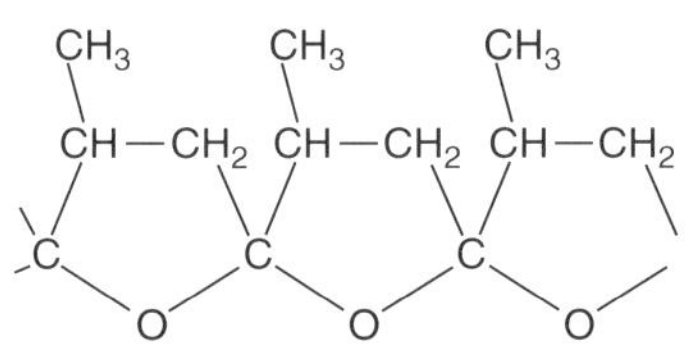

have been found to be present in the copolymer chains formed. This very peculiar aspect of the copolymerisation was paid special attention only when the first propylene/carbon monoxide copolymer with exclusively poly(spiroketal) structural units, poly[spiro-2,5-(3-methyltetrahydrofuran)], was obtained [490].

Different mechanistic interpretations of the formation of an alternating propylene/carbon monoxide copolymer of poly(spiroketal) structure were considered [107, 478, 480, 481, 489]. Any reasonable proposal, however, needs to take into account the nature of the end groups in the copolymer chains. To date this has not been possible owing to the low solubility of the copolymer in solvents other than hexafluoroisopropanol; however, this solvent, probably because of its acidic nature, causes transformation of the poly(spiroketal) structure into an isomeric poly(ketone) structure [489]. The formation of a cyclic polymeric structure could be favoured by minor entropy loss due to the intramolecularity of the process [480,481] and by the peculiar conformational situation of the poly(ketone) structure [491].

Under most conditions, only the simple poly(propylene ketone) is formed in propylene/carbon monoxide alternating copolymerisation. Isomerisation of poly(ketone) to poly(spiroketal) can occur, and it may be assisted by cationic palladium species and protonic acids. It must be emphasised that a low reaction temperature favours the formation of a spiroketal structure [107]. At a temperature above the ceiling temperature, the poly(spiroketal) depolymerises to the more flexible and entropically favoured poly(ketone) [481].

Palladium-based catalysts bearing chiral ligands have also been found to be capable of the stereospecific copolymerisation of allylbenzene [492] and its derivatives [493] with carbon monoxide; the formed copolymers appeared to be of an alternating, isotactic structure.

3.10 Cyclopolymerisation of Non-conjugated α,ω-Diolefins

Non-conjugated α,ω-diolefins subjected to polymerisation with Ziegler–Natta catalysts, such as $TiCl_4$–$AlEt_3$ and $TiCl_4$–$Al(i\text{-}Bu)_3$, were found in the 1950s to undergo polymerisation, which generally leads to the formation of polymers containing cyclic repeating units [2, 46, 494]. For instance, using heterogeneous titanium-based catalysts, monomers such as 1,5-hexadiene and 1,6-heptadiene were cyclopolymerised to crystalline, benzene-soluble polymers, the chains of which contained cyclic rings:

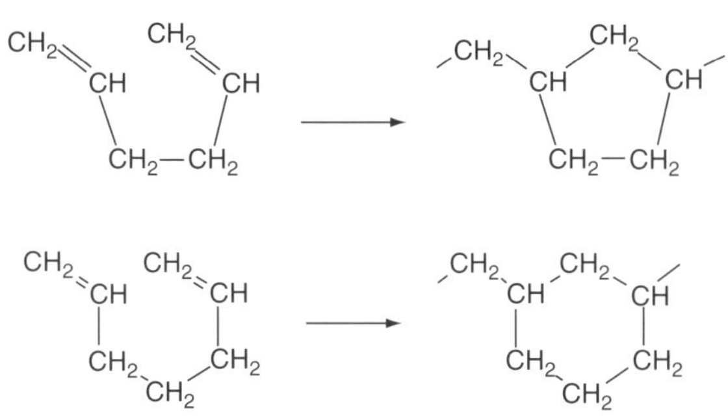

In all cases, linear polymers were obtained with chains that consisted predominantly of 1-methylene-3-cyclopentyl or 1-methylene-3-cyclohexyl units, respectively, and were characterised by a chain planar zigzag conformation [446, 495, 496].

Cyclopolymerisation may also concern the formation of polymers containing bicyclic repeating units. This is the case of the cyclopolymerisation of 3-vinyl-1,5-hexadiene, which yields polymers with 1-methylene-3-(2,5-methanocyclohexyl) units [494]:

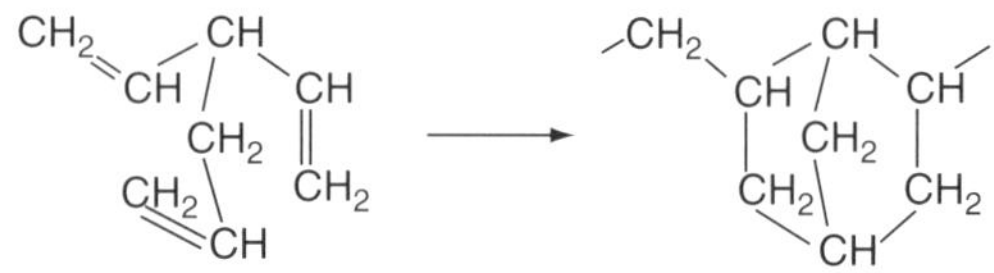

Cyclopolymerisation leading to polymers with monocyclic units in the main chain proceeds in two steps: the first step involves a 1,2-insertion of the coordinated α,ω-diolefin via one olefinic bond, and the second step, which is a ring-closing reaction, involves an intramolecular insertion of the other olefinic bond undergoing coordination; scheme (89) presents both steps for 1,5-hexadiene cyclopolymerisation leading to a cycloaliphatic polymer with poly(methylene-1,3-cyclopentane) structural units [30,450,497]:

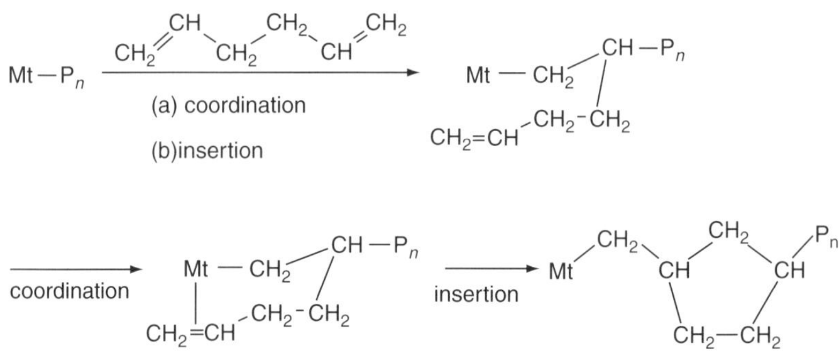

(89)

In the case of the formation of a cycloaliphatic polymer with bicyclic units in the main chain, such as the polymer with poly[methylene-1,3-(2,5-methanocyclohexane)] structural units obtained in the cyclopolymerisation of 3-vinyl-1,5-hexadiene, the intramolecular insertion, analogous to that presented in scheme (89), is followed by another intramolecular insertion involving a vinyl substituting group:

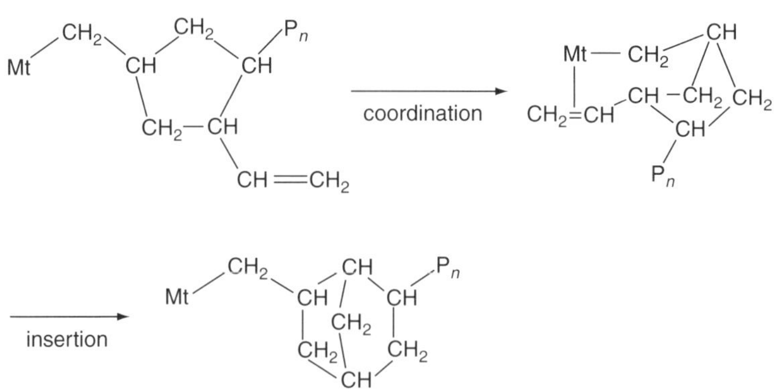

(90)

In view of the observed inactivity of α,α-disubstituted olefins towards polymerisation with Ziegler–Natta catalysts, it is interesting to note that α,ω-diolefins substituted at unsaturated carbon atoms, such as e.g. 2,5-dimethyl-1,5-hexadiene, also undergo cyclopolymerisation, analogously to unsubstituted parent monomers [2,446]. This can be interpreted in terms of a reaction pathway analogous to that shown in scheme (89). The insertions in the cyclopolymerisation appear to be facilitated by the nature of such a process.

Over the past few years, homogeneous Ziegler–Natta and related metallocene-based catalysts have been applied successfully for the cyclopolymerisation of α,ω-diolefins such as 1,5-hexadiene, 2-methyl-1,5-hexadiene, 1,6-heptadiene and 1,7-octadiene by Waymouth *et al.* [497–506], which has advanced remarkably the development of cycloaliphatic polymers. High molecular weight, highly stereoregular cycloaliphatic polymers, poly(methylene-1,3-cykloalkane)s, of various types of stereoisomerism have been obtained with the use of zirconocene-based catalysts. Titanocenes and hafnocenes, although typically of diminished activity, appeared also to be effective for cyclopolymerisation. For the activation of group 4 metallocenes, $[Al(Me)O]_x$, $B(C_6F_5)_3$ and $[PhNHMe_2]^+$ $[B(C_6F_5)_4]^-$ have been applied as cyclopolymerisation catalysts [497].

3.10.1 Stereoisomerism of Cycloaliphatic Polymers

Cycloaliphatic polymers such as poly(methylene-1,3-cykloalkane)s

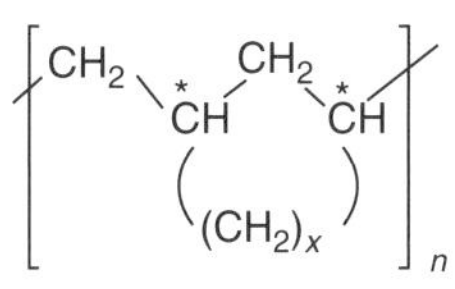

are capable of exhibiting stereoisomerism because of the presence of the two C* atoms at which the polymer main chain enters and leaves each ring. Four stereoregular structures are possible in such a case; Figure 3.48 shows the stereoisomerism of the 1,5-hexadiene cyclopolymer, poly(methylene-1,3-cyclopentane) [497].

The microstructure of the discussed cycloaliphatic polymers concerns the *cis–trans* geometrical isomerism of the rings and the relative stereochemistry between the rings. A modified Bovey *m–r* nomenclature [507] provides a useful description of the microstructure of poly(methylene-1,3-cycloalkane)s, where capital letters (*M* for *mesogenic*, *R* for *racemic*) denote the stereochemistry of the rings and lower case letters (*m* and *r*) denote the relative stereochemistry between the rings [503]. Therefore, *cis*-isotactic, *trans*-isotactic, *cis*-syndiotactic and *trans*-syndiotactic cyclopolymers may be formed. As in many other cases, ^{13}C NMR spectroscopy reveals information about both the tacticity of the polymer and the ratio of *cis* to *trans* rings.

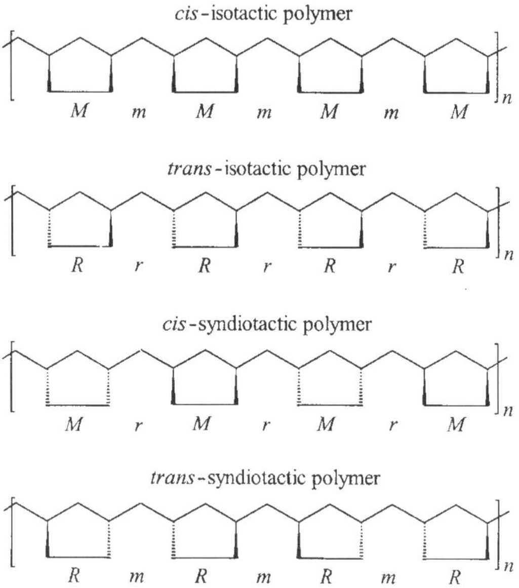

Figure 3.48 Stereoisomerism of poly(methylene-1,3-cyclopentane)

It is important to note that high molecular weight *trans*-isotactic poly(methylene-1,3-cyclopentane) contains no mirror or mirror glide planes of symmetry and is thus chiral by virtue of its main chain stereochemistry (it exhibits optical activity); this is in contrast to high molecular weight polypropylene and other poly(α-olefin)s, which contain an effective mirror plane perpendicular to the molecular axis in the middle of the molecule and are thus achiral [30,497].

3.10.2 Stereoregulation Mechanism

It has been reported [497] for cyclopolymerisations with metallocene-based catalysts that the enantioface selectivity of the insertion of the first olefinic bond in the α,ω-diolefin determines the relative stereochemistry between the rings (the occurrence of m or r diads), i.e. the tacticity of the cyclopolymer; the diastereoselectivity of the subsequent cyclisation involving the remaining olefinic bond determines, on the other hand, the relative stereochemistry in the rings (the occurrence of M or R diads), i.e. the *cis–trans* geometrical isomerism of the rings.

Heterogeneous Ziegler–Natta catalysts were found [446,495,496] to polymerise α,ω-diolefins, yielding polymers with predominantly occurring *cis*-(1-methylene-3-cyclopentyl) units. The *cis–trans* diastereoselectivity, which reflects the catalyst structure, can hardly be controlled with heterogeneous catalysts, but it is strongly influenced by substituents at the cyclopentadienyl ligands in homogeneous catalysts [498]. Cyclopolymerisation of 1,5-cycloxexadiene in the presence of the Cp_2MtX_2–$[Al(Me)O]_x$ catalyst (Mt=Ti, Zr, Hf;X=Cl, Me) leads to atactic polymers with predominant *trans* rings, whereas cyclopolymerisation in the presence of a more hindered catalyst such as $Cp^*{}_2MtX_2$–$[Al(Me)O]_x$ yields atactic polymers containing predominantly *cis* rings [497]. Conformational modelling concerning ligand effects on the diastereoselectivity of the cyclisation indicates that, between two cyclopentadienyl ligands, the lowest energy conformation for cyclisation is a chair-like conformation [504]. For this conformation, placement of the growing polymer chain in a pseudo-equatorial position results in a *trans* ring closing. Increasing the steric hindrance of the ligands results in non-bonded interactions between the incipient ring and the ligands, restricting the allowed conformations of the ring. The steric bulk of two pentamethylcyclopentadienyl rings disfavours the chair-like conformer relative to a boat-like conformer; consequently, placement of the growing polymer chain in a pseudo-equatorial position obliges *cis* ring cyclisation (Figure 3.49) [497].

A notable observation in the cyclopolymerisation of 1,5-hexadiene is that catalysts having fully substituted cyclopentadienyl ligands, irrespective of whether they are obtained from *ansa*-metallocene (chiral or achiral) or non-bridged metallocene precursors, show a high selectivity for the cyclisation; the more sterically congested the active site becomes, the smaller is the percentage of *trans* rings and the greater percentage of *cis* rings in the cyclopolymers produced. This assumes, in some oversimplification, that the diastereoselectivity is completely independent of the chirality of the catalyst [497].

The *cis*/*trans* ratio of the cyclic structures strongly influences the melting points of the cyclopolymers obtained; polymers with predominantly occurring *trans* rings are waxes exhibiting melting points not exceeding 70 °C, while polymers containing more than 90% *cis* cyclic units are crystalline and melt at 189 °C [30].

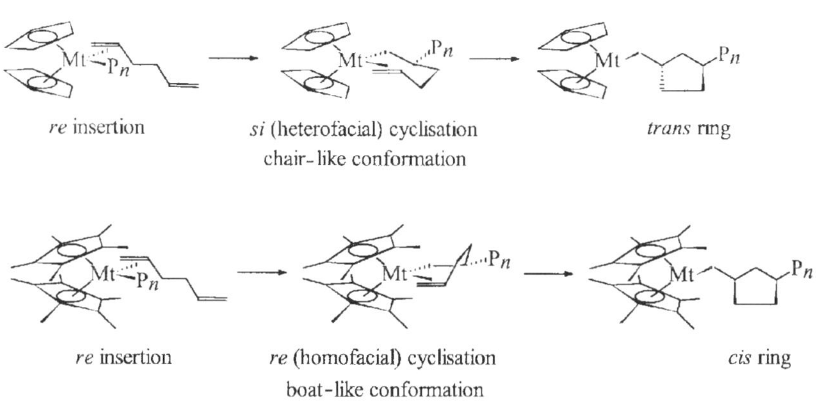

Figure 3.49 Diastereoselectivity of the cyclisation step in the cyclopolymerisation of 1,5-hexadiene with metallocene-based catalysts

The cyclopolymerisation of unsymmetrical α,ω-diolefins such as 2-methyl-1,5-hexadiene in the presence of catalysts such as $Cp_2{}^*ZrMe_2$–$[Al(Me)O]_x$, $[Cp_2{}^*ZrMe]^+[B(C_6F_5)_4]^-$ or $[Cp_2{}^*ZrMe]^+[MeB(C_6F_5)_3]^-$ yields highly regioregular cyclopolymers [501]. The perfectly head-to-tail linked monomeric units in the formed poly[methylene-1,3-(1-methylcyclopentane)] arises from the chemoselective insertion of the less hindered terminus of 2-methyl-1,5-hexadiene into the active Mt–P_n bond, followed by cyclisation involving the insertion of the disubstituted olefinic bond (Figure 3.50) [497]. The insertion of the disubstituted olefinic bond is made easier by its intramolecular nature.

The cyclopolymerisation of 1,5-hexadiene leads to polymers of substantially higher molecular weights than the polymerisation of 1-hexene with the same catalysts [498], undoubtedly owing to some steric hindrance of the H_β atom transfer that usually terminates the growth of a polymer chain [30].

The symmetry properties of cycloaliphatic polymers are such that polymers with certain microstructures, e.g. *trans*-isotactic poly(methylene-1,3-cyclopentane), are chiral; therefore, the cyclopolymerisation of α,ω-diolefins presents new opportunities in enantioselective polymerisation. The cyclopolymerisation of 1,5-hexadiene carried out with isospecific as well as *trans* selective catalysts of C_2 symmetry, such as methylaluminoxane-activated resolved (1*R*)-(Thind $CH_2)_2$Zr 1,1′-bi-2-naphtholate, yielded optically active *trans*-isotactic poly(-methylene-1,3-cyclopentane). The cyclopolymerisation with the (1*S*) enantiomer of the catalyst gave an enantiomeric polymer [505]. On the basis of analysis of ^{13}C NMR spectra, the degree of enantioface selectivity for this cyclopolymerisation was estimated to be of 91% [503,505].

The cyclopolymerisation of 1,5-hexadiene, 1,6-heptadiene and 1,7-octadiene with the *rac*.-(Thind$CH_2)_2$Zr 1,1′-bi-2-naphtholate–$[Al(Me)O]_x$ catalyst at room temperature yields poly(methylene-1,3-cycloalkane)s containing 67, 50 and 22% *trans* rings respectively; this testifies that the diastereoselectivity of the ring-closing step is quite dependent on the pendant olefin length [506]. Such a dependence has been interpreted in terms of competition between enantiomorphic site control and conformational control in the cyclisation step (Figure 3.51) [497].

Under an exclusively enantiomorphic site control mechanism, the symmetry of the catalyst would mandate homofacial insertion and cyclisation to yield *cis*

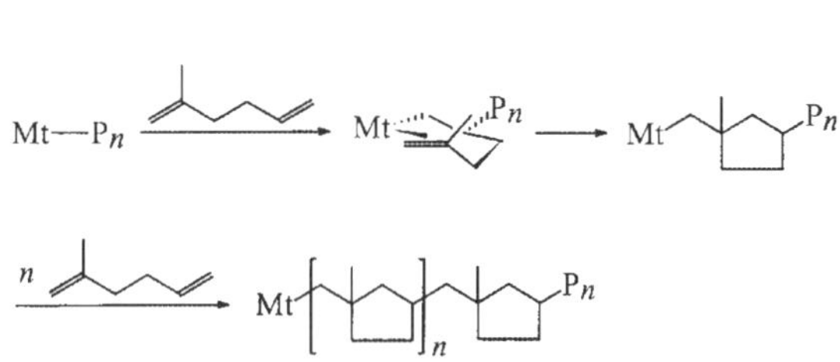

Figure 3.50 Cyclopolymerisation of 2-methyl-1,5-hexadiene with metallocene-based catalysts

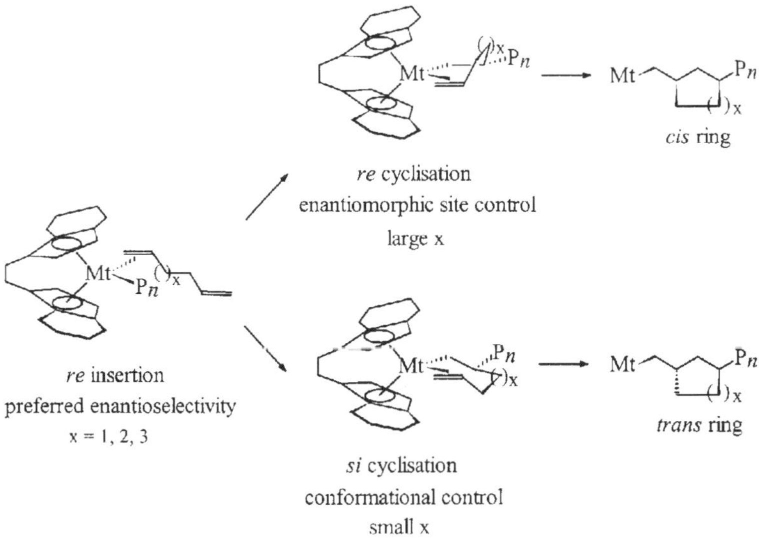

Figure 3.51 Enantiomorphic site control versus conformational control in the cyclisation step during cyclopolymerisation of α, ω-diolefins: 1,5-hexadiene (x=1), 1,6-heptadiene (x=2) and 1,7-octadiene (x=3) in the presence of a metallocene-based catalyst of C_2 symmetry

rings. On the other hand, a pure conformational control mechanism would imply that the conformational preference of the incipient ring determines the diastereoselectivity. With increasing length of the monomer molecule, the added degrees of freedom of the incipient ring should diminish the conformational preferences of the ring. In the limit of large x, the α, ω-diolefin would behave as two independent olefinic monomers, and for an isospecific catalyst one would expect a high *cis* selectivity, as observed for 1,7-octadiene cyclopolymerisation [497].

3.11 Polymerisation of Functionalised Olefins

Ziegler–Natta catalysts are commonly known as unique catalysts for polymerisations and copolymerisations of ethylene and α-olefins [39–41]. Although these catalysts, especially their new metallocene generations, greatly expand the range and versatility of technically feasible types of polyolefin material [43], relatively little attention has been paid to homopolymers and copolymers with olefins of polar vinyl monomers, which might be obtained using Ziegler–Natta and related coordination catalysts.

As with poly(α-olefin)s, the properties of polymers derived from polar monomers may be highly dependent on the stereoregularity of the polymer backbone. Moreover, copolymers of polar monomers with olefins may exhibit some advantageous properties with respect to those characteristic of non-polar polyolefins. The lack of reactive groups in polyolefins significantly limits many of their end uses, especially where adhesion, paintability, printability or compatibility is required.

In general, there are two ways to functionalise polyolefins: via direct copolymerisation of an olefin with a polar monomer and via chemical modification of preformed polymers [508–514].

Although direct copolymerisation of an olefin and polar monomer has succeeded, the achievements have been rather scant. Only a limited number of polymerisation systems with polar monomers could yield copolymers via coordination polymerisation involving monomer insertion into an active metal–carbon bond.

Ziegler–Natta catalysts are characterised by low activities in both olefin/polar monomer copolymerisation and polar monomer homopolymerisation. Homopolymers of polar monomers appeared in most cases to exhibit properties that were very similar to those of homopolymers obtained by radically initiated polymerisations; this testifies to the radical mechanism that operates in such polymerisations [28].

3.11.1 Coordination Homopolymerisation and Copolymerisation of Functionalised α-Olefins with Ethylene and α-Olefins

One of the long-standing challenges is to develop catalysts for incorporation of polar monomers via coordination polymerisation involving the insertion mechanism. Since most titanium-based catalysts are strong Lewis acids, the addition of polar monomers, which are Lewis bases, results in these monomers competing successfully with olefin monomers for vacant coordination sites; as a consequence, the catalysts become severely poisoned with respect to coordination polymerisation via the insertion mechanism. Moreover, small amounts of acidic hydrogen are sufficient to quench olefin polymerisation by cleaving the transition metal–alkyl bonds [450]. Although heteroatoms usually hinder the insertion reaction of C=C bonds into Mt–C bonds in coordination catalysts that polymerise olefins, some heteroatom-containing monomers have been homopolymerised and/or copolymerised with olefins according to the insertion mechanism, using heterogeneous and homogeneous, metallocene-based, Ziegler–Natta catalysts. This has been accomplished by minimising the interaction between the monomer heteroatom and the catalyst by applying various methods allowing active sites to retain their susceptibility for the insertion reaction: the heteroatom in the monomer is separated from the C=C bond by more than one methylene group, the heteroatom in the monomer and the Mt–C bond in the catalyst are sterically shielded, the electron-donating ability of the heteroatom in the monomer is lowered by attaching particular groups to it, the heteroatom has low reactivity, the heteroatom is complexed with the catalyst activator, and the labile hydrogen atom as in the hydroxyl group reacts with the catalyst activator or another compound [2,241].

An interesting new approach concerning the latter involves the copolymerisation of olefins with borane-containing monomers, such as ω-alkenylboranes, in the presence of heterogeneous and homogeneous Ziegler–Natta catalysts; the resulting copolymers are then converted to various functional polymers in post-polymerisation processes [515–518]. This may be illustrated by the copolymerisation of ethylene and α-olefins such as propylene, 1-butene or 1-octene with 5-hexenyl-9-borabicyclo[3.3.1]nonane, leading to polymers containing various functional groups, e.g. containing OH or I functions [518]:

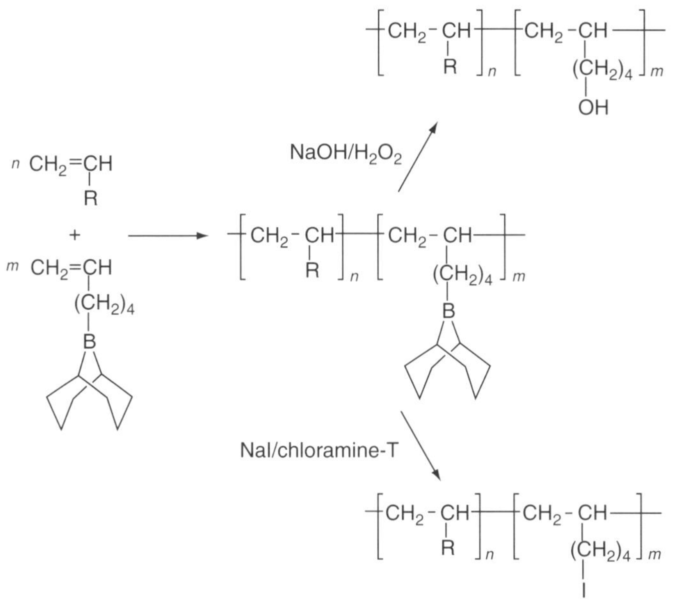

(91)

The use of borane-containing monomers clearly presents an effective and general approach in the functionalisation of polyolefins, which has the following advantages: stability of the borane moiety to coordination catalysts, solubility of borane compounds in hydrocarbon solvents (such as hexane and toluene) used as the polymerisation medium, and versatility of borane groups, which can be transformed to a remarkable variety of functionalities as well as to free radicals for graft-form polymerisations. The functionalised polymers are very effective interfacial modifiers in improving the adhesion between polyolefin and substrates and the compatibility in polyolefin blends and composites [518].

Advances towards the goal of polymerising polar monomers with coordination catalysts have been made with metallocene-based catalysts, especially aluminium-free cationic metallocene catalysts such as $[Cp^*{}_2ZrMe]^+[B(C_6F_5)_4]^-$ or $[Cp^*{}_2ZrMe]^+[MeB(C_6F_5)_3]^-$. Waymouth *et al.* [500] found these catalysts to be considerably more tolerant of polar functional groups than either heterogeneous or homogeneous Ziegler–Natta catalysts (i.e. those activated by aluminium alkyls or methylaluminoxane respectively). Functionalised olefinic monomers, however, must be carefully chosen in the case of both heterogeneous and homogeneous catalysts to avoid catalyst deactivation [30].

Another approach towards functionalised polyolefins is the generation of olefin polymers capped by polar end groups. Polymers of this kind are principally accessible through living polymerisations. Diblock copolymers of olefin or α, ω-diolefin with polar monomer can be obtained via a two-step polymerisation sequence [30,497]. Also, graft copolymers containing polymer segments grafted on to functionalised polyolefins are being developed [519–521].

Functionalised α-olefins capable of undergoing insertion polymerisation with Ziegler–Natta catalysts are, in principle, monomers in which the heteroatom (X) does not electronically interact with the double bond to be polymerised; in such monomers, the heteroatom is separated from the double bond: $CH_2{=}CH{-}(CH_2)_x{-}X$ [326,384,518,522–528]. Monomers with the heteroatom directly bound to the double bond, i.e. those of the $CH_2{=}CH{-}X$ type, may also undergo polymerization, but when the heteroatom is silicon or tin (X= Si, Sn) [522–526]. Representative examples of the insertion polymerisation of functionalised α-olefins and their copolymerisation with ethylene and α-olefins in the presence of heterogeneous Ziegler–Natta catalysts are shown in Table 3.7 [2,241,326,384,518,522–528].

Among the nitrogen-containing functionalised α-olefins, only tertiary amines have been polymerised and copolymerised (involving the insertion mechanism), with the exception of allyl amines. However, unsaturated amines with two methyl or ethyl substituents at the nitrogen atom, such as 5-*N*,*N*-dimethylamino-1-pentene and 5-*N*, *N*-diethylamino-1-pentene, cannot be polymerised to high polymers; these monomers can be polymerised using Ziegler–Natta catalysts only after complexing with a Lewis acid, e.g. diethylaluminium chloride. The polymerisation of functionalised α-olefins containing tertiary amine groups is influenced by the nature of both the monomer and the alkyl group bound to the Al atom in the catalyst activator. Polymerisation is favoured when both the nitrogen and the aluminium atoms are shielded by bulky alkyl groups [523,526].

Among the oxygen-containing functionalised α-olefins, those monomers that contain $O{-}SiMe_3$ groups undergo insertion polymerisation; the complexing ability of the O atom in these monomers is lower owing to both the steric hindrance caused by the $SiMe_3$ group and the presence of d–π interactions between the Si atom and the O atom. After hydrolysis with solutions of mineral acids in alcohols, the polymers obtained yield the respective polyalcohols [523,526]. Hydroxyl group-containing monomers can be transformed into respective boron-substituted [518] or aluminium-substituted [528] derivatives, which are polymerisable with a Ziegler–Natta catalyst; protecting groups can be removed from the obtained polymers and thus hydroxyl functionalities may be easily recovered.

Table 3.7 Insertion homopolymerisation and copolymerisation of functionalised α-olefins with ethylene and α-olefins in the presence of heterogeneous Ziegler–Natta catalysts

Monomer	Catalyst	Reference
$CH_2{=}CH{-}SiH_2Et$	$VCl_3{=}AlEt_3$	[522]
$CH_2{=}CH{-}SiMe_3^a$	$TiCl_4{-}AlEt_3$	[523]
$CH_2{=}CH{-}CH_2{-}SiH_3$	$TiCl_4{-}AlEt_3$	[524,525]
$CH_2{=}CH{-}CH_2{-}SiMe_3$	$TiCl_4{-}AlEt_3$	[524,525]
$CH_2{=}CH{-}(CH_2{-}SiMe_3$	$TiCl_4{-}AlEt_3$	[524,525]
$CH_2{=}CH{-}(CH_2)_x{-}N(i\text{-}Pr)_2; x{=}2, 3, 5, 9$	$\delta\text{-}TiCl_3.\frac{1}{3}AlCl_3{-}Al(i\text{-}Bu)_2Cl$	[526]
$CH_2{=}CH{-}(CH_2)_3{-}N(i\text{-}Pr)_2$	$\delta\text{-}TiCl_3{-}Al(i\text{-}Hx)_2Cl$	[326]
$CH_2{=}CH{-}(CH_2)_3{-}N(i\text{-}Bu)_2$	$\delta\text{-}TiCl_3{-}Al(i\text{-}Hx)_2Cl$	[384]
$CH_2{=}CH{-}(CH_2)_3{-}NMe_2{}^b$	$\delta\text{-}TiCl_3{-}AlEt_2Cl$	[526]
$CH_2{=}CH{-}(CH_2)_3{-}O{-}NEt_2{}^b$	$\delta\text{-}TiCl_3{-}AlEt_2Cl$	[526]
$CH_2{=}CH{-}(CH_2)_3{-}O{-}SiMe_3{}^c$	$\delta\text{-}TiCl_3.\frac{1}{3}AlCl_3{-}Al(i\text{-}Bu)_2Cl$	[526]
$CH_2{=}CH{-}(CH_2)_3{-}O{-}SiMe_3{}^c$	$\delta\text{-}TiCl_3{-}Al(i\text{-}Hx)_3$	[523]
$CH_2{=}CH{-}(CH_2)_9{-}O{-}SiMe_3{}^c$	$\delta\text{-}TiCl_3{-}Al(i\text{-}Hx)_3$	[523]
$CH_2{=}CH{-}(CH_2)_3{-}N(i\text{-}Pr){-}SiMe_3{}^d$	$\delta\text{-}TiCl_3 \cdot \frac{1}{3}AlCl_3{-}Al(i\text{-}Bu)_2Cl$	[526]
$CH_2{=}CH{-}(CH_2)_9{-}I$	$TiCl_3{-}AlEt_2Cl$	[527]
$CH_2{=}CH{-}(CH_2)_4{-}B(C_8H_{15})/CH_2{=}CH{-}R^e$	$TiCl_3{-}AlEt_2Cl$	[518]
$CH_2{=}CH{-}(CH_2)_x{-}C_6H_2(O{-}t\text{-}Bu)_2{-}OAlEt_2/CH_2{=}CH{-}CH_3; x{=}2{-}5^f$	$MgCl_2/TiCl_4{-}AlEt_3/Ph_2Si(OMe)_2$	[528]

[a] Polymer was obtained in a very low yield.
[b] Complexed with a stoichiometric amount of $AlEt_2Cl$.
[c] Polyalcohol was obtained after hydrolysis of the polymer formed.
[d] 5-N-Isopropylamino-1-pentene was obtained after hydrolysis of the polymer formed.
[e] Copolymerisation; R=H, Me, Et, Hx; $B(C_8H_{15})$ = 9-borabicyclo [3.3.1]nonane; in order to recover the OH functionality in the copolymer, the protecting borane-containing group was removed by $NaOH/H_2O_2$ treatment.
[f] Copolymerisation; functionalised α-olefin was derived from the reaction of 4-(ω-alkenyl)-2,6-di-(t-butyl)phenol with triethylaluminium (1:1 molar ratio); in order to recover the OH functionality in the copolymer, the protecting $AlEt_2$ group was removed by HCl/ethanol treatment.

Halogen-containing monomers can undergo insertion polymerisation only when the halogen atom is situated far away from the C=C bond; this concerns ω-halo-α-olefins [527].

A variety of functionalised α-olefins have been polymerised successfully via the insertion mechanism in the presence of borane-activated as well as aluminoxane-activated zirconocene catalysts [500,529]. Methylaluminoxane-activated zirconocene-based Ziegler–Natta catalysts have also been successfully applied for promoting the copolymerisation of functionalised α-olefins with ethylene, propylene and higher α-olefins [518,530,531]. Moreover, functionalised α, ω-diolefins have been cyclopolymerised to the corresponding functionalised poly(methylenecycloalkane)s using cationic metallocenes as catalysts [500]. Respective examples are presented in Table 3.8 [500,518,529–531].

Relative to heterogeneous Ziegler–Natta catalysts, metallocene-based catalysts offer a superior way to polymerise functionalised α-olefins. However,

Table 3.8 Insertion homopolymerisation and copolymerisation of functionalised α-olefins with ethylene and α-olefins in the presence of homogeneous metallocene catalysts

Monomer	Catalyst	Reference
$CH_2=CH–CH_2–O–SiMe_3$	*rac.*-$(IndCH_2)_2ZrMe_2–B(C_6F_5)_3$	[529]
$CH_2=CH–CH_2–N(i\text{-}Pr)_2$	*rac.*-$(IndCH_2)_2ZrMe_2–B(C_6F_5)_3$	[529]
$CH_2=CH–CH_2–PPh_2$	*rac.*-$(IndCH_2)_2ZrMe_2–B(C_6F_5)_3$	[529]
$CH_2=CH–CH_2–CH(CH_3)–O–Si(t\text{-}Bu)Me_2$	$[Cp_2{}^*ZrMe]^+[B(C_6F_5)_4]^-$	[500]
$CH_2=CH–(CH_2)_3–N(i\text{-}Pr)_2$	$[Cp_2{}^*ZrMe]^+[B(C_6F_5)_4]^-$	[500]
$CH_2=CH–(CH_2)_4–B(C_8H_{15})/CH_2=CH–R^{a,b}$	$Cp_2ZrCl_2–[Al(Me)O]_x$	[518]
$CH_2=CH–(CH_2)_4–B(C_8H_{15})/CH_2=CH–R^{a,b}$	*rac.*-$(IndCH_2)_2ZrMe_2 – [Al(Me)O]_x$	[518]
$CH_2=CH–(CH_2)_9–OAl<$ / $CH_2=CH–R^{a,c}$	$(BuCp)_2ZrCl_2 – [Al(Me)O]_x$	[530]
$CH_2=CH–(CH_2)_9–OAl<$ / $CH_2=CH–R^{a,c}$	*rac.*-$(IndCH_2)_2ZrMe_2 – [Al(Me)O]_x$	[530]
$CH_2=CH–(CH_2)_9–OAl<$ / $CH_2=CH–R^{a,c}$	*rac.*-$Me_2Si(Ind)_2ZrCl_2–[Al(Me)O]_x$	[531]
$(CH_2=CH–CH_2)_2CH–O–SiMe_3{}^d$	$[Cp^*{}_2ZrMe]^+[B(C_6F_5)_4]^-$	[500]

[a] Copolymerisation.
[b] $B(C_8H_{15})$ = 9-borabicyclo[3.3.1]nonane.
[c] $CH_2=CH–(CH_2)_9–OAl \Leftarrow CH_2=CH–(CH_2)_9–OH$ pretreated with $[Al(Me)O]_x$.
[d] Cyclopolymerisation; treatment of the resulting cyclopolymer with aqueous HCl yields poly[methylene-3,5-(1-hydroxy)cyclohexane].

several obstacles remain to be overcome, the most serious being the decay of the catalyst with the addition of the polar monomer [531].

The structure of the catalyst plays an important role in enchainment of the functionalised α-olefin. For instance, a bridged catalyst such as *rac.*-$Me_2Si(Ind)_2ZrCl_2–[Al(Me)O]_x$ seems to favour the polymerisation of 10-undecen-1-ol by comparison with a non-bridged zirconocene-based catalyst [531]. Another example shows that catalysts such as [*rac.*-$(ThindCH_2)_2ZrMe]^+[X]^-$ are active for the polymerisation of 5-*N*,*N*-diisopropylamino-1-pentene but not for that of 4-*tert*-butyldimethylsiloxy-1-pentene; the polymer of the former monomer obtained in this manner appeared to be isotactic [30].

3.11.2 Coordination Group-transfer Polymerisation of Alkyl (Meth)acrylates

Neutral organolanthanide metallocenes [532–536] and cationic zirconocene organic derivatives [537] appear to be effective catalysts for the polymerisation of polar vinyl monomers in which the heteroatom is conjugated with the double bond, such as acrylates [$CH_2=CH–C(OR)=O$] and methacrylates [$CH_2=C(Me)–C(OR)=O$].

An achiral organolanthanide, bis(pentamethylcyclopentadienyl)samarium hydride dimer $[(Cp^*{}_2SmH)_2]$, was shown efficiently to catalyse the living, highly syndiospecific polymerisation of methyl methacrylate [532–536]. The complexation of the catalyst with the carbonyl group of the monomer, followed by the 1,4-addition of the Sm–H functionality to the complexed monomer to generate a metal–enolate species, has been proposed. Furthermore, an eight-membered ring intermediate has been postulated to be formed, which should stabilise the enol chain end, also allowing monomer enchainment. The chain terminal binds to the metal in an enol form, while the penultimate methyl methacrylate unit complexes with the metal at its carbonyl group [532]:

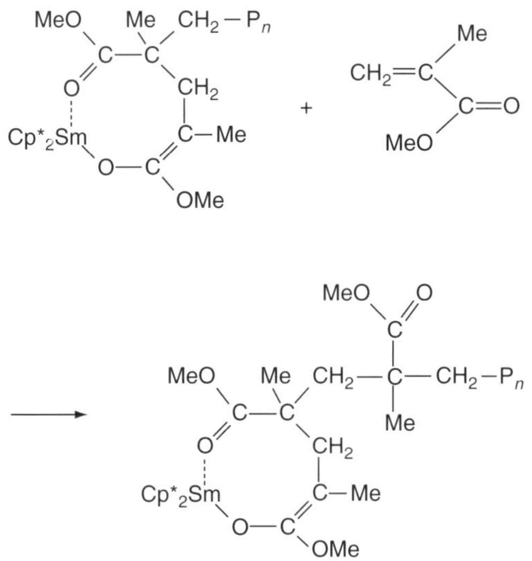

(92)

By using chiral organolanthanide *ansa*-metallocenes for methyl methacrylate polymerisation, highly stereoregular poly(methyl methacrylate)s were obtained; a syndiotactic or isotactic polymer could be synthesised, depending on the kind of metallocene catalyst [536].

The applicability of organolanthanide metallocenes as polymerisation catalysts can also be seen from the results of the block copolymerisation of ethylene and methyl methacrylate. The persistence of the lanthanide–alkyl bond has been utilised to prepare ethylene copolymers with polar poly(methyl methacrylate) blocks. For this purpose, ethylene is introduced as the first monomer into the polymerisation system with the samarocene catalyst, and then methyl methacrylate is polymerised, which leads to block copolymer formation [532–534]:

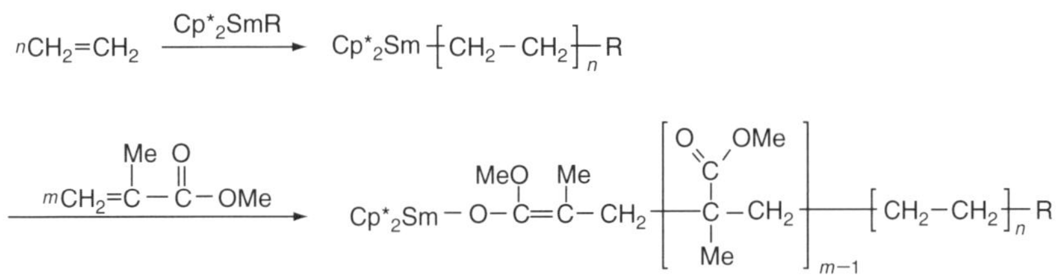

(93)

The application of achiral cationic zirconocene compounds for methyl methacrylate polymerisation, e.g. a mixture of $[Cp^*{}_2ZrMe(THF)]^+[BPh_4]^-$ and $Cp^*{}_2ZrMe_2$ in methylene chloride solution, leads to the formation of syndiotactic poly(methyl methacrylate). The species responsible for propagation are believed to be the bimetallic ones, involving cationic zirconium enolate and neutral zirconocene, which facilitates the process. Propagation is postulated to occur via the Michael reaction between the coordinating monomer and the cationic enolate [537]:

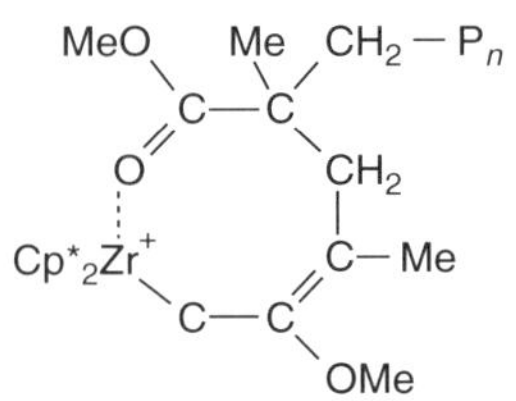

3.11.3 Radical Homopolymerisation and Copolymerisation of Polar Monomers with Olefins in the Presence of Modified Ziegler–Natta Catalysts

In the case of polar monomers, in which heteroatoms are directly bound to the double bond ($CH_2{=}CH{-}X$) or are conjugated with the double bond [$CH_2{=}CH{-}C(Z){=}X$, $CH_2{=}CH{-}C{\equiv}X$], polymerisation promoted by Ziegler–Natta catalysts takes place, but in principle no insertion mechanism operates during propagation. Active sites of Ziegler–Natta catalysts, which may be capable of promoting the radical polymerisation and/or copolymerisation of heteroatom-containing monomers with olefins, are formed in some systems, depending on the kind of monomers and catalysts used.

Monomers in which the heteroatom is directly bound to the olefinic carbon atom ($CH_2{=}CH{-}X$), such as vinyl chloride, can be polymerised with modified

Ziegler–Natta catalysts by a radical mechanism [538–542]. Dehydrochlorination is a problem in this case [543], although it can be reduced by using a modified activator such as dialkylaluminium alkoxide or substituted dialkylaluminium alkoxide like $Et_2AlOCH_2CH_2NR_2$ [538], or by the addition of a Lewis base such as tetrahydrofuran to the polymerisation system [539–542]. However, the essential role of the Lewis base present in a vinyl chloride polymerisation system with a Ziegler–Natta catalyst, e.g. $VOCl_3$–$AlEt_3$ or $VOCl_3$–$AlEt_2Cl$, is to accelerate the generation of free radicals capable of initiating the polymerisation [542]:

$$VOCl_3 + THF \cdot AlEt_3 \longrightarrow VOCl_2 + THF \cdot AlEt_2Cl + Et^{\cdot} \qquad (94)$$

It is to be noted in this connection that alkyl radicals normally formed during reduction of the transition metal compound in Ziegler–Natta systems [scheme (7)] do not initiate the radical polymerisation of olefins, in contrast to that of polar monomers. Most of the modified Ziegler–Natta catalysts for polar monomer polymerisation are characterised by low activities and lack of stereospecificity, producing polymers with properties that are very similar to those of polymers obtained by more conventional procedures for radical polymerisation [28].

Polar monomers in which a heteroatom is conjugated with the olefinic double bond [CH_2=CH–C(Z)=X, CH_2=CH–C≡X] undergo radical copolymerisation with olefins when initiated with Ziegler–Natta catalysts [544,545]. One of the possible reaction schemes of the formation of initiating radicals in the system acrylonitrile/$VOCl_3$–$AlEtCl_2$ is as follows [546]:

$$VOCl_3 + CH_2{=}CH{-}C{\equiv}N \cdot AlEtCl_2 \longrightarrow VOCl_2 + CH_2{=}CH{-}C{\equiv}N \cdot AlCl_3 + Et^{\cdot} \qquad (95)$$

However, the copolymerisation of polar vinyl monomers and olefins can also proceed in systems not containing a transition metal compound, i.e. in the presence of the organoaluminium compound alone. Organoaluminium compounds, e.g. $AlEtCl_2$, promote the free radical copolymerisation of polar monomers such as acrylonitrile with electron-donating monomers such as propylene [547–549]. The formation of a complex between a polar monomer and an organoaluminium compound, which a Lewis acid is, causes an increase in the electron-accepting properties of the complexed monomer and hence an increase in its susceptibility for copolymerisation with the olefin as the electron-donating monomer; the copolymerisation tends to produce an alternating copolymer. Such a copolymerisation in the system acrylonitrile/propylene–ethylaluminium dichloride is presented by the scheme [547–549]

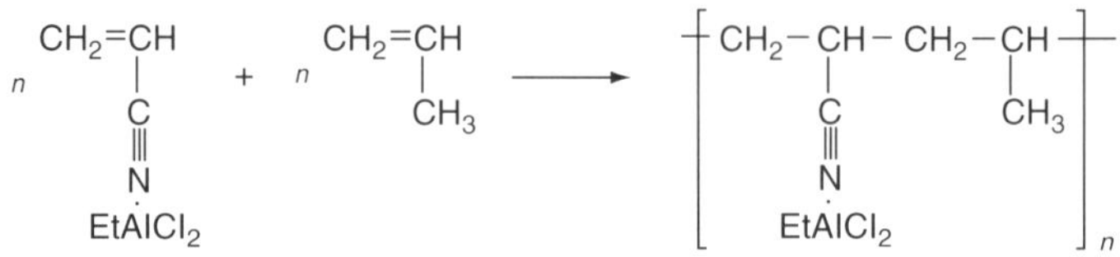

(96)

Organoaluminium compounds also produce free radicals by homolytic cleavage of the Al–C bond in the complexed polar monomer [scheme (97)], and thus they initiate the copolymerisation [549,550]:

$$CH_2{=}CH{-}C{\equiv}N \cdot AlEtCl_2 \longrightarrow CH_2{=}CH{-}C{\equiv}N \cdot \dot{A}lEtCl_2 + Et^{\cdot} \qquad (97)$$

It should be emphasised that spontaneous copolymerisation such as that presented by scheme (96), initiated by radicals formed according to scheme (97), is accelerated by the addition of a transition metal compound into the system (the formation of a modified Ziegler–Natta catalyst), since initiating radicals are more readily formed according to scheme (95) than to scheme (97).

3.12 Industrial Polymerisation Processes

Knowledge of the coordination polymerisation of olefins would not be complete without consideration of the types of process used in industry for polyolefin manufacture. Problems encountered in production influence developments in the area of catalysis in olefin polymerisation, an improvement in a catalyst being defined as leading to a reduction in the cost of making the polymer or giving better product properties. Therefore, the principal types of polyolefin production involving coordination catalysts of various types are dealt with briefly. Since modern polyolefin production processes offer a versatile range of polymers, the main commercially available olefin polymerisation products and their typical uses are also considered.

3.12.1 Polymerisation Methods

The technologies that have been developed for the production of polyolefins, olefin homopolymers and copolymers are slurry, solution and gas-phase polymerisation; bulk polymerisation of propylene in the liquid monomer as a special case of the slurry process has also emerged. The fundamental differences in the various olefin polymerisation processes reflect the different approaches that have been devised to remove the substantial heat of polymerisation. In addition, processes can be operated in a batch or a continuous mode. In the batch process the reagents are loaded into a polymerisation vessel, the polymer forms and the vessel is emptied before a new charge of reagents is introduced. In the continuous process, the catalyst precursor, activator and other necessary additives are continually introduced into the reaction vessel, and the polymer is removed. Continuous use of reactors without downtimes for charging and emptying makes economic sense [28,37,43,51,551,552].

3.12.1.1 Slurry Polymerisation

The slurry process is the oldest and still widely used method for manufacturing polymers of ethylene, propylene and higher α-olefins. In this process, the monomer dissolves in the polymerisation medium (hydrocarbon diluent) and forms a solid polymer as a suspension containing ca 40 wt-% of the polymer; the polymerisation occurs below the melting point of the polymer. In slurry polymerisation, the temperature ranges from 70 to 90 °C, with the ethylene pressure varying between 7 and 30 atm. The polymerisation time is 1–4 h and the polymer yield is 95–98 %. The polymer is obtained in the form of fine particles in the diluent and can be separated by filtration. Removal of the catalyst residues from the polymer can be achieved by the addition of alcohol (isopropanol, methanol), followed by recovery and extraction of the catalyst residues. The polymer is freed from diluent by centrifuging and then dried. In the case of polypropylene manufacture, the atactic fraction remains in the diluent [28,37].

The polymer molecular weight may be controlled by the addition of hydrogen to the reaction medium; the molecular weight distribution is regulated by varying the catalyst design or by carrying out polymerisation in several steps under varying conditions. The best conditions for polyolefin production are created when using stirred reaction vessels or loop reactors [37].

In some processes the polymerisation is carried out in a series of cascade reactors to allow variation in the hydrogen concentration through the operating steps in order to control the molecular weight distribution in the polymers formed.

A flow scheme of polypropylene production involving the slurry process with the removal of catalyst residues from the polymer is presented in Figure 3.52 [51].

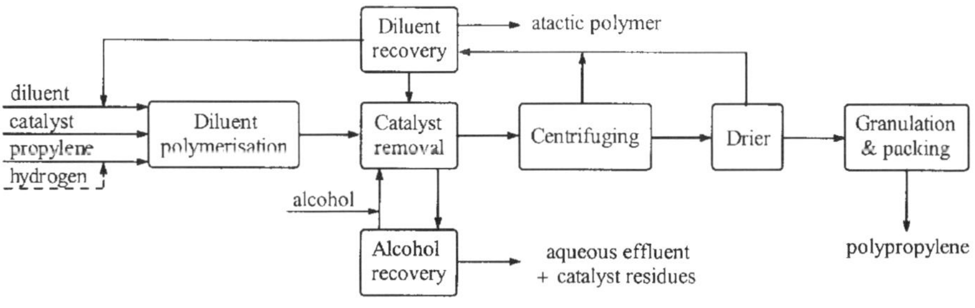

Figure 3.52 Flow scheme of polypropylene production using the slurry polymerisation process (with the removal of catalyst residues)

The development of supported catalysts has permitted the elimination of the expensive catalyst removal stage. Therefore, in slurry processes taking advantage of highly active catalysts, the diluent is recovered after centrifugation and recycled without purification (Figure 3.53) [51].

3.12.1.2 Bulk Polymerisation

A special case of the slurry process is bulk polymerisation, in which liquid propylene, with its poorer solvent power, is used instead of the inert diluent. In order to maintain the monomer in the liquid phase, a bulk process typically involves polymerisation at 55–80 °C under pressure in the range 20–30 atm. Vaporisation of propylene plays a significant role in the removal of the heat of polymerisation. Stirred autoclaves, incorporating evaporative cooling, and loop reactors provide good heat transfer conditions. The polymer formed is continuously removed from the reactor and separated from the unreacted monomer, which is purified to remove propylene oligomers (dimer, trimer, etc.) and other impurities before recycling. Hydrocarbon diluent and isopropyl alcohol are then added to remove the catalyst residues; atactic fraction of the polymer is again extracted in the diluent and isotactic polypropylene is finally dried and granulated [43,51]. A flow scheme of polypropylene production involving such a bulk process is presented in Figure 3.54 [51].

The catalyst may also be solubilised with polar complexing agents to permit extraction from the polymer by countercurrent washing with liquid propylene (Figure 3.55) [51].

The slurry bulk process is slightly more expensive than the slurry diluent process, the claimed advantage being low catalyst residues in the polypropylene produced [43,51]. The bulk process may be simplified by not removing the atactic polymer fraction (Figure 3.56) [51].

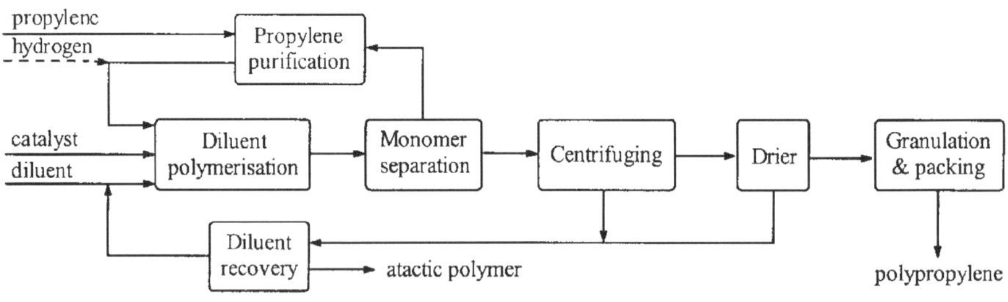

Figure 3.53 Flow scheme of polypropylene production using the slurry polymerisation process

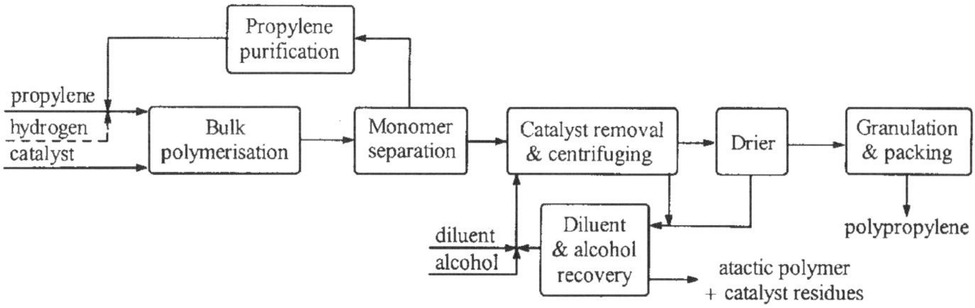

Figure 3.54 Flow scheme of polypropylene production using the bulk polymerisation process (with the removal of catalyst residues)

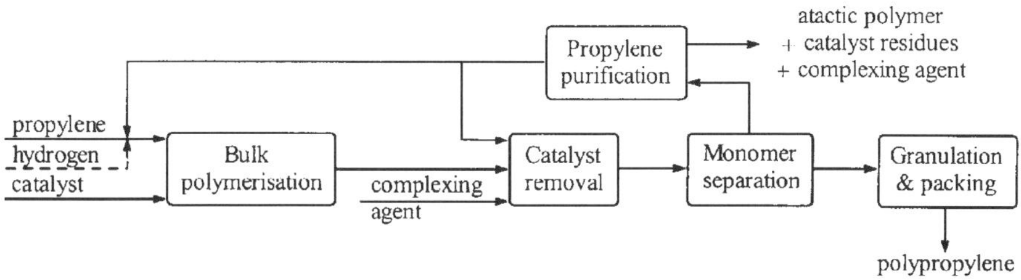

Figure 3.55 Flow scheme of polypropylene production using the bulk polymerisation process

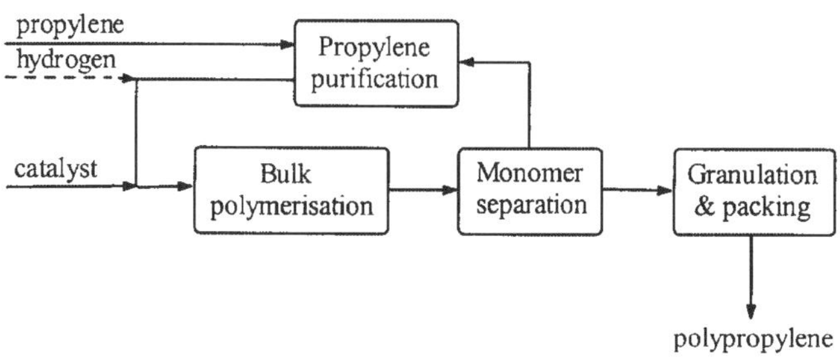

Figure 3.56 Flow scheme of polypropylene production using the bulk polymerisation process (without the removal of the atactic fraction)

3.12.1.3 Solution Polymerisation

The solution process has been developed mainly for the manufacture of polyethylene. The term 'solution' should not be taken literally since the polymer formed is often present in a molten state owing to maintenance of a high polymerisation temperature. In the solution process, a solvent such as cyclohexane, containing dissolved ethylene under a pressure of up to 100 atm, is heated; after addition of the catalyst, very rapid ethylene polymerisation occurs. The polymerisation temperature is maintained in the range 130–300 °C by indirect colling of the reactor with water or feeding the polymerisation vessel with cooled monomer solution; temperature control is also achieved via the ethylene pressure, which can be varied between ca 7 and 100 atm. The solution process allows high rates of polymerisation with a relatively compact stirred reactor. Residence times are of the order of a few minutes (2–10 min), as opposed to 1–4 h for the slurry process. Typically, 18–25 wt-% of polyethylene accumulates in solution; the polymer solution is discharged from the reactor, treated with a deactivating agent and the mixture is then filtered using filter aids such as alumina or passes through a bed of alumina where the deactivated catalyst residues are adsorbed. Finally, the purified polyethylene solution is introduced into the flash tank where most of the solvent is vaporised and recycled. The residual solvent in the polymer (ca 5 wt-% of the total) is removed in a devolatilising extruder during pelletisation. Compounding and pelletisation reduce the residual solvent in the product to a concentration of less than 0.05 wt-% [28,37,43,51,551].

In contrast to polyethylene production, solution polymerisation at high temperature is rarely applied for isotactic polypropylene, but some special-purpose polypropylene grades are manufactured (Figure 3.57) [51]. However, the solution process, which yields isotactic polypropylene with a very low level of impurities, is characterised by high overall costs. The solution process is being used to make atactic polypropylene, to which it seems more suited [43].

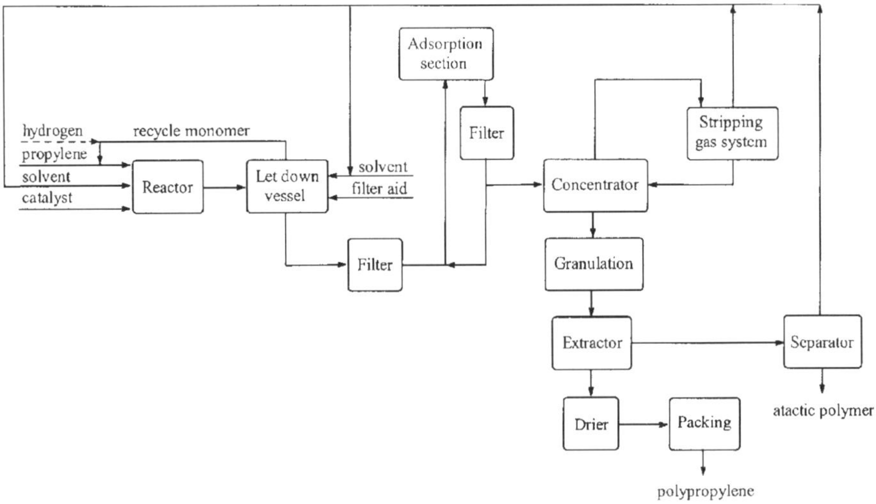

Figure 3.57 Flow scheme of polypropylene production using the high-temperature solution polymerisation process

3.12.1.4 Gas-phase Polymerisation

Gas-phase polymerisation is the newest technology introduced for the manufacture of polyolefins. It was developed for the production of polyethylene with third-generation Ziegler–Natta catalysts in the late 1960s. In the gas-phase process, the catalyst, which is supported on a bed of polymer, converts the monomer directly into a solid polymer. Since practically all the titanium chloride molecules are available to take part in the polymerisation, fewer are needed. Consequently, the yield of polymer formed per gram of titanium is so high that no extraction procedures are necessary. The elimination of the deashing part of the industrial process had obvious economic and environmental advantages and resulted in the explosive growth of polyolefins in the global plastics marketplace [49]. Although an innovative technology, the process for high-density polyethylene did not offer clear economic advantages over the established slurry process. In the case of linear low-density polyethylene, the fluidised-bed process is competitive with the solution and high-pressure processes, with lower capital and operating costs. In the case of polypropylene, industry uses two different methods for carrying out propylene polymerisation, depending on the chosen method of heat removal. Use is made of fluidised-bed systems, as in the case of polyethylene manufacture, and of mechanically agitated dry powder beds with evaporative cooling in vertical and horizontal autoclaves; propylene is recycled by removal as condensed vapour and reintroduction into the monomer feed [43]. This has the effect of removing the majority of the heat of polymerisation. The flow scheme of polypropylene manufacture involving the stirred-bed gas-phase polymerisation of propylene is shown in Figure 3.58 [51].

When fourth-generation Ziegler–Natta catalysts are used, the granulation step is avoided; the performance range of the catalyst is expanded to allow control of the morphological properties of the polymer (suitable shape, particle size distribution and compactness), apart from allowing control of its molecular structure (molecular weight distribution, branching and steric purity) and

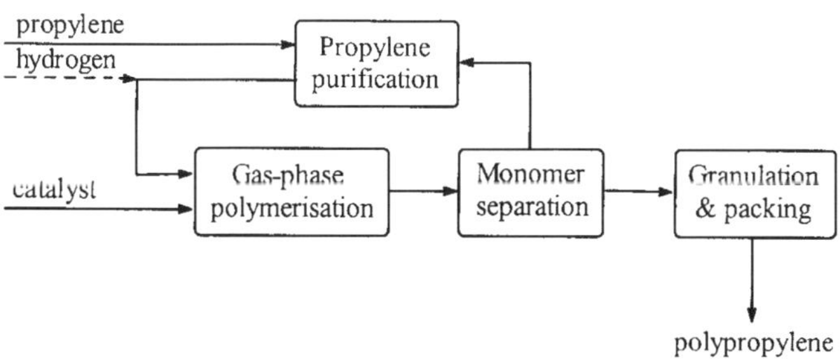

Figure 3.58 Flow scheme of polypropylene production using the gas-phase polymerisation process

control of the incorporation of other monomers to give random and heterophase copolymers with reproducible structures [49].

A single reactor system is used to make olefin homopolymers and random copolymers. Two reactors are operated in series for the production of block copolymers (impact copolymers). An inert conveying gas (nitrogen) is used to maintain the fluidised bed in the reactor for impact copolymerisation [43,51].

The development of the gas-phase technology represents a major advance for the commercial production of polyolefins. A gas-phase process avoids the problem of the high cost remaining in the high-mileage slurry and solution processes (associated with recycling of diluent or solvent and drying of the polymer).

The gas-phase polymerisation of ethylene is typically carried out at 85–100 °C under a pressure of 20–25 atm. The ethylene monomer circulates, thus removing the heat of polymerisation and fluidising the catalyst bed. In order to keep the temperature of the process at values below 100 °C, gas conversion is maintained at 2–3 per pass [37]. In the production of polypropylene by the gas-phase process, usually the polymerisation temperature is lower and ranges from about 50 to 85 °C, and the pressure range from ca 15 to 40 atm [553].

3.12.2 Polymerisation Catalysts

Catalysts used for the commercial production of olefin polymers are Ziegler–Natta catalysts and Phillips catalysts. Only a few catalysts have been utilised in industrial olefin polymerisation processes. High-density polyethylene and linear low-density polyethylene are both produced either with Phillips or with Ziegler–Natta catalysts. Improvements in the catalyst performance have allowed development of slurry and gas-phase processes in addition to the solution process utilising Phillips catalysts. Slurry processes involving polymerisation with Ziegler–Natta catalysts have undergone a considerable simplification after the introduction of highly active supported catalysts. The use of these catalysts has made it possible to eliminate one of the burdensome stages of the slurry process, which is the removal of catalyst residues from the polymer. Supported Ziegler–Natta catalysts are now also employed in the gas-phase process. Ethylene/propylene-based copolymers have been manufactured in solution (and slurry) processes by using soluble vanadium-based Ziegler–Natta catalysts since the 1960s [241]. However, processes involving high-productivity catalysts in a fluidised bed, in which no solvent is used and also no catalyst removal step is required, have been introduced into industrial production recently [553].

From 1991 onwards, polyethylenes produced by means of metallocene catalysts, supported on carriers, have appeared on the market. Plans to start other polyolefin manufacturing processes involving metallocene single-site catalysts (polypropylenes, ethylene/propylene and ethylene/propylene/non-conjugated diene rubbers, olefin copolymers with cyclic olefins) have been announced [554].

The recent achievements in catalysis and new polymerisation technologies make commercially available all the main polyolefins and polyolefin alloys, emphasising economic production processes while protecting the environment. However, high-activity $MgCl_2$-supported Ziegler–Natta catalysts will probably dominate polyolefin production for the next decade [555]. The life cycles of polyolefin catalysts, including those anticipated, are shown in Figure 3.59 [38].

In addition, it is worth mentioning that Ziegler–Natta catalysts, including metallocene-based catalysts, comprise several potentially hazardous components, many of which are sensitive to oxygen and moisture. Some metal alkyls, for instance, are pyrophoric (i.e. can combust spontaneously). Most manufacturing processes for olefin polymerisation catalysts comprise many solvent-consuming operations that generate effluent organic waste streams. Moreover, both the procatalysts and their activators tend to be expensive. Also, catalyst components are sensitive to many poisons, so must be handled and stored with care under dry nitrogen blanketing, with the total exclusion of air [555]. It should be stated, however, that the newest Ziegler–Natta catalysts (third, fourth and metallocene generations) are very efficient and are thus used in small amounts in relation to the amounts of polymers produced with their use.

3.12.3 Polymerisation Products

Among the large number of olefin polymers that may be obtained by coordination polymerisation, several are produced commercially on an enormous scale: high-density polyethylene (HDPE), linear low-density polyethylene (LLDPE),

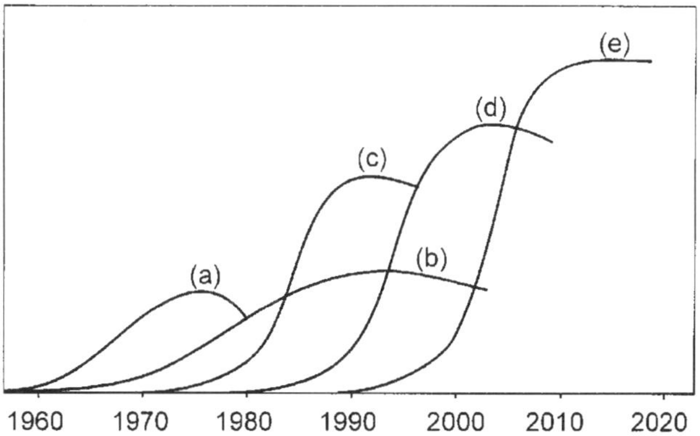

Figure 3.59 Life cycles of catalysts for olefin coordination polymerisation: (a) early-generation Ziegler–Natta catalysts for ethylene and propylene polymerisation; (b) Phillips catalysts for ethylene polymerisation; (c) fourth-generation Ziegler–Natta catalysts for ethylene polymerisation; (d) fourth-generation Ziegler–Natta catalysts for propylene polymerisation; (e) metallocene-based catalysts for olefin polymerisation leading to polymers of various stereoregularity

Table 3.9 Commercially available olefin homopolymers and copolymers, produced by coordination polymerisation, and their typical uses

Polymer	Typical use
Plastics	
High-density polyethylene	Bottles, drums, pipes, conduits, sheets, films, wire and cable insulations
Linear low-density polyethylene[a]	Blending with low-density polyethylene, films, packaging, bottles
Isotactic polypropylene	Automobile and appliance parts, ropes, cordages, webbing, carpeting, films
Isotactic poly(1-butene)	Films, pipes
Isotactic poly(4-methyl-1-pentene)[b]	Packaging, medical supplies, lighting
Ethylene/isotactic propylene block (compact) copolymers (polyallomers)	Food packaging, automotive trim, toys, bottles, films, heat-sterilisable containers
Elastomers	
Atactic polypropylene	Asphalt blends, sealants, adhesives, cable coatings
Ethylene/propylene copolymer	Impact modifiers for isotactic polypropylene
Ethylene/propylene/diene copolymer[c]	Wire and cable insulations, weather stripping, tyre sidewalls, hoses, seals

[a] Ethylene copolymers with 1-butene, 1-hexene or 1-octene.
[b] Usually obtained by copolymerisation of 4-methyl-1-pentene with small amounts of 1-pentene.
[c] Non-conjugated diene, e.g. 1,4-hexadiene, dicyclopentadiene, 5-ethylidene-2-norbornene.

isotactic polypropylene (it-PP), ethylene/propylene rubbers (EPR) and ethylene/propylene/non-conjugated diene rubbers (EPDM). Some other olefin polymers are produced in smaller quantities, such as isotactic poly(1-butene) and isotactic poly(4-methyl-1-pentene) [241].

The commercial production of high-density polyethylene started almost at the same time in late 1956 by Phillips using a chromium-based catalyst in a medium-pressure process and by Hoechst using a Ziegler catalyst in a low-pressure process. Polypropylene production began in Montecatini and Hercules plants in 1957. Poly(1-butene) and poly(4-methyl-1-pentene) have been produced in small commercial quantities since about 1965. The commercial production of ethylene/propylene-based rubbers started in 1960 [241].

Table 3.9 lists the main commercially available olefin polymers synthesised by coordination polymerisation and their typical uses [556].

3.13 Addendum: Recent Achievements in Polymerisation with Main Group Metal-based Catalysts

Recently, cationic aluminium complexes stabilised by appropriate monoanionic bidentate ligands capable of forming large chelate rings, e.g. the (*N*, *N*-diisopropylamino)troponiminate ligand, have found catalytic applications in olefin and functionalised olefin polymerisations [557]. Although such catalysts are not based on transition metal compounds, they exhibit the proper combination of electrophilic aluminium centre and active Al–C bond, which helps promote the insertion of the coordinating monomer [558].

Jordan *et al.* [557] found that three-coordinate aluminium alkyls or hydride

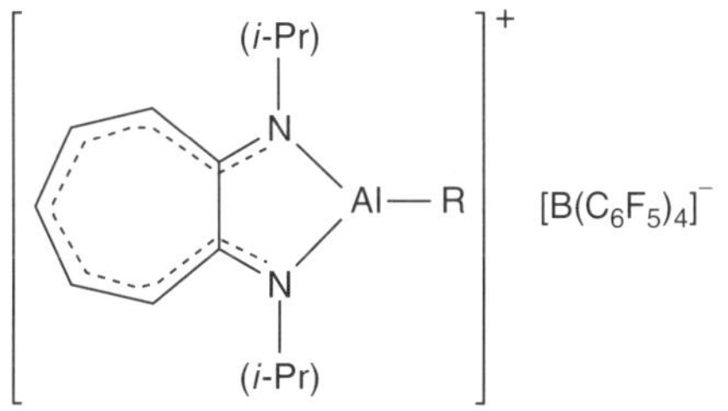

polymerised ethylene in a toluene medium under mild conditions (80–100 °C, 1–5 atm) to a high molecular weight polymer ($\bar{M}_n = 106\,500$; $\bar{M}_n = 2.4$; melting point 137.8 °C) in moderately low yields of ca 0.1 kg of polyethylene per gram Al/h atm.

The dinuclear hydride cation

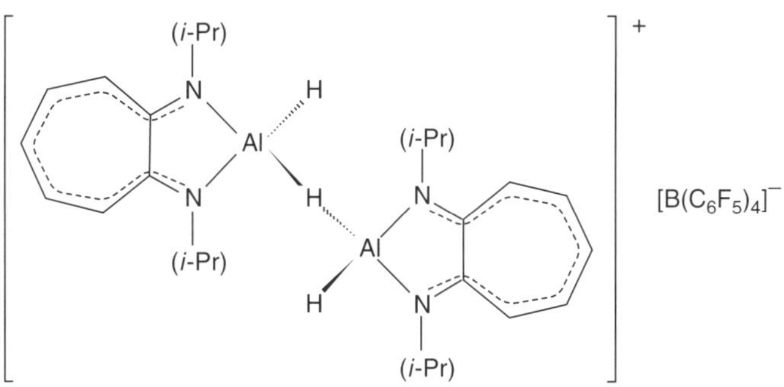

polymerised methyl methacrylate in a chlorobenzene or toluene medium at ambient temperature to a high molecular weight, predominantly syndiotactic polymer ($\bar{M}_n$ =187 000–228 000; $\bar{M}_w/\bar{M}_n = 1.8$) in high yields (80–100 %) [557]. Let us recall, in this connection, the syndiospecific polymerisation of methyl methacrylate with dinuclear cationic zirconocene complexes [537,559].

Polymerisation and Copolymerisation of Olefins in the Presence of Catalysts Containing Ni or Pd and α-Diimine Ligands

Very recently, McLain *et al.* [560] and Brookhart *et al.* [561] from du Pont de Nemours and Company invented catalysts that consist of transition metal compounds and α-diimine ligands, and sometimes other cocatalysts. Processes of polymerisation of ethylene, acyclic olefins and/or selected cyclic olefins, and optionally selected olefinic esters or carboxylic acids, carried out in the presence of these catalysts yield polymers useful as elastomers, moulding resins, in adhesives, etc. Since some of the polymerisations exhibit characteristics of living polymerisations, block copolymers can be readily made:

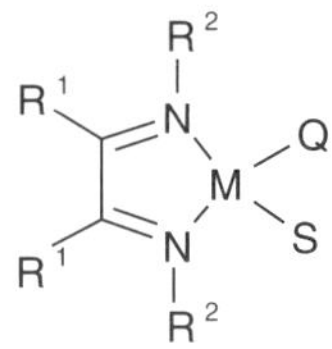

(I)

Coordination catalysts of formula (**I**), where M is Pd, Q is an alkyl, S is a halogen, R^1 is H or an alkyl (1–4 carbon atoms), preferably a methyl group, and R^2 is a hydrocarbyl (1–4 carbon atoms), preferably a methyl group, may be made by the reaction of the corresponding 1,5-cyclooctadiene (COD) Pd complex with the appropriate diimine. When M is Ni, (**I**) can be made by the displacement of another ligand, such as a dialkylether or a polyether such as 1,2-dimethoxyethane, by an appropriate diimine:

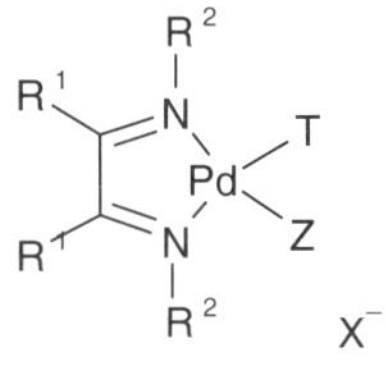

(II)

Catalysts of formula (**II**), where T is hydrogen or a hydrocarbyl not containing olefinic or acetylenic bonds, RC(=O)- or ROC(=O)-(preferably methyl), Z is a neutral Lewis base in which the donating atom is nitrogen, sulphur or oxygen (preferably R_2O or RCN) and X^- is BAF^- {BAF = tetrakis[3,5-bis(trifluoromethyl)phenyl]borate}, may be made by reacting a compound of formula (**I**), where Q is an alkyl and S is a halogen, with about one equivalent of an alkali metal salt, particularly the sodium salt, of HBAF, in the presence of a coordinating ligand, particularly a nitrile such as acetonitrile. When X^- is an anion such as BAF^-, SbF_6^- or BF_4^-, the same starting palladium compound can be reacted with the silver salt AgX:

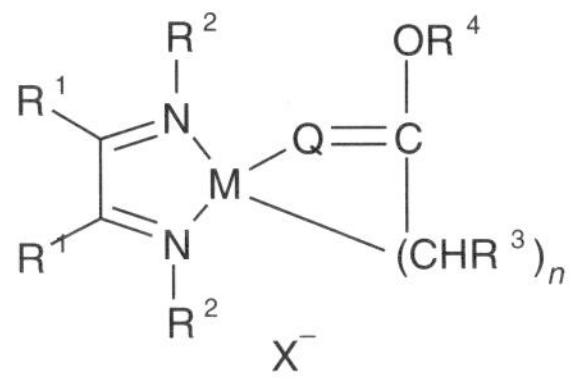

(III)

Compounds of formula **(III)**, where M is Ni(II) or Pd(II), can be made in the reaction of a compound of formula **(II)** with an acrylate ester $CH_2 = CHCO_2R^4$. This reaction is carried out in a non-coordinating solvent such as methylene chloride, preferably using a greater than 1–50-fold excess of the acrylate ester. In a preferred reaction, Q is methyl, and R^4 is an alkyl containing 1–4 carbon atoms, preferably a methyl group.

The nickel and palladium compounds described above are useful in processes for polymerising various olefins, and optionally also for copolymerising olefinic esters, carboxylic acids or other functional olefins with these olefins. When **(I)** is used as a catalyst, a neutral Lewis acid or a cationic Lewis or Bronsted acid whose counterion is a weakly coordinating anion is also present as part of the catalyst system. The neutral Lewis acid is originally uncharged (i.e. not ionic). Suitable neutral Lewis acids include SbF_5, Ar_3B and BF_3. By a cationic Lewis acid is meant a cation with a positive charge such as Ag^+, H^+ and Na^+.

In those instances where **(I)** (and similar catalysts that require the presence of a neutral Lewis acid or a cationic Lewis or Bronsted acid) does not contain an alkyl or hydride group already bonded to the metal (i.e. neither Q or S is an alkyl or hydride), the neutral Lewis acid or a cationic Lewis or Bronsted acid also alkylates or adds a hydride to the metal, i.e. causes an alkyl group or hydride to become bonded to the metal atom.

A preferred neutral Lewis acid, which can alkylate the metal, is a selected alkyl aluminium compound, such as R_3Al, R_2AlCl, $RAlCl_2$ and alkylaluminoxanes. Suitable alkyl aluminium compounds include methylaluminoxane (which is an oligomer with the general formula $[MeAlO]_n$), $(C_2H_5)_2AlCl$, $C_2H_5AlCl_2$ and $[(CH_3)_2CHCH_2]_3Al$. Metal hydrides such as $NaBH_4$ may be used to bond hydride groups to the metal M.

Preferred olefins in the polymerisation are one or more of ethylene, propylene, 1-butene, 2-butene, 1-hexene, 1-octene, 1-pentene, 1-tetradecene, norbornene and cyclopentene, with ethylene, propylene and cyclopentene. Other monomers that may be used with these catalysts (when it is a Pd(II) complex) to form copolymers with olefins and selected cycloolefins are carbon monoxide (CO) and vinyl ketones of the general formula $H_2C{=}CHC(O)R$. Carbon monoxide forms alternating copolymers with the various olefins and cycloolefins.

It is characteristic that homopolypropylene obtained using these catalysts exhibits a glass transition temperature of $-30\,^{\circ}C$ or less, and contains at least about 50 branches per 1000 methylene groups. In addition, some of the ethylene homopolymers have an exceptionally low density, less than about 0.86 g/mL.

In coordination polymerisation, α-olefins of the formula $CH_2{=}CH(CH_2)_aH$ may be inserted into the growing polymer chain in a 1,2 or 2,1 manner. Normally, these insertion steps lead to 1,2-enchainment or 2,1-enchainment of the monomer. Both of these steps form a $-(CH_2)_aH$ branch. However, with some catalysts herein, some of the initial product of 1,2-insertion can be rearranged by migration of the coordinated metal atom to the end of the last inserted monomer before insertion of additional monomer, similarly to that presented in schemes (69) and (70). This results in ω,2-enchainment and the formation of a methyl branch:

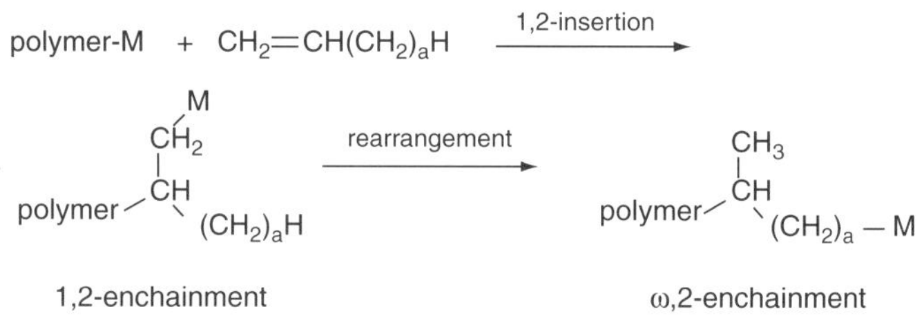

It is also known that, with certain other catalysts, some of the initial product of 2,1-insertion can be rearranged in a similar manner by migration of the coordinated metal atom to the end of the last inserted monomer. This results in ω,2-enchainment and no branch is formed:

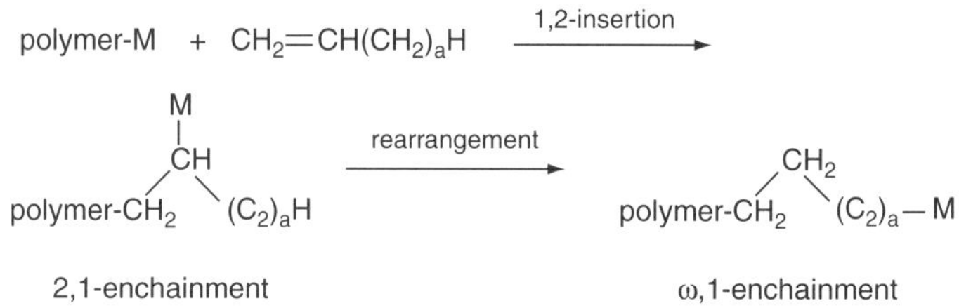

Because of the relatively large amounts of ω,1-enchainment that may be obtained using some of the polymerisation catalysts reported herein, novel polymers can be made, including homopolypropylene (PP). In some of the PPs obtained, the structure–$CH(-)CH_2CH_2CH_2(CH_2)_aCH_2CH_2CH_2CH(-)$– may be found.

Homopolymers of α-olefins such as propylene exhibit unusual properties compared with their 'normal' homopolymers. A homopolypropylene would usually have about 1000 methyl groups per 1000 methylene groups. Polypropylenes made using these catalysts typically have about half that many methyl groups, and in addition have some longer chain branches. Other α-olefins often field polymers whose microstructure is analogous to these polypropylenes when the above catalysts are used for the polymerisation.

In addition, these catalysts may have a bidentate ligand where coordination to the transition metal is through two different nitrogen atoms or through a nitrogen atom and a phosphorus atom, these nitrogen atoms being part of the bidentate ligand. It is believed that some of these compounds are effective polymerisation catalysts at least partly because the bidentate ligands have sufficient steric bulk on both sides of the coordination plane (of the square planar complex). If such a complex contains a bidentate ligand of appropriate steric bulk, it is believed that it produces polyethylene with a degree of polymerisation of at least about 10 or more.

It is also believed that the polymerisation catalysts herein are effective because the unpolymerised olefinic monomer can only slowly displace from the complex a coordinated olefin formed by β-hydride elimination from the growing polymer chain, which is attached to the transition metal. The displacement can occur by associative exchange. An increase in the steric bulk of the ligand slows the rate of associative exchange and allows polymer chain growth.

Preparation of Linear α-Olefins

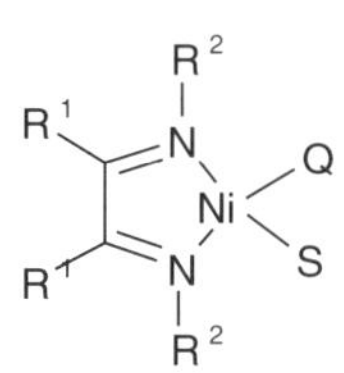

In the catalyst presented above, R^2 are hydrocarbyl or substituted hydrocarbyl groups. These groups greatly determine whether the α-olefins of this process are made, or whether higher polymeric materials, i.e. materials containing over 25 ethylene units, are coproduced or produced almost exclusively. If R^2 are highly sterically hindered about the nickel atom, the tendency is to produce a higher polymeric material. For instance, when the R^2 are both 2,6-diisopropylphenyl, mostly a higher polymeric material is produced. However, when the R^2 are both phenyl, mostly the α-olefins of this process are produced.

The α-olefins may also be formed by contact of ethylene with one of the catalysts:

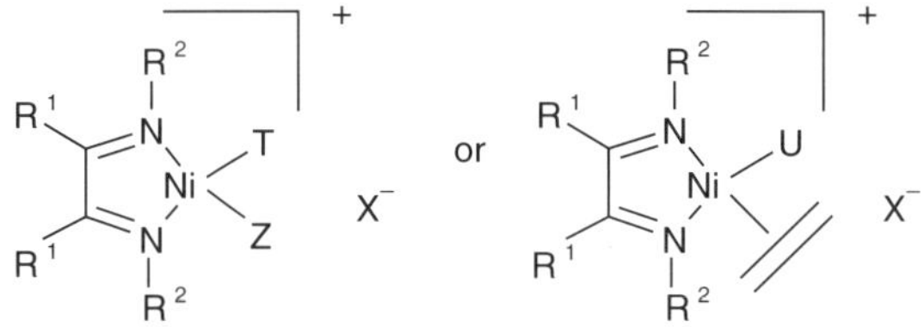

where U is an *n*-alkyl containing up to 38 carbon atoms, and X is a non-coordinating anion. Suitable groups for Z include dialkyl ethers such as diethyl ether, and alkyl nitriles such as acetonitrile.

In general, α-olefins can be made by this process using a Ni[II] complex of an α-diimine as a catalyst, where the Ni[II] complex is obtained using Ni[0], Ni[I] or Ni[II] precursors.

Polyketones Obtained by Copolymerisation of CO and α-Olefins in the Presence of Catalysts Containing Pd and Bidentate Phosphine Ligands

Dossett [562] from BP Chemicals Ltd has invented a catalyst composition prepared by reacting together a group 8 source metal such as palladium with a bidentate phosphine ligand having at least two phosphorus atoms joined by a bridging group of the formula $-(N)_x-(P)_yN-$ (where x is 0 or 1), and a promoter. The promoter can be a weakly or non-coordinating anion. Alternatively, it can be a boron hydrocarbyl compound or an aluminoxane. The catalysts can be used in the preparation of polyketones as interpolymers of olefins and carbon monoxide by polymerising a mixture of one or more olefins and carbon monoxide in the presence of such catalyst compositions. Particular examples of such catalysts are $[Pd[Ph_2PN(Me)N(Me)PPh_2]Cl_2]$ and $[Pd[Ph_2PN(Me)N(Me)PPh_2](PhCN)_2\}[BF_4]_2$.

Stewart *et al.* [563], also from BP Chemicals Ltd, used a bidentate phosphine ligand having at least two phosphorous atoms joined by another bridging group of the formula $-NR^2(CX)NR^2-$, where X = O, S or Se, and each R^2 is the same or different and is a hydrogen or hydrocarbyl group. An example of this bidentate phosphine ligand is $(Ph_2P)N(Me)CO(Ph)N(PPh_2)$.

As regards the promoter, this can be a source of an anion, which is either non-coordinating or weakly coordinating. Such anions are suitably the conjugate bases of strong acids having, for example, a pK_a of less than 6 and preferably less than 2 (e.g. $HBF_4, HPF_6, HSbF_6$, *p*-toluenesulphonic acid). Alternatively, the promoter can be a boron hydrocarbyl compound, for example a boron alkyl or boron aryl compound. The promoter can also be an aluminoxane.

References

1. Corradini, P., Busico, V. and Guerra, G., 'Monoalkene Polymerization: Stereospecificity', in *Comprehensive Polymer Science*, Pergamon Press, Oxford, 1989, Vol. 4, pp. 29–50.
2. Pasquon, I., Porri, L. and Giannini, U., 'Stereoregular linear polymers', in *Encyclopedia of Polymer Science and Engineering*, Wiley-Interscience, John Wiley & Sons, New York, 1989, Vol. 15, pp. 632–733.

3. Randall, J. C., 'Microstructure', in *Encyclopedia of Polymer Science and Engineering*, Wiley-Interscience, John Wiley & Sons, New York, 1987, Vol. 15, pp. 795–824.
4. Natta, G., *Experientia Suppl.*, **7**, 21 (1957).
5. Natta, G. and Corradini, P., *Nuovo Cimento Suppl.*, **15**(10), 9 (1960).
6. Danusso, F., Moraglio, G., Ghiglia, W., Motta, L. and Talamini, G., *Chim. Ind. (Milan)*, **41**, 748 (1959).
7. Natta, G., Pasquon, I., Zambelli, A. and Gatti, G., *J. Polym. Sci.*, **51**, 387 (1961).
8. Ciardelli, F., 'Optical Activity', in *Comprehensive Polymer Science*, Pergamon Press, Oxford, 1989, Vol. 1, pp. 561–571.
9. Natta, G., Pasquon, I., Corradini, P., Peraldo, M., Pegoraro, M. and Zambelli, A., *Atti Accad. Naz. Lincei. Cl. Sci. Fis. Mat. Nat. Rend.*, **28**(8), 539 (1960).
10. Natta, G., Pasquon, I. and Zambelli, A., *J. Am. Chem. Soc.*, **84**, 1488 (1962).
11. Natta, G., Peraldo, M. and Allegra, G., *Makromol. Chem.*, **75**, 215 (1964).
12. Bovey, F. A. and Tiers, G. V. D., *J. Polym. Sci.*, **44**, 173 (1960).
13. Bovey, F. A., 'Isomerism in Polymer Chains', in *High Resolution NMR of Macromolecules*, Academic Press, New York, 1972, pp. 53–64.
14. Bovey, F. A., 'The Study of Conformations of Vinyl Polymer Chains and Model compounds by NMR', in *High Resolution NMR of Macromolecules*, Academic Press, New York, 1972, pp. 182–204.
15. Brandrup, J., 'Stereochemical Definitions and Notations Relating to Polymers', *Polymer Handbook*, Wiley-Interscience, John Wiley & Sons, New York, 1989, pp. I/43–61.
16. Farina, M., *Macromol. Symp.*, **89**, 489 (1995).
17. Natta, G., Pino, P. and Mazzanti, G., *Gazz. Chim. Ital.*, **87**, 528 (1957).
18. Farina, M., Peraldo, M. and Natta, G., *Angew. Chem., Int. Ed. Engl.*, **4**, 107 (1965).
19. Mislow, K. and Bickart, P., *Israel J. Chem.*, **15**, 1 (1976/1977).
20. Wolfsgruber, C., Zannoni, G., Rigamonti, E. and Zambelli, A., *Makromol. Chem.*, **176**, 2765 (1975).
21. Doi, Y., Suzuki, E. and Keii, T., *Makromol. Chem., Rapid Commun.*, **2**, 293 (1981).
22. Ewen, J. A., *J. Am. Chem. Soc.*, **106**, 6355 (1984).
23. Ewen, J. A., Jones, R. L., Razavi, A. and Ferrara, J. D., *J. Am. Chem. Soc.*, **110**, 6255 (1988).
24. Farina, M., Di Silvestro, G., Sozzani, P. and Savaré, B., *Macromolecules*, **18**, 923 (1985).
25. Liberman, R. B. and Barbé, P. C., 'Propylene Polymers', in *Encyclopedia of Polymer Science and Engineering*, Wiley-Interscience, John Wiley & Sons, New York, 1988, Vol. 13, pp. 464–531.
26. Natta, G., Farina, M. and Peraldo, M., *Chim. Ind. (Milan)*, **42**, 255 (1960).
27. Zambelli, A., Giongo, M. G. and Natta, G., *Makromol. Chem.*, **112**, 183 (1968).
28. Tait, P. J. T., 'Monoalkene Polymerization: Ziegler–Natta and Transition Metal Catalysts', in *Comprehensive Polymer Science*, Pergamon Press, Oxford, 1989, Vol. 4, pp. 1–25.
29. Kaminsky, W., *Catal. Today*, **20**, 257 (1994).
30. Brintzinger, H. H., Fischer, D., Mülhaupt, R., Rieger, B. and Waymouth, R. M., *Angew. Chem., Int. Ed. Engl.*, **34**, 1143 (1995).
31. Coughlin, E. B. and Bercaw, J. E., *J. Am. Chem. Soc.*, **114**, 7606 (1992).
32. Turner, H. W., Eur. Pat. Appl. 277 004 (to Exxon Chem. Inc.) (1988); *Chem. Abstr.*, **110**, 58290a (1988).
33. Hlatky, G. G., Turner, H. W. and Eckmann, R. R., *J. Am. Chem. Soc.*, **111**, 2728 (1989).
34. Herfert, N. and Fink, G., *Makromol. Chem., Rapid Commun.*, **14**, 91 (1993).

35. Chien, J. C. W., Tsai, W. M. and Rausch, M. D., *J. Am. Chem. Soc.*, **113**, 8570 (1991).
36. Richardson, K., *Chem. in Britain*, **30**, 87 (1994).
37. Kaminsky, W., 'Polyolefins', in *Handbook of Polymer Synthesis*, M. Dekker, New York, 1992, Part A, pp. 1–76.
38. Galli, P., *Macromol. Symp.*, **89**, 13 (1995).
39. Ziegler, K., Holzkamp, E., Breil, H. and Martin, H., *Angew. Chem.*, **67**, 541 (1955).
40. Natta, G., *J. Polym. Sci.*, **34**, 21 (1959).
41. Natta, G., Pino, P., Corradini, P., Danusso, F., Mantica, E., Mazzanti, G. and Moraglio, G., *J. Am. Chem. Soc.*, **77**, 1708 (1955).
42. Natta, G., Corradini, P. and Allegra, G., *J. Polym. Sci.*, **51**, 399 (1961).
43. Whiteley, K. S., Heggs, T. G., Koch, H., Mawer, R. L., Goldbach, G. and Immel, W., 'Polyolefins', in *Ullman's Encyclopedia of Industrial Chemistry*, VCH Publishers, Inc., Weinheim, 1992, Vol. A21, pp. 487–577.
44. Tornqvist, E. G. M., Richardson, J. T., Wilchinsky, Z. W. and Looney, R. W., *J. Catal.*, **8**, 189 (1967).
45. Wilchinsky, Z. W., Looney, R. W. and Tornqvist, E. G. M., *J. Catal.*, **28**, 352 (1973).
46. Kissin, Y. V. and Bleach, D. L., 'Olefin Polymers', in *Encyclopedia of Polymer Science and Engineering*, Wiley-Interscience, John Wiley & Sons, New York, 1987, Vol. 10, pp. 395–408.
47. Endo, K., Fujii, K. and Otsu, T., *J. Polym. Sci., A, Polym. Chem.*, **29**, 1991 (1991).
48. Kollar, R., Schnecko, H. and Kern, W., *Makromol. Chem.*, **142**, 21 (1971).
49. Galli, P. and Haylock, J. C., *Makromol. Chem., Macromol. Symp.*, **63**, 19 (1992).
50. Diedrich, B., *Appl. Polym. Symp.*, **26**, 1 (1975).
51. Gavens, P. D., Bottrill, M., Kelland, J. W. and McMeeking, J., 'Ziegler–Natta Catalysis', *Comprehensive Organometallic Chemistry*, Pergamon Press, Oxford, 1982, Vol. 3, pp. 475–547.
52. Karol, F. J., *Catal. Rev.–Sci. Eng.*, **26**, 557 (1984).
53. Chien, J. C. W., 'Most Advanced Magnesium Chloride Supported Ziegler–Natta Catalyst', in *Catalysis in Polymer Synthesis*, ACS Symp. Ser. 496, Washington, DC, 1992, pp. 25–55.
54. Sobota, P., *Macromol. Symp.*, **89**, 63 (1995).
55. Baulin, A. A., Novikova, Ye. E., Mal'kova, G. Ya., Maksinov, V. L., Vyshinskaya, L. I. and Ivanchev, S. S., *Polym. Sci. USSR, Engl. Transl.*, **22**, 205 (1980).
56. Kabanov, V. A., Smetanyuk, V. I., Popov, V. G., Martynova, M. A. and Ulyanova, M. V., *Polym. Sci. USSR, Engl. Transl.*, **22**, 372 (1980).
57. Kabanov, V. A., Ivanchev, S. S., Smetanyuk, V. I., Baulin, A. A., Martynova, M. A. and Kopylov, V. M., *Polym. Sci. USSR, Engl. Transl.*, **22**, 382 (1980).
58. Barbé, P. C., Cecchin, G. and Noristi, L., *Adv. Polym. Sci.*, **81**, 1 (1987).
59. Galli, P., Barbé, P. C., Guidetti, G., Zannetti, R., Martorana, A., Marigo, A., Bergozza, M. and Fichera, A., *Eur. Polym. J.*, **19**, 19 (1983).
60. Kashiwa, N., *Polym. J.*, **12**, 603 (1980).
61. Dusseault, J. J. A. and Hsu, C. C., *J. Macromol. Sci. – Rev. Macromol. Chem. Phys. C*, **33**, 103 (1993).
62. Busico, V., Corradini, P., De Martino, L., Proto, A., Savino, V. and Albizzati, E., *Makromol. Chem.*, **186**, 1279 (1985).
63. Sergeev, S. A., Bukatov, G. D. and Zakharov, V. A., *Makromol. Chem.*, **184**, 2421 (1983).
64. Galli, P., Barbé, P. C. and Noristi, L., *Angew. Makromol. Chem.*, **120**, 73 (1984).
65. Mackie, P., Berger, M. N., Grieveson, B. M. and Lawson, D., *J. Polym. Sci., B*, **5**, 493 (1967).

66. Galli, P. and Haylock, J. C., *Prog. Polym. Sci.*, **16**, 443 (1991).
67. Albizzati, E., Giannini, U., Morini, G., Galimberti, M., Berino, L. and Scordamaglia, R., *Macromol. Symp.*, **89**, 73 (1995).
68. Albizzati, E., Giannini, U., Collina, G., Noristi, L. and Resconi, L., 'Catalysts and Polymerizations', in *Polypropylene Handbook*, Hanser Publishers, Munich, 1996, pp. 11–111.
69. Shiono, T., Uchino, H. and Soga, K., *Polym. Bull.*, **21**, 19 (1989).
70. Soga, K. and Park, J. R., in *Catalytic Olefin Polymerization*, Elsevier, New York, 1990, Vol. 56, pp. 131–138.
71. Zucchini, U., Cuffiani, I. and Pennini, G., *Makromol. Chem.*, **5**, 567 (1984).
72. Cozewith, C. and Ver Strate, G., *Macromolecules*, **4**, 482 (1971).
73. Makino, K., Tsuda, K. and Takai, M., *Polym. Bull.*, **27**, 41 (1991).
74. Zambelli, A., Natta, G., Pasquon, I. and Signorini, I., *J. Polym. Sci., C*, **16**, 2485 (1967).
75. Doi, Y., Ueki, S. and Keii, T., *Macromolecules*, **12**, 814 (1979).
76. Doi, Y., Ueki, S. and Keii, T., *Makromol. Chem.*, **180**, 1359 (1979).
77. Dall'Asta, G., Mazzanti, G., Natta, G. and Porri, L., *Makromol. Chem.*, **56**, 224 (1962).
78. Natta, G., Dall'Asta, G., Mazzanti, G. and Motroni, G., *Makromol. Chem.*, **69**, 163 (1963).
79. Natta, G., Porri, L. and Mazzei, A., *Chim. Ind. (Milan)*, **41**, 116 (1959).
80. Natta, G., Porri, L. and Carbonaro, A., *Atti Accad. Naz. Lincei, Cl. Sci. Fis. Mat. Nat. Rend.*, **31**(8), 189 (1961).
81. Natta, G., Zambelli, A., Pasquon, I. and Ciampelli, F., *Makromol. Chem.*, **79**, 161 (1964).
82. Natta, G., Dall'Asta, G., Mazzanti, G., Pasquon, I., Valvassori, A. and Zambelli, A., *J. Am. Chem. Soc.*, **83**, 3343 (1961).
83. Natta, G., Dall'Asta, G., Mazzanti, G. and Ciampelli, F., *Kolloid Z.*, **182**, 50 (1962).
84. Furukawa, J., Hirai, R. and Nakariva, N., *J. Polym. Sci., B*, **7**, 671 (1969).
85. Furukawa, J., Tsuruki, S. and Keiji, J., *J. Polym. Sci., A-1*, **11**, 2999 (1973).
86. Wieder, W. and Witte, J., *J. Appl. Polym. Sci.*, **26**, 2503 (1981).
87. Natta, G., *J. Am. Chem. Soc.*, **77**, 1708 (1955).
88. Natta, G., Pino, P., Mazzanti, G. and Lanzo, R., *Chim. Ind. (Milan).*, **39**, 1032 (1957).
89. Breslow, D. S. and Newburg, N. R., *J. Am. Chem. Soc.*, **79**, 5072 (1957).
90. Andersen, A., Cordes, H.-G., Herwig, J., Kaminsky, W., Merck, A., Mottweiler, R., Pein, J., Sinn, H. and Vollmer, H.-J., *Angew. Chem., Int. Ed. Engl.*, **15**, 630 (1976).
91. Sinn, H., Kaminsky, W., Vollmer, H.-J. and Woldt, R., *Angew. Chem.*, **92**, 396 (1980).
92. Reichert, K. H. and Meyer, K. R., *Makromol. Chem.*, **169**, 163 (1973).
93. Long, W. P. and Breslow, D. S., *Liebigs Ann. Chem.*, **1975**, 463 (1975).
94. Sinn, H. and Kaminsky, W., *Adv. Organomet. Chem.*, **18**, 99 (1980).
95. Kaminsky, W., Miri, M., Sinn, H. and Woldt, R., *Makromol. Chem., Rapid Commun.*, **4**, 417 (1983).
96. Sinn, H., *Macromol. Symp.*, **97**, 27 (1995).
97. Mason, M. R., Smith, J. M., Bott S. G. and Barron, A. R., *J. Am. Chem. Soc.*, **115**, 4971 (1993).
98. Barron, A. R., *Macromol. Symp.*, **97**, 15 (1993).
99. Belov, G. P. and Korneev, N. N., *Macromol. Symp.*, **97**, 63 (1995).
100. Kaminsky, W., *Angew. Makromol. Chem.*, **145/146**, 149 (1986).

101. Kaminsky, W., Külper, K., Brintzinger, H. H. and Wild, F. R. W. P., *Angew. Chem., Int. Ed. Engl.*, **24**, 507 (1985).
102. Herfert, N. and Fink. G., *Makromol. Chem.*, **193**, 1359 (1992).
103. Herwig, J. and Kaminsky, W., *Polym. Bull.*, **9**, 464 (1983).
104. De Boer, H. J. R. and Royan, B. W., *J. Mol. Catal.*, **90**, 171 (1994).
105. Horton, A. D., *Trends Polym. Sci.*, **2**, 158 (1994).
106. Chien, J. C. W. and Sugimoto, R., *J. Polym. Sci., A, Polym. Chem.*, **29**, 459 (1991).
107. Chien, J. C. W., 'Advances in Ziegler Catalysts', in *Ziegler Catalysts*, Springer-Verlag, Berlin, 1995, pp. 199–216.
108. Koga, N., Yoshida, T. and Morokuma, K., 'Theoretical Studies on Olefin Polymerisation using Group IV Metallocene Catalysts', in *Ziegler Catalysts*, Springer-Verlag, Berlin, 1995, pp. 275–289.
109. Jordan, R. F., *Adv. Organomet. Chem.*, **32**, 325 (1991).
110. Yang, X., Stern, C. L. and Marks, T. J., *Organometallics*, **10**, 840 (1991).
111. Yang, X., Stern, C. L. and Marks, T. J., *J. Am. Chem. Soc.*, **113**, 3623 (1991).
112. Ewen, J. A., Elder, M. J., Jones, R. L., Haspeslagh, L., Atwood, J. L., Bott, S. G. and Robinson, K., *Makromol. Chem., Macromol. Symp.*, **48/49**, 253 (1991).
113. Ewen, J. A. and Elder, M. J., *Makromol. Chem., Macromol. Symp.*, **66**, 179 (1993).
114. Corradini, P., Busico, V. and Cipullo, R., *Makromol. Chem., Rapid Commun.*, **13**, 21 (1992).
115. Cam, D. and Giannini, U., *Makromol. Chem.*, **193**, 1049 (1992).
116. Mallin, D. T., Rausch, M. D., Lin, Y.-G., Dong, S. and Chien, J. C. W., *J. Am. Chem. Soc.*, **112**, 2030 (1990).
117. Zambelli, A., Pellecchia, C. and Oliva, L., *Makromol. Chem., Macromol. Symp.*, **48/49**, 297 (1991).
118. Antberg, M., Dolle, V., Haftka, S., Rohrmann J., Spaleck, W., Winter, A. and Zimmermann, H. J., *Makromol. Chem., Macromol. Symp.*, **48/49**, 333 (1991).
119. Erker, G., Nolte, R., Aul, R., Wilker, S., Krüger, C. and Noe, R., *J. Am. Chem. Soc.*, **113**, 7594 (1991).
120. Chien, J. C. W., *Makromol. Chem., Macromol. Symp.*, **63**, 209 (1992).
121. Herfert, N. and Fink. G., *Makromol. Chem., Macromol. Symp.*, **66**, 157 (1993).
122. Farina, M., *Trends Polym. Sci.*, **2**, 85 (1994).
123. D'yachkovskii, F. S. and Tsvetkova, V. I., *Kinetics Catal.*, **35**, 487 (1994).
124. Ewen, J. A., *Macromol. Symp.*, **89**, 181 (1995).
125. Spaleck, W., Aulbach, M., Bachmann, B., Küber, F. and Winter, A., *Macromol. Symp.*, **89**, 237 (1995).
126. Galimberti, M., Martini, E., Piemontesi, F., Sartori, F., Camurati, I. and Albizzati, E., *Macromol. Symp.*, **89**, 259 (1995).
127. Razavi, A., Peters, L., Nafpliotis, L., Vereecke, D., Den Dauw, K., Atwood, J. L. and Thewald U., *Macromol. Symp.*, **89**, 345 (1995).
128. Coates, G. W. and Waymouth, R. M., *Science*, **267**, 217 (1995).
129. Hauptman, E., Waymouth, R. M. and Ziller, J. W., *J. Am. Chem. Soc.*, **117**, 11586 (1995).
130. Hoveyda, A. H. and Morken, J. P., *Angew. Chem., Int. Ed. Engl.*, **35**, 1262 (1996).
131. Herrmann, W. A., Rohrmann, J., Herdtweck, E., Spaleck, W. and Winter, A., *Angew. Chem., Int. Ed. Engl.*, **28**, 1511 (1989).
132. Mise, T., Miya, S. and Yamazaki, H., *Chem. Lett.*, **1989**, 1853 (1989).
133. Cheng, H. N. and Ewen, J. A., *Makromol. Chem.*, **190**, 1931 (1989).
134. Spaleck, W., Antberg, M., Dolle, V., Klein, R., Rohrmann J. and Winter, A., *New J. Chem.*, **14**, 499 (1990).

135. Kaminsky, W. and Buschermöhle, M., in *Recent Advances in Mechanistic and Synthetic Aspects of Polymerization*, D. Reidel, Dordrecht, 1987, p. 503.
136. Jolly, C. A. and Marynick, D. S., *J. Am. Chem. Soc.*, **111**, 7968 (1989).
137. Yu, Z. and Chien, J. C. W., *J. Polym. Sci., A, Polym. Chem.*, **33**, 1085 (1997).
138. Röll, W., Brintzinger, H. H., Rieger, B. and Zolk, R., *Angew. Chem., Int. Ed. Engl.*, **29**, 279 (1990).
139. Spaleck, W., Antberg, M., Aulbach, M., Bachmann, B., Dolle, V., Haftka, S., Küber, F., Rohrmann, J. and Winter, A., 'New Isotactic Polypropylenes via Metallocene Catalysts', in *Ziegler Catalysts*, Springer-Verlag, Berlin, 1995, pp. 83–97.
140. Ewen, J. A., Haspeslagh, L., Atwood, J. L. and Zhang, H., *J. Am. Chem. Soc.*, **109**, 6544 (1987).
141. Chien, J. C. W. and He, D., *J. Polym. Sci., A, Polym. Chem.*, **29**, 1585 (1991).
142. Herfert, N. and Fink, G., *Makromol. Chem.*, **193**, 773 (1992).
143. Razavi, A., Vereecke, D., Peters, L., Den Dauw, K., Nafpliotis, L. and Atwood, J. L., 'Manipulation of the Ligand Structure as an Effective and Versatile Tool for Modification of Active Site Properties in Homogeneous Ziegler–Natta Catalyst Systems', in *Ziegler Catalysts*, Springer-Verlag, Berlin, 1995, pp. 111–147.
144. Schnutenhaus, H. and Brintzinger H. H., *Angew. Chem.*, **91**, 837 (1979).
145. Katz, T. J. and Pesti, J., *J. Am. Chem. Soc.*, **104**, 346 (1982).
146. Ewen, J. A. and Elder, M. J., 'Isospecific Pseudo-helical Zirconocenium Catalysts', in *Ziegler Catalysts*, Springer-Verlag, Berlin, 1995, pp. 99–109.
147. Kravchenko, R. L., Waymouth, R. M. and Masood, Md. A., *Polym. Prepr. Am. Chem. Soc., Div. Polym. Chem.*, **37**(2), 475 (1996).
148. Stevens, J. C., Timmers, F. J., Wilson, D. R., Schmidt, D. F., Nickias, P. N., Rosen, R. K., Knight, G. W. and Lay, S. Y., Eur. Pat. Appl. 0416 815 A2, US Pat Appl. 0401 345 (to Dow Chemical Co.) (1990); *Chem. Abstr.*, **115**, 93163 m (1990).
149. Canich, J. A. M., Eur. Pat. Appl. 0420 436 (to Exxon Chem. Inc.) (1990).
150. Shapiro, P. J., Bunel, E., Schaefer, W. P. and Bercaw, J. E., *Organometallics*, **9**, 867 (1990).
151. Shapiro, P. J., Cotter, W. D., Schaefer, W. P., Labinger, J. A. and Bercaw, J. E., *J. Am. Chem. Soc.*, **116**, 4623 (1994).
152. Flores, J. C., Chien, J. C. W. and Rausch, M. D., *Organometallics*, **13**, 4140 (1994).
153. Canich, J. A. M., US Pat. 5 026 798 (to Exxon Chem. Inc.) (1991); *Chem. Abstr.*, **118**, 60284 (1994).
154. Turner, H. W., Hlatky, G. G. and Canich, J. A. M., PCT Int. Appl. WO 93 19 103 (to Exxon Chem. Inc.) (1993); *Chem. Abstr.*, **120**, 271442 (1994).
155. McKnight, A. L., Straus, D. A., Masood, Md. A. and Waymouth, R. M., *Polym. Prepr. Am. Chem. Soc., Div. Polym. Chem.*, **37**(2), 474 (1996).
156. Canich, J. A. M. and Turner, H. W., PCT Int. Appl. WO 92 12 612 (to Exxon Chem. Inc.) (1992).
157. Kempe, R., Oberthür, M., Hillebrand, G., Spannenberg, A. and Fuhrmann, H., *Polimery*, **42**, (1997).
158. Tjaden, E. B. and Jordan, R. F., *Macromol. Symp.*, **89**, 231 (1995).
159. Van der Linden, A., Schaverien, C. J., Meijboom, N., Ganter, C. and Orpen, A. G., *J. Am. Chem. Soc.*, **117**, 3008 (1995).
160. Kuran, W. and Sobota, P., unpublished data.
161. Oliva, L., Longo, P. and Pellecchia, C., *Makromol. Chem., Rapid Commun.*, **9**, 51 (1988).
162. Pellecchia C., Pappalardo, D. and van Beek, J. A. M., *Macromol. Symp.*, **89**, 335 (1995).

163. Bochmann, M. and Wilson, L. M., *J. Chem. Soc., Chem. Commun.*, **1986**, 1610 (1986).
164. Bochmann, M. and Jaggar, A. J., *J. Organomet. Chem.*, **424**, C5 (1992).
165. Bochmann, M. and Lancaster, S. J., *J. Organomet. Chem.*, **434**, C1 (1992).
166. Exhuis, J. J. W., Tan, Y. Y. and Teuben, J. H., *J. Mol. Catal.*, **62**, 277 (1990).
167. Exhuis, J. J. W., Tan, Y. Y., Meetsma, A., Teuben, J. H., Renkema, J. and Evens, G. G., *Organometallics*, **11**, 362 (1992).
168. Taube, J. and Krukowka, L., *J. Organomet. Chem.*, **347**, C9 (1988).
169. Hlatky, G. G., Eckman, R. R. and Turner, H. W., *Organometallics*, **11**, 1413 (1992).
170. Pellecchia, C., Proto, A., Longo, P. and Zambelli, A., *Makromol. Chem., Rapid Commun.*, **13**, 277 (1992).
171. Herfert, N. and Fink, G., *Makromol. Chem., Rapid Commun.*, **14**, 91 (1993).
172. Jordan, R. F., Dasher, W. E. and Echols, S. F., *J. Am. Chem. Soc.*, **108**, 1718 (1986).
173. Jordan, R. F., Bradley, P., Baenziger, N. C. and La Pointe, R. E., *J. Am. Chem. Soc.*, **112**, 1289 (1990).
174. Crowther, D. J., Borkowsky, S. L., Swenson, D., Meyer, T. Y. and Jordan, R. F., *Organometallics*, **12**, 2897 (1993).
175. Ballard, D. G. H., Courtis, A., Holton, J., McMeeking, J. and Pearce, R., *J. Chem. Soc., Chem. Commun.*, **1978**, 994 (1978).
176. Watson, P. L., *J. Am. Chem. Soc.*, **104**, 337 (1982).
177. Watson, P. L. and Roe, D. C., *J. Am. Chem. Soc.*, **104**, 6471 (1982).
178. Parkin, G., Bunel, E., Burger, B. J., Trimmer, M. S., van Asselt, A. and Bercaw, J. E., *J. Mol. Catal.*, **41**, 21 (1987).
179. Burger, B. J., Thompson, M. E., Cotter, W. D. and Bercaw, J. E., *J. Am. Chem. Soc.*, **112**, 1566 (1990).
180. Jeske, G., Lauke, H., Mauermann, H., Sweptson, P. N., Schumann, H. and Marks, T. J., *J. Am. Chem. Soc.*, **107**, 8091 (1985).
181. Keim, W., Kowaldt, F. H., Goddard, R. and Krüger, C., *Angew. Chem.*, **90**, 493 (1978).
182. Keim, W., Appel, R., Storeck, A., Krüger, C. and Goddard, R., *Angew. Chem. Int. Ed. Engl.*, **20**, 116 (1981).
183. Fink, G., Möhring, V., Heinrichs, A. and Denger, Ch., 'Migratory Nickel(0)–Phosphorane Catalyst', in *Catalysis in Polymer Synthesis*, ACS Symp. Ser. 496, Washington, DC, 1992, pp. 87–103.
184. Siedle, A. R., Newmark, R. A., Lamanna, W. M. and Schroepfer, J. N., *Polyhedron*, **9**, 301 (1990).
185. Horton, A. D. and Frijns, J. H. G., *Angew. Chem., Int. Ed. Engl.*, **30**, 1152 (1991).
186. Hlatky, G. G., Upton, D. J. and Turner, H. W., US Pat. Appl. 459 921 (1990); *Chem. Abstr.*, **115**, 256897v (1991).
187. Burger, B. J., Cotter, W. D., Coughlin, E. B., Chacon, S. T., Hajela, S., Herzog, T. A., Köhn, R., Mitchell, J., Piers, W. E., Shapiro, P. J. and Bercaw, J. E., 'Olefin Polymerisation with Single Component Organoscandium and Organoyttrium Catalysts', in *Ziegler Catalysts*, Springer-Verlag, Berlin, 1995, pp. 317–331.
188. Giannini, U., Zucchini, U. and Albizzati, E., *J. Polym. Sci., Polym. Lett. Ed.*, **8**, 405 (1970).
189. Pellecchia, C., Grassi, A. and Immirzi, A., *J. Am. Chem. Soc.*, **115**, 1160 (1993).
190. Pellecchia, C., Grassi, A. and Zambelli, A., *J. Mol. Catal.*, **82**, 57 (1993).
191. Möhring, V. M. and Fink, G., *Angew. Chem., Int. Ed. Engl.*, **24**, 1001 (1985).
192. Endo, K., Ueda, R. and Otsu, T., *J. Polym. Sci., A, Polym. Chem.*, **29**, 807 (1991).
193. Endo, K., Ueda, R. and Otsu, T., *J. Polym. Sci., A, Polym. Chem.*, **29**, 843 (1991).

194. Endo, K., Ueda, R. and Otsu, T., *Makromol. Chem.*, **193**, 539 (1992).
195. Endo, K. and Otsu, T., *J. Polym. Sci., A, Polym. Chem.*, **33**, 79 (1995).
196. Kaminaka, M. and Soga, K., *Makromol. Chem., Rapid Commun.*, **12**, 367 (1991).
197. Kaminaka, M. and Soga, K., *Polym. Commun.*, **33**, 1105 (1992).
198. Soga, K. and Kaminaka, M., *Makromol. Chem., Rapid Commun.*, **13**, 221 (1992).
199. Soga, K. and Kaminaka, M., *Makromol. Chem., Rapid Commun.*, **15**, 593 (1994).
200. Soga, K. and Kaminaka, M., *Macromol. Chem. Phys.*, **195**, 1369 (1994).
201. Soga, K. and Kaminaka, M., *Makromol. Chem.*, **194**, 1745 (1993).
202. Chien, J. C. W. and He, D., *J. Polym. Sci., A, Polym. Chem.*, **29**, 1603 (1991).
203. Collins, S., Kelly, W. M. and Holden, D. A., *Macromolecules*, **25**, 1780 (1992).
204. Kaminsky, W. and Renner, F., *Makromol. Chem., Rapid Commun.*, **14**, 239 (1993).
205. Kaminsky, W., *Macromol. Symp.*, **89**, 203 (1995).
206. Soga, K., Kim, H. J. and Shiono, T., *Macromol. Rapid Commun.*, **15**, 139 (1994).
207. Soga, K., *Macromol. Symp.*, **89**, 249 (1995).
208. Soga, K., Arai, T., Nozawa, H. and Uozumi, T., *Macromol. Symp.*, **97**, 53 (1995).
209. Soga, K., Kaminaka, M., Kim, H. J. and Shiono, T., 'Heterogeneous Metallocene Catalysts', in *Ziegler Catalysts*, Springer-Verlag, Berlin, 1995, pp. 333–342.
210. Soga, K., Kim, H. J. and Shiono, T., *Makromol. Chem., Rapid Commun.*, **14**, 765 (1993).
211. Otto, E., *Kunststoffe*, **83**, 188 (1993).
212. Kaminsky, W., in *Transition Metal Catalyzed Polymerization: Alkenes and Dienes*, Harwood Academic Publishers, New York, 1981, p. 225.
213. Lee, D.-H., Yoon, K.-B. and Huh, W.-S., *Macromol. Symp.*, **97**, 185 (1995).
214. Karol, F. J., Karapkina, G. L., Wu, C., Dow, A. W., Johnson, R. L. and Carrick, W. L., *J. Polym. Sci., A*, **10**, 2621 (1972).
215. Karol, F. J., Wu, C., Beichle, W. T. and Marschin, N. J., *J. Catal.*, **60**, 68 (1979).
216. Ballard, D. G. H., *Adv. Catal.*, **23**, 263 (1973).
217. Ballard, D. G. H., Jones, E., Wyatt, J. R., Murray, R. T. and Robinson, P. A., *Polymer*, **15**, 169 (1974).
218. Yermakov, Yu, I., *Catal. Rev.*, **13**, 77 (1976).
219. Yermakov, Yu. I. and Zakharov, V. A., *Adv. Catal.*, **24**, 173 (1975).
220. Hogan, J. P. and Banks, R. L., US Pat. 2 825 721 (to Phillips Petroleum Co.) (1954); *Chem. Abstr.*, **52**, 8621i (1958).
221. Clark, A., Hogan, J. P., Banks, R. L. and Lanning, W. C., *Ind. Eng. Chem.*, **48**, 1152 (1956).
222. McDaniel, M. P. and Welch, M. B., *J. Catal.*, **82**, 98 (1987).
223. Hogan, J. P., *J. Polym. Sci., A-1*, **8**, 2637 (1970).
224. McDaniel, M. P., *Adv. Catal.*, **33**, 47 (1985).
225. Groeneveld, C., Wittgen, P. M., Swinnen, H. D. M., Wernsen, A. and Schuit, G. C. A., *J. Catal.*, **83**, 346 (1983).
226. Baker, L. M. and Carrick, W. L., *J. Org. Chem.*, **35**, 774 (1970).
227. Rebenstorf, B., *Z. Anorg. Allg. Chem.*, **513**, 103 (1984).
228. Myers, D. L. and Lunsford, J. H., *J. Catal.*, **92**, 260 (1985).
229. Rebenstorf, B., *J. Mol. Catal.*, **38**, 355 (1986).
230. Lunsford, J. H., Fu, S.-L. and Myers, D. L., *J. Catal.*, **111**, 231 (1988).
231. Rebenstorf, B., *Z. Anorg. Allg. Chem.*, **571**, 148 (1989).
232. Rebenstorf, B., *J. Polym. Sci., A, Polym. Chem.*, **29**, 1949 (1991).
233. Krauss, H. I. and Hums, E. Z., *Z. Naturforsch., B Anorg. Chem., Org. Chem.*, **34B**, 1628 (1979).
234. Krauss, H. I. and Hums, E. Z., *Z. Naturforsch., B Anorg. Chem., Org. Chem.*, **35B**, 848 (1980).

235. Krauss, H. I. and Hums, E. Z., *Z. Naturforsch., B Anorg. Chem., Org. Chem.*, **38B**, 1412 (1983).
236. Natta, G., *Chim. Ind. (Milan).*, **37**, 888 (1955).
237. Hogan, J. P. and Witt, R. R., *Prepr. Am. Chem. Soc., Div. Petrol. Chem.*, **24**(2), 377 (1979).
238. Carrick, W. C., *J. Polym. Sci., A*, **10**, 2609 (1972).
239. Zletz, A., US Pat. 2 692 257 (to Standard Oil of Indiana) (1951); *Chem. Abstr.*, **49**, 2777d (1955).
240. Tait, P. J. T. and Watkins, N. D., 'Monoalkene Polymerisation: Mechanisms', in *Comprehensive Polymer Science*, Pergamon Press, Oxford, 1989, Vol. 4, pp. 533–573.
241. Pino, P., Giannini, U. and Porri, L., 'Insertion Polymerisation', in *Encyclopedia of Polymer Science and Engineering*, Wiley-Interscience, John Wiley & Sons, New York, 1987, Vol. 8, pp. 147–220.
242. Burfield, D. R., *Polymer*, **25**, 1645 (1984).
243. Fink, G., Herfert, P. and Montag, P., 'The Relationship Between Kinetics and Mechanism', in *Ziegler Catalysts*, Springer-Verlag, Berlin, 1995, pp. 159–179.
244. Hungenberg, K. D., Kerth, J., Langhauser, F., Marczinke, B. and Schlund, R., 'Gas Phase Polymerization of α-Olefins with Ziegler–Natta and Metallocene Catalysts: a Comparison', in *Ziegler Catalysts*, Springer-Verlag, Berlin, 1995, pp. 363–386.
245. Kashiwa, N. and Yoshitake, J., *Makromol. Chem.*, **183**, 1133 (1984).
246. Goodall, J., *J. Chem. Educ.*, **63**, 191 (1986).
247. Simonazzi, T., Cecchin, G. and Mazzullo, S., *Prog. Polym. Sci.*, **16**, 303 (1991).
248. Pino, P. and Rotzinger, B., *Makromol. Chem. Suppl.*, **7**, 41 (1984).
249. Boor Jr, J., *J. Polym. Sci., C*, **1**, 237 (1963).
250. Kissin, Y. V., *Adv. Polym. Sci.*, **15**, 91 (1974).
251. Burfield, D. R., McKenzie, I. D. and Tait, P. J. T., *Polymer*, **13**, 302 (1972).
252. Zakharov, V. A., Bukatov, G. D., Chumaevskii, N. B. and Yermakov, Yu, I., *Makromol. Chem.*, **178**, 967 (1977).
253. Busico, V., Cipullo, R. and Corradini, P., *Makromol. Chem. Rapid Commun.*, **14**, 97 (1993).
254. Rieger, B. and Chien, J. C. W., *Polym. Bull.*, **21**, 159 (1989).
255. Chen, Y. X., Rausch, M. D. and Chien, J. C. W., *Organometallics*, **13**, 748 (1994).
256. Stehling, U., Diebold, J., Kirsten, R., Röll, W., Brintzinger, H. H., Jüngling, S., Mülhaupt, R. and Langhauser, F., *Organometallics*, **13**, 964 (1994).
257. Soga, K., Shiono, T., Takemura, S., Takedi, S. and Kaminsky, W., *Makromol. Chem., Rapid Commun.*, **8**, 305 (1987).
258. Grassi, A., Zambelli, A., Resconi, L., Albizzati, E. and Mazzocchi, R., *Macromolecules*, **21**, 617 (1988).
259. Tsutsui, T., Mizuno, A. and Kashiwa, N., *Makromol. Chem.*, **190**, 1177 (1989).
260. Tsutsui, T., Kashiwa, N. and Mizuno, A., *Makromol. Chem., Rapid Commun.*, **11**, 565 (1990).
261. Asakara, T., Nakayama, N., Demura, M. and Asano, A., *Macromolecules*, **25**, 4876 (1992).
262. Tsutsui, T., Mizuno, A. and Kashiwa, N., *Polymer*, **30**, 428 (1989).
263. Spaleck, W., Küber, F., Winter, A., Rohrmann, J., Bachmann, B., Antberg, M., Dolle, V. and Paulus, E. F., *Organometallics*, **13**, 954 (1994).
264. Chien, J. C. W. and He, D., *J. Polym. Sci., A, Polym. Chem.*, **29**, 1595 (1991).
265. Aladyshev, A. M., Tsvetkova, V. L. and D'yachkovskii, F. S., *Vysokomol. Soed., Ser. A*, **33**, 865 (1991).
266. Toyota, A., Tsutsui, T. and Kashiwa, N., *J. Mol. Catal.*, **56**, 237 (1989).

267. Kaminsky, W. and Lücker, H., *Makromol. Chem., Rapid Commun.*, **5**, 225 (1989).
268. Cossee, P., *J. Catal.*, **3**, 80 (1964).
269. Mülhaupt, R., Klabunde, U. and Ittel, S., *J. Chem. Soc., Chem. Commun.*, **1985**, 1745 (1985).
270. Keii, T., Suzuki, E., Tamura, M., Murata, M. and Doi, Y., *Makromol. Chem.*, **183**, 2285 (1982).
271. Zakharov, V. A., Chumaevskii, N. B., Bukatov, G. D. and Yermakov, Y. I., *Makromol. Chem.*, **177**, 763 (1976).
272. Kawamura-Kuribayashi, H., Koga, N. and Morokuma, K., *J. Am. Chem. Soc.*, **114**, 8687 (1992).
273. Castonguay, L. A. and Rappé, A. K., *J. Am. Chem. Soc.*, **114**, 5832 (1992).
274. Prosenc, M.-H., Janiak, C. and Brintzinger, H. H., *Organometallics*, **11**, 4036 (1992).
275. Miyazawa, T. and Ideguchi, T., *J. Polym. Sci., B*, **1**, 389 (1963).
276. Cossee, P., *Tetrahedron Lett.*, **17**, 12 (1960).
277. Arlman, E. J. and Cossee, P., *J. Catal.*, **3**, 99 (1964).
278. Arlman, E. J., *J. Catal.*, **3**, 89 (1964).
279. Corradini, P., Barone, V., Fusco, R. and Guerra, G., *Eur. Polym. J.*, **15**, 133 (1979).
280. Corradini, P., Barone, V., Fusco, R. and Guerra, G., *J. Catal.*, **77**, 32 (1982).
281. Corradini, P., Barone, V. and Guerra, G., *Macromolecules*, **15**, 1242 (1982).
282. Corradini, P., Guerra, G. and Villani, V., *Macromolecules*, **18**, 1401 (1985).
283. Corradini, P., Barone, V., Fusco, R. and Guerra, G., *Gazz. Chim. Ital.*, **113**, 601 (1983).
284. Ystenses, M., *J. Catal.*, **129**, 383 (1991).
285. Rodriguez, L. A. M. and Van Looy, H. M., *J. Polym. Sci., A-1*, **4**, 1951 (1966).
286. Rodriguez, L. A. M. and Van Looy, H. M., *J. Polym. Sci., A-1*, **4**, 1971 (1966).
287. Doi, Y., Suzuki, E. and Keii, T., 'Alkenes and Dienes', in *Transition Metal Catalyzed Polymerizations, M.M.I. Press Symp.*, Harwood, New York, 1983, Ser. 4, Part A, pp. 47–82.
288. Pino, P., Guastalla, G., Rotzinger, B. and Mülhaupt, R., in *Proc. M.M.I. Int. Symp. on Transition Metal Catalyzed Polymerization*, Harwood, New York, 1981, part A, p. 435.
289. Doi, Y., *Makromol. Chem., Rapid Commun.*, **3**, 635 (1982).
290. Albizzati, E., Giannini, U., Morini, G., Smith, C. A. and Zeigler, R. C., 'Advances in Propylene Polymerization with $MgCl_2$-supported Catalysts', in *Ziegler Catalysts*, Springer-Verlag, Berlin, 1995, pp. 413–425.
291. Corradini, P., Guerra, G., Fusco, R. and Barone, V., *Eur. Polym. J.*, **16**, 835 (1980).
292. Sobota, P. and Utko, J., *Polym. Commun.*, **29**, 144 (1988).
293. Sobota, P. and Szafert, S., *Inorg. Chem.*, **1996**, 1778 (1996).
294. Zefirova, A. K. and Shilov, A. E., *Dokl. Akad. Nauk SSSR*, **136**, 599 (1961).
295. Shilov, A. E., Shilova, A. K. and Bobkov, B. N., *Vysokomol. Soed.*, **4**, 1688 (1962).
296. Shilov, A. E. and D'yachkovskii, F. S., *Zh. Fiz. Khim.*, **41**, 2515 (1967).
297. D'yachkovskii, F. S., Shilova, A. K. and Shilov, A. E., *J. Polym. Sci., C*, **16**, 2333 (1967).
298. Farina, M. and Puppi, C., *J. Mol. Catal.*, **82**, 3 (1993).
299. L'vovskii, V. E., Fushman, E. A. and D'yachkovskii, F. S., *J. Mol. Catal.*, **10**, 43 (1981).
300. Locatelli, P., *Trends Polym. Sci.*, **2**, 87 (1994).
301. Soga, K., Shiono, T. and Doi, Y., *Makromol. Chem.*, **189**, 1531 (1988).

302. Kashiwa, N., Yoshitake, J. and Toyota, A., *Polym. Bull.*, **19**, 333 (1988).
303. Corradini, P. and Guerra, G., *Prog. Polym. Sci.*, **16**, 239 (1991).
304. Ludlum, D. B., Anderson, A. W. and Ashby, C. E., *J. Am. Chem. Soc.*, **80**, 1380 (1959).
305. Spencer, M. D., Morse, A. M., Wilson, S. R. and Girolami, G. S., *J. Am. Chem. Soc.*, **115**, 2057 (1993).
306. Eisch, J. J., Pombrik, S. I., Shi, X. and Wu, S.-C., *Macromol. Symp.*, **89**, 221 (1995).
307. Fellman, J. D., Rupprecht, G. A. and Schrock, R. R., *J. Am. Chem. Soc.*, **101**, 5099 (1979).
308. Turner, H. W., Schrock, R. R., Fellman, J. D. and Holmes, S. J., *J. Am. Chem. Soc.*, **105**, 4942 (1983).
309. McKinney, R. J., *J. Chem. Soc., Chem. Commun.*, **1980**, 490 (1980).
310. Ivin, K. J., Rooney, J. J. and Stewart, C. D., *J. Chem. Soc., Chem. Commun.*, **1978**, 603 (1978).
311. Ivin, K. J., Rooney, J. J., Stewart, C. D. and Green, M. L. H., *J. Chem. Soc., Chem. Commun.*, **1978**, 604 (1978).
312. Green, M. L. H., *Pure Appl. Chem.*, **30**, 27 (1978).
313. Krauss, H. I. and Xing, Q., 'Organochromium Species in Phillips Catalysis', in *Proc. 2nd International School on Molecular Catalysis 'Organometallics and Catalysis'*, Poznan-Kiekrz, Poland, 1995, Abstracts, p. L-9.
314. Zambelli, A., Natta, G. and Pasquon, I., *J. Polym. Sci., C*, **4**, 411 (1963).
315. Zambelli, A., Pasquon, I., Signorini, R. and Natta, G., *Makromol. Chem.*, **112**, 160 (1968).
316. Natta, G., Mazzanti, G., Crespi, G. and Moraglio, G., *Chim. Ind. (Milan)*, **39**, 275 (1957).
317. Boor Jr, J. and Youngman, E. A., *J. Polym. Sci., A-1*, **4**, 1861 (1966).
318. Youngman, E. A. and Boor Jr, I., *Macromol. Rev.*, **2**, 33 (1967).
319. Lehr, M. H., *Macromolecules*, **1**, 178 (1968).
320. Lehr, M. H. and Corman, C. J., *Macromolecules*, **2**, 217 (1969).
321. Porri, L., Di Corato, A. and Natta, G., *Eur. Polym. J.*, **5**, 1 (1969).
322. Racanelli, P. and Porri, L., *Eur. Polym. J.*, **6**, 751 (1970).
323. Doi, Y., Kinoshiia, J., Morinaga, A. and Keii, T., *J. Polym. Sci., A-1*, **13**, 249 (1975).
324. Natta, G., Zambelli, A., Lanzi, S., Pasquon, I., Mognaschi, E. R., Segre, A. L. and Centola, P., *Makromol. Chem.*, **81**, 161 (1965).
325. Emde, H., *Angew. Makromol. Chem.*, **60**, 1 (1977).
326. Gumboldt, A., Helberg, J. and Schleitzer, G., *Makromol. Chem.*, **101**, 229 (1967).
327. Zambelli, A. and Allegra, G., *Macromolecules*, **13**, 42 (1980).
328. Corradini, P., Guerra, G. and Pucciariello, R., *Macromolecules*, **18**, 2030 (1985).
329. Tolman, C. A., *Chem. Soc. Rev.*, **1**, 337 (1972).
330. Reichert, K. H., Berthold, J. and Dornow, V., *Makromol. Chem.*, **121**, 258 (1969).
331. Reichert, K. H. and Schubert, E., *Makromol. Chem.*, **123**, 58 (1969).
332. Meyer, K. and Reichert, K. H., *Angew. Makromol. Chem.*, **12**, 175 (1970).
333. Reichert, K. H., *Angew. Makromol. Chem.*, **13**, 177 (1970).
334. Schnell, D. and Fink, G., *Angew. Makromol. Chem.*, **39**, 131 (1974).
335. Fink, G., Rottler, R. and Kreiter, C. G., *Angew. Makromol. Chem.*, **96**, 1 (1981).
336. Fink, G., Fenzl, W. and Mynott, R., *Z. Naturforsch. B*, **40**, 158 (1985).
337. Mynott, R., Fink, G. and Fenzl, W., *Angew. Makromol. Chem.*, **154**, 1 (1987).
338. Eisch, J. J., Piotrowski, A. M., Brownstein, S. K., Gabe, E. J. and Lee, F. L., *J. Am. Chem. Soc.*, **107**, 7219 (1985).
339. Eisch, J. J. and Boleslawski, M. P., *J. Organomet. Chem.*, **334**, C1 (1987).

340. Eisch, J. J., Caldwell, K. R., Werner, S. and Krüger, C., *Organometallics*, **10**, 3417 (1991).
341. Eisch, J. J., Pombrik, S. I. and Zheng, G. X., *Organometallics*, **12**, 3856 (1993).
342. D'yachkovskii, F. S., Grigoryan, E. A. and Babkina, O. N., *Int. J. Chem. Kinet.*, **13**, 603 (1981).
343. Laverty, D. T. and Rooney, J. J., *J. Chem. Soc., Faraday Trans.*, **79**, 89 (1983).
344. Brookhart, M. and Green, M. L. H., *J. Organomet. Chem.*, **250**, 395 (1983).
345. Corradini, P., Guerra, G., Cavallo, L., Moscardi, G. and Vacatello, M., 'Models for the Explanation of the Stereospecific Behaviour of Ziegler–Natta Catalysts', in *Ziegler Catalysts*, Springer-Verlag, Berlin, 1995, pp. 237–249.
346. Clawson, L., Soto, J., Buchwald, S. L. and Steigerwald, M. R., *J. Am. Chem. Soc.*, **107**, 3377 (1985).
347. Brookhart, M., Green, M. L. H. and Wong, L., *Prog. Inorg. Chem.*, **36**, 1 (1988).
348. Burger, B. J., Thompson, M. E., Cotter, W. D. and Bercaw, J. E., *J. Am. Chem. Soc.*, **112**, 1566 (1990).
349. Piers, W. E. and Bercaw, J. E., *J. Am. Chem. Soc.*, **112**, 9406 (1990).
350. Krauledat, H. and Brintzinger, H. H., *Angew. Chem., Int. Ed. Engl.*, **29**, 1412 (1990).
351. Woo, T. K., Fan, L. and Ziegler, T., *Organometallics*, **13**, 432 (1994).
352. Fan, L., Harrison, D., Woo, T. K. and Ziegler, T., *Organometallics*, **14**, 2018 (1995).
353. Obara, S., Koga, N. and Morokuma, K., *J. Organomet. Chem.*, **270**, C33 (1984).
354. Koga, N., Obara, S. and Morokuma, K., *J. Am. Chem. Soc.*, **106**, 4625 (1984).
355. Eisenstein, O. and Jean, Y., *J. Am. Chem. Soc.*, **107**, 1177 (1985).
356. Guo, Z., Svenson, D. C. and Jordan, R. F., *Organometallics*, **13**, 1424 (1994).
357. Guerra, G., Cavallo. L., Moscardi, G., Vacatello, M. and Corradini, P., *J. Am. Chem. Soc.*, **116**, 2988 (1994).
358. Woo, T. K., Fan, L. and Ziegler, T., 'A Combined Density Functional and Molecular Mechanics Study on Olefin Polymerisation by Metallocene Catalysts', in *Ziegler Catalysts*, Springer-Verlag, Berlin, 1995, pp. 291–315.
359. Harlan, C. J., Bott, S. G. and Barron, A. R., *J. Am. Chem. Soc.*, **117**, 6465 (1995).
360. Tritto, I., Sacchi, M. C., Locatelli, P. and Li, S. X., *Macromol. Symp.*, **97**, 101 (1995).
361. Brookhart, M., Volpe Jr, A. F. and Lincoln, D. M., *J. Am. Chem. Soc.*, **112**, 5634 (1990).
362. Corradini, P., Paiaro, G. and Panunzi, A., *J. Polym. Sci., C*, **16**, 2905 (1967).
363. Hanson, K. R., *J. Am. Chem. Soc.*, **88**, 2731 (1966).
364. Allegra, G., *Makromol. Chem.*, **145**, 235 (1971).
365. Corradini, P., Guerra, G., Vacatello, M. and Villani, V., *Gazz. Chim. Ital.*, **118**, 173 (1988).
366. Natta, G., *J. Inorg. Nucl. Chem.*, **8**, 589 (1958).
367. Locatelli, P., Tritto, I. and Sacchi, M. C., *Makromol. Chem., Rapid Commun.*, **5**, 495 (1984).
368. Zambelli, A., Sacchi, M. C., Locatelli, P. and Zannoni, G., *Macromolecules*, **15**, 211 (1982).
369. Locatelli, P., Sacchi, M. C., Tritto, I., Zannoni, G., Zambelli, A. and Piscitelli, V., *Macromolecules*, **18**, 627 (1985).
370. Crain Jr, W. O., Zambelli, A. and Roberts, J. D., *Macromolecules*, **4**, 330 (1971).
371. Pino, P., Lorenzi, G. P. and Lardicci, L., *Chim. Ind. (Milan)*, **42**, 712 (1960).
372. Pino, P., Ciardelli, F., Lorenzi, G. P. and Natta, G., *J. Am. Chem. Soc.*, **84**, 1487 (1962).
373. Pino, P., Ciardelli, F. and Lorenzi, G. P., *J. Polym. Sci., C*, **4**, 21 (1963).

374. Pino, P., *Adv. Polym. Sci.*, **4**, 393 (1965).
375. Pino, P., Fochi, G., Piccolo, O. and Giannini, U., *J. Am. Chem. Soc.*, **104**, 7381 (1982).
376. Chiellini, E. and Marchetti, M., *Makromol. Chem.*, **169**, 59 (1973).
377. Zambelli, A., Ammendola, P., Sacchi, M. C., Locatelli, P. and Zannoni, G., *Macromolecules*, **16**, 341 (1983).
378. Zambelli, A., Ammendola, P., Longo, P. and Grassi, A., *Gazz. Chim. Ital.*, **117**, 579 (1987).
379. Zambelli, A., Proto, A. and Longo, P., 'Stereochemistry of Polymerisation of Some α-olefins in the Presence of Ziegler-type Catalysts', in *Ziegler Catalysts*, Springer-Verlag, Berlin, 1995, pp. 217–235.
380. Ciardelli, F., Locatelli, P., Marchetti, M. and Zambelli, A., *Makromol. Chem.*, **175**, 923 (1974).
381. Pierono, O., Stigliani, G. and Ciardelli, F., *Chim. Ind. (Milan)*, **52**, 289 (1970).
382. Ciardelli, F., Carlini, C., Montagnoli, G., Lardicci, L. and Pino, P., *Chim. Ind. (Milan)*, **50**, 860 (1968).
383. Lazzaroni, L., Salvadori, P. and Pino, P., *J. Chem. Soc., Chem. Commun*, **18**, 1164 (1970).
384. Carlini, C., Nocci, R. and Ciardelli, F., *J. Polym. Sci., Polym. Chem. Ed.*, **15**, 767 (1977).
385. Pino, P. and Mülhaupt, R., *Angew. Chem., Int. Ed. Engl.*, **19**, 857 (1980).
386. Zambelli, A., Sacchi, M. C. and Locatelli, P., *Macromolecules*, **12**, 1051 (1979).
387. Takegami, Y. and Suzuki, T., *Bull. Chem. Soc. Jpn.*, **42**, 848 (1969).
388. Doi, Y., *Macromolecules*, **12**, 248 (1979).
389. Zambelli, A., Wolfsgrubber, C., Zannoni, G. and Bovey, F. A., *Macromolecules*, **7**, 750 (1974).
390. Zambelli, A., Bajo, G. and Rigamonti, E., *Makromol. Chem.*, **179**, 1249 (1978).
391. Asakura, T., Ando, I., Nishioka, A., Doi, Y. and Keii, T., *Makromol. Chem.*, **178**, 791 (1977).
392. Bovey, F. A., Sacchi, M. C. and Zambelli, A., *Macromolecules*, **7**, 752 (1974).
393. Doi, Y., Nozawa, F. and Soga, K., *Makromol. Chem.*, **186**, 2529 (1985).
394. Zambelli, A., Lety, A., Tosi, A. and Pasquon, I., *Makromol. Chem.*, **115**, 73 (1968).
395. Zambelli, A. and Tosi, C., *Adv. Polym. Sci.*, **15**, 31 (1974).
396. Zambelli, A., Tosi, C. and Sacchi, M. C., *Macromolecules*, **5**, 649 (1972).
397. Locatelli, P., Immirzi, A. and Zambelli, A., *Makromol. Chem.*, **176**, 1121 (1975).
398. Sacchi, M. C., Rigamonti, E. and Locatelli, P., *Makromol. Chem.*, **182**, 2881 (1981).
399. Shista, C., Hatorn, M. R. and Marks, T. J., *J. Am. Chem. Soc.*, **114**, 1112 (1992).
400. Ammendola, P., Pellecchia, C., Longo, P. and Zambelli, A., *Gazz. Chim. Ital.*, **117**, 65 (1987).
401. Collins, S., Gauthier, W. J., Holden, D. A., Kuntz, B. A., Taylor, N. J. and Ward, D. G., *Organometallics*, **10**, 2061 (1991).
402. Guerra, G., Corradini, P., Cavallo, L. and Vacatello, M., *Macromol. Symp.*, **89**, 307 (1995).
403. Wild, F. R. W. P., Zsolnai, L., Huttner, G. and Brintzinger, H. H., *J. Organomet. Chem.*, **232**, 233 (1982).
404. Wild, F. R. W. P., Wasiucionek, M., Huttner, G. and Brintzinger, H. H., *J. Organomet. Chem.*, **288**, 63 (1985).
405. Longo, P., Grassi, A., Pellecchia, L. and Zambelli, A., *Macromolecules*, **20**, 1015 (1987).
406. Cavallo, L., Guerra, G., Oliva, L., Vacatello, M. and Corradini, P., *Polym. Commun.*, **30**, 16 (1989).

407. Pino, P., Cioni, P. and Wei, J., *J. Am. Chem. Soc.*, **109**, 6189 (1987).
408. Kaminsky, W., Ahlers, A. and Möller-Lindenhof, N., *Angew. Chem., Int. Ed. Engl.*, **28**, 1216 (1989).
409. Sacchi, M. C., Barsties, E., Tritto, I., Locatelli, P., Brintzinger, H. H. and Stehling, U., *Macromolecules*, **30**, 1267 (1997).
410. Busico, V., Cipullo, R., Talarico, G., Segre, A. L. and Chadwick, J. C., *Macromolecules*, **30**, 4786 (1997).
411. Shelden, R. A., Fueno, T., Tsunetsugu, T. and Furukawa, J., *J. Polym. Sci., B*, **3**, 23 (1965).
412. Doi, Y. and Asakura, T., *Makromol. Chem.*, **176**, 507 (1975).
413. Angermund, K., Hanuschik, A. and Nolte, M., 'Forcefield Calculations on Zirconocene Compounds', in *Ziegler Catalysts*, Springer-Verlag, Berlin, 1995, pp. 251–274.
414. Babu, G. N., Newmark, R. A., Cheng, H. N., Llinas, G. H. and Chien, J. C. W., *Macromolecules*, **25**, 7400 (1992).
415. Weiss, H., Ehrig, M. and Ahlrichs, R., *J. Am. Chem. Soc.*, **116**, 4919 (1994).
416. Meier, R. J., van Doremaele, G. H. J., Iarlori, S. and Buda, F., *J. Am. Chem. Soc.*, **116**, 7274 (1994).
417. Hortmann, K. and Brintzinger, H. H., *New J. Chem.*, **16**, 51 (1992).
418. Razavi, A., Peters, L. and Nafpliotis, L., *J. Mol. Catal. A: Chem.*, **115**, 129 (1997).
419. Van der Leek, Y., Angermund, K., Reffke, M., Kleinschmidt, R., Goretzki, R. and Fink, G., *Chem. Eur. J.*, **3**, 585 (1997).
420. Haselwander, T., Beck, S. and Brintzinger H. H., 'Binuclear Titanocene and Zirconocene Cations with μ-Cl and $\mu - CH_3$ Bridges in Metallocene-based Ziegler–Natta Catalyst Systems – solution-NMR studies', in *Ziegler Catalysts*, Springer-Verlag, Berlin, 1995, pp. 181–197.
421. Bowey, F. A. and Kwei, T. K., 'Microstructure and Chain Conformation of Macromolecules', in *Macromolecules: an Introduction to Polymer Science*, Academic Press, New York, 1979, pp. 207–271.
422. Tonelli, A. E., *Macromolecules*, **11**, 565 (1978).
423. Schilling, F. C and Tonelli, A. E., *Macromolecules*, **13**, 270 (1980).
424. Bowey, F. A., 'NMR Investigations on the Mechanism of Propagation in Vinyl Polymerisation', in *High Resolution NMR of Macromolecules*, Academic Press, New York, 1972, pp. 146–181.
425. Stehling, F. C. and Knox, J. R., *Macromolecules*, **8**, 595 (1975).
426. Zambelli, A., Locatelli, P., Bajo, G. and Bovey, F. A., *Macromolecules*, **8**, 687 (1975).
427. Tonelli, A. E. and Schilling, F. C., *Acc. Chem. Res.*, **14**, 233 (1981).
428. Tonelli, A. E., *Macromolecules*, **11**, 634 (1978).
429. Asanuma, T., Nishimori, Y., Ito, N., Uchikawa, M. and Shiomura, T., *Polym. Bull.*, **25**, 567 (1991).
430. Zambelli, A., Ammendola, P. and Proto, A., *Macromolecules*, **22**, 2126 (1989).
431. Oliva, L., Di Serio, M., Peduto, N. and Santacesaria, E., *Macromol. Chem. Phys.*, **195**, 217 (1994).
432. Campbell, T. W. and Haven, A. C., *J. Appl. Polym. Sci.*, **1**, 73 (1959).
433. Endo, K. and Otsu, T., *J. Polym. Sci., Polym. Chem. Ed.*, **17**, 1453 (1979).
434. De Boer, H. J. R. and Royan, B. W., *J. Mol. Catal.*, **90**, 171 (1994).
435. Chien, J. C. W., Vizzini, J. C. and Kaminsky, W., *Makromol. Chem., Rapid Commun.*, **13**, 479 (1992).
436. Zambelli, A., Grassi, A., Galimberti, M. and Perego, G., *Makromol. Chem., Rapid Commun.*, **13**, 269 (1992).

437. Zambelli, A., Grassi, A., Galimberti, M. and Perego, G., *Makromol. Chem. Rapid Commun.*, **13**, 467 (1992).
438. Nakano, M. and Novak, M. B., *Polym. Prepr. Am. Chem. Soc. Div. Polym. Chem.*, **37**(2), 200 (1996).
439. Baker Jr, W. P., *J. Polym. Sci., A*, **1**, 655 (1963).
440. Havinga, R. and Schors, A., *J. Macromol. Sci. – Chem. A*, **2**, 1 (1968).
441. Van den Enk, J. E. and van der Ploeg, H. J., *J. Polym. Sci., A-1*, **9**, 2403 (1971).
442. Otsuka, S., Mori, K., Siminoe, T. and Imaizumi, F., *Eur. Polym. J.*, **3**, 73 (1967).
443. Porri, L., Rossi, R. and Ingrosso, G., *Tetrahedron Lett.*, **1971**, 1083 (1971).
444. Endo, K., Ueda, R. and Otsu, T., *Macromolecules*, **24**, 6849 (1991).
445. Endo, K., Okayama, S. and Otsu, T., *Eur. Polym. J.*, **28**, 153 (1992).
446. Marvel, C. S. and Stille, J. K., *J. Am. Chem. Soc.*, **80**, 1740 (1958).
447. Kennedy, J. P. and Otsu, T., *Adv. Polym. Sci.*, **7**, 369 (1970).
448. Tait, P. J. T. and Berry, I. G., 'Monoalkene Polymerisation: Copolymerisation', in *Comprehensive Polymer Science*, Pergamon Press, Oxford, 1989, Vol. 4, pp. 575–584.
449. Cherdron, H. and Brekner, M., *Macromol. Symp.*, **89**, 543 (1995).
450. Mülhaupt, R., 'Novel Polyolefin Materials and Processes: Overview and Prospects', in *Ziegler Catalysts*, Springer-Verlag, Berlin, 1995, pp. 35–55.
451. Natta, G., Valvassori, A., Mazzanti, G., and Sartori, G., *Chim. Ind. (Milan)*, **40**, 896 (1958).
452. Kissin, Y. V., *Macromol. Symp.*, **89**, 113 (1995).
453. Valvassori, A. and Sartori, G., *Adv. Polym. Sci.*, **5**, 24 (1967).
454. Natta, G., Mazzanti, G., Valvassori, A., Sartori, G. and Barbagallo, A., *J. Polym. Sci.*, **51**, 429 (1961).
455. Natta, G., Mazzanti, G., Valvassori, A., Sartori, G. and Fiumani, O., *J. Polym. Sci.*, **50**, 911 (1960).
456. Corbelli, L., Milani, F. and Fabbri, R., *Kunststoffe*, **34**, 11 (1981).
457. Busico, V., Corradini, P., Fontana, P. and Savino, V., *Makromol. Chem., Rapid Commun.*, **6**, 743 (1985).
458. Yang, S.-L., Xu, Z.-K. and Feng, L.-X., *Makromol. Chem., Macromol. Symp.*, **63**, 233 (1992).
459. Wang, J.-G., Chen, H. and Huang, B.-T., *Makromol. Chem.*, **194**, 1807 (1993).
460. Sugano, T., Gotoh, Y., Fujita, T., Uozumi, T. and Soga, K., *Makromol. Chem.*, **193**, 43 (1992).
461. Zambelli, A., Grassi, A., Galimberti, M., Mazzocchi, R. and Piemontesi, F., *Makromol. Chem., Rapid Commun.*, **12**, 523 (1991).
462. Chien, J. C. W. and He, D., *J. Polym. Sci., A, Polym. Chem.*, **29**, 1609 (1991).
463. Katayama, H., Shiraishi, H., Hino, T., Ogane, T. and Imai, A., *Macromol. Symp.*, **97**, 109 (1995).
464. Endo, K. and Otsu, T., *J. Polym. Sci., Polym. Chem. Ed.*, **17**, 1453 (1979).
465. Endo, K. and Otsu, T., *J. Polym. Sci., Polym. Chem. Ed.*, **24**, 1505 (1986).
466. Endo, K., Fujii, K. and Otsu, T., *Makromol. Chem., Rapid Commun.*, **12**, 409 (1991).
467. Dall'Asta, G. and Mazzanti, G., *Makromol. Chem.*, **61**, 178 (1963).
468. Kaminsky, W., *Macromol. Chem. Phys.*, **197**, 3907 (1996).
469. Kaminsky, W., Bark, A. and Arndt, M., *Makromol. Chem., Macromol. Symp.*, **47**, 83 (1991).
470. Kaminsky, W. and Spiehl, R., *Makromol. Chem.*, **190**, 515 (1989).
471. Kaminsky, W., Engehausen, R. and Kopf, J., *Angew. Chem., Int. Ed. Engl.*, **34**, 2273 (1995).

472. Kaminsky, W. and Noll, A., 'Polymerization of Phenyl Substituted Cyclic Olefins with Metallocene/Aluminoxane Catalysts', in *Ziegler Catalysts*, Springer-Verlag, Berlin, 1995, pp. 149–158.
473. Klabunde, U., Tulip, T. H., Roe, D. C. and Ittel, S. D., *J. Organomet. Chem.*, **334**, 141 (1987).
474. Klabunde, U. and Ittel, S. D., *J. Mol. Catal.*, **41**, 123 (1987).
475. Sen, A. and Lai, T.-W., *J. Am. Chem. Soc.*, **104**, 3520 (1982).
476. Drent, E., van Broekhoven, J. A. M. and Doyle, M. J., *J. Organomet. Chem.*, **417**, 235 (1991).
477. Starkweather, H., 'Olefin–Carbon Monoxide Copolymers', in *Encyclopedia of Polymer Science and Engineering*, Wile-Interscience, John Wiley & Sons, New York, 1987, Vol. 10, pp. 369–373.
478. Sen, A., *Acc. Chem. Res.*, **26**, 303 (1993).
479. Sen, A., *Adv. Polym. Sci.*, **73/74**, 125 (1986).
480. Wong, P. K., van Doorn, J. A., Drent, E., Sudmeijer, O. and Stil, H. A., *Ind. Eng. Chem. Res.*, **32**, 986 (1993).
481. Drent, E., van Broekhoven, J. A. M., Doyle, M. J. and Wong, P. K., 'Palladium Catalyzed Copolymerization of Carbon Monoxide with Olefins to Alternating Polyketones and Polyspiroketals', in *Ziegler Catalysts*, Springer-Verlag, Berlin, 1995, pp. 481–496.
482. Valli, V. L. K. and Alper, H., *J. Polym. Sci., A, Polym. Chem.*, **33**, 1715 (1995).
483. Belov, G. P., Golodkov, O. N. and Dzhabieva, Z. M., *Macromol. Symp.*, **89**, 455 (1995).
484. Rix, F. C. and Brookhart, M., *J. Am. Chem. Soc.*, **117**, 1137 (1995).
485. Batistini, A., Consiglio, G. and Suter, U. W., *Angew. Chem., Int. Ed. Engl.*, **31**, 303 (1992).
486. Jiang, Z., Dahlen, G. M., Houseknecht, K. and Sen, A., *Macromolecules*, **25**, 2999 (1992).
487. Jiang, Z., Sanganeria, S. and Sen, A., *J. Polym. Sci., A, Polym. Chem.*, **32**, 841 (1994).
488. Xu, F. Y., Zhao, A. X. and Chien, J. C. W., *Makromol. Chem.*, **194**, 2579 (1993).
489. Amevor, E., Bronco, S., Consiglio, G. and Di Benedetto, S., *Macromol. Symp.*, **89**, 443 (1995).
490. Batistini, A. and Consiglio, G., *Organometallics*, **11**, 3604 (1992).
491. Bronco, G., Consiglio, G., Hutter, R., Batistini, A. and Suter, U. W., *Macromolecules*, **27**, 4436 (1994).
492. Jiang, Z. and Sen, A., *J. Am. Chem. Soc.*, **117**, 4455 (1995).
493. Sen, A. and Murtuza, S., *Polym. Prepr. Am. Chem. Soc., Div. Polym. Chem.*, **37**(2), 194 (1996).
494. Butler, G. B., 'Cyclopolymerisation', in *Encyclopedia of Polymer Science and Engineering*, Wile-Interscience, John Wiley & Sons, New York, 1986, Vol. 6, pp. 523–583.
495. Makowski, H. S., Shim, B. K. C. and Wilchinsky, Z. W., *J. Polym. Sci., A*, **2**, 1549 (1964).
496. Cheng, H. N. and Khasat, N. P., *J. Appl. Polym. Sci.*, **35**, 825 (1988).
497. Miller, S. A. and Waymouth, R. M., 'Stereo- and Enantioselective Polymerisation of Olefins with Homogeneous Ziegler–Natta catalysts', in *Ziegler Catalysts*, Springer-Verlag, Berlin, 1995, pp. 441–454.
498. Resconi, L. and Waymouth, R. M., *J. Am. Chem. Soc.*, **112**, 4953 (1990).
499. Coates, G. W. and Waymouth, R. M., *J. Am. Chem. Soc.*, **113**, 6270 (1991).
500. Kesti, M. R., Coates, G. W. and Waymouth, R. M., *J. Am. Chem. Soc.*, **114**, 9679 (1992).

501. Kesti, M. R. and Waymouth, R. M., *J. Am. Chem. Soc.*, **114**, 3565 (1992).
502. Coates, G. W. and Waymouth, R. M., *Polym. Mater. Sci. Eng.*, **67**, 92 (1991).
503. Coates, G. W. and Waymouth, R. M., *J. Am. Chem. Soc.*, **115**, 91 (1993).
504. Cavallo, L., Guerra, G., Corradini, P., Resconi, L. and Waymouth, R. M., *Macromolecules*, **26**, 260 (1993).
505. Resconi, L., Coates, G. W., Mogstad, A. and Waymouth, R. M., *J. Macromol. Sci.–Chem. A*, **28**, 1225 (1991).
506. Coates, G. W. and Waymouth, M. R., *J. Mol. Catal.*, **76**, 189 (1992).
507. Frisch, H. L., Mallows, C. L. and Bovey, F. A., *J. Chem. Phys.*, **45**, 1565 (1966).
508. Deng, H., Shiono, T. and Soga, K., *Macromolecules*, **28**, 3067 (1995).
509. Mülhaupt, R., Duschek, T. and Rieger, B., *Makromol. Chem., Macromol. Symp.*, **48/49**, 317 (1991).
510. Shiono, T., Kurosawa, H., Ishida, O. and Soga, K., *Macromolecules*, **26**, 2085 (1993).
511. Mogstad, A. and Waymouth, R. M., *Macromolecules*, **27**, 2313 (1994).
512. Shiono, T., Kurosawa, H. and Soga, K., *Macromolecules*, **27**, 2635 (1994).
513. Chung, T. C., Raate, M., Berluche, E. and Schulz, D. N., *Macromolecules*, **21**, 1903 (1988).
514. Chung, T. C. and Rhubright, D., *J. Polym. Sci., A*, **31**, 2759 (1993).
515. Chung, T. C., *Macromolecules*, **21**, 865 (1988).
516. Ramakrishnan, S., Berluche, E. and Chung, T. C., *Macromolecules*, **23**, 378 (1990).
517. Chung, T. C., Lu, H. L. and Li, C. L., *Macromolecules*, **27**, 7533 (1994).
518. Chung, T. C., *Macromol. Symp.*, **89**, 151 (1995).
519. Chung, T. C. and Jiang, G. J., *Macromolecules*, **25**, 4816 (1992).
520. Chung, T. C., Jiang, G. J. and Rhubright, D., *Macromolecules*, **26**, 3467 (1993).
521. Chung, T. C. and Rhubright, D., *Macromolecules*, **27**, 1313 (1994).
522. Carbonaro, A., Greco, A. and Bassi, I. W., *Eur. Polym. J.*, **4**, 445 (1968).
523. Giannini, U., Brückner, G., Pellino, E. and Cassata, A., *J. Polym. Sci., B*, **5**, 527 (1967).
524. Natta, G., Mazzanti, G., Longi, P. and Bernardini, F., *J. Polym. Sci.*, **31**, 181 (1958).
525. Natta, G., Mazzanti, G., Longi, P. and Bernardini, F., *Chim. Ind. (Milan)*, **40**, 813 (1958).
526. Giannini, U., Brückner, G., Pellino, E. and Cassata, A., *J. Polym. Sci., C*, **22**, 157 (1968).
527. Clark, K. J. and Powell T., *Polymer*, **6**, 531 (1965).
528. Wilén, C.-E., Auer, M. and Näsman, J. H., *Polymer*, **33**, 5049 (1992).
529. Mogstad, A., Kesti, M. R., Coates, G. W. and Waymouth, M. R., *Polym. Prepr. Am. Chem. Soc., Div. Polym. Chem.*, **94**(1), 211 (1993).
530. Aaltonen, P. and Löfgren, B., *Macromolecules*, **28**, 5353 (1995).
531. Aaltonen, P., Fink, G., Löfgren, B. and Seppälä, J., *Macromolecules*, **29**, 5255 (1996).
532. Yasuda, H., Yamamoto, H., Yokota, K., Miyake, S. and Nakamura, A., *J. Am. Chem. Soc.*, **114**, 4908 (1992).
533. Yasuda, H., Furo, M., Yamamoto, H., Nakamura, A., Miyake, S. and Kibino, N., *Macromolecules*, **25**, 5115 (1992).
534. Yasuda, H., Yamamoto, H., Yamashita, M., Yokota, K., Nakamura, A., Miyake, S., Kai, Y. and Tanehisa, N., *Macromolecules*, **26**, 7134 (1993).
535. Jiang, T., Shen, Q., Lin, Y. and Lin S., *J. Organomet. Chem.*, **450**, 121 (1993).
536. Giardello, M. A., Yamamoto, Y., Brard, L. and Marks, T. J., *Macromolecules*, **26**, 3276 (1993).

537. Collins, S. and Ward, D. G., *J. Am. Chem. Soc.*, **114**, 5460 (1992).
538. Higashi, H., Watabe, K. and Namikawa, S., *J. Polym. Sci., B*, **5**, 1125 (1967).
539. Mazurek, V. V., Nesterchuk, G. T. and Merkureva, A. V., *Vysokomol. Soed., Ser. A*, **11**, 611 (1969); *Chem. Abstr.*, **70**, 115602k (1969).
540. Suzuki, Y. and Saito, M., *J. Polym. Sci., A-1*, **9**, 3639 (1971).
541. Chesworth, A. G., Haszeldine, R. L. and Tait, P. J. T., *J. Polym. Sci., A-1*, **12**, 1703 (1974).
542. Florjanczyk, Z., Kuran, W., Sitkowska, J. and Ziólkowski, A., *J. Polym. Sci., A.*, **25**, 343 (1987).
543. Pasynkiewicz, S. and Kuran, W., *J. Organometal. Chem.*, 1968, **15**, 307 (1968).
544. Florjanczyk, Z., Kuran, W., Pasynkiewicz, S. and Langwald, N., *Makromol. Chem.*, **179**, 287 (1978).
545. Florjanczyk, Z., Kuran, W., Langwald, N. and Sitkowska, J., *Makromol. Chem.*, **184**, 2457 (1983).
546. Florjanczyk, Z., Kuran, W., Pasynkiewicz, S. and Kasprzak, E., *Makromol. Chem.*, **178**, 1915 (1977).
547. Hirooka, M., Yabuuchi, H., Iseki, J. and Nakai, Y., *J. Polym. Sci., A-1*, **6**, 1381 (1968).
548. Pasynkiewicz, S., Kuran, W. and Diem, T., *J. Polym. Sci., A-1*, **7**, 2411 (1969).
549. Kuran, W., Pasynkiewicz, S., Florjanczyk, Z. and Kowalski, A., *Makromol. Chem.*, **175**, 3411 (1974).
550. Kuran, W., Pasynkiewicz, S. and Florjanczyk, Z., *Makromol. Chem.*, **174**, 73 (1973).
551. Schiers, J. and Evens, G., 'Polyethylene', in *The Polymeric Materials Encyclopedia*, CRC Press, Inc., Boca Raton, 1996, Vol. 8, pp. 5965–5977.
552. Moore Jr, E. P., 'Polypropylene', in *The Polymeric Materials Encyclopedia*, CRC Press, Inc., Boca Raton, 1996, Vol. 9, pp. 6578–6588.
553. Karol, F. J., *Macromol. Symp.*, **89**, 563 (1995).
554. Thayer, A. M., *Chem. Eng. News*, **1995**, 104 (1995).
555. Brockmeier, N. F. and Arzoumanidis, G. G., *Chem. Eng.*, **90**, 230 (1996).
556. Stevens, M. P., *Polymer Chemistry – an Introduction*, Oxford University Press, New York–Oxford, 1990, pp. 271–295.
557. Ihara, E., Young, Jr, V. G. and Jordan, R. F., *J. Am. Chem. Soc.*, **120**, 8277 (1998).
558. Coles, M. P. and Jordan, R. F., *J. Am. Chem. Soc.*, **119**, 8125 (1997).
559. Li, Y., Ward, D. G., Reddy, S. S. and Collins, S., *Macromolecules*, **30**, 1875 (1997).
560. McLain, St. J. and Feldman, J., US Pat. 5 852 145 (to du Pont de Nemours and Co.) (1998).
561. Brookhart, M. S., Johnson, L. K., Killian, Ch. M., McCord, E. F., McLain, St. J., Kreutzer, K. A., Ittel, St. D. and Tempel, D. J., US Pat. 5 880 241 (to du Pont de Nemours and Co.) (1999); Brookhart, M. S., Johnson, L. K., Killian, Ch. M., Arthur, S. D., Feldman, J., McCord, E. F., McLain, St. J., Kreutzer, K. A., Bennett, A. M. A., Coughlin, E. B., Ittel, St. D., Parthasarathy, A., Wang, L. and Yang, Z.-Y., US Pat. 5 866 663 (to du Pont de Nemours and Co.) (1999); Brookhart, M. S., Johnson, L. K., Arthur, S. D., Feldman, J., Kreutzer, K. A., Bennett, A. M. A., Coughlin, E. B., Ittel, St. D., Parthasarathy, A. and Tempel, D. J., US Pat. 5 886 224 (to du Pont de Nemours and Co.) (1999); Brookhart, M. S., Johnson, L. K., Killian, Ch. M., Arthur, S. D., McCord, E. F. and McLain, St. J. US Pat. 5 891 963 (to du Pont de Nemours and Co.) (1999); Brookhart, M. S., Johnson, L. K., Killian, Ch. M., McCord, E. F. and McLain, St. J., US Pat. 5 916 989 (to du Pont de Nemours and Co.) (1999); McLain, S. J., and Feldman, J., US

Pat. 6 002 034 (to du Pont de Nemours and Co.) (1999); Brookhart, M. S., Johnson, L. K., Killian, Ch. M., Arthur, S. D., Feldman, J., McLain, St. J., Kreutzer, K. A., Bennett, A. M. A., Coughlin, E. B., Ittel, S. D., Parthasarathy, A. and Tempel, D. J., US Pat. 6 034 259 (to du Pont de Nemours and Co.) (2000); Brookhart, M. S., Johnson, L. K., Killian, Ch. M. and McLain, St. J., US Pat. 6 140 439 (to du Pont de Nemours and Co.) (2000); Brookhart, M. S., and Small, B. L., US Pat. 6 150 482 (to du Pont de Nemours and Co.) (2000).

562. Dossett, St. J., US Pat. 6 018 016 (to BP Chemicals Ltd) (2000); Dossett, St. J. US Pat. 6 140 460 (to BP Chemicals Ltd) (2000).

563. Stewart, N. J., Dossett, St. J. (1998) US Pat. 5 770 684 (to BP Chemicals Ltd.).

Further Reading

Jenkins, A. D. and Loening, K. L., 'Nomenclature', in *Comprehensive Polymer Science*, Pergamon Press, Oxford, 1989, Vol. 1, pp. 13–54.

Bovey, F. A., 'Structure of Chains by Solution NMR Spectroscopy', in *Comprehensive Polymer Science*, Pergamon Press, Oxford, 1989, Vol. 1, pp. 339–375.

Fueno, T., Shelden, R. A. and Furukawa, J., 'Probabilistic Considerations of the Tacticity of Optically Active Polymers', in *J. Polym. Sci., A*, **3**, 1279–1288 (1965).

Keii, T., Doi, Y. and Soga, K., 'Living Polymer Systems – Ziegler–Natta polymerisation', in *Encyclopedia of Polymer Science and Engineering*, Wiley-Interscience, John Wiley & Sons, New York, 1989, Vol. 18, pp. 437–445.

Soga, K., Uozumi, T., Park, J. R. and Shiono, T., 'Recent Development in Stereochemical Control of Heterogeneous Ziegler–Natta Catalysts', in *Makromol. Chem., Macromol. Symp.*, **63**, 219–231 (1992).

Fan, Z.-Q., Feng, L.-X. and Yang, S.-L., 'Distribution of Active Centers on $TiCl_4/MgCl_2$ Catalyst for Olefin Polymerization', in *J. Polym. Sci., A, Polym. Chem.*, **34**, 3329–3335 (1996).

Czaja, K. and Bialek, M., 'Vanadium-based Ziegler–Natta Catalyst Supported on $MgCl_2(THF)_2$ for Ethylene Polymerisation', *Macromol. Rapid Commun.*, **17**, 253–260 (1996).

Wang, J.-G., Zhang, W.-B. and Huang, B.-T., 'Some Newer Concepts in Ethylene/α-Olefin Copolymerisation on Heterogeneous Ziegler–Natta Catalysts', *Makromol. Chem., Macromol. Symp.*, **63**, 245–258 (1992).

Tian, J. and Huang, B., 'Soluble Olefin Polymerization Catalysts', in *The Polymeric Materials Encyclopedia*, CRC Press, Inc., Boca Raton, 1996, Vol. 6, pp. 4740–4749.

Tritto, I., Li, S. X., Sacchi, M. C. and Locatelli, P., 'Metallocene/Aluminoxane Catalysts', in *The Polymeric Materials Encyclopedia*, CRC Press, Inc., Boca Raton, 1996, Vol. 6, pp. 4160–4169.

Tait, P. J. T., 'Metallocene Catalysts', in *The Polymeric Materials Encyclopedia*, CRC Press, Inc., Boca Raton, 1996, Vol. 6, pp. 4169–4177.

Bochmann, M., 'Metallocene Catalysts', in *The Polymeric Materials Encyclopedia*, CRC Press, Inc., Boca Raton, 1996, Vol. 6, pp. 4177–4191.

Huang, B. and Tian, J., 'Metallocene Catalysts', in *The Polymeric Materials Encyclopedia*, CRC Press, Inc., Boca Raton, 1996, Vol. 6, pp. 4191–4201.

Kaminsky, W., Niedoba, S., Möller-Lindenhof, N. and Rabe, O., 'Isotactic Olefin Polymeriation with Optically Active Catalysts', *Catalysis in Polymer Synthesis*, ACS Symp. Ser. 496, Washington, DC, 1992, pp. 63–71.

Grubbs, R. H. and Coates, G. W., 'α-Agostic Interactions and Olefin Insertion in Metallocene Polymerisation Catalysts', *Acc. Chem. Res.*, **29**, 85–93 (1996).

Michelotti, M., Altomare, A. and Ciardelli, F., 'Ethylene/α-Olefins Cooligomerization versus Copolymerization by Zirconocene Catalysts', *Polymer*, **37**, 5011–5016 (1996).
Kravchenko, R., Masood, A. and Waymouth, R. M., 'Propylene Polymerization with Chiral and Achiral Unbridged 2-Arylindene Metallocenes', *Organometallics*, **16**, 3635–3639 (1997).
Kurokawa, H. and Sugano, T., 'α-Olefin Polymerization by Various Alumoxanes Containing Isobutyl Group and Metallocene', *Macromol. Symp.*, **97**, 143–149.
Yu, Z. and Chien, J. C. W., 'Molecular Mechanics Study of Zirconium Catalyzed Isospecific Polymerization of Propylene', *J. Polym. Sci., A, Polym. Chem.*, **33**, 125–135 (1995).
Pieters, P. J. J., van Beek, J. A. M. and van Tol, M. F. H., 'A Method for the Prediction of Metallocene-type Catalyst Activity in Olefin (Co)Polymerization Reactions', *Macromol. Rapid Commun.*, **16**, 463–467.
Resconi, L., Piemontesi, F., Franciscono, G., Abis, L. and Fiorani, T., 'Olefin Polymerization at Bis(pentamethylcyclopentadienyl)zirconium and -hafnium Centers: Chain-transfer Mechanisms', *J. Am. Chem. Soc.*, **114**, 1025–1032 (1992).
Jüngling, S., Mülhaupt, R., Stehling, U., Brintzinger, H.-H., Fischer, D. and Langhauser, F., 'The Role of Dormant Sites in Propene Polymerization using Methylalumoxane Activated Metallocene Catalysts', *Macromol. Symp.*, **97**, 205–216 (1995).
Kaminsky, W., 'How to Reduce the Ratio Methylaluminoxane/Metallocene', *Macromol. Symp.*, **97**, 79–89 (1995).
Tritto, I., Sacchi, M. C., Locatelli, P. and Li, S. X., 'Metallocene Ion Pairs', *Macromol. Symp.*, **89**, 289–298 (1995).
Siedle, A. R., Hanggi, B., Newmark, R. A., Mann, K. R. and Wilson, T., 'How Coordinating are Non-coordinating Anions', *Macromol. Symp.*, **89**, 299–305 (1995).
Jin, J., Uozumi, T. and Soga, K., 'Ethylene Polymerization Initiated by SiO_2-supported Neodymocene Catalysts', *Macromol. Rapid Commun.*, **16**, 317–322 (1995).
Damiani, D. E., Juan, A. and Ferreira, M. L., 'Olefin Polymerization Catalysts', *The Polymeric Materials Encyclopedia*, CRC Press, Inc., Boca Raton, 1996, Vol. 6, pp. 4727–4734.
Lee, D.-H., 'Olefin Polymerization Catalysts', in *The Polymeric Materials Encyclopedia*, CRC Press, Inc., Boca Raton, 1996, Vol. 6, pp. 4734–4740.
Soga, K. and Shiono, T., 'Ziegler–Natta Catalysts for Olefin Polymerizations', *Prog. Polym. Sci.*, **22**, 1503–1546 (1997).
Kuran, W., 'Stereoregulation Mechanism in Coordination Polymerization of Hydrocarbon Monomers. Part I: Polymerization of α-Olefins', *Polimery*, **39**, 570–578 (1997).
Brockmeier, N. F., 'Gas-phase Polymerization', in *Encyclopedia of Polymer Science and Engineering*, Wiley-Interscience, John Wiley & Sons, New York, 1987, Vol. 7, pp. 480–488.
McMillan, F. M., 'Polyolefins', in *The Polymeric Materials Encyclopedia*, CRC Press, Inc., Boca Raton, 1996, Vol. 8, pp. 6424–6428.
Ross, J. F. and Mac Adams, J. L., 'Polyethylene (Commercial)', in *The Polymeric Materials Encyclopedia*, CRC Press, Inc., Boca Raton, 1996, Vol. 8, pp. 5953–5965.
Doak, K. W., 'Ethylene polymers', in *Encyclopedia of Polymer Science and Engineering*, Wiley-Interscience, John Wiley & Sons, New York, 1986, Vol. 6, pp. 383–386.
James, D. E., 'Linear Low Density Polyethylene', in *Encyclopedia of Polymer Science and Engineering*, Wiley-Interscience, John Wiley & Sons, New York, 1986, Vol. 6, pp. 429–454.
Beach, D. L. and Kissin, Y. V., 'High Density Polyethylene', in *Encyclopedia of Polymer Science and Engineering*, Wiley-Interscience, John Wiley & Sons, New York, 1986, Vol. 6, pp. 454–490.

Coughlan, J. J. and Hug, D. P., 'Ultrahigh Molecular Weight Polyethylene', in *Encyclopedia of Polymer Science and Engineering*, Wiley-Interscience, John Wiley & Sons, New York, 1986, Vol. 6, pp. 490–494.

Ver Strate, G., 'Ethylene–Propylene Elastomers', in *Encyclopedia of Polymer Science and Engineering*, Wiley-Interscience, John Wiley & Sons, New York, 1986, Vol. 6, pp. 523–564.

Cecchin, G., 'In Situ Polyolefin Alloys', *Macromol. Symp.*, **78**, 213–228 (1994).

Covezzi, M., 'The Spherilene Process: Linear Polyethylenes', *Macromol. Symp.*, **89**, 577–586 (1995).

Rangwala, H. A., Dalla Lana, I. G., Szymura, J. A. and Fiedorow, R. M., 'Copolymerization of Ethylene and 1-Butene Using a Cr–Silica Catalyst in a Slurry Reactor', *J. Polym. Sci., A, Polym. Chem.*, **34**, 3379–3387 (1996).

Kuran, W., 'Polar Vinyl Monomer Polymerization and Copolymerization with Olefins Promoted by Organometallic Catalysts', *Polimery*, **42**, 604–609 (1997).

Endo, K., 'Monomer-isomerization polymerization', in *The Polymeric Materials Encyclopedia*, CRC Press, Inc., Boca Raton, 1996, Vol. 6, pp. 4532–4536.

Fink, G., Möhrirg, V. M., Heinrichs, A., Denger, C., Schubbe, R. H. and Mühlenbrock, P. H., 'Olefin Polymerization by 2,ω-Linkage', in *The Polymeric Materials Encyclopedia*, CRC Press, Inc., Boca Raton, 1996, Vol. 6, pp. 4720–4727.

Endo, T. and Tomita, I., 'Allene Polymerization', in *The Polymeric Materials Encyclopedia*, CRC Press, Inc., Boca Raton, 1996, Vol. 1, pp. 161–166.

Garbassi, F. and Sommazzi, A., 'Olefin–Carbon Monoxide Copolymers', in *The Polymeric Materials Encyclopedia*, CRC Press, Inc., Boca Raton, 1996, Vol. 6, pp. 4701–4714.

Belov, G. P., 'Alternating Olefin–Carbon Monoxide Copolymers', in *The Polymeric Materials Encyclopedia*, CRC Press, Inc., Boca Raton, 1996, Vol. 6, pp. 4714–4720.

Koide, Y., Bott, S. G. and Barron, A. R., 'Alumoxanes as Cocatalysts in the Palladium-catalyzed Copolymerization of Carbon Monoxide and Ethylene: Genesis of a Structure–Activity Relationship', *Organometallics*, **15**, 2213–2226 (1996).

Problems

1. Name and characterise coordination catalysts for the polymerisation of ethylene and α-olefins.
2. Explain why only coordination catalysts are useful for polymerising α-olefins.
3. Characterise bonds forming a π complex between the olefin and transition metal.
4. Write down equations describing primary and secondary insertion of an α-olefin into the metal–carbon bond. Explain what the polymerisation regioselectivity is.
5. How can one explain the appearance of short and long branchings in high-density polyethylene produced by Ziegler–Natta catalysts. Why are short branchings of frequent occurrence, whereas long branchings are rare?
6. Explain why heterogeneous Ziegler–Natta catalysts produce polyolefins of high polydispersity but homogeneous catalysts produce polyolefins of relatively low polydispersity.
7. Explain how molecular hydrogen performs a control of polymer molecular weight in olefin polymerisation with Ziegler–Natta catalysts.
8. Explain why 1,4-hexadiene, but not 1,5-hexadiene, is used for obtaining ethylene/propylene/non-conjugated diene vulcanisable terpolymers.
9. Explain why small amounts of water, when present in a metallocene–trialkylaluminium system, accelerate olefin polymerisation.

10. Compare the electronic structure of group 4 and group 3 metallocene catalysts for olefin polymerisation.
11. Which are the essential factors determining the stereoregulation mechanism in the polymerisation of α-olefins with coordination catalysts?
12. Give the structure and stereochemical designation of all possible poly(α-olefin)s that could, in principle, be obtained by stereospecific polymerisation. Are they all likely to be obtainable in practice (with heterogeneous and homogeneous Ziegler–Natta catalysts)?
13. What structure will the polymers have formed in the polymerisation of 4-methyl-1-pentene with the following catalysts: δ-$TiCl_3$–$AlEt_3$, Cp_2ZrCl_2–$[Al(Me)O]_x$, *rac.*-$Me_2Si(Ind)_2ZrCl_2$–$[Al(Me)O]_x$ and $Me_2C(Cp)(Flu)ZrCl_2$–$[Al(Me)O]_x$. Give reasons why.
14. Show by adopted Fischer projections the various (if any) tactic (ditactic) polymers that might possibly be obtained from each of the olefins:

$CH_2{=}CH(^2H)$	$CH_2{=}CHCH_3$
$CH_2{=}C(^2H)_2$	$CH_2{=}C(CH_3)_2$
cis-,trans-$CH(^2H){=}CH(^2H)$	$CH(CH_3){=}C(CH_3)$
$CH_2{=}CHCH_2(^2H)$	$CH_2{=}C(CH_3)CH_2CH_3$
cis-,trans-$CH(^2H){=}CHCH_2(^2H)$	$CH_2{=}CHCH(CH_3)CH_3$
$CH_2 = C(^2H){=}CH_3$	$CH_2{=}CHCH_2CH(CH_3)CH_3$
cis-,trans-$CH(^2H){=}CHCH_3$	$CH(CH_3){=}CHCH_2CH_3$

15. Give examples of homogeneous chirotopic catalysts.
16. Give examples of homogeneous chirotopic catalysts that do not exist in a mesogenic form.
17. A polymer of which structure will be formed in 3-methyl-1-pentene polymerisation with the δ-$TiCl_3$–$Zn[CH(CH_3)CH_2CH_3]$ catalyst?
18. In which of the three groups (isotactic, syndiotactic or hemiisotactic) one may include syndioisostereoblock polypropylene?
19. How can one explain the occurrence of steric defects in tactic poly(α-olefin)s?
20. Explain why high-resolution nuclear magnetic resonance is the most convenient method for determining the chain microstructure in poly(α-olefin)s. Consider how ^1H and ^{13}C NMR spectroscopy can provide stereochemical information concerning α-olefin polymer chains on the diad level (*m*, *r*) and the triad level (*mm*, *rr*, *mr*).
21. Explain why β-olefins, which do not homopolymerise (without isomerisation) in the presence of Ziegler–Natta catalysts, undergo copolymerisation with ethylene in the presence of these catalysts.
22. Give reasons why ethylene and α-olefins undergo copolymerisation with carbon monoxide in the presence of coordination catalysts.
23. Explain why the cyclopolymerisation of 1,4-pentadiene has not been realised, in contrast to that of 1,5-hexadiene, 1,6-heptadiene and 1,7-octadiene.
24. Name and characterise methods for obtaining α-olefin copolymers with polar vinyl monomers in the presence of coordination catalysts.

4 Coordination Polymerisation of Vinylaromatic Monomers

Commercial polystyrene, one of the most representative general-purpose resins, is an amorphous atactic polymer with no melting point (softening temperature ca 100 °C, glass transition temperature ca 70–100 °C). Since the discovery of Ziegler–Natta catalysts, many efforts to produce stereoregular polystyrenes have been made.

The polymerisation of styrene, which is an exceptionally versatile monomer, in the presence of various Ziegler–Natta and related coordination catalysts produces high molecular weight polystyrenes, both highly isotactic polymers [1–4] and highly syndiotactic polymers [5–10].

Isotactic polystyrene is a partially crystalline polymer (it contains ca 30 % crystalline domains); the melting point of the crystalline zones of isotactic polystyrene ranges from ca 230 to 240 °C (glass transition temperature ca 87–97 °C). The chain conformation of crystalline isotactic polystyrene is a helical conformation with three monomeric units per helix turn. It was originally assumed that isotactic polystyrene, with its much better heat deformation resistance compared with the atactic amorphous polymer, would be a useful technical material, primarily as an admixture in atactic polystyrene. Test products appeared on the market towards the end of the 1960s [11]. The rate of crystallisation of isotactic polystyrene is, however, so low that this property cannot be exploited industrially [12–14]. However, isotactic polystyrene is still considered as a potentially valuable material because of its high heat resistance and good dielectric properties. On the other hand, low crystallinity, a low rate of crystallisation and high britlleness represent the principal obstacles to commercial application of isotactic polystyrene as a general-purpose plastic [15].

In contrast to isotactic polystyrene, syndiotactic polystyrene crystallises very rapidly (it contains ca 72 % crystalline domains) with a crystallisation rate similar to that of poly(ethylene terephthalate). The chain conformation of crystalline syndiotactic polystyrene is a planar zigzag conformation. However, the melting point of the syndiotactic polystyrene ranges from ca 270 to 275 °C (glass transition temperature ca 93–96 °C) so that the processing of this material is problematic; in spite of this, a range of processing methods, including precision moulding, has been developed. To date, no industrial applications of syndiotactic polystyrene are known. However, syndiotactic polystyrene, which exhibits a much higher crystallisation rate and crystallinity and enhanced mechanical properties, especially at elevated temperatures, by comparison with the isotactic isomer and excellent chemical and solvent resistance, is a potentially useful material as a low-cost engineering plastic (e.g. resistant fibres, films, containers, circuit boards) [14,15].

A large number of ring-substituted styrenes (with alkyl or electron withdrawing substituents) have been polymerised with heterogeneous Ziegler–Natta catalysts to isotactic polymers [16–20]. Some ring-substituted styrenes have also been polymerised with homogeneous Ziegler–Natta catalysts to syndiotactic polymers [21].

4.1 Isospecific Polymerisation of Vinylaromatic Monomers with Coordination Catalysts

The isospecific polymerisation of styrene was first reported by Natta *et al.* [1,2]. They obtained isotactic polystyrene, employing for the polymerisation a heterogeneous catalyst derived from titanium tetrachloride and an alkylaluminium compound.

Subsequent investigations revealed that, in principle, styrene undergoes isospecific polymerisation in the presence of heterogeneous Ziegler–Natta catalysts [1–4]. Although polystyrene of isotactic structure was also prepared with the use of homogeneous nickel-based coordination catalysts, it appeared to be of low molecular weight [22,23].

4.1.1 Polymerisation with Heterogeneous Ziegler–Natta Catalysts

Vinylaromatic monomers can be polymerised to isotactic polymers with some heterogeneous Ziegler–Natta catalysts that are active in the isospecific polymerisation of α-olefins [2,18–20,24–29]. The molecular weights of these polymers are usually very high, even of the order of several millions. Catalysts most widely used for the isospecific polymerisation of styrene are $TiCl_4$–$AlEt_3$, α-$TiCl_3$–$AlEt_3$, δ-$TiCl_3$–$AlEt_3$, VCl_3–$AlEt_3$, and supported catalysts such as $MgCl_2$/ethyl benzoate/$TiCl_4$–$AlEt_3$, $MgCl_2$/$TiCl_3$–$AlEt_3$, Mg(OH)Cl/Ti$(OBu)_4$–$AlMe_3$ and $TiCl_3$–Cp_2TiMe_2 [1,30–32]. Polymerisations are carried out usually at temperatures in the range 40–80 °C, in bulk or in solvents such as aliphatic or cycloaliphatic hydrocarbons (hexane, cyclohexane) and aromatic hydrocarbons (benzene, toluene).

It may be interesting to note that the isospecific polymerisation of vinylaromatic monomers is simpler than that of α-olefins. For example, it is known that propylene polymerisation with catalysts derived from $TiCl_4$ and $AlEt_3$ is much less stereospecific than polymerisation with the violet $TiCl_3$-based catalysts. Such a difference is not observed, however, in the polymerisation of styrene; in both cases, namely with the $TiCl_4$–$AlEt_3$ catalyst and the $TiCl_3$–$AlEt_3$ catalyst, highly isotactic polystyrene can be obtained. This indicates that no exact parallel exists between the catalyst requirements for the isospecific polymerisation of styrene (and its ring-substituted analogues) and these for the isospecific polymerisation of α-olefins.

Although the isospecific polymerisation of styrene monomers has much less steric demands for the Ziegler–Natta catalysts than that of α-olefins, it proceeds with much lower propagation rate constants by comparison with the polymerisation of α-olefins; for example, on a molar basis, styrene is ca 100 times less reactive than propylene in the polymerisation [30,33]. Also, compare the relatively slow polymerisation of styrene and other vinylaromatic monomers with the relatively fast polymerisation of vinylcyclohexane [20,31,34–36].

The mechanism of the isospecific polymerisation of styrene and its ring-substituted analogues in the presence of heterogeneous Ziegler–Natta catalysts is the same as for α-olefins. As regards the mode of monomer insertion, it has been reported as primary (1,2) insertion [37,38].

$$\mathrm{Mt-P}_n + \overset{(1)}{\mathrm{CH_2}}=\overset{(2)}{\underset{\mathrm{Ph}}{\mathrm{CH}}} \longrightarrow \mathrm{Mt}-\overset{(1)}{\mathrm{CH_2}}-\overset{(2)}{\underset{\mathrm{Ph}}{\mathrm{CH}}}-\mathrm{P}_n \qquad (1)$$

The steric control is attributed to the presence of chiral centres on the catalyst particle surface [39].

Reactivities of ring-substituted styrenes depend on the type and position of the substituents [18,19,27,40]. Table 4.1 lists the relative reactivities of different vinylaromatic monomers in polymerisation with the heterogeneous $TiCl_4$–$AlEt_3$ catalyst [41].

Except for a few cases originating from steric factors (e.g. for *ortho*-substituted styrenes), the relative reactivities of ring-substituted styrene follow the Hammett equation with a negative slope [19,27]. Therefore, monomer polymerisability increases with increasing electron density at the double bond undergoing the polymerisation. This may suggest the monomer coordination at the transition metal to be the rate determining step during polymerisation.

Styrene polymerisation with heterogeneous Ziegler–Natta catalysts activated by alkylaluminium compounds generally produces a mixture of isotactic and non-stereoregular polymer. For example, polystyrene produced with the

Table 4.1 Relative reactivity of vinylaromatic monomers in coordination polymerisation with heterogeneous Ziegler–Natta catalyst[a,b]

Monomer	Relative reactivity[c]	Monomer	Relative reactivity[c]
Styrene	1	*o*-Fluorostyrene	0.20
o-Methylstyrene	0.10	*m*-Fluorostyrene	0.50
m-Methylstyrene	0.43	*p*-Fluorostyrene	0.74
p-Methylstyrene	1.20	2-Methyl-4-fluorostyrene	0.10
2,4-Dimethylstyrene	0.10	*o*-Chlorostyrene	0
2,5-Dimethylstyrene	0.04	*m*-Chlorostyrene	0.40
3,4-Dimethylstyrene	0.48	*p*-Chlorostyrene	0.47
3,5-Dimethylstyrene	0.23	*p*-Bromostyrene	0.45
2,4,6-Trimethylstyrene	0	1-Vinylnaphthalene	0.22
p-Ethylstyrene	1.10	2-Vinylnaphthalene	0.67
p-Isopropylstyrene	0.49	1-Vinyl-4-chloronaphthalene	0.20
p-Cyclohexylstyrene	0.50	4-Vinyldiphenyl	0.73
		9-Vinylphenanthrene	0
		9-Vinylanthrene	0

[a] Results obtained with the $TiCl_4$–$AlEt_3$ catalyst (molar ratio Al/Ti = 3) at 60 °C.
[b] According to Refs 18 and 24.
[c] Based on reactivity of styrene as 1.
Reproduced by permission of John Wiley & Sons, Inc from Ref. 41. Copyright 1989 Wiley New York

α-$TiCl_3$–$AlEt_3$ catalyst contains ca 70 % of a highly isotactic and crystalline fraction insoluble in boiling toluene with an isotactic triad content higher than 95 %, ca 10 % of a fraction of medium isotacticity (insoluble in ketones) and ca 20 % of an amorphous, atactic fraction soluble in boiling methyl ethyl ketone [30]. It was suggested that atactic polystyrene could be produced by cationic polymerisation involving acidic species in the catalysts. The yield of atactic polystyrene increases with decreasing Al/Ti ratio below a certain level (e.g. 2.25–2.50 for the $TiCl_4$–$AlEt_3$ catalyst) [25]. The properties of this atactic polystyrene are identical to those of the polymer obtained with $TiCl_4$ only, which promotes the cationic polymerisation of styrene. It is to be mentioned in this connection that vinylaromatic monomers containing bulky substituents at the polymerising double bond, such as acenaphthylene, form stereoregular polymers of an identical structure by polymerisation with a $TiCl_4$-based Ziegler–Natta catalyst such as $TiCl_4$–$Al(i\text{-}Bu)_2H$ as well as with $TiCl_4$ alone [42].

4.1.2 Polymerisation with Homogeneous Nickel-complex Catalysts

Styrene undergoes polymerisation with the cationic organonickel(II) complex η^3-methallyl (η^4-cycloocta-1,5-diene)nickel hexafluorophosphate $\{[(MeAll)(Cod)Ni]^+[PF]_6^-\}$ combined *in situ* with tricyclohexylphosphine. The product of such polymerisation is a styrene oligomer with $\bar{M}_n = 1900$, characterised by a

98% degree of isotacticity [22,23]. It is interesting to note that the $[(MeAll)(Cod)Ni]^+[PF_6]^-$ complex alone, used without any addition of $P(Chx)_3$, is very active to styrene oligomerisation but is non-sterospecific [23]. One should add in this connection that high molecular weight polystyrenes can also be obtained with homogeneous catalysts based on organonickel(II) complexes; for instance, a catalyst such as nickel bisacetylacetonate combined with triethylamine and activated with methylaluminoxane, $Ni(Acac)_2/NEt_3$–$[Al\ (Me)O]_x$, produces polystyrenes soluble to different extents in boiling ketones [43]. Both acetone-soluble and acetone-insoluble polymer fractions appeared to have a moderately isotactic structure. The number-average molecular weight of the insoluble fractions varied with polymerisation temperature from $\bar{M}_n = 5.4 \times 10^3$ at 80 °C to $\bar{M}_n = 4.1 \times 10^4$ at −50 °C. The polydispersity was lower compared with that of polystyrene obtained with heterogeneous Ziegler–Natta catalysts.

Styrene polymerisation in both isospecific and non-stereospecific homogeneous catalyst systems is highly regioselective and involves a secondary (2,1) insertion of the monomer:

(2)

The initiation step in polymerisation with a catalyst derived from the $[(MeAll)(Cod)Ni]^+[PF_6]^-$ complex and $P(Chx)_3$ involves the secondary insertion of styrene into a Ni–H bond in the cationic nickel hydride species formed *in situ*:

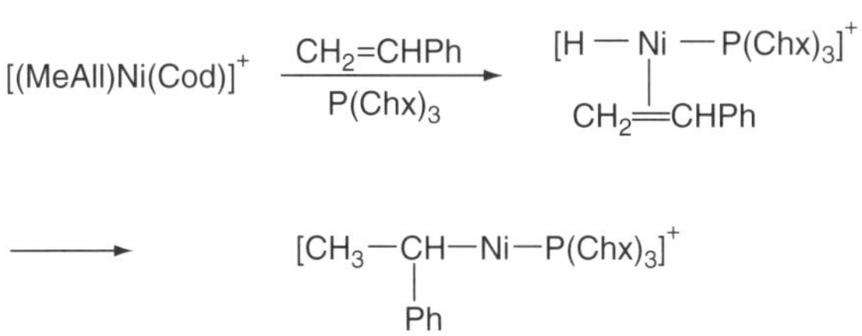

(3)

As already mentioned, the propagation proceeds via a secondary insertion of the monomer into a nickel–carbon bond:

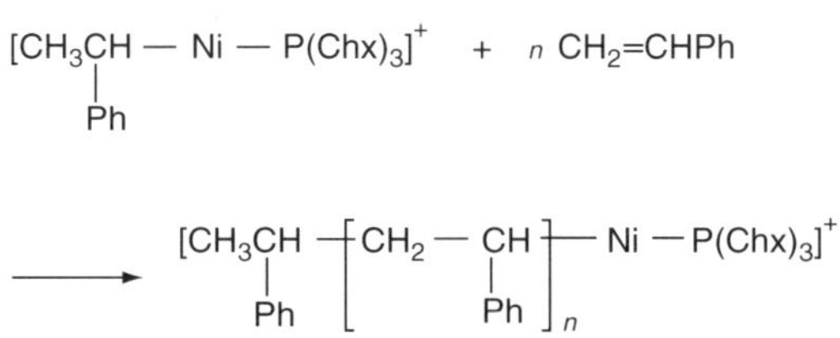

(4)

The secondary insertion of styrene involves species in which an interaction between the transition metal and the aromatic ring of the last inserted monomeric unit occurs [44]. This kind of interaction can be due to η^3-benzyl bond formation [45,46]:

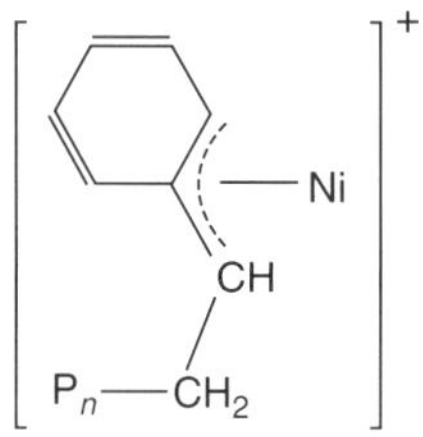

Secondary monomer insertions into a nickel–carbon bond and thus the occurrence of an η^3-benzyl bond at the propagating active site prevent a chain termination reaction, but, when only primary insertion takes place, chain termination occurs by β-hydrogen elimination [43]:

$$[P_n-CH_2-CH(Ph)-Ni-P(Chx)_3]^+ + CH_2{=}CHPh$$

$$\longrightarrow [P_n-CH_2-CH(Ph)-CH(Ph)-CH_2-Ni-P(Chx)_3]^+$$

$$\longrightarrow [H-Ni-P(Chx)_3]^+ + P_n-CH_2-CH(Ph)-C(Ph){=}CH_2 \qquad (5)$$

Although the mechanism of steric control that operates in the isospecific styrene polymerisation system with homogeneous nickel-based $[(MeAll)(Cod)Ni]^+[PF_6]^-/P(Chx)_3$ catalyst is not completely clear, it is evident that π-benzyl nickel species play an important role in determining the mode of the styrene insertion.

The contribution of a π-benzyl anchoring of the growing chain in the active species and the presence of the basic bulky $P(Chx)_3$ result in a very crowded nickel coordination sphere, which is believed to be responsible for the isospecificity of this homogeneous catalyst. Similarly, one can explain the increasing amount of stereoregular fraction in polystyrene formed with the $Ni(Acac)_2$–$[Al(Me)O]_x$ catalyst when it is modified by the addition of NEt_3 [47]. A ^{13}C NMR study of the 2-butanone-insoluble polystyrene fraction showed that the polymer consisted of blocks of m diads separated by single r diads or by a pair of r diads. Such a polymer microstructure could suggest that steric control arose from chiral sites in the catalytic nickel complex, but the configuration of these sites changed during propagation reaction.

4.2 Syndiospecific Polymerisation of Vinylaromatic Monomers with Coordination Catalysts

Syndiotactic polystyrene was first obtained only recently by Ishihara *et al.* [5] in polymerisation with a homogeneous catalyst derived from a transition metal compound such as monocyclopentadienyltitanium trichloride and methylaluminoxane in toluene. Since then, several authors have reported on the synthesis of syndiotactic polystyrene promoted by different catalysts based on metal hydrocarbyls such as benzyl compounds, half-sandwich metallocenes (e.g. monocyclopentadienyl, monopentamethylcyclopentadienyl and monoindenyl metal derivatives), metal alkoxides, metallocenes and some other compounds. These catalysts are commonly derived from titanium or zirconium compounds, either activated with methylaluminoxane or aluminium-free, such as those activated with tris(pentafluorophenyl)boron, and promote the syndiospecific polymerisation of styrene and substituted styrenes [5–10,21,48–70].

Representative examples of the syndiospecific polymerisation of styrene using catalysts based on various titanium compounds and methylaluminoxane are shown in Table 4.2 [6,52,53,56,58].

Table 4.2 Syndiospecific polymerisation of styrene in the presence of homogeneous catalysts based on a titanium compound and methylaluminoxane[a]

Titanium compound	[Al(Me)O] unit/Ti molar ratio	Styrene conversion (%)	Reference
$CpTiCl_2$	600	44.1	[6]
$CpTiCl_3$	600	99.2	[6]
Cp^*TiCl_3	900	100	[6]
$Me_4CpTi(O\text{-}i\text{-}Pr)_3$	500	3.3[b]	[52]
$IndTiCl_3$	4000	9.8	[53]
Cp_2TiCl_2	600	1.0	[6]
$Cp_2^*TiCl_2$	600	2.0	[6]
$TiCl_4$	800	4.1	[6]
$TiBr_4$	800	2.1	[6]
$Ti(OMe)_4$	800	3.8	[6]
$Ti(OEt)_4$	800	9.5	[6]
$Ti(OEt)_4$	290	4.0	[58]
$(Acac)_2TiCl_2$	800	0.4	[6]
$TiBz_4$	100	7.9[c]	[56]

[a] Polymerisation conditions: Ti compound 5×10^{-5} mol; styrene 2×10^{-1} mol; toluene 100 mL; temperature 50 °C; time 2 h.
[b] Polymerisation temperature 75 °C.
[c] Polymerisation syndiospecificity (acetone-insoluble polymer fraction) 93%.

It is now realised that almost all catalysts based on Ti(III) or Ti(IV) compounds and methylaluminoxane, soluble in aromatic solvents, could polymerise styrene into a highly syndiotactic polymer. The syndiotacticity measured by ^{13}C NMR spectroscopy can be greater than 98%. Syndiospecific polymerisation of styrene with homogeneous catalysts is characterised by a narrow molecular weight distribution ($\bar{M}_w/\bar{M}_n$ can reach a value of 2).

Also, divalent $TiPh_2$ activated with $[Al(Me)O]_x$ appeared to be a catalyst for syndiospecific styrene polymerisation [71]. Even δ-$TiCl_3$ or $(Acac)_3TiCl_3$, when activated with $[Al(Me)O]_x$, could yield a mixture of isotactic and syndiotactic polystyrenes. Some other catalysts, such as rare-earth coordination catalysts, have been successfully used to obtain syndiotactic-rich polystyrenes [72].

Monocyclopentadienyl titanium derivatives are the most active precursors for catalysts possessing high syndiospecific polymerisation activity for styrene and ring-substituted styrenes. The polymerisation activity of biscyclopentadienyl titanium compounds activated with methylaluminoxane is lower than that of other soluble titanium-based catalysts [73].

As regards a comparison of the relative effectiveness of titanium-and analogous zirconium-based catalysts in syndiospecific styrene polymerisation, the latter in general are less active than the former. Usually, polymer yields are lower, and a higher polymerisation temperature and reaction time as well as higher $[Al(Me)O]_x$/transition metal compound ratios are required. Among the few zirconium compounds examined, only tetrabenzylzirconium activated with methylaluminoxane has relatively higher syndiospecific activity [10,48,56].

As regards the polymerisation of ring-substituted styrenes, titanium-based catalysts containing a cyclopentadienyl ligand promote the syndiospecific polymerisation of such deactivated monomers as *p*-chlorostyrene, while catalysts not containing such ligands do not [55,73]. However, catalysts of both types promote the polymerisation of activated styrenes, such as *p*-methylstyrene [74]. It was recognised that there is a relation between the Hammett δ parameter of each substituent and reactivities in syndiospecific polymerisation. The reactivity is enhanced by electron-releasing substituents in an aromatic ring of the monomer; this means, for instance, that *p*-methylstyrene undergoes polymerisation faster than *p*-chlorostyrene [6,75]. Therefore, the polyinsertion of styrene might be defined as 'electrophilic', and the catalyst might be referred to as an 'electrophile' [55].

4.2.1 Polymerisation Regioselectivity and Stereospecificity

Styrene enchainment in the syndiospecific polymerisation of styrene with homogeneous catalysts, both containing and not containing a cyclopentadienyl or cyclopentadienyl-like ligand, is through a *cis* insertion [76,77]. This was evidenced by ^{1}H NMR analysis of copolymers of perdeuterostyrene and *trans*-1-(^{2}H) styrene. Styrene insertion has been reported as secondary (2,1) insertion [scheme (2)] [9,10,78].

^{13}C NMR analysis also showed the end groups $\cdots$–$CH(Ph)CH_3$ and $PhCH{=}CH$–$\cdots$ present in syndiotactic polystyrene in equal amounts [10]. Moreover, polymerisation in the presence of a catalyst with (^{13}C)-enriched triethylaluminium $[Al(^{13}CH_2CH_3)_3]$ as an activator yielded syndiotactic polystyrene, which contained some $\cdots$–$CH(Ph)CH_2\,^{13}CH_2CH_3$ end groups [9].

The structure of the observed end groups suggests that most of the polymer chains are initiated with the Mt–H species [scheme (6)] and that β-hydrogen elimination is the main chain termination reaction [scheme (7)] [57]:

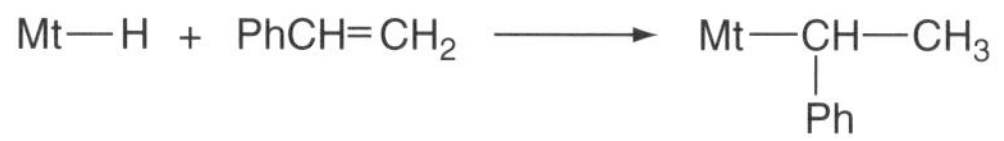

(6)

(7)

The stereocontrol mechanism that operates in styrene syndiospecific polymerisation is the chain end mechanism. ^{13}C NMR spectroscopic studies of the polymer microstructure revealed the presence of long sequences of *r* diads bridged by isolated *m* diads in polystyrene [7] which is diagnostic of steric ordering of the incoming monomer according to the configuration of the last monomeric unit of the growing chain. The electron donation from π electrons of the phenyl ring of the growing chain end and, eventually, that of the phenyl ring of the coordinating monomer to the titanium atom are considered to be important factors influencing the syndiospecific polymerisation of styrene [55].

4.2.2 Catalysts, Models of Active Sites and the Polymerisation Mechanism

Although there is no straightforward evidence concerning the structure of the true active species promoting the syndiospecific polymerisation of styrene and there is no one general mechanism that might be applicable for explaining all the experimental data concerning styrene polymerisation with homogeneous syndiospecific catalysts, some structures of active sites have been suggested reasonably and some features of the mechanism of syndiospecific polymerisation of styrene have been elucidated for various polymerisation systems.

4.2.2.1 Polymerisation with Metal Hydrocarbyls

Alkyltitanium compounds were found [79] to promote the polymerisation of styrene, leading, however, to an atactic polymer. Also, the polymerisation of styrene employing benzyl derivatives of group 4 metals, such as tetrabenzylzirconium, as a catalyst was reported [80] to produce atactic polystyrene. It appeared, however, that tetrabenzyl titanium or zirconium gave rise to moderately active syndiospecific catalysts when activated by methylaluminoxane [10,49].

Methylaluminoxane-free catalysts, such as cationic complexes derived from $TiBz_4$ and tris(pentafluorophenyl)borane $[B(C_6F_5)_3]$, appear to be much less active, poorly stereospecific catalysts for the polymerisation of styrene. Under analogous conditions, the use of *N*, *N*-dimethylanilinium tetrakis(pentafluorophenyl)borate $\{[Me_2N(Ph)H]^+[B(C_6F_5)_4]^-\}$ instead of $B(C_6F_5)_3$ does not yield active catalysts; only traces of syndiotactic polystyrene were obtained [70].

Although there is dispute about the exact oxidation state of titanium in the active species [Ti(III) or Ti(IV)], it was suggested, from the results of ESR measurements, that Ti(III) species form highly active sites for producing syndiotactic polystyrene in styrene polymerisation systems with the $TiBz_4$–$[Al(Me)O]_x$ catalyst [50]. The moderately low catalyst activity is attributable to the stability of the benzyl transition metal derivatives towards reduction.

A much lower stereospecificity of the $ZrBz_4$–$[Al(Me)O]_x$ catalyst for styrene polymerisation (ca 58% of the hot acetone-insoluble polymer fraction [56]) compared to that for the respective Ti-based catalyst should be noted. This can be explained in terms of the larger radius of zirconium than titanium, thus resulting in the impossibility of sufficiently effective chain end stereocontrol.

Zirconium-based cationic complexes derived from $ZrBz_4$ and $B(C_6F_5)_3$ or $[Me_2N(Ph)H]^+[B(C_6F_5)_4]^-$ do not provide active syndiospecific catalysts for styrene polymerisation [70].

4.2.2.2 Polymerisation with Half-sandwich Metallocenes

Catalyst precursors containing one cyclopentadienyl ligand attached to a transition metal atom (half-sandwich metallocenes) have been commonly used to obtain syndiotactic polystyrene. Several group 4 metal monocyclopentadienyl derivatives, $CpMtX_3$ (Mt = Ti, Zr; X = Cl, R, Ar, OR), were found to be highly active precursors for homogeneous syndiospecific methylaluminoxane-activated catalysts. The much higher activity of $CpMtX_3$–$[Al(Me)O]_x$ catalysts in the syndiospecific polymerisation of styrene by comparison with Mt Bz_4–$[Al(Me)O]_x$ catalysts was attributed entirely to the number of non-stereospecific and syndiospecific active species in the two systems, their respective catalytic species having the same intrinsic activity.

It may be worth recalling that catalysts such as $CpTiCl_3$–$[Al(Me)O]_x$ yield an atactic polymer when used for propylene polymerisation.

Cyclopentadienyltitanium alkoxides activated with methylaluminoxane are sensitive to various polymerisation conditions such as catalyst concentration, Al/Ti ratio, monomer concentration, dielectric constant of the reaction medium and polymerisation temperature.

As regards a comparison between the electrophilic properties of methylaluminoxane-activated titanium-based homogeneous catalysts for syndiospecific polymerisation of styrene, half-sandwich titanocene-based catalysts are stronger electrophiles than non-cyclopentadienyltitanium-based catalysts, the former catalysts thus being more active in the polymerisation [55].

One of the best catalysts for the syndiospecific polymerisation of styrene appeared to be that derived from $CpTiCl_3$ and methylaluminoxane. The polymerisation rate for this system decreases with increasing polymerisation time; such behaviour is very similar to that of other Ziegler–Natta catalysts. A maximum polymerisation rate is achieved at 50 °C [6]. Other catalysts such as $CpTi(OBu)_3$–$[Al(Me)O]_x$ are also reported to exhibit a very high activity and syndiospecificity in the polymerisation of styrene [50,51].

It was found that substituted cyclopentadienyltitanium trichloride, in a combination with methylaluminoxane, exhibited higher catalytic activity for syndiospecific polymerisation of styrene than $CpTiCl_3$ [52,53]. The efficiency of half-sandwich titanocenes as methylaluminoxane-activated precatalysts for the syndiospecific polymerisation of styrene increases in the following order: $CpTi(OMe)_3 \ll Me_4(Me_3Si)CpTi(OMe)_3 < Cp^*Ti(OMe)_3$. Thus, electron-donating substituents on the cyclopentadienyl ligand lead to increased catalyst activity and stability, stereospecificity and polymer $\bar{M}_w$.

The $Cp^*Ti(OR)_3$–$[Al(Me)O]_x$ system [52] appears to be one of the best catalysts. It has been suggested that ancillary ligands on titanium (Cl or OR) primarily influence the rate of formation of the active species but have little effect on the ability of the active site to assert stereocontrol of the propagation [81].

The optimum polymerisation temperature for the $Cp^*Ti(OR)_3$–$[Al(Me)O]_x$ catalyst to achieve its high catalytic activity and reasonable polymer molecular weight is 60–75 °C. Temperature greatly influences the rate of chain transfer and β-hydride elimination, providing a process tool for controlling the molecular weight. The Cp^* ligand with its electron-releasing methyl groups and greater steric bulk stabilises the active site towards β-hydride elimination, thus allowing for the formation of syndiotactic polystyrene of higher molecular weight. Bulk polymerisations of styrene yield polymers with higher molecular weights and catalyst efficiencies than those observed in slurry polymerisations [81].

ESR studies of $CpTiX_3$–$[Al(Me)O]_x$ and Cp^*TiX_3–$[Al(Me)O]_x$ catalysts indicate that Ti(III) species are formed by the reduction of Ti(IV) species in both the catalysts, to an extent of up to 40 and 85% respectively [81–84].

Among catalysts based on cyclopentadienyl-substituted half-sandwich titanocenes, the $IndTiCl_3$–$[Al(Me)O]_x$ catalyst in particular has extremely high activity and stereospecificity. It is relatively more sensitive to polymerisation conditions. A minimum concentration of about 50 mmol of methylaluminoxane was required to obtain the desired polymerisation activity. The activity also increases as the $[Al(Me)O]_x$ concentration increases, reaching a maximum at an Al/Ti molar ratio of 4000. For instance, the productivity of the $IndTiCl_3$–$[Al(Me)O]_x$ catalyst under optimum polymerisation conditions

($[IndTiCl_3] = 5 \times 10^{-7}$ mol/L, Al/Ti molar ratio 4000, temperature 50 °C) reached 37 000 kg of polystyrene per gram Ti/h (98.2% polymer fraction insoluble in hot methyl ethyl ketone) [53].

High-activity syndiospecific catalysts for styrene polymerisation have also been obtained using fluorinated half-sandwich titanocenes, such as $CpTiF_3$, $MeCpTiF_3$ and Cp^*TiF_3, as precatalysts, in a combination with methylaluminoxane. For instance, the productivity of the $CpTiF_3$–$[Al(Me)O]_x$ catalyst reaches 3000 kg of polystyrene per gram Ti/h ($\bar{M}_w = 100 \times 10^3$, $\bar{M}_w/\bar{M}_n = 2.0$) under the following polymerisation conditions: $[CpTiF_3] = 6.3 \times 10^{-4}$ mol/L, Al/Ti molar ratio 300, temperature 50 °C. Polystyrenes of the highest melting points can be obtained in polymerisations with the Cp^*TiF_3–$[Al(Me)O]_x$ catalyst. It exhibits a good catalytic activity, produces syndiotactic polystyrene characterised by a high molecular weight ($\bar{M}_w = 660 \times 10^3$), a very narrow molecular weight distribution ($\bar{M}_w/\bar{M}_n = 2.0$) and a high melting point (275 °C) and is stable up to a polymerisation temperature of 70 °C [59]; the chlorinated counterpart shows much less catalytic activity and produces a polymer of lower molecular weight.

The syndiospecific polymerisation of styrene has also been found [85] to proceed in the presence of the methylaluminoxane-activated trisiloxane-bridged dinuclear titanocene complex hexamethyltrisiloxanediylbis(cyclopentadienyltitanium trichloride) $[Cl_3TiCpSi(Me)_2OSi(Me)_2OSi(Me)_2CpTiCl_3]$, which also polymerised ethylene.

Methylaluminoxane-activated catalysts for syndiospecific styrene polymerisation exhibit higher activity than aluminium-free catalysts. In systems for the polymerisation of styrene with half-sandwich titanocene-based catalysts, as in polymerisation systems for α-olefins with metallocene-based catalysts, methylaluminoxane plays the role of maintaining the cationic species in the catalyst, acting as a non-coordinating anion [86]. Besides this function, methylaluminoxane also has the role of alkylating the transition metal compound when the Mt–C bond is not yet formed. Remember that it replaces one chlorine atom or an alkoxy group with an alkyl group and removes the second ligand from the transition metal, generating coordinatively unsaturated cationic species. Methylaluminoxane is always used in a large excess with respect to the transition metal compound in order to shift the equilibrium to the formation of cationic catalytic species, certainly acting also as a scavenger of impurities.

The different propagation rate constants for various $CpTix_3$-based catalysts could be explained, for instance, by considering the possibility of different ancillary ligands on the active species affecting the electrophilicity and possibly the intimate ion pair $\rightleftharpoons$ free ions equilibrium, if one assumes that the free cation may effectively promote the propagation [87].

Let us recall that the origin of stereospecificity in the syndiospecific polymerisation of styrene lies in a chain end stereochemical control mechanism [52,70]. Key features of the stereoregulation mechanism are stereorigid η^7 coordination of the growing chain end and diastereoselective η^2 coordination of the styrene molecule, imposed by direct interactions with the ancillary ligand on the metal in Ti(III) or Zr(III) chiral distorted pseudotetrahedral cationic species, which inverts its configuration after every syndiospecific insertion step [87]. The proposed structure for active species in the syndiospecific polymerisation of styrene with half-sandwich titanocene-based catalysts is shown in Figure 4.1 [88–90].

The insertion of η^2 coordinated styrene, by *cis* addition, would stepwise lead to η^7 bonding of the last incorporated monomeric unit and dissociation of the phenyl ring of the penultimate monomeric unit, thus providing the possibility of enantioselective η^2 coordination of another styrene molecule to give mirror related active species (Figure 4.2) [87].

The incorporation of the monomer is at the same time a chain migratory insertion and a nucleophilic substitution at the cationic metal centre by the phenyl group of the inserting monomer while the last monomeric unit of the propagating chain is leaving. However, each syndiospecific monomer enchainment involves the inversion of configuration at the metal centre. Thus, the stereochemical mechanism of syndiospecific styrene polymerisation implies steric control of monomer insertion by the chirality of the metal, which, in turn, is related to the configuration and conformation of the last monomeric unit of the growing chain (chain end stereocontrol) [87].

It is worth mentioning that the coordination of the mirror related styrene enantioface to give the diastereoisomer is hardly possible owing to the exceedingly high interactions with the Cp* ligand that occur when the conformation is

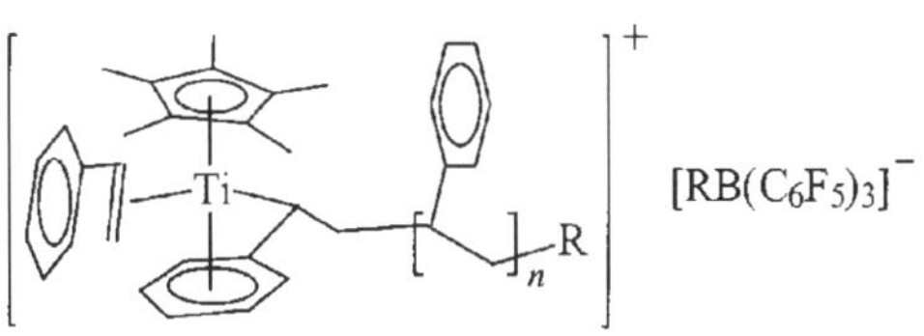

Figure 4.1 Proposed structure for the polymerising species in syndiospecific styrene polymerisation promoted by the $Cp^*TiR_3–B(C_6F_5)_3$ catalyst

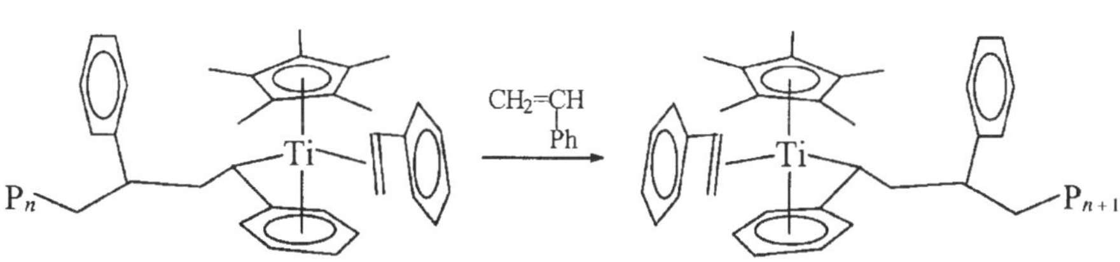

Figure 4.2 Inversion of the configuration of the active site after incorporation of the coordinated styrene molecule, η^7 coordination of the propagating chain end and η^2 coordination of another styrene molecule

that required by *cis* ligand migration and secondary (2,1) insertion. Considerable non-bonded interactions would also occur if the ancillary ligand were Cp instead of Cp* [87].

Considering the above stereochemical model for syndiospecific styrene polymerisation, one may conclude reasonably that η^4 coordination of the monomer at the active site could hardly be possible, and η^2 coordination would always be involved in the syndiospecific polymerisation of this monomer [87]. One should note that preliminary concepts concerning the stereoregulation mechanism of syndiospecific styrene polymerisation assumed the styrene monomer to undergo only η^4 coordination at the titanium centre, the propagating chain being anchored via a benzylic bond as an η^3 ligand at the titanium [44,55,70].

Homogeneous catalysts for the syndiospecific polymerisation of styrene have been extended to discrete cationic systems in the absence of methylaluminoxane. Such systems provide direct evidence for the cationic nature of the active species in the syndiospecific polymerisation of styrene. Catalysts derived from monocyclopentadienyl derivatives of titanium activated with $B(C_6F_5)_3$, $[Ph_3C]^+[B(C_6F_5)_4]^-$ or $[Et_3NH]^+[B(C_6F_5)_4]^-$ are capable of polymerising styrene in quite high yields [70,91]. Active catalysts were obtained by mixing Cp^*TiMe_3 or Cp^*TiBz_3 with $B(C_6F_5)_3$ in a molar ratio of 1:1. It is evident that the strong Lewis acid is able to extract an anionic ligand from the organometallic compound, generating a cationic species. The stability of such species should be due to interaction with the counterion, as demonstrated previously for zirconocenes [92].

The relatively high activity exhibited by some methylaluminoxane-free catalysts, such as e.g. Cp^*TiBz_3–$B(C_6F_5)_3$ [scheme (8)] and Cp^*TiMe_3–$[Ph_3C]^+[B(C_6F_5)_4]^-$ [scheme (9)], in the syndiospecific polymerisation of styrene might seem to favour the higher oxidation state of titanium [70]:

$$Cp^*TiBz_3 + B(C_6F_5)_3 \rightleftharpoons [Cp^*TiBz_2]^+ [BzB(C_6F_5)_3]^- \tag{8}$$

$$Cp^*TiMe_3 + [Ph_3C]^+ [B(C_6F_5)_4]^- \rightleftharpoons [Cp^*TiMe_2]^+ [B(C_6F_5)_4]^- + Ph_3CMe \tag{9}$$

However, NMR and ESR monitoring of the above catalysts showed that the initially formed [according to schemes (8) and (9)] Ti(IV) cationic species decompose stepwise, possibly leading to reduced Ti(III) species [87]:

$$[Cp^*TiR_2]^+ \longrightarrow [Cp^*TiR]^+ + R^\bullet \tag{10}$$

The rate of reduction of the titanium cationic species [scheme (10)] increases when styrene is added to the system [83,90,93].

As a consequence, it is not immediately apparent whether the active species are the Ti(IV) sites just mentioned or some decomposition product, possibly a Ti(III) site. However, evidence has been provided for an oxidation state of titanium present in the active species of +3 [50,87,94]. There are also findings for the $Cp^{*}TiCl_3$–$[Al(Me)O]_x$ catalyst, which, during its ageing, became active in syndiospecific styrene polymerisation while losing its catalytic activity for the polymerisation of ethylene. These findings suggest that the initially formed $[Cp^{*}TiMe_2]^+$ cation is only able to promote ethylene polymerisation, but the syndiospecific polymerisation of styrene is promoted by the $[Cp^{*}TiMe]^+$ cation [87,90].

The Ti(III) cationic active species has also been proved to be the true catalytic site in styrene polymerisation systems with Ti(III)-based catalysts such as $Cp^{*}Ti(OMe)_2$–$[Al(Me)O]_x$ and $Cp^{*}Ti(OMe)_2$–$[Me_2N(Ph)H]^+$ $[B(C_6F_5)_4]^-$ as well as with Ti(IV)-based catalysts such as $Cp^{*}Ti(OMe)_3$–$[A\ l(Me)O]_x$ and $Cp^{*}Ti(OMe)_3$–$[Me_2N(Ph)H]^+[B(C_6F_5)_4]^-$. The reduction of Ti(IV) species in systems with the $Cp^{*}Ti(OMe)_3$ precatalyst is facilitated by the addition of $Al(i\text{-}Bu)_3$ to the reaction system. The $Cp^{*}Ti(OMe)_2$-based catalysts, activated with either $[Al(Me)O]_x$ or $[Me_2N(Ph)H]^+[B(C_6F_5)_4]^-$, appeared to exhibit high activity in the syndiospecific polymerisation of styrene [95].

Half-sandwich zirconocene-based catalysts (e.g. those derived from the $CpZrCl_3$ precursor) show a remarkably low activity when compared with the titanium analogues. The lower activity of Zr-based catalysts might be due to the lower electrophilicity and lower concentration of catalytic active sites [71] as well as, at least in part, to the higher stability of the Zr(IV) species in comparison with the Ti(IV) species [55,57].

Half-sandwich group 3 metallocenes, such as $CpYCl_2 \cdot THF$, activated with alkylaluminioxane, have also been found to promote the polymerisation of styrene, but the polymer produced appeared to be atactic [96].

4.2.2.3 Polymerisation with Metal Alkoxides

Catalysts of the $Ti(OR)_4$–$[Al(Me)O]_x$ type show greatly inferior activity and syndiospecificity in the polymerisation of styrene by comparison with catalysts of the $CpTi(OR)_3$–$[Al(Me)O]_x$ type [54,70]. The activity and syndiospecificity of $Ti(OR)_4$–$[Al(Me)O]_x$ catalysts increases when the Al/Ti molar ratio in the polymerisation system is increased. The maximum activity of $Ti(OR)_4$–$[Al(Me)O]_x$ catalysts is observed at an Al/Ti molar ratio of ca 100 [54,55]. It is worth mentioning that, under the same polymerisation conditions, these catalysts yield syndiotactic polystyrene with a higher molecular weight than does the $CpTiCl_3$–$[Al(Me)O]_x$ catalyst [71].

Catalysts derived from alkoxytitanium chlorides and methylaluminoxane have also appeared to promote styrene polymerisation [58].

4.2.2.4 Polymerisation with Metallocenes

Metallocenes such as Cp_2TiCl_2 and Cp_2ZrCl_2 alone are capable of polymerising styrene to an atactic polymer (involving a free radical propagation mechanism) [97]. The same metallocenes activated with methylaluminoxane form active catalysts for the polymerisation of styrene; their productivity and syndiospecificity, however, are not very high. In contrast, when activated with aluminium alkyls, these metallocenes do not afford catalysts that might be active in the polymerisation of styrene [98,99].

The activity of metallocene-based catalysts is strongly influenced by steric factors connected with the structure of the ligands coordinating to the transition metal. As already mentioned, the origin of the stereospecificity in styrene polymerisation with homogeneous syndiospecific catalysts is believed to lie in a chain end stereochemical control mechanism [52,70]. The sterically demanding nature of the electronically preferred 2,1-insertion of the monomer into the titanium–carbon bond helps to explain the negligible activity of the more crowded metallocene precatalyst containing two cyclopentadienyl ligands. It has been found [60,61] that catalytic activities and selectivities of Cp'_2TiCl_2–$[Al(Me)O]_x$ catalysts towards styrene syndiospecific polymerisation were mostly affected by steric factors of the ligands coordinating to the titanium centre; a larger coordination gap aperture angle (smaller centroid–Ti–centroid angle) gave rise to higher activity and selectivity to syndiotactic polystyrene. Therefore, the efficiency of titanocene-based catalysts for the syndiospecific polymerisation of styrene increases for precursors in the following order: $Cp_2TiCl_2 < Cp^*_2TiCl_2 < (Cp^*CH_2)_2TiCl_2 < Me_2C(Cp)_2$ $TiCl_2Me_2Si\ (Cp)_2\ TiCl_2 < H_2C(Cp)_2TiCl_2$.

Mechanistic studies on *ansa*-titanocene–methylaluminoxane systems have revealed the formation of intermediate compounds [60,61]:

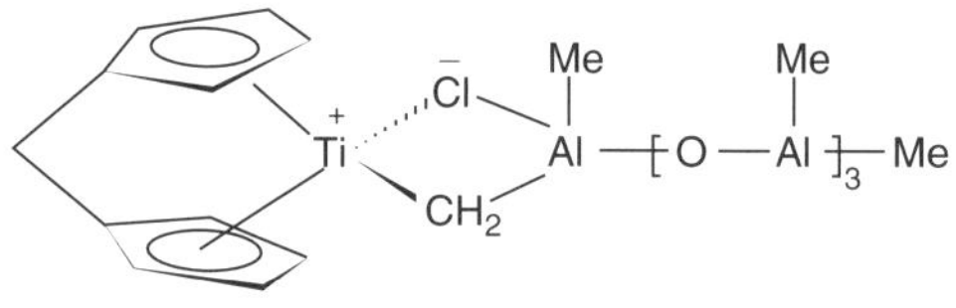

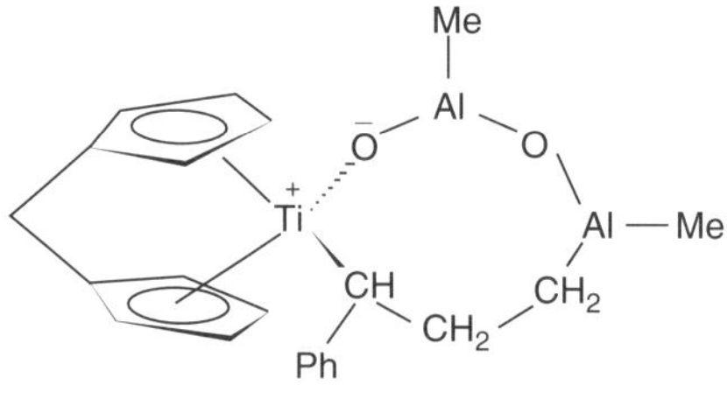

These compounds exhibited fairly high catalytic activity for the syndiospecific polymerisation of styrene.

It is worth mentioning that a *rac.-ansa*-titanocene–methylaluminoxane catalyst, such as *rac.*-$Ph_2C(Cp)(Ind)TiCl_2$–$[Al(Me)O]_x$, which yields an isotactic polymer in propylene polymerisation, promotes the syndiospecific polymerisation of styrene [73,100]. This is the first example where two different stereoregular polymers, isotactic and syndiotactic, can be obtained using the same catalyst in the case of two different monomers.

4.2.2.5 Polymerisation with Heterogeneous Catalysts

Heterogeneous non-supported or supported catalysts that have been used successfully for the syndiospecific polymerisation of styrene are presented in Tables 4.3 and 4.4 [62,63,66–69,101–103].

The polymerisation of styrene with hydrocarbon-insoluble δ-$TiCl_3$ activated by $[Al(Me)O]_x$ yields a mixture of highly isotactic and highly syndiotactic polymers. Supported catalysts with chlorine-containing magnesium compounds as carriers also produce a mixture of these two polystyrene stereoisomers; it is interesting that atactic polystyrene fraction is not present among the products. In general, the amount of syndiotactic polystyrene increases with increasing Ti/Al molar ratio [73]. In order to explain the nature of the active species responsible for the formation of isotactic and syndiotactic polystyrenes, $TiCl_3$ and $TiCl_4$ supported on $Mg(OH)_2$ were reacted with $[Al(Me)O]_x$ in a

Table 4.3 Polymerisation of styrene using various heterogeneous titanium-based catalysts activated with methylaluminoxane[a]

	Catalyst			
Precursor	Quantity (mmol)	[Al(Me)O] unit/ Ti molar ratio	Styrene conversion %	Polystyrence
$TiCl_4$	40.0	10	72.0	Isotactic
$TiCl_4$	5.0	40	0.4	Isotactic + syndiotactic
$TiCl_4$	0.2	500	0.7	Syndiotactic
δ-$TiCl_3$ [b]	1.0	100	8.2	Isotactic + syndiotactic
δ-$TiCl_3$ [b]	0.2	1000	2.0	Isotactic + syndiotactic
$Mg(OEt)_2$/EB[c]/$TiCl_4$	2.0	50	2.9	Isotactic + syndiotactic (84%) (16%)
$Mg(OEt)_2$/EB[c]/$TiCl_4$	0.2	500	1.1	Isotactic + Syndiotactic (12%) (88%)
$Mg(OEt)_2$/EB[c]/$TiCl_4$	0.2	1000	1.4	Isotactic + syndiotactic (10%) (90%)

[a] Polymerisation conditions: styrene 4.3×10^{-1} mol; toluene 100 mL; temperature 50 °C; time 2 h.
[b] Derived from $TiCl_4$ by reduction with metallic Al.
[c] Ethyl benzoate.

Table 4.4 Polymerisation of styrene using various heterogeneous titanium-based catalysts

Catalyst			
Precursor	Activator	Polystyrene	Reference
δ-$TiCl_3$	$[Al(Me)O]_x$	Isotactic + syndiotactic	[62,101]
$Mg(OH)_2/Ti(OBu)_4$	$[Al(Me)O]_x$	Syndiotactic	[68]
$Mg(OH)Cl/Ti(OBu)_4$	$[Al(Me)O]_x$	Isotactic + syndiotactic	[68]
$MgCl_2/Ti(OEt)_4$	$[Al(Me)O]_x$	Isotactic + syndiotactic	[66]
$MgCl_2/Ti(OBu)_4$	$[Al(Me)O]_x$	Isotactic + syndiotactic	[68]
$SiO_2/Ti(OBu)_4$	$[Al(Me)O]_x$	Syndiotactic	[67]
$Al_2O_3/CpTiCl_3$	$Al(i\text{-}Bu)_3$	Isotactic + syndiotactic	[69,102]
$Al_2O_3/MeCpTiCl_3$	$Al(i\text{-}Bu)_3$	Isotactic + syndiotactic	[69]
Al_2O_3/Cp^*TiCl_3	$Al(i\text{-}Bu)_3$	Isotactic + syndiotactic	[69]
CD^a/Cp^*TiCl_3	$[Al(Me)O]_x$	Syndiotactic	[103]

[a] α-Cyclodextrin.

toluene medium. The results demonstrated that, on reaction with methylaluminoxane, titanium compounds showed partial solubility. Soluble catalytic species were found to promote the syndiospecific polymerisation of styrene, while the species present in the separated insoluble part were found to promote isospecific polymerisation [62,63,73]. Moreover, the amount of 'soluble $TiCl_3$' in the polymerisation medium appeared to increase with an increasing amount of $[Al(Me)O]_x$ used. It is worth noting that the soluble $TiCl_3$.THF complex in combination with $[Al(Me)O]_x$ polymerised styrene to syndiotactic polystyrene [63].

In view of these results, it was suggested that syndiospecific and isospecific polymerisations with heterogeneous catalysts containing a chlorine atom in the titanium compound or in the support are promoted by homogeneous species formed in the polymerisation system and by heterogeneous species respectively.

In the case of chlorine-free catalysts such as $Mg(OH)_2/Ti(OBu)_4$–$[Al(Me)O]_x$ and $SiO_2/Ti(OBu)_4$–$[Al(Me)O]_x$, heterogeneous species are assumed to promote the syndiospecific polymerisation of styrene [67,68]. In a polymerisation system with the latter catalyst, best results were achieved when treating the carrier with $[Al(Me)O]_x$ prior to supporting $Ti(OBu)_4$ (no further activation with methylaluminoxane was needed). The polymerisation rate reaches a maximum at an Al/Ti molar ratio of 20; this is much lower than the value of the Al/Ti molar ratio required to reach the maximum polymerisation rate in the respective homogeneous system, i.e. the system without a carrier [54].

When supported on Al_2O_3 as a carrier, half-sandwich metallocenes such as $CpTiCl_3$ and Cp^*TiCl_3 also gave rise to suitable stereospecific catalysts that could even be activated by $Al(i\text{-}Bu)_3$ [69]. However, in contrast to the respective homogeneous catalysts (yielding syndiotactic polystyrene), polymerisation with these heterogeneous catalysts afforded isotactic and syndiotactic polystyrenes.

As Al_2O_3 contains considerable amounts of acidic centres on the surface, it was suggested that the syndiospecific polymerisation of styrene takes place via polyinsertion involving Ti–C cationic species. Non-cationic species with the Ti–C bond on the alumina surface are believed to promote isospecific polymerisation.

When α-cyclodextrin-supported Cp^*TiCl_3 is used as the catalyst precursor, $[Al(Me)O]_x$ should be used as an activator in order to obtain a suitable catalyst for the stereospecific polymerisation of styrene (polymerisation syndiospecificity ca 93%); with $AlMe_3$ as the activator, α-cyclodextrin-supported Cp^*TiCl_3 had less activity for styrene polymerisation even if α-cyclodextrin was treated with $[Al(Me)O]_x$ prior to supporting Cp^*TiCl_3 [103].

4.3 Copolymerisation

As mentioned earlier, heterogeneous Ziegler–Natta catalysts, which yield isotactic polystyrene, are also active in the polymerisation of ethylene and α-olefins [2]. Such catalysts also promote the polymerisation of ring-substituted styrenes [16–20] and conjugated dienes [104]. These monomers are also capable of copolymerising with each other. The substituted styrenes are copolymerised with styrene in the presence of isospecific catalysts, yielding random copolymers that are co-isotactic [28,105]. The reactivities of ring-substituted styrenes in their copolymerisation with styrene are governed by the electronic effects of the ring substituents (following the Hammett equation with a negative slope [19,27]), i.e. the copolymerisability increases with increasing electron density at the double bond undergoing copolymerisation, which is illustrated by Table 4.1 [18,41].

Similarly, the same catalysts that promote the syndiospecific polymerisation of styrene also polymerise ethylene and α-olefins [106,107], ring-substituted styrenes [6] and conjugated dienes [44,74,108–110]. These monomers can also be copolymerised with each other [111–114]. Substituted styrenes, which yield syndiotactic polymers by polymerisation run with syndiospecific catalysts, form copolymers with styrene; the polymerisation rate increases with increasing nucleophilicity of the comonomer. The random copolymers formed are co-syndiotactic [6,111,112].

Styrene and ring-substituted styrenes also form copolymers with carbon monoxide, similarly to ethylene and α-olefins. These styrene/carbon monoxide copolymers are of alternating, highly regioregular head-to-tail structure and are characterised by different stereoregularity (syndiotactic, isotactic) [115–117].

4.3.1 Copolymerisation with Olefins

Styrene, which can be treated formally as an α-olefin branched in the 3-position, forms copolymers with ethylene and α-olefins (as well as with β-olefins, involving isomerisation–copolymerisation). Both heterogeneous Ziegler–Natta catalysts and a single-site metallocene catalyst promote the copolymerisation.

4.3.1.1 Copolymerisation with Heterogeneous Ziegler–Natta Catalysts

Styrene undergoes copolymerisation with ethylene and various α-olefins in the presence of heterogeneous Ziegler–Natta catalysts. Its reactivity in the copolymerisation is quite low, which is illustrated by the values of the relative reactivity ratios, r_1 and r_2, presented in Table 4.5 [118]. One may note, however, a considerably high relative reactivity of styrene in copolymerisation with vinylcyclohexane. The copolymerisation of styrene with small amounts of α-olefin, such as 1-octene or 1-decene, yields copolymers of reduced crystallinity and thus reduced brittleness compared with the homopolymer of styrene.

Styrene/α-olefin copolymers containing a predominant amount of styrene units can be easily formed through the monomer isomerisation–copolymerisation of styrene and β-olefin such as *cis*-2-butene in the presence of heterogeneous Ziegler–Natta catalysts such as $TiCl_3$–$AlEt_3$. Styrene appeared to be a favourable comonomer for monomer isomerisation–copolymerisation with internal olefins, since only the isomerisation of β-olefin to α-olefin, and not the isomerisation of styrene (in contrast to the olefin), occurs in the presence of Ziegler–Natta catalysts [119]:

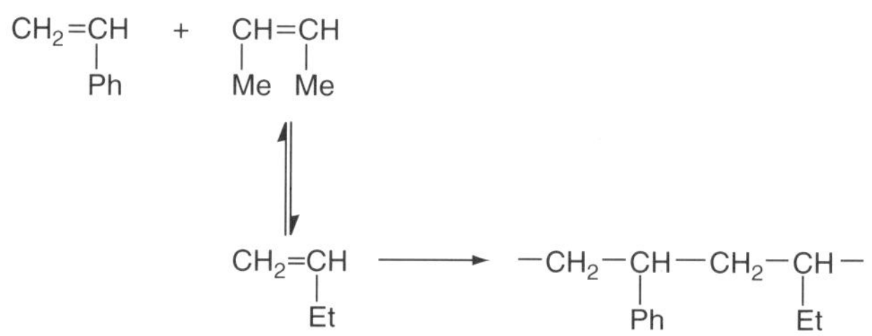

(11)

By adding $NiCl_2$ to the $TiCl_3$–$AlEt_3$ catalyst (Ni/Ti/Al molar ratio 1:1:3), the contents of 1-butene units in the copolymer were increased at the same comonomer feed as compared with the copolymerisation system without added

Table 4.5 Relative reactivity ratios for copolymerisation of styrene (r_1) and α-olefins (r_2) with heterogeneous Ziegler–Natta catalysts

α-olefin (M_2)	$r_1 = k_{11}/k_{12}$	$r_1 = k_{22}/k_{21}$
Ethylene	0.012	81
Propylene	0.12–0.20	8–20
1-Hexene	0.19	9.8
1-Heptene	0.61	5.7
4-Methyl-1-pentene	0.98	3.9
4-Methyl-1-hexene	1.80	1.3
5-Methyl-1-heptene	0.59	4.0
Vinylcyclohexane	2.10	0.18

$NiCl_2$. It is interesting that copolymers consisting almost entirely of styrene units are produced easily via monomer isomerisation–copolymerisation of styrene and *cis*-2-butene, contrary to the ordinary copolymerisation of styrene and 1-butene [119].

4.3.1.2 Copolymerisation with Single-site Catalysts

The remarkable capabilities of single-site catalysts for promoting polymerisation have led to a novel range of high molecular weight styrene/ethylene copolymers, referred to as 'pseudo-random' copolymers, which were hardly accessible with the use of heterogeneous Ziegler–Natta catalysts of the $TiCl_3$–AlR_3 type. These new copolymers have been shown to consist of comonomers linked randomly, with the exception that styrene units are always separated by at least one ethylene unit. Such copolymers are rubbery and are expected to compete with other elastomers.

Using half-sandwich metallocene-based catalysts such as constrained-geometry catalysts with substituted cyclopentadienylamidotitanium species such as $[Me_4CpSi(Me)_2N(t\text{-}Bu)]TiCl_2$ [120,121] as well as the aluminium-free $Me_4CpTiBz_3$–$B(C_6F_5)_3$ catalyst [122] for styrene/ethylene copolymerisation, one may achieve an incorporation of styrene at levels of up to 45 mol.-%, which is equivalent to 75 wt-%. However, the obtained copolymers have no head-to-tail styrene unit arrangement in chains and no tacticity in alternating styrene–ethylene chain segments.

On the other hand, an alternating styrene/ethylene copolymer has been formed in a copolymerisation system containing a non-cyclopentadienyltitanium-based catalyst, which was a thiobisphenolatotitanium-based catalyst. Copolymerisation run with 2,2′-thiobis(4-methyl-6-*t*-butylphenoxy)titanium bisisopropoxide $[(Tbp)Ti(O\text{-}i\text{-}Pr)_2]$–$[Al(Me)O]_x$ or 2,2′-thiobis(4-methyl-6-*t*-butylphenoxy)titanium dichloride $[(Tbp)TiCl_2]$–$[Al(Me)O]_x$ as a catalyst yielded an alternating head-to-tail styrene/ethylene copolymer. Apart from the copolymer, a styrene homopolymer of syndiotactic structure was formed in this copolymerisation system. The styrene/ethylene alternating copolymer surprisingly appeared to be isotactic [123].

Such an isotactic copolymer has also been obtained using some *ansa*-zirconocene-based catalysts such as *rac.*-$(IndCH_2)_2ZrCl_2$–$[Al(Me)O]_x$ (C_2 symmetric). Catalysts based on *ansa*-zirconocenes such as $Ph_2C(Cp)(Flu)ZrCl_2$ (C_s symmetric), $Me_2C(Cp)_2ZrCl_2$ (C_{2v} symmetric) and *rac.*-$Me_2C(MeCp)_2ZrCl_2$ (C_2 symmetric) yielded copolymers with predominant atactic styrene–ethylene units. These results suggest that the aromatic or fused ring substituent of the cyclopentadienyl ligand having C_2 symmetry (and not the aliphatic substituent) is be responsible for stereoregulation in styrene/ethylene alternating copolymerisation [124].

4.3.2 Copolymerisation with Carbon Monoxide

4.3.2.1 Stereoisomerism of Alternating Styrene/Carbon Monoxide Copolymers

Although regioregular head-to-tail alternating styrene/carbon monoxide copolymers, referred to as polyketones, are formed in the copolymerisation of a prochiral rather than a chiral monomer, which is styrene, they possess true chiral sites constituted by the tertiary carbon atoms in the main chain: $-CH_2C(O)-C^*H(Ph)-CH_2C(O)-$. Therefore, isotactic diads and syndiotactic diads constitute respective triads and longer stereosequences in copolymer chains. Stereoisomerism of regioregular head-to-tail styrene/carbon monoxide alternating copolymers $\{-[-CH(Ph)-CH_2-C(O)-]_n-\}$, which lies in the appearance of isotactic and syndiotactic poly(1-oxo-2-phenyltrimethylene)s, is illustrated in Figure 4.3.

Note that in the isotactic polymer molecule (the adapted Fischer projection shows tertiary carbon atoms of the same configuration) the successive phenyl substituents appear alternately in front of and behind the plane of the extended polymer backbone in the flat zigzag projection. However, in the syndiotactic polymer molecule (the adapted Fischer projection shows neighbouring tertiary carbon atoms of opposite configurations to each other) the successive phenyl substituents appear in front of or behind the plane of the extended polymer backbone in the flat zigzag projection. Such spatial placements of phenyl

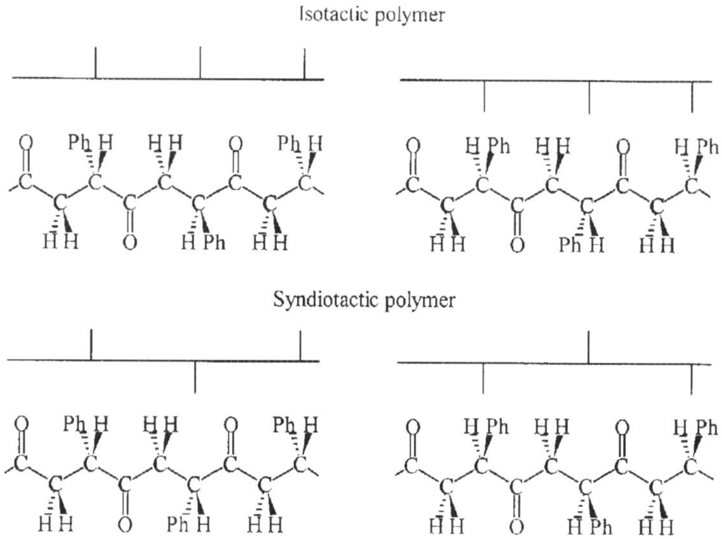

Figure 4.3 Stereoisomerism of poly(1-oxo-2-phenyltrimethylene) $\{-[-CH(Ph)-CH_2-C(O)-]_n-\}$. Isotactic and syndiotactic polymers

groups in the extended polymer backbone in flat zigzag projections for isotactic and syndiotactic poly(1-oxo-2-phenyltrimethylene)s is characteristic of tactic polymers containing an odd number of atoms in the monomeric units constituting the backbone of the macromolecule. Let us recall that this is in contrast to the spatial placements of substituents at tertiary carbon atoms in the extended polymer backbone in flat zigzag projections for isotactic and syndiotactic poly(α-olefin)s, which contain an even number of atoms in the monomeric units constituting the backbone of the macromolecule.

4.3.2.2 Stereospecific Alternating Copolymerisation of Styrene and Carbon Monoxide

The alternating copolymerisation of styrene and ring-substituted styrenes with carbon monoxide proceeds in the presence of palladium-based catalysts with chelating ligands. It gives isotactic or syndiotactic copolymers, poly(1-oxo-2-phenyltrimethylene)s, depending on the kind of ligand in the catalyst as well as on the kind of monomer applied for the copolymerisation. In the case of this copolymerisation, involving the monomer containing an aromatic ring in the molecule, a more complete stereocontrol of the polymer growth reaction is possible compared with that in the case of the copolymerisation of an aliphatic monomer (α-olefin) [125].

Palladium cationic systems containing bidentate nitrogen ligands such as 2,2′-bipyridine or 1,10-phenantroline yield a copolymer that is prevailingly syndiotactic [115,126,127]. The molecular weight of styrene/carbon monoxide copolymers is relatively low. The highest values were achieved when using a preformed catalyst, 1,10-phenantroline as the ligand, an anion from a very strong acid (trifluoromethanesulfonic acid), moderate levels of ethanol and the progressive addition of benzoquinone during the copolymerisation [128–130]. a high molecular weight styrene/carbon monoxide copolymer ($\bar{M}_w = 170 \times 10^3$) with prevailing syndiotactic diads is a highly crystalline polymer [131]. The absence of pairs of isotactic diads, i.e. isotactic triads, in copolymer chains suggests a chain end stereocontrol mechanism operating in the copolymerisation [116,132]. The degree of syndiotacticity can be slightly increased by using 4-nitro-1,10-phenantroline as the ligand [133]. Bipyridine or phenanthroline type ligands containing substituents possessing centres of chirality form catalysts that yield copolymers still exhibiting the same degree of syndiotacticicty as those without chiral ligands.

However, when (*S*,*S*)-3,3′-(2,3-butanediol)-2,2′-bipyridine or (*R*)-3,3′-(1,2-propanediol)-2,2′-bipyridine was used as the ligand [125], copolymers were obtained that had a higher content of isotactic triads. An effective control towards the isospecificity of copolymerisation (<98 % isotacticity in the copolymer) is fulfilled for the copolymerisation of ring-substituted styrene such as *p*-*t*-butylstyrene and carbon monoxide with catalysts containing cationic methylpalladium species [117]:

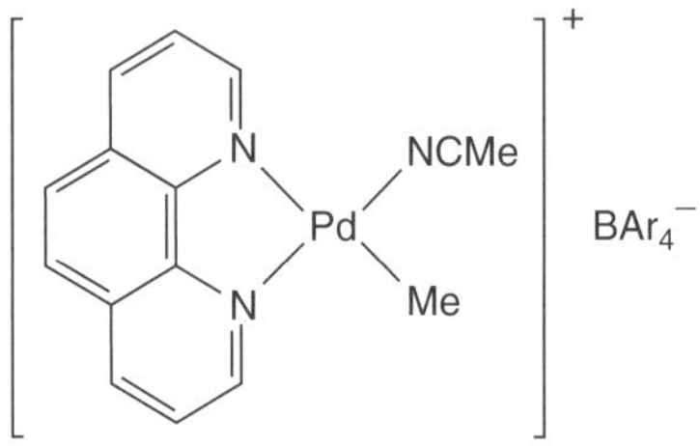

The copolymerisation of styrene and carbon monoxide with a Pd-based catalyst utilising separately the *R* and *S* enantiomers of 2-pyridinecarboxaldehyde-*N*-1-phenylethylideneimine yielded copolymers showing a high optical activity [117,134].

A valuable ligand for the highly isotactic alternating copolymerisation of styrene and *p*-methylstyrene with carbon monoxide appeared to be C_2-symmetric (4*S*, 4′*S*)-(–)-4,4′,5,5′-tetrahydro-4,4′-bis(1-methylethyl)-2,2′-bioxazole coordinated to Pd(II) in the cationic complex [135]:

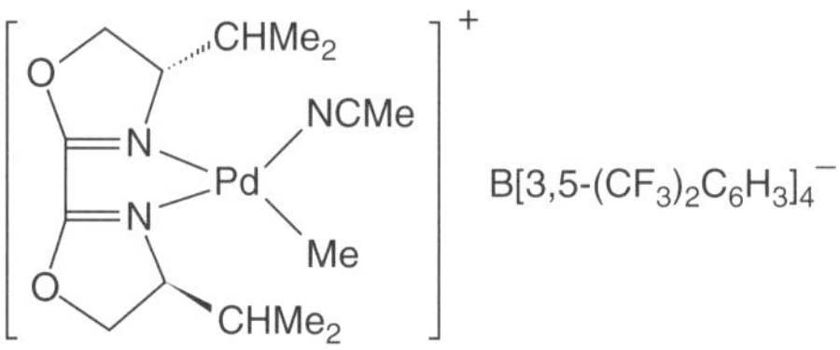

This copolymerisation run at room temperature and under low carbon monoxide pressure (1–4 atm) yields highly isotactic optically active polymers.

References

1. Natta, G., Pino, P., Corradini, P., Danusso, F., Mantica, E., Mazzanti, G. and Moraglio, G., *J. Am. Chem. Soc.*, **77**, 1708 (1955).
2. Natta, G. and Corradini, P., *Makromol. Chem.*, **16**, 77 (1955).
3. Overberger, C. and Mark, H., *J. Polym. Sci.*, **35**, 381 (1959).
4. Kern, R. J., Hurst, H. G. and Richard, W. J., *J. Polym. Sci.*, **45**, 195 (1960).
5. Ishihara, N., Seimiya, T., Kuramoto, M. and Uoi, M., *Macromolecules*, **19**, 2464 (1986).
6. Ishihara, N., Kuramoto, M. and Uoi, M., *Macromolecules*, **21**, 3356 (1988).
7. Ammendola, P., Pellecchia, C., Longo, P. and Zambelli, A., *Gazz. Chim. Ital.*, **117**, 65 (1987).
8. Grassi, A., Pellecchia, C., Longo, P. and Zambelli, A., *Gazz. Chim. Ital.*, **117**, 249 (1987).
9. Pellecchia, C., Longo, P., Grassi, A., Ammendola, P. and Zambelli, A., *Makromol. Chem., Rapid Commun.*, **8**, 277 (1987).
10. Zambelli, A., Longo, P., Pellecchia, C. and Grassi, A., *Macromolecules*, **20**, 2035 (1987).

11. Curie, J. A. and Dole, M. J., *J. Phys. Chem.*, **73**, 3384 (1969).
12. Overbergh, N., *J. Polym. Sci., Polym. Phys. Ed.*, **14**, 1177 (1976).
13. Usmani, A. M. and Sayler, I. O., *J. Elast. Plast.*, **12**, 90 (1980).
14. Cimmino, S., Di Pace, E., Martuscelli, E. and Silvestre, C., *Polymer*, **32**, 1080 (1991).
15. Silvestre, C. and Cimmino, S., 'Stereoregular Polystyrene', in *The Polymeric Materials Encyclopedia*, CRC Press, Inc., Boca Raton, 1996, Vol. 9, pp. 6820–6828.
16. Sianesi, D., Rampichini, M. and Danusso, F., *Chim. Ind. (Milan)*, **41**, 287 (1959).
17. Natta, G., *J. Polym. Sci.*, **16**, 143 (1955).
18. Natta, G., Danusso, F. and Sianesi, D., *Makromol. Chem.*, **28**, 253 (1958).
19. Danusso, F. and Sianesi, D., *Makromol. Chem.*, **30**, 238 (1959).
20. Heller, J. and Miller, D. B., *J. Polym. Sci., A-1*, **5**, 2323 (1967).
21. Abis, L., Albizzati, E., Conti, G., Giannini, U., Resconi, L. and Spera, S., *Makromol. Chem., Rapid Commun.*, **9**, 209 (1988).
22. Ascenso, J. R., Dias, A. R., Gomes, P. T., Ramao, C. C., Pham, Q., Neibecker, D. and Tkatchenko, I., *Macromolecules*, **22**, 1000 (1989).
23. Ascenso, J. R., Dias, A. R., Gomes, P. T., Ramao, C. C., Neibecker, D., Tkatchenko, I. and Revillon, A., *Makromol. Chem.*, **190**, 2773 (1989).
24. Natta, G., and Danusso, F., *Chim. Ind. (Milan)*, **40**, 445 (1958).
25. Danusso, F. and Sianesi, D., *Chim. Ind. (Milan)*, **40**, 450 (1958).
26. Danusso, F., Sianesi, D. and Calcagno, B., *Chim. Ind. (Milan)*, **40**, 628 (1958).
27. Danusso, F. and Sianesi, D., *Chim. Ind. (Milan)*, **44**, 493 (1962).
28. Natta, G., *Makromol. Chem.*, **35**, 93 (1960).
29. Ciardelli, F., Pieroni, O., Carlini, C. and Menicagli, C., *J. Polym. Sci., A-1*, **10**, 809 (1972).
30. Kissin, Y. V., in *Isospecific Polymerization of Olefins with Heterogeneous Ziegler–Natta Catalysts*, Springer-Verlag, New York, 1985.
31. Soga, K. and Yanagihara, H., *Makromol. Chem., Rapid Commun.*, **9**, 23 (1988).
32. Soga, K., Uozumi, T., Yanagihara, H. and Shiono, T., *Makromol. Chem., Rapid Commun.*, **11**, 229 (1990).
33. Natta, G., Danusso, F. and Pasquon, I., *Collect. Czech. Chem. Commun.*, **22**, 191 (1957).
34. Danusso, F., *Chim. Ind. (Milan)*, **44**, 611 (1962).
35. Sianesi, D., Machi, A. and Danusso, F., *Chim. Ind. (Milan)*, **41**, 964 (1959).
36. Tanikawa, K., Kusabayashi, S., Kirata, H. and Mikawa, H., *J. Polym. Sci., B*, **6**, 275 (1968).
37. Benaboura, A., Deffieux, A. and Sigwalt, P., *Makromol. Chem.*, **188**, 21 (1987).
38. Ammendola, P., Tancredi, P. and Zambelli, A., *Macromolecules*, **19**, 307 (1986).
39. Aubert, J. H., *Polym. Prepr. Am. Chem. Soc., Div. Polym. Chem.*, **28**, 147 (1987).
40. Gunesin, B. Z. and Murray J. G., US Pat. 4 749 763 (to Mobil Oil Corp.) (1988); *Chem. Abstr.*, **109**, 150251r (1988).
41. Pasquon, I., Porri, L. and Giannini, U., 'Stereoregular Linear Polymers,' in *Encyclopedia of Polymer Science and Engineering*, Wiley-Interscience, John Wiley & Sons, New York, 1989, Vol. 15, pp. 632–733.
42. Kuran, W., *Roczn. Chem.* (*Ann. Soc. Chim. Polonorum*), **41**, 221 (1967); *Chem. Abstr.*, **67**, 32992z (1967).
43. Longo, P., Grassi, A., Oliva, L. and Ammendola, P., *Makromol. Chem.*, **191**, 237 (1990).
44. Porri, L., Giarrusso, A. and Ricci, G., *Makromol. Chem., Macromol. Symp.*, **48/49**, 239 (1991).
45. Cotton, F. A. and La Prade, M. D., *J. Am. Chem. Soc.*, **25**, 5418 (1968).
46. Carmona, E., Marin, J. M. and Poveda, M. L., *Organometallics*, **6**, 1757 (1987).

47. Dias, M. L. and Giarrusso, A., 'Stereospecific Polymerisation of Styrene Monomers with Ni Catalysts,' in Proceedings of International Symposium on *Synthetic, Structural and Industrial Aspects of Stereospecific Polymerisation*, Milan, Italy, 1994, Abstracts, P-71.
48. Soga, K., Yu, C. H. and Shino, T., *Makromol. Chem., Rapid Commun.*, **9**, 354 (1988).
49. Chien, J. C. W. and Salajka, Z., *J. Polym. Sci., A, Polym. Chem.*, **29**, 1243 (1991).
50. Chien, J. C. W., Salajka, Z. and Dong, S., *Macromolecules*, **25**, 3199 (1992).
51. Chien, J. C. W. and Salajka, Z., *J. Polym. Sci., A, Polym. Chem.*, **29**, 1253 (1991).
52. Kucht, A., Kucht, H., Barry, S., Chien, J. C. W. and Rausch, M. D., *Organometallics*, **12**, 3075 (1993).
53. Ready, T. E., Day, R. O., Chien, J. C. W. and Rausch, M. D., *Macromolecules*, **26**, 5822 (1993).
54. Oliva, L., Pellecchia, C., Cinquina, P. and Zambelli, A., *Macromolecules*, **22**, 1642 (1989).
55. Zambelli, A., Pellecchia, C., Oliva, L., Longo, P. and Grassi, A., *Makromol. Chem.*, **192**, 223 (1991).
56. Longo, P., Proto, A. and Oliva, L., *Macromol. Rapid Commun.*, **15**, 151 (1994).
57. Zambelli, A., Pellecchia, C. and Oliva, L., *Makromol. Chem., Macromol. Symp.*, **48/49**, 297 (1991).
58. Kaminsky, W. and Lenk, S., *Macromol. Chem. Phys.*, **195**, 2093 (1994).
59. Kaminsky, W., *Polimery*, **42**, 587 (1997).
60. Miyashita, A., Nabika, M. and Susuki, T., *Polym. Prepr. Jpn*, **43**, 1967 (1994).
61. Miyashita, A., Nabika, M. and Susuki, T., 'Mechanistic Study on Syndiotactic Polymerisation of Styrene Catalysed by Titanocene–Methylalumoxane Complexes,' in Proceedings of International Symposium on *Synthetic, Structural and Industrial Aspects of Stereospecific Polymerisation*, Milan, Italy, 1994, Abstracts, III-18.
62. Mani, R. and Burns, M., *Macromolecules*, **24**, 5476 (1991).
63. Dias, M. L., Giarrusso, A. and Porri, L., *Macromolecules*, **26**, 6664 (1993).
64. Contreras, J. M., Ayal, H. A. and Rabagliati, F. M., *Polym. Bull.*, **32**, 367 (1994).
65. Quyoum, R., Wang, Q., Tudoret, M. -J., Baird, M. C. and Gillis, D. J., *J. Am. Chem. Soc.*, **116**, 6435 (1994).
66. Dall'Occo, T., Sartori, F., Vecellio, G., Zucchini, U. and Maldotti, A., *Makromol. Chem.*, **194**, 151 (1993).
67. Soga, K. and Nakatani, H., *Macromolecules*, **23**, 957 (1990).
68. Soga, K. and Monoi, T., *Macromolecules*, **23**, 1558 (1990).
69. Soga, K., Koide, R. and Uozumi, T., *Makromol. Chem., Rapid Commun.*, **14**, 511 (1993).
70. Pellecchia, C., Longo, P., Proto, A. and Zambelli, A., *Makromol. Chem., Rapid Commun.*, **13**, 265 (1992).
71. Zambelli, A., Oliva, L. and Pellecchia, C., *Macromolecules*, **22**, 2129 (1989).
72. Yang, M., Cha, C. and Shen, Z., *Polym. J.*, **22**, 919 (1990).
73. Ishihara, N., *Macromol. Symp.*, **89**, 553 (1995).
74. Zambelli, A., Ammendola, P. and Proto, A., *Macromolecules*, **22**, 2126 (1989).
75. Soga, K. and Monoi, T., *Macromolecules*, **22**, 3823 (1989).
76. Longo, P., Grassi, A., Proto, A. and Ammendola, P., *Macromolecules*, **21**, 24 (1988).
77. Mitani, N., Ishihara, N., Seimiya, T., Ijitsu, T. and Takyu, T., *Polym. Prepr. Jpn.*, **37**, 24 (1988).
78. Mitani, N. and Ishihara, N., *Polym. Prepr. Jpn.*, **41**, 1800 (1992).
79. Chien, J. C. W., Wu, J. C. and Rausch, M. D., *J. Am. Chem. Soc.*, **103**, 1180 (1981).

80. Ballard, D. G. H., Dawkins, J. V., Key, J. M. and van Linden, P. W., *Makromol. Chem.*, **165**, 173 (1973).
81. Campbell Jr, R. E., Newman, T. H. and Malanga, M. T., *Macromol. Symp.*, **97**, 151 (1995).
82. Gillis, D. J., Tudoret, M. -J. and Baird, C., *J. Am. Chem. Soc.*, **115**, 2543 (1993).
83. Grassi, A., Pellecchia, C., Oliva, L. and Laschi, F., *Macromol. Chem. Phys.*, **196**, 1093 (1995).
84. Tomotsu, N., Kuramoto, M., Takeuchi, M. and Maezawa, H., *Metallocenes '96*, **1996**, 179 (1996).
85. Lee, D.-H., Yoon, K.-B., Lee, E.-H., Noh, S.-K., Byun, G.-G. and Lee, C.-S., *Macromol. Rapid. Commun.*, **16**, 265 (1995).
86. Garbassi, F., Gila, L. and Proto, A., *Polymer News*, **19**, 367 (1994).
87. Zambelli, A., Pellecchia, C. and Proto, A., *Macromol. Symp.*, **89**, 373 (1995).
88. Pellecchia, C., Grassi, A. and Zambelli, A., *J. Chem. Soc., Chem. Commun.*, **1993**, 947 (1993).
89. Pellecchia, C., Grassi, A. and Zambelli, A., *Organometallics*, **13**, 298 (1994).
90. Grassi, A. and Zambelli, A., *Polym. Prepr. Am. Chem. Soc., Div. Polym. Chem.*, **37**, (2), 533 (1996).
91. Pellecchia, C., Pappalardo, D., Oliva, L. and Zambelli, A., *J. Am. Chem. Soc.*, **117**, 6593 (1995).
92. Yang, X., Stern, C. L. and Marks, T. J., *J. Am. Chem. Soc.*, **113**, 3623 (1991).
93. Grassi, A., Zambelli, A. and Laschi, F., *Organometallics*, **15**, 480 (1996).
94. Bueschges, U. and Chien, J. C. W., *J. Polym. Sci., Polym. Chem. Ed.*, **27**, 1529 (1989).
95. Newman, T. H. and Malanga, M. T., *Polym. Prepr. Am. Chem. Soc., Div. Polym. Chem.*, **37**(2), 534 (1996).
96. Ito, H., Shirahama, H. and Yasuda, H., *Polym. Prepr. Jpn.*, **43**, 1730 (1994).
97. Kaeriyama, K., *Polymer*, **12**, 422 (1971).
98. Reichert, K. H., Berthold, J. and Dornow, V., *Makromol. Chem.*, **121**, 258 (1969).
99. Reichert, K. H. and Berthold, J., *Makromol. Chem.*, **124**, 103 (1969).
100. Green, M. L. H. and Ishihara, N., *J. Chem. Soc., Dalton Trans.*, **1994**, 657 (1994).
101. Zambelli, A. and Ammendola, P., *Prog. Polym. Sci.*, **16**, 203 (1991).
102. Yim, J.-H., Chu, K.-J., Choi, K.-W. and Ihm, S.-K., *Eur. Polym. J.*, **32**, 1381 (1996).
103. Lee, D.-H., Yoon, K.-B. and Huh, W.-S., *Macromol. Symp.*, **97**, 185 (1995).
104. Porri, L. and Giarrusso, A., 'Conjugated Diene Polymerisation', in *Comprehensive Polymer Science*, Pergamon Press, Oxford, Vol. 4, pp. 53–108.
105. Allegra, G. and Bassi, I. W., *Adv. Polym. Sci.*, **15**, 91 (1969).
106. Giannetti, E., Nicoletti, G. M. and Mazzocchi, R., *J. Polym. Sci., Polym. Chem. Ed.*, **23**, 2117 (1985).
107. Oliva, L., Longo, P. and Pellecchia, C., *Makromol. Chem., Rapid Commun.*, **9**, 51 (1988).
108. Oliva, L., Longo, P., Grassi, A. and Pellecchia, C., *Makromol. Chem., Rapid Commun.*, **11**, 519 (1990).
109. Ricci, G., Italia, S., Giarrusso, A. and Porri, L., *J. Organomet. Chem.*, **451**, 67 (1993).
110. Ricci, G., Italia, S. and Porri, L., *Macromolecules*, **27**, 868 (1994).
111. Grassi, A., Lomgo, P., Proto, A. and Zambelli, A., *Macromolecules*, **22**, 104 (1989).
112. Soga, K., Nakatani, H. and Monoi, T., *Macromolecules*, **23**, 953 (1990).
113. Pellecchia, C., Proto, A. and Zambelli, A., *Macromolecules*, **25**, 4450 (1992).

114. Zambelli, A., Proto, A., Longo, P. and Oliva, L., *Macromol. Chem. Phys.*, **195**, 2623 (1994).
115. Corradini, P., De Rosa, C., Panunzi, A., Petrucci, G. and Pino, P., *Chimia*, **44**, 52 (1990).
116. Barasacchi, M., Consiglio, G., Medici, L., Petrucci, G. and Suter, U. W., *Angew. Chem., Int. Ed. Engl.*, **30**, 989 (1991).
117. Brookhart, M., Wagner, M. I., Bolavoine, G. G. A. and Haddou, H. A., *J. Am. Chem. Soc.*, **116**, 3641 (1994).
118. Kissin, Y. V., *Adv. Polym. Sci.*, **15**, 91 (1974).
119. Endo, K. and Otsu, T., *J. Polym. Sci., A, Polym. Chem.*, **33**, 79 (1995).
120. Stevens, J. C., Timmers, F. J., Wilson, D. R., Schmidt, G. F., Nickias, P. N., Rosen, R. K., Knight, G. W. and Lai, S.-Y., Eur. Pat. Appl. 0 416 815 (1990); Jpn Pat. Appl. 7 053 618 (to Dow Chemical Company) (1990).
121. Arai, T., Ohtsu, T. and Suzuki, S., *Polym. Prepr. Am. Chem. Soc., Div. Polym. Chem.*, **38**(2), 349 (1997).
122. Pellecchia, C., Papplardo, D., D'Arco, M. and Zambelli, A., *Macromolecules*, **29**, 1158 (1996).
123. Kakugo, M., Miyatake, M. and Mizunuma, K., *Stud. Surf. Sci. Catal.*, **56**, 517 (1990).
124. Inoue, N. and Shiomura, T., *Polym. Prepr. Jpn.*, **42**, 2292 (1993).
125. Amevor, E., Bronco, S., Consiglio, G. and Di Benedetto, S., *Macromol. Symp.*, **89**, 443 (1995).
126. De Rosa, C. and Corradini, P., *Eur. Polym. J.*, **29**, 163 (1993).
127. Brookhart, M., Rix, F. C., DeSimone, J. M. and Barborak, J. C., *J. Am. Chem. Soc.*, **114**, 5894 (1992).
128. Drent, E., US Pat. 4 786 714 (to Koninklijke/Shell) (1988).
129. Klingensmith, G. B., US Pat. 4 916 208 (to Koninklijke/Shell) (1990).
130. Koster, R. A. and Birk, R. H., *Polym. Prepr. Am. Chem. Soc., Div. Polym. Chem.*, **37**(2), 525 (1996).
131. Sen, A., *Acc. Chem. Res.*, **26**, 303 (1993).
132. Busico, V., Corradini, P., Landriani, L. and Trifuoggi, M., *Makromol. Chem. Rapid Commun.*, **44**, 52 (1990).
133. Sen, A. and Jiang, Z., *Macromolecules*, **26**, 911 (1993).
134. Jiang, Z., Adams, S.E. and Sen, A., *Macromolecules*, **27**, 2694 (1994).
135. Bartolini, S., Carfagna, C. and Musco, A., *Macromol. Rapid Commun.*, **16**, 9 (1995).

Further Reading

Dias, M. L., 'Styrene Stereospecific Polymerisation', in *The Polymeric Materials Encyclopedia*, CRC Press, Inc., Boca Raton, 1996, Vol. 10, pp. 8014–8024.

Kuramoto, M., 'Syndiotactic polystyrene', in *The Polymeric Materials Encyclopedia*, CRC Press, Inc., Boca Raton, 1996, Vol. 9, pp. 6828–6838.

Kissin, Y. V., 'Stereoregular Styrene Polymers', in *Encyclopedia of Polymer Science and Engineering*, Wiley-Interscience, John Wiley & Sons, New York, 1989, Vol. 15, pp. 763–770.

Chien, J. C. W., 'Advances in Ziegler Catalysts. Syndiotactic Styrene Polymerisation', in *Ziegler Catalysts*, Springer-Verlag, Berlin, 1995, pp. 209–210.

Huang, B. and Tian, J., 'Metallocene Catalysts (Group 4 Elements). Homopolymerisation of Styrene and Styrene/Ethylene Copolymerisation', in *The Polymeric Materials Encyclopedia*, CRC Press, Inc., Boca Raton, 1996, Vol. 6 pp. 4191–4201.

Soga, K., Arai, T., Nozawa, H. and Uozumi, T., 'Recent Development in Heterogeneous Metallocene Catalysts', *Macromol. Symp.*, **97**, 53–62 (1995).
Brintzinger, H.-H., Fischer, D., Mlhaupt, R., Rieger, B. and Waymouth, R. M., 'Stereospecific Olefin Polymerisation with Chiral Metallocene Catalysts. Activities of *ansa*-Zirconocene Catalysts', *Angew. Chem., Int. Ed. Engl.*, **34**, 1156–1157 (1995).
Horton, A. D., 'Metallocene Catalysis: Polymers by Design?' *Trends Polym. Sci.*, **2**, 158–166 (1994).
Wünsch, J. R., 'Syndiotaktisches Polystyrol', in *Polystyrol*, Carl Hanser Publishers, Munich, 1996, pp. 82–104.

Problems

1. Name and characterise coordination catalysts for the isospecific and syndiospecific polymerisation of styrene.
2. Name and characterise bonds anchoring the polystyrene growing chain at the transition metal atom, depending on the kind of catalyst.
3. Explain the role of methylaluminoxane as an activator for the catalyst precursor in styrene polymerisation systems.
4. Discuss the mechanisms of isospecific and syndiospecific polymerisation of styrene.
5. Explain why the polymerisability of ring-substituted styrenes increases with increasing electron density at the double bond undergoing polymerisation.
6. How can one obtain styrene-rich styrene/α-olefin copolymers using heterogeneous Ziegler–Natta catalysts? Give an example.
7. What are the advantages of half-sandwich metallocene-based catalysts as compared with heterogeneous Ziegler–Natta catalysts in styrene polymerisation? What are the possible consequences of this for developing industrial processes?
8. Name the alternating head-to-tail copolymer of styrene and carbon monoxide. Draw adapted Fischer projections and planar zigzag projections for both stereoisomers of this copolymer.

5 Coordination Polymerisation of Conjugated Dienes

Conjugated dienes, namely 1,3-butadiene and isoprene, have represented and still represent the most widely used standard materials for the production of synthetic rubber. Methods for the polymerisations of conjugated dienes have been known for a long time. Butadiene rubber, produced since the 1930s by the polymerisation of butadiene with metallic sodium, was the first synthetic rubber made for general purpose. However, it had poor low-temperature resistance owing to the occurrence of various isomeric monomeric units in the polymeric chains. In connection with the development of emulsion polymerisation processes for producing styrene/butadiene rubbers, emulsion homopolymerisation of butadiene has also been developed and the production of butadiene rubber by this method began in Poland in the late 1930s. Emulsion polybutadiene does not have such good properties as other types of butadiene rubber afforded by solution processes (anionic polymerisation initiated with alkyllithium or with Alfin catalysts), but it is cheaper [1].

Using Alfin catalysts, butadiene polymers were already obtained in the 1940s. The Alfin catalytic system consists of three components (formed *in statu nascendi* from alkyl chloride, metallic sodium, alcohol and olefin): sodium salt of secondary alcohol (e.g. sodium isopropoxide), alkenylsodium (e.g. allylsodium) and finely dispersed sodium chloride (the name Alfin originates from *al*cohol + ole*fin*) [2,3]. Since the molecular weight of polybutadiene obtained with Alfin catalysts is very high (it can reach a value of a few millions), 1,4-dihydronaphthalene is often added to the polymerisation system for the regulation of molecular weight [1].

Metal alkyls, preferably of alkali metals such as alkyllithium, exhibit quite good effectiveness in isoprene polymerisation [4–6].

However, no method of polymerisation known before 1954 allowed one to obtain polymers with a high regularity of structure from the most common conjugated dienes. A true breakthrough in the development of conjugated diene rubbers took place after the discovery of stereospecific polymerisation with transition metal-based coordination catalysts. From the late 1950s, a rapid development of industrial production of solution types of polybutadiene by means of polymerisation with Ziegler–Natta catalysts was observed.

The polymerisation of conjugated dienes with coordination catalysts is characterised by two distinct features compared with other methods used for the polymerisation of these monomers. Firstly, it shows a high degree of chemoselectivity, i.e. it can yield polymers with chains made up almost exclusively of one type of monomeric unit (1,4, 1,2 or eventually 3,4). Secondly, it shows a high degree of stereoselectivity, i.e. it can produce polymers characterised by a high configurational regularity when the monomeric units contain sites of stereoisomerism (internal double bonds and eventually tertiary carbon atoms).

It is worth noting that no other method of polymerisation of conjugated dienes so far known shows the high degree of chemo- and stereoselectivity of coordination catalysts. Radically initiated polymerisations give polymers that are amorphous by X-ray at room temperature, since they consist of monomeric units of different types, in variable proportions. Cationically initiated polymerisations give regular, cross linked polymers and thus with a lower degree of unsaturation than expected. A higher chemo- and stereoselectivity is displayed by alkali metal-based catalysts. The Alfin catalysts yield polybutadiene consisting of ca 70 % *trans*-1,4 monomeric units, which exhibits crystallinity at room temperature owing to the presence in each chain of relatively long sequences made up of *trans*-1,4 units only. A good degree of chemoselectivity is shown by lithium alkyls. They yield polybutadiene consisting of more than 90 % 1,2 monomeric units by polymerisation run in ethers, and of ca 93 % 1,4 monomeric units by polymerisation in hydrocarbons. However, both these polymers are amorphous, the former because it lacks configurational ordering at the tertiary carbon atoms, the latter because it possesses randomly distributed *cis*-1,4 and *trans*-1,4 monomeric units. However, lithium alkyls show a good degree of chemo- and stereoselectivity in the polymerisation of isoprene, producing polyisoprene consisting of ca 93 % *cis*-1,4 monomeric units [7]. The properties of all these polymers do not match, however, the properties of respective conjugated diene polymers afforded by polymerisation with transition metal-based coordination catalysts.

5.1 Stereoisomerism of Conjugated Diene Polymers

Isomerism in conjugated diene polymers is a considerably more complex phenomenon than that encountered in α-olefin polymers.

1,3-Butadiene and symmetric internally disubstituted butadienes (of the $CH_2{=}CR{-}CR{=}CH_2$ type) can give polymers containing 1,4 or 1,2 monomeric units, each of which, independently of its structure, possesses one site of stereoisomerism. Therefore, four stereoregular isomers are foreseeable for these polymers: *cis*-1,4, *trans*-1,4 [poly(1,2-dialkyl-*cis*-1-butenylene), poly(1,2-dialkyl-*trans*-1-butenylene)] $\{-[-C(R){=}C(R)-CH_2-CH_2-]_n-$ or $-[-CH_2-C(R){=}C(R)-CH_2-]_n\}$, 1,2-isotactic and 1,2-syndiotactic. Polymers afforded by isoprene and other unsymmetric internally monosubstituted butadienes, i.e. monomers of the $CH_2{=}CR{-}CH{=}CH_2$ type, can also be formed via 3,4-monomer enchainment and thus exist additionally as 3,4-isotactic and 3,4-syndiotactic stereoisomers.

Stereoisomerism of polymers derived from 1,3-pentadiene and other unsymmetric terminally monosubstituted butadienes ($CH_2{=}CH{-}CH{=}CHR$) is more complex. Stereoregular polymers can be formed from these monomers via their 1,4-, 1,2- or 3,4-concatenation; each polymer contains two sites of stereoisomerism in the monomeric unit:

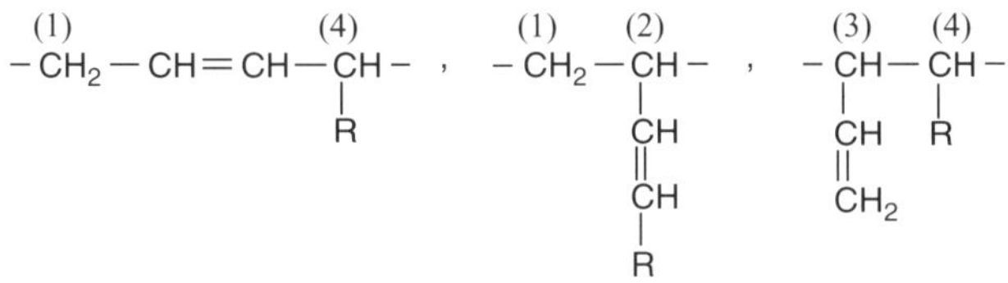

The monomeric units formed via 1,4-polymerisation of monomers of the $CH_2{=}CH{-}CH{=}CHR$ type display stereoisomerism at the double bond (*cis–trans*) and at the tertiary carbon atom which can assume two opposite configurations. Hence, the obtained 1,4-polymers can appear as *cis*-1,4-isotactic, *cis*-1,4-syndiotactic, *trans*-1,4-isotactic and *trans*-1,4-syndiotactic isomers. One should note that the tertiary carbon atom constitutes a true chiral site in the polymer chain in this case, although the polymer chain is formed by the polymerisation of a prochiral, and not chiral, monomer. The isotactic polymers can assume two enantiomeric forms; isotactic polymers consisting predominantly of one enantiomeric form have actually been obtained by 1,3-pentadiene polymerisation using an optically active catalyst [8,9]. The 1,2 monomeric units in polymers afforded by monomers of the $CH_2{=}CH{-}CH{=}CHR$ type contain a double bond in the side group, hence two types of isotactic and two types of syndiotactic polymer made up of these units are possible, each type differing for the configuration of the double bonds, *cis* or *trans*. The 3,4 monomeric units contain two tertiary carbon atoms, and thus three types of stereoregular polymer derived from the $CH_2{=}CH{-}CH{=}CHR$ monomers with this structure are possible: *erythro*-diisotactic, *threo*-diisotactic and disyndiotactic. In summary, a monomer of the $CH_2{=}CH{-}CH{=}CHR$ type can give, in principle, eleven types of stereoregular polymer.

A similar complex isomerism is also displayed by unsymmetric internally and terminally disubstituted butadienes (of the $CH_2{=}CH{-}CR{=}CHR$ and $CH_2{=}CR{-}CH{=}CHR$ type).

Stereoregular polymers that can be afforded by 2,4-hexadiene and other symmetric terminally disubstituted butadienes (of the CHR=CH–CH=CHR type) exhibit still more complex stereoisomerism, since each monomeric unit in these polymers possesses three sites of isomerism. The formation of these polymers involves 1,2- and 1,4-polymerisation. The 1,2-polymers derived from the CHR=CH–CH=CHR monomers exhibit the same type of stereoisomerism as polymers with a 3,4 structure obtained from monomers of the CH_2=CH–CH=CHR type. However, owing to the presence of the R substituent at the double bond in the side group of the polymer derived from a monomer of the CHR=CH–CH=CHR type, two types of *erythro*-diisotactic, *threo*-diisotactic and disyndiotactic polymer are foreseeable, each type with either *cis* or *trans* configuration of the double bond, as in the 1,2-polymer derived from a monomer of the CH_2=CH–CH=CHR type. Thus, six stereoisomeric forms of 1,2-polymer are possible for the CHR=CH–CH=CHR monomer. The 1,4 monomeric units in the polymers formed by the polymerisation of CHR=CH–CH=CHR monomers contain one double bond (in either *cis* or *trans* configuration) and two tertiary carbon atoms and therefore can exist as two sets of enantiomers, *erythro* and *threo*:

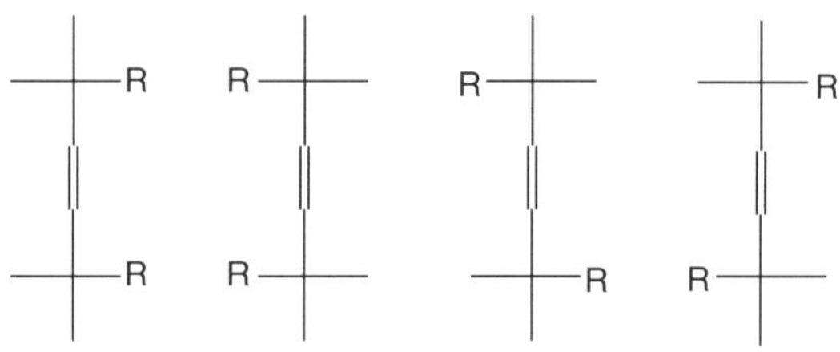

erythro *threo*

As a consequence, eight stereoisomeric forms are foreseeable for the 1,4-polymers of the CHR=CH–CH=CHR monomer: *cis*-1,4-(or *trans*-1,4-) *erythro*-diisotactic, *cis*-1,4-(or *trans*-1,4-) *threo*-diisotactic, *cis*-1,4-(or *trans*-1,4-) *erythro*-disyndiotactic, and *cis*-1,4-(or *trans*-1,4-) *threo*-disyndiotactic. For instance, Figures 5.1 and 5.2 illustrate the stereoisomerism of *cis*-1,4- and *trans*-1, 4-polymers of CHR=CH–CH=CHR monomers respectively.

In this case, the occurrence of two types of stereoisomeric form of disyndiotactic polymers, *erythro* and *threo*, should be emphasised {let us recall that there were no differences in the stereostructure (*erythro* or *threo*) of disyndiotactic deuterated poly (α-olefins), e.g. polymers of β-monodeuterated α-olefins, $-[-CH(R)-CH(^2H)-]_n-$, when omitting different chain end groups}.

One should also pay attention in this case to the appearance of true chiral sites in the polymer chain, although it is formed by the polymerisation of a prochiral, and not chiral, monomer. In summary, a monomer of the CHR=CH–CH=CHR type can give, in principle, 14 types of stereoregular polymer [10].

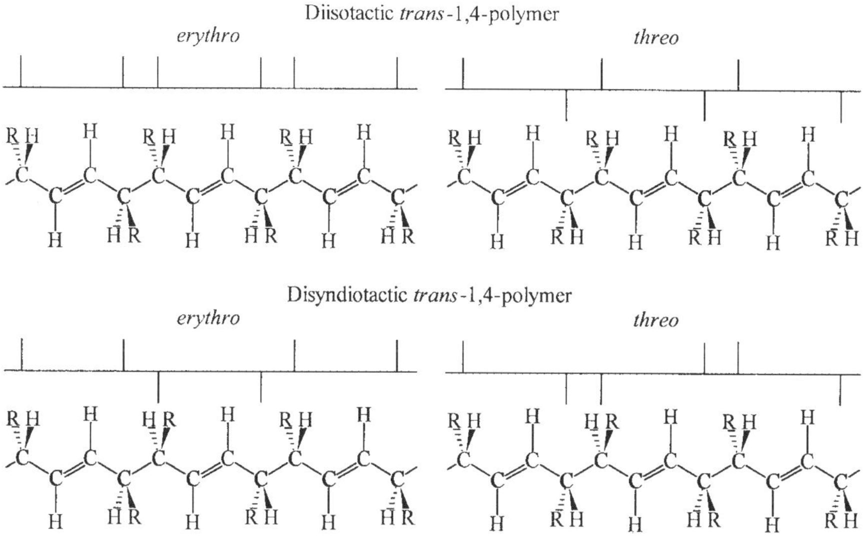

Figure 5.1 Stereoisomerism of *cis*-1,4 polymers of 1,4-disubstituted butadienes, poly(3,4-dialkyl-*cis*-1-butenylene) $\{[-CH{=}CH{-}CH(R){-}CH(R)-]_n-$ or $-[-CH(R){-}CH{=}CH{-}CH(R)-]_n-\}$. Diisotactic and disyndiotactic *erythro* and *threo* polymers; only one of the enantiomorphic forms of the polymers is shown

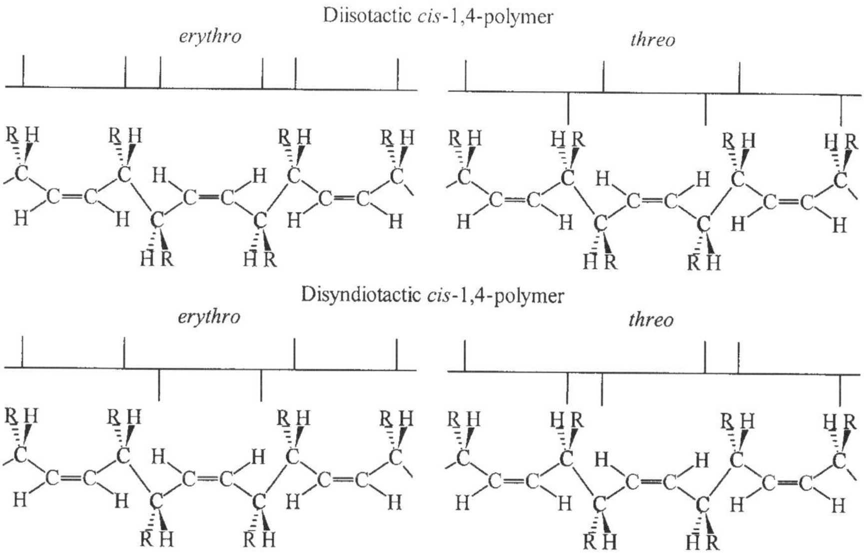

Figure 5.2 Stereoisomerism of *trans*-1,4 polymers of 1,4-disubstituted butadienes, poly(3,4-dialkyl-*trans*-1-butenylene) $\{-[-CH{=}CH{-}CH(R){-}CH(R)-]_n-$ or $-[-CH(R){-}CH{=}CH{-}CH(R)-]_n-\}$. Diisotactic and disyndiotactic *erythro* and *threo* polymers; only one of the enantiomorphic forms of the polymers is shown

A comparison between the theoretically foreseeable possibilities and the results actually obtained shows the progress that stereospecific polymerisation has made in the field of conjugated dienes. Namely, stereoregular polymers have been obtained from each of the above types. By the use of coordination catalysts, it is possible to obtain from 1,3-butadiene all four foreseeable stereoregular polymers. All these polymers have been obtained with a high degree of steric purity and hence are highly crystalline or crystallisable. Moreover, polymers consisting predominantly of only two types of monomeric unit practically in a 1:1 ratio, alternating along the polymer chain, such as *cis*-1,4/*trans*-1,4 and *cis*-1,4/1,2, were also obtained. These polymers have been referred to as equibinary polymers: eb-*cis*-1,4/*trans*-1,4 and eb-*cis*-1,4/1,2 polybutadienes [7].

Some of the polybutadienes obtained with transition metal-based coordination catalysts have practical significance: the most important is *cis*-1,4-polybutadiene, which exhibits excellent elastomeric properties. As regards isoprene polymers, two highly stereoregular polyisoprenes, a *cis*-1,4 polymer (very similar to natural rubber) and a *trans*-1,4-polymer (of equal structure to that of gutta percha or balata) have been obtained with coordination catalysts. Various polymers of mixed 3,4 structure, amorphous by X-ray, were also obtained [7].

The coordination polymerisation of 1,3-pentadiene is of particular interest owing to the complex phenomena of stereoisomerism displayed by its polymers. This monomer and other monomers of the $CH_2{=}CH{-}CH{=}CHR$ type exist in two isomeric forms – 'entgegen' (*E*) and 'zusammen' (*Z*):

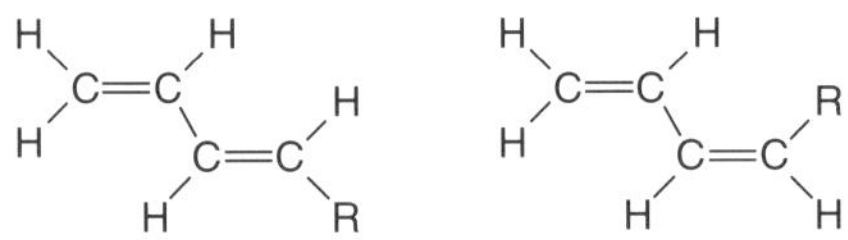

(*E*) (*Z*)

Both isomers are polymerised by some coordination catalysts, whereas with others only the (*E*) isomer is polymerised. Of the various possible stereoregular poly(1,3-pentadiene)s, only five have been prepared so far: *cis*-1,4-isotactic [11,12], *trans*-1,4-isotactic [13], *cis*-1,4-syndiotactic [14,15], 1,2-*cis*-syndiotactic [16] and 1,2-*trans*-syndiotactic [17] polymers. In addition to these stereoregular polymers, amorphous polymers of various types, consisting of different monomeric units in variable proportions, have been obtained [7].

Also, the coordination polymerisation of 2,4-hexadiene and 1,4-deuterated 1,3-butadienes is of particular interest owing to isomerism phenomena shown by their polymers. 2,4-Hexadiene and other monomers of the $CHR{=}CH{-}CH{=}CHR$ type exist in three isomeric forms – (*E,E*), (*Z,Z*) and (*E,Z*):

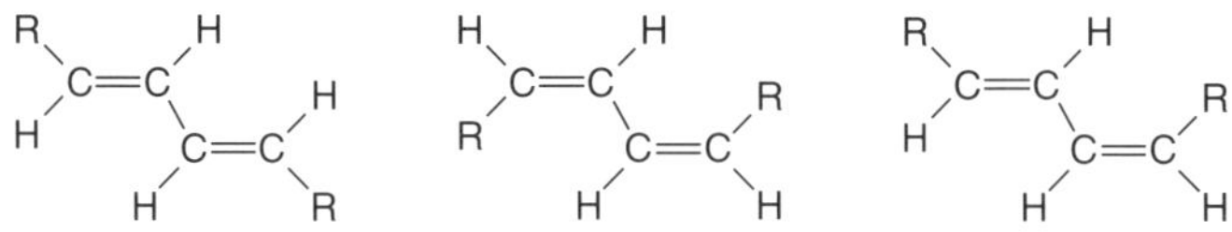

Crystalline polymers with a *trans*-1,4-*threo*-diisotactic structure were obtained from (*E*,*E*)-2,4-hexadiene [18,19], while amorphous polymers were yielded by the (*E*,*Z*)-isomer. The (*Z*,*Z*)-isomer is not polymerised with any of the coordination catalysts used. It is interesting that poly(2,4-hexadiene) with a *trans* structure is formed in the presence of catalysts that give predominantly *cis*-1,4 polymers from butadiene, isoprene and other simple conjugated dienes. 1,4-Deuterated 1,3-butadiene, such as (*Z*,*Z*)-$CH(^2H)$=CH–CH=$CH(^2H)$, was polymerised with coordination catalysts, producing *trans*-1,4 stereoregular polymers with some catalysts, whereas with others *cis*-1,4 stereoregular polymers of this monomer were obtained [20]. By using various catalysts, several stereoregular polymers with a 1,4 structure were also obtained from the two deuterated 1,3-pentadienes (*E*,*E*)-$CH(^2H)$=CH–CH=$CHCH_3$ and (*E*,*Z*)-$CH(^2H)$=CH–CH=$CHCH_3$, namely *trans*-1,4-*threo*-diisotactic, *cis*-1,4-*threo*-disyndiotactic and *cis*-1,4-*erythro*-diisotactic polymers from the former monomer and a *trans*-1,4-*erythro*-diisotactic polymer from the latter monomer [21,22].

As regards polymers of other conjugated dienes (higher alkyl 1,4-disubstituted 1,3-butadienes), it has already been mentioned that several stereoisomers of such polymers have been obtained [7].

5.2 Polymerisation Catalysts

The coordination polymerisation of conjugated dienes began in 1954, soon after the first results on the polymerisation of propylene had been obtained. The first stereoregular conjugated diene polymers, *cis*-1,4-polyisoprene [23] and then *trans*-1,4-polyisoprene, along with stereoregular polymers of 1,3-butadiene and 1,3-pentadiene, i.e. *cis*-1,4- and *trans*-1,4-polymers as well as 1,2-syndiotactic and 1,2-isotactic polymers [23–26], were derived in the Goodrich-Gulf Co. and the Montecatini SpA laboratories from polymerisations with catalysts that had been found to be effective for the polymerisation of ethylene and propylene, i.e. $TiCl_4$ and $TiCl_3$ [or $Ti(OR)_4$] in a combination with aluminium alkyls.

At the beginning of the 1960s it was found that some simple derivatives of transition metals such as $RhCl_3 \cdot 3H_2O$ and $Rh(NO_3)_3 \cdot 2H_2O$ polymerised butadiene to *trans*-1,4-polymers in protic solvents such as water and ethanol [27,28]. Also, other catalysts consisting of transition metal salts (monometallic or bimetallic) and not containing a preformed metal–carbon bond appeared to be active in butadiene polymerisation in non-polar hydrocarbon media yielding *cis*-1,4- or *trans*-1,4-polymers as well as 1,2-polymers [29–35].

Soon after, it was reported that allylnickel halides could polymerise butadiene to crystalline polymers consisting of *cis*-1,4 or *trans*-1,4 units [36] and that trisallylchromium polymerised it predominantly to 1,2-polybutadiene but bisallylcobalt iodide polymerised it predominantly to *cis*-1,4-polybutadiene [37].

In the late 1980s and in the 1990s, homogeneous metallocene-based Ziegler–Natta and related aluminium-free catalysts as well as other non-metallocene-based homogeneous single-site catalysts, which are active in the syndiospecific polymerisation of styrene, were found also to promote the polymerisation of conjugated dienes [16,38–43].

Following extensive work carried out over the last decades and continuing up to present, a vast number of catalysts have been discovered that are yielding stereoregular polymers of conjugated dienes. The most common coordination catalysts that have been employed in the polymerisation of conjugated dienes are usually a combination of transition metal compounds of groups 4 to 8 of the periodic system or of the lanthanide or actinide series with aluminium alkyls or hydrides, and thus they fit the definition of Ziegler–Natta catalysts. Such catalysts may be heterogeneous or soluble in hydrocarbons used as the polymerisation medium. Conjugated dienes can be polymerised with several types of Ziegler–Natta catalyst, either heterogeneous or homogeneous (soluble aluminium alkyl- or hydride-activated transition metal catalysts and methylaluminoxane-activated metallocene-based and related non-metallocene-based single-site catalysts). The Ziegler–Natta catalysts derived from Ti-, Co-, Ni- and Nd-based catalyst precursors and organometallic or metal hydride activators are at present the most important with regard to both activity and stereospecificity, and they are the only ones (heterogeneous and hydrocarbon soluble) now used for the industrial manufacture of conjugated diene (butadiene and isoprene) polymers [7].

The single-site homogeneous Ziegler–Natta catalysts and their aluminium-free analogues, e.g. such as those activated with tris(pentafluorophenyl)borane or *N*, *N*-dimethylanilinium tetrakis(pentafluorophenyl)borate, are of interest for two essential reasons. Firstly, they are active not only for the polymerisation of simple conjugated dienes such as butadiene and isoprene (yielding the respective *cis*-1,4 polymers [39]) but also for the polymerisation of monomers that are not polymerised by other soluble catalysts such as (*Z*)-1,3-pentadiene and 4-methyl-1,3-pentadiene [38–42]; therefore, studies of such polymerisation systems may supply additional information on the influence of monomer structure on stereospecificity. Secondly, a comparison of conjugated diene polymerisation in these systems with the polymerisation of styrene using the same catalysts may give still more information concerning the stereoregulation mechanism [43]. The interest in homogeneous single-site catalysts for conjugated diene polymerisation, however, is not as pronounced as in the case of α-olefin polymerisation and even in the case of styrene polymerisation.

Although Ziegler–Natta catalysts are the most advantageous for performing conjugated diene polymerisation, catalysts including allylmetal derivatives of transition metals have been developed. Such catalysts, consisting of an allylmetal derivative single component and in a combination with a Lewis acid or Lewis base, are of interest because they are good models for active site structure and therefore provide information that is useful for elucidating the polymerisation mechanism. Some of them, however, exhibit productivity and stereospecificity comparable with that of Ziegler–Natta catalysts [7].

As regards catalysts not containing preformed metal–carbon bonds such as rhodium salts active in butadiene polymerisation in aqueous emulsions or in protic solvents as well as other catalysts of this type (used in non-polar hydrocarbon media), there is a theoretical rather than practical interest paid to such catalysts. However, a few of them have activity and stereospecificity comparable with Ziegler–Natta catalysts [27–35].

5.2.1 Ziegler–Natta Catalysts

5.2.1.1 Aluminium Alkyl or Hydride-activated Catalysts

Catalysts derived from transition metal compounds and aluminium alkyl or hydride comprise a vast number, some of which are soluble whereas others are insoluble in the polymerisation medium (aromatic and/or aliphatic hydrocarbons). For instance, trialkylaluminium-activated $TiCl_4$-, $TiCl_3$-, VCl_4- and VCl_3-based Ziegler–Natta catalysts are insoluble in the polymerisation diluent, while trialkylaluminium-activated $Ti(OR)_4$-, $V(Acac)_3$- and $Cr(Acac)_3$-based catalysts are soluble in the polymerisation medium. Derivatives of various transition metals, in particular Ti, V, Cr, Mo, Co and Ni as well as Ce, Pr, Nd, Gd and Yb (lanthanide series) and U (actinide series), can be used for the preparation of the catalysts. Also, various derivatives of transition metals can be applied in order to prepare the catalysts, but the most common are halides, alcoholates, acetylacetonates and carboxylates. In some cases, the metal halides are used in a combination with Lewis bases as catalyst precursors. Aluminium alkyls, such as AlR_3, AlR_2Cl and $AlRCl_2$ ($R = Et$, *i*-Bu in most cases), are preferably used as precatalyst activators; however, aluminium hydrides, usually complexed with ethers or amines (e.g. $AlHCl_2 \cdot OEt_2$, $AlHCl_2 \cdot NMe_3$, $AlH_2\ Cl \cdot NMe_3$, AlH_2NMe_2, $AlH_2Cl \cdot NMe_3$), and oligo(aminoalane)s of the formula $[Al(H)N(R)]_x$ also appeared to be effective. Other metal alkyls such as LiR, MgR_2, CaR_2 and ZnR_2 can be applied in some cases, but they are of minor interest [7].

Titanium-based Ziegler–Natta catalysts for the polymerisation of conjugated dienes are usually obtained using $TiCl_4$, $TiCl_3$, TiI_4 or $Ti(OR)_4$ as catalyst precursors. Of the $TiCl_4$-based catalysts, only those containing β-$TiCl_3$, i.e. $TiCl_4$–AlR_3 (1:1) and $TiCl_4$–AlH_xCl_{3-x} (0.9–1.5:1) catalysts, are of interest, since they have been successfully applied for the synthesis of *cis*-1,4-polyisoprene

[23]. In these catalysts, the active species containing Ti–C or Ti–H bonds are located on the surface of the β-$TiCl_3$ being formed under conditions applied for the catalyst preparation. As for other heterogeneous Ziegler–Natta catalysts without an inert support, only ca 1% of the Ti present in the system forms active centres [44,45]. Effective catalysts are also obtained using separately prepared β-$TiCl_3$ with AlR_3 or AlR_2Cl as activators [46,47]. TiI_4–AlR_3 catalysts produce polybutadiene containing more *cis*-1,4 monomeric units than polybutadiene obtained with the corresponding $TiCl_4$-based catalyst, but hardly polymerise isoprene [48].

In contrast to $TiCl_4$, TiI_4 is only slightly soluble in hydrocarbon solvents, and therefore alternative catalysts based on soluble iodine-containing precursors, such as $TiCl_2I_2$, or iodine-containing activators, such as e.g. $AlEt_2I$, have been developed [49–52]. These catalysts lead to practically the same contents of *cis*-1,4 monomeric units in polybutadiene as the TiI_4–AlR_3 catalyst. The equivalence of all the above catalysts with regard to stereospecificity is attributable to the fact that iodine invariably ends up bonding to titanium to form the active species. It is interesting that ca 5% of the Ti introduced with $TiCl_2I_2$ into the polymerisation system forms active centres [44]. Although the catalyst-forming reaction in such iodine-containing systems gives rise to a precipitate, the polymerisation proved to proceed in a homogeneous phase [48,51].

Catalysts of the $Ti(OR)_4$–AlR_3 type are soluble in hydrocarbon solvents and polymerise butadiene and isoprene into polymers having predominant 1,2 and 3,4 structures respectively. The catalyst-forming reaction consists of an alkylation of $Ti(OR)_4$ with the formation of titanium species in a lower oxidation state. The catalyst activity increases with increasing Ti(III) concentration [53]. In the $Ti(OBu)_4$–$Al(i\text{-}Bu)_3$ catalyst, less than 1% of the Ti present in the system is active in the polymerisation [54]. When alkylaluminium chlorides ($AlEt_2Cl$, $AlEt_{1.5}Cl_{1.5}$ or $AlEtCl_2$) are used as activators for $Ti(OR)_4$ in an Al/Ti molar ratio higher than 4, an exchange reaction between Cl and OR groups occurs and a precipitate consisting of β-$TiCl_3$ is formed; such heterogeneous catalysts therefore have the same stereospecificity as $TiCl_4$–AlR_3 (1:1) catalysts, and thus *cis*-1,4-polyisoprene is provided [55,56].

Heterogeneous vanadium-based catalysts derived from VCl_3, $VOCl_3$ or VCl_4 and AlR_3 or AlR_2Cl have been developed for the polymerisation of conjugated dienes affording polymers with predominant *trans*-1,4 monomeric units [57,58]. VCl_3, which is insoluble in hydrocarbon solvents and apparently does not react with AlR_3 or AlR_2Cl, gives catalysts with surface V–R bonds characterised by relatively high stereospecificity but low productivity. Supporting VCl_3 on an inert carrier such as kaolin, clay, etc., leads to more efficient catalysts [59]. $VOCl_3$ and VCl_4, which are soluble in hydrocarbons, give a precipitate on reaction with AlR_3 or AlR_2Cl. Such heterogeneous catalysts contain vanadium in a lower oxidation state, V(III), like VCl_3-based catalysts [58].

There are also known soluble vanadium-based catalysts. Those obtained from soluble compounds such as $V(Acac)_3$ or $VO(OR)_3$ as precursors being activated with AlR_3 have been used for the polymerisation of butadiene to a 1,2-syndiotactic polymer. Catalysts derived from the reaction of soluble vanadium precursors and alkylaluminium chlorides are also soluble in hydrocarbon solvents but have a different stereospecificity, since they yield polybutadiene consisting predominantly of *trans*-1,4 monomeric units [60]. In soluble $V(Acac)_3$- and $VO(OR)_3$-based catalysts, when prepared with an Al/V molar ratio higher than 5, chlorine ends up eventually on vanadium and the catalyst contains the $VRCl_2$ species complexed with the aluminium alkyl [61]. This is the reason why the $VCl_3.3THF–AlR_2Cl$ catalyst is practically equivalent to the $V(Acac)_3–AlR_2Cl$ catalyst. The V–C bond is rather unstable at ambient temperature and homolytic cleavage of this bond leads to V(II) inactive species; thus, such catalysts are preferably used at lowered temperature. However, it should be noted that, in the presence of chlorine donors capable of reoxidising V(II) to V(III), such as Cl_3CCOOH, polymerisation could be carried out even at elevated temperature [7].

Catalysts based on cobalt compounds and alkylaluminium chlorides ($AlEt_2Cl$, $AlEt_{1.5}Cl_{1.5}$ or $AlEtCl_2$) readily dissolve in hydrocarbon solvents and are effective for the *cis*-1,4-polymerisation of butadiene. A variety of Co compounds have been used as catalyst precursors, such as cobalt chlorides complexed with Lewis bases (e.g. pyridine), cobalt acetylacetonates, cobalt carboxylates and cobalt tetrachloroaluminate $[Co(AlCl_4)_2]$ as well as complexes of Co(I) and Co(0) [62–67]. Water is often added to polymerisation systems with homogeneous catalysts based on a cobalt compound and aluminium alkyl [68]. An alkylaluminoxane is formed in the reaction of R_2AlCl with H_2O; this turns out to be an effective activator for the *cis*-1,4-polymerisation of butadiene. Cobalt-based catalysts supported on inorganic or organic carriers are highly active in the polymerisation of butadiene, leading to polymers with high contents of *cis*-1,4 monomeric units [69,70].

Soluble nickel-based catalysts derived from reactions of nickel compounds such as $Ni(Acac)_2$, $Ni(OCOR)_2$ or $NiCl_2.2Py$ with alkylaluminium chlorides are as useful as cobalt-based catalysts for the *cis*-1,4-polymerisation of butadiene. However, the Ni-based catalysts are less stereospecific and afford polymers characterised by lower molecular weights [71]. Butadiene polymers characterised by higher molecular weights and high contents of *cis*-1,4 monomeric units can be obtained using fluorine-containing catalysts such as $Ni(OCOR)_2–AlR_3–BF_3.OEt_2$ and $Ni(OCOR)_2–AlR_3–HF$ [72].

Chromium- and molybdenum-based catalysts yield predominantly 1,2-polybutadiene. Several soluble chromium compounds such as $Cr(Acac)_3$ give apparently hydrocarbon-soluble catalysts when activated with AlR_3. Also, soluble catalysts are formed from soluble molybdenum compounds such as $Mo(Acac)_3$ or $MoO_2(Acac)_2$ on reaction with $AlEt_3$ in the presence of ethyl acetate or sulphides. ESR studies of $Mo(OR)_xCl_{4-x}$-based catalysts showed the Mo(III) species to be present in the catalysts. It should be mentioned, however, that catalysts based on group 6 metals have not been extensively studied [7].

Catalysts derived from compounds of lanthanide and actinide metals, which possess vacant f orbitals, are used for the preparation of conjugated diene polymers with high contents of *cis*-1,4 monomeric units. Catalysts based on cerium, neodymium and uranium derivatives are especially effective in conjugated diene polymerisation [73–75]. At the beginning of the 1960s, catalysts based on uranium were developed. For instance, homogeneous $U(All)_4$–$AlEt_2Cl$ and $U(OR)_4$–$AlEtCl_2$ catalysts are highly active and afford polybutadiene with a content of *cis*-1,4 units not less than 99% [74–76]. Despite their very high activity and stereospecificity for the *cis*-1,4-polymerisation of butadiene and isoprene, the uranium-based catalysts have not attained industrial utilisation because of the problems connected with the radioactive residue remaining in the polymer. Cerium-based catalysts, which were the first catalysts from the lanthanide and actinide series, also have some disadvantages owing to the fact that cerium residues remaining in the polymer catalyse its oxidation. Therefore, in the second half of the 1970s, interest was focused on neodymium-based catalysts, which do not present the disadvantages of U- and Ce-based catalysts and are more active than those based on other elements of the series (Pr, Gd, Tb, etc.) [77–79]. Nd-based catalysts were initially obtained as binary systems and are now widely used as ternary systems [80]. Ternary catalysts are derived from soluble Nd compounds, such as carboxylates (naphthenoate, wersatate, octanoate) and higher alcoholates, alkylaluminium chloride and a branched aluminium alkyl such as $Al(i\text{-}Bu)_3$ or $Al(i\text{-}Bu)_2H$ in hydrocarbon solvents. For instance, $Nd(OCOR)_3$–$AlEt_2Cl$–$Al(i\text{-}Bu)_3$ (1:2.7–3:20–30) catalysts are prepared by firstly reacting the Nd salt of carboxylic acid with $AlEt_2Cl$ in a hydrocarbon solvent and then adding $Al(i\text{-}Bu)_3$ to the precipitate formed. Ternary catalysts such as $Nd(OCOR)_3$–$t-$BuCl–$Al(i\text{-}Bu)_2H$ have also been prepared [81–83]. Binary catalysts are derived from an Nd compound, usually complexed with a Lewis base, and $Al(i\text{-}Bu)_3$ in a hydrocarbon medium. The catalysts prepared from $NdCl_3$ are practically insoluble in hydrocarbon solvents. More soluble catalysts can be obtained using $Nd(OCOR)_xCl_{3-x}$ or $Nd(OR)_xCl_{3-x}$ (R = higher alkyl). The presence of chlorine in the catalyst is crucial for the formation of a *cis*-1,4-polymer of conjugated diene. Catalysts prepared from $Nd(OCOR)_3$ or $Nd(OR)_3$ and AlR_3 afford, in fact, polybutadiene with *trans*-1,4 monomeric units [84]. Determinations of the number of active centres in a binary $NdCl_3.EtOH$–AlR_3 catalyst and a ternary $Nd(OCOR)_3$–$AlEt_2Cl$–$Al(i\text{-}Bu)_3$ catalyst have revealed 5–7 % of the Nd introduced into the polymerisation system to constitute the active sites at room temperature [78]. The efficiency of the catalyst depends on the extent of alkylation of the neodymium compound, which is determined by the nature of the aluminium alkyl activator, and on the Al/Nd ratio, as well as on the stability of the Nd–C bond, which depends on temperature [78,81,85]. In the case of conjugated diene polymerisation with Nd-based catalysts, saturated aliphatic hydrocarbons are favoured as the polymerisation medium because of competitive coordination to the active site by both aromatic ring electrons (when an aromatic solvent used as the medium) and the diene monomer [86,87]. A great advantage of the Nd-based catalysts is that polymerisation with such catalysts can be carried out at temperatures in a wide range, from −70 °C to 130 °C [88]. However, a decrease in the polybutadiene *cis*-1,4 content was observed as polymerisation temperature was increased [89]; but the content of 1,2 monomeric units in polybutadiene formed remained very low (<1 %). Since polybutadiene with high content of *cis*-1,4 monomeric units (>98 %) can be obtained in the presence of Nd-based catalysts even at an elevated temperature of 20–50 °C, such a process is advantageous for industrial utilisation as compared with processes involving butadiene polymerisations at a lowered temperature with Ti-, Co- and Ni-based Ziegler–Natta catalysts.

The microstructure of representative conjugated diene polymers obtained by polymerisation with various aluminium alkyl or hydride-activated Ziegler–Natta catalysts is presented in Table 5.1 [7,12,26,41,49,57,58,62,73,77,81,83, 89–106] and Table 5.2 [7,12,13,17–19,41,57,77,81,83,91,92,100,104,105, 107–113].

The versatility of Ziegler–Natta catalysts regarding their stereoregulation ability in the polymerisation of conjugated dienes, depending on the kind of transition metal and its ligands, the kind of activator as well as the kind of polymerised monomer, should be emphasised. One may note that conjugated dienes can undergo polymerisation with several types of Ziegler–Natta catalyst, either insoluble or soluble in the reaction medium [7,114]. In polymerisation with heterogeneous Ziegler–Natta catalysts, stereospecificity depends essentially on catalyst structure, but in general it is only slightly influenced by the kind of monomer. For instance, the VCl_3–$AlEt_3$ catalyst yields polymers with a *trans*-1,4 structure from all the most common conjugated dienes, while the $TiCl_4$–$AlEt_3$ (1:1) catalyst produces predominantly polymers of *cis*-1,4 structure [7]. In contrast, stereospecificity is strongly dependent on the kind of monomer in polymerisations with soluble Ziegler–Natta catalysts. For instance, the $Ti(OR)_4$–$AlEt_3$ catalyst yields a 1,2-syndiotactic polymer from butadiene, while it affords a *cis*-1,4-isotactic polymer from (*E*)-1,3-pentadiene [12,91]; the $Co(OCOR)_2$–$AlEt_2Cl$–H_2O catalyst gives a *cis*-1,4-polymer from butadiene, while it yields a *trans*-1,4-polymer from 2,4-hexadiene [7,18].

5.2.1.2 Homogeneous Single-site Catalysts

Homogeneous single-site Ziegler–Natta catalysts consisting of a half-sandwich metallocene such as $CpTiCl_3$, $CpTi(OR)_3$, $CpTiCl_2 \cdot 2THF$, $[CpTiCl_2]_x$ and $CpVCl_2 \cdot 2PEt_3$ in a combination with methylaluminoxane, $CpMtX_n$–$[Al(Me)O]_x$, and catalysts consisting of tetraalkoxytitanium combined with methylaluminoxane, such as $Ti(OBu)_4$–$[Al(Me)O]_x$, are active for the polymerisation of several types of conjugated diene [16,38–43,115,116]. These catalysts have provided data that permit better information on the relationship of monomer structure and polymerisation chemoselectivity and stereospecificity.

Table 5.1 Polymerisation of 1,3-butadiene with aluminium alkyl or hydride-activated Ziegler–Natta catalysts

	Polymer microstructure			
Catalyst (component molar ratio)	*cis*-1,4 (%)	*trans*-1,4 (%)	1,2[a] (%)	Reference
$TiCl_4$–AlR_3 (1:1)	65–70			[90]
$TiCl_4$–$AlEt_3$ (1:1.2)	50	45		[7]
$TiCl_4$–$AlEt_3$ (1:0.5)		91		[12,91]
β-$TiCl_3$–AlR_2Cl (1:1)	65–70			[92]
γ-$TiCl_3$–$AlEt_3$ (1:1)		92–95		[92]
TiI_4–$Al(i\text{-}Bu)_3$ (1:4)	95	2	3	[49,93,94]
TiI_4–$LiAlH_4$ (1:3)	85			[7]
$TiCl_4$–AlI_3–$AlHCl_2.OEt_2$ (1:1.3:5.5)	93	2.5	4.5	[95]
$Ti(OBu)_4$–$AlEtCl_2$ (1:10)		93–94		[7]
$Ti(OR)_4$–AlR_3 (1:7)			95	[91]
$Ti(OBu)_4$–AlR_3			90–100	[96]
VCl_3–$AlEt_3$		98–99		[57,97]
$VCl_3.3THF$–$AlEt_2Cl$ (1:50)		99–100		[98]
$VOCl_3$–AlR_3		High		[57,58,97]
$V(Acac)_3$–$AlEt_3$			90–95	[26]
$Cr(Acac)_3$–$AlEt_3$ (1:3)			High	[26]
$Cr(Acac)_3$–$AlEt_3$ (1:10)			High[b]	[26]
$Cr(PhCN)_6$–$AlEt_3$ (1:2.5)			100	[12,26,91]
$Cr(PhCN)_6$–$AlEt_3$ (1:5)			100[b]	[12,26,91]
$Cr(CO)_5$.–Py–$AlEt_3$ (1:5)			High[b]	[26]
$Mo(Acac)_3$–AlR_3 (1:5)			80	[7]
$Mo(OR)_2Cl_2$–AlR_3 (1:10)	> 49	< 1	> 49	[7]
$CoCl_2.2Py$–$AlEt_2Cl$–H_2O	98	1	1	[99]
$CoCl_2.2Py$–AlR_3			High	[100]
$CoCl_2.2Py$–AlR_3–H_2O			High	[101]
$Co(Acac)_2$–CS_2–AlR_3 (1:3:60)			99–100	[41]
$Co(Acac)_3$–$AlEt_3$			90–95	[92]
$Co(Acac)_3$–CS_2–AlR_3			High	[102,103]
$Ni(Acac)_2$–$AlEt_2Cl$	90			[62]
$Ni(OCOR)_2$–$BF_3.OEt_2$–$AlEt_3$ (1:17:15)	97			[104,105]
$Rh(Acac)_3$–AlR_3		98		[106]
$Ce(OCOR)_3$–$AlEt_{1.5}$–$Cl_{1.5}$–AlR_3	97			[73]
$Pr(OCOCCl_3)_3AlEt_2Cl$–$Al(i\text{-}Bu)_3$ (1:2:25)	98			[83]
$NdCl_3.xPy$–$Al(i\text{-}Bu)_3$ (1:30)	97	2.7	0.3	[89]
$Nd(OCOCCl_3)_3$–$AlEt_2Cl$–$Al(i\text{-}Bu)_3$ (1:2:25)	97–99			[77,81,83]
$Nd(OCOR)_3$–t-BuCl–$Al(i\text{-}Bu)_2H$ (1:3:20)	> 98			[83]
$U(OR)_4$–$AlEtCl_2$	99			[92]

[a] The 1,2-polymers have mainly a syndiotactic structure, the isotactic polymers having been obtained only with mixed-structure polymers, from the $Cr(Acac)_3$–$AlEt_3$ (1:10), $Cr(PhCN)_6$–$AlEt_3$ (1:5) and $Cr(CO)_5$.Py–$AlEt_3$ (1:5) catalysts; in general, structures are variable and the highest-purity crystalline polymers are separated by fractionation.

[b] Isotactic polymer.

Table 5.2 Polymerisation of isoprene, 1,3-pentadiene and 2,4-hexadiene with aluminium alkyl-activated Ziegler–Natta catalysts

	Polymer microstructure			
Catalyst (component molar ratio)	*cis*-1,4 (%)	*trans*-1,4 (%)	1,2[a] (%)	Reference
$TiCl_4$–AlR_3 (1:1)	96–98			[92]
$TiCl_4$–AlR_3 (Al/Ti < 1)		95[b]		[12,91]
$TiCl_4$–AlR_3–NR_3	100[b]			[107]
$TiCl_4$–AlR_3 (1:1)	65–70[c]			[92]
α-$TiCl_3$–AlR_3		91[b]		[108]
β-$TiCl_3$–AlR_3 (1:1)	95[b]			[92]
γ-$TiCl_3$–AlR_3		High[c,d]		[41]
$Ti(OR)_4$–AlR_3 (1:7)	85–90[c,d]		<90[b]	[12,91]
$Ti(OR)_4$–AlR_2Cl	96–97[b]			[109]
VCl_3–$AlEt_3$		99[b]		[92,110]
VCl_3–AlR_3		80–90[c,d]		[13,41,57]
VCl_4–Al(*i*-Bu)$_3$		98[b]		[57]
$Fe(OCOR)_3$–BuNCS–$AlEt_3$			High[c,e]	[105]
$CoCl_2$.2Py–AlR_2Cl	95[b]			[111]
$Co(Acac)_3$–$AlEt_2Cl$ (1:600)			90[c,e]	[17,105]
$Co(Acac)_2$–$AlEt_2Cl$–H_2O (1:600:300)	98[c,e]			[41,105]
$Co(Acac)_2$–$AlEt_2Cl$		High[f,g]		18,19,41]
$Ni(OCOR)_2$–BF_3.PhOH–$AlEt_3$ (1:1.5:10)	high[c,e]			[41,105]
$Ni(OCOR)_2$–BF_3.OEt_2–$AlEt_3$	High[c,e]			[104,105]
$NdCl_3$.*x*Py–Al(*i*-Bu)$_3$ (1:30)			95–96[b]	[7]
$NdCl_3$.*x*Py–Al(*i*-Bu)$_3$	96–97[b]			[100]
$Nd(OCOR)_3$–$AlEt_2Cl$–Al(*i*-Bu)$_3$	95[b], 95[c,d]	High[f,g]		[18,19,41,77,81, 83,112,113]
$Nd(OCOR)_3$–$AlEt_2Cl$–THF	97[b]			[110]

[a] The 1,2 structure mixed with the predominant 3,4 structure in the case of polyisoprene.
[b] Polyisoprene.
[c] Poly[(*E*)-1,3-pentadiene].
[d] Isotactic polymer.
[e] Syndiotactic polymer.
[f] Poly[(*E,E*)-2,4-hexadiene].
[g] *threo*-Diisotactic polymer.

The $CpTiCl_3$- and $CpTi(OR)_3$-based catalysts afford polymers with a *cis*-1,4-isotactic structure from 1,3-butadiene, 2,3-dimethyl-1,3-butadiene, (*E*)-2-methyl-1,3-pentadiene and (*Z*)-1,3-pentadiene, and mixed (equibinary) *cis*-1,4/1,2-polymers from 1,3-butadiene and (*E*)-1,3-pentadiene, while they give a 1,2-syndiotactic polymer from 4-methyl-1,3-pentadiene and a mixed 1,2/*trans*-1,4 polymer from (*E,E*)-2,4-hexadiene. The catalysts derived from $CpTiCl_2$.2THF or $[CpTiCl_2]_x$ and $[Al(Me)O]_x$ are less active than the $CpTiCl_3$–$[Al(Me)O]_x$ catalyst but give the same type of polymers. It is interesting that a change in stereospecificity with temperature was observed in the polymerisation of (*Z*)-1,3-pentadiene with $CpTiCl_3$- and $CpTi(OR)_3$-based catalysts; a *cis*-1,4-isotactic polymer containing minor amounts of 1,2 units was obtained in polymerisation run at 20 °C, but a 1,2-syndiotactic polymer was yielded in polymerisation at −30 °C. As regards the $Ti(OBu)_4$–$[Al(Me)O]_x$ catalyst, it exhibits lower activity but similar stereospecificity in polymerisations of butadiene and 4-methyl-1,3-pentadiene as compared with the $CpTi(OBu)_3$–$[Al(Me)O]_x$ catalyst. In (*Z*)-1,3-pentadiene polymerisation with the $Ti(OBu)_4$–$[Al(Me)O]_x$ catalyst, irregular polymers are formed and consist of a mixture of *cis*-1,4,*trans*-1,4 and 1,2 units, the former being predominant, irrespective of the polymerisation temperature. However, a highly stereoregular polymer, consisting of almost 100 % *trans*-1,4 units, has been obtained from (*E*,*E*)-2,4-hexadiene using the $Ti(OBu)_4$–$[Al(Me)O]_x$ catalyst. Table 5.3 summarises data concerning the polymerisation of some conjugated dienes in the presence of homogeneous catalysts [43].

A particularly high reactivity is exhibited by 4-methyl-1,3-pentadiene in polymerisation with the homogeneous catalysts discussed; this monomer is 50–100 times more reactive than butadiene [38,39,43,117].

Table 5.3 Polymerisation of some conjugated dienes with homogeneous catalysts[a, b]

Conjugated diene monomer	Catalyst	Polymer microstructure *cis*-1,4 (%)	*trans*-1,4 (%)	1,2 (%)
1,3-Butadiene	$CpTiCl_3$–$[Al(Me)O]_x$	80	2	18
1,3-Butadiene	$CpTi(OBu)_3$–$[Al(Me)O]_x$	82	2	16
1,3-Butadiene	$Ti(OBu)_4$–$[Al(Me)O]_x$	81	4	15
(*E*)-1,3-Pentadiene	$CpTiCl_3$–$[Al(Me)O]_x$	43		57
(*E*)-1,3-Pentadiene	$CpTi(OBu)_3$–$[Al(Me)O]_x$	55		45
(*E*)-1,3-Pentadiene	$Ti(OBu)_4$–$[Al(Me)O]_x$	59	28	13
(*Z*)-1,3-Pentadiene	$CpTiCl_3$–$[Al(Me)O]_x$	>99		
(*Z*)-1,3-Pentadiene	$CpTi(OBu)_3$–$[Al(Me)O]_x$	74		26
(*Z*)-1,3-Pentadiene	$CpTi(OBu)_3$–$[Al(Me)O]_x$			>99[c]
2,3-Dimethyl-1,3-butadiene	$CpTiCl_3$–$[Al(Me)O]_x$	>99		
(*E*)-2-Methyl-1,3-pentadiene	$CpTiCl_3$–$[Al(Me)O]_x$	>99		
(*E*)-2-Methyl-1,3-pentadiene	$CpTi(OBu)_3$–$[Al(Me)O]_x$	>99		
(*E*)-2-Methyl-1,3-pentadiene	$Ti(OBu)_4$–$[Al(Me)O]_x$		>90	
4-Methyl-1,3-pentadiene	$CpTiCl_3$–$[Al(Me)O]_x$			99
4-Methyl-1,3-pentadiene	$CpTi(OBu)_3$–$[Al(Me)O]_x$			>99
4-Methyl-1,3-pentadiene	$Ti(OBu)_4$–$[Al(Me)O]_x$			100
(*E*,*E*)-2,4-Hexadiene	$CpTiCl_3$–$[Al(Me)O]_x$		30	70
(*E*,*E*)-2,4-Hexadiene	$Ti(OBu)_4$–$[Al(Me)O]_x$		>99	

[a] Polymerisation conditions: monomer 2 mL; solvent (toluene) 16 mL; $CpTiCl_3$ 1×10^{-5} mol; [Al(Me)O] unit/Ti molar ratio 1000; temperature 20 °C.
[b] According to Ref. 43.
[c] Polymerisation temperature −30 °C.

Let us recall that half-sandwich metallocene-based catalysts, either activated with methylaluminoxane or aluminium free, such as those activated with tris(perfluorophenyl)boron, also promote the syndiospecific polymerisation of styrene.

5.2.2 Supported Half-sandwich Metallocene Catalysts

Conjugated dienes have been polymerised using supported half-sandwich metallocene catalysts. For instance, catalysts derived by supporting $CpTiCl_3$ on alumina–silica gels, containing–O–Ti(Cp)Cl_2 species, displayed activity in isoprene polymerisation without the addition of any other activator. Depending on the alumina–silica gel composition, the kind of polymerisation medium and the temperature, these catalysts exhibited various activities and selectivities; polyisoprenes with a predominant 3,4 structure and mixed 1,2/*trans*-1,4 structure were obtained [118,119].

It is interesting that the presence of the alumina-silica carrier in the catalyst is sufficient enough for the activation of the titanocene for the polymerisation, similarly like the presence of methylaluminoxane in the titanocene homogeneous polymerisation system; $CpTiCl_3$ alone does not exhibit catalytic activity in conjugated diene polymerisation.

5.2.3 Catalysts Based on η^3-Allyl Derivatives of Transition Metals

Catalysts of the π-allylic type for the polymerisation of conjugated dienes are based in particular on complexes of the following metals: Ti, Zr, Nb, Cr, Mo, W, Co, Ni, Ru, Rh, Nd and U. Usually they include simple π-allyl complexes of such metals as Ti, Nb, Cr, Mo, W, Ni, Co and Rh as well as more complex systems derived from the reaction of the π-allyl complexes with Lewis bases or acids as well as other electron donors or acceptors. Simple π-allyl type catalysts are soluble in hydrocarbon diluents used as the polymerisation medium [e.g. Ni(All)X, where X = Cl, Br, I, $OCOCF_3$; Cr$(All)_3$]. However, bimetallic catalysts such as Ni(All)Cl–$TiCl_4$ form heterogeneous systems in hydrocarbon diluents applied for polymerisation. Although some catalysts of the π-allylic type exhibit activity and stereospecificity comparable with those of Ziegler–Natta catalysts, they have not been used in industrial polymerisation processes for manufacturing polymers of conjugated dienes. However, catalysts based on π-allyl (η^3) derivatives of transition metals serve as model catalysts useful for elucidating mechanistic problems of conjugated diene polymerisation.

Table 5.4 [33,36,37,116,120–134] and Table 5.5 [7,74,81,103,120,122,126, 135–147] report the microstructure of 1,3-butadiene polymers derived from polymerisation with some representative catalysts of the π-allylic type (η^3-allyl, η^3-crotyl, η^3-methallyl and higher η^3-hydrocarbyl derivatives of transition metals).

Most π-allylic catalysts contain an η^3-allyl or η^3-crotyl ligand, but they also may contain an η^3-methallyl ligand:

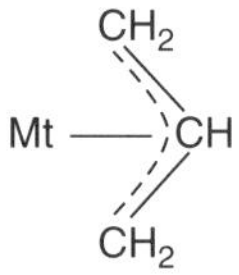

Mt-[η^3-(C_3H_5)]

Mt-(π-allyl)

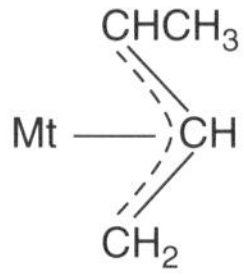

Mt-[η^3-(C_4H_7)]

Mt-(π-crotyl)

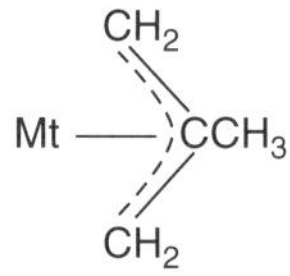

Mt-[η^3-(C_4H_7)]

Mt-(π-methallyl)

Table 5.4 Polymerisation of 1,3-butadiene with π-allylnickel catalysts

	Polymer microstructure			
Catalyst[a, b]	*cis*-1,4 (%)	*trans*-1,4 (%)	1,2 (%)	Reference
Ni(η^3-C_4H_7)Cl	92	6	2	[36,120,121]
Ni(η^3-C_4H_7)I	<4	93–99	<3	[36,116,121,122]
Ni(η^3-C_3H_5)$OCOCCl_3$	94	4	2	[122]
Ni(η^3-C_3H_5)$OCOCF_3$	96–97	<2	1–2	[120,123,124]
Ni(η^3-$C_{12}H_{19}$)$OCOCF_3$	98	2		[120]
Ni(η^3-C_3H_5)$OC_6H_2(NO_2)_3$	94–97	3–4	<2	[120,125]
Ni(η^3-C_3H_5)$OC_6H_2Br_3$		96–97	3–4	[122,126]
Ni(η^3-C_3H_5)$OSO_2C_6H_4CH_3$		96	4	[33,125]
[Ni(η^3-C_3H_5).P$(OPh)_3$]PF_6	5	94	1	[127]
[Ni(η^3-C_3H_5).$SbPPh_3$]PF_6	88	9	3	[127]
Ni$(\eta^3$-$C_3H_5)_2$–Cl_3CCOOH	95	4	1	[126]
Ni(η^3-C_4H_7)$OCOCF_3$–F_3CCOOH	50	49	1	[128]
Ni(η^3-C_3H_5)$OCOCF_3$–EtOH		96	4	[128]
Ni(η^3-$C_{12}H_{19}$)$OCOCF_3$–EtOH		96	4	[128]
Ni(η^3-C_3H_5)$OCOCF_3$–$PhNO_2$	50	50		[124]
Ni(η^3-C_3H_5)$OCOCF_3$–P$(OPh)_3$		99		[124]
Ni(η^3-C_3H_5)$OCOCF_3$–arene(C_6H_6)	49	49	2	[124,128]
Ni(η^3-C_4H_7)Cl–Cl_3CCHO	95	4	1	[122]
Ni(η^3-C_4H_7)Cl–THF	54	41	5	[120,126]
Ni(η^3-C_4H_7)I–*p*-fluoranil	97	2	1	[122]
Ni$(\eta^3$-$C_4H_7)_2$[c]–*p*-chloranil	94	3	3	[129]
Ni$(\eta^3$-$C_4H_7)_2$–NiI_2		95	5	[126]
Ni$(\eta^3$-$C_3H_5)_2$–$SnCl_2$	52	46	2	[130]
Ni$(\eta^3$-$C_3H_5)_2$–SnI_4		95	5	[126,130]
Ni(C_8H_{12})–HI[d]		High		[131,132]
Ni(C_8H_{12})–F_3CCOOH[d]	High			[129,132]
Ni($C_{12}H_{18}$)–HI[e]	100			[37,131,133]
Ni($C_{12}H_{18}$)–F_3CCOOH[e]	92	4	4	[37,131,133]
Ni($C_{12}H_{18}$)–ClF_2CCOOH[e]	50	50		[37,129,134]

[a] Eventual catalyst association is not shown

[b] η^3-C_3H_5 = π-allyl; η^3-C_4H_7 = π-crotyl; η^3-C_8H_{13} = π-(octa-2,6-diene-1-yl); η^3-$C_{12}H_{19}$ = π-(dodeca-2,6,10-triene-1-yl)

[c] η^3-C_4H_7 = π-methallyl

[d] Mono (π-allyl) η^3-C_8H_{13} nickel species are formed from the nickel bis (π-allylic) Ni(C_8H_{12}) species in the polymerisation system [ref. 132]

[e] Mono (π-allyl) η^3-$C_{12}H_{19}$ nickel species are formed from the nickel bis (π-allylic) Ni($C_{12}H_{18}$) species in the polymerisation system [ref. 37]

Table 5.5 Polymerisation of 1,3-butadiene with catalysts based on π-allyl derivatives of transition metals

	Polymer microstructure			
Catalyst[a, b]	*cis*-1,4 (%)	*trans*-1,4 (%)	1,2 (%)	Reference
$Nb(\eta^3\text{-}C_3H_5)_3$	1	2	97	[135]
$Nb(\eta^3\text{-}C_4H_7)_3$			100	[122,136]
$Nb(\eta^3\text{-}C_4H_7)_2Cl$	91	5–5.5	3.5–4	[122,136]
$Cr(\eta^3\text{-}C_3H_5)_3$		10	90	[137]
$Cr(\eta^3\text{-}C_3H_5)_2Cl$	90	5	5	[120]
$Cr(\eta^3\text{-}C_3H_5)_2OCOCCl_3$	93	4	3	[136,138,139]
$Cr(\eta^3\text{-}C_4H_7)_2OCOCCl_3$	93	4	3	[122]
$Cr(\eta^3\text{-}C_3H_5)_3$–$Cl_3CCOOH$	93			[137]
$Cr(\eta^3\text{-}C_3H_5)_3$–$0.5O_2$		92.5–93	<7.5	[137,140]
$Cr(\eta^3\text{-}C_4H_7)_3$–$TiI_4$	90	6	4	[126,137]
$Cr(\eta^3\text{-}C_4H_7)_3$–$NiBr_2$		95	5	[126]
$Mo(\eta^3\text{-}C_3H_5)_4$		2	98	[141]
$Mo(\eta^3\text{-}C_3H_5)_3Cl$			90	[142]
$Mo(\eta^3\text{-}C_3H_5)_4$–$I_2$		2	98	[141]
$Mo(\eta^3\text{-}C_3H_5)_4$–$CH_2 = CHCH_2I$		0.5	99.5	[122]
$Mo(\eta^3\text{-}C_4H_7)_4$–$Cl_3CCOCl$		0.5	99.5	[143,144]
$Co(\eta^3\text{-}C_8H_{13})(C_4H_6)$	1	1	98	[7]
$Co(\eta^3\text{-}C_8H_{13})(C_4H_6)$–$CS_2$			100[c]	[103]
$Co(\eta^3\text{-}C_3H_5)_2I$	High			[37]
$Co(\eta^3\text{-}C_4H_7)_3$–$2HCl$	91.5	1.5	7	[122]
$Co(\eta^3\text{-}C_4H_7)_3$–$I_2$	45	12	43	[137,145]
$Rh(\eta^3\text{-}C_4H_7)_3$		94	6	[122,137]
$Nd(\eta^3\text{-}C_3H_5)_3$.DOX	3	93	4	[146]
$Nd(\eta^3\text{-}C_3H_5)_3$.DOX–$AlEtCl_2$[d]	94	5	1	[146]
$Li[Nd(\eta^3\text{-}C_4H_7)_4]$.DOX	2.6	92.8	4.6	[81]
$U(\eta^3\text{-}C_3H_5)_3Cl$	99	0.7	0.3	[74,147]
$U(\eta^3\text{-}C_3H_5)_3Br$	98.5	1.0	0.5	[147]
$U(\eta^3\text{-}C_3H_5)_3I$	98.5	1.2	0.3	[147]

[a] Eventual catalyst association is not shown.
[b] $\eta^3\text{-}C_3H_5 = \pi$-allyl; $\eta^3\text{-}C_4H_7 = \pi$-crotyl; $\eta^3\text{-}C_8H_{13} = \pi$-(octa-2,6-diene-1-yl).
[c] Syndiotactic polymer.
[d] Ziegler–Natta catalyst by definition.

Higher η^3-hydrocarbyl derivatives of transition metals can be obtained by partial protolysis of the bis(π-allyl)metal compounds formed from cyclo-octa-1,5-diene metal complexes and 1,3-butadiene with one equivalent of the Brønsted acid in the presence of the proper ligand in a polar solvent at lowered temperature (e.g. $-40\,^{\circ}C$). For instance, Ni–$[\eta^3\text{-}(C_8H_{13})]$ and Ni–$[\eta^3\text{-}(C_{12}H_{19})]$ derivatives have been yielded by the reaction of $Ni(C_8H_{12})_2$ $[(Ni(Cod)_2]$ with 1,3-butadiene (C_4H_6), followed by treatment with a Brønsted acid, according to scheme (1) [132] and scheme (2) [37] respectively:

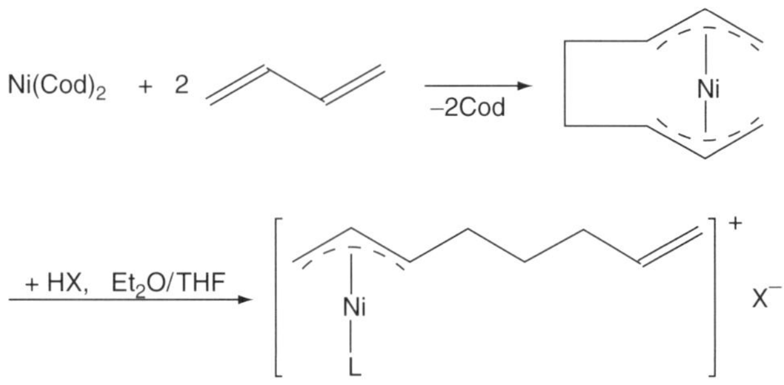

(1)

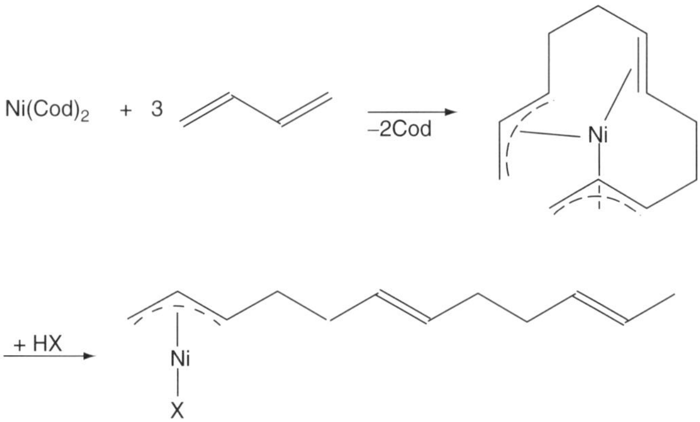

(2)

The activity and stereospecificity of π-allylic catalysts for conjugated diene polymerisation depend both on the kind of metal and on the nature of the ligand attached to this metal. For instance, $Cr(All)_3$ [137] and $Co(\eta^3\text{-}C_8H_{13})(C_4H_6)\text{–}CS_2$ [103] catalysts yield 1,2-polybutadiene, while $Cr(All)_2Cl$ [120], $Cr(All)_2I$ [134] and $U(All)_3Cl$ [147] catalysts produce *cis*-1,4-polybutadiene, but an $Nd(All)_3$.DOX catalyst gives *trans*-1,4-polybutadiene [146] and a $Co(\eta^3\text{-}C_4H_7)_3\text{–}I_2$ catalyst yields eb-*cis*-1,4/1,2-polybutadiene [137,145] (Table 5.5).

Catalysts of the Ni(All)X type give 1,4-polybutadienes, but the polymer microstructure and catalyst activity depend strongly on the nature of the X ligand. Polymers with a high content of *cis*-1,4 monomeric units are obtained with π-allylic complexes having electron-withdrawing groups such as Cl [120,121] or anionic ligands such as F_3CCOO^- and Cl_3CCOO^- [120–124]. However, even the nature of the electron-withdrawing group or anionic ligand can result in changes in the microstructure of the formed polymer. For instance, *cis*-1,4-polybutadienes are obtained with η^3-allylnickel chloride [36,120,121] and with η^3-allylnickel picrate [36,120,125], whereas *trans*-1,4-polybutadienes are yielded by η^3-crotylnickel iodide [36,116,121,122] and by η^3-allylnickel-2,4,6-tribromophenolate [122,126]. It is interesting that the addition of an electron acceptor such as *p*-fluoranil to η^3-crotylnickel iodide changes the catalyst stereospecificity, and *cis*-1,4-polybutadiene is formed with such a catalyst [122]. The addition of electron donors or Lewis bases to η^3-allylnickel halides or η^3-allylnickel trihaloacetates causes a decrease in the content of *cis*-1,4 monomeric units and an increase in the content of *trans*-1,4 units in polybutadienes formed with such catalysts. For example, the addition of an electron donor such as benzene to η^3-allylnickel trifluoroacetate (which alone gives predominantly *cis*-1,4-polybutadiene [123,124]) results in the formation of eb-*cis*-1,4/*trans*-1,4-polybutadiene [124,128]. Moreover, η^3-crotylnickel chloride and η^3-allylnickel trifluoroacetate, which yield *cis*-1,4-polybutadienes in hydrocarbon solvents [36,120–124], give eb-*cis*-1,4/*trans*-1,4 [120,126] and *trans*-1,4-polybutadienes [128] in EtOH or THF as solvents; also, the addition of $P(OPh)_3$ to $Ni(All)OCOCF_3$ results in the formation of *trans*-1,4-polybutadiene [124]. The addition of Lewis acids to bis(η^3-allyl)metal catalysts generally results in an increase in the catalyst activity; the *trans*-1,4 structure of polymers is favoured in some bimetallic systems of this kind, but *cis*-1,4-polymers can also be formed, depending on the ligand attached to the transition metal atom and the substituent at the metal of the Lewis acid [126,130] (Table 5.4).

Catalysts based on π-allylic derivatives of transition metals supported on alumina, silica or silica–alumina gels exhibit generally enhanced activity by comparison with their unsupported counterparts, while the stereospecificity depends on the nature of the catalyst carrier. For instance, $Cr(All)_3$, which predominantly produces 1,2-polybutadiene [137], becomes a stereospecific catalyst for the formation of *trans*-1,4-polybutadiene when supported on silica or silica–alumina gel and for the formation of *cis*-1,4-polybutadiene when supported on alumina [148]. However, an increase in the content of *cis*-1,4 monomeric units in polybutadiene with increasing silica concentration in π-allylnickel–alumina–silica catalysts has been observed [149].

5.2.4 Catalysts Based on Transition Metal Salts without an Organometallic or Metal Hydride Activator

Various transition metal-based catalysts not containing preformed metal–carbon bonds have been developed for the polymerisation of conjugated dienes [27–35, 150–158]. These catalysts include monometallic precursors such as Rh, Co and Ni salts and bimetallic precursors such as $CoCl_2$–$AlCl_3$. Some of them are soluble in a polymerisation medium, e.g. $Rh(NO_3)_3$ in protic solvents (ROH, H_2O) [27,150–154] and $CoCl_2$–$AlCl_3$ in aprotic solvents [155–157], and some others are insoluble in a polymerisation diluent, e.g. $TiCl_4$–$Ni(PCl_3)_4$ [158].

Rhodium salts such as $Rh(NO_3)_3.2H_2O$ and $RhCl_3.3H_2O$ exhibit catalytic activity for polymerisation of butadiene in protic solvents or aqueous emulsions and yield *trans*-1,4-polybutadiene ($\geq$99%) [27,28,150–154]. Propagation active species are proved to be formed in such systems by the insertion of butadiene into the rhodium–hydrogen bond existing in aqueous and related solutions of rhodium salts [154].

Several bimetallic catalysts give *cis*-1,4-polybutadiene, e.g. $TiCl_4$–$Ni(PCl)_3)_4$ [158], $CoCl_2$–$AlCl_3$ and $CoCl_2$–$AlCl_3$–Al [155–157]. It is assumed that in such systems an active metal–carbon bond is formed upon insertion of the diene monomer into the transition metal–chlorine bond for species containing metal in a higher oxidation state or through an oxidative addition of the monomer to a species containing metal in a low oxidation state.

Some bimetallic catalysts of various stereospecificity are based on transition metal carbonyls and metal halides; for instance, $Ni(CO)_4$–VCl_4 and $Ni(CO)_4$–WCl_6 catalysts produce *cis*-1,4-polybutadiene, but the $Co_2(CO)_8$ –$MoCl_5$ catalyst yields 1,2-polybutadiene [35].

5.3 Polymerisation Mechanism and Stereochemistry

The polymerisation of conjugated dienes with transition metal-based catalysts is an insertion polymerisation, with the growing chain end being in the form of an η^3-allylic ligand attached to the metal (after the insertion of the coordinated monomer) or in the form of an η'-allylic ligand (prior to the insertion of the coordinating monomer). Polymer growth then occurs by the repetitive insertion of the monomer into the metal–carbon bond, alternately changing its nature from π-allyl to σ-allyl species, and vice versa, in each propagation step [scheme (3) in Chapter 2) [7].

Note the similarity between the polymerisation of conjugated dienes and monoalkenes; several monoalkene/1,3-dialkene copolymers can be prepared e.g. ethylene/butadiene [159] and propylene/butadiene [160] alternating copolymers in the presence of transition metal (V, Ti) chloride or oxychloride–triisobutylaluminium Ziegler–Natta catalysts. However, the copolymerisation of conjugated dienes and monoolefins does not proceed as readily as the copolymerisation involving only the diene monomers. This is due to the fact that π-allylic bonds, appearing in diene polymerisation, are more stable than respective σ bonds; ethylene or α-olefin thus is inserted less readily into the metal–carbon π bond than into the metal–carbon σ bond. The appearance of active species of the π-allylic type, which is the most distinctive feature of conjugated diene polymerisation, also results in poor catalytic activity of highly active conjugated diene polymerisation catalysts in the polymerisation of ethylene and α-olefins (e.g. soluble $CoCl_2 \cdot 2Py$–$AlEt_2Cl$–H_2O) [99] and $Ni(OCOR)_2$ –$BF_3.OEt_2$–$AlEt_3$ [105] Ziegler–Natta catalysts). On the other hand, heterogeneous Ziegler–Natta catalysts such as $TiCl_4$–$AlEt_3$, $TiCl_3$–$AlEt_3$ and VCl_3–$AlEt_3$, which are highly active for the polymerisation of ethylene and α-olefins, are also markedly active in the polymerisation of conjugated dienes [7] (and in the isospecific polymerisation of styrene [161]).

In contrast to heterogeneous Ziegler–Natta catalysts, homogeneous catalysts based on biscyclopentadienyl derivatives of group 4 transition metals, which contain cationic metallocene species of formally d^0 14-electronic structure, hardly promote the polymerisation of conjugated dienes, since the diene can act as a donor of four electrons rather than of two electrons as in monoolefin polymerisation (let us recall that the polymerisation of conjugated dienes is catalysed by half-sandwich metallocene-based catalysts). However, it has been reported [162] that statistical copolymers of ethylene and butadiene were obtained with the Cp_2ZrCl_2–$[Al(Me)O]_x$ catalyst.

5.3.1 Mechanistic–Kinetic Aspects of Polymerisation

The initial step of the coordination polymerisation of conjugated dienes has been elucidated by ^{1}H NMR studies on systems consisting of simple π-allylic nickel derivatives such as $[Ni(\eta^3\text{-crotyl})I]_2$ and 1,3-butadiene. By applying a non-deuterated $[Ni(\eta^3\text{-}C_4H_7)I]_2$ catalyst and perdeuterated monomer (C_4D_6) as well as perdeuterated catalyst $[Ni(\eta^3\text{-}C_4D_7)I]_2$ and non-deuterated monomer (C_4H_6), it was proved that the insertion of the first monomer molecule occurred between nickel and the crotyl group [163]. More detailed information about the mode of insertion of the conjugated diene into a metal–carbon bond of the π-allylic type has been supplied by ^{1}H NMR examination of a reaction system containing 1,3-butadiene and $[Pd(\eta^3\text{-}C_4H_7)Cl]_2$; the reaction in such a system is slow enough to be followed by NMR spectroscopy. The first step consists of the coordination of butadiene to palladium, causing the formation of η^3-allylic species that are in equilibrium with η^3-allylic species:

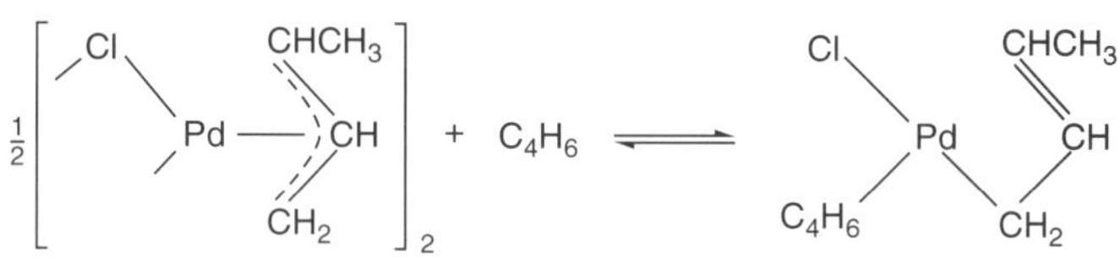

(3)

In the second step, the actual insertion of coordinating butadiene occurs into the Pd–(η^1–C_4H_7) bond; it involves the migration of the allylic group to the coordinated monomer. The insertion results in the formation of an η^3-butenyl group [164,165].

The propagation reaction in conjugated diene polymerisation occurs by an analogous insertion, which has been demonstrated by ^{1}H NMR analysis of the polymerisation of partially deuterated butadiene (CD_2=CH–CH=CD_2) with a simple π-crotyl derivative of nickel, $[Ni(\eta^3\text{-}C_4D_6H)I]_2$ [166].

As regards chain termination reactions, several processes can bring about termination of the growing macromolecule. These processes have been studied mostly for the polymerisation of butadiene or isoprene with Ziegler–Natta catalysts to *cis*-1,4-polymers because of the commercial significance of such polymers. Depending on the catalyst used and the polymerisation conditions, termination reactions consist in chain transfer with the monomer and/or with the aluminium alkyl activator, transfer with the suitable chain transfer agents introduced into the polymerisation system and spontaneous terminations [7].

Transfer to the monomer can occur in various polymerisation systems to a greater or lesser extent. In the polymerisation of butadiene or isoprene with several Ziegler–Natta catalysts, e.g. $Co(Acac)_2$–$AlEt_2Cl$–H_2O [167], $Ti(OR)_4$–$AlEt_3$ [167], $TiCl_4$–AlR_3 [169,170] and $Ni(OCOR)_2$–$AlEt_2Cl$ [71], as well as with aluminium alkyl-free monometallic Rh catalysts [171], a transfer of this type is the main termination process; such a transfer is also the main process regulating the molecular weight of the polymer formed.

Termination by reaction of the growing chain with an alkylaluminium compound used as the catalyst activator has been observed in several types of polymerisation and is often the main termination process in butadiene polymerisation systems with such catalysts as $Nd(OCOR)_3$–$AlEt_2Cl$–$Al(i\text{-}Bu)_3$ [71,89] and TiI_4–$Al(i\text{-}Bu)_3$–$O(i\text{-}Pr)_2$ [48], and in isoprene polymerisation system with the VCl_3–AlR_3 catalyst [172]. In particular, in butadiene polymerisation systems with neodymium-based catalysts, the molecular weight of the formed polymer (*cis*-1,4-polybutadiene) decreases with increasing $Al(i\text{-}Bu)_3/Nd(OCOR)_3$ molar ratio. Transfer with aluminium alkyls appears to be a very useful practical method now available for regulating polymer molecular weight with Nd-based catalysts. The transfer reaction involving aluminium alkyl in butadiene polymerisation systems with neodymium-based Ziegler–Natta catalysts can be represented as follows [7]:

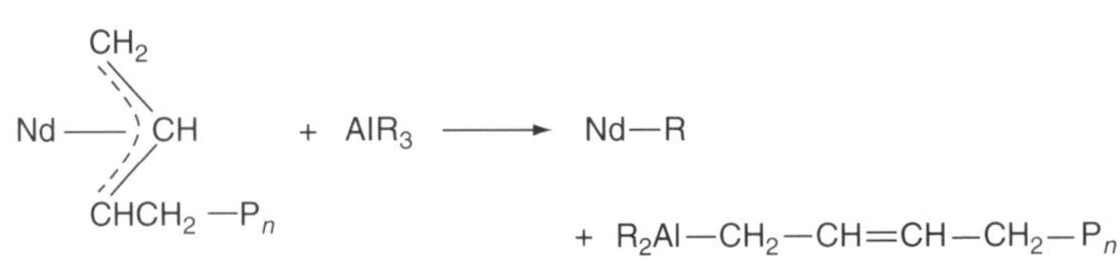

(4)

Chain transfer agents, which are introduced into the polymerisation system in order to regulate the molecular weight of the formed polymer, can be of various nature; cumulated dienes (allenes such as 1,2-butadiene) are particularly efficient even at low concentration, but ethylene, α-olefins (e.g. propylene) and also hydrogen can be used to reduce the polymer molecular weight [173,174]. The mechanism of transfer with cumulated dienes has not been elucidated. Ethylene, the reaction of which in a butadiene polymerisation system with the $CoCl_2 \cdot 2Py$–$AlEt_2Cl$ catalyst follows scheme (5), is the most effective among the olefins:

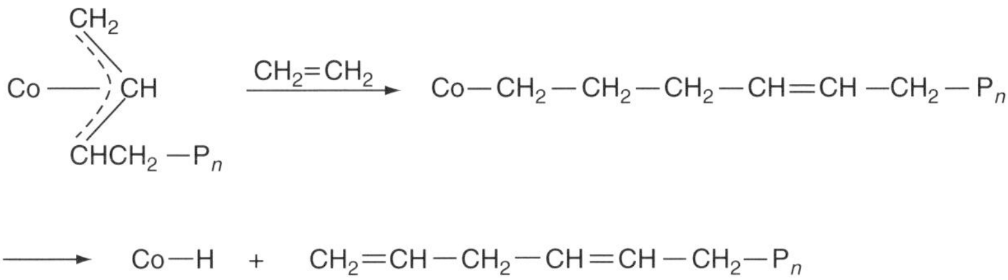

(5)

Reaction with hydrogen gives rise to the terminated polymer chain and to the Mt–H species, which reforms a new η^3-allyl group on reaction with the monomer [7]:

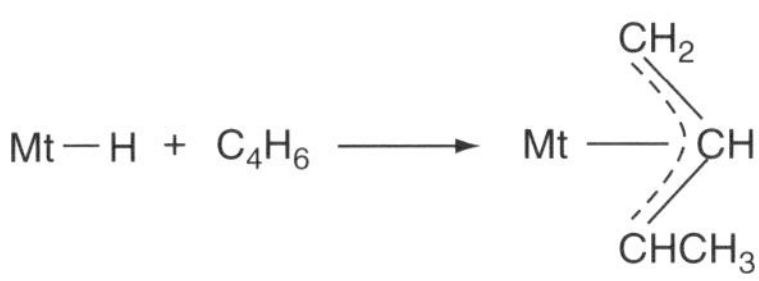

(6)

Spontaneous transfer reactions in conjugated diene polymerisation systems are more complex than those in monoalkene polymerisation systems. Two types of chain termination reaction can occur in principle in polymerisation systems containing conjugated diene. The first type, mononuclear termination, consists in a hydrogen abstraction from the growing chain with the formation of an Mt–H bond [scheme (7)] which reforms an Mt–[η^3-(All)] bond on reaction with the monomer:

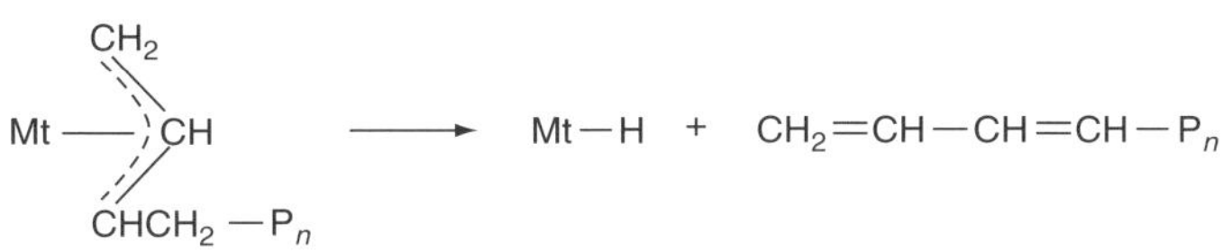

(7)

The second type, bimolecular termination, involves homolytic cleavage of the Mt–[η^3-(All)] bond, leading to a polymer chain bearing a free radical, while the metal assumes the lower oxidation state:

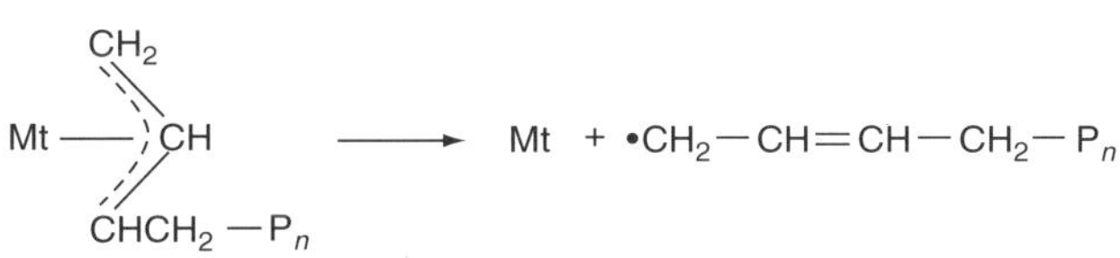

(8)

The radical chain will form a terminated macromolecule either by coupling with another radical or by disproportionation, or by hydrogen abstraction from the polymerisation diluent. Note that monomolecular termination, leading to Mt–H species according to scheme (7), which is characteristic of systems with Co- and Ni-based catalysts of the π-allylic type [40,85,175,176], does not terminate the kinetic chain of the polymerisation. On the other hand, bimolecular termination leading to reduced metal species according to scheme (8), which occurs in some systems with V-based heterogeneous Ziegler–Natta catalysts [172] and with Ti-based soluble Ziegler–Natta catalysts [48], terminates the kinetic chain of the polymerisation; a rapid drop in catalyst activity is observed in such systems [7].

As regards the kinetics of conjugated diene polymerisation in terms of the dependence of the rate of polymerisation on the monomer and catalyst concentrations, most studies show that the polymerisation rate is first order with respect to both the monomer and the catalyst concentration. For instance, the rate of polymerisation of isoprene with the $TiCl_4$–$AlEt_3$ (1:1) catalyst, leading to a *cis*-1,4-polymer, is represented by [177]

$$R_p = k_p \times [M]^1 \times [MtX_n]^1 \times [A]^1$$

For the polymerisation of butadiene to a *cis*-1,4-polymer in the presence of the $Co(Acac)_3$–$AlEt_2Cl$–H_2O catalyst, the polymerisation rate is expressed as follows [167]:

$$R_p = k_p \times [M]^1 \times [MtX_n]^1$$

The rate is independent of the concentration of the activator, $AlEt_2Cl$, ($R_p \sim [A]^0$) as long as a low concentration of the cobalt procatalyst and a fixed concentration of H_2O are maintained.

As regards the polymerisation of butadiene and isoprene to *trans*-1,4-polymers, rather scant data are available in the literature; the rate of polymerisation was found to be first order with respect to the monomer concentration in systems with the $VOCl_3$–$Al(i\text{-}Bu)_3$ catalyst [178].

Kinetic investigations of butadiene polymerisation to 1,2-polymers with various Ziegler–Natta catalysts such as $Ti(OBu)_4$–$AlEt_3$ [168] showed that the polymerisation rate is first order with respect to the monomer and catalyst concentration.

For conjugated diene polymerisations in systems with various π-allylic catalysts, a first-order dependence of the polymerisation rate on the monomer concentration has been observed in principle. However, the order with respect to the catalyst reflects the different degrees of association of the π-allylic complexes of transition metals under polymerisation conditions; a first-order dependence was found for systems with $[Ni(\eta^3\text{-All})OCOCF_3]_2$ and $Ni(\eta^3\text{-All})X \cdot LA$ catalysts [126,176,179], whereas an order of 0.5 was found for systems with $[Pd(\eta^3\text{-}C_4H_7)Cl]_2$ and $[Ni(\eta^3\text{-All})X]_2$ catalysts [124,126,180].

5.3.2 Regiospecificity and Chemoselectivity of the Propagation Reaction

In the coordination polymerisation of butadiene, the last inserted monomeric unit, which is the butenyl group attached to the metal atom as an η^3-allylic ligand, can assume one of two forms, *anti* and *syn*:

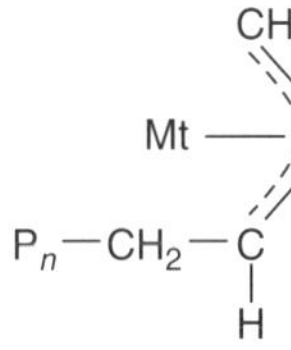

anti

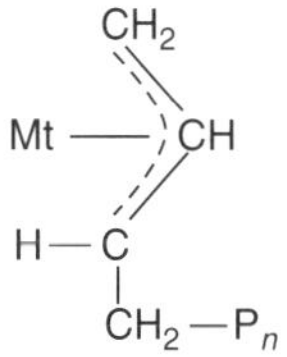

syn

In the case of the polymerisation of non-symmetrically substituted 1,3-butadienes, the most representative of which are isoprene [CH_2=C(CH_3)–CH=CH_2] and 1,3-pentadiene (CH_2=CH–CH=CH–CH_3), two different positional isomers can be formed instead. Isoprene can give two isomeric η^3-allylic groups on insertion into a metal–carbon bond, each of which can exist in the *anti* and *syn* form (only the syn form is shown):

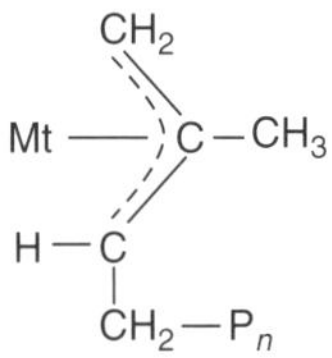

syn

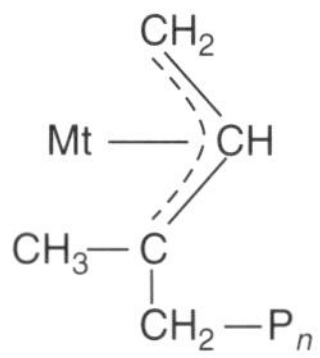

syn

For 1,3-pentadiene, in principle the following η^3-allylic groups can exist (again, only the *syn* form is shown):

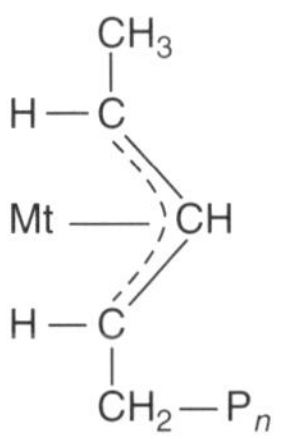

syn

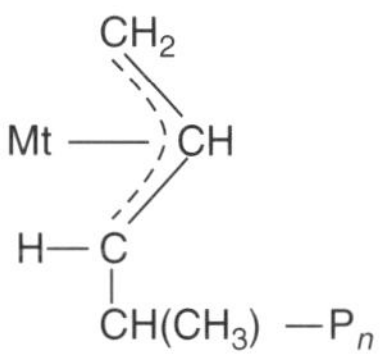

syn

On the basis of ^{1}H NMR investigations it was shown that, in polymerisation with most stereospecific catalysts, isoprene gives, on insertion, predominantly the η^3-[$CH_2C(Me)CHCH_2$–P_n] group, while 1,3-pentadiene yields the η^3-[$CH_2CHCHCH(Me)$–P_n] group [7]. Therefore, polymerisation involving such subsequent insertions is highly regiospecific and thus leads to highly regioregular polymers.

The chemoselectivity of the polymerisation, i.e. the formation of 1,4 as against 1,2 monomeric units in resulting polymers of conjugated dienes, depends on whether enchainment of the incoming monomer occurs at the C_1 or at the C_3 atom of the last inserted monomeric unit. Enchainment via the Mt–C_1 bond (in both the *anti* and *syn* forms) gives rise to the formation of a 1,4 monomeric unit (*cis* or *trans* respectively), whereas enchainment via the Mt–C_3 bond (in both the *anti* and *syn* forms) gives rise to a 1,2 unit. Both cases of the reaction in the butadiene polymerisation system are illustrated below:

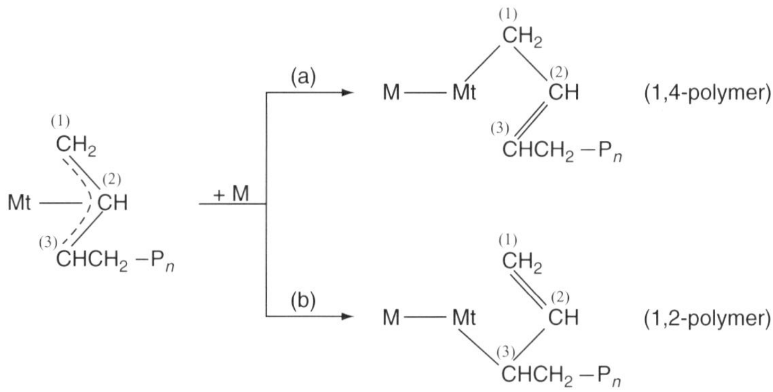

(9)

The factors that determine the reaction of the η^3-butenyl group at the C_1 atom [scheme (9a)] or at the C_3 atom [scheme (9b)] with various catalysts are both steric and electronic in nature, but they have not been fully elucidated. However, the behaviour of some non-symmetrically substituted 1,3-butadienes, e.g. 4-methyl-1,3-pentadiene, in polymerisation with various Ziegler–Natta catalysts can be interpreted on the basis of steric effects [181]. It is obvious that, in the polymerisation of non-symmetrically substituted butadiene, the type of monomeric unit that results from reaction at the C_3 atom will depend on the type of π-allylic ligand. Thus, the η^3–[$CH_2C(Me)CHCH_2$–P_n] group will afford a 3,4-isoprene unit, whereas the η^3-[$CH_2CHC(Me)CH_2$–P_n] group will give a 1,2-isoprene unit in the polymer chain formed. Analogously, the η^3-[$CH(Me)CHCHCH_2$–P_n] group will afford a 1,2-pentadiene unit and the η^3-[$CH_2CHCHCH(Me)$–P_n] group a 3,4-pentadiene unit. It has been found that only 1,2-polymers of 4-methyl-1,3-pentadiene [$CH_2{=}CH{-}CH{=}C(CH_3)_2$], with the exclusion of 1,4 monomeric units, were formed by polymerisation in the presence of several Ziegler–Natta catalysts, even in the presence of catalysts that produce 1,4-polymers from butadiene and isoprene (e.g. $TiCl_4$–AlR_3 and α-$TiCl_3$–AlR_3) [46,181,182]. Let us note that the C_1 atom in butenyl ligands formed from butadiene and isoprene, η^3-[$CH_2CHCHCH_2$–P_n], η^3-[$CH_2C(Me)CHCH_2$–P_n] and η^3-[$CH_2CHC(Me)CH_2$–P_n], is less substituted and thus more reactive than the C_3 atom; hence, 1,4-polymers can be formed more readily than 1,2- or 3,4-polymers. However, in the butenyl ligand formed from 4-methyl-1,3-pentadiene, η^3-[$C(Me)_2CHCHCH_2$–P_n], the C_3 atom is less substituted than the C_1 atom and hence more reactive; therefore a 1,2-polymer is formed in this case.

Considering the above, one cannot attribute the formation of polybutadiene with a 1,2 structure of monomeric units only to steric factors in terms of steric hindrances at the C_1 and C_3 atoms of the butenyl group; the C_3 atom in this group is more substituted than the C_1 atom. Instead, 1,2-polybutadiene formation has been interpreted (Figure 5.3) [181] on the basis of observations made for polymerisation with soluble Ziegler–Natta catalysts derived from Ti-, V-, Cr-, Mo- and Co-based precursors and trialkylaluminium as a catalyst activator. These observations are as follows: the non-1,2 monomeric units appearing in formed polybutadienes are predominantly of *cis*-1,4 structure; polymers of butadiene homologues yielded in systems with some of the catalysts mentioned are characterised by a high content of *cis*-1,4 units or by a mixed *cis*-1,4/1,2 structure; polybutadienes of eb-*cis*-1,4/1,2 structure are formed with Mo- and Co-based catalysts.

According to schemes presented in Figure 5.3, a *cis*-1,4 unit or 1,2 unit is formed, depending on whether the coordinating monomer molecule reacts via its $C_{4'}$ atom at the C_1 atom of the butenyl group or reacts via its $C_{4'}$ atom at the C_3 atom of the butenyl group. Small variations in the position of the incoming monomer with respect to the last inserted monomeric unit, due to the nature of metal substituents or the nature of the monomer itself, may cause preferential formation of one or other type of monomeric unit. Therefore, the formation of

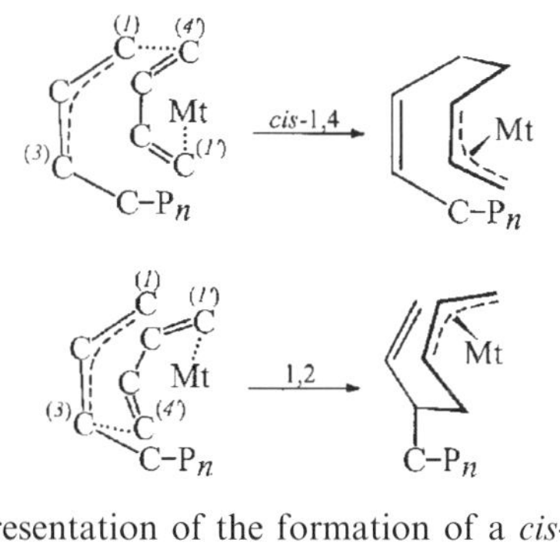

Figure 5.3 Schematic presentation of the formation of a *cis*-1,4 diad (isotactic in the case of the terminally monosubstututed dienes) versus a 1,2 diad (syndiotactic) in the polymerisation of conjugated dienes. The last inserted monomeric unit (as an η^3-allylic ligand) is below the plane, the incoming coordinating monomer is above the plane and the transition metal atom is on the plane

equibinary *cis*-1,4/1,2-polybutadiene in some systems, e.g. with the $Ti(OR)_4$–$AlEt_3$catalyst, is obvious [183–185]; the two modes of reaction presented in Figure 5.3 are equally probable in such cases [7,41]. The schemes in Figure 5.3 also account for the formation of 1,2-polybutadiene [91] and *cis*-1,4-polypentadiene [12] in polymerisation with the same catalyst. However, the interpretation given in Figure 5.3 is not valid in general. There are cases in which the formation of 1,2 units depends on other factors, not yet fully clarified [7].

An interesting example of high chemoselectivity in the coordination polymerisation of conjugated dienes is the polymerisation of 1,3-cyclohexadiene in the presence of Ziegler–Natta [186] and π-allylnickel [187] catalysts, which leads to polymers consisting mainly of 1,4 monomeric units.

5.3.3 cis–trans Isomerism in 1,4-Propagation Reactions

The mode of monomer enchainment via 1,4-insertion, i.e. the formation of *cis*-1,4 or *trans*-1,4 monomeric units in the polymers formed, is determined by the structure, *anti* or *syn*, of the last inserted monomeric unit in the growing chain. A plausible scheme for the formation of *cis*-1,4 and *trans*-1,4 monomeric units for the polymerisation of 1,3-butadiene is presented below [7]:

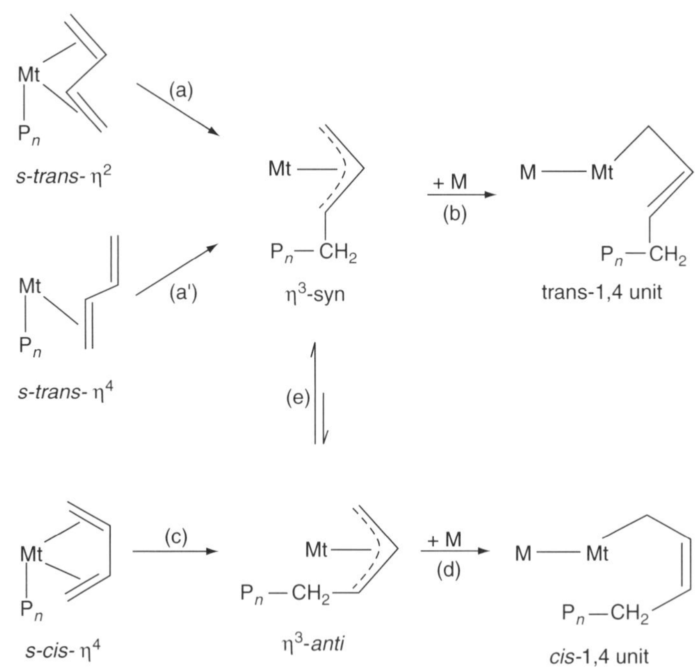

(10)

The *anti* form of η^3-butenyl group gives rise to the formation of a *cis*-1,4 monomeric unit, whereas the *syn* form leads to a *trans*-1,4 unit. It is believed that the appearance of the η^3-butenyl group in *anti* or *syn* form is related to the mode of monomer coordination to the metal atom in the catalyst species.

A conjugated diene can coordinate to a transition metal by only one double bond, as an *s-trans*-η^2 ligand, or with the two double bonds, as an *s-cis*-η^4 or as an *s-trans*-η^4 ligand [188]. A coordinated *transoid* monomer (as an *s-trans*-η^2 or an *s-trans*-η^4 ligand) is inserted into the metal–carbon bond, acquiring the *syn*-η^3-allylic structure of the growing chain end. On the other hand, when a *cisoid* monomer coordinates to a metal (as an *s-cis*-η^4 ligand), an *anti*-η^3-allylic structure is formed.

The *anti* and *syn* forms of the π-allylic ligand are in equilibrium. If no bulky substituent is present at the C_2 atom of the butenyl group, the equilibrium at ambient temperature is completely shifted towards the *syn* form which is thermodynamically much more stable than the *anti* form [148,189]. Therefore, *trans*-1,4 monomeric units can be generated, either involving a coordinated *transoid* monomer [pathways (a)–(b) and (a′)–(b), scheme (10)] or involving a coordinated *cisoid* monomer [pathway (c)–(e)–(b), scheme (10)], if the rate of *anti* → *syn* isomerisation [pathway (e), scheme (10)] is greater than that of insertion. When the rate of this isomerisation is lower that that of insertion, *cis*-1,4 monomeric units are obtained from the coordinated *cisoid* monomer [pathway (c)–(d), scheme (10)].

An *anti* → *syn* isomerisation was demonstrated to occur in several polymerisation systems. For instance, $Co(Acac)_2$–$AlEt_2Cl$ and $Nd(OCOR)_3$–$AlEt_2Cl$–$Al(i\text{-}Bu)_3$ catalysts polymerise butadiene to yield a *cis*-1,4-polymer. The same catalysts produce a *trans*-1,4-polymer when 2,4-hexadiene is subjected to polymerisation [19]. The polymerisation of 2,4-hexadiene proceeds at a lower rate than that of 1,3-butadiene, probably owing to the steric effects exerted by the methyl substituents. Since the insertion step is slow, the *anti*-η^3-allylic ligand has enough time to undergo isomerisation to the more stable *syn* form before the monomer concatenation. Such an interpretation is also in agreement with results obtained from the copolymerisation of 1,3-butadiene and 2,4-hexadiene with Co- and Nd-based catalysts [190]. In copolymers whose contents of butadiene are high (> 90 mol.-%), the hexadiene units are isolated between the butadiene units and have a *cis*-1,4 structure. Insertion of butadiene into the Mt–[η^3-CH(Me)CHCHCH(Me)–P_n] bond formed between the metal and chain-terminating hexadiene unit is faster than that of hexadiene into the same bond, and hence the growing chain end with a hexadiene terminal cannot be isomerised from the *anti* to *syn* form before insertion occurs. As a result, the *trans*-1,4 content of hexadiene units in the copolymer decreases [7].

The *anti* → *syn* isomerisation is also responsible for a decrease in the content of *cis*-1,4 monomeric units and the simultaneous increase in the *trans*-1,4 content of polybutadienes derived from polymerisation with several Ti-, Co- and Ni-based catalysts, at very low monomer concentration. Under polymerisation conditions, the rate of propagation is low and hence the isomerisation takes place more readily than at higher monomer concentrations [191,192].

It should be noted that *trans*-1,4-polymers formed by *anti* $\rightarrow$ *syn* isomerisation, i.e. via the pathway (c)–(e)–(b) [scheme (10)] when the monomer is coordinated to a metal atom as an *s-cis*-η^4 ligand, are produced by various catalysts, especially by soluble ones. It is also worth noting that polymers with a *trans*-1,4 structure of monomeric units are formed by heterogeneous catalysts via the pathway (a)–(b) or (a′)–(b) [scheme (10)] from the (Z)-isomers of conjugated dienes such as $CH_2{=}CH{-}CH{=}CHR$, for which the *s-trans*-η^2 or *s-trans*-η^4 coordination is preferred and the *s-cis*-η^4 coordination is unfavoured [13,193].

The reactions presented in scheme (10) also account for effects exerted by the addition of Lewis bases or acids (as well as other electron donors and acceptors) to the polymerisation system on the microstructure of the polymers formed. As shown in Tables 5.4 and 5.5, some catalysts that are highly stereospecific for the formation of *cis*-1,4-polybutadiene yield *trans*-1,4-polybutadiene (or eb-*cis*-1,4/*trans*-1,4-polybutadiene) after the addition of a Lewis base or other electron donor to the catalyst system. A plausible explanation of the observed phenomena is that the added component occupies a coordination site at the transition metal, thus forcing the incoming monomer molecule to coordinate as an *s-trans*-η^2 ligand. When the additional catalyst component has a basicity comparable with that of the monomer, a competitive monomer/Lewis base (electron donor) coordination takes place, as shown below [7]:

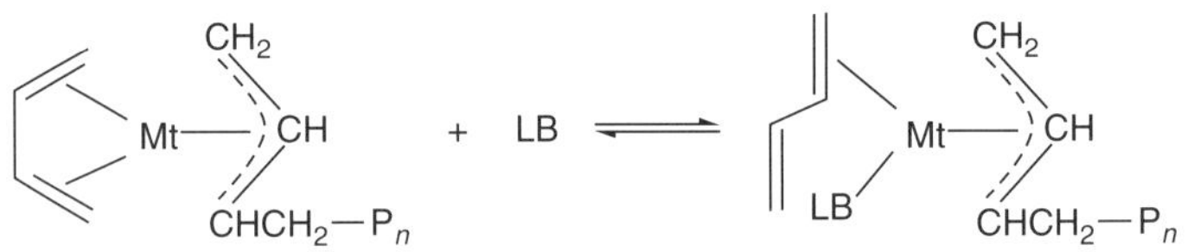

(11)

Catalyst complexation with a Lewis base or other electron donor may affect the polymer microstructure in different ways. If the added component occupies one coordination site, a monomer coordinates to another site of the active species with one double bond, i.e. as an *s-trans*-η^2 ligand, which gives rise to the formation of *trans*-1,4 monomeric units via the pathway (a)–(b) [scheme (10)]. Depending on the lifetimes of metal species complexed with the monomer and with the Lewis base or the other donor [scheme (11)], mixed *cis*-1,4/*trans*-1,4-polybutadienes or an eb-*cis*-1, 1,4/*trans*-1,4-polymer can be formed. One should mention in this connection that equibinary *cis*-1,4/*trans*-1,4-butadiene polymers can also be formed in systems without the addition of a Lewis base or other electron donor; in such cases, the equilibrium of the *anti-syn* isomerisation is not shifted and there are equal probabilities for the reaction pathways involving coordination of a *transoid* monomer and a *cisoid* monomer [7].

Reactions proceeding according to scheme (10) also account for the effect of Lewis acids or other electron acceptors on catalysts of the π-allylic type. The formation of ionic or charge transfer complexes favours the coordination of the conjugated diene as an *s-cis*-η^4 ligand, and hence the formation of *cis*-1,4 monomeric units, if an *anti* form is predominant [7]. The formation of an ionic complex is shown below, as an example, for π-allylnickel chloride and aluminium trichloride [194].

$$[Ni–(\eta^3\text{-All})Cl]_2 + Al_2Cl_6 \longrightarrow 2\,[Ni–(\eta^3\text{-All})]^+AlCl_4^- \qquad (12)$$

5.3.4 Isospecificity and Syndiospecificity of the Propagation Reaction

A striking feature of conjugated diene polymerisation is its capability to yield stereoregular 1,4-polymers with an isotactic or a syndiotactic structure when terminally disubstituted 1,3-butadiene, containing only deuterium atoms as substituents, and also terminally monodeuterated 1,3-pentadiene are applied as monomers. As already mentioned, the polymerisations of (*Z,Z*)-1,4-$(^2H)_2$-1,3-butadiene and (*E,E*)-1-(^2H)-1,3-pentadiene were found [20–22] to yield *trans*-1,4-*threo*-diisotactic polymers (with the VCl_3–$AlEt_3$ catalyst), *cis*-1,4-*threo*-disyndiotactic polymers [with the $Co(Acac)_2$–$AlEt_2Cl$–H_2O catalyst] and *cis*-1,4-*erythro*-diisotactic polymers [with the $Nd(OCOR)_3$–$AlEt_2Cl$–$Al(i\text{-}Bu)_3$ catalyst]. However, the polymerisation of (*E,Z*)-1-(^2H)-1,3-pentadiene was reported [21] to produce a *trans*-1, 4-*erythro*-diisotactic polymer (with the VCl_3–$AlEt_3$ catalyst).

It must be emphasised that no stereoregular polymer has so far been obtained from deuterated ethylene [$CH_2{=}CH(^2H)$, $CH(^2H){=}CH(^2H)$], evidently because deuterium atoms do not exert a steric effect. Thus, the stereoregulation in the polymerisation of conjugated dienes must come from different origins than those that have been considered for α-olefin polymerisation, i.e. non-bonding interactions between the alkyl group at the double bond of the incoming monomer (α-olefin) and the transition metal-bonded monomeric unit of the growing polymer chain.

The origin of the stereoregulation in conjugated diene coordination polymerisation was recognised by Porri *et al.* [7,41] who explained the different modes of formation of the isotactic and the syndiotactic polymers in the presence of various transition metal-based catalysts.

Note that the Mt–(η^3-butenyl) species, *anti* or *syn*, constitute a chiral centre, irrespective of the substituents. The chirality of such species is connected with the presence of two asymmetric carbon atoms, C_2 and C_3, in the butenyl group. In the case of the polymerisation of 1,3-butadiene, the enantiomorphic

Mt–(η^3-butenyl) active species can be presented for *anti* and *syn* forms as follows:

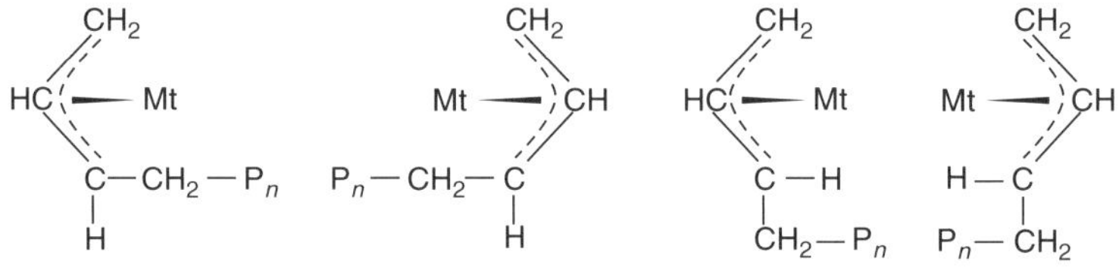

A chiral Mt–(η^3-butenyl) species will give, depending on its structure and thus on the orientation of the incoming monomer, a new Mt–(η^3-butenyl) species of the same chirality as the previous one (and hence an isotactic diad) or of the opposite chirality (and hence a syndiotactic diad); it is obvious that the tacticity may concern only 1,2-polymers of non-substituted or substituted butadiene and 1,4-polymers of terminally monosubstituted and symmetrically disubstituted butadiene. The mode of the formation of the butenyl group of the same or opposite chirality with respect to the preceding butenyl group is shown, for 1,3-butadiene insertion, in Figure 5.4 [7].

An arrangement as in Figure 5.4a gives, after monomer insertion, a new butenyl group of the same chirality as the previous one, while an arrangement as in Figure 5.4b will give a butenyl group of the opposite chirality. It should be emphasised that neither the monomer nor the butenyl group must bear any substituent, since the chirality of the discussed species originates from the presence of two mentioned earlier asymmetric carbon atoms, C_2 and C_3, in the butenyl group.

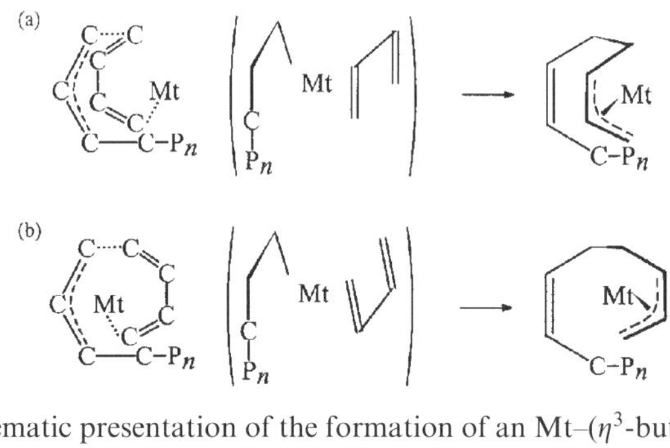

Figure 5.4 Schematic presentation of the formation of an Mt–(η^3-butenyl) group (a) of the same chirality and (b) of opposite chirality to the preceding one (in the *anti* form), depending on the orientation of the incoming monomer (1,3-butadiene). The monomer is above the plane, the η^3-butenyl group is below and Mt is on the plane of the figure. In brackets, the same situation is shown, but observed along an axis perpendicular to the plane that contains the η^3-butenyl group, from left to right

5.3.4.1 Isotactic and Syndiotactic cis-1,4-Polydiene Formation

A representative example of the formation of isotactic and syndiotactic *cis*-1,4-polymers of conjugated dienes is the polymerisation of terminally substituted butadienes such as e.g. 1,3-pentadiene in the presence of Ziegler–Natta catalysts based on Ti or Nd and Co or Ni precursors respectively [22,195]. The formation of isotactic and syndiotactic *cis*-1,4-polypentadiene has been proposed as in the schemes in Figure 5.4 [7,41,43]. The butenyl group exists in an *anti* form and the monomer coordinates as an *s-cis*-η^4 ligand, since a *cis*-1,4-polymer is obtained. An isotactic polymer is formed in the case of an arrangement as in Figure 5.4a, while a syndiotactic polymer is formed when ligands are arranged as in Figure 5.4b.

The validity of the proposed mechanism is borne out by the results of the polymerisation of (*E,E*)-1-(^{2}H)-1,3-pentadiene with $Nd(OCOR)_3$–$AlEt_2Cl$–$Al(i\text{-}Bu)_3$ and $Co(Acac)_2$–$AlEt_2Cl$–H_2O catalysts [20–22,195]. The Nd-based catalyst gave a *cis*-1,4-*erythro*-diisotactic polymer, while the Co-based catalyst yielded a *cis*-1,4-*threo*-disyndiotactic polymer. The formation of such polymers is shown by the schemes in Figure 5.5 [7,41].

The different orientation of the monomer in systems with Ziegler–Natta catalysts based on Ti or Nd and Co or Ni precursors has been suggested [41] to be determined by the presence in the isospecific catalysts (Ti- and Nd-based ones) of anionic ligands bound to the transition metal and the absence of any anionic ligand in the syndiospecific catalysts (Co- and Ni-based ones). In the

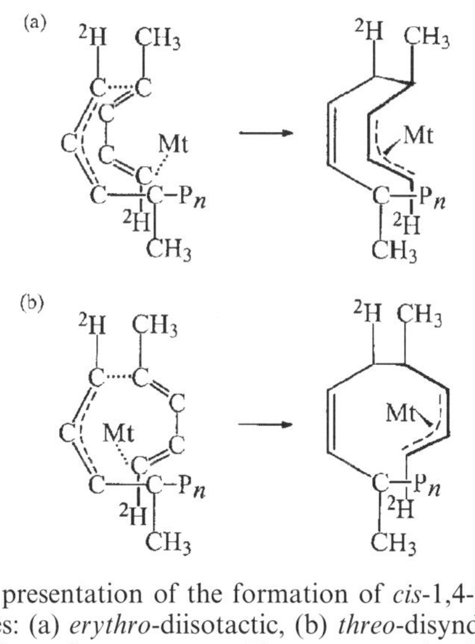

Figure 5.5 Schematic presentation of the formation of *cis*-1,4-polymers of terminally disubstituted butadienes: (a) *erythro*-diisotactic, (b) *threo*-disyndiotactic, depending on the orientation of the incoming monomer. The monomer is above the plane, the η^3-butenyl group is below and Mt is on the plane of the figure

presence of the ligand attached to the transition metal, the incoming monomer should be forced, for steric reasons, to assume the orientation in Figure 5.4a rather than that in Figure 5.4b. On the other hand, in the absence of any anionic ligand at the transition metal, i.e. in the case of cationic metal species, the orientation of the incoming monomer seems to be essentially determined by the interaction between the last inserted monomeric unit and the coordinating monomer; the orientation as in Figure 5.4b presumably minimises such an interaction.

An implication of the above mechanism of stereoregulation is that only some types of *cis*-1,4-stereoregular polymer can be obtained from terminally symmetrically disubstituted butadienes. Considering the stereoregulation mechanism operating in conjugated diene polymerisation systems, which is shown in Figures 5.4 and 5.5, it is obvious that the (*E*,*E*)-CHR=CH–CH=CHR monomer will give only *threo*-disyndiotactic and *erythro*-diisotactic *cis*-1,4-polymers, and not *threo*-diisotactic and *erythro*-diisyndiotactic polymers; for example, the stereochemistry of the formation of *cis*-1,4-*threo* monomeric units, via the *anti* form of the η^3-butenyl group being formed from the *s-cis*-η^4 complex, is presented in Figure 5.6 [7].

5.3.4.2 Isotactic and Syndiotactic 1,2-Polydiene Formation

One can see in Figure 5.3 that the formation of a *cis*-1,4 unit gives rise to a butenyl group of the same chirality as the preceding one; a *cis*-1,4 isotactic polymer is thus obtained. However, the formation of a 1,2 unit leads to a butenyl group of the opposite chirality as the previous one and hence to a syndiotactic polymer.

The formation of syndiotactic 1,2-polybutadiene [91,96] (and isotactic *cis*-1,4-polypentadiene [12,91]) is preferred with catalysts containing anionic ligands (e.g. with the $Ti(OR)_4$–AlR_3 catalyst). This can be interpreted in terms of the mode of monomer coordination that is shown in Figure 5.4a; reaction of the η^3-butenyl group via its C_1 atom with the coordinating monomer will give a *cis*-1,4-isotactic polymer, while reaction of the η^3-butenyl group

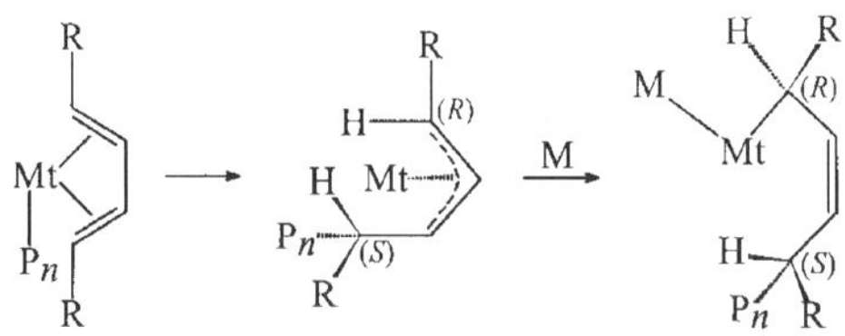

Figure 5.6 Stereochemistry of the formation of *cis*-1,4-*threo* monomeric units in the polymerisation of terminally symmetrically disubstituted conjugated dienes such as (*E*,*E*)-CHR=CH–CH=CHR

via its C_3 atom will result in the formation of a 1,2-syndiotactic polymer (Figure 5.3).

Isotactic 1,2-polybutadiene has so far been obtained only with soluble catalysts derived from Cr-based precursors and $AlEt_3$ as an activator at high Al/Cr molar ratios ($\geq$ 5) [12,26,91]. It is interesting that syndiotactic 1,2-polybutadiene is formed instead with Cr-based catalysts when applying catalyst components at a low Al/Cr molar ratio [26]. This has been interpreted [41] in terms of the presence or absence of an anionic ligand at the transition metal atom (Cr). In the catalysts prepared at lower Al/Cr ratios, some ligands remain bonded to the Cr atom, hence creating a situation of the coordinating monomer as in Figure 5.4a, which leads to 1,2-syndiotactic polybutadiene. At higher Al/Cr ratios, all the anionic ligands are presumably removed from the Cr atom, hence creating an arrangement of the coordinating monomer as in Figure 5.4b. Therefore, reaction of the monomer coordinating as in Figure 5.4b at the C_3 atom of the η^3-butenyl group leads to a 1,2-isotactic polymer (whereas reaction at the C_1 atom of the η^3-butenyl group leads to a *cis*-1,4-syndiotactic polymer).

The polymers obtained from unsymmetrically terminally disubstituted butadiene such as 4-methyl-1,3-pentadiene are made up of 1,2 monomeric units only, irrespective of the catalyst used; this is due to the presence of two methyl substituents at the C_4 atom in the monomer. Two stereoregular polymers have so far been obtained from 4-methyl-1,3-pentadiene, one with a 1,2-isotactic structure and one with a 1,2-syndiotactic structure. The isotactic polymer has been yielded by heterogeneous Ziegler–Natta catalysts, e.g. $TiCl_4$–$AlEt_3$ and $\alpha - TiCl_3$– [$AlEt_3$ [182]. The factors that determine the orientation of the coordinating monomer in this case are not, however, completely clear [41].

Syndiotactic 1,2-poly(4-methyl-1,3-pentadiene) has been formed by polymerisation with homogeneous catalysts, e.g. $TiBz_4$–$[Al(Me)O]_x$ and $CpTiCl_3$–$[Al(Me)O]_x$ [41,43]. The coordination of the monomer as an *s-trans*-η^2 ligand rather than an *s-cis*-η^4 ligand at the Ti atom has been postulated to be involved in the polymerisation. The *s-cis*-η^4 monomer coordination is less favoured for steric reasons in the case of 4-methyl-1,3-pentadiene. A possible scheme for the formation of the 1,2-syndiotactic polymer from this monomer is presented in Figure 5.7 [41,43].

The orientation of the coordinating monomer is determined by steric interaction with the last inserted monomeric unit. Note that, with the orientation shown in Figure 5.7, the coordinating monomer will give, after insertion, an η^3-butenyl group of opposite chirality with respect to the preceding one, thus leading to a 1,2-syndiotactic polymer.

It is interesting that the polymerisation of 4-methyl-1,3-pentadiene proceeds faster at $-20\,^\circ C$ than at $20\,^\circ C$. It has been suggested [43] that at a temperature below $0\,^\circ C$ this monomer coordinates as an *s-trans*-η^2 ligand. At a higher temperature it can probably coordinate either as an *s-trans*-η^2 ligand or as an *s-cis*-η^4 ligand, but monomeric units in the polymer are derived mainly from the

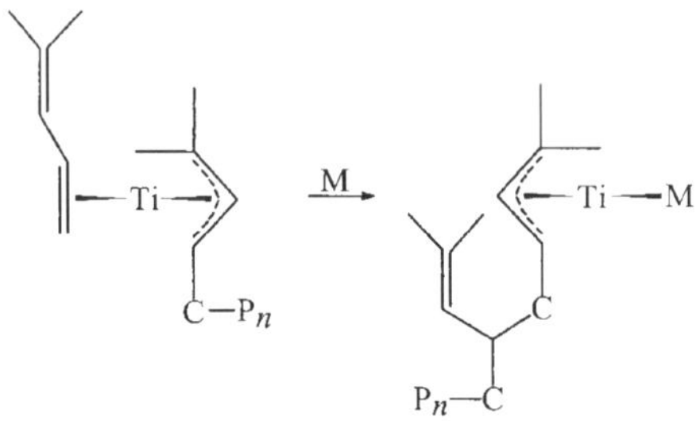

Figure 5.7 Schematic presentation of the formation of 1,2-syndiotactic poly(4-methyl-1,3-pentadiene)

s-trans-η^2-coordinated monomer, because the enchainment of such a monomerproceeds more rapidly than that of the *s-cis*-η^4-coordinated monomer. Occasional insertion of *s-cis*-η^4-coordinated 4-methyl-1,3-pentadiene may cause the appearance of steric defects, which are responsible for lower crystallinity of 1,2 poly(4-methyl-1,3-pentadiene) derived at a temperature higher than 20 °C.

It should be mentioned that the polymerisation of conjugated dienes with half-sandwich metallocene catalysts such as $CpTiCl_3$–$[Al(Me)O]_x$ affords polymers of lower stereoregularity than that of polymers formed using conventional transition metal catalysts. However, studies of polymerisations of conjugated diene monomers with metallocene catalysts seem to be very promising in order to resolve a challenging problem, which is to develop catalysts that achieve perfect stereoregularities in the polymerisation [42,115]. It is very probable that the well-defined molecular structure of these catalysts and the possibility of their easy modification will allow the synthesis of perfectly stereoregular polymers of conjugated dienes.

5.3.4.3 Isotactic and Syndiotactic trans-1,4-Polydiene Formation

As mentioned earlier, several stereoregular *trans*-1,4-polymers have been obtained from terminally substituted butadienes [20–22]. For instance, a symmetrically disubstituted butadiene such as (*E*,*E*)-CHR=CH–CH=CHR undergoes stereospecific polymerisation to yield *threo*-diisotactic and *erythro*-disyndiotactic *trans*-1,4-polymers, but not *erythro*-diisotactic and *threo*-disyndiotactic polymers [7]. One may note in this connection that only some types of stereoregular polymer can be formed from conjugated dienes, which is dueo stereo chemical implications of the stereoregulation mechanism operating in the polymerisation of these monomers. Figure 5.8 shows, for example, the stereochemistry of the formation of *trans*-1,4-*erythro* monomeric units via the *syn* form of the η^3-butenyl group formed from the *s-trans*-η^2 complex [7].

Let us recall that a *syn* form of the η^3-butenyl group can also be obtained by *anti* → *syn* isomerisation [scheme (10)]. The stereochemistry of this isomerisation is presented in Figure 5.9 [7].

One may note that the rearrangement of the *anti* form of an η^3-butenyl group occurs through a rotation around the C_2–C_3 bond in the intermediate η^1-allylic species and a reattachmnent of the double bond to the metal atom, with the other face, to yield a new η^3-butenyl group (*syn* form); the inversion of the configuration at the C_3 atom from *anti* to *syn* is accompanied with the inversion of the chirality at the C_1 atom.

Figure 5.10 shows schematically [41] the formation of *trans*-1,4-*threo*-diisotactic poly(2,4-hexadiene) by the *anti* → *syn* isomerisation of the last polymerised unit in a polymerisation run with the $Co(Acac)_2$–$AlEt_2Cl$ or Nd $(OCOR)_3$–Al Et_2Cl–Al(i-Bu)$_3$ catalyst [18,19].

The 2,4-hexadiene monomer coordinates as a *s–cis–*η^4 ligand, as does butadiene or isoprene, to give an *anti* form of the η^3-butenyl group after insertion. The orientation of the coordinating monomer is determined, in this case, by the interaction of the monomer itself with the last polymerised unit. As already mentioned above, the insertion of a new coordinating 2,4-hexadiene monomer is slow; this is due to the steric effect exerted by the methyl group. As a consequence, the last polymerised unit in an *anti* form is rearranged to the more stable *syn* form, prior to the insertion of the next incoming monomer, and hence a *trans*-1,4-*threo* monomeric unit is formed.

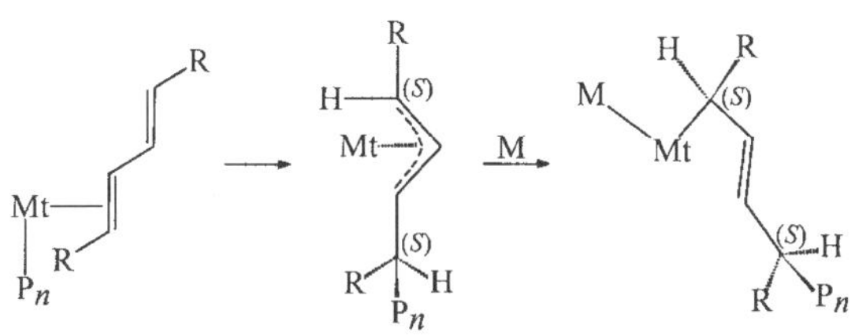

Figure 5.8 Stereochemistry of the formation of *trans*-1,4-*erythro* monomeric units in the polymerisation of terminally symmetrically disubstituted conjugated dienes such as (*E*,*E*)-CHR=CH–CH=CHR

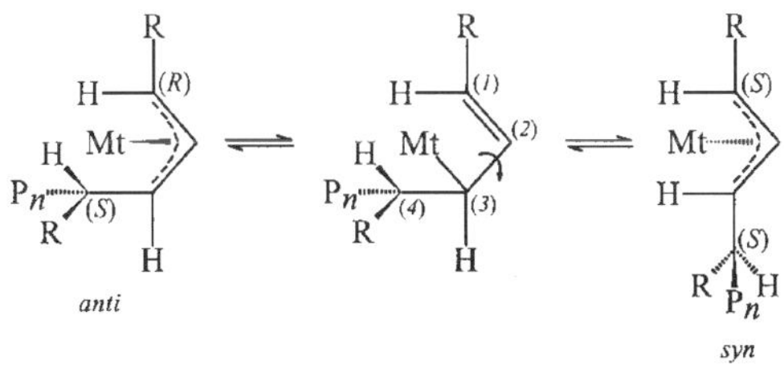

Figure 5.9 Stereochemistry of the *anti* → *syn* isomerisation of an η^3-butenyl group in the polymerisation of terminally symmetrically disubstituted conjugated dienes

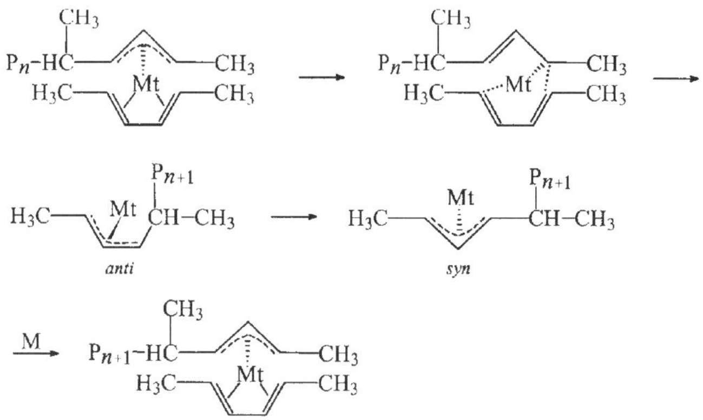

Figure 5.10 Schematic presentation of the formation of *trans*-1,4-*threo*-1,4-diisotactic poly[(*E*,*E*)-2,4-hexadiene] involving an *anti* → *syn* isomerisation

5.4 Copolymerisation

In general, copolymerisation of conjugated dienes with coordination catalysts is difficult because of the strong selectivity of the catalysts towards monomers of the particular type. The use of suitable catalysts, however, makes it possible to obtain copolymers of conjugated dienes with various monomers. Obtained copolymers can be random or alternating, depending on the kind of comonomers and catalysts.

5.4.1 Copolymerisation of 1,3-Butadiene with Higher Conjugated Dienes

Copolymers of butadiene with other conjugated dienes formed in the presence of coordination catalysts are characterised by a random distribution of comonomers [7].

In the copolymerisation of butadiene and isoprene with Ti-based catalysts, both monomeric units of the copolymers obtained are essentially of a *cis*-1,4 structure; the microstructure of monomeric units in the copolymers does not differ substantially from that in the homopolymers [196–198]. Nd-based catalysts provide butadiene/isoprene copolymers with more than 95% *cis*-1,4 monomeric units [89,199,200]. On the other hand, Co-based catalysts give copolymers in which the structure of the monomeric units depends markedly on copolymer composition [19,201,202]. Similarly, the structure of the monomeric units depends on copolymer composition in copolymers of butadiene and 2,3-dimethylbutadiene obtained by copolymerisation with Co-based catalysts [201,203].

The copolymerisation of butadiene and 1,3-pentadiene in the presence of V-based catalysts yields copolymers with a *trans*-1,4 structure of the butadiene units and a mixed *trans*-1,4/1,2 structure of the pentadiene units (in a ratio of 0.60–0.65/0.35–0.40 respectively) [61,204,205].

A good example illustrating the dependence of the structure of monomeric units on the copolymer composition is the already mentioned copolymerisation of butadiene with 2,4-hexadiene in the presence of $Co(Acac)_2$–$AlEt_2Cl$–H_2O and $Nd(OCOR)_3$–$AlEt_2Cl$–$Al(i\text{-}Bu)_3$ catalysts [190]. Let us recall that this copolymerisation yields copolymers containing hexadiene units with a *cis*-1,4 structure when they are isolated by butadiene units (in copolymers containing more than 90 mol.-% butadiene units) and with a *trans*-1,4 structure in the hexadiene homosequences. This has been attributed to the fact that 2,4-hexadiene insertion is more rapid in the case of a butadiene-terminated growing polymer chain (relatively low steric hindrances) than in the case of a hexadiene-terminated growing chain, so that *anti* → *syn* isomerisation of the η^3-allylic species (responsible for the formation of the *trans*-1,4 structure) does not occur.

5.4.2 Copolymerisation of Conjugated Dienes with Ethylene and α-Olefins

Copolymerisation of conjugated dienes and alkenes can lead to random or alternating copolymers, depending on the applied catalyst and copolymerisation conditions [7].

Random copolymers of butadiene, isoprene and/or pentadiene with ethylene and/or propylene have been obtained in the presence of various catalysts, mainly based on Ti or V compounds [206,207]. Statistical butadiene/ethylene copolymers can also be formed with zirconocene catalysts [162].

Of particular interest are, however, alternating copolymers of conjugated dienes (mainly butadiene and isoprene) and alkenes (mainly ethylene and propylene) [159,160,208–220]. Suitable catalysts for the alternating copolymerisation of conjugated dienes and alkenes are those derived from precursors based on Ti and V compounds. The butadiene units in the copolymers formed exhibit a predominant *trans*-1,4 structure, but their content usually does not exceed ca 85%. However, strictly alternating copolymers obtained with the bis(2,2-dimethylpropyloxy)vanadium oxychloride–triisobutylaluminium catalyst, $VO(ONp)_2Cl$–$Al(i\text{-}Bu)_3$, are highly stereoregular with more than 98% *trans*-1,4-butadiene units [210,211].

A scheme for the alternating copolymerisation of conjugated dienes and α-olefins has been proposed in the case of isoprene/propylene copolymerisation with V-based Ziegler–Natta catalysts [209]:

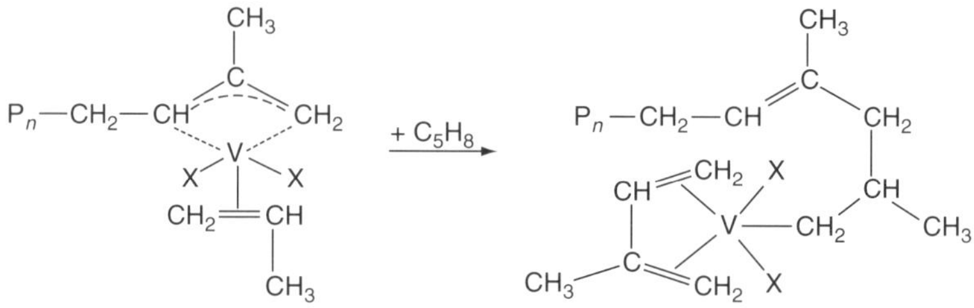 (13)

The insertion of the coordinating diene monomer proceeds via an η^3-allylic species. The metal atom with the attached copolymer chain terminated by such a species gives its two coordination sites to be occupied by the η^3-allylic group. This allows the next coordination of only one additional α-olefin. After insertion of the coordinating α-olefin, two coordination sites at the metal atom are again available, and they are preferably complexed by another conjugated diene molecule.

The first crystalline highly alternating butadiene/ethylene copolymers contained predominant *trans*-1,4-butadiene units. These copolymers were isolated from crude products of copolymerisation at lowered temperature (−25 °C) with soluble Ziegler–Natta catalysts containing weak Lewis bases, e.g. VCl_4–Al$(i\text{-Bu})_2$Cl–Al$(i\text{-Bu})_3$–PhOMe (1:2:2:2) [159]. Later on, highly alternating butadiene/ethylene copolymers were obtained with the $TiCl_4$–AlR_3 (1:2) catalysts at a temperature of 25 °C. The addition of Lewis bases such as amines to the polymerisation system enhanced the degree of alternation of the copolymers formed but reduced the activity of the catalysts. In this case, the alternating butadiene/ethylene copolymers contained mainly *cis*-1,4-butadiene units [208].

Alternating butadiene copolymers with propylene or higher α-olefins (1-butene, 1-pentene, 3-methyl-1-butene, 4-methyl-1-pentene, 1-dodecene) have been derived from copolymerisations run with Ziegler–Natta catalysts based on vanadium compounds, e.g. $VOCl_3$, VO$(ONp)_2$Cl, VCl_4 and VO$(Acac)_2$, as precursors and AlR_3 or AlR_2Cl as activators, in hexane or toluene at lowered temperatures ranging from −30 to −50 °C [160,210–218]. It is worth mentioning that the presence of halogen atoms in the catalysts (originated from the precursor or activator) is essential for the catalyst activity. Butadiene units in copolymers obtained with V-based catalysts exhibited prevailingly a *trans*-1,4 structure. By contrast, in copolymers obtained with Ti-based catalysts, *cis*-1,4-butadiene units were also present in small amounts.

Alternating copolymers of butadiene and propylene exhibit good elastomeric properties [213,221].

5.4.3 Copolymerisation of 1,3-Butadiene with Styrene

Although butadiene/styrene copolymers (manufactured by free radical polymerisation processes) are elastomers of great technical importance, rather little attention has been paid to the coordination copolymerisation of these monomers. Only recently has the copolymerisation of butadiene and styrene been found [222–224] to proceed in the presence of Ziegler–Natta catalysts derived from lanthanide-based precursors containing trichloroacetate ligands and organoaluminium activators, such as $Ln(OCOCCl_3)_3$–$AlEt_2Cl$–$Al(i\text{-}Bu)_3$ (1:2:25 molar ratio). It is interesting that the copolymerisation could be carried out in a hexane medium at elevated temperature of 50 °C. The high molecular weight butadiene/styrene copolymers obtained using equimolar amounts of comonomers in the feed contained 50–80 mol.-% butadiene units. The relative monomer reactivity ratios were evaluated as $r_1 = 3.1$ and $r_2 = 0.3$ for butadiene and styrene respectively (in copolymerisation with a Gd-based catalyst) [223].

The copolymers were characterised by the occurrence of butadiene units with a predominant *cis*-1,4 structure (73–79 %) and with *trans*-1,4 (8–18 %) and 1,2 (7–14 %) structures. The contents of *cis*-1,4-butadiene units in copolymers thus appeared to be lower than those in butadiene homopolymers yielded by $Ln(OCOCCl_3)_3$–$AlEt_2Cl$–$Al(i\text{-}Bu)_3$ catalysts under similar conditions. Moreover, spectroscopic ^{13}C NMR diad analysis of the butadiene/styrene copolymer obtained by copolymerisation with an Nd-based catalyst (the copolymer contained ca 78 mol.-% butadiene units) indicated the *cis*-1,4 fraction of butadiene units to be low in the styrene–butadiene diad sequence (*cis*-1,4/*trans*-1,4/1,2 ratios of 0.28/0.44/0.28) and high in the butadiene-butadiene diad sequence (*cis*-1,4/*trans*-1,4/1,2 ratios of 0.83: 0.15: 0.02). Therefore, the microstructure of the inserted butadiene unit was suggested to be influenced by the penultimate monomeric units; butadiene enchainment as *cis*-1,4 monomeric units was suggested to be induced by backbiting coordination. The proposed backbiting coordination model is shown in Figure 5.11 [223]. One can see in Figure 5.11 that the penultimate butadiene unit promotes the formation of *cis*-1,4-butadiene units, whereas the penultimate styrene unit interferes with butadiene *cis*-1,4-propagation and favours the formation of *trans*-1,4-butadiene units.

Furthermore, it was suggested that the microstructure of the butadiene unit of copolymers obtained with lanthanide-based catalysts was affected not only by the penultimate monomeric unit but also by the pen-penultimate unit,

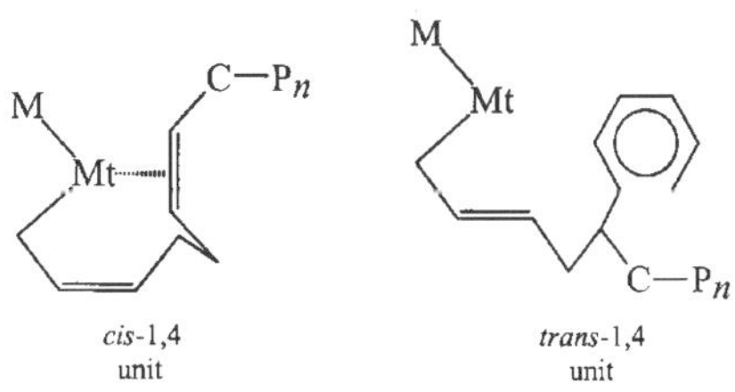

Figure 5.11 Schematic presentation of the effect of the kind of penultimate monomeric unit on the formation of a *cis*-1,4- or *trans*-1,4-butadiene unit in butadiene/styrene copolymerisation with a lanthanide-based catalyst

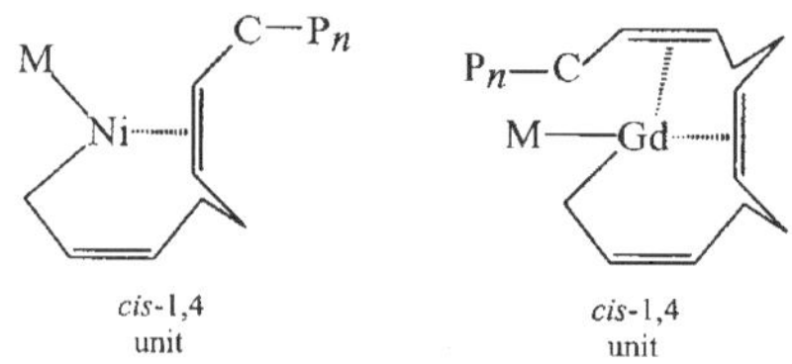

Figure 5.12 Schematic presentation of the backbiting coordination involving penultimate and pen-penultimate *cis*-1,4-butadiene units in butadiene/styrene copolymerisation with transition metal- and rare-earth metal-based catalysts

whereas that of the copolymers obtained with transition metal-based catalysts was affected only by the penultimate monomeric unit. Figure 5.12 shows the effects of the penultimate and pen-penultimate unit for microstructure control of the butadiene unit in copolymerisations run in the presence of Ni-based and Gd-based catalysts respectively [223].

The multidentated-type coordination in the case of the rare-earth metal-based catalysts may arise from 4f and 5d orbital features [225].

The order of activity of the $Ln(OCOCCl_3)_3$–$AlEt_2Cl$–$Al(i\text{-}Bu)_3$ catalysts in butadiene/styrene copolymerisation, which is $Nd > Gd > Pr > Dy \approx Yb$, appeared to be almost the same as that in butadiene homopolymerisation ($Nd > Pr > Gd \gg Dy \approx Yb$).

It has been suggested, therefore, that it is the kind of rare-earth metal that influences the reactivity of butadiene in homo- and copolymerisation rather than the kind of ligand [224].

Note that only the rare-earth metals with trichloroacetate ligands, i.e. $Ln(OCOCCl_3)_3$, were found to be suitable precursors to form (with organoaluminium activators) catalysts yielding homo- and copolymers of butadiene and styrene in relatively high yields (up to 67% under the copolymerisation conditions applied); catalysts with other acetate ligands (F_3CCOO, Cl_2CHCOO, $ClCH_2COO$ and CH_3COO) exhibited lower activity and hardly yielded copolymers of butadiene and styrene [223].

It has thus been suggested [223,224] that it is the kind of ligand rather than the kind of lanthanide that affects styrene reactivity in copolymerisation; the optimum ligand at the rare-earth metal atom, which appeared to be Cl_3CCOO, gave an adequate energy level of the catalyst, thus favouring a proper balance of the donation and back-donation processes between styrene and the rare-earth metal for the copolymerisation.

5.5 Industrial Polymerisation Processes

High molecular weight *cis*-1,4 polybutadiene is commercially produced using a solution butadiene polymerisation with aluminium alkyl-activated Ziegler–Natta catalysts based on Ti, Co and Ni precursors, and, more recently, catalysts based on lanthanides [226]. Nd-based catalysts, being the most active, are particularly favoured. The greater ionic radius and coordination capabilities of neodymium and other lanthanides when compared with transition metals not only provide high molecular weight stereoregular homopolymers in high yields but also make it possible to produce stereoregular copolymers of butadiene and other conjugated dienes [89].

Industrial polymerisation processes with the use of titanium-, cobalt- and nickel-based aluminium alkyl-activated Ziegler–Natta catalysts, which are employed for the manufacture of *cis*-1,4-polybutadiene, involve a solution polymerisation in low-boiling aromatic hydrocarbons such as toluene or in a mixture of aromatic and aliphatic hydrocarbons such as *n*-heptane or cyclohexane. The polymerisation is carried out in an anhydrous hydrocarbon solvent system. The proper ratio of butadiene monomer and solvent is blended and then completely dried in the tower, followed by molecular sieves. The alkylaluminium activator is added, the mixture is agitated and then the transition metal precatalyst is introduced. This blend then passes through a series of reactors in a cascade system in which highly exothermic polymerisation occurs. Therefore, the reaction vessels are cooled to slightly below room temperature.

In a polymerisation process run with the use of the $CoCl_2$–$Al(i\text{-}Bu)_3$ or $CoCl_2$–$AlEt_2Cl$ catalyst, the reaction temperature is lowered to ca 10 °C and the butadiene pressure is ca 3 atm. The residence time of the monomer is 5 h to achieve its 80 % conversion. In some processes in which the polymer molecular weight is regulated by the addition of hydrogen or another agent, the use of a series of reactors allows variation in the hydrogen concentration through the operating steps in order to control the molecular weight distribution in the polymers formed.

A flow scheme of *cis*-1,4-polybutadiene production involving polymerisation with cobalt-based Ziegler–Natta catalysts in a solution process with the removal of catalyst residues from the polymer is presented in Figure 5.13 [227].

The post-polymerisation mixture leaving the reactors contains about 15 % rubber solids dissolved in the solvent. This mixture is treated with alcohol to stop the catalyst action and is then blended in large holding tanks to ensure uniformity. Here, processing chemicals may also be added if desired, which makes the final product as easy to process as natural rubber. From the holding tanks, the mixture passes to a tank where catalyst residues are removed and dumped. Then, in a series of agitated vessels, water and steam are added to coagulate the rubber and to strip out unreacted monomer and solvent, which are recycled. The *cis*-1,4-polybutadiene elastomer is obtained as a crumb that is dewatered on a screen and then reslurried. This apparently redundant slurrying should be done in order to remove all traces of hydrocarbons before the crumb can be dried safely.

As regards *cis*-1,4-polyisoprene manufacture, this is done using the same apparatus and according to analogous procedure.

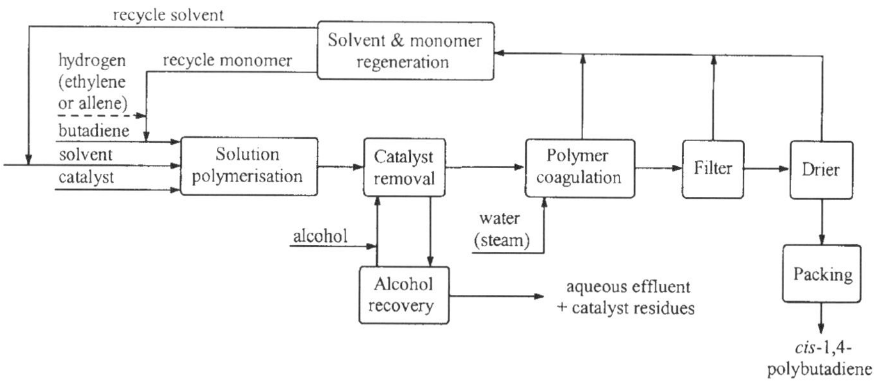

Figure 5.13 Flow scheme of polybutadiene production using the lowered-temperature solution polymerisation process

The development of Nd-based catalysts has permitted the industrial realisation of a polymerisation process for *cis*-1,4-polybutadiene in which temperature varies in the range 20–50 °C. In such a process, the polymerisation proceeds relatively fast; the average residence time of the monomer varies in the range 0.5–4 h to achieve 90 % butadiene conversion.

Nd-based Ziegler–Natta catalysts also make it possible to produce polybutadiene with especially high contents of *cis*-1,4 monomeric units, which improves its fracture resistance. This is due to the ability of high *cis*-1,4-polybutadiene to undergo strain-induced crystallisation which, acting as a reinforcing mechanism, enhances the fatigue resistance of polybutadiene-based elastomers. The polymer, obtained using Nd-based Ziegler–Natta catalysts, exhibits superior physical and processing properties compared with polymers produced by Ti-, Co- and Ni-based catalysts.

High *cis*-1,4-polybutadiene is manufactured on a large industrial scale and occupies a well-defined position in the elastomers market. It is employed mainly in the tyre industry, where it is blended with natural rubber and/or with styrene–butadiene rubber and applied in either sidewalls, threads or rims of tyres. It should be noted in this connection that natural rubber, in contrast to its synthetic counterpart, displays some physical properties that appear to be useful in the manufacture of tyres for heavy-duty machines. The fact is that some non-hydrocarbon substances appearing in natural rubber in small amounts (such as polypeptides) protect the high-dimensional tyre formed against collapsing prior to the vulcanisation process and thus enable a high-quality product to be obtained.

By contrast, *cis*-1,4-polyisoprene is produced in limited amounts, since it is not price competitive with natural rubber (owing to the relatively high costs of manufacturing the isoprene monomer). The same applies to *trans*-1,4-polyisoprene, which is more expensive than its natural counterparts – gutta percha and balata.

As regards high *trans*-1,4-polybutadiene, it has a few applications, especially as a blend with natural rubber. Syndiotactic 1,2-polybutadiene is a unique material that combines the properties of plastic and rubber. These properties lead to applications both as a thermoplastic resin and as a rubber. As regards isotactic 1,2-polybutadiene, one may note that its properties have not excited sufficient interest for commercial development.

Some of the most important applications of the mentioned polymers of conjugated dienes are listed in Table 5.6 [228].

Table 5.6 Commercially available polymers of butadiene and isoprene, produced by coordination polymerisation, and their typical uses

Polymer	Typical use
cis-1,4-Polybutadiene	Tyres, conveyer belts, hoses, wire and cable insulation, footwear, seals, mounts
cis-1,4-Polyisoprene	Tyres, footwear, adhesives, coated fabrics, hoses, mounts
trans-1,4-Polybutadiene	Tyres (blending with natural rubber)
trans-1,4-Polyisoprene	Protective coatings, adhesives
syndiotactic 1,2-Polybutadiene	Films, footwear, tubes, hoses, sponges, gloves

References

1. Meissner, B., Schätz, M. and Brajko, V., 'Synthetic Rubbers', in *Elastomers and Rubber Compounding Materials*, Elsevier Science, Amsterdam, 1989, pp. 76–299.
2. Morton, A. A., *Ind. Eng. Chem., Prod. Res. Dev.*, **42**, 1488 (1950).
3. Morton, A. A., Nelidow, I. and Schoenberg, E., *Rubber Chem. Technol.*, **30**, 426 (1957).
4. Adams, H. E., Bebb, R. L., Foreman, L. E. and Wakefield, L. B., *Rubber Chem. Technol.*, **45**, 1252 (1972).
5. Patterson, D. B. and Halasa, A. F., *Macromolecules*, **24**, 4489 (1991).
6. Arest-Yakubovich, A. A., *Macromol. Symp.*, **85**, 279 (1994).
7. Porri, L. and Giarrusso, A., 'Conjugated Diene Polymerisation', in *Comprehensive Polymer Science*, Pergamon Press, Oxford, 1989, Vol. 4, pp. 53–108.
8. Natta, G., Porri, L. and Valenti, S., *Makromol. Chem.*, **67**, 225 (1963).
9. Costa, G., Locatelli, P. and Zambelli, A., *Macromolecules*, **6**, 653 (1973).
10. Farina, M., *Top. Stereochem.*, **17**, 1 (1987).
11. Natta, G., Porri, L., Stoppa, G., Allegra, G. and Ciampelli, F., *J. Polym. Sci., B*, **1**, 67 (1963).
12. Natta, G., Porri, L., Carbonaro, A. and Stoppa, G., *Makromol. Chem.*, **77**, 114 (1964).
13. Natta, G., Porri, L., Corradini, P., Zanini, G. and Ciampelli, F., *J. Polym. Sci.*, **51**, 463 (1961).
14. Porri, L. and Carbonaro, A., *Makromol. Chem.*, **60**, 236 (1963).
15. Natta, G., Porri, L., Carbonaro, A., Ciampelli, F. and Allegra, G., *Makromol. Chem.*, **51**, 229 (1962).
16. Ricci, G., Italia, S. and Porri, L., *Macromolecules*, **27**, 868 (1994).

17. Porri, L., Di Corato, A. and Natta, G., *Eur. Polym. J.*, **5**, 1 (1969).
18. Destri, S., Gallazzi, M. C., Giarrusso, A. and Porri, L., *Makromol. Chem., Rapid Commun.*, **1**, 293 (1980).
19. Wang, F., Bolognesi, A., Immirzi, A. and Porri, L., *Makromol. Chem.*, **182**, 3617 (1981).
20. Porri, L. and Aglietto, M., *Makromol. Chem.*, **177**, 1465 (1976).
21. Destri, S., Gatti, G. and Porri, L., *Makromol. Chem., Rapid Commun.*, **2**, 605 (1981).
22. Destri, S., Bolognesi, A., Porri, L. and Wang, F., *Makromol. Chem., Rapid Commun.*, **3**, 187 (1982).
23. Horne Jr, S. E., Kiehl, J. P., Shipman, J. J., Folt, V. L., Gibbs, C. F., Wilson, E. A., Newton, E. B. and Reinhart, M. A., *Ind. Eng. Chem., Prod. Res. Dev.*, **48**, 784 (1956).
24. Natta, G., Corradini, P. and Porri, L., *Rend. Accad. Naz. Lincei*, **8**, 728 (1956).
25. Natta, G. and Porri, L., Ital. Pat. 538 453 (to Montecatini SpA) (1956); *Chem. Abstr.*, **52**, 5032 (1958).
26. Natta, G., Porri, L., Zanini, G. and Palvarini, A., *Chim. Ind. (Milan)*, **41**, 1163 (1959).
27. Rinehart, R. E., Smith, H. P., Witt, H. and Romeyn Jr, H., *J. Am. Chem. Soc.*, **83**, 4864 (1961).
28. Rinehart, R. E., Smith, H. P., Witt, H. and Romeyn Jr. H., *J. Am. Chem. Soc.*, **84**, 4145 (1962).
29. Porri, L., Gallazzi, M. C. and Vitulli, G., *J. Polym. Sci., B*, **5**, 629 (1967).
30. Dawans, F. and Teyssié, P., *J. Polym. Sci., B*, **3**, 1045 (1965).
31. Otsu, T. and Yamaguchi, M., *J. Polym. Sci., A-1*, **7**, 387 (1969).
32. Tinyakova, N. I., Alferov, A. V., Golenko, T. G., Dolgoplosk, B. A., Oreshkin, I. A., Sharaev, O. K., Chernenko, G. N. and Yakovlev, V. A., *J. Polym. Sci., C*, **16**, 2625 (1967).
33. Anderson, W. S., *J. Polym. Sci., A-1*, **5**, 429 (1967).
34. Wen, T. C., Chang, C. C., Chuang, Y. D., Chin, J. P. and Chang, C. T., *J. Am. Chem. Soc.*, **103**, 4576 (1981).
35. Babitskii, B. D., Kormer, V. A., Lapuk, I. M. and Skoblikova, V. I., *J. Polym. Sci., C*, **16**, 3219 (1968).
36. Porri, L., Natta, G. and Gallazzi, M. C., *Chim. Ind. (Milan)*, **46**, 428 (1964).
37. Wilke, G., Bogdanovic, B., Hardt, P., Heimbach, P., Keim, W., Kromer M., Oberkirch, W., Tanaka, K., Steinrucke, E., Walter, D. and Zimmermann, H., *Angew. Chem., Int. Ed. Engl.*, **5**, 151 (1966).
38. Zambelli, A., Ammendola, P. and Proto, A., *Macromolecules*, **22**, 2125 (1989).
39. Oliva, L., Longo, P., Grassi, A. and Zambelli, A., *Makromol. Chem., Rapid Commun.*, **11**, 519 (1990).
40. Ricci, G., Italia, S., Comitani, C. and Porri, L., *Polym. Commun.*, **17**, 514 (1991).
41. Porri, L., Giarrusso, A. and Ricci, G., *Makromol. Chem., Macromol. Symp.*, **48/49**, 239 (1991).
42. Ricci, G., Italia, S., Giarrusso, A. and Porri, L., *J. Organomet. Chem.*, **451**, 67 (1993).
43. Ricci, G., Porri, L. and Giarrusso, A., *Macromol. Symp.*, **89**, 383 (1995).
44. Rafikov, S. R., Monakov, Yu. B., Tolstikov, G. A., Marina, N. G., Minchenkova, N. Kh. and Sevel'eva, I. G., *J. Polym. Sci., Polym. Chem. Ed.*, **21**, 2697 (1983).
45. Minsker, K. S., Karpasa, M. M., Monakov, Yu. B. and Zaikov, G. E., *Eur. Polym. J.*, **21**, 973 (1985).
46. Natta, G., Porri, L. and Fiore, L., *Gazz. Chim. Ital.*, **89**, 761 (1959).
47. Smith, G. H. and Perry, D. C., *J. Polym. Sci., A-1*, **7**, 707 (1969).

48. Henderson, J. F., *J. Polym. Sci., C*, **4**, 233 (1963).
49. Mayer, P. H., *J. Polym. Sci., A*, **3**, 209 (1965).
50. Mayer, P. H. and Lehr, M. H., *J. Polym. Sci., A*, **3**, 217 (1965).
51. Bresler, S. L., Grechanovskii, G. A., Muzhai, A. and Poddubnyi, I. Ya., *Makromol. Chem.*, **133**, 111 (1970).
52. Marconi, W., Araldi, M., Beranger, A. and De Maldé, M., *Chim. Ind. (Milan)*, **45**, 522 (1963).
53. Ivanova, A. M., Lukmanova, R. Z. and Grechishnikov, Yu. G., *Khim. Vysokomol. Soed., Neftkhim.*, **1975**, 65 (1975); *Chem. Abstr.*, **85**, 63423 (1976).
54. Monakov, Yu. B., Marina, N. G., Savel'eva, I. G. and Rafikov, S. R., *Vysokomol. Soed., Ser. A*, **23**, 50 (1981); *Chem. Abstr.*, **94**, 140246 (1981).
55. Cucinella, S., Mazzei, A., Marconi, W. and Busetto, C., *J. Macromol. Sci. – Chem. A*, **4**, 1549 (1970).
56. Nakatomi, S., *Nippon Kagaku Kaishi*, **1976**, 29 (1976); *Chem. Abstr.*, **85**, 6105 (1976).
57. Natta, G., Porri, L., Corradini, P. and Morero, D., *Chim. Ind. (Milan)*, **40**, 362 (1958).
58. Natta, G., Porri, L. and Mazzei, A., *Chim. Ind. (Milan)*, **41**, 116 (1959).
59. Lasky, J. S., Garner, H. K. and Ewart, E. H., *Ind. Eng. Chem., Prod. Res. Dev.*, **1**, 82 (1962).
60. Froehlich, H. O. and Kaholdt, H., *Z. Chem.*, **15**, 365 (1975).
61. Porri, L., Carbonaro, A. and Ciampelli, F., *Makromol. Chem.*, **61**, 90 (1963).
62. Longiave, C., Castelli, R. and Croce, G. F., *Chim. Ind. (Milan)*, **43**, 625 (1961).
63. Dolgoplosk, B. A., Erussalimskii, B. L., Kropatcheva, E. N. and Tinyakova, E. I., *J. Polym. Sci.*, **58**, 1333 (1962).
64. Gippin, M., *Ind. Eng. Chem., Prod. Res. Dev.*, **4**, 160 (1965).
65. Byrikhin, V. S., Tverskoi, V. A., Potapov, S. S., Luzina, N. N. and Fedorov, N. P., *Polym. Sci. USSR*, **16**, 1222 (1974).
66. Volkov, L. A., Chernova, I. D., Radushnova, I. L., Robysheva, V. T. and Lysova, V. I., *Polym. Sci. USSR*, **18**, 1701 (1976).
67. Takahashi, A. and Kambara, S., *J. Polym. Sci., B*, **3**, 279 (1965).
68. Storr, A., Jones, K. and Laubengayer, A. W., *J. Am. Chem. Soc.*, **90**, 3173 (1978).
69. Golubeva, N. D., Pomagailo, A. D., Kuzaev, A. I., Ponomarev, A. I. and Dyachkovskii, F. S., *J. Polym. Sci., Polym. Symp.*, **68**, 33 (1980).
70. Soga, K. and Yamamoto, K., *Polym. Bull.*, **6**, 263 (1982).
71. Lee, D. H. and Hsu, C. C., *J. Appl. Polym. Sci.*, **25**, 2373 (1980).
72. Throckmorton, M. C. and Farson, M. S., *Rubber Chem. Technol.*, **45**, 268 (1972).
73. Throckmorton, M. C., *Kautsch. Gummi Kunstst.*, **22**, 293 (1969).
74. Lugli, G., Mazzei, A. and Poggio, S., *Makromol. Chem.*, **175**, 2021 (1974).
75. Bruzzone, M., Mazzei, A. and Giuliani, G., *Rubber Chem. Technol.*, **47**, 1175 (1974).
76. De Chirico, A., Lanzani, P. C., Poggi, E. and Bruzzone, M., *Makromol. Chem.*, **175**, 2029 (1974).
77. Shen, Z., Quyang, J., Wang, F., Hu, Z., Yu, F. and Qian, B., *J. Polym. Sci., Polym. Chem. Ed.*, **18**, 3345 (1980).
78. Monakov, Yu. B., Marina, N. G., Savel'eva, I. G., Zhiber, L. E., Kozlov, V. G. and Rafikov, R. S., *Dokl. Akad. Nauk SSSR*, **265**, 1431 (1982); *Chem. Abstr.*, **98**, 54523 (1983).
79. Kaita, S., Kobayashi, E., Sakakibara, S., Aoshima, S. and Furukawa, J., *J. Polym. Sci., A, Polym. Chem.*, **34**, 3431 (1996).
80. Marina, N. G., Monakov, Yu. B., Rafikov, S. R. and Gadeleva, K. H., *Polym. Sci. USSR*, **26**, 1251 (1984).

81. Mazzei, A., *Makromol. Chem., Suppl.*, **4**, 61 (1981).
82. Nickaf, J. B., Burford, R. P. and Chaplin, R. P., *J. Polym. Sci., A, Polym. Chem.*, **33**, 1125 (1995).
83. Wilson, D. J. and Jenkins, D. K. *Polym. Bull.*, **27**, 407 (1992).
84. Chigir, N. N., Sharaev, O. K., Tinyakova, E. I. and Dolgoplosk, B. A., *Vysokomol. Soed., Ser. B*, **25**, 47 (1983); *Chem. Abstr.*, **98**, 126695 (1983).
85. Ricci, G., Italia, S., Cabassi, F. and Porri, L., *Polym. Commun.*, **28**, 223 (1987).
86. Ricci, G., Boffa, G. and Porri, L., *Makromol. Chem., Rapid Commun.*, **7**, 355 (1986).
87. Li, Y. L. and Quyang, J., *Acta Polym. Sinica*, **1**, 39 (1988).
88. Li, X., Sun, Y. and Jin, Y., *Acta Chim. Sinica*, **44**, 1163 (1986).
89. Hsieh, H. and Yeh. H., *Rubber Chem. Technol.*, **58**, 117 (1985).
90. Natta, G., Porri, L. and Mazzei, A., *Chim. Ind. (Milan)*, **41**, 398 (1959).
91. Natta, G., Porri, L. and Carbonaro, A., *Makromol. Chem.*, **77**, 126 (1964).
92. Cooper, W. and Vaughan, G., *Prog. Polym. Sci.*, **1**, 128 (1967).
93. Saltman, W. M. and Link, T. H., *Ind. Eng. Chem., Prod. Res. Dev.*, **3**, 199 (1964).
94. Kraus, G., Short, J. and Thornton, V., *Rubber Plast. Age*, **38**, 880 (1957).
95. Marconi, W., Mazzei, A., Araldi, A. and De Maldé, M., *J. Polym. Sci., A*, **3**, 735 (1965).
96. Wilke, G., *J. Polym. Sci.*, **38**, 45 (1959).
97. Natta, G., Porri, L., Fiore, L. and Zanini, G., *Chim. Ind. (Milan)*, **40**, 116 (1958).
98. Natta, G., Porri, L. and Carbonaro, A., *Atti Accad. Naz. Lincei, Cl. Sci. Fis. Mat. Nat. Rend.*, **31**(8), 189 (1961).
99. Gippin, M., *Ind. Eng. Chem., Prod. Res. Dev.*, **1**, 32 (1962).
100. Longiave, C. and Castelli, R., *J. Polym. Sci., C*, **4**, 387 (1963).
101. Susa, E., *J. Polym. Sci., C*, **4**, 399 (1963).
102. Ashitaka, H., Ishikawa, H., Ueno, H. and Nagasaka, A., *J. Polym. Sci., Polym. Chem. Ed.*, **21**, 1853 (1983).
103. Ashitaka, H., Jinda, K. and Ueno, H., *J. Polym. Sci., Polym. Chem. Ed.*, **21**, 1989 (1983).
104. Sakata, R., Hosono, J., Onishi, A. and Ueda, K., *Makromol. Chem.*, **139**, 73 (1970).
105. Beebe, D. H., Gordon, C. E., Thudium, R. N., Throckmorton, M. C. and Hanlon, T. L., *J. Polym. Sci., Polym. Chem. Ed.*, **16**, 2285 (1978).
106. Zachoval, J. and Vernovic, B., *J. Polym. Sci., B*, **4**, 965 (1966).
107. Razuvaev, G. A., Minsker, K. S., Fedoseeva, G. T. and Savel'ev I. A., *Vysokomol. Soed.*, **1**, 1691 (1959); *Chem. Abstr.*, **54**, 19015g (1960).
108. Bawn, C. E. H., North, A. M. and Walker, J. S., *Polymer*, **5**, 419 (1964).
109. Cucinella, S., *Chim. Ind. (Milan)*, **59**, 696 (1977).
110. Brock, M. J. and Hackathorn, M. J., *Rubber Chem. Technol.*, **45**, 1303 (1972).
111. Dolgoplosk, B. A., Kropacheva, E. N., Khremnikova, E. K., Kuznetzova, E. I. and Golodova, K. G., *Dokl. Akad. Nauk SSSR*, **135**, 847 (1960); *Chem. Abstr.*, **55**, 13292f (1960).
112. Yoshimoto, T., Kamatsuki, K., Sakata, R., Yamamoto, K., Takeuchi, Y., Onishi, A. and Ueda, K., *Makromol. Chem.*, **139**, 61 (1970).
113. Panasenko, A. A., Odinakov, V. N., Monakov, Yu. B., Khalilov, L. M., Bezgina, A. S., Ignatyuk, V. K. and Rafikov, S. R., *Vysokomol. Soed., Ser. B*, **19**, 656 (1977).
114. Porri, L., Giarrusso, A. and Ricci, G., *Prog. Polym. Sci.*, **16**, 405 (1991).
115. Porri, L., Giarrusso, A. and Ricci, G., *Makromol. Chem., Macromol. Symp.*, **66**, 231 (1993).
116. Ricci, G. and Porri, L., *Polymer*, **38**, 4499 (1997).
117. Zambelli, A., Pellecchia, C. and Proto, A., *Macromol. Symp.*, **89**, 373 (1995).
118. Skupinski, W. and Walczuk, P., *Polimery*, **31**, 393 (1986).

119. Skupinski, W., Cieslowska-Glinska, I. and Wasilewski, A., *J. Mol. Catal.*, **33**, 129 (1985).
120. Dawans, F. and Teyssié, P., *Ind. Eng. Chem., Prod. Res. Develop.*, **10**, 261 (1971).
121. Matsumoto, T. and Furukawa, J., *J. Polym. Sci., B*, **5**, 935 (1967).
122. Dolgoplosk, B. A., *Polym. Sci. USSR*, **13**, 367 (1971).
123. Bourdauducq, B. P. and Dawans, F., *J. Polym. Sci., A-1*, **10**, 2527 (1972).
124. Maréchal, J. C., Dawans, F. and Teyssié, P., *J. Polym. Sci., A-1*, **8**, 1993 (1970).
125. Dawans, F., Durand, J. P. and Teyssié, P., *J. Polym. Sci., B*, **10**, 493 (1972).
126. Yakovlev, V. A., Dolgoplosk, B. A., Tinyakowa, E. I. and Yakovleva, O. N., *Vysokomol. Soed., Ser. A*, **11**, 1645 (1969); *Chem. Abstr.*, **71**, 102236q (1969).
127. Taube, R., Gehrke, J. P. and Schmidt, U., *J. Organomet. Chem.*, **292**, 287 (1985).
128. Durand, J. P., Dawans, F. and Teyssié, P., *J. Polym. Sci., A-1*, **8**, 979 (1970).
129. Durand, J. P. and Teyssié, P., *J. Polym. Sci., B*, **6**, 229 (1968).
130. Golenko, T. G., *Izv. Akad. Nauk., Ser. Khim.* **1968**, 2271.
131. Durand, J. P., Dawans, F. and Teyssié, P., *J. Polym. Sci., B*, **5**, 785 (1967).
132. Taube, R., Wache, S. and Sieler, J., *J. Organomet. Chem.*, **459**, 335 (1993).
133. Dawans, F. and Teyssié, P., *Compt. Rend. Acad. Sci. (Paris)*, **263C**, 1512 (1966).
134. Yakovlev, V. A., Dolgoplosk, B. A., Makovetskii, K. L. and Tinyakova, E. I., *Dokl. Akad. Nauk SSSR*, **187**, 354 (1969).
135. Witte, J., in *Houben Weyl Methoden der Organischen Chemie: Makromolekulare Stoffe*, Georg Thieme Verlag, Stuttgart, 1986, Vol. E 20/1, p. 94.
136. Oreshkin, I. A., Ostrovskaya, I. Ya., Yakovlev, V. A., Tinyakova, E. I. and Dolgoplosk, B. A., *Dokl. Akad. Nauk SSSR*, **173**, 1349 (1967).
137. Dolgoplosk, B. A. and Tinyakova, E. I., *Izv. Akad. Nauk SSSR, Ser. Khim.*, **1970**, 344 (1970).
138. Oreshkin, I. A., Chernenko, G. M., Tinyakova, E. I. and Dolgoplosk, B. A., *Dokl. Akad. Nauk SSSR*, **169**, 1102 (1966).
139. Dolgoplosk, B. A., Oreshkin, I. A., Tinyakova, E. I. and Yakovlev, V. A., *Izv. Akad. Nauk SSSR, Ser. Khim.*, **1967**, 2130 (1967).
140. Oreshkin, I. A., Tinyakova, E. I. and Dolgoplosk, B. A., *Polym. Sci. USSR*, **11**, 2096 (1969).
141. Dolgoplosk, B. A. and Tinyakova, E. I., *Polym. Sci. USSR*, **19**, 2805 (1977).
142. Jolly, P. W., Krueger, C., Romao, C. C. and Romao, M. J., *Organometallics*, **3**, 936 (1984).
143. Ter-Minasyan, R. I., Parenago, O. P., Frolov, V. M. and Dolgoplosk, B. A., *Dokl. Akad. Nauk SSSR*, **214**, 824 (1974).
144. Ter-Minasyan, R. I., Parenago, O. P., Chirkova, V. G., Frolov, V. M. and Dolgoplosk, B. A., *Kinet. Katal.*, **17**, 935 (1976).
145. Wilke, G., *Angew. Chem., Int. Ed. Engl.*, **2**, 105 (1963).
146. Taube, R., Windisch, H. and Maiwald, S., *Macromol. Symp.*, **89**, 393 (1995).
147. Witte, J., *Angew. Makromol. Chem.*, **94**, 119 (1981).
148. Dolgoplosk, B. A., Tinyakova, E. I., Sefanovskaya, N. N., Oreshkin, I. A. and Shmonina, V. L., *Eur. Polym. J.*, **10**, 605 (1974).
149. Skupinski, W., Zawartke, M. and Malinowski, S., *React. Kinet. Catal. Lett.*, **13**, 319 (1980).
150. Dauby, R., Dawans, F. and Teyssié, P., *J. Polym. Sci., C*, **16**, 1989 (1967).
151. Ochiai, E., Hirai, R. and Makishima, S., *J. Polym. Sci., B*, **4**, 1003 (1966).
152. Morton, M. and Das, B., *J. Polym. Sci., C*, **27**, 1 (1969).
153. Sokolov, V. N., Babitskii, B. D., Kormer, V. A. and Poddubnyi, I. Ya., *J. Polym. Sci., C*, **16**, 4351 (1967).
154. Kormer, V. A., Babitskii, B. D., Lobach, M. I. and Chesnokova, N. N., *J. Polym. Sci., C*, **16**, 4351 (1967).

155. Scott, H., Frost, R. E., Belt, R. F. and O'Reilly, D. E., *J. Polym. Sci., A*, **2**, 3233 (1964).
156. Scott, H., *J. Polym. Sci., B*, **4**, 105 (1966); *J. Polym. Sci., A*, **2**, 3233 (1964).
157. O'Reilly, D. E., Poole Jr, C. D., Belt, R. F. and Scott, H., *J. Polym. Sci., A*, **2**, 3257 (1964).
158. Jenkins, D. K., Timms, D. G. and Duck, W. E., *Polymer*, **7**, 419 (1966).
159. Natta, G., Zambelli, A., Pasquon, I. and Ciampelli, F., *Makromol. Chem.*, **79**, 161 (1964).
160. Kawasaki, A., Maryuma, I., Taniguchi, M., Hirai, R. and Furukawa, J., *J. Polym. Sci., Polym. Lett. Ed.*, **7**, 613 (1969).
161. Natta, G. and Danusso, F., *Chim. Ind. (Milan)*, **40**, 445 (1958).
162. Kaminsky, W. and Schlobom, M., *Makromol. Chem., Macromol. Symp.*, **4**, 103 (1986).
163. Lobach, M. I., Kormer, V. A., Tsereteli, I. Yu., Kondratenkov, G. A., Babitskii, B. D. and Klepikova, V. I., *J. Polym. Sci., Polym. Lett. Ed.*, **9**, 71 (1971).
164. Sokolov, V. N., Khvostik, G. M., Poddubnyi, I. Ya. and Kondratenkov, G. A., *J. Organomet. Chem.*, **29**, 313 (1971).
165. Hughes, R. P., Jack, T. and Powell, J., *J. Am. Chem. Soc.*, **94**, 7723 (1972).
166. Klepikova, V. I., Erussalimskii, G. B., Lobach, M. I., Churlaeva, L. A. and Kormer, V. A., *Macromolecules*, **9**, 217 (1976).
167. Bawn, C. E. H., *Rubber Plast. Age*, **46**, 510 (1965).
168. Dawes, D. H. and Winkler, C. A., *J. Polym. Sci., A*, **2**, 3029 (1964).
169. Gibbs, C. F., Horne Jr, S. E., Macey, J. H. and Tucker, H., *Rubber World*, **144**, 69 (1961).
170. Yamazaki, N., Suminoe, T. and Kambara, S., *Makromol. Chem.*, **65**, 157 (1963).
171. Zachoval, J., Mikes, F., Krepelka, J., Prouzova, O. and Pradova, O., *Eur. Polym. J.*, **8**, 397 (1972).
172. Cooper, W., Eaves, D. E., Owen, G. D. T. and Vaughan, G., *J. Polym. Sci., C*, **4**, 211 (1964).
173. Longiave, C., Castelli, R. and Ferraris, M., *Chim. Ind. (Milan)*, **44**, 725 (1962).
174. Dubini, M., Longiave, C. and Castelli, R., *Chim. Ind. (Milan)*, **45**, 923 (1963).
175. Thomassin, J. M., Walkiers, E., Warrin, R. and Teyssié, P., *J. Polym. Sci., Polym. Chem. Ed.*, **13**, 1147 (1975).
176. Harrod, J. T. and Wallace, L. R., *Macromolecules*, **5**, 686 (1972).
177. Saltman, W. M., Gibbs, W. E. and Lal, J., *J. Am. Chem. Soc.*, **80**, 5615 (1958).
178. Pantukh, B. I., Rozentsvet, V. A., Monakov, Yu. B. and Rafikov, S. R., *Acta Polym.*, **34**, 732 (1983).
179. Medvedev, S. S., *Russ. Chem. Rev.*, **37**, 834 (1968).
180. Pakuro, N. I., Zabolotskaya, E. V. and Medvedev, S. S., *Vysokomol. Soed., Ser. B*, **10**, 3 (1968); *Chem. Abstr.*, **68**, 69914x (1968).
181. Gallazzi, M. C., Giarrusso, A. and Porri, L., *Makromol. Chem., Rapid Commun.*, **2**, 59 (1981).
182. Porri, L. and Gallazzi, M. C., *Eur. Polym. J.*, **2**, 189 (1966).
183. Furukawa, J., Haga, K., Kobayashi, E., Iseda, Y., Yashimoto, T. and Sakamoto, K., *Polym. J.*, **3**, 371 (1971).
184. Furukawa, J., Kobayashi, E. and Kawagoe, T., *Polym. J.*, **5**, 231 (1973).
185. Zhang, Z. Y., Zhang, H. J., Ha, H. M. and Wu, Y., *J. Mol. Catal.*, **17**, 65 (1982).
186. Freg, D. A., Hasegawa, G. E. and Marvel, C. S., *J. Polym. Sci., A*, **1**, 2057 (1963).
187. Dolgoplosk, B. A., Beilin, S. I., Korshak, Y. V., Chernenko, G. M., Vardanyan, L. M. and Teterina, M. P., *Eur. Polym. J.*, **9**, 895 (1973).
188. Yasuda, H. and Nakamura, A., *Angew. Chem., Int. Ed. Engl.*, **26**, 723 (1987).

189. Lucas, J., van Leeuwen, P. W. N. M., Volger, H. C. and Kouwenhoven, A. P., *J. Organomet. Chem.*, **47**, 153 (1973).
190. Bolognesi, A., Destri, S., Zinam, Z. and Porri, L., *Makromol. Chem., Rapid Commun*, **5**, 679 (1984).
191. Dolgoplosk, B. A., Makovetskii, K. E., Red'kina, L. I., Soboleva, T. V., Tinyakova, E. I. and Yakovlev, V. A., *Dokl. Akad. Nauk SSSR*, **205**, 387 (1972); *Chem. Abstr.*, **77**, 140604 (1972).
192. Makovetskii, K. E. and Red'kina, L. I., *Dokl. Akad. Nauk SSSR*, **215**, 1380 (1974); *Chem. Abstr.*, **83**, 61116 (1975).
193. Natta, G., Porri, L. and Gallazzi, M. C., *Chim. Ind. (Milan)*, **46**, 1158 (1964).
194. Cooper, W., *Ind. Eng. Chem., Prod. Res. Dev.*, **9**, 457 (1970).
195. Porri, L., Gallazzi, M. C., Destri, S. and Bolognesi, A., *Makromol. Chem., Rapid Commun.*, **4**, 485 (1983).
196. Furukawa, J., Saegusa, T., Irako, K., Hirooka, N. and Narumiya, T., *Kogyo Kagaku Zasshi*, **65**, 2074 (1962); *Chem. Abstr.*, **58**, 11553 (1963).
197. Bresler, L. S., Dolgoplosk, B. A., Kolechkova, M. F. and Kropacheva, E. N., *Rubber Chem. Technol.*, **37**, 121 (1964).
198. Suminoe, T., Sasaki, K., Yamazaki, N. and Kambara, S., *Kobunshi Kagaku*, **21**, 9 (1964); *Chem. Abstr.*, **61**, 801 (1964).
199. Monakov, Yu. B., Marina, N. G., Duvakina, N. V. and Rafikov, S. R., *Dokl. Akad. Nauk SSSR*, **236**, 617 (1977); *Chem. Abstr.*, **87**, 185045 (1977).
200. Shen, Z., Song, X., Xiao, S., Yang, J. and Kan, X., *J. Appl. Polym. Sci.*, **28**, 1585 (1983).
201. Pasquon, I., Porri, L., Zambelli, A. and Ciampelli, F., *Chim. Ind. (Milan)*, **43**, 509 (1961).
202. Weber, H., Schleimer, B. and Winter, H., *Makromol. Chem.*, **101**, 320 (1967).
203. Cabassi, F., Italia, S., Giarrusso, A. and Porri, L., *Makromol. Chem.*, **187**, 913 (1986).
204. Natta, G., Porri, L., Carbonaro, A. and Lugli, G., *Makromol. Chem.*, **53**, 52 (1962).
205. Carbonaro, A., Zamboni, V., Novajra, G. and Dall'Asta, G., *Rubber Chem. Technol.*, **46**, 1274 (1973).
206. Furukawa, J., Kobayashi, E. and Haga, K., *J. Polym. Sci., Polym. Chem. Ed.*, **11**, 629 (1973).
207. Bruzzone, M., Carbonaro, A. and Corno, C., *Makromol. Chem.*, **179**, 2173 (1978).
208. Furukawa, J. and Hirai, R., *J. Polym. Sci., A-1*, **10**, 3027 (1972).
209. Furukawa, J. and Kobayashi, E., *Rubber Chem. Technol.*, **51**, 600 (1978).
210. Wieder, W., Krömer, H. and Witte, J., *J. Appl. Polym. Sci.*, **27**, 3639 (1982).
211. Jiao, S., Su, D., Yu, D. and Hu, L., *Chinese J. Polym. Sci.*, **6**, 135 (1988).
212. Furukawa, J., Hirai, R. and Nakaniwa, N., *J. Polym. Sci., Polym. Lett. Ed.*, **7**, 671 (1969).
213. Furukawa, J., *Angew. Makromol. Chem.*, **23**, 189 (1972).
214. Furukawa, J., Amano, H. and Hirai, R., *J. Polym. Sci., A-1*, **10**, 681 (1972).
215. Wieder, W. and Witte, J., *J. Appl. Polym. Sci.*, **26**, 2503 (1981).
216. Furukawa, J., Tsuruki, S. and Kiji, J., *J. Polym. Sci., A-1*, **11**, 2999 (1973).
217. Furukawa, J., *Pure Appl. Chem.*, **42**, 495 (1975).
218. Furukawa, J. and Hirai, R., *J. Polym. Sci., A-1*, **10**, 2139 (1972).
219. Khurshio, A., Toppare, L. and Akbulut, U., *Polym. Commun.*, **28**, 269 (1987).
220. Kuntz, J., Powers, K. W., Hsu, C. S. and Rose, K. D., *Makromol. Chem., Macromol. Symp.*, **13/14**, 337 (1988).
221. Furukawa, J., *J. Polym. Sci., Polym. Symp.*, **9**, 895 (1974).

222. Kobayashi, E., Kaita, S., Aoshima, S. and Furukawa, J., *J. Polym. Sci., A, Polym. Chem.*, **32**, 1195 (1994).
223. Kobayashi, E., Kaita, S., Aoshima, S. and Furukawa, J., *J. Polym. Sci., A, Polym. Chem.*, **33**, 2175 (1995).
224. Kaita, S., Kobayashi, E., Sakakibara, S., Aoshima, S. and Furukawa, J., *J. Polym. Sci., A, Polym. Chem.*, **34**, 3431 (1996).
225. Sabirov, Z. M. and Monakov, Yu. B., *Inorg. Chim. Acta*, **169**, 221 (1990).
226. Brydson, J. A., in *Rubbery Materials*, Elsevier, London, 1988, p. 127.
227. Price, F. C., *Chem. Eng.*, **1963**, 84 (1963).
228. Dreyfuss, P., 'Polybutadienes', in *The Polymeric Materials Encyclopedia*, CRC Press, Inc., Boca Raton, 1996, Vol. 8, pp. 5657–5663.

Further Reading

Yu, G. and Li, Y., 'Butadiene Polymerization', in *The Polymeric Materials Encyclopedia*, CRC Press, Inc., Boca Raton, 1996, Vol. 1, pp. 871–882.

Natta, G. and Porri, L., 'Elastomers by Coordinated Anionic Mechanism', in *Polymer Chemistry of Synthetic Elastomers; Diene Elastomers*, Wiley-Interscience, John Wiley & Sons, New York, 1969, Part II, pp. 597–678.

Pino, P., Giannini, U. and Porri, L., 'Insertion Polymerization', in *Encyclopedia of Polymer Science and Engineering*, Wiley-Interscience, John Wiley & Sons, New York, 1987, Vol. 8, pp. 147–220.

Pasquon, I., Porri, L. and Giannini, U., 'Stereoregular Linear Polymers', in *Encyclopedia of Polymer Science and Engineering*, Wiley-Interscience, John Wiley & Sons, New York, 1989, Vol. 15, pp. 632–763.

Kaminsky, W., 'Polymeric Dienes', in *Handbook of Polymer Synthesis*, M. Dekker, New York, 1992, Part A, pp. 385–431.

Watanabe, H. and Masuda, T. 'Diene Polymerisation', in *Catalysis in Precision Polymerisation*, John Wiley & Sons, Chichester–New York, 1997, pp. 55–66.

Gavens, P. D., Bottrill, M., Kelland, J. W. and McMeeking J., 'Ziegler–Natta catalysis', in *Comprehensive Organometallic Chemistry*, Pergamon Press, Oxford, 1982, Vol. 3, pp. 475–549.

Henderson, J. E., 'Polymeric Butadiene Derivatives', in *Encyclopedia of Polymer Science and Engineering*, Wiley-Interscience, John Wiley & Sons, New York, 1985, Vol. 2, pp. 515–536.

Tate, B. T. and Bethea, T. W., 'Butadiene Polymers', in *Encyclopedia of Polymer Science and Engineering*, Wiley-Interscience, John Wiley & Sons, New York, 1985, Vol. 2, pp. 537–590.

Burford, R. P., Nickaf, J. B. and Chaplin, R. P., 'Neodymium-catalyzed Butadiene Polymerization', in *The Polymeric Materials Encyclopedia*, CRC Press, Inc., Boca Raton, 1996, Vol. 1, pp. 882–888.

Halasa, A. F. and Hsu, W. L., 'Crystalline 3,4-Polyisoprene', in *The Polymeric Materials Encyclopedia*, CRC Press, Inc., Boca Raton, 1996, Vol. 8, pp. 6337–6340.

Furukawa, J., 'Alternating Copolymers', in *Encyclopedia of Polymer Science and Engineering*, Wiley-Interscience, John Wiley & Sons, New York, 1986, Vol. 4, pp. 233–261.

Gowie, J. M. G., 'Alternation in the Presence of Ziegler–Natta Catalysts', in *Comprehensive Polymer Science*, Pergamon Press, Oxford, 1989, Vol. 4, pp. 409–412.

Franta, I., 'Brief History of Rubber. Some General Properties of Rubber', in *Elastomers and Rubber Compounding Materials*, Elsevier Science, Amsterdam, 1989, pp. 19–30.

Franta, I. and Duchácek, V., 'Natural Rubber', in *Elastomers and Rubber Compounding Materials*, Elsevier Science, Amsterdam, 1989, pp. 31–64.

Zachoval, J. and Brajko, V., 'Synthetic Rubbers Production Methods', in *Elastomers and Rubber Compounding Materials*, Elsevier Science, Amsterdam, 1989, pp. 65–75.

Problems

1. Name and characterise coordination catalysts for the polymerisation of conjugated dienes. Give examples of representative catalysts.
2. Name and characterise bonds anchoring the polymer growing chain in the polymerisation of conjugated dienes.
3. Show by structural drawings and by adopted Fischer projections the various stereoregular polymers that might possibly be obtained from each of the monomers:

$$CH_2{=}CH{-}CH{=}CH_2$$

$$CH_2{=}CH{-}CH{=}CH{-}CH_3$$

$$CH_2{=}C(CH_3){-}CH{=}CH_2$$

$$CH_2{=}C(CH_3){-}C(CH_3){=}CH_2$$

4. Explain how ethylene performs a control of polymer molecular weight in butadiene polymerisation with Ziegler–Natta catalysts.
5. Explain why butadiene forms alternating copolymers with ethylene in copolymerisation with some Ziegler–Natta catalysts.
6. Discuss the formation of 1,2- and 1,4-polymers of conjugated dienes. Explain why the non-1,2 monomeric units appearing in polymers of butadiene and pentadiene, formed with $Ti(OR)_4$–AlR_3 catalysts, are predominantly of *cis*-1,4 structure.
7. Discuss the role of the mode of the monomer coordination in the formation of *cis*-1,4- or *trans*-1,4-polymers of conjugated dienes.
8. Explain why *s-cis*-η^4-coordinated butadiene can give a *trans*-1,4-polymer.
9. Discuss the dependence of the microstructure of 2,4-hexadiene monomeric units (*cis*-1,4 or *trans*-1,4) in 2,4-hexadiene/1,3-butadiene copolymers on the composition of these copolymers (distribution of comonomers along copolymer chains).
10. Which are the essential factors determining the stereoregulation mechanism in the polymerisation of conjugated dienes with coordination catalysts? Compare these factors with those determining the stereoregulation mechanism in the polymerisation of α-olefins.
11. Explain the capability of terminally symmetrically disubstituted 1,3-butadiene containing only deuterium atoms as substituents to yield stereoregular 1,4-polymers with a diisotactic or disyndiotactic structure.
12. Give the structure and stereochemical designation of all possible 1,4-polymers of terminally symmetrically disubstituted butadienes that could, in principle, be obtained by stereospecific polymerisation. Are they all likely to be obtainable in practice or can only some types of stereoregular polymer be obtained? Give reasons why.
13. Explain why polybutadienes obtained with rare-earth metal-based catalysts exhibit a higher degree of stereoregularity (higher contents of *cis*-1,4 monomeric units) than those derived from polymerisations in the presence of transition metal-based catalysts.
14. What are the advantages of Nd-based Ziegler–Natta catalysts over Ti-, Co- or Ni-based Ziegler–Natta catalysts in the manufacture of butadiene rubber (*cis*-1,4-polybutadiene).

6 Coordination Polymerisation of Cycloolefins

Internal cycloolefins can be homopolymerised in several ways, involving ring-preserving or ring-opening reactions, depending on the kind of monomer and the kind of catalyst used for polymerisation. The ring-retaining polymerisation of cycloolefins can produce 1,2- and 1,3-polymers (1,2- and 1,3-polycycloolefins – polycycloalkylenes) that exhibit stereoisomerism (tacticity) originating from the presence of chirotopic tertiary carbon atoms involved in the formation of cyclic structures in the polymer main chains. Alternatively, the ring-opening metathesis polymerisation of cycloolefins ('romp') occurs, giving polymers with double bonds in the polymer chains (polyalkenylenes – polyalkenamers) that exhibit *cis/trans* isomerism; in the case of polymers obtained from bicyclic or multicyclic olefins that contain both a cyclic ring and a double bond in their chains, tacticity as well as the *cis/trans* isomerism is displayed. Ring-closing polymerisation ('rcp') or cyclic non-conjugated diolefins (cyclopolymerisation), leading to polymers possessing multicyclic repeating units in their main chains, should also be mentioned.

It should also be noted that exocyclic olefins, the most representative of which are exomethylenecycloalkanes, can be homopolymerised by a ring-opening reaction.

6.1 Ring-retaining Polymerisation

Acyclic internal olefins (β-olefins), e.g. 2-butene, are not homopolymerised (unless they are isomerised to the corresponding α-olefins) in the presence of coordination catalysts, but undergo only a cross-propagation reaction with ethylene [scheme (76) in Chapter 3] giving alternating copolymers [1]. On the other hand, cyclic olefins, which also possess an internal double bond, have been successfully homopolymerised to yield 1,2-polycycloolefins, poly(1,2-cycloalkylene)s, or 1,3-polycycloolefins, poly (1,3-cycloalkylene)s, depending on the kind of monomer subjected to the polymerisation and the kind of catalyst applied. The first polycycloolefins were obtained by 1,2-insertion polymerisation, but only in the case of the polymerisation of cycloolefins having a highly strained ring, such as cyclobutene [2–4] and norbornene (bicyclo[2.2.1]-hept-2-ene) [5–7], in the presence of Ziegler–Natta catalysts; however, only low molecular weight polymers were formed from the latter monomer. Real progress in the polymerisation of cycloolefins was made by utilising homogeneous Pd-based catalysts [8–11] as well as metallocene catalysts [12–17]; norbornene was found to undergo 1,2-insertion polymerisation yielding a high molecular weight polymer, poly (2,3-bicyclo[2.2.1]hept-2-ene) [8–17], while cyclopentene appeared to undergo 1,3-insertion isomerisation–polymerisation leading to a high molecular weight polymer, poly(1,3-cyclopentylene) [12,15,18–20].

6.1.1 1,2-Insertion Polymerisation

6.1.1.1 *Stereoisomerism of Poly(1,2-cycloalkylene)s*

Polymers formed in the 1,2-insertion polymerisation of cycloolefins, poly(1,2-cycloalkylene)s, possess two chiral centres in monomeric units; they are constituted by the carbon atoms at which the polymer backbone enters and leaves each ring. Four possible stereoisomeric structures for such polymers with chiral centres arising from rings linking adjacent main chain carbon atoms can be predicted: *erythro*-diisotactic, *erythro*-disyndiotactic, *threo*-diisotactic and *threo*-disyndiotactic. Considering, however, that the common cycloolefins are *cis*-substituted at the double bond and that polymerisation involves a *cis* insertion of the monomer into a metal–carbon bond, only *erythro*-diisotactic and *erythro*-disyndiotactic poly(1,2-cycloalkylene)s, such as poly(1,2-cyclobutylene)s, have been obtained (Figure 6.1) [4].

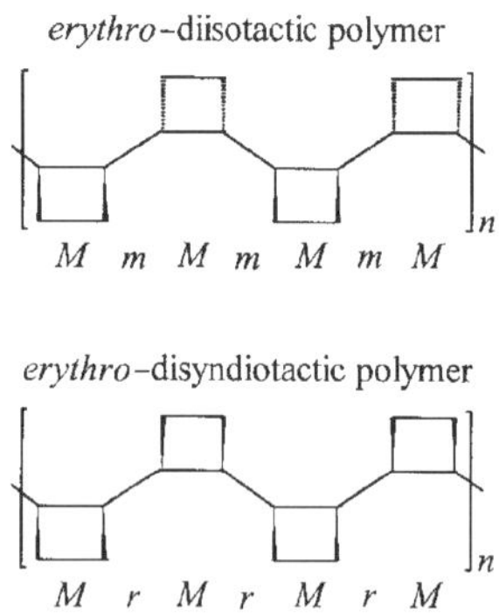

Figure 6.1 Stereoisomerism of poly(1,2-cyclobutylene)s. Diisotactic and disyndiotactic *erythro* polymers

For poly(1,2-cycloalkylene)s, the *erythro* structures are those in which there is a *cis* configuration of the polymer main chain bonds entering and leaving each ring; the *threo* structures possess a *trans* configuration of the polymer main chain bonds entering and leaving each ring.

In the case of poly(1,2-cycloalkylene)s containing symmetrical rings such as poly(1,2-cyclobutylene)s, the *erythro* and *threo* structures are also referred to as *meso* (*M*) and *racemo* (*R*) structures for denoting the stereochemistry of the rings; the relative stereochemistry between the rings is denoted by *meso* (*m*) and *racemo* (*r*) configurations corresponding to diisotactic and disyndiotactic structures respectively [21,22].

6.1.1.2 *Polymerisation Catalysts and Stereochemistry*

Cycloolefins, unlike acyclic internal olefins, undergo coordination polymerisation owing to the presence of ring strain. Loss of ring strain is an important contribution to the driving force of this polymerisation.

The most strained cycloolefins, which are substituted cyclopropenes, e.g. 3,3-dimethylcyclopropene or 3-methyl-3-ethylcyclopropene, appeared to be polymerised readily to respective substituted poly(1,2-cyclopropene)s in the presence of Pd-based catalysts containing very bulky non-labile ligands. Such catalysts are characterised by reduced activity in order to prevent ring opening of the cyclopropene monomer [23].

Poly(1,2-cyclobutylene) with an *erythro*-diisotactic structure was obtained by polymerisation in the presence of Cr- and some V-based Ziegler–Natta catalysts such as CrO_2Cl_2–$AlEt_2Cl$ and $Cr(Acac)_3$–$AlEt_2Cl$ (in toluene) [4], VCl_4–$AlEt_3$ (in heptane at lowered temperature) [3,4], $V(Acac)Cl_2$–$AlEt_2Cl$ (in toluene) [4] or other coordination catalysts such as Ni(All)Br–EtOH [24] and $RhCl_3$–H_2O [25]. Instead, poly(1,2-cyclobutylene) with an *erythro*-disyndiotactic structure was obtained using other V-based Ziegler–Natta catalysts such as $V(OBu)_3$–$AlEt_2Cl$ (in toluene) [4] and $V(Acac)_3$–$AlEt_2Cl$ (in toluene) [3,4].

High molecular weight polymers of norbornene, poly(2,3-bicyclo[2.2.1]hept-2-ene)s, have been obtained by double bond opening with certain electrophilic Pd(II) complexes, e.g. cationic palladium complexes such as $[Pd(MeCN)_4][BF_4]_2$ [8–10] as well as methylaluminoxane-activated metallocenes such as Cp_2ZrCl_2, *rac*.-$(IndCH_2)_2ZrCl_2$, *rac*.-$Me_2Si(Ind)_2ZrCl_2$ or $Ph_2C(Flu)(Cp)ZrCl_2$ [15].

In particular, C_2- and C_s-symmetrical zirconocenes, activated by methylaluminoxane, exhibit very high activities in the polymerisation [18]. Polymers of norbornene obtained with these catalysts are characterised by predominant *erythro*-disotactic and *erythro*-disyndiotactic structure respectively [15]. Active species formed in the *rac*.-$Me_2Si(Ind)_2ZrCl_2$–$[Al(Me)O]_x$ catalyst possess homotopic coordination sites for the incoming monomer, and hence pure enantiomorphic site steric control should lead to an *erythro*-diisotactic polymer. Instead, the $Ph_2C(Flu)(Cp)ZrCl_2$–$[Al(Me)O]_x$ catalyst forms an active species with enantiotopic sites for the coordination of the monomer. Thus, enantiomorphic site control with chain migration should form an *erythro*-disyndiotactic polymer. Since the C_{2v}-symmetrical Cp_2ZrCl_2–$[Al(Me)O]_x$ catalyst is achiral, the only stereocontrol possible can be performed by the polymer chain end [15].

In contrast to metallocene catalysts, Pd-based catalysts produce atactic polymers of norbornene [8].

The polymerisation of norbornene showed features of a 'living' process when $[Pd(RCN)_4][BF_4]_2$ complexes were used as catalysts. Poly(2,3-bicyclo[2.2.1]-hept-2-ene)s of high molecular weight (in the approximate range 10×10^3–100×10^3) with small polydispersities ($\bar{M}_w/\bar{M}_n < 1.10$), characterised by a relatively high glass transition temperature ($T_g = 300\,°C$), could be obtained when a solvent mixture of chlorobenzene and nitrobenzene was used at a reaction temperature of 0 °C [10].

It is worth noting that catalysts based on simple Co salts are also effective in polymerising norbornene to poly(2,3-bicyclo[2.2.1]hept-2-ene)s; interestingly, the polymers exhibited a very high molecular weight ($\bar{M}_w > 1600 \times 10^3$) and glass transition temperature ($T_g = 380\,°C$) while being readily soluble in simple hydrocarbons such as cyclohexane [26].

The polymerisation of norbornene occurs in the presence of cationic Pd(II) complexes with weakly nucleophilic ligands via the *cis* insertion involving predominantly the *exo* face of the monomer (the diastereotopic *endo* face of the monomer is much less reactive) [10]:

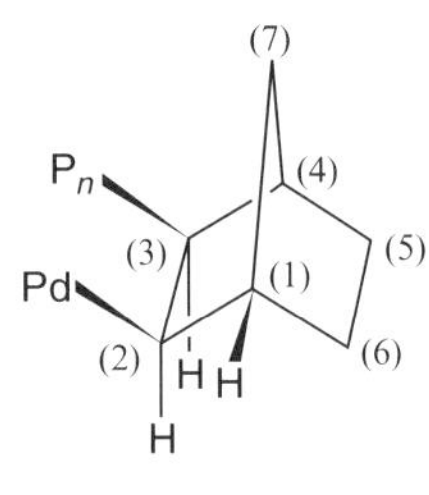

However, norbornene polymerisation run with methylaluminoxane-activated zirconocene catalysts shows exclusively *cis–exo* insertion [15].

Decomposition of the Pd–M–P_n species via β-hydrogen elimination is unfavourable, since both β-hydrogen atoms, H_3 and H_1, are not easily accessible for a Pd–H bond forming process. The suppression of β-hydrogen abstraction is a prerequisite for the monomer undergoing *cis*-insertion polymerisation, leading, in this case, to poly(2,3-bicyclo[2.2.1]hept-2-ene) [10].

At the end of considerations dealing with the insertion polymerisation of norbornene one should mention the Pd-catalysed polymerisation of other bicyclic olefins such as *exo*-8,9-dihydrodicyclopentadiene (tricyclo[5.2.1.0$^{2.6}$]

dec-8-ene) which appeared to give a high molecular weight polymer characterised by low polydispersity ($\bar{M}_w/\bar{M}_n$=1.3) in a 'living' process [10]:

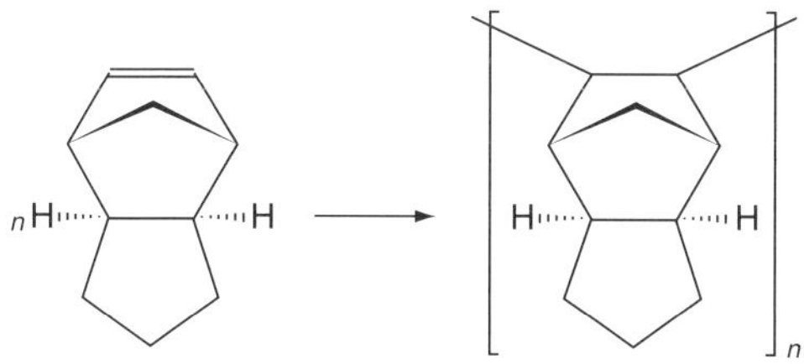

(1)

By contrast, *endo, exo*-1,4,5,8-dimethano-1,2,3,4,4a,5,8,8a-octahydronaphthalene was hardly polymerised to a high molecular weight polymer ('non-living' process) [10]:

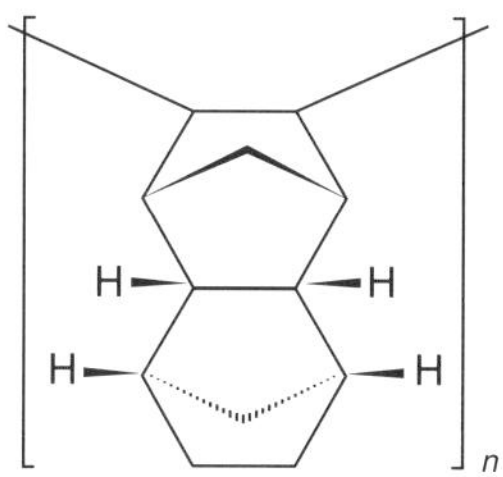

Similar polymerisation behaviour was exhibited by *exo*- and *endo*-dicyclopentadiene (tricyclo[5.2.1.0^{2,6}]deca-3,8-diene) [10]:

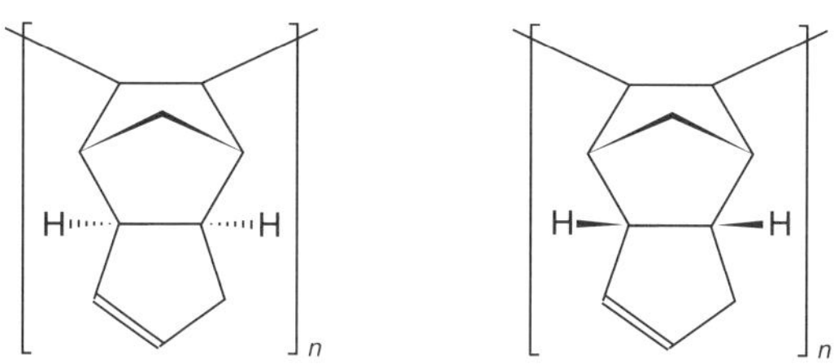

exo *endo*

It is interesting that the double bond of the less strained five-membered ring of both *exo*- and *endo*-monomers remains intact.

6.1.1.3 Copolymerisation of Cycloolefins and Carbon Monoxide

Analogously to ethylene–carbon monoxide copolymers, alternating copolymers between cycloolefins such as norbornene and carbon monoxide have been synthesised using cationic Pd(II) complexes modified by phosphorus ligands such as $[Pd(MeCN)_n(PPh)_{4-n}][BF_4]_2 (n = 1,2,3)$ [27]. General requirements for the catalyst are weakly coordinating anionic ligands. Cationic formally d^0 16-electron Pd species with an active Pd–C bond are the active catalyst sites. A square planar geometry would be expected for such a catalytic species. Monomer insertion into the Pd–C bond can occur only if the coordinating monomer molecule and polymer chain are bound in *cis*, and not in *trans*, configuration at the Pd central atom, and the remaining ligands are also in *cis* configuration [28]:

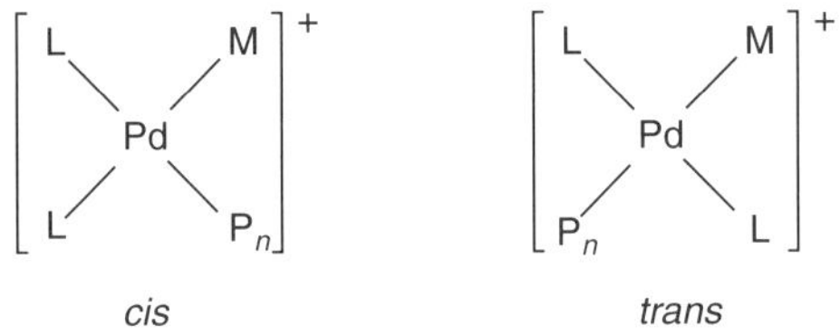

As in the case of homopolymerisation, no problem of isomerisation of propagating species exists in the copolymerisation of norbornene with carbon monoxide and hence 2,3-bicyclo[2.2.1]hept-2-ene units, as the only norbornene monomeric units appear in alternating copolymers with carbon monoxide [27]:

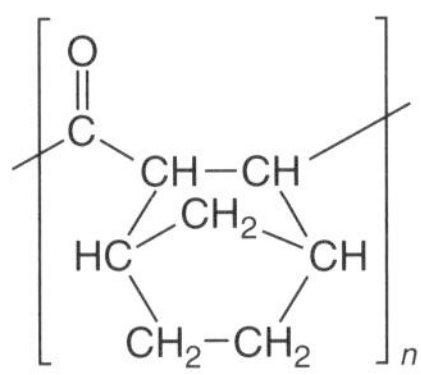

Such a structure of poly(carbon monoxide-*alt*-norbornene) is due to the lack of β-hydrogen abstraction from the norbornene monomeric unit bound to palladium.

Alternating copolymers of cycloolefins and carbon monoxide have also been obtained in the case of cyclopentene; the copolymerisations were run using cationic Pd(II) complexes modified by an achiral 1,3-bis(diphenylphosphino)-propane ligand or a chiral 2,4-bis(diphenylphosphino)pentane ligand.

Although the competing β-hydrogen abstraction is suppressed by the relatively fast insertion of carbon monoxide into the Pd–C bond, two different modes of cyclopentene enchainment, namely by 1,2- and 1,3-insertion, have been recognised in the copolymers formed [29]:

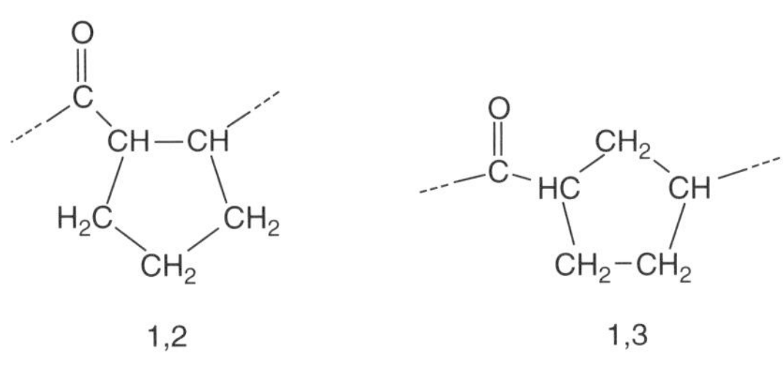

The 1,3-concatenation of cyclopentene is due to isomerisation involving β-hydrogen abstraction [20].

Depending upon the ligand applied, ratios of 1,2- to 1,3-enchainment of between 0.5:0.5 and 0.05:0.95 have been found. The competitive 1,2- and 1,3-monomer enchainments are influenced both by the nature of the ligand and the nature of the anion. It has been assumed that the prevailing tacticity of the copolymers corresponds to an isotactic structure [29,30].

6.1.2 Cyclopolymerisation of Non-conjugated Cyclodiolefins

The polymerisation of monocyclic non-conjugated diolefins, which takes place by the double bond opening with simultaneous transannular migration, leads to cyclopolymers having bicyclic repeating units present in the main chain. A typical example of such cyclopolymerisation is the polymerisation of 1,5-cyclooctadiene [31,32]:

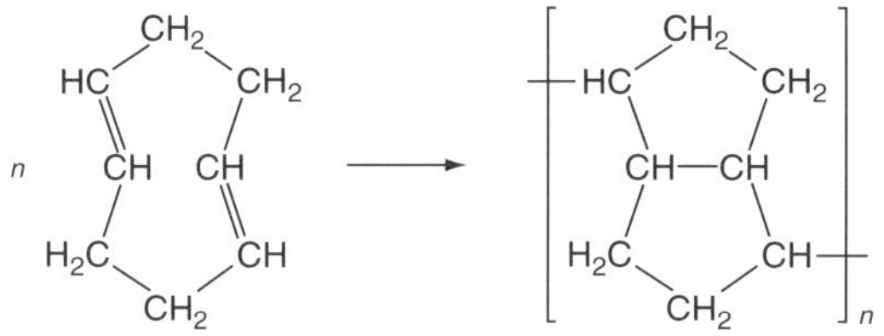

(2)

Ziegler–Natta catalysts such as $TiCl_4$–AlR_3 [31] and Cp_2MtCl_2–$AlEt_3$ (Mt = Ti, Zr, Hf) [32] appeared to be effective in cyclopolymerisation. It is noteworthy that *cis*, *cis*-1-methyl-1,5-cyclooctadiene was also cyclopolymerised in the presence of Ziegler–Natta catalysts, involving a pathway analogous to that in scheme (2) [33].

A cationic polymerisation mechanism, however, has been suggested [32] to operate in the systems discussed.

6.1.3 1,3-Insertion Isomerisation–Polymerisation

6.1.3.1 Stereoisomerism of Poly (1,3-cycloalkylene)s

Poly(1,3-cycloalkylene)s contain a ring incorporated via 1,3 atoms into the main chain. Each carbon atom at which the polymer main chain enters and leaves each ring constitutes a chiral centre, similarly to the case of poly(1,2-cycloalkylene)s. Hence, two chiral centres are present in a monomeric unit in the polymer backbone. The microstructure of poly(1,3-cycloalkylene)s concerns the *cis–trans* geometrical isomerism of the rings and the relative stereochemistry between the rings, denoted by *meso* (*m*) and *racemo* (*r*) assignments. Therefore, four possible stereoregular structures can be foreseen for poly(1,3-cycloalkylene)s: *cis*-isotactic, *cis*-syndiotactic, *trans*-isotactic and *trans*-syndiotactic.

However, *cis*-isotactic and *cis*-syndiotactic polymers have been obtained predominantly from common cycloolefins such as cyclopentene; the 1,3-insertion isomerisation–polymerisation of cycloalkenes such as cyclopentene farely yields polymers with a *trans* structure (usually in an amount not exceeding 3%) [15,19,20]. Figure 6.2 shows both *cis*-isomers of poly(1,3-cyclopentylene).

It is worth emphasising that the notations *cis* and *trans* are used in the case of poly(1,3-cyclopentylene)s, similarly to the case of other cycloaliphatic polymers such as poly(1-methylene-3-cyclopentylene), which is obtained by the cyclopolymerisation of cyclohexa-1,5-diene. Let us recall that *erythro–threo* nomenclature, which is used, by definition, for denoting relative configurations at two contiguous carbon atoms in the polymer backbone, cannot be applied in the case of polymers with 1,3-cycloalkylene rings in the main chain; chiral 1,3 carbon atoms in these rings are not contiguous [22].

6.1.3.2 Polymerisation Mechanism

The polymerisation of cyclopentene, affording predominantly poly(1,3-cyclopentylene), proceeds in the presence of homogeneous Ziegler–Natta catalysts based on non-bridged as well as stereorigid *ansa*-zirconocenes, such as Cp_2ZrCl_2–$[Al(Me)O]_x$, *rac*.-$(IndCH_2)_2ZrCl_2$–$[Al(Me)O]_x$, *rac*.-$Me_2Si(Ind)_2ZrCl_2$–$[Al(Me)O]_x$ and $Ph_2C(Flu)(Cp)ZrCl_2$–$[Al(Me)O]_x$ [12,15,18–20].

A plausible polymerisation mechanism that accounts for the 1,3-enchainment of cyclopentene, leading to both *cis*- and *trans*-poly (1,3-cyclopentylene)s, is outlined below [15,19,20]:

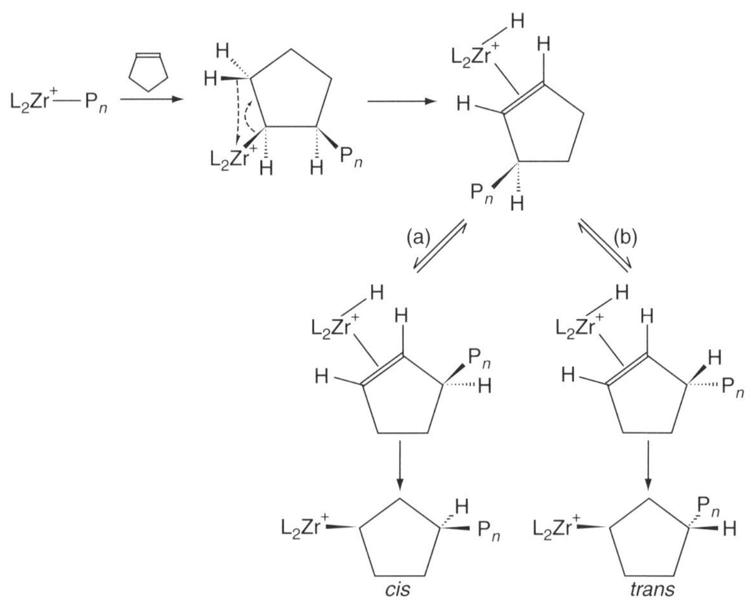

(3)

cis-isotactic polymer

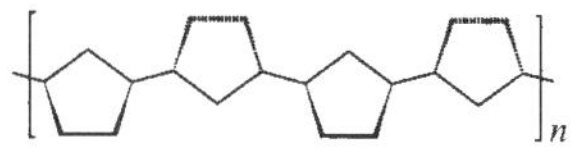

cis-syndiotactic polymer

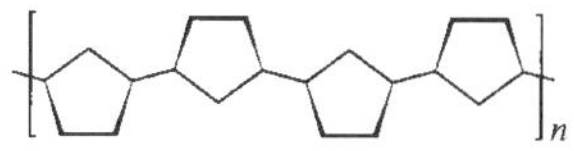

Figure 6.2 Stereoisomerism of poly(1,3-cyclopentylene)s. Isotactic and syndiotactic *cis* polymers

In the first step, a *cis* 1,2-insertion of cyclopentene takes place; the intermediate derived from this insertion is sterically too demanding to permit insertion of another monomer molecule and is isomerised before another cyclopentene molecule is inserted. Since only β-hydrogen atoms are available for the stabilisation of the active species (β-agostic interaction), β-hydride transfer, followed by monomer rotation or migration, and then reinsertion occur, leading to 1,3-linked polymer chains. The rotation around the Zr–olefin π bond results in the formation of *cis*-1,3-poly(cyclopentylene) [scheme (3a)] [20] while the π face migration of the olefin across the nodal plane of the double bond results in the formation of *trans*-1,3-poly(cyclopentylene) [scheme (3b)] [19].

The discussed mechanism is similar to that operating in the 1,3-enchainment of propylene using metallocene-based catalysts [scheme (67) in chapter 3] [34–39].

It is worthwhile mentioning that 1,3-homopolymerisation is possible using homogeneous metallocene catalysts but impossible with heterogeneous Ziegler–Natta catalysts which are not capable of the isomerisation illustrated by scheme (3).

The 1,3-insertion isomerisation–polymerisation characteristic of cyclopentene is not possible in the case of norbornene polymerisation, since no β-hydrogen atoms are available in norbornene to stabilise the product of its initial 1,2-insertion owing to the rigid structure of the monomeric unit [15].

The microstructure of poly(1,3-cyclopentylene)s prepared by metallocene catalysts depends on the kind of metallocene precursor and polymerisation conditions. Polymerisations conducted with methylaluminoxane-activated Cp_2ZrCl_2, *rac.*-$Me_2Si(Ind)_2ZrCl_2$ and $Ph_2C(Flu)(Cp)ZrCl_2$ featured mainly *cis*-1,3 structures. The $Ph_2C(Flu)(Cp)ZrCl_2$–$[Al(Me)O]_x$ catalyst is not highly stereospecific but favours the formation of *cis*-syndiotactic polymers [15]. Polymers yielded with *rac.*-$(IndCH_2)_2ZrCl_2$–$[Al(Me)O]_x$ and *rac.*-$Me_2Si(Ind)_2ZrCl_2$–$[Al(Me)O]_x$ catalysts are characterised by a *cis*-isotactic structure [15,20].

Note the chain end stereocontrol mechanism operating in cyclopentene polymerisation, which is similar to norbornene polymerisation but differs from the case of norbornene in the fact that in cyclopentene polymerisation the chiral centres of the growing polymer chain are located in the α and γ position [15].

6.2 Ring-opening Metathesis Polymerisation

The first ring-opening polymerisation of cycloolefin was reported shortly after discoveries by Ziegler and Natta; Anderson and Merckling [40] claimed that the polymerisation of norbornene could be promoted by the $TiCl_4$–MgEtBr catalyst. The structure of the product, however, was not recognised immediately, but is now known to have been poly(1-vinylene-3-cyclopentylene):

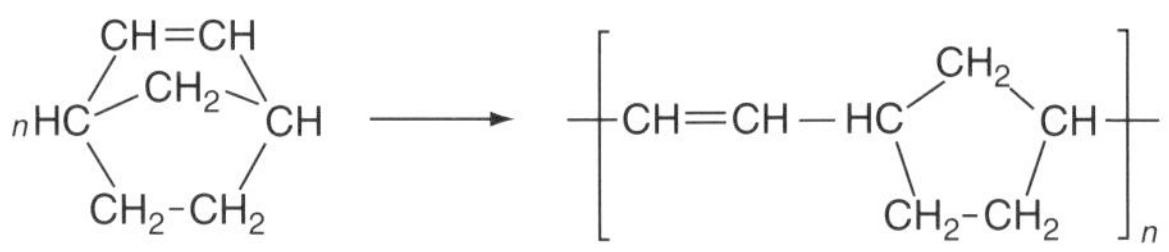

(4)

Subsequently, Eleuterio [41] filed a patent concerning the ring-opening polymerisation of cyclopentene leading to poly(1-pentenylene):

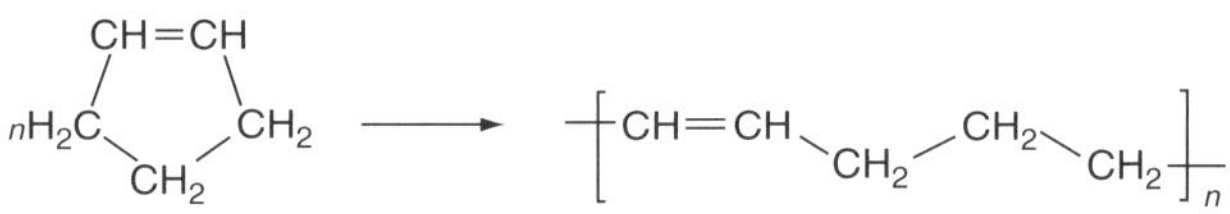

(5)

Later on, Calderon *et al.* [42,43] recognised that the ring-opening polymerisation of cyclic olefins is a special case of the more general alkene metathesis reaction, e.g. as for propylene:

$$2CH_3{-}CH{=}CH_2 \rightleftharpoons CH_3{-}CH{=}CH{-}CH_3 + CH_2{=}CH_2 \quad (6)$$

Confirmation that the ring-opening polymerisation of cycloolefins also proceeds by complete scission of the C=C bond (transalkylidenation) was provided soon after by Dall'Asta and Motroni [44].

These early results, along with a vast number of other data [45], establish that ring-opening metathesis polymerisation proceeds via a chain process [scheme (4) in Chapter 2] in which the structures of the active species fluctuate between metal alkylidenes (carbenes) and four-membered metallacycles (metallacyclobutane)s, a concept that was first introduced by Hérisson and Chauvin [46].

There exists a wide variety of cyclic olefins capable of being polymerised via a ring-opening metathesis reaction: From high-strained cycloolefins (cyclobutene and homologues, norbornene and homologues) up to low-strained (cyclopentene) and unstrained cycloolefins (cycloheptene, cyclooctene) [45].

In common with the polymerisation of acyclic olefins (α-olefins) by Ziegler–Natta catalysts, the ring-opening metathesis polymerisation of monocyclic and bicyclic olefins is promoted by alkylmetal-activated transition metal halides, and only a relatively small proportion of the transition metal atoms introduced into the system is converted into the active sites for the polymerisation. Also, as in the polymerisation of ethylene by Phillips catalysts, the metathesis polymerisation of cycloolefins can be promoted by alumina- or silica-supported catalysts based on transition metal oxides. However, several well-characterised metal–carbene complexes as well as related metallacyclobutanes appeared to catalyse the ring-opening metathesis polymerisation of cycloolefins, especially in living polymerisation systems [45,47]. It is worth mentioning that, in contrast to the polymerisation of α-olefins by metal alkyl- or hydride-activated transition metal chlorides as catalysts, several transition metal chlorides exhibit catalytic activity in cycloolefin polymerisation without the need to be pretreated with a metal alkyl or hydride as an activator.

It is worth noting that, contrary to α-olefin polymerisation systems, cycloolefin polymerisation systems are not restricted to dry, oxygen-free reaction conditions and hydrocarbon monomers. For example, iridium-based catalysts are efficient in the polymerisation of *exo*-bicyclo[2.2.1]hept-5-ene-2-carboxylic acid in aqueous or ethanolic media [48].

6.2.1 Stereoisomerism of Poly(alkenylene)s

Ring-opening polymerisation of cycloolefins gives rise to the occurrence of a number of types of stereoisomerism exhibited by the polymers formed. The first of these is related to the carbon–carbon double bond as it appears in the polymer main chain as a *cis* or *trans* structure and applies to all cyclic olefin monomers. The second arises in circumstances such as that found in the bicyclic olefin norbornene, where the two tertiary carbon atoms are of opposite chiralities. Hence, if the carbon atoms on either side of the vinylene group in the backbone of the polymer formed are of opposite chirality, a *meso* (*m*) diad is generated, whereas carbon atoms with the same chirality give a *racemo* (*r*) diad, and thus isotactic, syndiotactic and atactic polymers can be formed (let us recall that *m* and *r* assignments denote the relative stereochemistry between the rings). Figure 6.3 shows four possible stereoregular chains of the norbornene polymer poly(1-vinylene-3-cyclopentylene) [49].

Unfortunately, the structural feature of poly(1-vinylene-3-cyclopentylene) does not give rise to any detectable *m*–*r* splitting either in the ^{1}H or in the ^{13}C NMR spectra of the polymer, and thus no information on the tacticity of poly(1-vinylene-3-cyclopentylene) is available [50,51]. However, one can resolve this problem by using 1-substituted or 5-substituted norbornenes, such as 1-methylbicyclo[2.2.1]hept-2-ene, 5-methylbicyclo[2.2.1]hept-2-ene and 5,5-dimethylbicyclo[2.2.1]hept-2-ene, for the polymerisation. During the ring-opening metathesis polymerisation of these monomers, in addition to the isomerism associated with the arrangement of substituents about the double bond and the position of the cycloalkylene ring, there are possibilities of head-to-tail (h–t), head-to-head (h–h) and tail-to-tail (t–t) arrangement of the monomeric units along the polymer backbone. Note that these unsymmetrically substituted norbornenes are chiral; therefore, the polymerisation of one

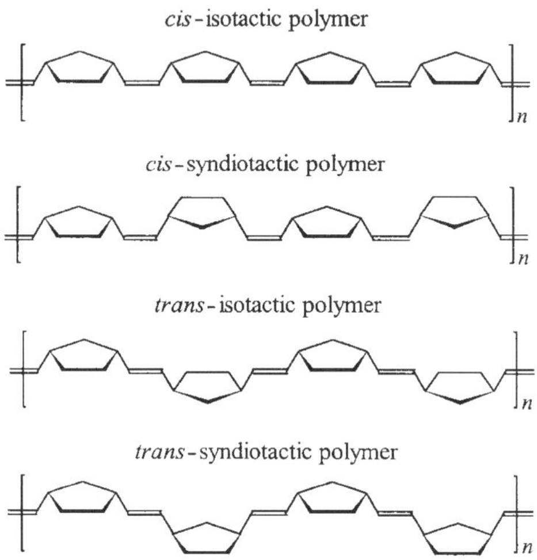

Figure 6.3 Stereoisomerism of poly(1-vinylene-3-cyclopentylene). Isotactic and syndiotactic *cis* and *trans* polymers

enantiomer will lead to a polymer with an isotactic structure along with an h–t arrangement of monomeric units, while a syndiotactic polymer will be formed along with h–h and t–t arrangement of the monomeric units [52]. These may be illustrated schematically for the case of 5,5-dimethylbicyclo[2.2.1]hept-2-ene polymerisation:

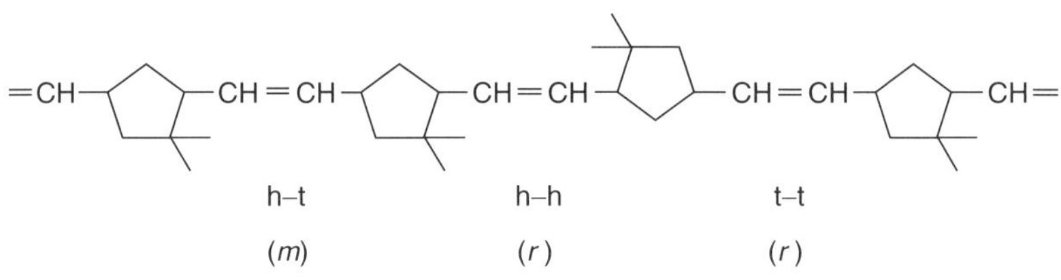

Hence, the tacticity effect is transformed into the regioselectivity effect in unsymmetrically substituted norbornene polymers, which can be detected by ^{13}C NMR spectroscopy. This approach also makes it possible to gain insight into the relationships between *cis* and *trans* structures of monomeric units in polymer chains and their tacticity. For instance, it shows that the all-*cis* polymer of racemic (±)-*exo*-5-methylbicyclo[2.2.1]hept-2-ene, formed with a catalyst that produces a polymer with a high *cis* content, has a syndiotactic structure (with an h–t arrangement of units), whereas the all-*trans* polymer, obtained with another catalyst, appeared to be atactic with a slight bias towards *m* diads (with h–h, t–t and h–t arrangements of the units) [50,53]. Moreover, polymerisation of racemic (±)-1-methylbicyclo[2.2.1]hept-2-ene yields an all-*cis* syndiotactic, h–t, polymer that is the alternating copolymer of the two enantiomers [54]:

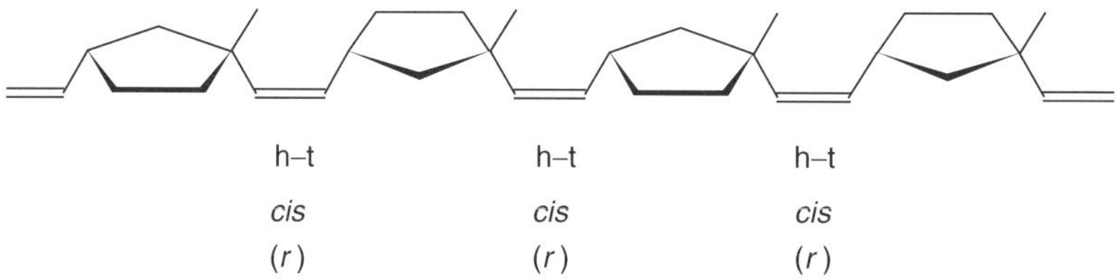

The above examples show that the ring-opening metathesis polymerisation of cycloolefins, even simple substituted bicyclic olefins, gives rise, in principle, to polymers with a very wide range of microstructures defined by the frequency and distribution of *cis* and *trans* vinylene units, *m* and *r* diads and h–h, t–t or h–t arrangements of cycloaliphatic units.

6.2.2 Polymerisation Catalysts and Active Sites

There is a wide variety of transition metal compounds, ranging from group 4 (Ti) to group 8 metals (Ir), that can be applied as catalysts or catalyst precursors for the ring-opening metathesis polymerisation of cycloolefins. However, the most commonly used are W, Mo, Re and Ru compounds; tungsten-based catalysts appeared to be the most effective. Other transition metal compounds such as Nb and Ta compounds have also often been used as catalysts, but especially for mechanistic studies [45].

The catalysts effective for ring-opening metathesis polymerisation can be either homogeneous or heterogeneous (including supported catalysts). Conventional catalysts are generally prepared from tungsten or molybdenum halide complexes and organometallic compounds of groups 1a to 4a (metal alkyl, allyl or aryl derivatives) in an organic solvent (or in monomer bulk). The addition of a third component, such as H_2O, ROH, PhOH and O_2, into a polymerisation system often causes a marked increase in catalyst activity. As with Ziegler–Natta catalysts, the active species can appear to be either soluble or separate as a solid phase. Typical soluble catalysts of this type are WCl_6–$AlEtCl_2$, WCl_6–$AlEtCl_2$–EtOH, WCl_6–$AlEtCl_2$–PhOH, WCl_6–$SnMe_4$, $WOCl_4$–$SnMe_4$ [55–57], $MoCl_5$–$AlEt_3$ and $Mo(NO)_2Cl_2(PPh_3)_2$–$AlEtCl_2$ [2,58,59]. Conventional, multicomponent catalysts are often used industrially because of their high activity and cost effectiveness. However, owing to the presence of a Lewis acid activator in the system, side reactions often compete with metathesis polymerisation. The supported catalysts are generally prepared by depositing the active metal as the oxide or carbonyl complexes on inert carriers such as alumina or silica, followed by several thermal and chemical treatments; typically, MoO_3/Al_2O_3, Re_2O_7/Al_2O_3, WO_3/SiO_2 and $W(CO)_6/SiO_2$ catalysts (retreated by metal alkyls or hydrides, e.g. $LiAlH_4$) are used [41,60]. However, supported heterogeneous catalysts have not found much application in the ring-opening metathesis polymerisation of cycloolefins, whereas soluble catalysts are very important.

Then, there are the deliberately prepared and well-defined metal alkylidene (carbene) complexes and metallacyclobutanes which, however, are relatively difficult to synthesise in satisfactory yields and generally exhibit much lower catalytic activity than metal alkyl-activated transition metal halide-based conventional catalysts. Homogeneous catalysts that already contain preformed alkylidene (carbene) or metallacyclobutane moieties are usually based on W, Mo, Ru, Nb, Ta and Ti compounds. Apart from containing well-defined active species, these catalysts offer the advantage of exclusively catalysing metathesis polymerisation without the competing side reactions associated with conventional catalysts [61], since they do not require a Lewis acid activator to be metathesis active. Another useful feature of well-defined homogeneous catalysts is the yielding of poly(alkenylene)s with a high *cis* content [62–64]. Catalysts of the discussed type, which exhibit moderate activity in cycloolefin polymerisation, can be further activated by the addition of a Lewis acid, e.g. $GaBr_3$ [65].

There are also catalysts that lack any apparent source of metal–carbon bonds. These catalysts include the aforementioned alumina- and silica-supported transition metal oxides (which, in principle, do not demand any activation by organometallic compounds), and also several group 6–8 transition metal chlorides (soluble in hydrocarbons or chlorohydrocarbons), most typically $RuCl_3$. Some of these transition metal halides require activation by a cocatalyst of the Lewis acid type (e.g. $AlCl_3$, $GaBr_3$, $TiCl_4$) [66,67]. Noble metal chlorides may be used in alcoholic solvents or in water containing emulsifiers [68].

The current general understanding of the mechanism operating in cycloolefin metathesis polymerisation leads us towards the acceptance of the structure of active sites in systems with catalysts belonging to the aforesaid three major groups as one that alternates between metal carbene complexes and metallacyclobutanes.

6.2.2.1 Catalysts Containing a Preformed Transition Metal-Carbon Bond

Various relatively stable catalysts comprising both metal carbene complexes and metallacyclobutanes have been prepared and shown to be active in the ring-opening metathesis polymerisation of cycloolefins. When utilised for promoting cycloolefin polymerisation, such catalysts essentially possess the requisite characteristics of living systems. These well-characterised homogeneous systems make it possible to follow the metathesis polymerisation pathway involving the interconversion of a metal carbene and metallacyclobutane, either of which may be the stable chain-carrying species [47].

Transition metal carbene complexes can be divided into two classes: electrophilic carbenes (Fischer carbene [69–71], Casey carbene [72,73]) and nucleophilic carbenes (Osborn carbene [74,75], Schrock carbene [76–79]):

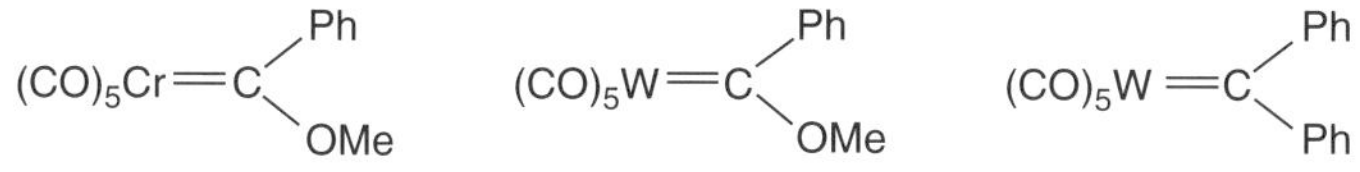

Fischer carbene Fischer carbene Casey carbene

electrophilic carbenes

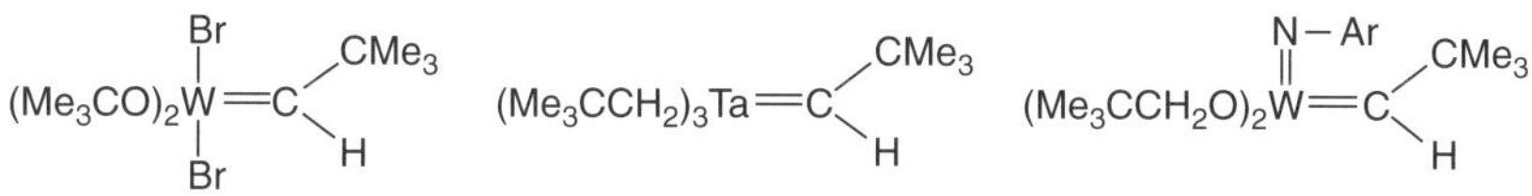

Osborn carbene Schrock carbene Schrock carbene

nucleophilic alkylidenes

Complexes of the Fischer and Casey type contain a low-valent transition metal bound to an acceptor ligand; as a consequence, the carbene carbon usually behaves as an electrophilic centre. The stability of the former complex is due to the partial compensation of the electron deficiency of the carbene carbon (which is sp^2-hybridised and possesses an empty p orbital) by the electron pair of the vicinal oxygen. The latter complex is stabilised by the partial conjugation of the π clouds of the phenyl rings with the free p orbital of the carbene carbon. Electrophilic complexes (with the carbene carbon atom comparable with that in organic carbonyl compounds) are especially susceptible to cyclopropanation of olefins and to adduct formation with a Lewis base. On the other hand, complexes of the Osborn and Schrock type contain a higher-valent transition metal and strong donor ligands; in these complexes (which may be regarded as transition metal-modified ylides) the carbene carbon behaves as a nucleophilic centre. Nucleophilic complexes are especially susceptible to adduct formation with a Lewis acid and to Wittig-type alkylation (carbonyl olefination reactions), as well as to olefin metathesis [80].

As mentioned above, the electrophilic metal carbene complexes are stabilised by the presence of heteroatoms or phenyl rings at the divalent carbon atom, while hydrogen or alkyl groups stabilise the nucleophilic complexes. Therefore, there is a distinction between 'carbenoids' and 'alkylidenes' when designing carbene ligands corresponding to the former or the latter class.

These two greatly different properties of metal carbene complexes reflect a crucial difference in the metal–carbon bonding. Single carbene moieties (present in Fischer and Casey complexes) form a donor–acceptor metal carbene bond involving a σ donor bond and a π re-donor bond. On the other hand, triplet carbene moieties (appearing in Osborn and Schrock complexes) form a covalent metal alkylidene double bond (σ, π) via spin pairing of the sp^2-hybridised carbon and metal σ electrons and the carbon pπ and metal dπ electrons respectively. Bonding in transition metal carbene and alkylidene complexes is presented schematically in Figure 6.4 [80].

Although this view is oversimplified and 'borderline' metal carbene complexes have been isolated, this approach is convenient for discussing the activity of metal carbene species in the ring-opening metathesis polymerisation of cycloolefins. Calculations have predicted [81,82] and recent results have shown [83] that, in some systems, 'metal alkylidene' reactivity is competitive with 'metal carbene' reactivity, i.e. olefin metathesis is competitive with olefin cyclopropanation.

Electrophilic metal carbene complexes such as $(CO)_5W{=}C(Ph)OMe$ generally exhibit poor activity as catalysts for metathesis polymerisation, and higher temperatures are required to bring about the polymerisation of high-strained cycloolefins such as norbornene or cyclobutene [84,85]. However, their activity can be enhanced by the addition of a Lewis acid such as $TiCl_4$ into the polymerisation system [86]. Electrophilic complexes such as $(CO)_5W{=}CPh_2$ also generally exhibit poor activity but they are more active than those mentioned above and enable the polymerisation of various cycloolefins [87,88].

Nucleophilic metal alkylidene complexes are more useful for promoting the metathesis polymerisation of cycloolefins than electrophilic metal carbenes. For instance, $Br_2(Me_3CCH_2O)_2W{=}CHCMe_3$ is a moderately active catalyst [75,89] that can be further activated by the addition of Lewis acids such as $GaBr_3$ to

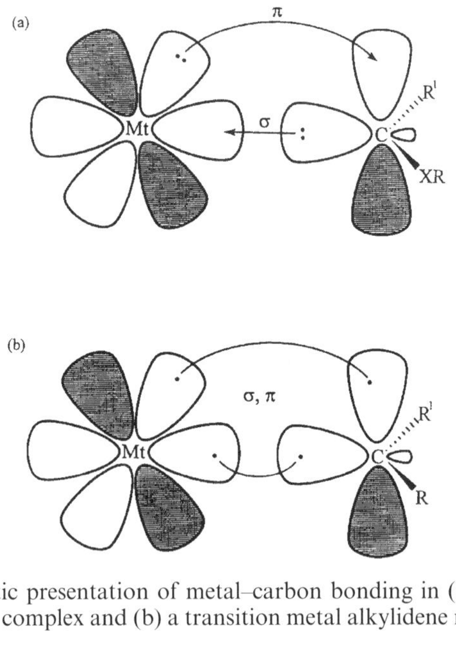

Figure 6.4 Schematic presentation of metal–carbon bonding in (a) a transition metal carbene electrophilic complex and (b) a transition metal alkylidene nucleophilic complex

form relatively highly active cationic metathesis polymerisation species with four-coordinate tungsten [65]:

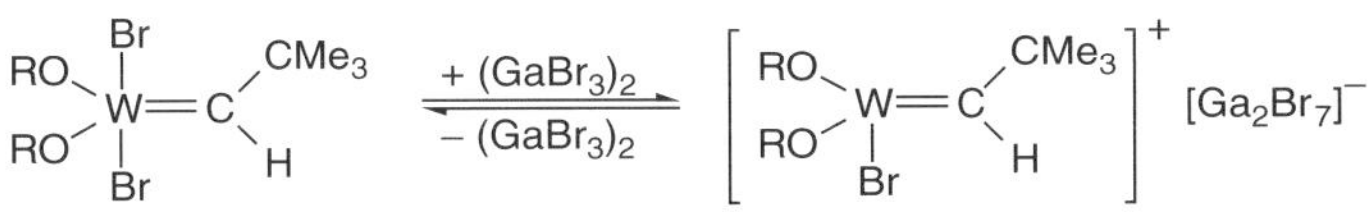

(7)

The ionic compound $[Br(Me_3CCH_2O)_2W{=}CHCMe_3]^+[Ga_2Br_7]^-$ displays catalytic activity increased by a factor of ca 3400 compared with the neutral tungsten alkylidene complex.

The most useful metal alkylidenes as cycloolefin metathesis polymerisation catalysts, however, are nucleophilic W and Mo complexes of the Schrock type [90]. The best-characterised complexes of this type are four-coordinate neopentylidene ($=CHCMe_3$) or neophylidene ($=CHCMe_2Ph$) complexes of tungsten [79,91] and molybdenum [62] that contain bulky ligands; the transition metal coordination sphere includes two bulky alkoxides and a bulky imido ligand, the most convenient of which is the 2,6-diisopropylphenylimido ligand. The four-coordinate transition metal allows a relatively small substrate molecule to attack the metal alkylidene complex to give a metallacyclobutane intermediate with a five-coordinate metal atom. By contrast, bulky alkoxide and imido ligands prevent decomposition reactions [49].

The nucleophilic ruthenium alkylidene complex $Cl_2(PPh_3)_2Ru{=}CHCH{=}CPh_2$ also appeared to catalyse the polymerisation of norbornene; the catalyst has been shown to be living with the norbornene monomer [63]. Another highly strained monomer, bicyclo[3.2.0]hept-6-ene, has been polymerised in a living system with this ruthenium alkylidene catalyst [92]:

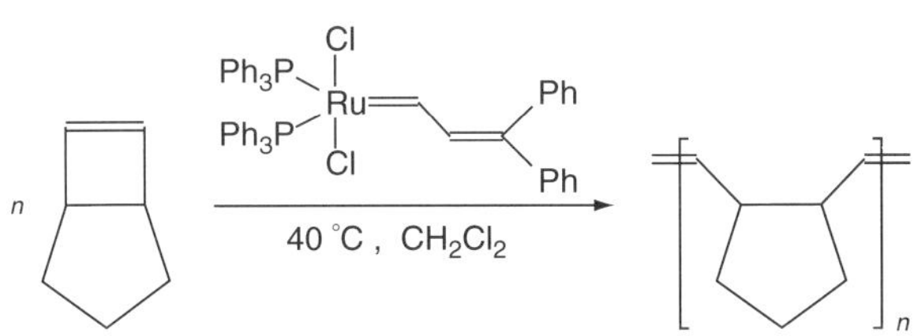

(8)

The activity of the ruthenium alkylidene complex can be greatly increased by exchanging triphenylphosphine ligands for the more electron-donating bulky tricyclohexylphosphine ligands; a complex with the latter ligands is active for the polymerisation of only slightly strained cyclopentene [93,94].

It is interesting that, to promote the ring-opening metathesis polymerisation of cycloolefins, metal carbyne complexes can also be used; in such a case, the carbyne complex is rearranged to form the actual metal carbene complex [scheme (9)] capable of initiating the polymerisation [95]:

$$Br(CO)_4W\equiv C-Ph \longrightarrow (CO)_4W=C(Ph)(Br) \quad (9)$$

It is worth realising in this connection that the chain carrying the carbene can also be formed by a direct reaction of the monomer with the metal carbyne complex, in spite of the fact that complete scission of the triple bond in this complex does not take place [96]:

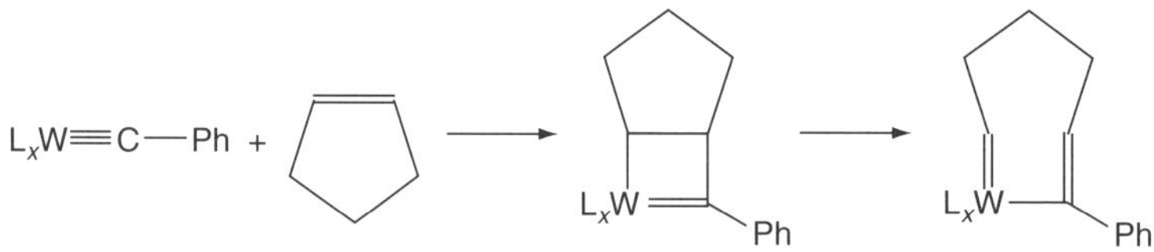

(10)

As regards stable metallacycle catalysts for cycloolefin polymerisation, catalysts based on Ti (titanacyclobutanes) and Ta (tantallacyclobutanes) are used [47,49]. The first demonstration of a living and well-controlled system concerned the polymerisation of norbornene in the presence of substituted biscyclopentadienyltitanacyclobutane in which four-coordinate titanium possesses formally a d^0 16-electron structure [97,98]. The initial titanacycle undergoes a reaction with norbornene at 20 °C to yield the trisubstituted metallacyclobutane:

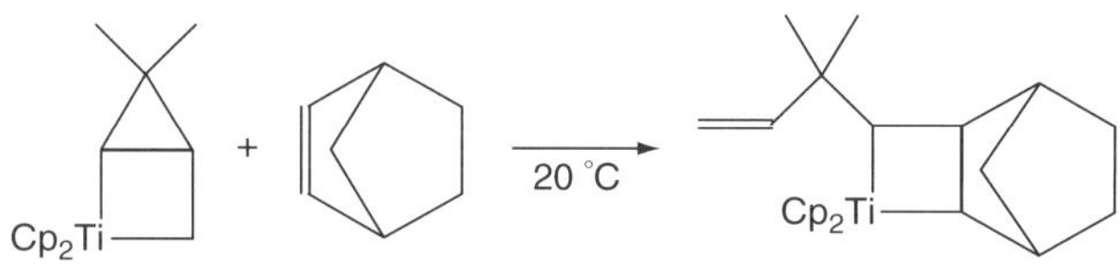

(11)

When heated at 65 °C in the presence of excess monomer, the formed metallacycle promotes ring-opening metathesis polymerisation:

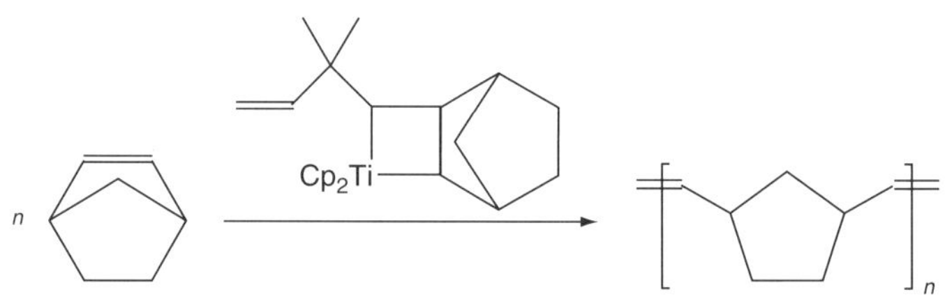

(12)

The metallacycle formed according to scheme (11) is in equilibrium with a small (unobservable) amount of titanium alkylidene complex formed by the opening of the titanacycle ring; the alkylidene complex is then trapped by norbornene to give a new titanacycle, and thus the polymer chain is propagated [49]:

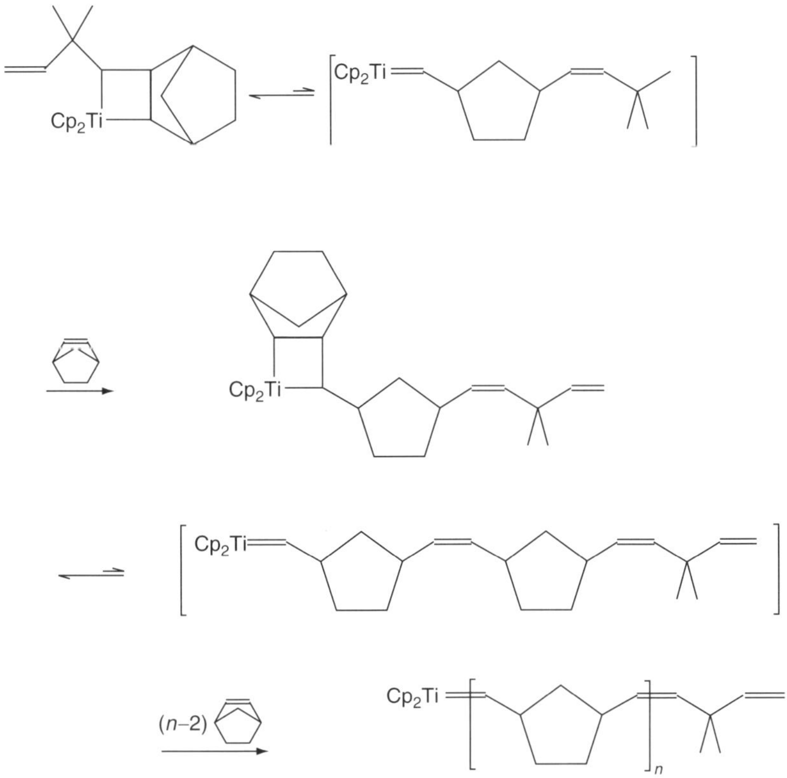

(13)

The titanacycle catalyst is also effective in promoting the polymerisation of slightly strained cycloolefins, namely cyclopentene, cycloheptene and cyclooctene. The poly(1-alkenylene)s obtained in these polymerisations, however, are characterised by a broadened molecular weight distribution (typically $\bar{M}_w/\bar{M}_n = 1.2\text{–}1.8$) compared with that for the norbornene polymer (typically $\bar{M}_w/\bar{M}_n = 1.08\text{–}1.2$) [99].

Analogously to the titanacycle catalyst, similar trisubstituted tantallacyclobutanes were prepared, but with the use of tantalum alkylidene for the reaction with norbornene [100–102]:

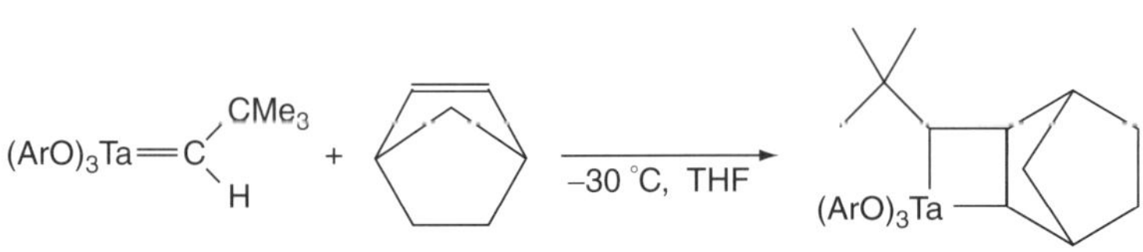

(14)

The tantallacycle catalyst (where Ar=2,6-diisopropylphenoxide) polymerised norbornene at 50 °C, similarly to the trisubstituted titanacyclobutane according to scheme (13) [102].

The stability of metal alkylidene (carbene) complexes and the corresponding metallacycles can be dependent on various factors, but it is worth noting that the kind of metal, the metal oxidation state and the ligands surrounding the metal are considered to be of essential significance. Although stable metal carbene complexes are usually obtained from W and Mo compounds whereas metallacycles are obtained from Ti compounds, systems have been found in which both the metal alkylidene complex and its precursor metallacyclobutane can be detected at lowered temperature by NMR spectroscopy [45].

6.2.2.2 Metal Hydrocarbyl or Hydride-activated Catalysts

Although the formation of transition metal carbene complexes in cycloolefin metathesis polymerisation systems with multicomponent catalysts is not completely understood, the reaction between the two main components of the catalytic systems first occurs in the same way as in typical Ziegler–Natta systems to give an organometallic intermediate possessing a transition metal–carbon σ bond. In the case of tungsten-based catalysts, these alkyltungsten intermediates with a W–C σ bond are active enough to promote routine addition polymerisation of some acyclic olefins (ethylene, propylene) to yield low molecular weight wax-like products. However, the alkyltungsten compounds are unstable and, unlike the metal–carbon σ bond in common Ziegler–Natta catalysts, have a tendency to decompose by α-hydrogen elimination to give highly reactive metal carbene species [103]. The formation of such a species when using the WCl_6–$AlEtCl_2$ catalyst for the metathesis polymerisation of cycloolefins can be schematically shown as follows:

$$WCl_6 + AlEtCl_2 \xrightarrow[-AlCl_3]{} Cl_5W-CH_2CH_3 \xrightarrow[-HCl]{} Cl_4W{=}CHCH_3 \qquad (15)$$

The metal carbene species can be formed in a similar way in systems with the WCl_6–$SnMe_4$ catalyst in which, however, the disproportionation of a methylated tungsten derivative takes place [104]:

$$2WCl_6 + 2SnMe_4 \xrightarrow[-2SnMe_3Cl]{} 2Cl_5W-CH_3 \longrightarrow Cl_4W{=}CH_2 + WCl_6 + CH_4 \qquad (16)$$

The disproportionation reaction is obviously not a possible route to the active species in the case of a system with the WCl_6–$SnPh_4$ catalyst; initially formed Cl_5 W–Ph must undergo other reactions leading to metal carbene species, such as reaction of the W–C bond with the monomer [105]. The pathway has also been proposed for systems with other catalysts, e.g. WCl_6–$Si(All)_4$ [scheme (17)], which yields a large amount of tungsten carbene species [106]:

$$WCl_6 + 2Si(CH_2CH{=}CH_2)_4 \xrightarrow[-2Si(CH_2CH{=}CH_2)_3Cl]{} Cl_4W(CH_2CH{=}CH_2)_2 \longrightarrow Cl_4W{=}CHCH{=}CH_2 + CH_3CH{=}CH_2 \qquad (17)$$

Metal hydrocarbyl or hydride-activated catalysts are in principle susceptible to deactivation by reactions with traces of moisture and other protic compounds; however, the activity of soluble catalysts can be enhanced in some instances by the addition of a protic compound as the third component [107]. It is also worthy noting that soluble catalyst precursors can give, after activation with organometallic compounds, soluble or finely divided heterogeneous catalysts.

6.2.2.3 *Catalysts without a Metal Hydrocarbyl or Hydride Activator*

Catalysts for the ring-opening metathesis polymerisation of cycloolefins that either do not contain a preformed metal–carbon bond or must not be pretreated with an organometallic (or metal hydride) activator include in principle most supported heterogeneous catalysts and a number of catalysts prepared from soluble precursors, e.g. WCl_6, $MoCl_5$, $ReCl_5$ and $RuCl_3$. Active sites can be formed via the reaction of the Mt–Cl bond in the precursor molecule with the coordinating cycloolefin [103,108]. However, the most general routes leading to transition metal carbenoid active sites are those involving the formation of metal hydride and metallacycle species [109].

Metal hydride species can undergo an insertion reaction with an olefinic double bond, resulting in the formation of a metal alkyl species which then generates the requisite carbenoids by α-hydride abstraction [110–112]:

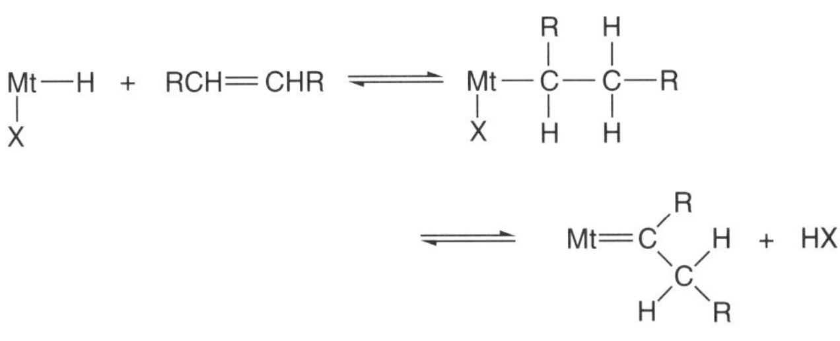

(18)

It has been found [113,114] that there are numerous sources of original hydride in various polymerisation systems. In supported catalysts, the carrier, which contains surface hydroxyl groups, can serve as the addendum. In other cases the hydride can be originated from water, alcohol or phenol if they are required for the activation [115].

Metallacyclobutane species (which then afford initiating carbenoids) can be formed in polymerisation systems with supported catalysts by monomer reaction via a metal π-allyl complex and metal hydride. A reaction of this type, where the olefin is 2,8-decadiene, is shown by the following scheme ($R = H_3CCH{=}CH{-}(CH_2)_4{-}$) [116]:

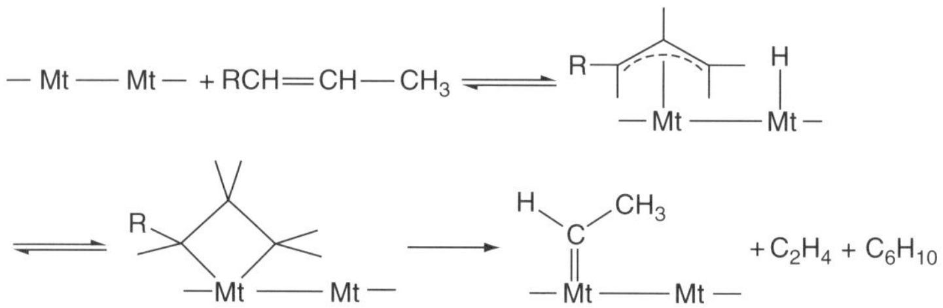

(19)

In the case of polymerisation in the presence of heterogeneous catalysts with the transition metal in a low oxidation state, metallacyclobutane species can also be formed by the intentional addition of cyclopropane to the system [117]:

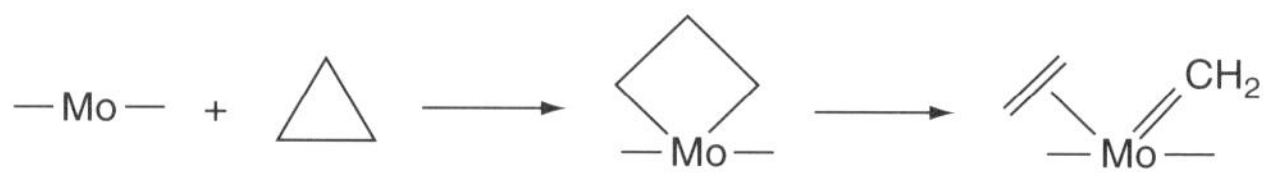

(20)

It is interesting that catalysts containing highly oxidised species can undergo a reduction reaction to produce a metal carbene complex and an aldehyde on reaction with the olefinic monomer [118]:

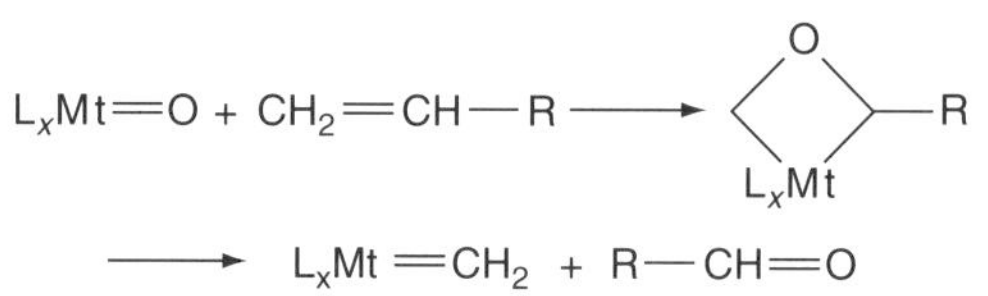

(21)

6.2.3 Mechanistic–Kinetic and Thermodynamic Aspects of Polymerisation

If one considers a general scheme of the ring-opening metathesis polymerisation of cycloolefins [scheme (22)], it involves complexation of the monomer molecule at the free coordination site, followed by an attack of the carbene carbon atom on the complexed monomer [46]:

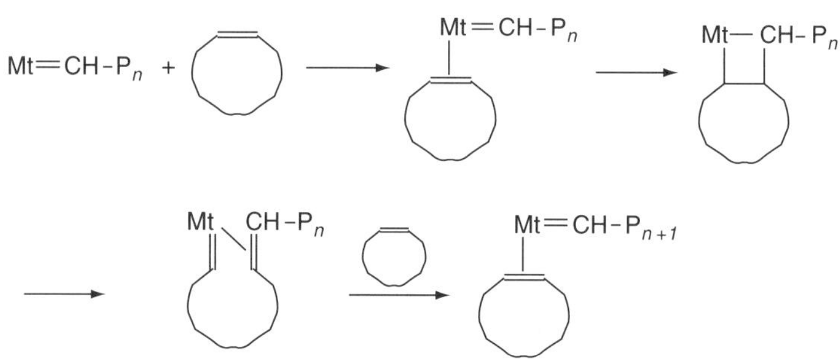

(22)

The attack of the carbene on the coordinated cycloolefin molecule involves a monomer complexed in the *cis* position with respect to the carbene carbon atom to yield the metallacycle; coordination of the monomer in the *trans* position leads instead to an inactive complex [119]. The metallacycle then undergoes a rupture either degeneratively to yield starting species or productively to yield a growing polymer chain with reformatted metal alkylidene species and carbon–carbon double bond. A new double bond thus formed in the growing polymer chain decoordinates, which releases a vacancy at the transition metal atom to be complexed by the next cycloolefin molecule. However, more detailed studies on metathesis polymerisation suggest that it can proceed via two different pathways as regards the mode of in-chain double bond coordination and decoordination at the active site: the pathway involving the complexation of one in-chain C=C bond analogous to that proceeding according to scheme (22), and the pathway involving the complexation of two in-chain C=C bonds [schemes (23) and (24) respectively) [120]:

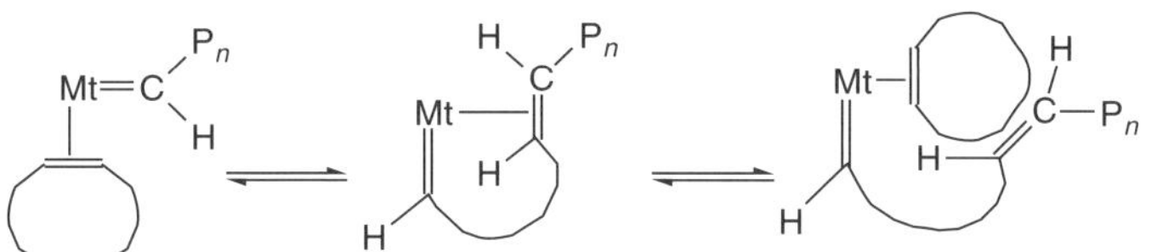

(23)

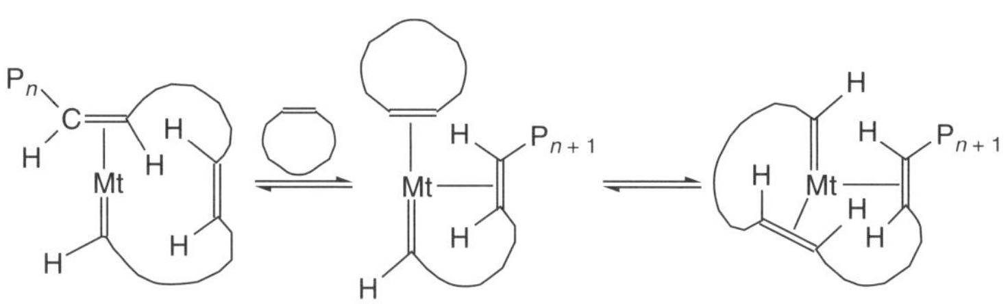

(24)

As regards metal alkylidene and metallacycle active sites participating in metathesis polymerisation, it should be emphasised that either the alkylidene or the metallacyclobutane can be the resting state of the catalyst, depending on the catalyst used for the polymerisation [99].

The propagation reaction consists in repetitive interaction of the cycloolefin double bond with a metal alkylidene species formed via the initiation reaction, and differs from this reaction only in the rate constant value. The rate of the propagation reaction should be first order with respect to the monomer concentration. However, the kinetics of polymerisation of cycloolefins is complicated by the fact that any active catalytic species in the polymerisation system will be able not only to coordinate the monomer double bond to lead to a propagation step but also to coordinate a C=C bond from the polymer chain. If the coordination of the latter C=C bond proceeds intramolecularly, then a new propagating metal alkylidene species is formed with elimination of an oligomeric cyclic molecule [121]:

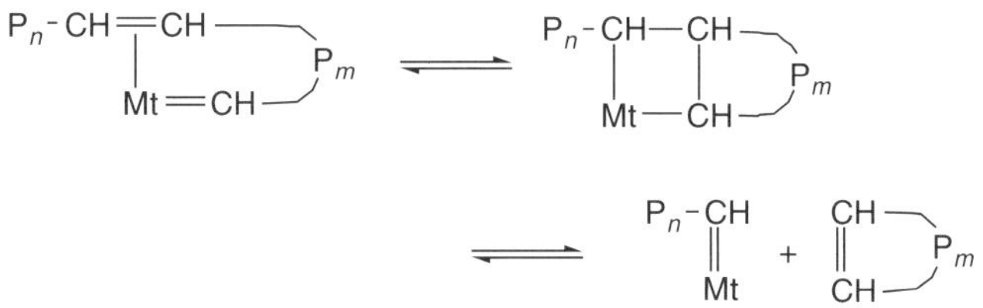

(25)

The occurrence of this intramolecular secondary metathesis reaction (backbiting reaction) results in a lowering of the molecular weight of the obtained polymer. However, the general consequence of backbiting reactions in metathesis polymerisation is the presence of cyclic oligomers 'contaminating' the high molecular weight poly(1-alkenylene)s.

On the other hand, intermolecular coordination of the in-chain C=C bond at the active site will lead to a secondary metathesis reaction according to the following scheme [121]:

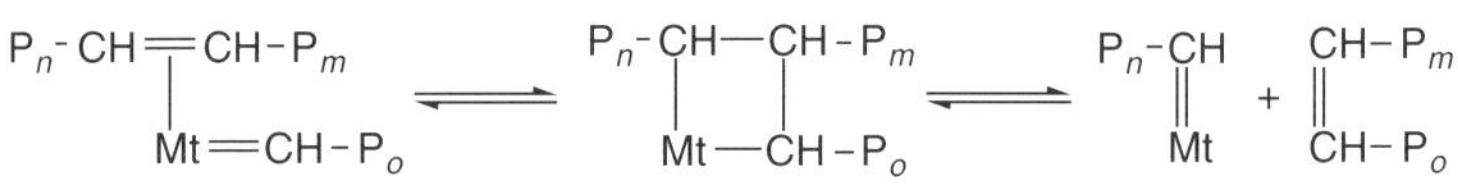

(26)

The occurrence of this chain transfer reaction results in a *cis–trans* isomerisation of double bonds in the polymer chains; however the *cis* or *trans* structure of these double bonds has no essential influence on their susceptibility to a backbiting reaction. An important implication of the intermolecular secondary metathesis reaction is, instead, the tendency of the molecular weight distribution in the resulting polymer to attain the equilibrium condition $\bar{M}_w/\bar{M}_n = 2$ [122].

Both the intramolecular and the intermolecular secondary metathesis reactions affect the polymerisation kinetics by decreasing the rate of polymerisation, because a fraction of the active sites that should be available as propagation species are involved in these non-productive metathesis reactions. The kinetics of polymerisation in the presence of metal alkyl-activated and related catalysts shows in some cases a tendency towards retardation, again due to gradual catalyst deactivation [123]. Moreover, several other specific reactions can influence the polymerisation. Among them, the addition of carbene species to an olefinic double bond, resulting in the formation of cyclopropane derivatives [108], and metallacycle decomposition via reductive elimination of cyclopropane [109] deserve attention.

The mole fraction of individual oligomers formed by secondary metathesis reactions decreases continuously with increasing degree of polymerisation, in agreement with the Jacobson–Stockmayer theory [124]. It is obvious that the total fraction of cyclic oligomers under equilibrium conditions depends on the thermodynamic parameters of the polymerisation which are influenced primarily by the structure of the monomer and polymer, their concentration and the polymerisation temperature and pressure. However, the ratio of cyclic oligomer fraction to linear polymer fraction formed within a short polymerisation time is a kinetic problem [125].

The ring strain energy is the principal factor considered in the polymerisation of cycloolefins [126,127]. The metathesis polymerisation proceeds by the cleavage of the ring to form a linear molecule, which is accompanied with the release of the associated ring strain energy. The propagation–depropagation equilibrium relating to the ring-opening metathesis polymerisation of cycloolefins is expressed as follows:

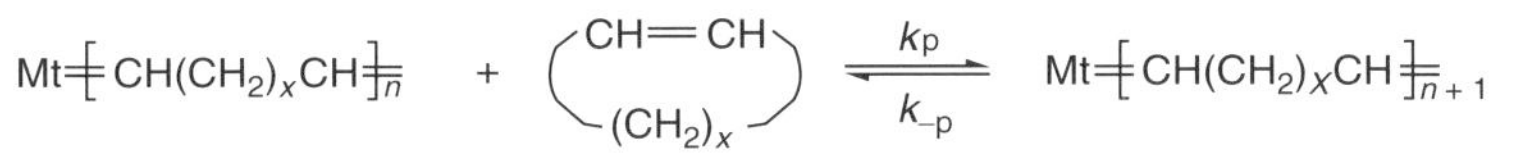

(27)

According to the principles of kinetic and thermodynamic equilibria, the following dependences are valid for the discussed system [121]:

$$K_{\mathrm{p}} = k_{\mathrm{p}}/k_{-\mathrm{p}}, \qquad k_{\mathrm{p}} \times [\mathrm{M}]_{\mathrm{eq}} = k_{-\mathrm{p}}, \qquad K_{\mathrm{p}} = 1/[\mathrm{M}]_{\mathrm{eq}},$$
$$\Delta G = \Delta H - T\Delta S = -RT\ln K_{\mathrm{p}} = RT\ln[\mathrm{M}]_{\mathrm{eq}}$$

Highly strained rings, free of steric hindrance around the carbon–carbon double bond, offer, for both thermodynamic and kinetic reasons, the best chance to obtain a high molecular weight linear polymer almost free of cyclic oligomers [125]. Therefore, the choice of norbornene as the monomer for studies of many polymerisations, especially in living systems, is not coincidental. Norbornene belongs to unique monomers, since it possesses a very reactive C=C bond, which is then converted into a sterically hindered unreactive C=C bond in the polymer. The ring strain of this monomer (27.2 kcal/mol) is essential for driving the reaction in the forward direction by suppressing depolymerisation through a backbiting reaction of the penultimate C=C bond in the growing polymer chain. Moreover, chain transfer and isomerisation reactions, which might result from backbiting into the other acyclic C=C bonds in the polymer backbone, are inhibited by the steric bulk provided by the adjacent cyclopentane rings [99].

6.2.4 Copolymerisation

Copolymerisation, which is often carried out in order to gain a better insight into the nature of polymerisation initiating and/or propagating species and to modify the properties of the polymeric products formed, has also been satisfactorily carried out in the case of cycloolefins. The ring-opening metathesis copolymerisation of these monomers can proceed in various systems involving both high- and low-strain cyclic olefins. Depending on the kind of monomer and catalyst applied for the reaction, as well as the conditions and mode of carrying out the copolymerisation, either random or block copolymers can be obtained. An important goal of such a copolymerisation is the preparation of copolymers that have a very narrow molecular weight distribution [45,49,125].

Metathesis copolymerisations may be controlled in particular in systems in which well-defined homogeneous catalysts are used as catalysts; the living nature of these catalysts makes it possible to control the formation of block copolymers. Certain metathesis catalysts generate propagating species that are stable enough to be employed as living species in block copolymer preparation [75,97]. For this purpose it is necessary to operate with polymerisation systems where secondary metathesis reactions do not present a serious complication. Norbornene and related bicyclic olefins offer the best possibilities in this regard [45]. By changing the nature of the propagating species during the polymerisation, the block ring-opening metathesis copolymerisation of two cycloolefins of two different reactivities can be accomplished. Most conveniently, it can be realised in systems with two highly strained cycloolefins by the monomer sequential technique; by successive addition of monomers into a polymerisation system with the $Cl_2(PPh_3)_2Ru{=}CHCH{=}CPh_2$ catalyst, diblock and triblock copolymers of norbornene and bicyclo[3.2.0]hept-6-ene have been obtained [92]. Since the latter comonomer is more reactive than the former, it was observed in the polymerisation of an equimolar mixture of these monomers by a ruthenium complex that the propagation of norbornene did not begin until the polymerisation of bicyclo[3.2.0]hept-6-ene was complete. This led to the formation of a block copolymer in the system with a mixture of the above-mentioned monomers [128].

The copolymerisation of highly strained cycloolefins such as norbornene and relatively low-strained cyclic monomers such as cyclopentene leads to random copolymers in the presence of WCl_6–$SnPh_4$ [129] and $(p\text{-Me-}C_6H_4\text{-}SO_3)_2Ru(H_2O)_6$ [128] catalysts. Since the homopolymerisation of cyclopentene with the latter catalyst was found hardly to proceed, it was postulated [128] that the initiation step involved the reaction of the ruthenium complex with norbornene to generate a highly active catalyst, which then copolymerised the two monomers indiscriminately to yield random copolymers. Once initiated, the high activity of the propagating species derived from the $(p\text{-Me–}C_6H_4\text{–}SO_3)_2Ru(H_2O)_6$ complex is demonstrated by the lack of discrimination in the copolymerisation of two monomers of different reactivity.

6.2.5 Stereochemical Aspects of Polymerisation

Cycloolefins can be ring-open polymerised with a high degree of stereospecificity. Many poly(1-alkenylene)s prepared by metathesis polymerisation of cycloolefins are stereoregular polymers with presumably a *cis* or *trans* configuration of the C=C bonds. Although the mechanism that controls the configuration of these bonds during the ring-opening metathesis polymerisation of cycloolefins is still not completely clear, the major controlling factors are considered to be connected with steric interactions between the cycloolefin and metal carbene complex and in the metallacyclobutane intermediate [109].

As mentioned above, the metathesis polymerisation has been suggested [120] to proceed by two different pathways, the difference being concerned with the mode of coordination of the in-chain carbon–carbon double bonds. The most common pathway [scheme (23)] exhibits low stereoselectivity and leads to equilibrated products, while the second pathway [scheme (24)] is highly *cis* selective. The *cis* selectivity results from a highly chelated intermediate that requires two C=C bonds to be complexed to the metal at the active site. The steric demand for such a complexation requires *cis* C=C bonds [scheme (24)]. It is worth mentioning that a low-strained cycloolefin such as cyclopentene produces poly(1-pentenylene) of high *cis* content (ca 98%) with the WF_6–$AlEtCl_2$ catalyst [120]. Let us also note that catalysts that produce *cis*-poly(1-pentenylene) can be used satisfactorily to polymerise other cycloolefins to highly *cis*-polymers [84]. With a number of other catalysts, one in-chain C=C bond at most is complexed to the metal at the active site. The lack of chelation relaxes the stereochemical demands on the double bond and allows non-chelating in-chain C=C bonds to compete with a polymerisable (potentially chelating) cycloolefin, which results in secondary metathesis reactions [109].

General schemes accounting for the changes in polymer structure that are brought about by changes in the structure of the catalyst have been proposed. The kind of transition metal, its oxidation state and the ligand environment of the metal are decisive for determination of the steric control in the propagation steps [121].

To gain more insight into the stereoregulation mechanism operating in the ring-opening metathesis polymerisation of cycloolefins, let us firstly have a look at the structure of the propagating species in systems with unsymmetrically substituted bicyclic olefins, which are most convenient for mechanistic studies. One can note that, in the case of the polymerisation of 1-methylnorbornene, the simplest representative of such monomers, there are propagating species of two types, one in which the 1-methyl substituent is adjacent to the carbene carbon atom (P_h) and the second in which it is distant from the carbene (P_t):

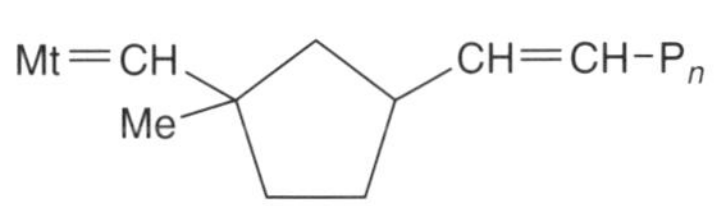

P_h

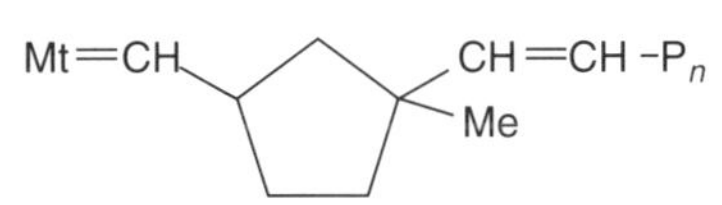

P_t

If it is supposed that the metal centre is chiral, then each of the P_h and P_t propagating species may appear in two enantiomorphic forms, Δ and Λ (the assumption is that the configuration of active sites $P_{h(\Delta)}$ and $P_{h(\Lambda)}$ or $P_{t(\Delta)}$ and

$P_{t(\Lambda)}$ remains intact between successive propagation steps, i.e. there is no ligand migration or rotation about the Mt=C bond) [121]:

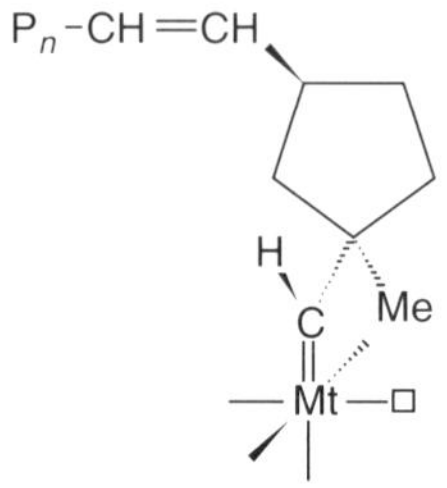

$P_{h(\Delta)}$

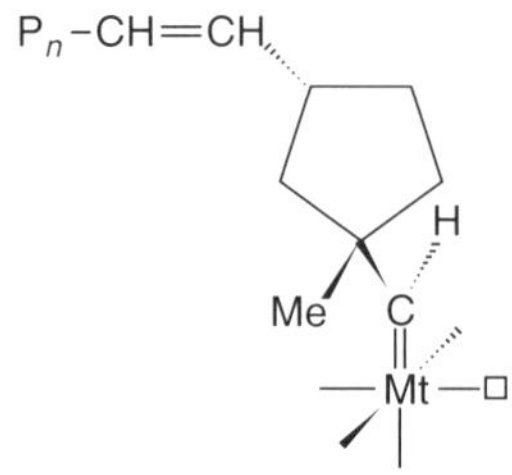

$P_{h(\Lambda)}$

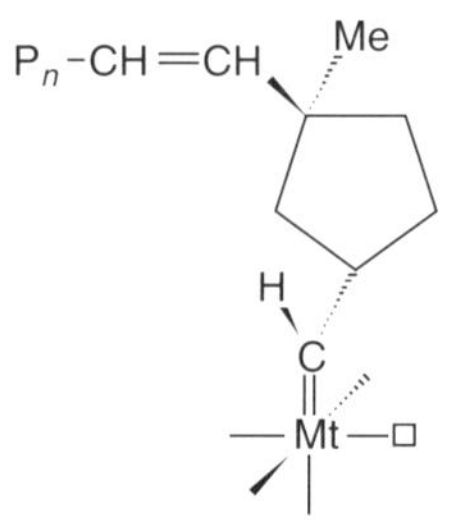

$P_{t(\Delta)}$

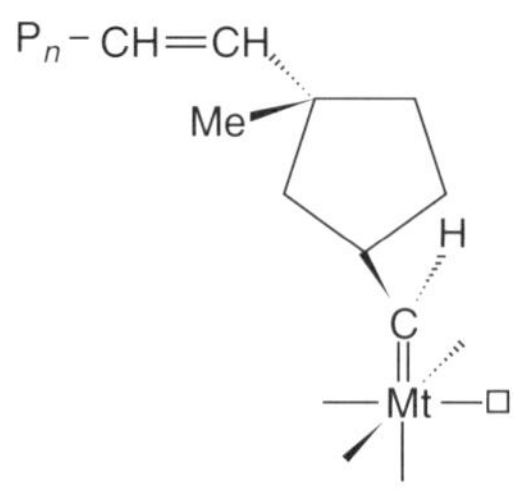

$P_{t(\Lambda)}$

Let us also note that 1-methylnorbornene exists in two enantiomeric forms, *R* and *S*:

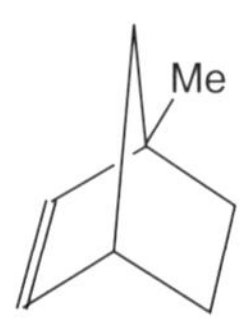

(*R*)

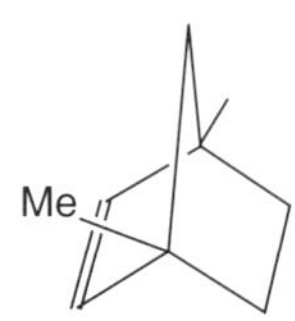

(*S*)

The interpretation of stereoisomerism exhibited by cycloolefin polymers formed via ring-opening metathesis polymerisation can be given in terms of the enantiomorphism of the propagating species with an octahedral symmetry around the transition metal atom. The enantiomeric substituted bicyclic olefin can coordinate at the enantiomorphic site in one of two ways so as to form *cis* or *trans* double bonds in the propagation step; the formed metallacyclobutanes in which the growing polymer chain is *cis* to the added monomer give rise to *cis* C=C bonds in the polymer chain, while those having the growing chain *trans* to the metallacycle ring give rise to *trans* C=C bonds in the polymer chain. The formation of a *cis* C=C bond results in a propagating species of opposite chirality, whereas the formation of a *trans* C=C bond gives a species of the same chirality. Therefore, an h–t all-*cis* polymer should be syndiotactic [schemes (28) and (29)] and an h–t all-*trans* polymer should be isotactic [schemes (30) and (31)] [45,121]:

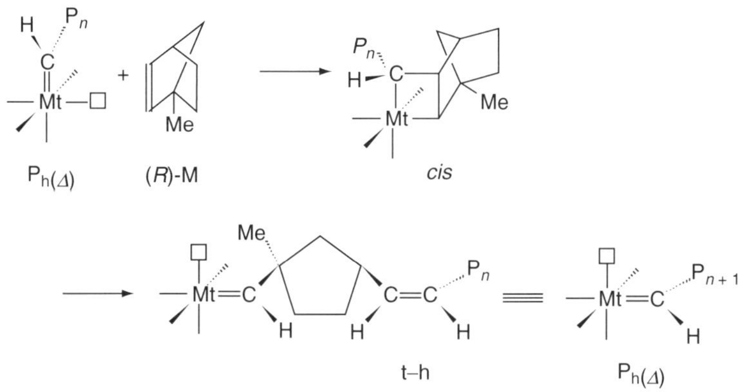

(28)

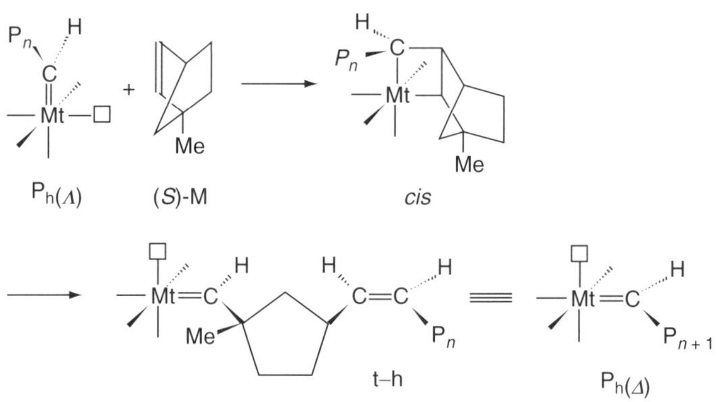

(29)

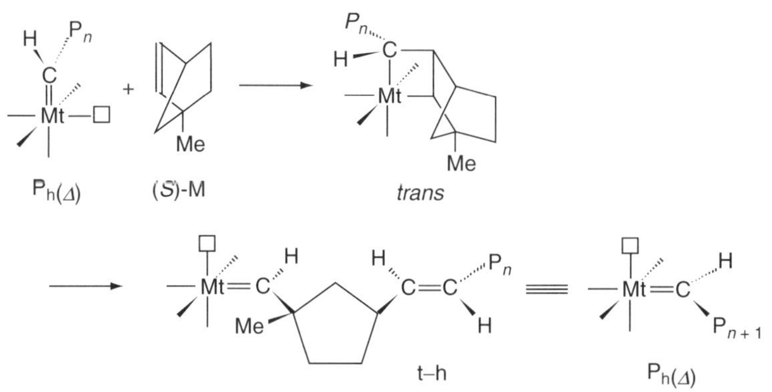

(30)

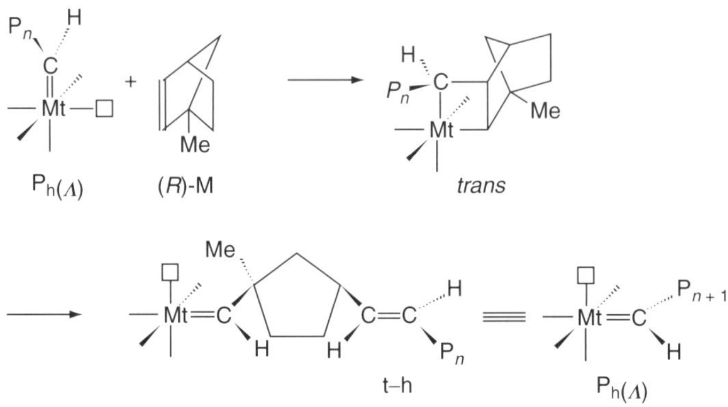

(31)

The mechanism shown in schemes (28) and (29) predicts both the enantiomers to be present in the polymerisation system if propagation leading to an h–t all-*cis* syndiotactic polymer is to proceed. Presupposing that 'like reacts with like' and considering that the propagating species must alternate in structure between the *Δ* and *Λ* configurations, it is clear why both enantiomers have to participate in the propagation. It has been observed, in fact, that catalysts that polymerise racemic 1-methylnorbornene to a high-*cis* h–t syndiotactic polymer will not polymerise a pure enantiomer [54]. One would expect that some of the catalyst active sites (of the same chirality) will react in one step with their conjugate enantiomer, but such reaction propagation comes to a halt because the other enantiomer, essential for maintaining the propagation, is absent. On the other hand, if either form of propagating species P_h is to react to give the *trans* C=C bond by an h–t reaction, then it must do so by reacting unfavourably with the non-conjugate enantiomer; i.e. if the $P_{h(\Delta)}$ site is to form a *trans* C=C bond, it must to do so by reacting with (*S*)–M, and the $P_{h(\Lambda)}$ site must react with (*R*)-M to form a *trans* C=C bond. Such reactions lead to h–t all-*trans* isotactic polymers with retention of the configuration of the catalyst active sites [schemes (30) and (31)]. However, the predominance of propagation via 'like' enantiomers reacting together in active species–monomer pairs predicts that *trans* C=C bonds, if they arise in the polymerisation system, should be associated with other reactions of propagating species involving h–h and t–t arrangements of monomeric units rather than those reactions that would lead to h–t structures with retention of the configuration of the active sites. There are suggestions, supported by experimental evidence, that h–h sequence formation [schemes (32 and 33)] is usually preceded by an h–t arrangement of monomeric units with *trans* double bonds, followed immediately by a t–t reaction that may be either *cis* [schemes (34 and 35)] or *trans* [schemes (36 and 37)] generating [45, 121].

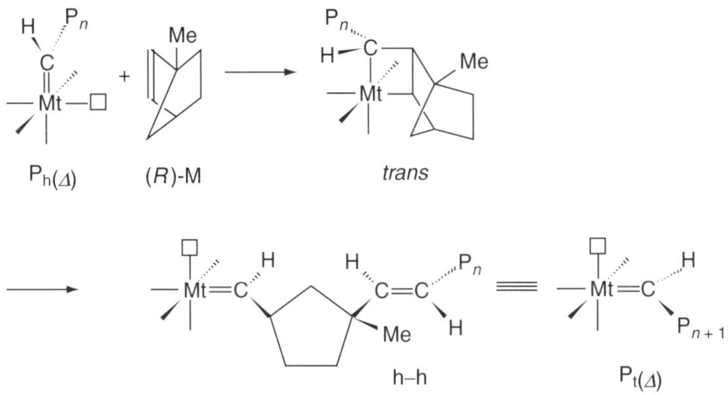

(32)

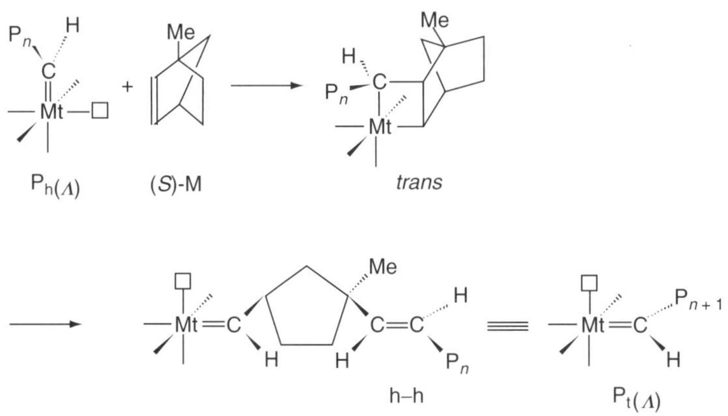

(33)

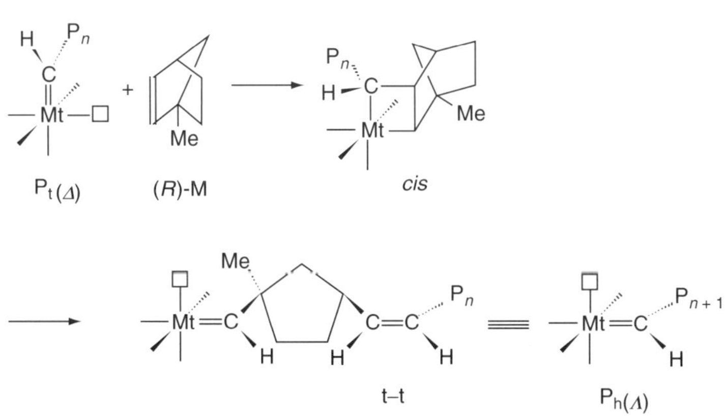

(34)

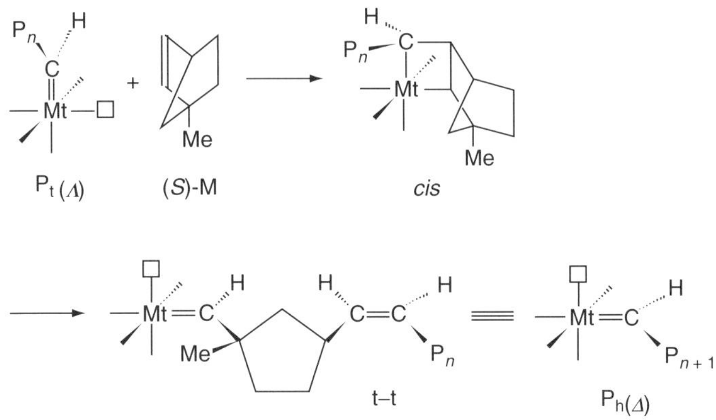

(35)

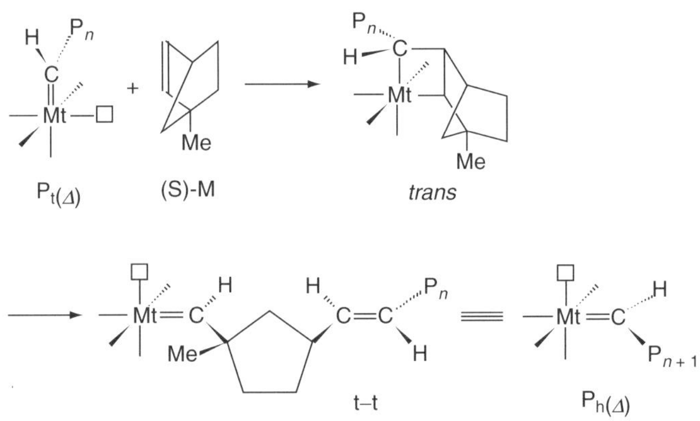

(36)

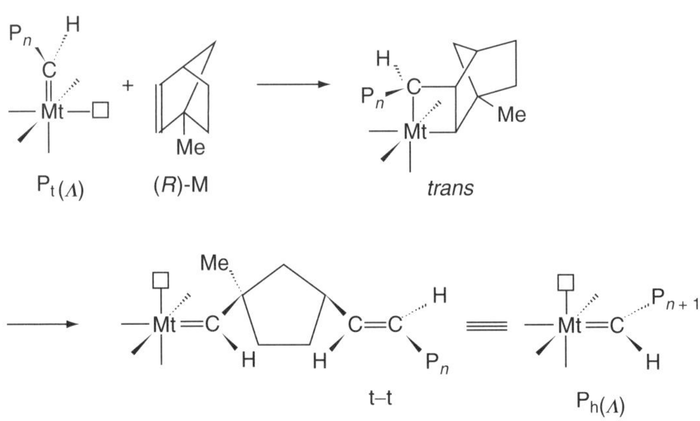

(37)

As regards metathesis polymerisation leading to *cis*/*trans* polymers, it is obvious that the *cis* junctions must be associated with *r* diads and the *trans* junctions with *m* diads [130]. In the case of the polymerisation affording fully atactic polymers, the active sites may be either achiral or chiral but epimerised. In the polymerisation leading to polymers exhibiting intermediate tacticity, partial epimerisation may occur between propagation steps [131].

6.2.6 Polymerisation of Cyclic Polyenes

Among cyclic polyenes, cyclic dienes, trienes and tetraenes have been ring-open polymerised via the metathesis reaction. Representative of the cyclodienes most commonly used for polymerisation are 1,5-cyclooctadiene, norbornadiene (bicyclo[2.2.1]hept-2,5-diene) and dicyclopentadiene as mono-, bi- and tricyclic diolefins respectively. Cycloocta-1,5-diene metathesis polymerisation is another approach to the preparation of 1,4-polybutadiene:

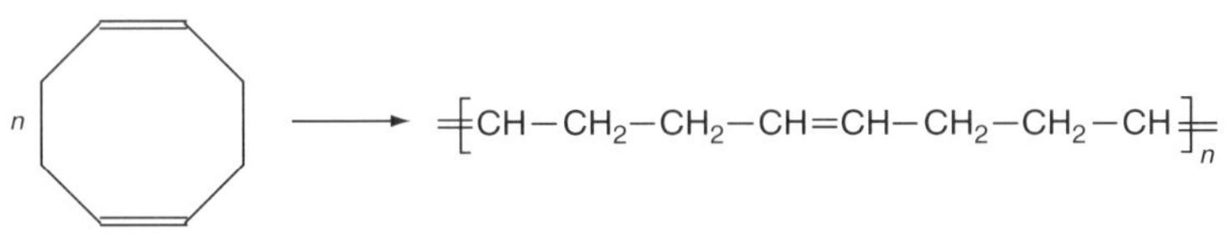

(38)

The polymerisation of *cis,cis* and *cis,trans* isomers of 1,5-cyclooctadiene can be run with usual soluble tungsten-based catalysts, e.g. such as WCl_6–$AlEtCl_2$, WCl_6–$Si(All)_4$ and WCl_6–$SnMe_4$ [106, 132, 133], and supported catalysts such as Re_2O_7/Al_2O_3 [134]. In addition to 1,5-cyclooctadiene, several substituted *cis*-1,5-cyclooctadienes were polymerised with metathesis catalysts [135–137]. For example, the polymerisation of 1-methylcycloocta-1,5-diene in the presence of a W-based catalyst yields a polymer that has the structure of an alternating copolymer of butadiene and isoprene; the unsubstituted carbon–carbon double bond is opened during the polymerisation [137]. It is worth noting that the polymerisation of 1,5-cyclodecadiene with the usual metathesis catalysts such as WCl_6–$AlEtCl_2$–EtOH leads to a polymer whose structure corresponds to that of the ethylene/butadiene (1:2 molar ratio) copolymer [138].

Norbornadiene, like norbornene, can be ring-open polymerised by a broad variety of metathesis catalysts, but crosslinking is difficult to prevent owing to the double bond of the cyclopentene ring present in the repeating unit [139]:

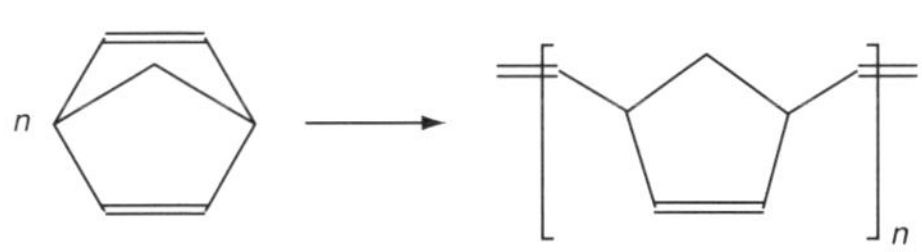

(39)

Nevertheless, such catalysts as WCl_6–$SnMe_4$ allow, under specific conditions, norbornadiene polymerisation to linear, non-crosslinked polymers [140]. When the carbon atom at one double bond of norbornadiene is substituted, selective polymerisation via the unsubstituted double bond is feasible without any subsequent crosslinking of the polymer formed [141].

The metathesis polymerisation of dicyclopentadiene (*endo*, *exo*) yields a linear polymer when carried out with single-component homogeneous metathesis catalysts of the metallacyclobutane or metal alkylidene type in diluted toluene solutions [142, 143].

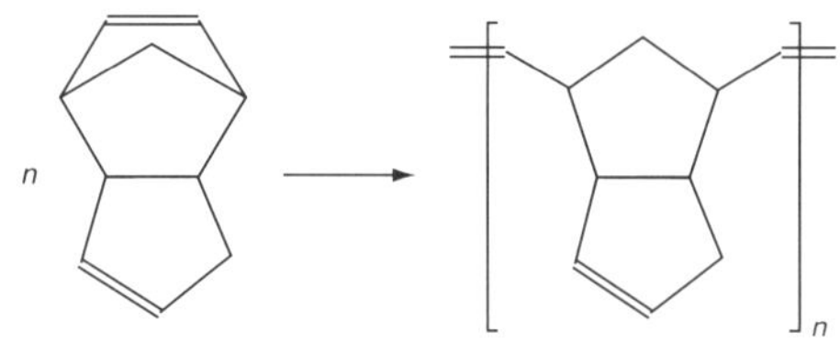

(40)

The polymer displays narrow molecular weight distributions characteristic of living polymerisation systems. Among catalysts exhibiting an ability to perform such controlled polymerisations of dicyclopentadiene according to scheme (40), titanacycles and Mo or W alkylidenes are representative [142, 143]:

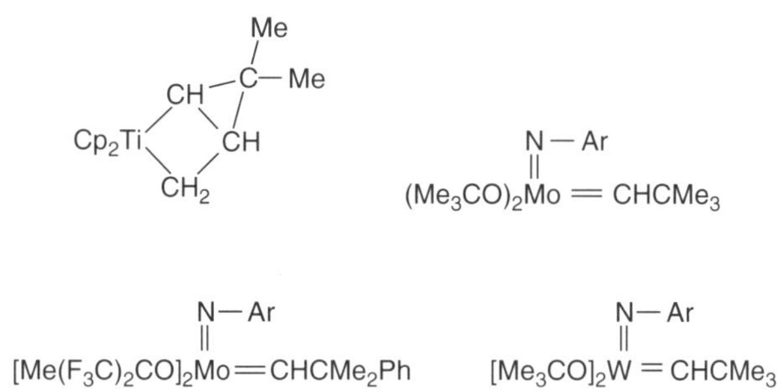

Ar = 2,6-diisopropylphenyl

The polymerisation of dicyclopentadiene is plagued by the same problem as that of norbornadiene, namely crosslinking. In polymerisations with homogeneous well-defined catalysts such as those mentioned above, the solution concentration determines whether a soluble or an insoluble polymer will be formed. The formation of an insoluble polymer is attributed to crosslinking either via a ring-opening reaction of the pendant cyclopentene rings in the initially formed linear polymer by a metathesis mechanism [scheme (41)] [142] or (at least partially) via a ring-retaining olefinic addition reaction of the pendant cyclopentene rings [scheme (42)], promoted by the heat released during the metathesis polymerisation [143]:

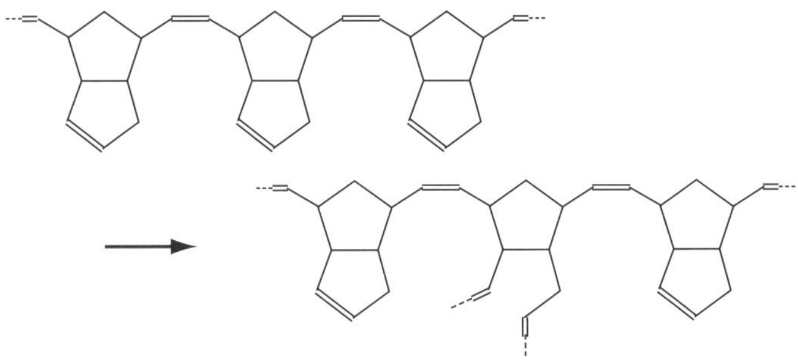

(41)

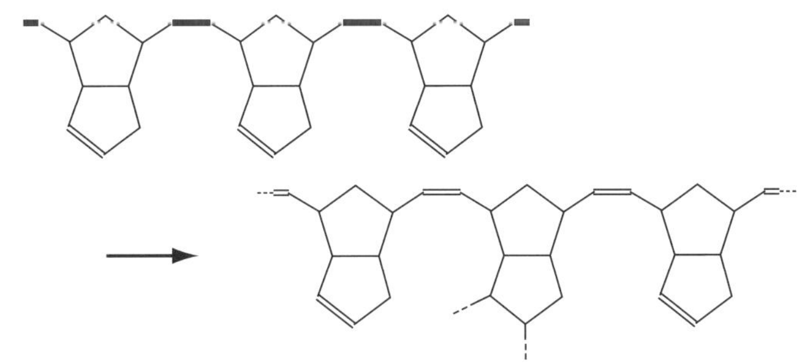

(42)

If the heat is removed from the polymerisation system through dilution or by cooling the solution, a soluble linear polymer of dicyclopentadiene remains the final product [143].

Polymerisations of dicyclopentadiene with conventional metathesis catalysts such as WCl_6, $MoCl_5$, $ReCl_5$, $NbCl_5$–$SnMe_4$, WCl_6–$SnMe_4$, WCl_6–$AlEtCl_2$–EtOH, (2,4,6-$Me_3C_6H_3$)$W(CO)_5$–$AlEtCl_2$, $(PhO)_xWOCl_{4-x}$-SnR_3H ($x = 1, 2, 3, 4$), carried out in hydrocarbon or chlorohydrocarbon solvents or without solvent in bulk, lead to insoluble crosslinked polymers [144–147]. Such processes are attracting increasing interest from the commercial point of view and have been utilised for industrial production; poly(dicyclopentadiene)s, both crosslinked elastomers and thermosetting resins, are the most technically important polymers made today via metathesis polymerisation [125].

[2.2]Paracyclophane-1,9-diene is an interesting cyclic diene monomer with a very strained ring. It has been easily polymerised with a tungsten carbene complex [148] to low molecular weight poly(*p*-phenylene vinylene) [148]:

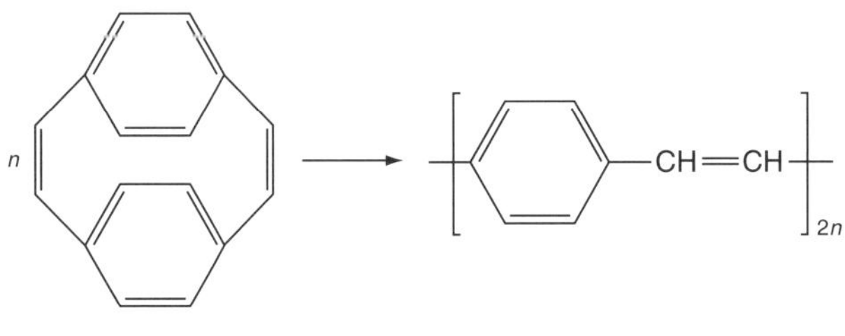

(43)

As regards the metathesis polymerisation of cyclic trienes, it has been carried out in an attempt to find alternative routes for preparing soluble and meltable precursors of polyacetylene [149, 150]. Hence, several substituted or unsubstituted tricyclic or other polycyclic trienes were subjected to polymerisation in the presence of metathesis catalysts such as WCl_6–$SnMe_4$ [151–154] and the tungsten neopentylidene complex $[Me(F_3C)_2CO]_2W(=NAr)(=CHCMe_3)$ [155]. A successful solution of the problem is outlined below [125,150]:

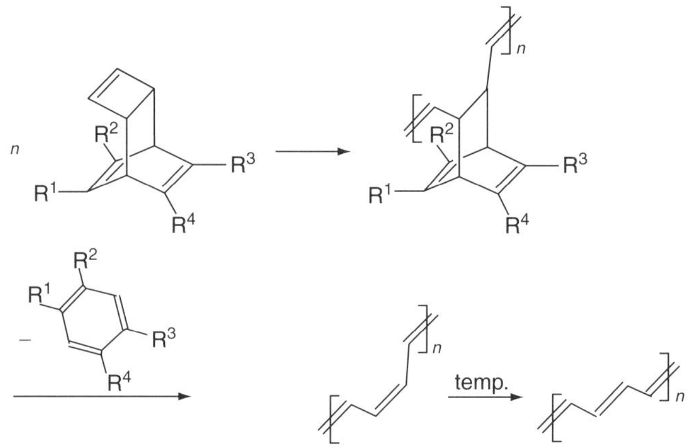

(44)

$R^1 = R^2 = CF_3$, COOMe; R^1–R^2 =–CH=CH–CH=CH–; $R^3 = R^4 =$ H

The policyclic triene monomer undergoes metathesis polymerisation exclusively by the cyclobutene double bond under mild conditions (in toluene solution at 20 °C) to give a soluble precursor polymer. This polyacetylene precursor can be purified and characterised prior to its conversion at elevated temperature, via thermally initiated symmetry-allowed elimination (retro Diels–Alder reaction), to polyacetylene (a heat treatment of the product also results in isomerisation of the initial *cis* form to a more stable *trans* form) [150].

An interesting instance of the ring-opening metathesis polymerisation of cyclic trienes is the polymerisation of the novel conjugated diene 3,4-diisopropylidenecyclobutane in the presence of a bis(cyclopentadienyl)titanacyclobutane derivative as a catalyst, which affords a linear cross-conjugated polymer [156]:

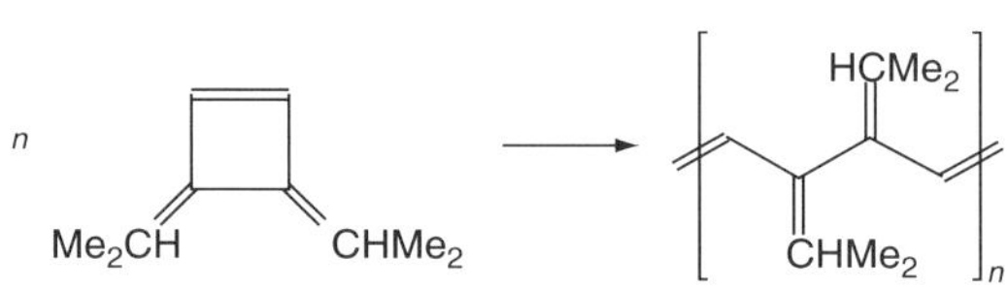

(45)

An alternative approach to the preparation of polyacetylene is provided by the ring-opening polymerisation of 1,3,5,7-cyclooctatetraene in the presence of a metathesis catalyst, e.g. WCl_6–$AlEt_2Cl$–epichlorohydrin ($Cl_4W[OCH(CH_2Cl)_2$ –$AlEt_2Cl$) [157–163]:

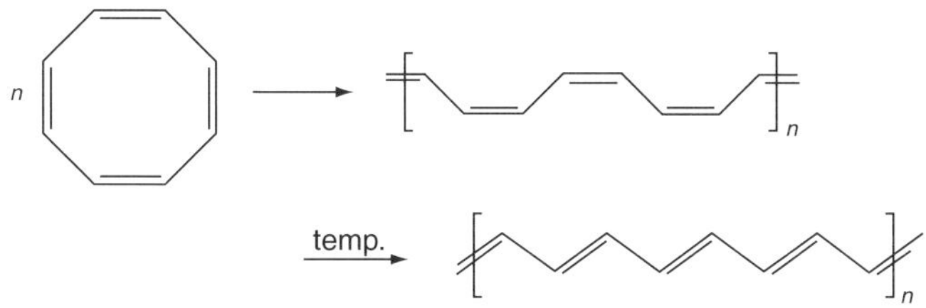

(46)

Although polyacetylene formed after thermal treatment has a *trans* structure, it undergoes a reversible photoinduced *cis–trans* isomerisation that can be a consequence of the formation of cyclic oligomers owing to secondary metathesis (backbiting) reactions occurring during cyclooctatetraene polymerisation [164]. It is worth mentioning in this connection that, during cyclooctatetraene polymerisation in dilute solution, benzene is formed irreversibly by a backbiting secondary metathesis reaction; by contrast, benzene formation is negligible in bulk metathesis polymerisation [159].

An interesting extension of cyclooctatetraene polymerisation is the polymerisation of trimethylsilyl-substituted cyclooctatetraene to give soluble polyacetylene substituted with one trimethylsilyl group every eight carbon atoms [162]. Such materials have been used to construct solar cells [165].

6.3 Ring-opening Polymerisation of Exocyclic Olefins

An interesting example of the polymerisation of exocyclic olefins is the ring-opening polymerisation of exomethylenecycloalkanes such as exomethylenecyclobutane (isomeric to cyclopentene) with the bis(1,2-dimethylcyclopentadienyl)-dimethylzirconium $[(Me_2Cp)_2ZrMe_2]–B(C_6F_5)_3$ catalyst [166]. This insertion isomerisation–polymerisation, in contrast to the ring-retaining and ring-opening polymerisation of internal cycloolefins such as cyclopentene, yields a polymer with 2-methylene-1-buten-4-yl units rather than 1-methylenecyclobutan-1-yl units. The latter units, which are initially formed by the insertion of the monomer into a Zr–C bond in cationic species $[(Me_2Cp)_2Zr–Me]^+$ $[B(Me)(C)_6F_5)_3]^-$, rapidly undergo a rearrangement to the former units; this rearrangement involves β-alkyl transfer:

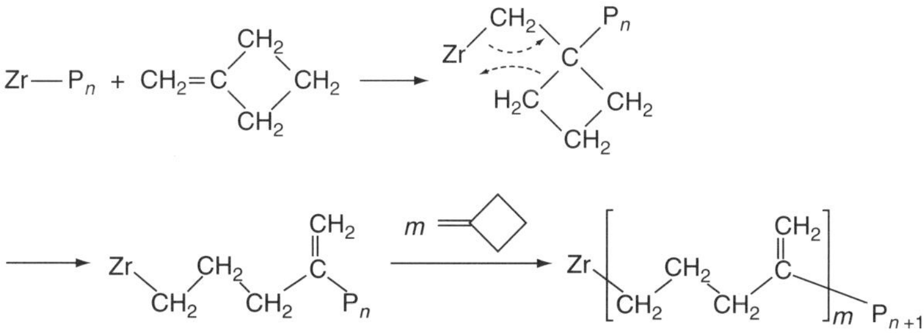

(47)

Apparently, the insertion of a β,β-disubstituted olefin (which exomethylenecyclobutane is) into a Zr–alkyl bond with a quaternary C_β atom is so slow that the rearrangement shown in scheme (47) is the preferred process.

The polymerisation of exomethylenecyclobutane may present interesting arguments to explain the lack of polymerisability of β,β-disubstituted acyclic olefins such as isobutylene according to the insertion mechanism in the presence of coordination catalysts. Probably, also in this case, β-methyl transfer in species that may initially be formed by 1,2-insertion [scheme (48)] occurs much faster than the next monomer insertion:

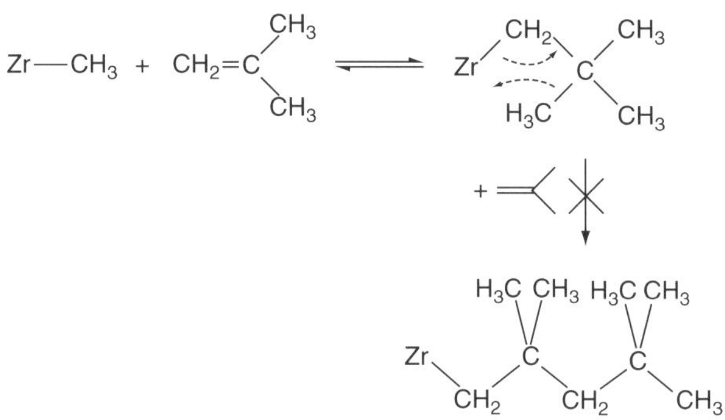

(48)

It can be seen from scheme (48) that such a transfer, if it occurs, reduces the chance of the polyisobutylene chain growth.

6.4 Industrial Polymerisation Processes

Both the ring-retaining and ring-opening polymerisations of cycloolefins lead to polymers that are of high potential utility as polymeric resins or elastomers. The polymers formed by ring-retaining homopolymerisation of cycloolefins are not processable owing to their high melting point and their insolubility in common organic solvents. However, when cycloolefins are copolymerised with ethylene, phase transition temperatures are lowered. This facilitates the melt processing of such polymers. Cycloolefin copolymers characterised by a low ethylene content represent a new class of thermoplastic amorphous materials. They exhibit heat resistance to ca 200 °C, high rigidity, toughness and environmental stability combined with low water uptake and excellent optical properties. These amorphous, transparent cycloaliphatic polymers have been used successfully for manufacturing optical fibres, lenses and disc coatings. Novel high distortion temperature polyolefins and their blends offer attractive potential for the development of engineering resins [167–171].

The products of the ring-opening metathesis polymerisation of cycloolefins, poly(1-alkenylene)s, are known as polyalkenamers according to the nomenclature proposed by Huggins [172], some of which have industrial significance. Polyalkenamers have made an increasingly important impact in the advanced materials field. When prepared from monocyclic olefins, they are usually crystalline substances with rather low melting points depending on the configuration (*cis* or *trans*) of their double bonds [173–176]. Metathesis polymerisations of monocyclic olefins that have been developed commercially comprise cyclopentene and cyclooctene. In commercial polypentenamer and polyoctenamer, the double bond configuration is primarily *trans*. Although the synthesis of polypentenamer was developed in the early 1970s, ostensibly for the tyre market, problems have delayed high-mileage production, since the technology makes it more expensive than traditional well-established polybutadiene and polyisoprene materials. However, polyoctenamer manufacture has been introduced successfully into industry [150,177,178].

Polymers derived from the ring-opening metathesis polymerisation of bicyclic olefins are highly moulded rigid substances. Poly(1-vinylene-3-cyclopentylene) with a high *trans* content of the double bonds and molecular weights of over 10^6, produced from norbornene, is a tough thermoplastic material, but becomes elastomeric after absorption of oil (up to 5 times it own mass) [150,177].

The metathesis polymerisation of dicyclopentadiene, an inexpensive monomer (commercially available cyclopentadiene dimer produced by a Diels–Alder addition reaction containing ca 95 % *endo* and ca 5 % *exo* form), leads to a polymer that may be transformed into a technically useful elastomer [144–146, 179] and thermosetting resin [180,181]. The polymerisation has characteristics that make it readily adaptable to the reaction injection moulding ('rim') process [182]. The main feature of this process comes from the fact that the polymerisation is carried out directly in the mould of the desired final product. The active metathesis catalyst is formed when two separate reactants, a precatalyst (tungsten-based) component and an activator (aluminium-based) component, are combined. Monomer streams containing one respective component are mixed directly just before entering the mould, and the polymerisation into a partly crosslinked material takes place directly in this mould (Figure 6.5) [147,168,183–186].

Polydicyclopentadiene obtained in such a process exhibits a glass transition temperature well above 100 °C, high stiffness, high impact strength, low water uptake and low dielectric constant. Recently, the process has been improved by using monomer-soluble trisalkylammoniummolybdates, activated by alkylaluminium alkyls, as catalysts. Owing to the improved solubility of the catalyst in the dicyclopentadiene monomer, the addition of toluene (for dissolving the catalyst) is eliminated, thus enhancing the monomer conversion and improving the thermal and mechanical properties of the product [183].

Commercially available polymers, produced from cycloolefins by ring-retaining polymerisation and ring-opening metathesis polymerisation of cycloolefins, and their typical uses are listed in Table 6.1 [12,14,144–147, 150,167–171,177–186].

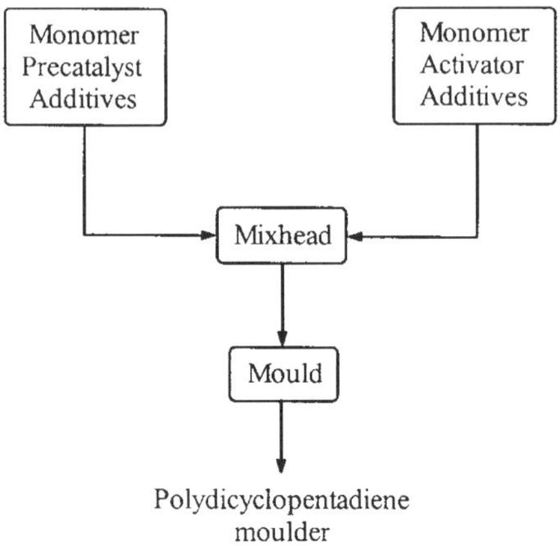

Figure 6.5 Flow scheme of polydicyclopentadiene production via ring-opening metathesis polymerisation in the reaction injection moulding process

Table 6.1 Commerically available polymers of cycloolefins, produced by coordination polymerisation, and their typical uses

Polymer	Typical use
Cycloolefin/ethylene copolymers[a]	Engineering plastics especially for optical applications (compact discs, optical fibres, lenses)
Polydicyclopentadiene[b]	Special-purpose elastomers
Polydicyclopentadiene[c]	Structural plastics
Poly(1-octenylene)[d]	Blending with other more common elastomers
Poly(1-vinylene-3-cyclopentylene)[e]	Moulding compounds, engine mounts, car bumper guards, noise and vibration damping applications

[a] Cyclopentene and norbornene random copolymers with ethylene formed via ring-retaining polymerisation; according to Refs 12, and 14 169 to 171.
[b] Dicyclopentadiene polymer formed via ring-opening metathesis polymerisation (elastomer); according to Refs 144–146 and 179.
[c] Dicyclopentadiene polymer formed via ring-opening metathesis polymerisation (thermosetting resin); according to Refs 180 to 186.
[d] Cyclooctene polymer (polyoctenamer) with a *trans*-structure, formed via ring-opening metathesis polymerisation; according to Refs 150, 177 and 178.
[e] Norbornene polymer (polynorbornenamer) with a *trans*-structure, formed via ring-opening metathesis polymerisation; according to Refs 150 and 177.

References

1. Natta, G., Dall'Asta, G., Mazzanti, G., Pasquon, I., Valvassori, A. and Zambelli, A., *J. Am. Chem. Soc.*, **83**, 3343 (1961).
2. Natta, G., Dall'Asta, G. and Mazzanti, G., *Angew. Chem., Int. Ed. Engl.*, **3**, 723 (1964).

3. Dall'Asta, G., Mazzanti, G., Natta, G. and Porri, L., *Makromol. Chem.*, **56**, 224 (1962).
4. Natta, G., Dall'Asta, G., Mazzanti, G. and Motroni, G., *Makromol. Chem.*, **69**, 163 (1963).
5. Truett, W. L., Johnson, D. R., Robinson, I. M. and Montague, B. A., *J. Am. Chem. Soc.*, **82**, 2337 (1960).
6. Sartori, G., Ciampelli, F. and Cameli, F., *Chim. Ind. (Milan)*, **45**, 1479 (1963).
7. Tsujino, T., Saegusa, T. and Furukawa, J., *Makromol. Chem.*, **85**, 71 (1965).
8. Mehler, C. and Risse, W., *Makromol. Chem., Rapid Commun.*, **12**, 255 (1991).
9. Mehler, C. and Risse, W., *Macromolecules*, **25**, 4226 (1992).
10. Melia, J., Connor, E., Rush, S., Breunig, S., Mehler, C. and Risse, W., *Macromol. Symp.*, **89**, 433 (1995).
11. Seehof, N., Mehler, C., Breunig, S. and Risse, W., *J. Mol. Catal.*, **76**, 219 (1992).
12. Kaminsky, W., Bark, A. and Arndt, M., *Makromol. Chem., Macromol. Symp.*, **47**, 83 (1991).
13. Kaminsky, W. and Bark, A., *Polym. Int.*, **28**, 251 (1992).
14. Kaminsky, W. and Noll, A., *Polym. Bull.*, **31**, 175 (1993).
15. Arndt, M. and Kaminsky, W., *Macromol. Symp.*, **97**, 225 (1995).
16. Arndt, M. and Kaminsky, W., *Macromol. Symp.*, **95**, 167 (1995).
17. Arndt, M., Engehausen, R., Kaminsky, W. and Zoumis, K., *J. Mol. Catal., A, Chem.*, **101**, 171 (1995).
18. Kaminsky, W. and Spiehl, R., *Makromol. Chem.*, **190**, 515 (1989).
19. Kelly, W. M., Taylor, N. J. and Collins, S., *Macromolecules*, **27**, 4477 (1994).
20. Collins, S. and Kelly, W. M., *Macromolecules*, **25**, 233 (1992).
21. Frisch, H. L., Mallows, C. L. and Bovey, F. A., *J. Chem. Phys.*, **45**, 1565 (1966).
22. Brandrup, J., 'Stereochemical Definitions and Notations Relating to Polymers', in *Polymer Handbook*, Wiley-Interscience, John Wiley & Sons, New York, 1989, pp. I/43–61.
23. Rush, S., Reinmuth, A. and Risse, W., in proceeding of International Symposium on *Olefin Metathesis and Related Chemistry*, Durham, UK, Abstracts, P50.
24. Porri, L., Natta, G. and Gallazzi, M. C., *Chim. Ind. (Milan)*, **46**, 428 (1964).
25. Natta, G., Dall'Asta, G. and Motroni, G., *J. Polym. Sci., B*, **2**, 349 (1964).
26. Goodall, B. L., McIntosh III, L. H. and Rhodes, L. F., *Macromol. Symp.*, **89**, 421 (1995).
27. Sen, A. and Lai, T. W., *J. Am. Chem. Soc.*, **104**, 3520 (1982).
28. Drent, E., van Broeckhoven, J. A. M. and Doyle, M. J., *J. Organomet. Chem.*, **417**, 235 (1991).
29. Amevor, E., Bronco, S., Consiglio, G. and Di Benedetto, S., *Macromol. Symp.*, **89**, 443 (1995).
30. Amevor, E. and Consiglio, G., *Chimia*, **47**, 283 (1993).
31. Reichel, B., Marvel, C. S. and Greenley, R. Z., *J. Polym. Sci., A*, **1**, 2935 (1963).
32. Bokaris, E. P., Siskos, M. G. and Zarkadis, A. K., *Eur. Polym. J.*, **12**, 1441 (1992).
33. Valvassori, A., Sartori, G., Turba, V. and Lachi, M. P., *J. Polym. Sci., C*, **16**, 23 (1967).
34. Soga, K., Shiono, T., Takemura, S. and Kaminsky, W., *Makromol. Chem., Rapid Commun.*, **8**, 305 (1987).
35. Grassi, A., Zambelli, A., Resconi, L., Albizzati, E. and Mazzochi, R., *Macromolecules*, **21**, 617 (1988).
36. Tsutsui, T., Mizuno, A., and Kashiwa, N., *Makromol. Chem.*, **190**, 1177 (1989).
37. Rieger, B. and Chien, J. C. W., *Polym. Bull.*, **21**, 159 (1989).
38. Rieger, B., Mu, X., Mallin, D. T., Rausch, M. D. and Chien, J. C. W., *Macromolecules*, **23**, 3559 (1990).

39. Asakara, T., Nakayama, N., Demura, M. and Asano, A., *Macromolecules*, **25**, 4876 (1992).
40. Anderson, A. W. and Merckling, N. G., US Pat. 2 721 189 (to Du Pont de Nemours) (1955); *Chem. Abstr.*, **50**, 3008 (1956).
41. Eleuterio, H. S., US Pat. 3 074 918 (to Du Pont de Nemours) (1957); *Chem. Abstr.*, **55**, 16005 (1961).
42. Calderon, N., Chen, N. Y. and Scott, K. W., *Tetrahedron Lett.*, **1967**, 3327 (1967).
43. Calderon, N., Ofstead, E. A., Ward, J. P., Judy, W. A. and Scott, K. W., *J. Am. Chem. Soc.*, **90**, 4133 (1968).
44. Dall'Asta, G. and Motroni, G., *Eur. Polym. J.*, **7**, 707 (1971).
45. Ivin, K. J., 'Metathesis Polymerisation,' in *Encyclopedia of Polymer Science and Engineering*, Wiley-Interscience, John Wiley & Sons, New York, 1987, Vol. 9, pp. 634–669.
46. Hérisson, J. L. and Chauvin, T., *Makromol. Chem.*, **141**, 161 (1970).
47. Grubbs, R. and Novak, B. M., 'Living Polymer Systems – Olefin Metathesis', in *Encyclopedia of Polymer Science and Engineering*, Wiley-Interscience, John Wiley & Sons, New York, 1989, Supplement Vol., pp. 420–429.
48. Feast, W. J., Gibson, V. C., Kosravi, E., Mashall, E. L. and Wilson, B., 'Recent Developments in the Synthesis of Functionalised Polymers via Living Ring-opening Metathesis Polymerisation', in *Ziegler Catalysts*, Springer-Verlag, Berlin, 1995, pp. 469–480.
49. Schrock, R. R., 'Ring-opening Metathesis Polymerisation', in *Ring-opening Polymerisation*, Hanser Publishers, Munich, 1993, pp. 129–156.
50. Ivin, K. J., Laverty, D. T. and Rooney, J. J., *Makromol. Chem.*, **178**, 1545 (1977).
51. Ivin, K. J., Lapienis, G. and Rooney, J. J., *J. Chem. Soc., Chem. Commun.*, **1979**, 1068 (1979).
52. Ivin, K. J., Laverty, D. T., Rooney, J. J. and Watt, P., *Recl. Trav. Chim. Pays-Bas*, **96**, M54 (1977).
53. Ivin, K. J., *Pure Appl. Chem.*, **52**, 1907 (1980).
54. Hamilton, J. G., Ivin, K. J., Rooney, J. J. and Waring, L. C., *J. Chem. Soc., Chem. Commun.*, **1983**, 159 (1983).
55. Dall'Asta, G. and Manetti, R., *Eur. Polym. J.*, **4**, 145 (1968).
56. Höcker, H., Reimann, W., Reif, L. and Riebel, K., *J. Mol. Catal.*, **8**, 191 (1980).
57. Denisova, T. T., Syatkovskii, A. I. and Skuratova, T. B., *Polym. Sci. USSR, A*, **25**, 798 (1983).
58. Natta, G., Dall'Asta, G., Bassi, I. W. and Carella, G., *Makromol. Chem.*, **91**, 87 (1966).
59. Laroche, C., Laval, J. P., Lattes, A., Leconte, M., Quignard, F. and Basset, J. M., *J. Chem. Soc., Chem. Commun.*, **1983**, 220 (1983).
60. Turner, L. and Bradshaw, C. P. C., Brit. Pat. 1 105 565 (to British Petroleum Co.) (1965); *Chem. Abstr.*, **68**, 87758 (1968).
61. Lindmark-Hamberg, M. and Wagener, K. B., *Macromolecules*, **20**, 2949 (1987).
62. Schrock, R. R., Murdzek, J. S., Bazan, G. C., Robbins, J., Dimare, M. and O'Regan, M., *J. Am. Chem. Soc.*, **112**, 3875 (1990).
63. Nguyen, S. T., Johnson, L. K. and Grubbs, R. H., *J. Am. Chem. Soc.*, **114**, 3975 (1992).
64. Novak, B. M. and Grubbs, R. H., *J. Am. Chem. Soc.*, **110**, 7542 (1988).
65. Kress, J. and Osborn, J. A., *J. Am. Chem. Soc.*, **105**, 6346 (1983).
66. Natta, G., Dall'Asta, G. and Porri, L., *Makromol. Chem.*, **81**, 253 (1965).
67. Michelotti, F. W. and Keaveney, W. P., *J. Polym. Sci., A*, **3**, 895 (1965).
68. Rinehart, R. E. and Smith, H. R., *J. Polym. Sci., B*, **3**, 1049 (1965).
69. Fischer, E. O. and Maasbol, A., *Angew. Chem., Int. Ed. Engl.*, **3**, 580 (1964).

70. Fischer, E. O. and Dotz, K. H., *Chem. Ber.*, **105**, 3966 (1972).
71. Fischer, E. O., *J. Organomet. Chem.*, **14**, 1 (1976).
72. Casey, C. P. and Burkhardt, T. J., *J. Am. Chem. Soc.*, **95**, 5833 (1973).
73. Casey, C. P. and Burkhardt, T. J., *J. Am. Chem. Soc.*, **96**, 7808 (1974).
74. Kress, J., Wesolek, M. and Osborn, J. A., *J. Chem. Soc., Chem. Commun.*, **1982**, 514 (1982).
75. Kress, J., Osborn, J. A., Greene, R. M. E., Ivin, K. J. and Rooney, J. J., *J. Chem. Soc., Chem. Commun.*, **1985**, 874 (1985).
76. Wengrovius, J., Schrock, R. R., Churchill, M. R., Missert, J. R. and Young, W. J., *J. Am. Chem. Soc.*, **102**, 4515 (1980).
77. Schaverien, C. J., Dewan, J. C. and Schrock, R. R., *J. Am. Chem. Soc.*, **108**, 2771 (1986).
78. Schrock, R. R., Feldman, J., Cannizzo, L. F. and Grubbs, R. H., *Macromolecules*, **20**, 1169 (1987).
79. Schrock, R. R., DePue, R. T., Feldman, J., Schaverien, C. J., Dewan, J. C. and Liu, A. H., *J. Am. Chem. Soc.*, **110**, 1423 (1988).
80. Demonceau, A. and Noels, A. F., *Education Adv. Chem.*, **1**, 25 (1993).
81. Carter, E. A. and Goddard III, W. A., *J. Am. Chem. Soc.*, **108**, 2180 (1986).
82. Carter, E. A. and Goddard III, W. A., *J. Am. Chem. Soc.*, **108**, 4746 (1986).
83. Noels, A. F., Demonceau, A., Carlier, E., Hubert, A. J., Márquez-Silva, R.-L. and Sánchez-Delgado, R. A., *J. Chem. Soc., Chem. Commun.*, **1988**, 783 (1988).
84. Katz, T. J. and Acton, L., *Tetrahedron Lett.*, **1976**, 4251 (1976).
85. Katz, T. J., *Adv. Organomet. Chem.*, **16**, 283 (1986).
86. Chauvin, Y., Commereuc, D. and Zaborowski, G., *Makromol. Chem.*, **179**, 1285 (1978).
87. Katz, T. J., McGinnis, J. and Altus, C., *J. Am. Chem. Soc.*, **91**, 606 (1976).
88. Katz, T. J., Lee, S. J. and Shippey, M. A., *J. Mol. Catal.*, **8**, 219 (1980).
89. Novak, B. M. and Grubbs, R. H., *J. Am. Chem. Soc.*, **110**, 960 (1988).
90. Feldman, J. and Schrock, R. R., *Prog. Inorg. Chem.*, **39**, 1 (1991).
91. Schrock, R. R., DePue, R. T., Feldman, J., Yap, K. B., Yang, D. C., Davis, W. M., Park, L. Y., DiMare, M., Schofield, M., Anhaus, J., Walborsky, E., Evitt, E., Krüger, C. and Betz, P., *Organometallics*, **9**, 2262 (1990).
92. Wu, Z., Benedicto, A. D. and Grubbs, R. H., *Macromolecules*, **26**, 4975 (1993).
93. Nguyen, S. T., Grubbs, R. H. and Ziller, J. W., *J. Am. Chem. Soc.*, **115**, 9858 (1993).
94. Lynn, D. M., Kanaoka, S. and Grubbs, R. H., *J. Am. Chem. Soc.*, **118**, 784 (1996).
95. Katz, T. J., Ho, T. H., Shih, N.-Y., Ying, Y.-C. and Stuart, V. I. W., *J. Am. Chem. Soc.*, **106**, 2659 (1984).
96. Fischer, E. O. and Wagner, W. R., *J. Organomet. Chem.*, **116**, C21 (1976).
97. Gilliom, L. R. and Grubbs, R. H., *J. Am. Chem. Soc.*, **108**, 733 (1986).
98. Gilliom, L. R. and Grubbs, R. H., *Organometallics*, **5**, 721 (1986).
99. Novak, B. M. and Perrott, M. G., 'Ring-opening Metathesis Polymerisation', *Catalysis in Precision Polymerisation*, John Wiley & Sons, Chichester, 1997, pp. 100–112.
100. Wallace, K. C., Dewan, J. C. and Schrock, R. R., *Organometallics*, **5**, 2162 (1986).
101. Wallace, K. C. and Schrock, R. R., *Macromolecules*, **20**, 448 (1987).
102. Wallace, K. C., Liu, A. H., Dewan, J. C. and Schrock, R. R., *J. Am. Chem. Soc.*, **110**, 4964 (1988).
103. Dolgoplosk, B. A., Oreshkin, I. A. and Makovetsky, K. L., Tinyakova, E. I., Ostrovskaya, I. Ya., Kershenbaum, I. L. and Chernenko, G. M., *J. Organomet. Chem.*, **128**, 339 (1977).

104. Grubbs, R. H. and Hoppin, C. R., *J. Chem. Soc., Chem. Commun.*, **1977**, 634 (1977).
105. Bencze, L., Marko, L., Optiz, R. and Thiele, K. H., *Hung. J. Ind. Chem.*, **4**, 15 (1976).
106. Oreshkin, I. A., Red'kina, L. I., Kershenbaum, J. L., Chernenko, G. M., Makovetsky, K. L., Tinyakova, E. I. and Dolgoplosk, B. A., *Eur. Polym. J.*, **13**, 447 (1977).
107. Kupper, F. W. and Streck, R., *Makromol. Chem.*, **175**, 2055 (1974).
108. Amass, A. J. and Zurimendi, J. A., *J. Mol. Catal.*, **8**, 243 (1980).
109. Grubbs, R. H., 'Alkene and Alkyne Metathesis Reactions', in *Comprehensive Organometallic Chemistry*, Pergamon Press, Oxford, 1982, Vol. 8, pp. 499–551.
110. Calderon, N., Lawrence, J. P. and Ofstead, E. A., *Adv. Organomet. Chem.*, **17**, 449 (1979).
111. Banks, R. L., *Chem. Tech.*, **1979**, 494 (1979).
112. Grubbs, R. H., *Prog. Inorg. Chem.*, **24**, 1 (1978).
113. Laverty, D. T., McKervey, M. A., Rooney, J. J. and Stewart, A., *J. Chem. Soc., Chem. Commun.*, **1976**, 193 (1976).
114. Laverty, D. T., Rooney, J. J. and Stewart, A., *J. Catal.*, **45**, 110 (1976).
115. Muetterties, E. L. and Band, E., *J. Am. Chem. Soc.*, **102**, 6572 (1980).
116. Grubbs, R. H. and Swetnick, S. J., *J. Mol. Catal.*, **8**, 25 (1980).
117. Lombardo, E. A., Jacano, M. L. and Hall, W. K., *J. Catal.*, **51**, 243 (1978).
118. Rappe, A. K. and Goddard III, W. A., *J. Am. Chem. Soc.*, **102**, 5114 (1980).
119. Doherty, M., Siove, A., Parlier, A., Rudler, H. and Fontanille, M., *Makromol. Chem., Macromol. Symp.*, **6**, 33 (1986).
120. Ofstead, E. A., Lawrence, J. P., Senyek, M. L. and Calderon, N., *J. Mol. Catal.*, **8**, 227 (1980).
121. Amass, A. J., 'Metathesis Polymerisation: Chemistry', in *Comprehensive Polymer Science*, Pergamon Press, Oxford, 1989, Vol. 4, pp. 109–134.
122. Amass, A. J. and Zurimendi, J. A., *Polymer*, **23**, 211 (1982).
123. Amass, A. J. and Tuck, C. N., *Eur. Polym. J.*, **14**, 817 (1978).
124. Jacobson, H. and Stockmayer, W. H., *J. Chem. Phys.*, **18**, 1600 (1950).
125. Kricheldorf, H. R., 'Metathesis Polymerisation of Cycloolefins', in *Handbook of Polymer Synthesis*, M. Dekker, New York, 1992, Part A, pp. 433–479.
126. Dainton, F. S. and Ivin, K. J., *Q. Rev., Chem. Soc.*, **12**, 61 (1958).
127. Lebedev, B. and Smirnova, N., *Macromol. Chem. Phys.*, **195**, 35 (1994).
128. Hillmyer, M. A., Benedicto, A. D., Nguyen, S. T., Wu, Z. and Grubbs, R. H., *Macromol. Symp.*, **89**, 411 (1995).
129. Ivin, K. J., O'Donnell, J. H., Rooney, J. J. and Stewart, C. D., *Makromol. Chem.*, **180**, 1975 (1979).
130. Devine, G. I., Ho, H. T., Ivin, K. J., Mohamed, M. A. and Rooney, J. J., *J. Chem. Soc., Chem. Commun.*, **1982**, 1229 (1982).
131. Ho, H. T., Ivin, K. J. and Rooney, J. J., *J. Mol. Catal.*, **15**, 245 (1982).
132. Ivin, K. J., Rooney, J. J., Bencze, L., Hamilton, J. G., Lam, L. M., Lapienis, G., Reddy, B. S. R. and Ho, T. H., *Pure Appl. Chem.*, **54**, 447 (1982).
133. Zerpner, D., Holtrup, W. and Streck, R., *J. Mol. Catal.*, **36**, 115 (1986).
134. Sato, H., Tanaka, Y. and Taketami, T., *Makromol. Chem.*, **178**, 1993 (1977).
135. Syatkovskii, A. I., Denisova, T. T., Abramenko, Y. L., Khatchaturov, A. S. and Babitsky, B. D., *Polymer*, **22**, 1554 (1981).
136. Tlenkopachev, M. A., Avdeikina, E. G., Korshak, Y. V., Bondarenko, G. N., Dolgoplosk, B. A. and Kutepov, D. L., *Dokl. Akad. Nauk SSSR*, **268**, 133 (1983).
137. Khatchaturov, A. S., Abramenko Y. L. and Syatkovskii, A. I., *Polym. Sci. USSR*, **24**, 775 (1982).
138. Furukawa, J. and Mizoe, Y., *J. Polym. Sci., Polym. Lett. Ed.*, **11**, 263 (1973).

139. Ivin, K. J., Laverty, T. and Rooney, J. J., *Makromol. Chem.*, **179**, 253 (1978).
140. Reif, L. and Höcker, H., *Makromol. Chem., Rapid Commun.*, **2**, 745 (1981).
141. Feast, W. J. and Wilson, B., *J. Mol. Catal.*, **8**, 277 (1980).
142. Fisher, R. A. and Grubbs, R. H., *Makromol. Chem., Macromol. Symp.*, **63**, 271 (1992).
143. Davidson, T. A. and Wagener, K. B., *J. Mol. Catal. A, Chem.*, **133**, 67.
144. Marshall, P. R. and Ridgewell, B. J., *Eur. Polym. J.*, **5**, 29 (1969).
145. Rinehart, R. E., *J. Polym. Sci., A*, **7**, 27 (1969).
146. Hamilton. J. G., Ivin, K. J. and Rooney, J. J., *J. Mol. Catal.*, **36**, 115 (1986).
147. Bell, A., 'Dicyclopentadiene Polymerisation using Well-characterized Tungsten Phenoxide Complexes', in *Catalysis in Polymer Synthesis*, ACS Symp. Ser. 496, Washington, DC, 1992, pp. 121–133.
148. Grubbs, R. H. and Johnson, L. K., *J. Am. Chem. Soc.*, **112**, 5384 (1990).
149. Thorn-Csányi, E. and Höhnk, H.-D., *J. Mol. Catal.*, **76**, 101 (1992).
150. Feast, W. J., 'Metathesis Polymerization: Applications', in *Comprehensive Polymer Science*, Pergamon Press, Oxford, 1989, Vol. 4, pp. 135–142.
151. Edwards, J. H., Feast, W. J. and Bott, D. C., *Polymer*, **25**, 359 (1984).
152. Edwards, J. H. and Feast, W. J., *Polym. Commun.*, **21**, 595 (1980).
153. Bott, D. C., Chai, C. K., Edwards, J. H., Feast, W. J., Friend, R. H. and Horton, M. E., *J. Phys., C*, **3**, 143 (1983).
154. Bott, D. C., *Synth. Met.*, **14**, 245 (1986).
155. Schrock, R. R., *Polym. Mater. Sci. Eng.*, **58**, 92 (1988); *Chem. Abstr.*, **109**, 170958e (1988).
156. Swager, T. M. and Grubbs, R. H., *J. Am. Chem. Soc.*, **108**, 2771 (1986).
157. Korshak, Yu. V., Korshak, V. V., Kanischka, G. and Höcker, H., *Makromol. Chem., Rapid Commun.*, **6**, 685 (1985).
158. Tlenkopachev, M. A., Korshak, V. V., Orlov, V. A. and Korshak, V. V., *Dokl. Akad. Nauk SSSR*, **291**, 409 (1986).
159. Klavetter, F. L. and Grubbs, R. H., *J. Am. Chem. Soc.*, **110**, 7807 (1988).
160. Klavetter, F. L. and Grubbs, R. H., *Polym. Mater. Sci. Eng.*, **58**, 855 (1988).
161. Klavetter, F. L. and Grubbs, R. H., *Polym. Mater. Sci. Eng.*, **59**, 586 (1988).
162. Ginsburg, E. J., Gorman, C. B., Marder, S. R. and Grubbs, R. H., *J. Am. Chem. Soc.*, **111**, 7621 (1989).
163. Klavetter, F. L. and Grubbs, R. H., *Synth. Met.*, **28**, D99 (1989).
164. Berdyugin, V. V., Burshtein, K. Ya., Shorygin, P. P., Korshak, Yu. V., Orlov, A. V. and Tlenkopachev, M. A., *Dokl. Akad. Nauk SSSR*, **312**, 410 (1990).
165. Sailor, M. J., Ginsburg, E. J., Gorman, C. B., Kumar, A., Grubbs, R. H. and Lewis, N. S., *Science*, **249**, 1146 (1990).
166. Yang, X., Jia, L. and Marks, T. J., *J. Am. Chem. Soc.*, **115**, 3392 (1993).
167. Moriya, A., Ishimoto, A. and Takahashi, M., Eur. Pat. Appl. 486 365 (to Mitsui Petrochemicals) (1991).
168. Benedict, G. M., Goodall, B. L., Marchant, N. S. and Rhodes, L. F., *New J. Chem.*, **18**, 105 (1994).
169. Kashiwa, H. and Toyota, A., *Chem. Econ. Engng*, **18**, 14 (1986).
170. Cherdron, H., Brekner, M.-J. and Osan, F., *Angew. Makromol. Chem.*, **223**, 121 (1994).
171. Cherdron, H. and Brekner, M.-J., *Macromol. Symp.*, **89**, 543 (1995).
172. Huggins, M. L., *J. Polym. Sci.*, **8**, 257 (1952).
173. Haas, F, Nützel, K., Pampus, G. and Theisen, D., *Rubber Chem. Technol.*, **43**, 1116 (1970).
174. Natta, G., Dall'Asta, G., Bassi, J. W. and Carella, G., *Makromol. Chem.*, **91**, 87 (1966).

175. Minchak, R. J. and Tucker, H., *Polym. Prepr. Am. Chem. Soc., Div. Polym. Chem.*, **13**, 885 (1972).
176. Dall'Asta, G., *Rubber Chem. Technol.*, **47**, 511 (1974).
177. Streck, R., *Chemtech.*, **13**, 758 (1983).
178. Streck, R., *J. Mol. Catal.*, **4**, 305 (1988).
179. Matejka, L., Houtman, C. and Macosko, C. W., *J. Appl. Polym. Sci.*, **30**, 2787 (1985).
180. Breslow, D. S., *Prog. Polym. Sci.*, **18**, 1141 (1993).
181. Fitzer, E., *Angew. Chem., Int. Ed. Engl.*, **19**, 375 (1980).
182. Macosko, C. W., in *RIM – Fundamentals of Reaction Injection Molding*, Hanser Publishers, New York, 1989.
183. Goodall, B. L., Kroenke, W. J., Minchak, R. J. and Rhodes, L. F., *J. Appl. Polym. Sci.*, **47**, 607 (1993).
184. Breslow, D. S., *Chemtech.*, **20**, 540 (1990).
185. Klosiewicz, D. W., US Pat. 4 400 340 (to Hercules) (1983).
186. Minchak, R. J. US Pat. 4 002 815 (to Goodrich) (1977).

Further Reading

Kaminsky, W., 'Zirconocene Catalysts for Olefin Polymerisation', *Catal. Today*, **20**, 257–271 (1994).

Kaminsky, W. and Noll. A., 'Polymerization of Phenyl Substituted Cyclic Olefins with Metallocene/Aluminoxane Catalysts', in *Ziegler Catalysts*, Springer-Verlag, Berlin, 1995, pp. 149–158.

Kaminsky, W., 'New Polymers by Metallocene Catalysis', *Macromol. Chem. Phys.*, **197**, 3907–3945 (1996).

Huang, B. and Tian, J., 'Metallocene Catalysts', in *The Polymeric Materials Encyclopedia*, CRC Press, Inc., Boca Raton, 1996, Vol. 6, pp. 4191–4201.

Kaminsky, W. and Arndt, M., 'Metallocenes for Polymer Catalysis', *Adv. Polym. Sci.*, **127**, 143–187 (1997).

Arndt, M., 'New Polyolefins', in *Handbook of Thermoplastics*, M. Dekker, Inc., New York, 1997, pp. 39–56.

Kaminsky, W., 'Highly Active Metallocene Catalysts for Olefin Polymerisation', *J. Chem. Soc., Dalton Trans.*, **1998**, 1413–1418 (1998).

Höcker, H., 'Ring-opening Polymerisation of Cycloolefins by Means of Metathesis Catalysts: Kinetic and Thermodynamic Effects on the Product Distribution', *Makromol. Chem. Macromol. Symp.*, **6**, 47–52 (1986).

Eleuterio, H. S., 'Scientific Discovery and Technological Innovation: an Electic Odyssey into Olefin Metathesis Chemistry', *J. Macromol. Sci. – Chem. A*, **28**, 907–915 (1991).

Korshak, Yu. V., 'Metathesis Polymerization of Cycloolefins', in *The Polymeric Materials Encyclopedia*, CRC Press, Inc., Boca Raton, 1996, Vol. 6, pp. 4250–4263.

Karlen, T., Ludi, A., Mühlebach, A., Bernhard, P. and Pharisa, C., 'Photoinduced Ring-opening Metathesis Polymerisation of Strained Bicyclic Olefins with Ruthenium Complexes', *J. Polym. Sci., A, Polym. Chem.*, **33**, 1665–1674 (1995).

Thorn-Csányi, E. and Kraxner, P., 'Metathesis Preparation of *p*-Phenylene Vinylene Oligomers, Homo- and Copolymers', in *The Polymeric Materials Encyclopedia*, CRC Press, Inc., Boca Raton, 1996, Vol. 7, pp. 5055–5067.

Stelzer, F., Schimetta, M. and Leising, G., 'Poly(cyclopentadienylene-vinylene)', in *The Polymeric Materials Encyclopedia*, CRC Press, Inc., Boca Raton, 1996, Vol. 8, pp. 5715–5725.

Problems

1. Name and characterise the main types of coordination polymerisation of cycloolefins and give examples of representative catalysts for each type.
2. Name and characterise bonds anchoring the growing chain of cycloolefin polymers at the transition metal atom, depending on the kind of catalyst.
3. Give the structure and stereochemical designation of all possible cycloolefin polymers that might, in principle, be obtained by stereospecific ring-retaining polymerisation.
4. Explain why ring-retaining polymerisation of cyclopentene with a zirconocene catalyst yields poly(1,3-cyclopentylene) but not poly (1,2-cyclopentylene).
5. Explain why ring-retaining polymerisation of norbornene with a zirconocene catalyst proceeds via 1,2-insertion, leading to poly(2,3-nonorbornene), which is in contrast to the polymerisation of cyclopentene with this catalyst.
6. Give the structure and stereochemical designation of all possible cycloolefin polymers that might, in principle, be obtained by stereospecific ring-opening metathesis polymerisation.
7. Show by structural drawings the various (if any) stereoregular polymers that might be possibly obtained from each of the following cycloolefins:

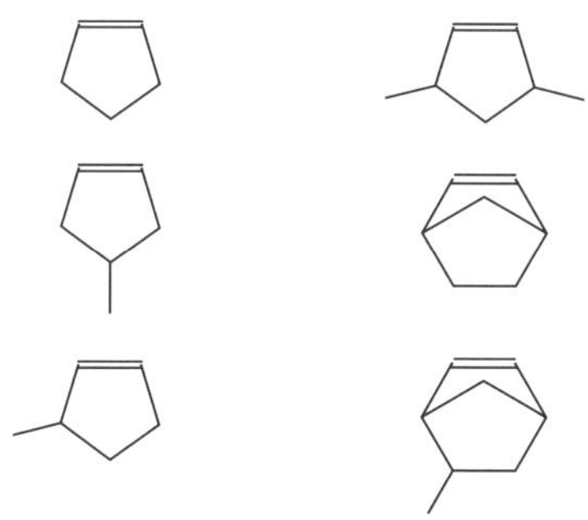

8. What structure will be possessed by polymers formed in the polymerisation of norbornene with the following catalysts: $TiCl_4$–MgEtBr, WCl_6–$AlEt_2Cl$, *rac.*-$Me_2Si(Ind)_2ZrCl_2$–$[Al(Me)O]_x$, $ReCl_5$, $[Pd(MeCN)_4][BF_4]_2$, $(ArO)_3Ta{=}CHCMe_3$.

9. Arrange the following monomers according to their polymerisability with metathesis catalysts: cyclobutene, cyclopentene, cyclooctene, norbornene, dicyclopentadiene.
10. Which cycloolefins (monocyclic, bicyclic or policyclic) will undergo secondary metathesis reactions during ring-opening metathesis polymerisation? Give reasons why.
11. Explain why ring-opening metathesis polymerisation of cycloolefins is accompanied with backbiting reactions leading to cyclic oligomers more readily in dilute solution than in bulk.
12. Indicate cycloolefin monomers that will be polymerised by metathesis catalysts to polymers with a structure equivalent to polybutadiene, an alternating ethylene/butadiene copolymer, an alternating butadiene / isoprene copolymer and polyacetylene.

7 Coordination Polymerisation of Alkynes

Acetylene was first polymerised to polyacetylene (polyvinylene) according to scheme (1) by Ziegler, who patented a process employing new catalysts prepared from transition metal salts and metal alkyls [1]:

$$n\ \mathrm{CH{\equiv}CH} \longrightarrow \mathrm{-\!\!\left[CH{=}CH\right]\!\!-}_n \qquad (1)$$

Thereafter, Natta *et al.* [2,3] polymerised acetylene in the presence of soluble $Ti(OBu)_4$–$AlEt_3$ catalysts to crystalline polyacetylene with a predominant *trans* structure of the conjugated carbon–carbon double bonds. A number of other Ziegler–Natta catalysts, including soluble and insoluble ones, were subsequently demonstrated to be capable of promoting acetylene polymerisation to high molecular weight polyacetylene [4–9]. It was found that the Ti $(OBu)_4$–$AlEt_3$ (1:4) catalyst was the most efficient Ziegler–Natta catalyst for acetylene polymerisation [10]. Polyacetylene yielded by Ziegler–Natta catalysts appeared to have an increasingly high *cis* content of the conjugated double bond sequences as the polymerisation temperature was decreased, and thus an increasingly high *trans* content with increasing polymerisation temperature [2,3,10]. As regards the conformation of these acetylene polymers, let us note that two conformations, *cisoid* and *transoid*, are possible for each polyacetylene isomer:

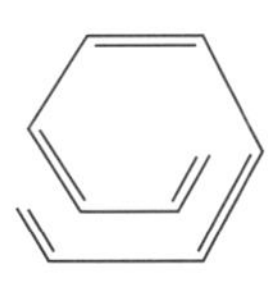

cis-cisoid

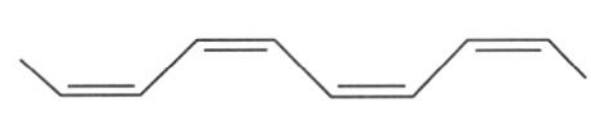

cis-transoid

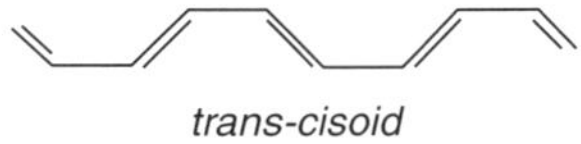

trans-cisoid

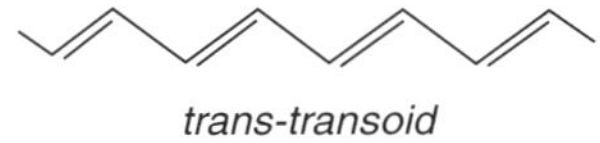

trans-transoid

Several monosubstituted acetylenes (terminal alkynes) with a small (not presenting steric hindrances) substituent, R–C≡CH (where R=primary or secondary alkyl group), have also been polymerised using Ziegler–Natta catalysts [11–15].

Acetylenic monomers also appeared to undergo polymerisation with conventional olefin metathesis catalysts. This relates to monosubstituted highly branched alkylacetylenes and arylacetylenes as well as disubstituted acetylenes (internal alkynes) [16–18]. It has been demonstrated that acetylene itself may also be polymerised using olefin metathesis catalysts [19,20]. The polymerisation of alkynes [scheme (2)] involves a metathesis reaction [scheme (5) of Chapter 2] analogously to that of cycloolefins [21]:

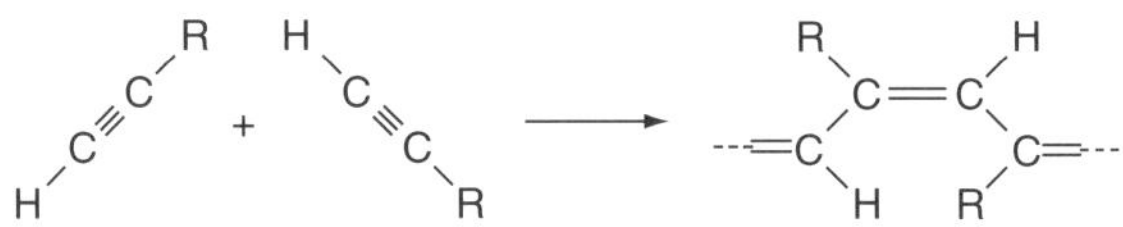

(2)

The mechanism of polymerisation of alkynes with metathesis catalysts requires that the original triple bond of the acetylenic monomer becomes a single bond in the polymer [scheme (5) of Chapter 2], in contrast to the insertion mechanism of acetylene polymerisation with Ziegler–Natta catalysts, where the triple bond becomes a double bond [scheme (1)]. Ideas about the mechanism of metathesis polymerisation of cycloolefins suggested that isolable metal carbenes might promote the polymerisation of cycloolefins suggested that isolable metal carbenes might promote the polymerisation of alkynes, as indeed turned out to be true, as several metal carbenes were found [22–24] to cause alkyne polymerisation.

Whether the insertion or metathesis mechanism operates in the polymerisation depends on the kind of acetylenic monomer and the catalyst applied for the reaction. Various monosubstituted acetylenes, not only hydrocarbons but also heteroatom-containing compounds, are polymerised when suitable catalysts are used. Moreover, both sterically unhindered and encumbered alkynes undergo polymerisation with appropriate catalysts. Let us consider that sterically unencumbered acetylenic monomers, which *n*-alkylacetylenes are, readily undergo polymerisation with Ziegler–Natta catalysts, while acetylenes containing bulky substituents such as *t*-butylacetylene and *o*-substituted phenylacetylenes polymerise with metathesis catalysts; unsubstituted phenylacetylene is also polymerised in the presence of Rh-based catalysts to yield stereoregular polymers. A variety of disubstituted acetylenic monomers also undergo polymerisation; the catalysts effective for this polymerisation, however, are virtually restricted to group 5 and 6 transition metal-based metathesis catalysts. One should emphasise that disubstituted monomers with less steric hindrance, such as di(*n*-alkyl)acetylenes, tend to be polymerised with Mo- and W-based catalysts, whereas the sterically encumbered counterparts, e.g. 1-(trimethylsilyl)-1-

propyne and diphenylacetylene, are polymerised only with Nb- and Ta-based catalysts [25].

Apart from the discussed monoalkynes, acetylenic monomers capable of producing high molecular weight linear polymers also include α, ω-dialkynes. These monomers undergo cyclohomopolymerisation to yield soluble, long conjugated polymers with ring-closed structures, as well as cyclisation copolymerisation with heterocumulenes such as carbon dioxide and isocyanates to yield soluble polymers containing heterocyclic moieties in the polymer backbone [25].

Polyacetylene has promise as a polymeric electroconducting material [26–31]. The simplest way to obtain this polymer is by polymerisation of acetylene, although alternative methods involving the metathesis polymerisation of cyclic polyenes are also effective.

Polyacetylene appeared to be insoluble in all solvents tested [10,32]. Note, however, that units of polyacetylene in soluble form have been obtained by using graft or block copolymerisation methods, such as grafting polyacetylene to soluble polymers [33–37], grafting soluble polymeric chains on the main backbone of polyacetylene [38] and diblock copolymerisation [39–42]. In contrast to polyacetylene, polymers that can be obtained from substituted acetylenes are soluble in common solvents.

7.1 Insertion Polymerisation

7.1.1 Polymerisation of Monoalkynes

Acetylene is most frequently polymerised with the soluble $Ti(OBu)_4$–$AlEt_3$ Ziegler–Natta catalyst (Shirakawa catalyst) [10,11]; typical polymerisation conditions are as follows: $[Ti(OBu)_4] = 0.25M$, $[AlEt_3] = 1.0M$, toluene medium, temperature $-78\,^{\circ}C$, acetylene pressure 600 torr (note the rather high catalyst concentration). Other soluble or insoluble Ziegler–Natta catalysts have also been used to polymerise acetylene and acetylenic compounds [4–9,12–15,43–54]. Interesting examples of the metallocene-catalysed polymerisation of alkynes are the polymerisations of phenylacetylene with the Cp_2TiCl_2–AlMe_2Cl [52,53] and Cp_2HfCl_2–AlR_xCl_{3-x} [54] catalysts, in which monomer insertion is postulated to occur into a Mt–C active bond in group 4 metallocene cationic species. Coordination catalysts based on lanthanides have also been successfully applied for acetylene polymerisation; rare-earth metal compounds such as lanthanide naphthenates and 2-ethylhexyl phosphonates activated with trialkylaluminium $[AlEt_3, Al(i-Bu)_3]$ yield a *cis*-rich (87–95%) crystalline polyacetylene at $30\,^{\circ}C$ [55–58]. The $Ln(naphthenate)_3$–$AlEt_3$ and $Ln(naphthenate)_3$–$AlEt_3$–EtOH catalysts also appeared to polymerise phenylacetylene [59] and terminal alkynes such as 3-methyl-1-butyne, 1-pentyne, 3-methyl-1-pentyne, 4-methyl-1-pentyne and 1-hexyne to yield polymers with a high content of *cis* configuration [60,61].

Interesting features in the polymerisation of acetylenic monomers are displayed by rhodium-based catalysts; they may be applied in metal akyl or hydride-activated systems, e.g. $RhCl_3$–$LiBH_4$ [49] and $[(Cod)Rh]^+[[BPh_4]^-$ –$SiEt_3H$ [62], or be used without an organometallic or hydride activator, e.g. $[(Cod)RhCl]_2$ [63–66], $(Ph_3P)_2Rh(C{\equiv}CPh)(Nbd)$ (Nbd=2,5-norbornadiene) [67] and $[(Nbd)_2RhCl]_2$ [68–73]. The polymerisation of phenylacetylene with the $(Ph_3P)_2Rh(C{\equiv}CPh)(Nbd)$ catalyst, in which a stereoregular all-*cis* polymer is formed, has all the features of living polymerisation [67]. Rhodium-catalysed polymerisation proceeds in various solvents such as benzene, tetrahydrofuran, ethanol and triethylamine [69,73]. Monomers that undergo polymerisation with rhodium catalysts are restricted to acetylene and non-crowded or moderately crowded terminal alkynes such as phenylacetylene and alkyl propiolates. The polymerisation of phenylacetylene with the $[(Nbd)RhCl]_2$– NEt_3 catalyst yields polymers whose molecular weight reaches ca 1×10^6 [69], which is noteworthy because such high molecular weight poly(phenylacetylene) cannot be obtained with Mo- and W-based metathesis catalysts. It is also to be noted that polymers of *cis-transoid* configuration are formed in the discussed rhodium-mediated polymerisations.

An original method for acetylene polymerisation is that using bis(ylide)nickel(II) catalysts with a highly active Ni–Ph bond [74,75]:

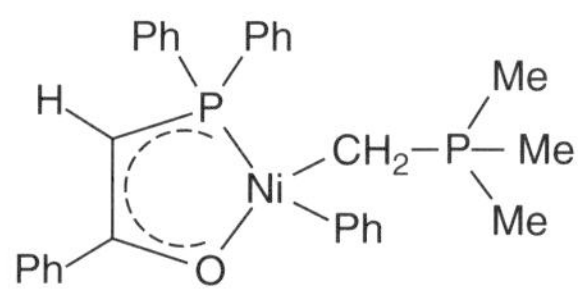

The *cis–trans* content of polyacetylene formed with coordination catalysts depends on the polymerisation temperature; the increasing *trans* content with increasing polymerisation temperature has been explained in terms of a thermally induced *cis–trans* isomerisation which occurs before crystallisation of the chain segment [10,76–78].

The active species formed in $Ti(OBu)_4$–$AlEt_3$ Ziegler–Natta catalysts probably involve a Ti(III)–Et bond resulting from alkylation and reduction of $Ti(OBu)_4$ by $AlEt_3$ [79]. The monomer insertion into the Ti–C bond occurs by *cis*-addition leading to a *cis-transoid* structure of the polymer at low temperature [78], which has been demonstrated in the case of the polymerisation of phenylacetylene with the Cp_2TiCl_2–$AlMe_2Cl$ catalyst [52,53]. A schematic representation of such an insertion mechanism for acetylene polymerisation in the presence of Ti-based Ziegler–Natta catalysts is shown in below [78,80–82]:

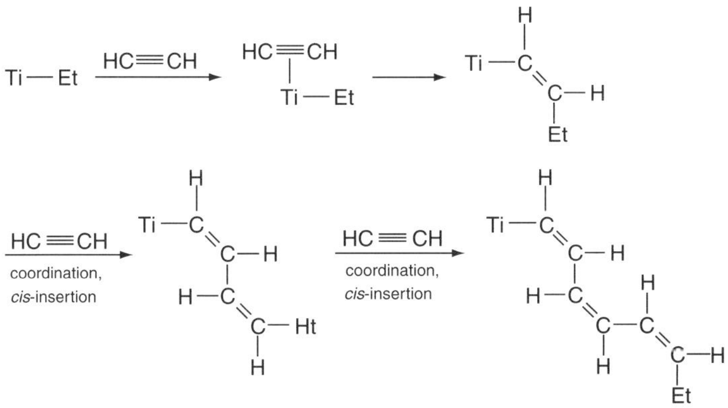

(3)

In an alkyne polymerisation system with Ziegler–Natta catalysts, chain transfer and termination reactions, similar to those postulated for olefin polymerisation, have been suggested to take place [25]. A possible chain transfer reaction is the formation of the Ti–H species from the propagating chain end by β-hydride elimination:

$$\text{Ti—CH=CH—P}_n \longrightarrow \text{Ti—H} + \text{CH}\equiv\text{C—P}_n \qquad (4)$$

With some catalysts, selectivity to linear polymers is low owing to cyclooligomerisation (forming aromatic rings) and other secondary reactions accompanying the polymerisation; cyclotrimerisation, one of these side reactions leading to benzene, occurs by backbiting of the propagating chain end assuming the *cis-cisoidal* conformation [80,83].

It is worth noting that suitable olefins added to the polymerisation system can act as chain transfer agents during the metathesis polymerisation of acetylenic monomers; for instance, trimethylvinylsilane has been found [84] to be an effective chain transfer agent in the polymerisation of phenylacetylene in the presence of the WCl_6–$SnPh_4$ catalyst.

7.1.2 Cyclopolymerisation of α, ω-Dialkynes

7.1.2.1 Homopolymerisation

The cyclohomopolymerisation of 1,6-heptadiyne by using Shirakawa [85] and related [86] catalysts is a representative insertion polymerisation of acetylenic monomers with Ziegler–Natta catalysts:

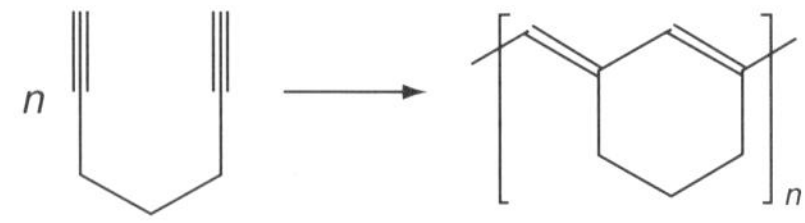

(5)

A characteristic feature of such a polymerisation is that the insertion of the triple bond in the monomer is followed by the insertion of the triple bond in the pendant group, but not in the new monomer molecule, which results in ring closure.

Cyclopolymerisation of 1,6-heptadiyne derivatives has also been carried out in the presence of Pd-based catalysts [87–89].

7.1.2.2 Copolymerisation with Heterocumulenes

An extremely interesting extension of the cyclopolymerisation of α, ω-diacetylenes is the new Ni(0)-catalysed dialkyne cycloaddition copolymerisation with heterocumulenes (O=C=O, RN=C=O, RN=C=NR), unsaturated compounds ($RC\equiv N$) and carbene-type compounds ($C\equiv O$, $C\equiv NR$) [90]. The 1:1 cycloaddition copolymerisation of internal dialkynes $R{-}C\equiv C{-}(CH_2)_x{-}C\equiv C{-}R$ (R=Et;$x > 6, x < 2$), e.g. 3,11-tetradecadiyne, with carbon dioxide in the presence of the $(R_3P)_2Ni(Cod)_2$ complex (R=n-C_8H_{17}) led to high molecular weight poly(2-pyrone) [91]:

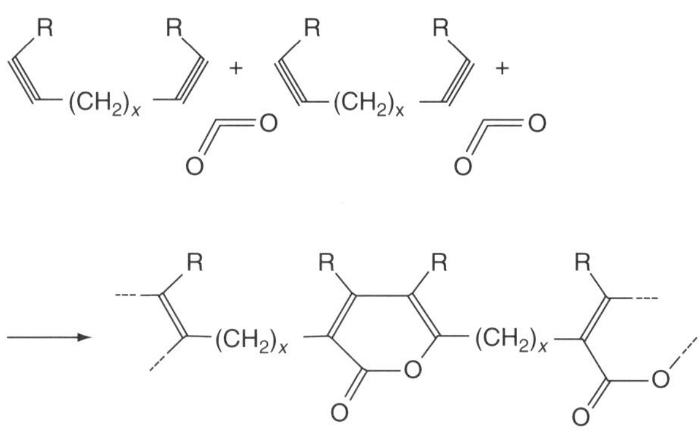

(6)

The dialkyne/carbon dioxide copolymerisation is controlled by the relative rate of inter- and intramolecular cyclisations of the dialkyne; the former is favoured when the number of methylene groups in the monomer $R{-}C\equiv C{-}(CH_2)_x{-}C\equiv C{-}R$ is equal to 3, 4 or 5 ($x = 3$–5), but the intermolecular cyclisation of the dialkyne is favoured to effect 1:1 cycloaddition copolymerisation of the dialkyne and CO_2 to a poly(2-pyrone) when the number x has other values [91–96].

The 1:1 cycloaddition copolymerisation of internal dialkynes, R–C≡C–$(CH_2)_x$–C≡C–R, (R=Me, Et; $x = 2, 4, 6$), e.g. 3,11-tetradecadiyne, 3,9-dodecadiyne and 2,6-octadiyne, with isocyanates $R^1N{=}C{=}O$ (R^1 = Chx, n-C_8H_{17}, Ph) in the presence of the $(R_3P)_2Ni(Cod)_2$ complex (R = n-C_8H_{17}) led to high molecular weight poly(2-pyridone) [97]:

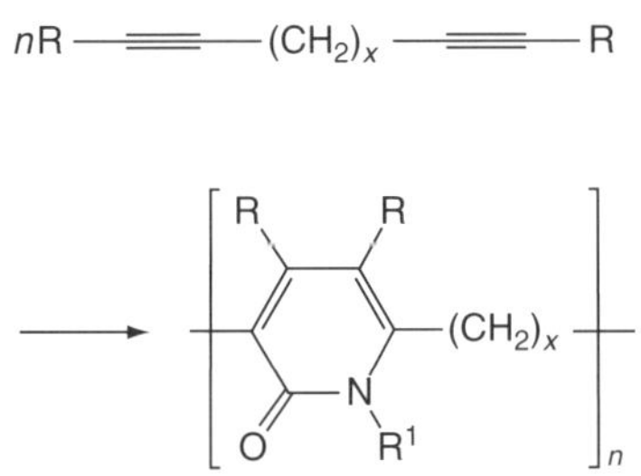

(7)

Analogously, aromatic high molecular weight poly(pyridone)s have been obtained using terminal and internal aromatic diynes, e.g. 1,4-diethynylbenzene [98] and 1,4-bis(phenylethynyl)benzene [99] respectively, for 1:1 cycloaddition copolymerisation with isocyanates.

By the use of cyclic dialkynes for Ni(0)-catalysed cycloaddition copolymerisation with heterocumulenes such as carbon dioxide and isocyanates, respective ladder poly(2-pyrone)s and poly(2-pyridone)s have been obtained [100]. For instance, 1,7-cyclotridecadiyne, 1,7-cyclotetradecadiyne and 1,8-cyclopentadecadiyne undergo a cycloadditon copolymerisation with carbon dioxide and isocyanates to yield ladder poly(2-pyrone)s [91,101] and poly(2-pyridone)s [102] respectively. The 1:1 cycloaddition copolymerisation of 1,8-cyclopentadecadiyne with carbon dioxide and n-octyl isocyanate, $R^1N{=}C{=}O$ ($R^1 = n$-C_8H_{17}), in the presence of the $(R_3P)_2Ni(Cod)_2$ complex (R = n-C_8H_{17}), leading to high molecular weight soluble polymers, is presented by schemes (8) and (9) respectively:

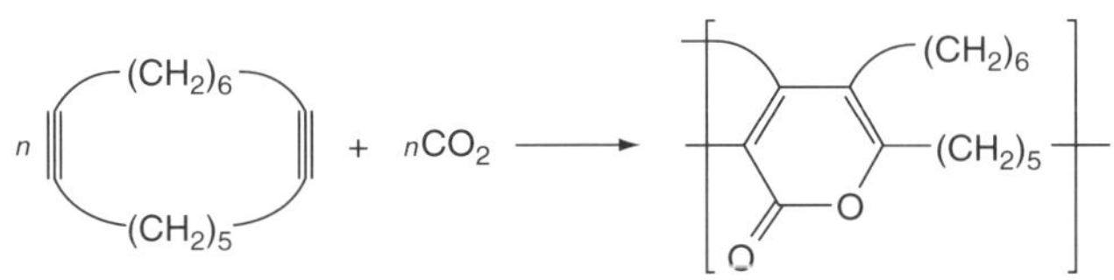

(8)

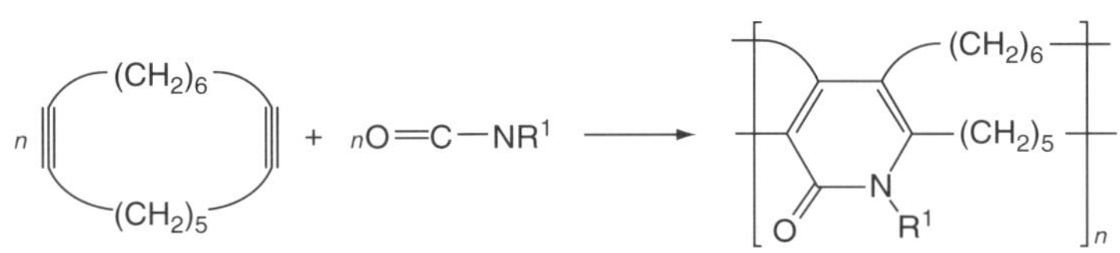

(9)

The ladder copolymers are preferentially formed (by the intermolecular reaction) because the formation of a tricyclic heterocycle by intramolecular cyclisation is difficult for steric reasons [100].

The mechanism of ring formation from monoalkyne and heterocumulenes, catalysed by Ni(0) complexes, L_x Ni(0), has been proposed to involve one-step cycloaddition; scheme (10) [103] and scheme (11) [104, 105] show the formation of the 2-pyrone ring in the alkyne reaction with carbon dioxide and the 2-pyridone ring in the alkyne reaction with isocyanate respectively:

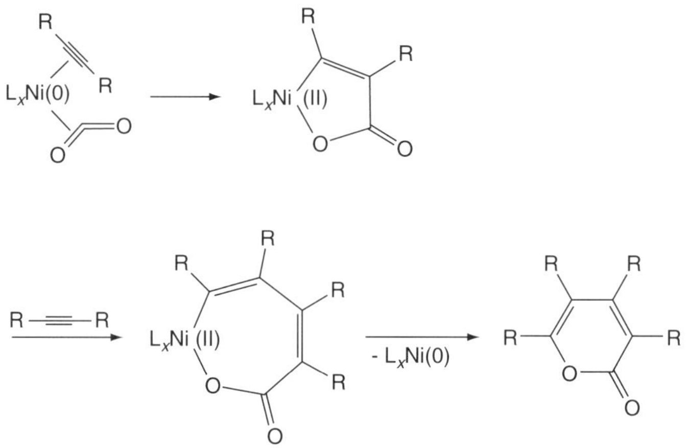

(10)

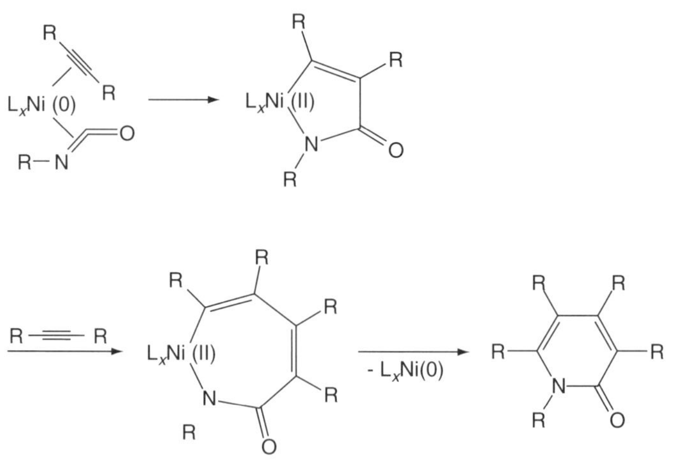

(11)

It is worth noting that these reactions require a change in the oxidation state of the metal (Ni) in the catalytic cycle.

7.2 Metathesis Polymerisation

7.2.1 Polymerisation of Monoalkynes

Metathesis polymerisation of alkynes proceeds with olefin metathesis catalysts, both the conventional (e.g. $TaCl_5$, $TaCl_5$–$SnBu_4$, $MoCl_5$, $MoCl_5$–$SnPh_4$, $MoCl_5$–LiBu, $Mo(CO)_6$–PhOH, WCl_6, WCl_6–$AlEtCl_2$, WCl_6–$SnPh_4$) and well-defined metal carbenes (e.g. Fischer carbene $(CO)_5W{=}C(OMe)Ph$), Casey carbene $(CO)_5W{=}CPh_2$ and Rudler carbene $(CO)_4\overline{W{=}C(OMe)CH_2CH_2{=}CH_2})$ [15–20, 22–24, 106–134].

As rcgards mctal carbcnc such as tri[*t*-butoxy(ncopcntylidync)tungstcn, $(Me_3CO)_3W{\equiv}CCMe_3$ (Schrock carbyne [135–137]), they appeared, however, to catalyse the transalkylidynation of acetylenes, but not the metathesis polymerisation. Also, heterogeneous catalysts that function as olefin metathesis catalysts, such as WO_3/SiO_2 [138] and MoO_3/SiO_2 [139], will yield transalkylidynation products of 1-alkynes. However, most terminal acetylenes are trimerised to aromatics with these catalysts.

By contrast, the Fischer carbyne *trans*-bromotetracarbonyl(phenylmethylidene)tungsten, $Br(CO)_4W{\equiv}CPh$ [140–142], will polymerise monosubstituted acetylenes by a metathesis mechanism [143] but will not give transalkylidynation products. The Fischer carbyne and related carbynes are effective catalysts for the metathesis polymerisation of not only monosubstituted acetylenes, whose polymerisation can be brought about by a number of catalysts [108, 109, 118, 120–125, 128–134], but also of acetylenes polymerised previously by only a few or no catalysts, i.e. even disubstituted acetylenes [112–115, 126, 127, 144–148] and unsubstituted acetylene [19, 20, 42, 119].

It was proved that metal carbynes are sources of metal carbenes [e.g. scheme (9) in Chapter 6] promoting the polymerisation of acetylenic monomers. Therefore, related metal carbynes and carbenes appeared to catalyse the polymerisation of alkynes in the same way as regards the identity of the products, in particular as regards stereochemistry. For the terminal and internal alkynes, the Fischer carbyne acts much like the Casey and Fischer metal carbenes. The Fischer carbyne also promotes acetylene polymerisation, and it does this where the Fischer carbene fails and the Casey carbene is much less effective [22,143].

The fact that the Schrock carbyne, $(Me_3CO)_3W{\equiv}CCMe_3$, catalyses the transalkylidynation of alkynes whereas the Fischer carbyne, $Br(CO)_4W{\equiv}CPh$, polymerises them may be explained in terms of coordination unsaturation of the former carbyne and saturation of the latter. If the coordination unsaturation of tungsten is required for the reaction to occur, the Schrock metal alkylidyne can react directly, but the Fischer carbyne $Br(CO)_4W{\equiv}CPh$ has to lose a ligand from the metal. If that ligand is bromine and if it is rearranged to the adjacent carbon atom [scheme (9) in Chapter 6], the coordination unsaturation appears but at the cost of the tungsten carbyne transforming into a tungsten carbene. Thus, only polymerisation by metal carbene species, and not the transalkylidynation of alkynes requiring metal carbyne species, proceeds with the Fischer carbyne, in contrast to the Schrock carbyne [143].

Interesting evidence supporting the mechanism of polymerisation of acetylenes via carbene species is provided by the block and random copolymerisation of acetylenic monomers with cycloolefins. For instance, block copolymers of acetylene and cyclopentene with the WCl_6–$AlEtCl_2$ catalyst [41] and block copolymers of acetylene and norbornene with the $(Me_3CO)_2W(\!=\!NAr)\!=\!CHMe_3$ catalyst [42] have been obtained; moreover, random copolymers of phenylacetylene and norbornene with the WCl_6 catalyst have also been obtained [149, 150].

Since the Schrock carbyne is active in the transalkylidynation reaction, the possibility of ring-opening polymerisation of cycloalkynes by acetylene transalkylidynation catalysts (metal alkylidyne complexes) has been evaluated [151]. Unfortunately, cyclooctyne is one of the few relatively stable strained cyclic acetylenes, but it is not strained enough to react selectively with the catalyst and yield a polymer according to the scheme

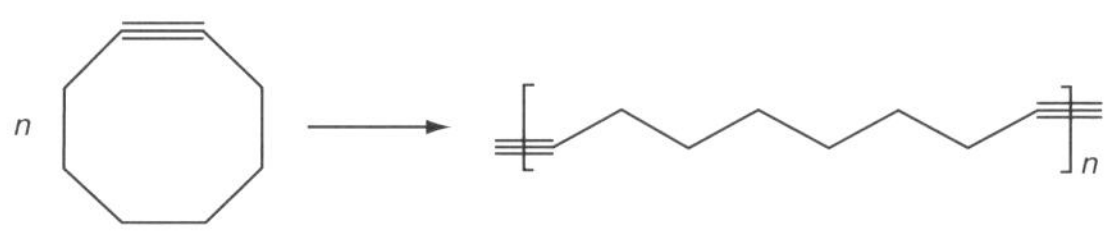

(12)

As in the case of the ring-opening metathesis polymerisation of cycloolefins, an important matter is the control of polymerisation to prepare acetylenic polymers having precise structures. A living polymerisation is of practical importance in the synthesis of monodisperse polymers, such as terminally functionalised polymers and block copolymers. The metathesis catalysts that promote the living polymerisation of acetylene [42] and acetylenic monomers include $MoOCl_4$–$SnBu_4$–EtOH, $NbCl_5$ and Ta, Mo and W alkylidenes [84, 133, 152, 153].

7.2.2 Cyclopolymerisation of α,ω-Dialkynes

The cyclopolymerisation of α,ω-diacetylenes with metathesis catalysts concerns 1,6-heptadiyne and its 4,4-disubstituted derivatives [87, 106, 154–163] and 1,7-octadiyne [164].

The cyclopolymerisation of 1,6-heptadiyne derivatives such as diethyl dipropargyl malonate by a Schrock carbene, $[Me(CF_3)_2CO]_2Mo(\!=\!NAr)\!=\!CHMe_3$, via a metathesis mechanism, proceeds in a living fashion to provide a conjugated polymer having both five-and six-membered rings, which is shown schematically below [154, 155]:

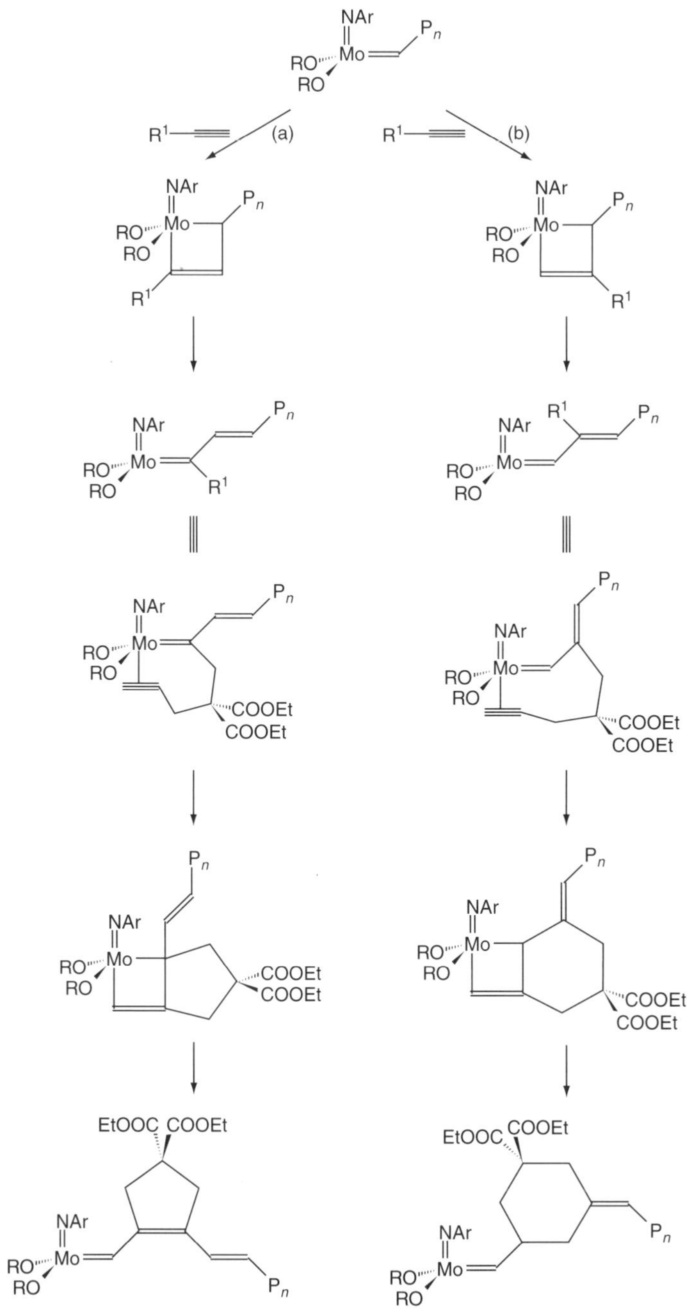

(13)

Ar = 2,6-(i-Pr)$_2$C$_6$H$_3$; R = Me(CF$_3$)$_2$C ; R^1 = CH≡CCH$_2$C(COOEt)$_2$CH$_2$

The ring size is controlled by the extent of 2,1-addition versus 1,2-addition of the acetylenic group in the monomer to metal alkylidene, the two modes that are presumably giving rise to a five-membered ring [scheme (13a)] and a six-membered ring [scheme (13b)] respectively. Note the tail-to-tail (t–t) structure of the polymer chain sequences containing five-membered rings (1-vinylene-2-cyclopent-1-enylene units) and the head-to-tail (h–t) structure of the polymer chain sequences containing six-membered rings (3-methylidene-1-cyclohex-1-enylene units). The prevailing formation of cyclopolymers containing five-membered rings over six-membered rings has been observed in systems with conventional metathesis catalysts as well as defined metal alkylidene catalysts [155].

The living nature of the discussed cyclopolymerisation of the 1,6-heptadiyne derivative in the presence of Schrock carbene as the catalyst has been demonstrated by the synthesis of a block copolymer with 2,3,-dicarbomethoxynorbornadiene [25].

Interesting examples of the cyclopolymerisation are the homopolymerisation of 1,7-octadiyne and its copolymerisation with 1-hexyne promoted by the $Mo(CO)_6$–*m* – Cl–C_6H_4–OH catalyst [164] which lead to cyclopolymers containing seven-membered rings in the polymer backbone

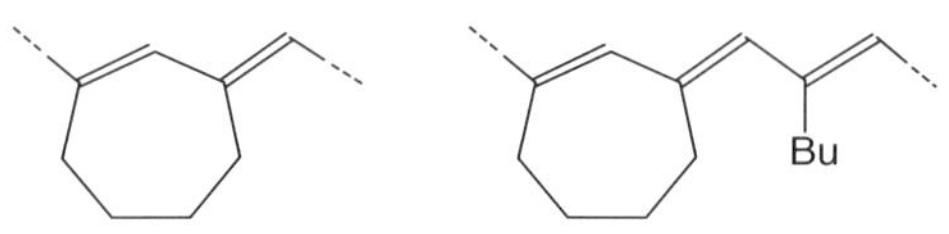

References

1. Ziegler, K., *Angew. Chem.*, **67**, 541 (1955).
2. Natta, G., Mazzanti, G. and Pino, P., *Angew. Chem.*, **69**, 686 (1957).
3. Natta, G., Mazzanti, G. and Corradini, P., *Atti Accad. Naz. Lincei, Cl Sci. Fis. Mat. Nat. Rend*, **25**(8), 3 (1958).
4. Luttinger, L. B., *Chem. Ind. (Lond.)*, **1960**, 1135 (1960).
5. Korshak, V. V., *Dokl. Akad. Nauk SSSR*, **136**, 1342 (1961).
6. Matkovskii, P. E. and Zavorkhin, N. D., *Izv. Akad. Nauk Kaz. SSR, Ser. Khim.*, **15**, 70 (1965).
7. Nasirov, T. M., Krentsel, B. A. and Davydov, B. E., *Izv. Akad. Nauk SSSR, Ser. Khim.*, **1965**, 1009 (1965).
8. Nicolescu, I. V. and Angelescu, E., *J. Polym. Sci., A*, **3**, 1227 (1965).
9. Higashimura, K. and Oiwa, M., *Bull. Chem. Soc. Jpn.*, **69**, 109 (1966).
10. Ito, T., Shirakawa, K. and Ikeda, S., *J. Polym. Sci., Polym. Chem. Ed.*, **12**, 11 (1974).
11. Shirakawa, H. and Ikeda, S., *Polym. J.*, **2**, 231 (1971).
12. Meriwether, L. S., Colthup, E. C., Kennerly, G. W. and Reusch, R. N., *J. Org. Chem.*, **26**, 5155 (1961).
13. Trepka, W. J. and Sonnenfeld, R. J., *J. Polym. Sci., A-1*, **8**, 2721 (1970).
14. Ciardelli, F., Lanzillo, S. and Pieroni, O., *Macromolecules*, **7**, 174 (1974).

15. Masuda, T., Kawasaki, M., Okano, Y. and Higashimura, T., *Polym. J.*, **14**, 371 (1982).
16. Masuda, T., Hasegawa, K. and Higashimura, T., *Macromolecules*, **7**, 728 (1974).
17. Woon, P. S. and Farona, M. F., *J. Polym. Sci., Polym. Chem. Ed.*, **12**, 1749 (1974).
18. Navarro, F. R. and Farona, M. F., *J. Polym. Sci., Polym. Chem. Ed.*, **14**, 2335 (1976).
19. Shimamura, K., Karasz, F. E., Hirsch, J. A. and Chien, J. C. W., *Makromol. Chem., Rapid Commun.*, **2**, 473 (1981).
20. Aldissi, M., Linaya, C., Sledz, J., Schue, F., Giral, L., Fabre, J. M. and Rolland, M., *Polymer*, **23**, 243 (1982).
21. Masuda, T., Sasaki, N. and Higashimura, T., *Macromolecules*, **8**, 717 (1975).
22. Katz, T. J. and Lee, S. J., *J. Am. Chem. Soc.*, **102**, 422 (1980).
23. Katz, T. J. and Lee, S. J., *J. Am. Chem. Soc.*, **102**, 7940 (1980).
24. Katz, T. J., Lee, S. J. and Shippey, M. A., *J. Mol. Catal.*, **8**, 219 (1980).
25. Masuda, T., 'Acetylene Polymerization', in *Catalysis in Precision Polymerisation*, John Wiley & Sons, Chichester–New York, 1997, pp. 67–97.
26. Frommer, J. E. and Chance, R. R., 'Electrically Conductive Polymers', in *Encyclopedia of Polymer Science and Engineering*, Wiley-Interscience, John Wiley & Sons, New York, 1986, Vol. 5, pp. 462–507.
27. Etemad, S., Pron, A., Heeger, A. J., MacDiarmid, A. G., Mele, E. J. and Rice, M. J., *Phys. Rev. B*, **23**, 5137 (1981).
28. Ikehata, S., Kaufer, J., Woerner, T., Pron, A., Druy, M. A., Sivak, A., Heeger, A. J. and MacDiarmid, A. G., *Phys. Rev. Lett.*, **45**, 1123 (1980).
29. Kletter, M. J., Woerner, T., Pron, A., MacDiarmid, A. G., Heeger, A. J. and Park, Y. W., *J. Chem. Soc., Chem. Commun.*, **1980**, 426 (1980).
30. Weinberger, B. R., Ehrenfreund, E., Pron, A., Heeger, A. J. and MacDiarmid, A. G., *J. Chem. Phys.*, **72**, 4749 (1980).
31. Weinberger, B. R., Kaufer, J., Heeger, A. J., Pron, A. and MacDiarmid, A. G., *Phys. Rev. B*, **20**, 223 (1979).
32. Pez, G. P., *J. Am. Chem. Soc.*, **98**, 8072 (1976).
33. Bates, F. S. and Baker, G. L., *Macromolecules*, **16**, 704 (1983).
34. Baker, G. L. and Bates, F. S., *Macromolecules*, **17**, 2169 (1984).
35. Bates, F. S. and Baker, G. L., *Macromolecules*, **16**, 1013 (1983).
36. Destri, S., Catellani, M. and Bolognesi, A., *Makromol. Chem., Rapid Commun.*, **5**, 353 (1984).
37. Bolognesi, A., Catellani, M., Porzio, W. and Destri, S., *Polymer*, **27**, 1128 (1986).
38. Kminek, I. and Trekoval, J., *Makromol. Chem., Rapid Commun.*, **14**, 479 (1984).
39. Galvin, M. E. and Wnek, G. E., *Mol. Cryst. Liq. Cryst.*, **117**, 33 (1985).
40. Aldissi, M., *Synth. Met.*, **13**, 87 (1986).
41. Farren, T. R., Amass, A. J., Beevers, M. S. and Stowell, J. A., *Makromol. Chem.*, **188**, 2535 (1987).
42. Schlund, R., Schrock, R. R. and Crowe, W. E., *J. Am. Chem. Soc.*, **111**, 8004 (1989).
43. Luttinger, L. B., *J. Org. Chem.*, **27**, 1591 (1962).
44. Luttinger, L. B. and Colthup, E. C., *J. Org. Chem.*, **27**, 3752 (1962).
45. Naarmann, H. and Theophilou, N., *Synth. Met.*, **22**, 1 (1987).
46. Naarmann, H., *Synth. Met.*, **17**, 223 (1987).
47. Oh, S. Y., Akagi, K. and Shirakawa, H., *Synth. Met.*, **32**, 245 (1989).
48. Terlemezyan, L. and Mikhailov, M., *Makromol. Chem., Rapid Commun.*, **3**, 613 (1982).
49. Kern, R. J., *J. Polym. Sci., A-1*, **7**, 621 (1969).
50. Yokota, K., Ohtubo, M., Hirabayashi, T. and Imai, Y., *Polym. J.*, **25**, 1079 (1993).

51. Biyani, B., Campagna, A. J., Daruwalla, D., Srivastava, C. M. and Ehrlich, P., *J. Macromol. Sci. – Chem. A*, **9**, 327 (1975).
52. D'yachkovskii, F. S., Yarovitskii, P. A. and Bystrov, V. F., *Vysokomolek. Soed.*, **6**, 659 (1964).
53. Katz, T. J., Hacker, S. M., Kendrik, R. D. and Yannoni, C. S., *J. Am. Chem. Soc.*, **107**, 2182 (1985).
54. Siskos, M. G., Bokaris, E. P., Zardakis, A. K. and Kyriakakou, G., *Eur. Polym. J.*, **12**, 1127 (1992).
55. Shen, Z., Yang, M., Shi, M. and Cao, Y., *J. Polym. Sci., Polym. Chem. Ed.*, **20**, 411 (1982).
56. Cao, Y., Qian, R., Wang, F. and Zhao, X., *Makromol. Chem., Rapid Commun.*, **3**, 687 (1982).
57. Shen, Z., Wang, F. and Cao, Y., *Inorg. Chim. Acta*, **110**, 55 (1985).
58. Hu, X., Wang, F., Zhao, X. and Yan, D., *Chin. J. Polym. Sci.*, **5**, 221 (1987).
59. Shen, Z. and Farona, M. F., *J. Polym. Sci., Polym. Chem. Ed.*, **22**, 1009 (1984).
60. Shen, Z. and Farona, M. F., *Polym. Bull.*, **10**, 298 (1983).
61. Mullagaliev, I. R. and Mudarisova, R. K., *Izv. Akad. Nauk SSSR, Ser. Khim.*, **7**, 1687 (1988).
62. Goldberg, Y. and Alper, H., *J. Chem. Soc., Chem. Commun.*, **1994**, 1209 (1994).
63. Furlani, A., Napoletano, S., Russo, M. V. and Feast, W. J., *Polym. Bull.*, **16**, 311 (1986).
64. Furlani, A., Licoccia, S. and Russo, M. V., *J. Polym. Sci., A*, **24**, 991 (1986).
65. Furlani, A., Napoletano, S., Russo, M. V., Camus, A. and Marish, N., *J. Polym. Sci., A*, **27**, 75 (1989).
66. Haupt, H. J. and Ortmann, U., *Z. Anorg. Allg. Chem.*, **619**, 1209 (1993).
67. Kishimoto, Y., Eckerle, P., Miyatake, T., Ikariya, T. and Noyori, R., *J. Am. Chem. Soc.*, **116**, 12131 (1994).
68. Tabata, M., Yang, W. and Yokota, K., *Polym. J.*, **22**, 1105 (1990).
69. Yang, W., Tabata, M., Kobayashi, S., Yokota, K. and Shimizu, A., *Polym. J.*, **23**, 1135 (1991).
70. Lindgren, M., Lee, H. S., Yang, W., Tabata, M. and Yokota, K., *Polymer*, **32**, 1531 (1991).
71. Tabata, M., Yang, W. and Yokota, K., *J. Polym. Sci., A*, **32**, 1113 (1994).
72. Tabata, M., Tamura, H., Yokota, K., Nozaki, Y., Hoshina, T., Minakawa, H. and Kodaira, K., *Macromolecules*, **27**, 6234 (1994).
73. Tabata, M., Inaba, Y., Yokota, K. and Nozaki, Y., *J. Macromol. Sci. – Pure Appl. Chem. A*, **31**, 465 (1994).
74. Ostoja-Starzewski, K. A. and Witte, J., *Angew. Chem., Int. Ed. Engl.*, **24**, 599 (1985).
75. Ostoja-Starzewski, K. A. and Witte, J., *Angew. Chem.*, **100**, 861 (1988).
76. Fukui, K. and Inagaki, S., *J. Am. Chem. Soc.*, **97**, 4445 (1975).
77. Ito, T., Shirakawa, H. and Ikeda, S., *J. Polym. Sci., Polym. Chem. Ed.*, **13**, 1943 (1975).
78. Shen, M. A., Karasz, F. A. and Chien, J. C. W., *J. Polym. Sci., Polym. Chem. Ed.*, **21**, 2787 (1983).
79. Chien, J. C. W., Karasz, F. E., Wnek, G. E., MacDiarmid, A. G. and Heeger, A. J., *J. Polym. Sci., Polym. Chem. Ed.*, **18**, 45 (1980).
80. Shirakawa, H. and Ikeda, S., *J. Polym. Sci., Polym. Chem. Ed.*, **12**, 929 (1974).
81. Berlin, D. A. and Cherkashin, M. I., *Vysokomol. Soed., Ser. A*, **13**, 2298 (1971).
82. Simionescu, C., Dumitrescu, S., Negulescu, I., Percec, V., Grigoras, M., Diaconu, I., Leanca, L. and Gorash, L., *Vysokomol. Soed., Ser. A*, **16**, 790 (1974).
83. Ikeda, S. and Tamaki, A., *J. Polym. Sci., B*, **4**, 605 (1966).

84. Kouzai, H., Masuda, T. and Higashimura, T., *Macromolecules*, **25**, 7096 (1992).
85. Gibson, H. W., Bailey, F. C., Epstein, A. J., Rommelmann, H., Kaplan, S., Harbour, J., Yang, X.-Q., Tanner, D. B. and Pochan, J. M., *J. Am. Chem. Soc.*, **105**, 4417 (1983).
86. Stille, J. K. and Frey, D. A., *J. Am. Chem. Soc.*, **83**, 1697 (1961).
87. Jin, S. H., Kang, S. W., Park, J. G., Lee, J. C. and Choi, K. S., *J. Macromol. Sci. – Pure Appl. Chem. A*, **32**, 455 (1995).
88. Akopyan, L. A., Ambartsumyan, G. V., Ovakimyan, E. V. and Matsoyan, S. G., *Vysokomol. Soed., Ser. A*, **19**, 271 (1977).
89. Akopyan, L. A., Ambartsumyan, G. V., Grigoryan, S. G. and Matsoyan, S. G., *Vysokomol. Soed., Ser. A*, **19**, 1068 (1977).
90. Tsuda, T., 'Transition Metal Catalyzed Diyne Cycloaddition Copolymerization', in *The Polymeric Materials Encyclopedia*, CRC Press, Boca Raton, Vol. 3, pp. 1905–1915.
91. Tsuda, T., Maruta, K. and Kitaike, Y., *J. Am. Chem. Soc.*, **114**, 1498 (1992).
92. Tsuda, T. and Maruta, K., *Macromolecules*, **25**, 6102 (1992).
93. Tsuda, T., Ooi, O. and Maruta, K., *Macromolecules*, **26**, 4840 (1993).
94. Tsuda, T., Kitaike, Y. and Ooi, O., *Macromolecules*, **26**, 4956 (1993).
95. Tsuda, T. and Hokazono, H., Y., *Macromolecules*, **27**, 1289 (1994).
96. Tsuda, T., *Gazz. Chim. Ital.*, **125**, 101 (1995).
97. Tsuda, T. and Hokazono, H., Y., *Macromolecules*, **26**, 1796 (1993).
98. Tsuda, T. and Tobisawa, A., *Macromolecules*, **28**, 1360 (1995).
99. Tsuda, T. and Tobisawa, A., *Macromolecules*, **27**, 5943 (1994).
100. Tsuda, T., 'Cycloaddition Copolymerization of Cyclic Diynes – Ladder Polymers', in *The Polymeric Materials Encyclopedia*, CRC Press, Inc., Boca Raton, 1996, Vol. 5, pp. 3525–3530.
101. Tsuda, T., *Macromolecules*, **28**, 1312 (1995).
102. Tsuda, T. and Hokazono, H. Y., *Macromolecules*, **26**, 5528 (1993).
103. Hoberg, H., *J. Organomet. Chem.*, **266**, 203 (1984).
104. Hoberg, H. and Oster, B. N., *Synthesis*, **1982**, 324 (1982).
105. Hoberg, H. and Oster, B. N., *J. Organomet. Chem.*, **252**, 359 (1983).
106. Schrock, R. R., Lour, S. F., Zanetti, N. C. and Fox, H. H., *Organometallics*, **13**, 3396 (1994).
107. Krouse, S. A. and Schrock, R. R., *Macromolecules*, **21**, 1885 (1988).
108. Masuda, T., Okano, Y., Kuwane, Y. and Higashimura, T., *Polym. J.*, **12**, 907 (1980).
109. Okano, Y., Masuda, T. and Higashimura, T., *Polym. J.*, **14**, 477 (1982).
110. Ho, H. T., Ivin, K. J. and Rooney, J. J., *J. Chem. Soc., Faraday Trans. 1*, **78**, 2227 (1982).
111. Masuda, T. and Higashimura, T., *Macromolecules*, **12**, 9 (1979).
112. Higashimura, T., Deng, Y.-X. and Masuda, T., *Macromolecules*, **15**, 234 (1982).
113. Masuda, T., Takahashi, A. and Higashimura, T., *J. Chem. Soc., Chem. Commun.*, **1982**, 1297 (1982).
114. Masuda, T., Isobe, E., Higashimura, T. and Takada, K., *J. Am. Chem. Soc.*, **105**, 7473 (1983).
115. Masuda, T., Ohtori, T. and Higashimura, T., *Am. Chem. Soc. Polym. Prepr., Div. Polym. Chem.*, **20**, 1043 (1979).
116. Hasegawa, K., Masuda, T. and Higashimura, T., *Macromolecules*, **8**, 255 (1975).
117. Amass, A. J., Beevers, M. S., Farona, T. R. and Stowell, J. A., *Makromol. Chem., Rapid Commun.*, **8**, 119 (1987).
118. Gal, Y.-S., Jung, B., Kim, J.-H., Lee, W.-C. and Choi, S.-K., *J. Polym. Sci., A, Polym. Chem.*, **33**, 307 (1995).

119. Theophilou, N., Munardi, A., Aznar, R., Sledz, G., Schue, F. and Naarmann H., *Eur. Polym. J.*, **23**, 15 (1987).
120. Ofstead, E. A., Lawrence, J. R., Senyek, M. L. and Calderon, N., *J. Mol. Catal.*, **8**, 227 (1980).
121. Oreshkin, I. A., Redkina, L. I., Kershenbaum, I. L., Chernenko, G. M., Makovetskii, K. L., Tinyakova, E. I. and Dolgoplosk, B. A., *Eur. Polym. J.*, **13**, 447 (1977).
122. Liaw, D.-J. and Lin, C.-L., *Polym. Int.*, **36**, 29 (1995).
123. Liaw, D.-J. and Lin, C.-L., *J. Polym. Sci., Polym. Chem. Ed.*, **31**, 3151 (1993).
124. Liaw, D.-J., Leu, S.-D., Lin, C.-L. and Lin, C.-F., *Polym. J.*, **24**, 889 (1992).
125. Liaw, D.-J., Soum, A., Fontanille, M., Parlier, A. and Rudler, H., *Makromol. Chem., Rapid Commun.*, **6**, 309 (1985).
126. Masuda, T., Niki, A., Isobe, E. and Higashimura, T., *Macromolecules*, **18**, 2109 (1985).
127. Niki, A., Masuda, T. and Higashimura, T., *J. Polym. Sci., A*, **25**, 1553 (1987).
128. Vosloo, H. C. M. and Du Plessis, J. A. K., *Polym. Bull.*, **30**, 273 (1993).
129. Tamura, K., Masuda, T. and Higashimura, T., *Polym. Bull.*, **30**, 537 (1993).
130. Tamura, K., Masuda, T. and Higashimura, T., *Polym. Bull.*, **32**, 289 (1995).
131. Ganesamoorthy, S. and Sundararajan, G., *Macromolecules*, **25**, 2060 (1992).
132. Kiyashina, Z. S., Pomagailo, A. D., Kuzaev, A. I., Lagodsinskaya, G. V. and D'yachkovskii, F. S., *Vysokomol. Soed., Ser. A*, **21**, 1796 (1979).
133. Nakano, M., Masuda, T. and Higashimura, T., *Macromolecules*, **27**, 1344 (1994).
134. Nakayama, Y., Mashima, K. and Nakamura, A., *Macromolecules*, **26**, 6267 (1993).
135. Wengrovius, J. H., Sancho, J. and Schrock, R. R., *J. Am. Chem. Soc.*, **103**, 3932 (1981).
136. Sancho, J. and Schrock, R. R., *J. Mol. Catal.*, **15**, 75 (1982).
137. Leigh, G. F., Rahman, M. T. and Wilson, E. R., *J. Chem. Soc., Chem. Commun.*, **1982**, 541 (1982).
138. Moulijn, J. A., Reitsma, H. J. and Boelhouwer, C., *J. Catal.*, **25**, 434 (1972).
139. Mortreux, A. and Blanchard, M., *Bull. Soc. Chim. France*, **1972**, 1641 (1972).
140. Fischer, E. O., Kreis, G., Kreiter, C. G., Muller, J., Huttner, G. and Lorenz, H., *Angew. Chem., Int. Ed. Engl.*, **12**, 564 (1973).
141. Fischer, E. O. and Kreis, G., *Chem. Ber.*, **109**, 1673 (1976).
142. Fischer, E. O., Schubert, U. and Fischer, H., *Inorg. Synth.*, **19**, 172 (1979).
143. Katz, T. J., Ho, T. H., Shih, N.-Y., Ying, Y.-C. and Van Stuart, I. W., *J. Aqm. Chem. Soc.*, **106**, 2659 (1984).
144. Masuda, T., Kawai, H. and Higashimura, T., *Polym. J.*, **11**, 813 (1979).
145. Masuda, T., Kuwane, Y. and Higashimura, T., *Polym. J.*, **13**, 301 (1981).
146. Masuda, T., Kawasaki, K., Okano, Y. and Higashimura, T., *Polym. J.*, **14**, 371 (1982).
147. Masuda, T. and Higashimura, T., *Adv. Polym. Sci.*, **81**, 121 (1987).
148. Cotton, F. A., Hall, W. T., Cann, K. J. and Karol, F. J., *Macromolecules*, **14**, 233 (1981).
149. Makio, H., Masuda, T. and Higashimura, T., *Polymer*, **34**, 1490 (1993).
150. Makio, H., Masuda, T. and Higashimura, T., *Polymer*, **34**, 2218 (1993).
151. Krouse, S. A. and Schrock, R. R., *Macromolecules*, **22**, 2569 (1989).
152. Masuda, T., Fujimori, J., Rahman, M. Z. A. and Higashimura, T., *Macromol. Chem. Phys.*, **196**, 1769 (1995).
153. Wallace, K. C., Leu, A. H., Davis, W. M. and Schrock, R. R., *Organometallics*, **8**, 644 (1989).
154. Fox, H. H. and Schrock, R. R., *Organometallics*, **11**, 2763 (1992).

155. Fox, H. H., Wolf, M. O., O'Dell, R., Lin, B. L., Schrock, R. R. and Wrghton, M. S., *J. Am. Chem. Soc.*, **116**, 2827 (1994).
156. Saunders, R. S., Cohen, R. E. and Schrock, R. R., *Macromolecules*, **24**, 5599 (1991).
157. Ahn, H. K., Kim, Y. H., Jin, S. H. and Choi, S. K., *Polym. Bull.*, **29**, 625 (1992).
158. Park, J. W., Lee, J. H., Cho, H. N. and Choi, S. K., *Macromolecules*, **26**, 1191 (1993).
159. Koo, K. M., Han, S. H., Kang, Y. S., Kim, U. Y. and Choi, S. K., *Macromolecules*, **26**, 2485 (1993).
160. Kim, Y. H., Kwon, S. K. and Choi, S. K., *Bull. Korean Chem. Soc.*, **13**, 459 (1992).
161. Choi, S. K., *Makromol. Chem., Macromol. Symp.*, **33**, 145 (1990).
162. Gal, Y. S. and Choi, S. K., *J. Polym. Sci., A, Polym. Chem.*, **31**, 345 (1993).
163. Kim, Y. H., Kwon, S. K., Gal. Y. S. and Choi, S. K., *J. Macromol. Sci. – Pure Appl. Chem. A*, **29**, 589 (1992).
164. Vosloo, H. C. M. and Du Plessis, J. A. K., *J. Appl. Polym. Sci., Appl. Polym. Symp.*, **48**, 561 (1991).

Further Reading

Grubbs, R. H., 'Alkene and Alkyne Metathesis Reactions', in *Comprehensive Organometallic Chemistry*, Pergamon Press, Oxford, 1982, Vol. 8, pp. 499–550.

Gibson, H. W. and Pochan, J. M., 'Acetylenic Polymers', in *Encyclopedia of Polymer Science and Engineering*, Wiley-Interscience, John Wiley & Sons, New York, 1985, Vol. 1. pp. 87–130.

Bolognesi, A., Catellani, M. and Destri, S., 'Polymerization of Acetylene', in *Comprehensive Polymer Science*, Pergamon Press, Oxford, 1989, Vol. 4, pp. 143–153.

Costa, G., 'Polymerisation of Mono- and Di-substituted Acetylenes', in *Comprehensive Polymer Science*, Pergamon Press, Oxford, 1989, Vol. 4, pp. 155–161.

Amass, A. J., 'Metathesis Polymerization: Chemistry', in *Comprehensive Polymer Science*, Pergamon Press, Oxford, 1989, Vol. 4, pp. 109–134.

Furlani, A. and Russo, M. V., 'Polyphenylacetylene', in *The Polymeric Materials Encyclopedia*, CRC Press, Inc., Boca Raton, 1996, Vol. 8, pp. 6481–6492.

Kishimoto, Y., Noyori, R., Eckerle, P., Miyatake, P. and Ikariya, T., 'Stereospecific Living Polymerisation of Phenylacetylene with Rh Complexes', in *The Polymeric Materials Encyclopedia*, CRC Press, Inc., Boca Raton, 1996, Vol. 7, pp. 5051–5055.

Problems

1. Name and characterise the main types of coordination polymerisation of alkynes and give representative catalysts for each type.
2. Show by structural drawings the stereoregular polymers that are obtained from terminal acetylenes, $R-C{\equiv}CH$; indicate various chain conformations for each stereoisomer.
3. Give reasons why pristine polyacetylenes formed with coordination catalysts of various types at low temperature are of the *cis* configuration.
4. Give an example of the copolymerisation of alkyne and cycloalkene and explain why it can proceed.

5. Give reasons why the cyclopolymerisation of α, ω-dialkyne such as 1,6-heptadiyne proceeds via both five-membered and six-membered ring closure.
6. Explain why the metal carbyne-mediated transalkylidynation reaction leads to polymer chain propagation when cycloalkyne is applied as the monomer.

8 Coordination Polycondensation

Step growth reactions involving the homo- and heteropolycondensation of various bifunctional unsaturated monomers in the presence of transition metal-based coordination catalysts have appeared to be a very useful synthetic tool for the preparation of low and high molecular weight polymers with an unsaturation in the polymer backbone. These reactions lead to unsaturated hydrocarbon and non-hydrocarbon polymers where polymeric chains are formed by carbon–carbon and carbon–heteroatom coupling respectively.

Essentially, two types of unsaturated bifunctional compound display the capability of being condensed in high yield, via a transition metal-catalysed carbon–carbon coupling reaction, into their requisite generic high molecular weight polymers plus a small molecule: non-conjugated acyclic dialkenes and haloaromatic derivatives.

Acyclic dienes undergo metathesis polycondensation (acyclic diene metathesis – 'admet') to poly(alkenylene)s in the presence of W, Mo and Ru alkylidene complexes as catalysts [1]:

$$n\mathrm{CH_2{=}CH{-}R{-}CH{=}CH_2} \rightleftharpoons \mathrm{CH_2{=}\!\left[CH{-}R{-}CH\right]_n\!{=}CH_2} + (n-1)\mathrm{CH_2{=}CH_2} \qquad (1)$$

A unique aspect of acyclic diene metathesis polycondensation is its ability to produce new polymer backbones by strategic monomer design [1].

Haloarene derivatives used for coordination polycondensation are primarily represented by halostyrenes and haloarylacetylenes which undergo self-coupling to poly(arylene vinylene)s [scheme (2)] and poly(aryleneacetylene)s [scheme (3)] respectively, and by dihaloarenes which, mainly in a combination with alkenes or divinylarenes and acetylenes or diacetylenes, undergo cross-coupling to poly(arylene vinylene)s [schemes (4) and (5)] and poly(arylene acetylene)s [schemes (6) and (7)] respectively [2]:

$$n\mathrm{X{-}Ar{-}CH{=}CH_2} \longrightarrow \mathrm{X\!\left[Ar{-}CH{=}CH\right]_n\!H} + (n-1)\mathrm{HX} \qquad (2)$$

$$n\,\mathrm{X{-}Ar{-}C{\equiv}CH} \longrightarrow \mathrm{X{+}Ar{-}C{\equiv}C{+}_{n}H} + (n-1)\mathrm{HX} \tag{3}$$

$$n\,\mathrm{X{-}Ar{-}X} + n\mathrm{CH_2{=}CH_2} \longrightarrow \mathrm{X{+}Ar{-}CH{=}CH{+}_{n}H} + (2n-1)\mathrm{HX} \tag{4}$$

$$n\,\mathrm{X{-}Ar^1{-}X} + n\mathrm{CH_2{=}CH{-}Ar^2{-}CH{=}CH_2} \longrightarrow$$

$$\mathrm{X{+}Ar^1{-}CH{=}CH{-}Ar^2{-}CH{=}CH{+}_{n}H} + (2n-1)\mathrm{HX} \tag{5}$$

$$n\,\mathrm{X{-}Ar{-}X} + n\,\mathrm{CH{\equiv}CH} \longrightarrow \mathrm{X{+}Ar{-}C{\equiv}C{+}_{n}H} + (2n-1)\mathrm{HX} \tag{6}$$

$$n\,\mathrm{X{-}Ar^1{-}X} + n\mathrm{CH{\equiv}C{-}Ar^2{-}C{\equiv}CH} \longrightarrow$$

$$\mathrm{X{+}Ar^1{-}C{\equiv}C{-}Ar^2{-}C{\equiv}C{+}_{n}H} + (2n-1)\mathrm{HX} \tag{7}$$

Haloarene derivatives have been self- and cross-coupled with organometallic compounds. Among these compounds, boron-substituted 9-borabicyclo[3.3.1]-nonyl (Bbn) derivatives, which are obtained by the hydroboration of olefinic compounds, deserve attention. For example, transition metal-catalysed polycondensation involving the self-coupling of monomers containing both aryl halide and alkylboronate functions and that involving cross-coupling of two corresponding bifunctional monomers, which leads to poly(arylene alkylene)s, are shown in schemes (8) and (9) respectively [3]:

$$n\mathrm{X{-}Ar{-}CH{=}CH_2} + n\mathrm{Bbn{-}H} \longrightarrow n\mathrm{X{-}Ar{-}CH_2{-}CH_2{-}Bbn}$$

$$\longrightarrow \mathrm{X{+}Ar{-}CH_2{-}CH_2{+}_{n}Bbn} + (n-1)\mathrm{Bbn{-}X} \tag{8}$$

$$n\mathrm{CH_2{=}CH{-}(CH_2)_4{-}CH{=}CH_2} + 2n\mathrm{Bbn{-}H} \longrightarrow n\mathrm{Bbn{-}(CH_2)_8{-}Bbn}$$

$$\xrightarrow{n\mathrm{X{-}Ar{-}X}} \mathrm{X{+}Ar{-}(CH_2)_8{+}_{n}Bbn} + (2n-1)\mathrm{Bbn{-}X} \tag{9}$$

Bbn = −B 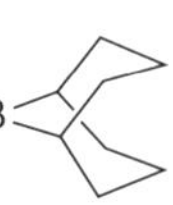

The boron-containing chain ends can be functionalised by oxidation with either H_2O_2/NaOH or I_2/NaOH.

On the other hand, aromatic monomers containing a boronic acid function apart from a halide function, i.e. haloarene boronic acids, undergo coordination homopolycondensation in the presence of transition metal-based catalysts, which results in the formation of poly(arylene)s [2]:

$$nX-Ar-B(OH)_2 \longrightarrow X\text{-}[Ar]_n\text{-}B(OH)_2 + (n-1)X-B(OH)_2 \qquad (10)$$

The transition metal-catalysed cross-coupling of arene diboronic acids with dihaloarenes also yields poly(arylene)s [2]:

$$nX-Ar^1-X + n(HO)_2B-Ar^2-B(OH)_2 \longrightarrow X\text{-}[Ar^1-Ar^2]_n\text{-}B(OH)_2 + (2n-1)X-B(OH)_2 \qquad (11)$$

It is worth noting that arene boronic acids and arene diboronic acids are synthesised from the corresponding halo- and dihaloarenes.

Poly(arylene)s are also formed in analogous coordination heteropolycondensation of dihaloarenes with di(alkylmetal)arenes, especially di(tributyltin)arenes [2]:

$$nX-Ar^1-X + nBu_3Sn-Ar^2-SnBu_3 \longrightarrow X\text{-}[Ar^1-Ar^2]_n\text{-}SnBu_3 + (2n-1)X-SnBu_3 \qquad (12)$$

The coordination homopolycondensation of metallated haloarenes also leads to poly(arylene)s [2]:

$$nX-Ar-X + nZn \longrightarrow nX-Ar-ZnX \longrightarrow X\text{-}[Ar]_n\text{-}ZnX + (n-1)ZnX_2 \qquad (13)$$

Catalysts for coordination polycondensation involving haloarene derivatives are usually based on Pd complexes, although Ni complexes can be employed in some instances.

Also, the coordination polycondensation of dihaloarenes via reductive coupling with metals or metal complexes, catalysed by transition metal complexes, leads to polyarylenes [2,3]:

$$nX-Ar-X + nZn \longrightarrow [Ar]_n + nZnX_2 \qquad (14)$$

Transition metal-catalysed polycondensations involving the formation of carbon–carbon bonds are an important subject of current research in polymer chemistry, owing to the unique thermal, electrical and optical properties of the resulting materials [2].

Transition metal-catalysed carbon-heteroatom coupling reactions primarily comprise the carbonylation polycondensation of dibromoarenes with aromatic diamines and bisphenols in the presence of carbon monoxide which leads to polyamides [scheme (15)] and polyesters [scheme (16)] respectively [4]:

$$n\mathrm{X-Ar^1-X} + n\,\mathrm{H_2N-Ar^2-NH_2} + 2n\mathrm{CO} \longrightarrow \left[\mathrm{Ar^1-\overset{\displaystyle O}{\overset{\|}{C}}-NH-Ar^2-NH-\overset{\displaystyle O}{\overset{\|}{C}}}\right]_n + 2n\,\mathrm{HX} \qquad (15)$$

$$n\mathrm{X-Ar^1-X} + n\mathrm{HO-Ar^2-OH} + 2n\mathrm{CO} \longrightarrow \left[\mathrm{Ar^1-\overset{\displaystyle O}{\overset{\|}{C}}-O-Ar^2-O-\overset{\displaystyle O}{\overset{\|}{C}}}\right]_n + 2n\mathrm{HX} \qquad (16)$$

The transition metal-catalysed polycondensation of dibromoalkanes with diacetylene arenes in the presence of carbon dioxide, which leads to polyesters [scheme (17)], is an instance where the carbon–heteroatom bond is formed by coupling of the carboxylated species [5]:

$$n\mathrm{Br-R-Br} + n\mathrm{CH{\equiv}C-Ar-C{\equiv}CH} + 2n\mathrm{CO_2} \longrightarrow \left[\mathrm{C{\equiv}C-Ar-C{\equiv}C-O-\overset{\displaystyle O}{\overset{\|}{C}}-R-\overset{\displaystyle O}{\overset{\|}{C}}-O}\right]_n + 2n\mathrm{HBr} \qquad (17)$$

Carbon–heteroatom coupling reactions including carbonylation and carboxylation polycondensations, promoted by transition metal catalysts, are becoming a promising route for various types of new condensation polymer.

The carbonylation oxidative polycondensation of bisphenol, 2,2-bis(4-hydroxyphenyl)propane, with transition metal-based catalysts, which yields the respective aromatic polycarbonate, is of high potential interest [6]:

$$n\mathrm{HO-Ar-OH} + n\mathrm{CO} + n/2\,\mathrm{O_2} \longrightarrow \left[\mathrm{Ar-O-\overset{\displaystyle O}{\overset{\|}{C}}-O}\right]_n + n\mathrm{H_2O} \qquad (18)$$

Also of interest is the oxidative polycondensation of 2,6-disubstituted phenols with transition metal catalysts such as palladium complexes, which produces aromatic polyethers [2].

8.1 Acyclic Diene Metathesis

Terminal dienes (α, ω-dialkenes), $CH_2{=}CH{-}(CH_2)_x{-}CH{=}CH_2$ ($x = 1$–7), are the preferred monomers in metathesis polycondensation [scheme (1)] for both entropic and steric reasons. Although acyclic diene metathesis occurs efficiently with no obvious driving force except for the entropic increase associated with an equilibrium process, the removal of the small-molecule byproduct from the reaction causes a shift in the productive metathesis equilibrium towards high molecular weight poly(1-alkenylene) formation. The metathesis of terminal diene units yields ethylene as a small molecule, which is initially removed from the reaction by a simple change in the state of matter to the gaseous phase and can then be completely removed from the polycondensation system under reduced pressure [1].

Let us emphasise that the driving force for acyclic diene metathesis, which is a step-growth condensation polymerisation, is the release and removal of a small condensate molecule. The polycondensation is performed preferably under bulk conditions (no solvent used), since acyclic diene metathesis is thermally neutral and there is no need to remove the heat of the reaction, in contrast to exothermic cyclic olefin ring-opening metathesis polymerisation.

Internal dienes, $RCH{=}CH{-}(CH_2)_x{-}CH{=}CHR$, can also undergo metathesis polycondensation, currently forming poly(1-alkenylene) and the appropriate alkene small molecule [scheme (19)] but at the expense of reduced activity and rate [1]:

$$n\mathrm{RCH{=}CH{-}(CH_2)}x\mathrm{{-}CH{=}CHR} \rightleftharpoons \mathrm{RCH{=}}\left[\mathrm{CH{-}(CH_2)}x\mathrm{{-}CH}\right]_n\mathrm{{=}CHR} + (n{-}1)\,\mathrm{RCH{=}CHR} \qquad (19)$$

It should be noted that when an R group, e.g. a methyl group, is present at the internal unsaturated carbon atom in a terminal diene molecule, as in the $CH_2{=}C(Me){-}(CH_2)_x{-}C(Me){=}CH_2$ monomer, the cyclic diene metathesis ceases [7,8]. The steric effect exerted by the R substituent can even be important at the α position to the double bond in the monomer. Sterically encumbering this position hinders polymer formation [9].

Also, divinylarenes such as 1,4-divinylbenzene can undergo metathesis polycondensation according to scheme (1) [10–12].

Acyclic α, ω-diene metathesis is a method for the synthesis of poly(1-alkenylene)s, which are also available from the ring-opening metathesis polymerisation of cycloolefins [13–15].

Metathetical polycondensation of acyclic dienes has not been successful with conventional catalysts used for the ring-opening metathesis polymerisation of cycloolefins, which is due to the fact that Lewis acids are usually present, and produce deleterious side reactions [13,16,17]. Only Lewis acid-free, well-defined catalysts have been successfully applied for acyclic diene metathesis polycondensation; the key success has been to choose catalysts that obviate other pathways not involving the metathesis mechanism [18–20]. It was Wagener *et al.* [16,21] who first were able to convert an acyclic α, ω-diene (1,9-decadiene), by using an acid-free metal alkylidene catalyst, to a high molecular weight polymer [poly(1-octenylene)] with a number-average molecular weight $\bar{M}_n = 50\,000$. The polycondensation was carried out in bulk to maximise the monomer concentration in this equilibrium reaction, with monomer to catalyst ratios of ca 1000:1; conversions occurred over a period of hours, leading to polymers characterised by polydispersity approaching a value of 2.0 [14].

It may be interesting that unsaturated ketones can also be subjected to metathetical polycondensation with the WCl_6 catalyst [22,23]:

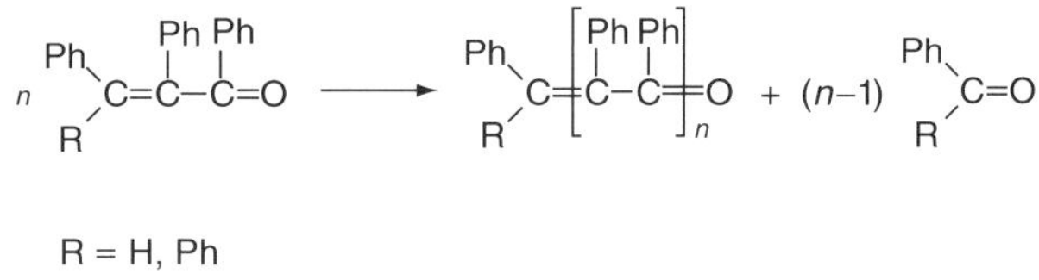

(20)

However, only a low molecular weight poly(1,2-diphenylacetylene) was obtained.

8.1.1 Polycondensation Catalysts

Lewis acid-free catalysts for acyclic diene metathesis obviate the formation of carbocations, which in turn completely eliminates competing reactions, mostly involving cationic oligomerisation via 1-alkene bonds. Thus, metathesis polycondensation occurs quantitatively to yield high molecular weight poly(1-alkenylene)s with vinyl end groups and ethylene as a byproduct.

Efficient catalysts capable of promoting metathesis condensation polymerisation are based on tungsten, molybdenum and ruthenium complexes [1]. Usually, the Schrock alkylidenes, especially their tungsten version, which have been shown to be the fastest metathesis polycondensation catalysts known to date, are employed [14,18–21,24,25]:

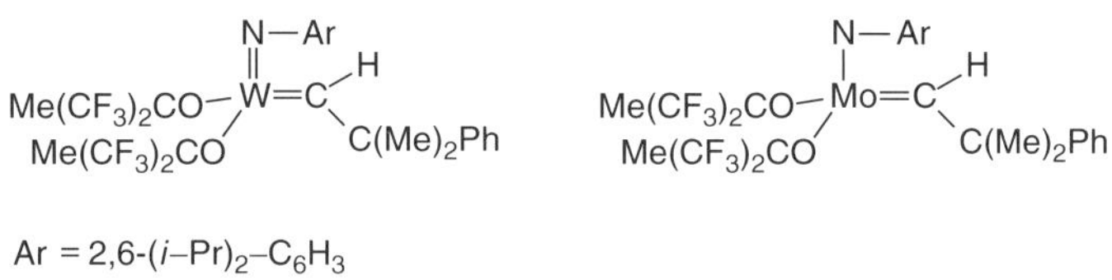

It should be noted that these catalysts ensure rapid conversions of any type of monomer susceptible to metathesis polycondensation, even of heteroatom-containing (functionalised) α,ω-dienes [1].

The ruthenium-based catalyst provided by Grubbs *et al.* [19] also promotes acyclic diene metathesis polycondensation, although with higher concentrations being required to achieve reasonable reaction rates [24,25]:

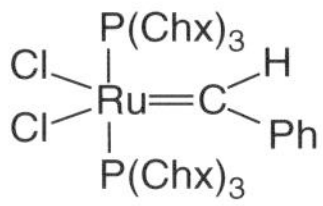

Ruthenium-based catalysts are relatively easy to synthesise and are tolerant of alcohol functionality in the acyclic diene monomer [26].

It has been demonstrated recently that conventional metathesis catalysts, used as a combination of aryloxytungsten-based precursor and tetramethyltin, tetrabutyltin or tributyltin hydride activator, are capable of promoting acyclic diene metathesis polycondensation [27]:

Ar = 2,6-$(Br)_2$–C_6H_3, 2,6-$(Ph)_2$–C_6H_3

These conventional catalysts are formed by the alkylation of the tungsten-based precursor by an alkyltin activator which is followed by α-hydrogen elimination to the tungsten alkylidene derivative [28]. It has been demonstrated [29] that the properly functionalised acyclic diene monomer employed for the polycondensation can even serve as an activator for the conventional tungsten-based precursors. For instance, bis(4-pentenyl)dibutylstannate $\{[CH_2=CH(CH_2)_3]_2SnBu_2\}$ undergoes metathesis condensation polymerisation to the respective high molecular weight polymer, $=[=CH(CH_2)_3–Sn(Bu)_2–(CH_2)_3CH=]_n=$, in a system with the $W(OAr)_2Cl_4$ or $W(O)(OAr)_2Cl_2$ catalyst, without the need for an added cocatalyst [30].

8.1.2 Polycondensation Mechanism

The peculiarity of acyclic diene metathesis is that it proceeds as a typical equilibrium condensation polymerisation process but is governed by a mechanism characterised by the occurrence of metal carbene and metallacyclobutane species; these species are found both in the ring-opening metathesis polymerisation of cycloolefins ('romp') and the ring-closing metathesis of acyclic dienes ('rcm') [31,32]. It has been conclusively demonstrated [14] that a metallacyclobutane ring is generated during acyclic diene metathesis polycondensation. If the metal alkylidene $L_xMt{=}CHR^1$ is used as the catalyst, the initial step of the reaction involves the formation of a π complex between one of the olefin groups of the diene monomer and the metal, followed by the collapse to a corresponding metallacyclobutane ring. In productive metathesis, this ring undergoes a cleavage in such a manner as to place the metal centre on the end of the monomer [1]:

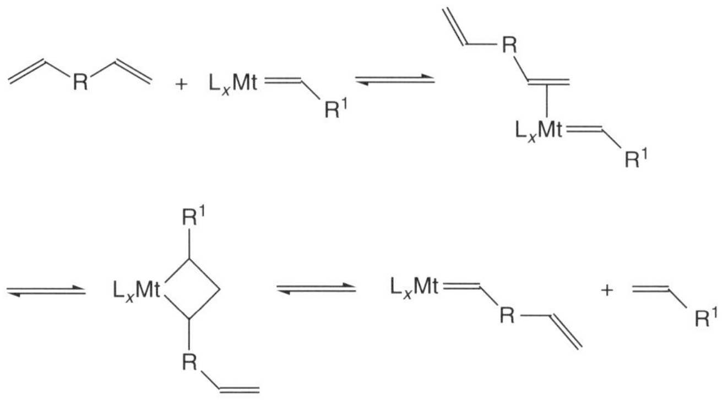

(21)

The new metal alkylidene species $L_xMt = CHRCH = CH_2$ initiates the polycondensation cycle, which is illustrated in Figure 8.1 [1].

The polycondensation cycle is started in an analogous way to that shown in scheme (21); one of the double bonds in a diene or in a polymer chain end forms a π complex with the metal, which leads to another metallacyclobutane ring. In the next step, this ring collapses to form an internal olefin within a polymer chain and generates the true catalyst for the metathesis polycondensation, a methylidene, $L_xMt = CH_2$. The cycle continues by the formation of a metallacyclobutane ring, again via the same π complexation as that described above. Cleavage of the latter metallacyclobutane ring evolves ethylene, and the cycle is repeated, growing the polymer in a stepwise manner with every cycle [1].

The participation of metallacyclobutane rings in metathesis polycondensation is the reason why the aforementioned sterically encumbered acyclic diene monomers hardly undergo reaction [7–9]; in such cases, steric influences at particular positions in the monomer hinder the formation of the necessary metallacyclobutane ring, thus inhibiting polymer formation [1].

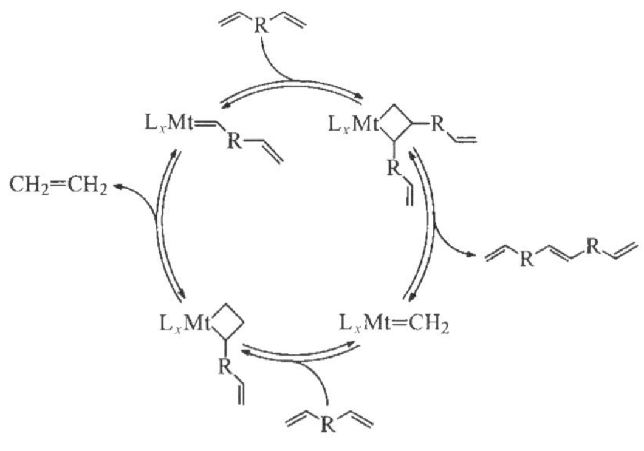

Figure 8.1 Schematic presentation of the mechanism for acyclic diene metathesis condensation polymerisation

Note that the acyclic diene metathesis reaction [scheme (22)] and the ring-closing metathesis reaction [scheme (23)] are inter- and intramolecular displays of the same mechanistic event [1]:

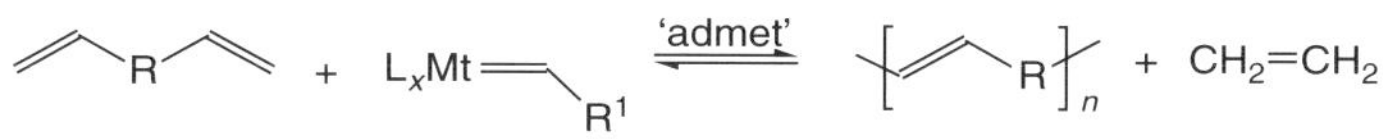

(22)

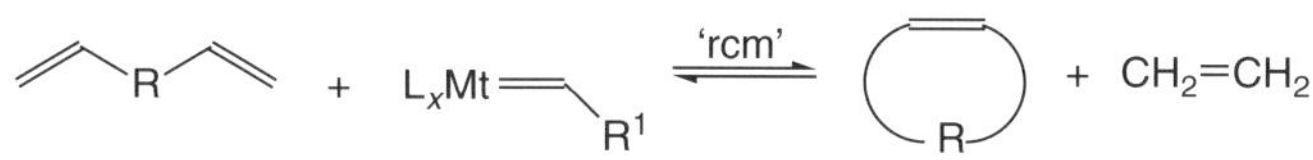

(23)

In the ring-closing metathesis reaction, intramolecular metathesis closes a ring to form a small cyclic molecule with concurrent loss of a small molecule (ethylene). Conversely, in the case of the acyclic diene metathesis reaction, macromolecules are formed by successive intermolecular condensation of two olefinic molecules [1].

Using 1,5-hexadiene, it was shown that, depending upon whether molybdenum- or ruthenium-based catalysts are employed, a change in mechanism appears to occur. In the presence of Schrock's molybdenum catalyst, 1,5-hexadiene produces principally linear poly(1-butenylene) [scheme (24)] [33], but with Grubbs's ruthenium catalyst the primary product is the cyclic dimer 1,5-cyclooctadiene [scheme (25)] [25,33]:

(24)

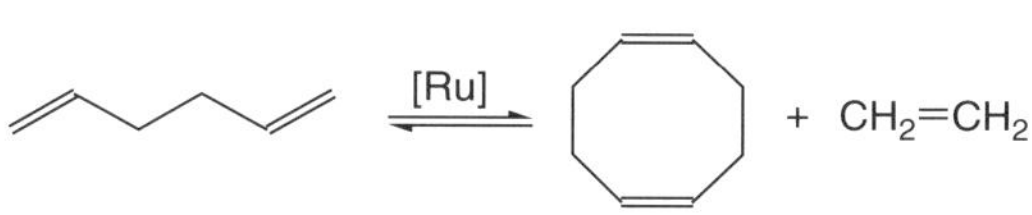

(25)

This change in mechanism has been attributed to the formation of a π chelate complex via coordination of the second olefinic bond present in the diene at the Ru atom, which then favours the intramolecular ring-closing reaction over intermolecular condensation polymerisation [1].

It is worth noting that the polycondensation of hydrocarbon monomers is completely reversible [34]. Unsaturated polymers such as 1,4-poly(butadiene) can be converted to diene monomers via depolymerisation with ethylene [35–37]:

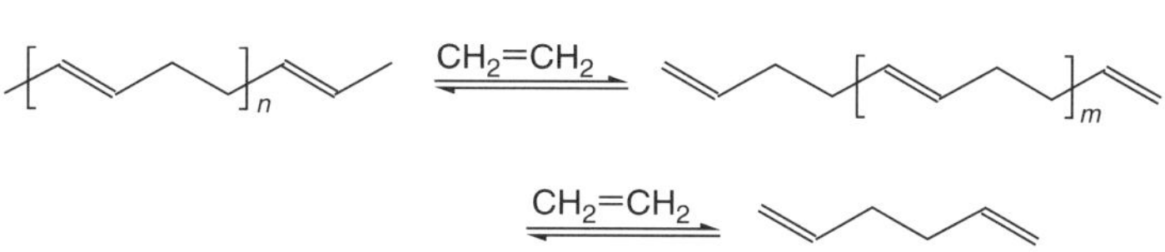

(26)

It is possible not only to achieve high mass conversions of polybutadiene to 1,5-hexadiene but also to create telechelic oligomers in this manner [38–40]. Catalyst selection plays an important role here; ruthenium-based catalysts appeared to be best in bringing about clean conversions of high molecular weight unsaturated polymers to their telechelic oligomers [1].

8.1.3 Tailored Hydrocarbon and Functionalised Polymers

Catalyst and monomer development in acyclic diene metathesis remains a subject of interest, the goal being to obtain macromolecules with well-defined backbone structures and architectures by easily accessible and less expensive means. By the application of an appropriate design of monomers and a careful choice of catalysts, a variety of non-functionalised and functionalised dienes have been polymerised via metathesis condensation to high molecular weight polymers.

It is now possible to produce both conjugated [41] and saturated [42] hydrocarbon polymers (e.g. strictly linear polyethylene [42]) by the metathesis polycondensation of acyclic dienes. In the latter case the initially formed unsaturated polymer is subjected to hydrogenation using various reducing agents. However, in order to extend the utility of metathesis condensation polymerisation, Wagener *et al.* [43] have devised a novel hydrogenation catalyst based on a ruthenium complex which permits direct conversion of the unsaturated polymer formed via metathesis polycondensation to its saturated analogue; the homogeneous ruthenium-based catalyst in its native state drives the metathesis condensation polymerisation. Then, silica gel is introduced directly into the reaction vessel, thereby converting the ruthenium catalyst into a heterogeneous hydrogenation catalyst. Exhaustive hydrogenation of the initially formed poly(1-alkenylene) can then be performed to produce a completely saturated polymer. Typical examples of very pure hydrocarbon polymers formed by acyclic diene metathesis include poly(1,4-phenylene vinylene) [12], 1,4-polybutadiene [44] and poly(1-octenylene) [45], polymers which are of general interest. Another instance is the possibility of obtaining polymers with a structure identical to perfectly alternating butadiene/isoprene copolymers from a single monomer by choosing the appropriate triene monomer [46]:

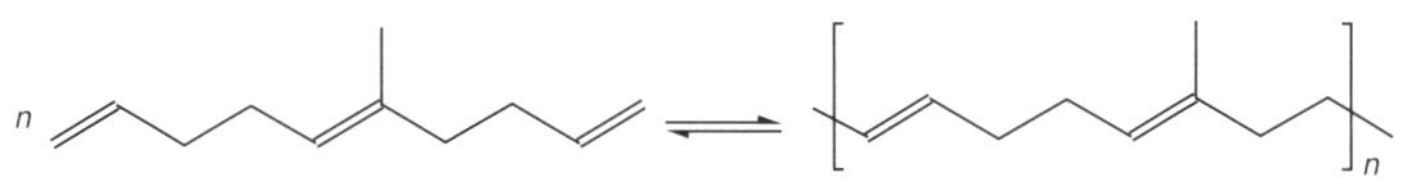

(27)

Also, a series of well-defined polyolefins, including perfectly branched polyethylene and ethylene/α-olefin copolymers, have been synthesised via acyclic diene metathesis polycondensation [scheme (28)] [47]; these well-defined polyolefins have been designed to model the crystallisation of polyethylene and olefin copolymers.

$$n\mathrm{CH_2{=}CH{-}(CH_2)_x{-}\overset{\displaystyle R}{\overset{|}{C}H}{-}(CH_2)_x{-}CH{=}CH_2}$$

$$\xrightarrow[-n\mathrm{CH_2\ CH_2}]{} \left[\mathrm{CH{=}CH{-}(CH_2)_x{-}\overset{\displaystyle R}{\overset{|}{C}H}{-}(CH_2)_x}\right]_n$$

$$\xrightarrow{\mathrm{H_2}} \left[\mathrm{CH_2{-}CH_2{-}(CH_2)_x{-}\overset{\displaystyle R}{\overset{|}{C}H}{-}(CH_2)_x}\right]_n \qquad (28)$$

The technique illustrated by scheme (28) gives control of the number of methylene groups between branch points as well as the length and identity of the branch itself [1].

By applying monomers containing functional groups for acyclic diene metathesis polycondensation, a wide variety of functionalised, unsaturated and saturated polymers have been obtained. The choice of catalyst as well as the position of the functional group within the monomer makes a significant impact, regarding both the rate of polycondensation and the molecular weight of the polymers obtained. The optimum conditions for metathesis polycondensation occur by positioning the functional group at least two methylene units distant from the metathesising olefin within the monomer unit. If the functional group is any closer, an intramolecular complex can be formed between the lone electron pair within the functional group and metal [15,33]:

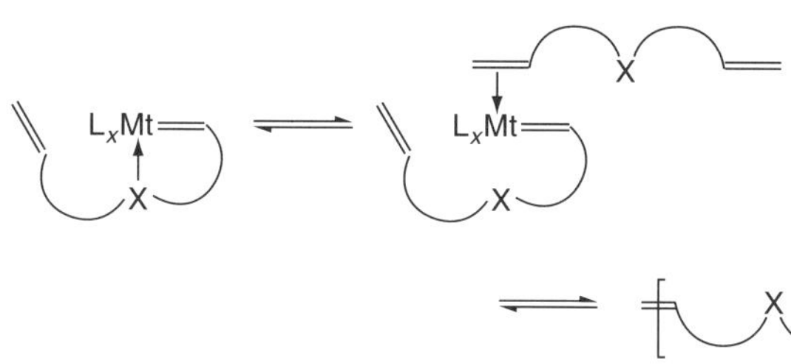

(29)

If this complex is sufficiently stable, then no further reaction occurs, and the polycondensation is obviated. However, as mentioned above, by using controlled monomer design, a variety of functionalised polymers (typical by with $\bar{M}_n = 10\,000$–$30\,000$ and $\bar{M}_w/\bar{M}_n = 2.0$) can be synthesised: polyethers [15,48–50], poly(thioether)s [48], polyesters [51,52], polycarbonates [53], polyketones [54], polysiloxanes [55–58], poly(carbosiloxane)s [59], poly(carbosilane)s [60], poly(carbodichlorosilane)s [61] and polymers with a conjugated π system [62].

It has also been possible to design a series of segmented functionalised copolymers (with polyether or polysiloxane segments) using α,ω-dienyl telechelic oligomers for metathesis polycondensation [63].

Telechelic dienes for metathesis polycondensation, containing functional groups such as those in alcohols, esters, carboxylic acids and imides, can be obtained via acyclic diene metathesis depolymerisation [64,65]. They can then be used in further reactions to create hydrophobic polyurethanes and other special-purpose polymers [1].

8.2 Polycondensation via Carbon–Carbon Coupling Reactions Involving Haloarene Derivatives

8.2.1 Aryl–Vinyl Coupling

The transition metal-catalysed coupling of aryl halides with alkenes (alkenylation of aromatic halides) is referred to as the Heck reaction [66–75]. Heitz *et al.* [76–82] were the first to utilise the Heck reaction to synthesise high molecular weight poly(arylene vinylene)s by the self-coupling of bromostyrene [scheme (2)] or the cross-coupling of dibromoarene with ethylene [scheme (4)] or divinylarene [scheme (5)].

Heck-type step-growth condensation polymerisation involves mainly palladium-based catalysts, although nickel-based catalysts are also effective. It is worth noting that this polycondensation requires a change in the oxidation state of the metal (e.g. Pd) [schemes (30) and (31)] [71], which is in contrast to chain growth polymerisation, such as ethylene/carbon monoxide alternating copolymerisation promoted by Pd-based catalysts [schemes (82) and (83) in Chapter 3], for which the preservation of the oxidation state of palladium, Pd(II), is typical [83–85]:

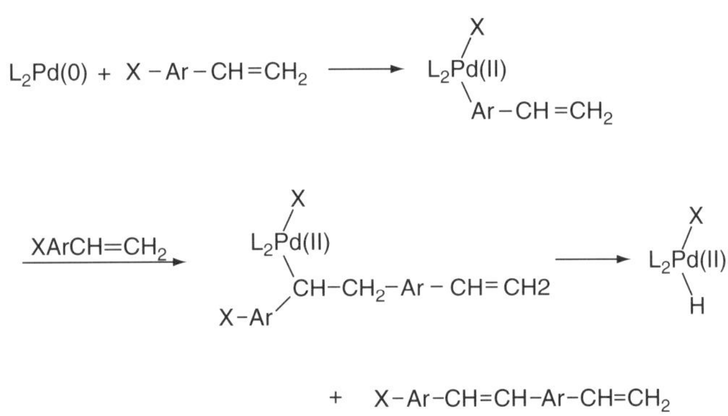

(30)

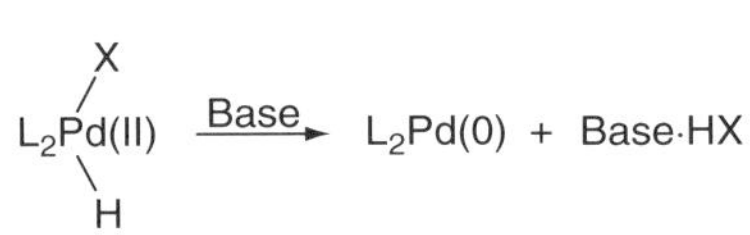

(31)

The first step in catalytic reactions of the Heck type is the oxidative addition of the organic halide to Pd(0) species to form an intermediate organopalladium halide constituting the Pd(II) species. This is followed by insertion of the olefinic bond and subsequent β-hydrogen elimination [scheme (30)]. The catalyst is recycled by the reaction of the Pd(II)-hydride species with a base [scheme (31)]. It is worth noting here that palladium species, $L_2(X)Pd{-}ArCH{=}CH_2$, do not propagate the chain growth polymerisation of the $CH_2{=}CHArX$ monomer via its olefinic bond in the discussed process.

Palladium(0)-catalysed coupling reactions of haloarenes with alkenes, leading to carbon–carbon bond formation between unsaturated species containing sp^2-hybridised carbon atoms, follow a similar mechanistic scheme; as already stated, the general features of the catalytic cycle involve an oxidative addition–alkene insertion–reductive elimination sequence. The reaction is initiated by the oxidative addition of electrophile to the zero-valent metal [86]. The most widely used are diverse Pd(0) complexes, usually with weak donor ligands such as tertiary phosphines. A coordinatively unsaturated Pd(0) complex with a formally d^0 14-electron structure has meanwhile been proven to be a catalytically active species. This complex is most often generated *in situ* [87–91].

Tetrakis(triphenylphosphine)palladium(0), $Pd(PPh_3)_4$, which exists in equilibrium with tris(triphenylphosphine)palladium(0), $Pd(PPh_3)_3$, and free triphenylphosphine in solution are frequently employed. The loss of the second phosphine ligand [92,93] leads to the catalytically active bis(triphenylphosphine)palladium(0), $Pd(PPh_3)_2$. A number of sterically encumbered diphosphine-palladium(0) complexes are known [94–96], but the $Pd(PPh_3)_2$ complex is exceptional. However, Pd(II) complexes [e.g. bis(triphenylphosphine)palladium dichloride, $Pd(PPh_3)_2Cl_2$], which readily undergo reduction in the reaction medium, are more commonly employed as catalysts for convenience, as they are inherently stable in air. A '$Pd(PPh_3)_2$' species generated by treatment of $Pd(PPh_3)_2Cl_2$ with proper reductants constitutes the active sites [97–99].

If a Pd(II) complex is used as the catalyst (PdX_2), the first step of the reaction involves the reduction of Pd(II) species to Pd(0) species (PdL_2) by the olefin [100]:

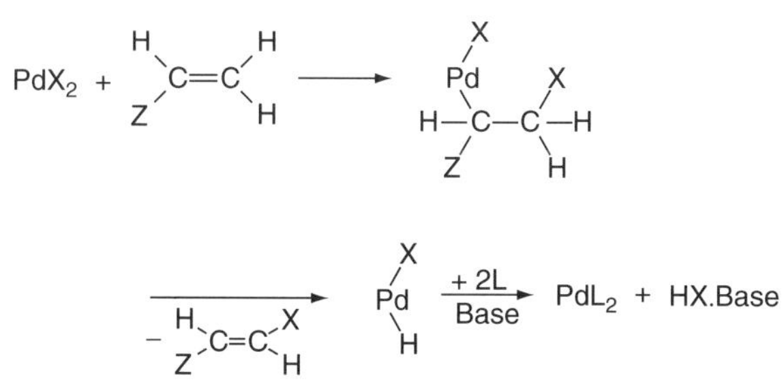

(32)

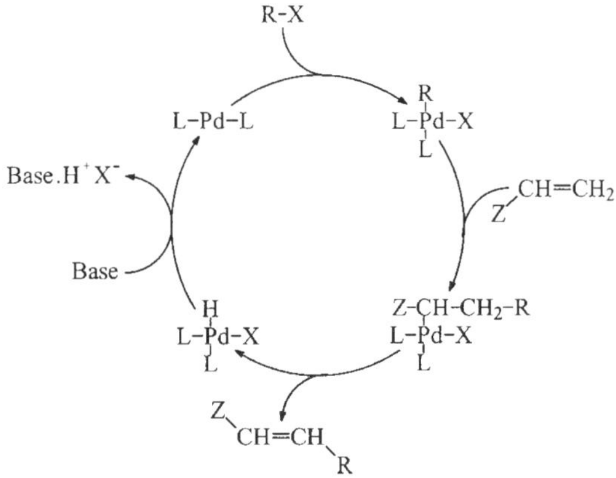

Figure 8.2 Schematic presentation of the mechanism for the Heck reaction

The general catalytic cycle in Heck-type polycondensation consisting of the oxidative addition of aryl halide, olefin insertion and reductive elimination is shown in Figure 8.2 [100–104].

For the crucial stage of the oxidative addition of an electrophile to the zero-valent palladium, double synergistic interaction takes place:

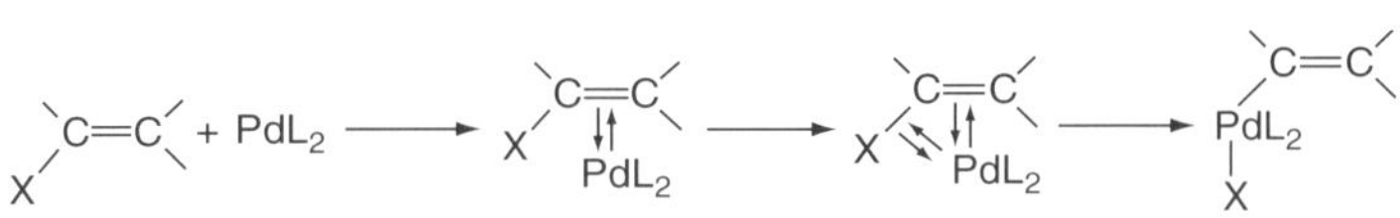

(33)

The π-complexed Pd(0) species are thought to undergo concerted synergistic interaction with a carbon–halogen bond, proceeding with retention of the configuration [102–105]. In the next stage, which follows the insertion of an olefinic bond into the Pd–C bond, β-hydride elimination takes place, but only after an internal rotation around the former double bond [scheme (34)], as it requires at least one β-hydrogen to be oriented synperiplanar with respect to the halopalladium moiety [105]:

(34)

The subsequent *syn*-elimination, yielding a 1,2-substituted alkene and hydridopalladium halide, is, however, reversible, and therefore the thermodynamically more stable (*E*)-alkene, (*E*)-RCH=CHZ, is generally produced when the coupling reaction is performed with a terminal olefin. It is worth mentioning that the formation of this 1,2-disubstituted alkene is accompanied with the formation of 1,1-disubstituted alkene, $CH_2 = C(Z)R$, the amount of which depends on the kind of catalyst and the reaction conditions. a lower reaction temperature provides for a lower ratio of 1,1-to 1,2-disubstituted olefins, i.e. for higher regioselectivity of the Heck-type polycondensation [101]. Finally, reductive elimination of HX from the hydridopalladium halide, aided by the added base, regenerates the active catalyst species and thereby completes the catalytic cycle.

As regards the rate determining step of the Heck coupling reaction, the oxidative addition has been assumed to be this step, although it has not been proven in all details.

The application of Heck coupling polycondensation is not limited to the synthesis of poly(arylene vinylene)s via the alkenylation of haloarenes in simple monomer systems but includes a variety of self- and cross-coupling reactions involving reactants with various functionalities. For instance, the polycondensation of diiodoarene with bis(acrylamido)arene by the $Pd(OOCCH_3)_2$–P $(o\text{-}C_6H_4\text{–}CH_3)_3$ catalyst yields respective aromatic polycinnamamide [106]:

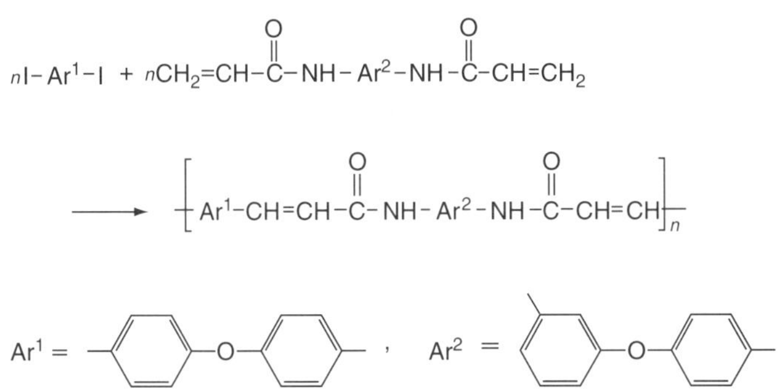

(35)

The polycinnamamide obtained according to scheme (35) is a high molecular weight polymer. However, when the corresponding dibromoarene, $BrC_6H_4OC_6H_4Br$, was used instead of diiodoarene, only the polycinnamamide of low molecular weight was formed. Polymers with a cinnamoyl backbone were also prepared by Heck coupling of dihaloarenes and acrylates [107].

Other examples of functionalised poly(arylene vinylene)s comprise polymers with alkyl, aryl, fluoroalkyl, nitro, alkoxy, porphyrin and metalloporphyrin moieties [100].

8.2.2 Aryl–Alkynyl Coupling

The transition metal-catalysed coupling reaction of aryl halides with alkynes (alkynylation of aryl halides) is frequently also considered as a Heck reaction [68, 108–111]. Polymers containing arylacetylene and diarylacetylene units in the backbone have been obtained by the self-coupling of ethynylaryl halides [scheme (3)] and the cross-coupling of aromatic dihalides with acetylene [scheme (6)] or diethynylbenzenes [scheme (7)] [112–121]. The majority of these crystalline polymers are of low molecular weight owing to their insolubility [3]. However, soluble high molecular weight polymers were obtained in polycondensations of respective substrates substituted with alkoxy groups containing long aliphatic chains [122].

The reactivity of the aryl halide decreases in the halide order I > Br, with chlorides failing to react. The reactivity of substituted aryl halides increases upon going from electron-donating substituents through unsubstituted aryl halides to electron-withdrawing substituents. Both reactivity patterns of ArI > ArBr and activation of aryl halides with electron-withdrawing substituents follow the general reactivity of aryl halides in Pd(0) oxidative addition reactions [3].

Aryl–acetylene coupling polycondensation is performed mainly in the presence of Pd-based catalysts under normal Heck reaction conditions (temperature in the region of 50–100 °C). However, the most effective coupling reaction is achieved by combining Pd(0) catalysts with a Cu(I) compound, thereby allowing ambient reaction temperature [109,110,123–129]. Usually, various combinations of reagents containing Pd(II) complexes that generate Pd(0) species can be used; for instance, $Pd(PPh_3)_2Cl_2$ can be reduced to the active $Pd(PPh_3)_2$ complex by reaction with an acetylenic compound [3]:

$$\begin{aligned} &Pd(PPh_3)_2Cl_2 + HC{\equiv}CAr \xrightarrow[-2HCl]{} Pd(PPh_3)_2(C{\equiv}CAr)_2 \\ &\longrightarrow Pd(PPh_3)_2 + ArC{\equiv}C{-}C{\equiv}CAr \end{aligned} \tag{36}$$

The general catalytic cycle in the Pd(0)/Cu(I)-promoted aryl–acetylene coupling reaction is shown in Figure 8.3 [3].

The $Pd(PPh_3)_2$ complex (PdL_2) has two vacant coordination sites and easily undergoes a Cu(I)-promoted oxidative addition with aryl halide (Ar–X) to generate a Pd(II) complex. This is followed by the substitution of the halide

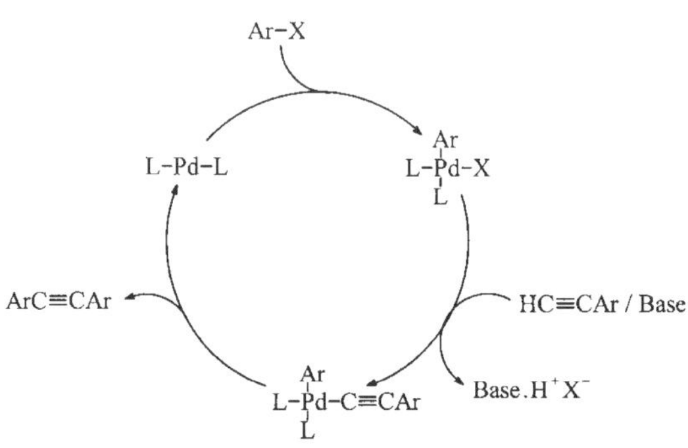

Figure 8.3 Schematic presentation of the mechanism for the aryl–alkynyl coupling reaction

with an acetylide, and reductive elimination of the diarylacetylene to regenerate Pd(0) active sites. A base is required to deprotonate acetylene [3]. In the absence of a base, the coordination of acetylene to Pd(II) species would be through the triple bond rather than through the acetylide carbon bond, thereby allowing oligomerisation of the acetylene derivative to occur in a chain growth manner [130].

It is worth mentioning that acetylenes used for the coupling reaction with aromatic halides can be applied as acetylene adducts; this method makes it possible to avoid complications with the purification and dosage of acetylene and to perform a one-pot synthesis [131,132]. When 2-methyl-3-butyne-2-ol is used as the monoprotected acetylene (with acetone), the synthesis is carried out in a liquid–liquid phase transfer system and requires, apart from the Pd(0)/Cu(I) catalyst, an inorganic base such as NaOH or KOH. For instance, polycondensation by the coupling of dihaloarene ($BrAr^2Br$) and the product derived from the coupling of another dihaloarene (IAr^1I) and 2-methyl-3-butyne-2-ol, subsequently deprotected, leads to the respective poly(arylene acetylene) [133].

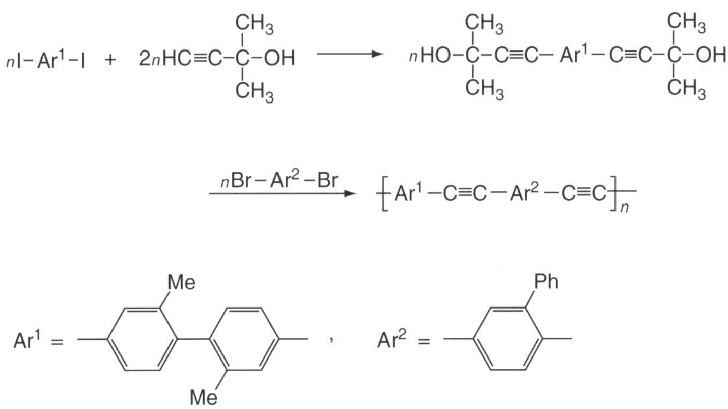

(37)

In an analogous way, various poly(arylene alkynylene)s were obtained [120,134].

It should be noted that functionalised diacetylene monomers have also been used for coupling reactions with haloarenes. By reacting diacetylene-substituted *p*-aminoanilines [135] and diethynyl(methyl)(*n*-octyl)silane [136] with diiodoarenes, polyamides and polysilanes have been prepared, respectively.

In addition to aryl–alkynyl coupling reactions involving dihaloarenes and acetylenes or dialkynylarenes which lead to poly(arylene acetylene)s, it is worth noting that these polymers have also been obtained by the Pd(0)-catalysed heteropolycondensation of dialkoxy-substituted dibromoarenes with bis(tributylstannyl)acetylenes [122].

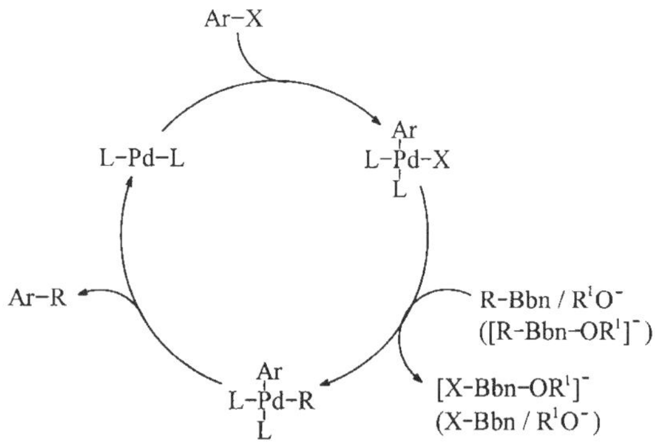

Figure 8.4 Schematic presentation of the mechanism for the aryl–alkyl coupling reaction

8.2.3 Aryl–Alkyl Coupling

Monomers that contain haloarene and alkylboronate functions undergo an aryl–alkyl bond-forming coupling in the presence of palladium-based catalysts [schemes (8) and (9)] [3,137]. The mechanism of this coupling is outlined in Figure 8.4.

The aryl halide first oxidatively adds to Pd(0) species and a Pd(II) complex is generated. Following activation of the alkylboronate (R–Bbn) with (R^1O^-) base, the alkyl group undergoes transmetallation from boron to palladium to generate another Pd(II) complex, which reductively eliminates the aryl–alkyl coupled product with regeneration of the Pd(0) species [3]. It is to be mentioned that the discussed phase transfer Pd(0)-catalysed polycondensation of monomers containing aryl halide and alkylboronate functions is accompanied with side reactions to some extent.

8.2.4 Aryl–Aryl Coupling

Higher molecular weight poly(arylene)s containing solubilising alkyl substituents were prepared by Pd-catalysed self-coupling of bromoalkylboronic acids [scheme (10)] as well as by cross-coupling of dihaloarenes with arene diboronic acids [scheme (11)] in a mixture of benzene and aqueous sodium carbonate [138–140]. These reactions also tolerate functional groups such as ether, carbonyl and N-alkyl groups in dihaloarenes when coupled with arene diboronic acids [140–144]. The Pd(0)-catalysed coupling of haloarenes and aromatic boronic acids, known as Suzuki coupling [145–147], was introduced for the synthesis of a wide variety of linear polyphenylenes. Since the discussed polycondensation proceeds regiospecifically under mild conditions, it appears to be an ideal way for the synthesis of poly(arylene)s containing various functions.

Polycondensation of this type was also introduced for the synthesis of hyperbranched polyphenylenes [138,148].

Electrophilic haloarene function and nucleophilic areneboronic acid function undergo carbon–carbon coupling only in the presence of Pd(0) catalytic species. The general catalytic cycle for this aryl–aryl coupling, involving an oxidative addition–transmetallation–reductive elimination sequence, follows the mechanistic scheme shown in Figure 8.4 for aryl–alkyl coupling.

Various aromatic dibromides have been cross-coupled with difunctional tributyltin aromatic compounds in the presence of palladium-based catalysts to yield poly(arylene)s [scheme (12)] [149–151]. The mechanistic pathway of this coupling, known as Stille coupling [152], follows an oxidative addition–transmetallation–reductive elimination sequence.

The coordination homopolymerisation of metallated haloarenes [scheme (13)] by coupling reactions of haloarene and metallated arene moieties in the presence of palladium-based catalysts have been widely used for the synthesis of poly(arylene)s. The generally accepted mechanism involves a catalytic cycle similar to that taking place in polycondensation systems containing aromatic halide and organotin function [2].

The dehalogenating polycondensation of dihaloarenes to poly(arylene)s, involving reductive coupling with metals such as magnesium or zinc [scheme (14)], proceeds in the presence of palladium- or nickel-based catalysts [153–156]. The generally accepted mechanism of Ni(0)-catalysed polycondensation involves Ni(I) and Ni(III) species, with the Ni(II) species being reactive [157,158]. Excess zinc is required to reduce Ni(II) to Ni(0) in the first step of the reaction, and to reduce unreactive Ni(III) species formed throughout the catalytic cycle. The initially generated Ni(0) species undergoes an oxidative addition with the haloarene to generate an Ar–Ni(II) complex. This is reduced by zinc to the Ar–Ni(I) complex, which is more reactive towards oxidative addition than the Ar–Ni(II) complex. The Ar–Ni(I) complex undergoes oxidative addition with a second aryl halide to generate an Ar–Ni(III)–Ar complex, which reductively eliminates the Ar–Ar product, resulting in a Ni(I) complex. The regenerated Ni(I) complex is again reactive towards oxidative addition with an aryl halide, or it can be reduced to Ni(0) species by zinc [3].

8.3 Polycondensation via Carbenoid Coupling Reactions of Bis(dichloromethyl)arenes

In connection with nickel-catalysed reductive coupling reactions of dihaloarenes, leading to poly(arylene)s, the copper-catalysed reductive carbenoid coupling reactions involving substituted bis(dichloromethyl)arenes and metals or other reductants should be mentioned. The reductive carbenoid coupling of bis(phenyldichloromethyl)arene with zinc in the presence of Cu-based catalysts is shown below:

$$n\mathrm{Cl_2C(Ph)\text{-}Ar\text{-}C(Ph)Cl_2} + 2n\mathrm{Zn} \longrightarrow \mathrm{-[Ar\text{-}C(Ph){=}C(Ph)]_n\text{-}} + 2n\mathrm{ZnCl_2} \qquad (38)$$

In this case, the respective substituted poly(arylene vinylene) is formed [159–161].

8.4 Polycondensation via Carbon–Heteroatom Coupling Reactions

8.4.1 Carbonylative Coupling

Palladium- or nickel-catalysed carbonylation polycondensation of dibromoarenes with bifunctional nucleophilic monomers such as aromatic diamines and bisphenols in the presence of carbon monoxide appeared to be a new promising method for the synthesis of aromatic polyamides [scheme (15)] and polyesters [scheme (16)] respectively [162–170]. The first successful example was the synthesis of high molecular weight polyaramide according to scheme (15) ($Ar^1 = m\text{-}C_6H_4\text{-}$, $Ar^2 = p\text{-}C_6H_4\text{-}O\text{-}C_6H_4\text{-}$) [162].

The general catalytic cycle for this carbonylation coupling reaction is analogous to direct carbon–heteroatom coupling [scheme (39)] except that carbon monoxide insertion takes place after the oxidative addition step and prior to the nucleophilic attack of the amine [scheme (40)]:

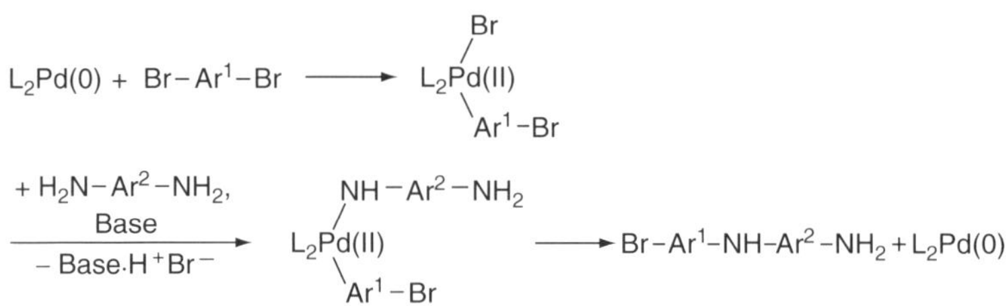
(39)

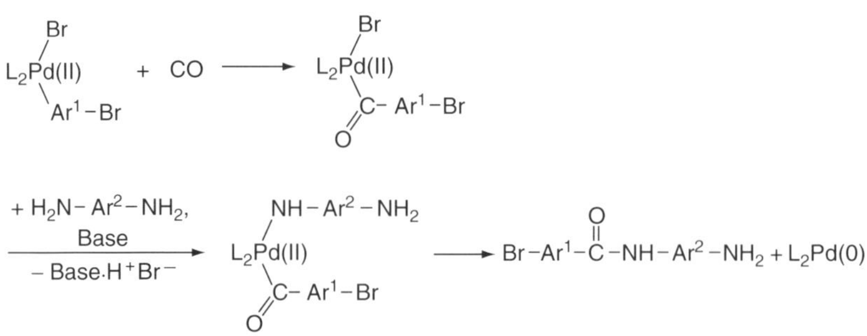
(40)

Coordination carbonylation polycondensation has been extended from the synthesis of polyamides [scheme (15)] and polyarylates [scheme (16)] to reactions using other nucleophilic monomers that, with dihaloarenes and carbon monoxide, yield poly(imide–amide)s, poly(acylhydrazide)s, and poly(benzoxazole)s [165,170,171].

It is preferable for relatively inexpensive Ni-based catalysts to be used instead of expensive Pd-based catalysts in the carbonylative polycondensation. However, in the case of carbonylative coupling of dibromoarenes with aromatic diamines, yielding polyaramides, the use of nickel-based catalysts is not so successful for the synthesis of high molecular weight polymers, compared with the method using palladium-based catalysts [166].

8.4.2 Carboxylative Coupling

Copper complex-catalysed carboxylation polycondensation of dibromoalkanes with arenediyls in the presence of carbon dioxide [scheme (17)], although producing low molecular weight polyesters in moderate yields, also represents pioneering efforts in the use of carbon dioxide as a non-petroleum-based carbon source [5].

8.5 Oxidation Carbonylative Polycondensation of Bisphenol

As stated above, the carbonylation oxidative polycondensation of bisphenol in the presence of transition metal-based catalysts leads to aromatic polycarbonate [scheme (18)] [6]. The reaction of bisphenol (HOArOH, e.g. Ar = p-C_6H_4–CMe_2–C_6H_4–), carried out under CO and O_2 pressure in a chlorohydrocarbon solvent under anhydrous conditions, using a group 8 metal-based catalyst (e.g. a $PdBr_2$ complex) and a redox catalyst (e.g. Mn(II) (benzoinoxime)$_2$, L_xMn) in the presence of a base (e.g. 2,2,6,6,N-pentamethylpiperidine, R_3N), involves most probably the pathway shown schematically below:

$$n\text{PdBr}_2 + n\text{HO–Ar–OH} + n\text{CO} + 2n\text{R}_3\text{N} \longrightarrow \left[\text{Ar–O–}\overset{\overset{\displaystyle\text{O}}{\|}}{\text{C}}\text{–O}\right]_n + 2n\text{R}_3\text{NH}^+\,\text{Br}^- + n\text{Pd(0)} \quad (41)$$

The Pd(0) species undergo bromination to Pd(II) species, $PdBr_2$, via oxidation catalysed by a Mn(II) complex:

$$n\text{Pd(0)} + 2n\text{R}_3\text{NH}^+\,\text{Br}^- + n/2\text{O}_2 \longrightarrow n\text{PdBr}_2 + 2n\text{R}_3\text{N} + n\text{H}_2\text{O} \quad (42)$$

Note that the essential step of polycarbonate formation involves the reaction sequence shown below:

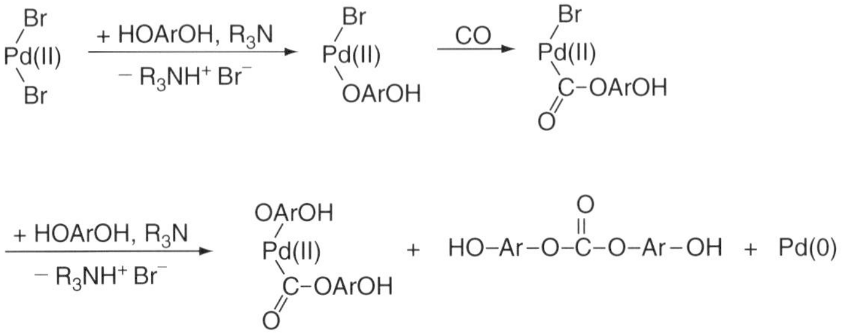

(43)

Although only low molecular weight polycarbonates in modest yield have been obtained by this method, it represents an interesting non-phosgene route to aromatic polycarbonates.

References

1. Tindall, D., Pawlow, J. H. and Wagener, K. B., Recent Advances in ADMET chemistry, pp. 183–198.
2. Kiehl, A. and Müllen, K., 'Polycondensation', in *Catalysis in Precision Polymerisation*, John Wiley & Sons, Chichester–New York, 1997, pp. 134–187.
3. Percec, V., Pugh, C., Cramer, E., Okita, S. and Weiss, R., *Makromol. Chem., Macromol. Symp.*, **54/55**, 113 (1992).
4. Imai, Y., *Makromol. Chem., Macromol. Symp.*, **54/55**, 151 (1992).
5. Oi, S., Fukue, Y., Nemoto, K. and Inoue, Y., *Macromolecules*, **29**, 2694 (1996).
6. Hallgren, J. E., US Pat. 4 201 721 (to General Electric Co.) (1980).
7. Konzelman, J. and Wagener, K. B., *Macromolecules*, **28**, 4686 (1995).
8. Wagener, K. B. and Konzelman, J., *Polym. Prepr. Am. Chem. Soc., Div. Polym. Chem.*, **32**(1), 375 (1991).
9. Konzelman, J. and Wagener, K. B., *Polym. Prepr. Am. Chem. Soc., Div. Polym. Chem.*, **33**(1), 1072 (1992).
10. Kumar, A. and Eichinger, B. E., *Makromol. Chem., Rapid Commun.*, **13**, 311 (1992).
11. Thorn-Csányi, E. and Höhnk, H.-D., *J. Mol. Catal.*, **76**, 101 (1992).
12. Wolf, A. and Wagener, K. B., *Polym. Prepr. Am. Chem. Soc., Div. Polym. Chem.*, **33**(1), 535 (1991).
13. Wagener, K. B., Nel, J. G., Konzelman, J. and Boncella, J. M., *Macromolecules*, **23**, 5155 (1990).
14. Wagener, K. B., Boncella, J. M. and Nel, J. G., *Macromolecules*, **24**, 2649 (1991).
15. Brzezinska, K. and Wagener, K. B., *Macromolecules*, **24**, 5273 (1991).
16. Lindmark-Hamberg, M. and Wagener, K. B., *Macromolecules*, **20**, 2949 (1987).
17. Wagener, K. B., Boncella, J. M., Nel, J. G., Duttweiler, R. P. and Hillmyer, M. A., *Makromol. Chem.*, **191**, 365 (1990).
18. Schrock, R. R., Murdzek, J. S., Bazan, G. C., Robbins, J., Di Mare, M. and O'Regan, M., *J. Am. Chem. Soc.*, **112**, 3875 (1990).

19. Schwab, P. F., Marcia, B., Ziller, J. W. and Grubbs, R. H., *Angew. Chem., Int. Ed. Engl.*, **34**, 2039 (1995).
20. Nugent, W. A., Feldman, J. and Calabrese, J. C., *J. Mol. Catal.*, **36**, 13 (1995).
21. Wagener, K. B., Nel, J. G., Duttweiler, R. P., Hillmyer, M. A., Boncella, J. M., Konzelman, J., Smith Jr, D. W., Puts, R. D. and Willoughby, L., *Rubber Chem. Technol.*, **64**(1), 83 (1991).
22. Höcker, H., Reimann, W., Riebel, K. and Szentivanyi, Z., *Makromol. Chem.*, **177**, 1707 (1976).
23. Schopov, I. and Mladenova, L., *Makromol. Chem., Rapid Commun.*, **6**, 659 (1985).
24. Brzezinska, K., Anderson, J. D. and Wagener, K. B., *Polym. Prepr. Am. Chem. Soc., Div. Polym. Chem.*, **37**(1), 327 (1996).
25. Brzezinska, K., Wolfe, P. S., Watson, M. D. and Wagener, K. B., *Macromol. Chem. Phys.*, **197**, 2065 (1996).
26. Valenti, D. J. and Wagener, K. B., *Polym. Prepr. Am. Chem. Soc., Div. Polym. Chem.*, **37**(1), 325 (1996).
27. Gomez, F. J. and Wagener, K. B., *PSME Prepr. Am. Chem. Soc., Div. Polym. Mat. Sci. Eng.*, **76**, 59 (1997).
28. Nugent, W. A., Feldman, J. and Calabrese, J. C., *J. Am. Chem. Soc.*, **117**, 8992 (1995).
29. Wolfe, P. S., Gomez, F. J. and Wagener, K. B., *PSME Prepr. Am. Chem. Soc., Div. Polym. Mat. Sci. Eng.*, **76**, 250 (1997).
30. Wolfe, P. S., Gomez, F. J. and Wagener, K. B., *Macromolecules*, **30**, 714 (1997).
31. Ofstead, E. A. and Wagener, K. B., 'Polymer Synthesis via Metathesis Chemistry', in *New Methods for Polymer Synthesis*, Plenum Press, New York, 1992, Ch. 8.
32. Ivin, K. J. and Mol, J. C., *Olefin Metathesis and Metathesis Polymerization*, Academic Press, San Diego–London, 1997.
33. Wagener, K. B., Brzezinska, K., Anderson, J. D., Younkin, T. R. and De Boer, W., *Macromolecules*, **30**, 7363 (1997).
34. Wagener, K. B., Nel, J. G., Smith Jr, D. W. and Boncella, J. M., *Polym. Prepr. Am. Chem. Soc., Div. Polym. Chem.*, **31**(2), 711 (1990).
35. Watson, M. D. and Wagener, K. B., *Polym. Prepr. Am. Chem. Soc., Div. Polym. Chem.*, **37**(1), 609 (1996).
36. Wagener, K. B., Puts, R. D. and Smith Jr, D. W. *Makromol. Chem., Rapid Commun.*, **12**, 419 (1991).
37. Wagener, K. B. and Puts, R. D., *Polym. Prepr. Am. Chem. Soc., Div. Polym. Chem.*, **32**(1), 379 (1991).
38. Marmo, J. C. and Wagener, K. B., *Macromolecules*, **28**, 2602 (1995).
39. Marmo, J. C. and Wagener, K. B., *Macromolecules*, **26**, 2137 (1993).
40. Wagener, K. B. and Marmo, J. C., *Macromol. Commun.*, **16**, 557 (1995).
41. Tao, D. and Wagener, K. B., *Macromolecules*, **27**, 1281 (1994).
42. O'Gara, J. E., Wagener, K. B. and Hahn, S. F., *Makromol. Chem., Rapid Commun.*, **14**, 657 (1993).
43. Wagener, K. B., Gomez, F. J. and Watson, M., 3rd International School on *Molecular Catalysis*, Lagow-Poznan, Poland, 1998, Abstracts, L 5.
44. Nel, J. G., Wagener, K. B. and Boncella, J. M., *Polym. Prepr. Am. Chem. Soc., Div. Polym. Chem.*, **30**(2), 130 (1989).
45. Nel, J. G., Wagener, K. B., Boncella, J. M. and Duttweiler, R. P., *Polym. Prepr. Am. Chem. Soc., Div. Polym. Chem.*, **30**(1), 283 (1989).
46. Konzelman, J. and Wagener, K. B., *Macromolecules*, **29**, 7657 (1996).
47. Wagener, K. B. and Valenti, D., *Macromolecules*, **30**, 6688 (1997).
48. O'Gara, J. E., Portmess, J. D. and Wagener, K. B., *Macromolecules*, **26**, 2837 (1993).

49. Wagener, K. B., Brzezinska, K. and Bauch, C. G., *Makromol. Chem., Rapid Commun.*, **13**, 75 (1992).
50. Brzezinska, K. and Wagener, K. B., *Macromolecules*, **25**, 2094 (1992).
51. Wagener, K. B., Patton, J. T. and Boncella, J. M., *Macromolecules*, **25**, 3862 (1992).
52. Bauch, C. G., Boncella, J. M. and Wagener, K. B., *Makromol. Chem., Rapid Commun.*, **12**, 413 (1991).
53. Wagener, K. B. and Patton, J. T., *Macromolecules*, **26**, 249 (1993).
54. Wagener, K. B., Patton, J. T., Forbes, M. D., Myers, T. L. and Maynard, H. D., *Polym. Int.*, **32**, 411 (1993).
55. Smith Jr, D. W. and Wagener, K. B., *Macromolecules*, **24**, 6073 (1991).
56. Smith Jr, D. W. and Wagener, K. B., *Macromolecules*, **26**, 3533 (1993).
57. Cummings, S. K., Anderson, J. and Wagener, K. B., *Polym. Prepr. Am. Chem. Soc., Div. Polym. Chem.*, 192 (1997).
58. Marciniec, B. and Lewandowski, M., *J. Inorg. Organomet. Polym.*, **5**, 647 (1995).
59. Smith Jr, D. W. and Wagener, K. B., *Macromolecules*, **26**, 1633 (1993).
60. Marciniec, B. and Lewandowski, M., *J. Polym. Sci., A, Polym. Chem.*, **34**, 1443 (1996).
61. Cummings, S. K., Smith Jr, D. W. and Wagener, K. B., *Macromol. Commun.*, **16**, 347 (1995).
62. Wagener, K. B. and Tao, D., *Macromolecules*, **27**, 1281 (1994).
63. Wagener, K. B., Brzezinska, K., Anderson, J. D. and Dilocker, S., *J. Polym. Sci., A, Polym. Chem.*, **35**, 3441 (1997).
64. Marmo, J. C. and Wagener, K. B., *Polym. Prepr. Am. Chem. Soc., Div. Polym. Chem.*, **35**(1), 817 (1994).
65. Viswanathan, T., Gomez, F. and Wagener, K. B., *J. Polym. Sci., A, Polym. Chem.*, **32**, 2469 (1994).
66. Mizoroki, T., Mori, K. and Ozaki, A., *Bull. Chem. Soc. Jpn*, **44**, 581 (1971).
67. Heck, R. F. and Nolley Jr, J. P. *J. Org. Chem.*, **37**, 2320 (1972).
68. Dieck, H. A. and Heck, R. F., *J. Organomet. Chem.*, **93**, 259 (1975).
69. Heck, R. F., *Pure Appl. Chem.*, **50**, 691 (1978).
70. Heck, R. F., *Acc. Chem. Res.*, **12**, 146 (1979).
71. Heck, R. F., *Organic React.*, **27**, 345 (1981).
72. De Meijre, A. and Meyer, F. E., *Angew. Chem., Int. Ed. Engl.*, **33**, 2379 (1994).
73. Cabri, W. and Candiani, I., *Acc. Chem. Res.*, **28**, 2 (1995).
74. Negishi, E., Copéret, C., Ma, S., Liou, S.-Y. and Liu, F., *Chem. Rev.*, **96**, 365 (1996).
75. Shibasaki, M., Boden, C. O. J. and Kojima, A., *Tetrahedron*, **53**, 7371 (1997).
76. Heitz, W., Brügging, W., Freund, L., Gailberger, M., Greiner, A., Jung, H., Kampschulte, U., Niessner, N., Osan, F., Schmidt, H. W. and Wicker, M., *Makromol. Chem.*, **189**, 119 (1988).
77. Greiner, A. and Heitz, W., *Makromol. Chem., Rapid Commun.*, **9**, 581 (1988).
78. Brenda, M., Greiner, A. and Heitz, W., *Makromol. Chem.*, **191**, 1083 (1990).
79. Greiner, A., Martelock, H., Noll, A., Siegfried, N. and Heitz, W., *Polymer*, **32**, 1857 (1991).
80. Heitz, W., *Polym. Prepr. Am. Chem. Soc., Div. Polym. Chem.*, **32**(1), 327 (1991).
81. Greiner, A. and Heitz, W., *Polym. Prepr. Am. Chem. Soc., Div. Polym. Chem.*, **32**(1), 333 (1991).
82. Martelock, H., Greiner, A. and Heitz, W., *Makromol. Chem.*, **192**, 967 (1991).
83. Sen, A., *Adv. Polym. Sci.*, **73/76**, 125 (1986).
84. Sen, A. and Lai, T.-W., *J. Am. Chem. Soc.*, **104**, 3520 (1982).
85. Sen, A. and Lai, T.-W., *J. Am. Chem. Soc.*, **103**, 4627 (1981).

86. Stille, J. K. and Lau, K. S. Y., *Acc. Chem. Res.*, **11**, 434 (1977).
87. Amatore, C., Carré, E., Jutand, A., M'Barki, M. A. and Meyer, G., *Organometallics*, **14**, 5605 (1995).
88. Amatore, C., Carré, E., Jutand, A. and M'Barki, M. A., *Organometallics*, **14**, 1818 (1995).
89. Amatore, C., Jutand, A., Khalil, F., M'Barki, M. A. and Mottier, L., *Organometallics*, **12**, 3168 (1993).
90. Brown, J. M., Pérez-Torrente, J. J., Allcock, N. W. and Clase, H. J., *Organometallics*, **14**, 207 (1995).
91. Brown, J. M. and Hii, K. K., *Angew. Chem., Int. Ed. Engl.*, **35**, 657 (1996).
92. Kuran, W. and Musco, A., *J. Organomet. Chem.*, **40**, C47 (1972).
93. Fauvarque, J. F., Pflüger, F. and Troupel, M., *J. Organomet. Chem.*, **208**, 419 (1981).
94. Musco, A., Kuran, W., Silvani, A. and Anker, M., *J. Chem. Soc., Chem. Commun.*, **1973**, 938 (1973).
95. Kuran, W. and Musco, A., *Inorg. Chim. Acta*, **12**, 187 (1975).
96. Otsuka, S., Yoshida, T., Matsumoto, M. and Nakatsu, K., *J. Am. Chem. Soc.*, **98**, 5850 (1976).
97. Amatore, C., Azzabi, M. and Jutand, A., *J. Am. Chem. Soc.*, **113**, 8375 (1991).
98. Paul, F., Patt, J. and Hartwig, J. F., *J. Am. Chem. Soc.*, **116**, 5969 (1994).
99. Hartwig, J. F. and Patt, J., *J. Am. Chem. Soc.*, **117**, 5373 (1995).
100. Yu, L. and Bao, Z., 'Poly(*p*-phenylene vinylene)s by the Heck Coupling Reaction', in *The Polymeric Materials Encyclopedia*, CRC Press, Inc., Boca Raton, 1996, Vol. 9, pp. 6532–6537.
101. Heitz, W. and Greiner, A., 'Palladium-catalyzed Synthesis (Monomers and Polymers)', in *The Polymeric Materials Encyclopedia*, CRC Press, Inc., Boca Raton, 1996, Vol. 7, pp. 4865–4871.
102. Amatore, C. and Pflüger, F., *Organometallics*, **9**, 2276 (1990).
103. Jutand, A. and Mosleh, A., *Organometallics*, **14**, 1810 (1995).
104. Cianfriglia, P., Narducci, V., Sterzo, C. L., Viola, E., Bocelli, G. and Kodenkandath, T. A., *Organometallics*, **15**, 5220 (1996).
105. Friestad, G. K. and Branchand, B. P., *Tetrahedron Lett.*, **36**, 7047 (1995).
106. Yoneyama, M., Tanaka, M., Kakimoto, M. and Imai, Y., *Macromolecules*, **22**, 4148 (1989).
107. Suzuki, M., Sho, K., Lim, J. C. and Saegusa, T., *Polym. Bull.*, **21**, 415 (1989).
108. Cassr, L., *J. Organomet. Chem.*, **93**, 253 (1975).
109. Sonogashira, K., Tohda, Y. and Hagihara, N., *Tetrahedron Lett.*, **50**, 4467 (1975).
110. Takahashi, S., Kuroyama, Y., Sonogashira, K. and Hagihara, N., *Synthesis*, **1980**, 627 (1980).
111. Ratovelomana, V. and Linstrumelle, G., *Synth. Commun.*, **11**, 917 (1981).
112. Sanechika, K., Yamamoto, T. and Yamamoto, A., *Bull. Chem. Soc. Jpn*, **57**, 752 (1984).
113. Harris, F. W., Pamidimukkala, A., Gupta, R., Das, S., Wu, T. and Mock, G., *J. Macromol. Sci. – Chem. A*, **21**, 1117 (1984).
114. Havens, S. J. and Hergenrother, P. M., *J. Polym. Sci., Polym. Lett. Ed.*, **23**, 587 (1985).
115. Trumbo, D. L. and Marvel, C. S., *J. Polym. Sci., Polym. Chem. Ed.*, **24**, 2231 (1986).
116. Trumbo, D. L. and Marvel, C. S., *J. Polym. Sci., Polym. Chem. Ed.*, **24**, 2311 (1986).
117. Trumbo, D. L. and Marvel, C. S., *J. Polym. Sci., Polym. Chem. Ed.*, **25**, 839 (1987).

118. Brown, I. M. and Wilbur, J. M., *Macromolecules*, **21**, 1859 (1988).
119. Kondo, K., Okuda, M. and Fujitani, T., *Macromolecules*, **26**, 7382 (1993).
120. Moroni, M. and Le Moigne, J., *Macromolecules*, **27**, 562 (1994).
121. Beginn, C., Grazulevicius, J. V. and Strohriegel, P., *Macromol. Chem. Phys.*, **195**, 2353 (1994).
122. Giesa, R. and Schulz, R. C., *Makromol. Chem.*, **191**, 857 (1990).
123. Ames, D. L., Bull, D. and Takundwa, C., *Synthesis*, **1981**, 364 (1981).
124. Trybulski, E. J., Reeder, E., Blount, J. F., Walser, A. and Fryer, R. I., *J. Org. Chem.*, **47**, 2441 (1982).
125. Bumagin, N. A., Ponomaryov, A. B. and Beletskaya, I. P., *Synthesis*, **1984**, 728 (1984).
126. Havens, S. J. and Hergenrother, P. M., *J. Org. Chem.*, **50**, 1763 (1985).
127. Neenan, T. X. and Whitesides, G. M., *J. Org. Chem.*, **53**, 2489 (1988).
128. Tao, W., Nesbitt, S. and Heck, R. F., *J. Org. Chem.*, **55**, 63 (1990).
129. Sabourin, E. T. and Onopchenko, A., *J. Org. Chem.*, **48**, 5135 (1983).
130. Simionescu, C. I., Percec, V. and Dimitrescu, S., *J. Polym. Sci., Polym. Chem. Ed.*, **15**, 2497 (1977).
131. Carpita, A., Lessi, A. and Rossi, R., *Synthesis*, **1984**, 571 (1984).
132. Pugh, C. and Percec, V., *J. Polym. Sci., Polym. Chem. Ed.*, **28**, 1101 (1990).
133. Solomin, V. A. and Heitz, W., *Macromol. Chem. Phys.*, **195**, 303 (1994).
134. Mangel, T., Eberhardt, A., Scherf, U., Bunz, U. H. F. and Müllen, K., *Macromol. Commun.*, **16**, 571 (1995).
135. Müller, W. T. and Ringsdorf, H., *Macromolecules*, **23**, 2825 (1990).
136. Corriú, R. J.-P., Douglas, W. E. and Yang, Z.-X., *Eur. Polym. J.*, **29**, 1563 (1993).
137. Cramer, E. and Percec, V., *J. Polym. Sci., Polym. Chem. Ed.*, **28**, 3029 (1990).
138. Rehahn, M., Schlüter, A.-D., Wegner, G. and Feast, W. J., *Polymer*, **30**, 1060 (1989).
139. Fahnenstich, K., Koch, K.-H. and Müllen K., *Makromol. Chem., Rapid Commun.*, **10**, 563 (1989).
140. Rehahn, M., Schlüter, A.-D. and Wegner, G., *Makromol. Chem., Rapid Commun.*, **11**, 535 (1990).
141. Scherf, U. and Müllen K., *Makromol. Chem., Rapid Commun.*, **12**, 489 (1991).
142. Scherf, U. and Müllen K., *Macromolecules*, **25**, 3546 (1992).
143. Wallow, T. I. and Novak, B. M., *Polym. Prepr. Am. Chem. Soc., Div. Polym. Chem.*, **33**(1), 908 (1992).
144. Kaufmann, T. and Lexy, H., *Chem. Ber.*, **114**, 3674 (1981).
145. Miyaura, N., Yanagi, T. and Suzuki, A., *Synth. Commun.*, **11**, 513 (1981).
146. Miller, R. B. and Dugar, S., *Organometallics*, **3**, 1261 (1984).
147. Miyaura, N. and Suzuki, A., *Chem. Rev.*, **9**, 2457 (1995).
148. Kim, Y. H. and Webster, O. W., *Polym. Prepr. Am. Chem. Soc., Div. Polym. Chem.*, **29**(2), 3102 (1988).
149. Bochman, M. and Kelly, K., *J. Chem. Soc., Chem. Commun.*, **1989**, 532 (1989).
150. Bochman, M., Kelly, K. and Lu, J., *J. Polym. Sci., Polym. Chem. Ed.*, **30**, 2511 (1992).
151. Bochman, M. and Lu, J., *J. Polym. Sci., Polym. Chem. Ed.*, **32**, 2493 (1994).
152. Stille, J. K., *Angew. Chem., Int. Ed. Engl.*, **25**, 508 (1986).
153. Colon, I. and Kwiatkowski, G. T., *J. Polym. Sci., Polym. Chem. Ed.*, **28**, 367 (1990).
154. Ueda, M. and Ichikawa, F., *Macromolecules*, **23**, 926 (1990).
155. Yamamoto, T., Osakada, K., Wakabayashi. T. and Yamamoto, A., *Makromol. Chem., Rapid Commun.*, **6**, 671 (1985).
156. Yamamoto, T., Kashiwazaki, A. and Kato, K., *Makromol. Chem.*, **190**, 1649 (1989).

157. Colon, I. and Kelsey, D. R., *J. Org. Chem.*, **51**, 2627 (1986).
158. Amatore, C. and Jutand, A., *Organometallics*, **7**, 2203 (1988).
159. Yoneyama, M., Konishi, T., Kakimoto, M. and Imai, Y., *Polym. Prepr. Jpn.*, **39**, 1799 (1990).
160. Hoerhold, H.-H., Gottschaldt, J. and Opfermann, J., *J. Prakt. Chem.*, **319**, 611 (1977).
161. Feast, W. J. and Millichamp, I. S., *Polym. Commun.*, **24**, 102 (1983).
162. Yoneyama, M., Kakimoto, M. and Imai, Y., *Macromolecules*, **21**, 1908 (1988).
163. Yoneyama, M., Kakimoto, M. and Imai, Y., *J. Polym. Sci., Polym. Chem. Ed.*, **27**, 1985 (1989).
164. Yoneyama, M., Kakimoto, M. and Imai, Y., *Macromolecules*, **22**, 2593 (1989).
165. Yoneyama, M., Kakimoto, M. and Imai, Y., *Macromolecules*, **22**, 4152 (1989).
166. Yoneyama, M., Konishi, T., Kakimoto, M. and Imai, Y., *Makromol. Chem., Rapid Commun.*, **11**, 381 (1990).
167. Imai, Y., *Polym. Prepr. Am. Chem. Soc., Div. Polym. Chem.*, **32**(1), 331 (1991).
168. Perry, R. J. and Turner, S. R., *Polym. Prepr. Am. Chem. Soc., Div. Polym. Chem.*, **32**(1), 335 (1991).
169. Perry, R. J. and Turner, S. R., *J. Macromol. Sci.–Chem. A*, **28**, 1213 (1991).
170. Perry, R. J., Turner, S. R. and Blevins, R. W., *Macromolecules*, **26**, 1509 (1993).
171. Perry, R. J. and Wilson, B. D., *Macromolecules*, **27**, 40 (1994).

Further Reading

Thorn-Csányi, E. and Kraxner, P., '*p*-Phenylene Vinylene Oligomers, Homo- and Copolymers (Metathesis Preparation)', in *The Polymeric Materials Encyclopedia*, CRC Press, Inc., Boca Raton, 1996, Vol. 7, pp. 5055–5067.

Korshak, Y. V., 'Metathesis Polymerization, Cycloolefins', in *The Polymeric Materials Encyclopedia*, CRC Press, Inc., Boca Raton, 1996, Vol. 7, pp. 4250–4236.

Ivin, K. J., 'Metathesis Polymerization', in *Encyclopedia of Polymer Science and Engineering*, Wiley-Interscience, John Wiley & Sons, New York, 1987, Vol. 9, pp. 634–669.

Nagishi, E. and Liu, F., 'Palladium- or Nickel-catalyzed Cross-coupling with Organometals Containing Zinc, Magnesium, Aluminium and Zirconium', in *Metal-catalyzed Cross-coupling Reactions*, Wiley-VCH, Weinheim, 1998, pp. 1–47.

Bräse, S. and de Meijre, A., 'Palladium-catalyzed Coupling of Organyl Halides to Alkenes – the Heck Reaction', in *Metal-catalyzed Cross-coupling Reactions*, Wiley-VCH, Weinheim, 1998, pp. 99–166.

Sonogashira, K., 'Cross-coupling Reactions to *sp* carbon atoms', in *Metal-catalyzed Cross-coupling Reactions*, Wiley-VCH, Weinheim, 1998, pp. 203–229.

Tsuji, J. and Mandai, T., 'Palladium-catalyzed Coupling Reactions of Propargylic Compounds', in *Metal-catalyzed Cross-coupling Reactions*, Wiley-VCH, Weinheim, 1998, pp. 455–489.

Problems

1. Explain why well-defined metal alkylidienes, but not conventional two- or three-component metathesis catalysts, are preferably used for acyclic diene metathesis polycondensation.
2. Give reasons why terminal dienes are the preferred monomers in acyclic diene metathesis polycondensation.

3. Indicate α, ω-diene monomers that will undergo metathesis condensation-polymerisation to the following polymers:

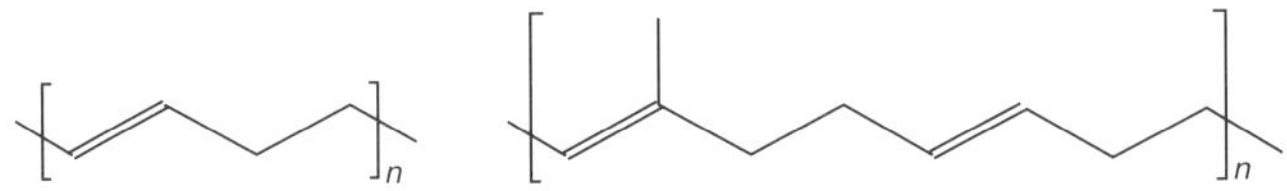

4. Indicate the main features differentiating acyclic diene metathesis polycondensation and single C–C bond-forming polycondensation with respect to the kind of monomers and catalyst action.
5. Indicate monomers and catalysts that will produce poly(1,4-phenylene vinylene) and poly(1,4-phenylene-1,6-hexenylene) via coordination polycondensation of various types.
6. Explain why carbon–heteroatom carbonylative coupling reactions are preferably carried out with palladium-based catalysts; give an example of the catalytic cycle in the carbonylative polycondensation of dihaloarene and bisphenol, leading to a polyester.

9 Coordination Polymerisation of Non-hydrocarbon (Heterocyclic and Heterounsaturated) Monomers

Polymerisation and copolymerisation of non-hydrocarbon, heteroatom-containing cyclic and acyclic monomers in the presence of coordination catalysts constitute a distinct group of coordination polymerisation processes that differ essentially from the coordination polymerisation of unsaturated hydrocarbon monomers in terms of the kinds of catalyst applicable and the polymerisation mechanism. While the coordination polymerisation and copolymerisation of hydrocarbon monomers involve π complex formation between the monomer and the metal atom, polymerisations of heterocyclic and heterounsaturated monomers take place by monomer coordination via σ bond formation between the heteroatom and the metal atom. Apart from the mode of monomer coordination, the polymerisation of heteroatom-containing monomers differs from that of hydrocarbon monomers in the mode of monomer enchainment which consists in the nucleophilic attack of the initiating group or polymer chain terminal bound to the metal on the coordinating monomer in the initiation or propagation step respectively [schemes (6) to (9) in chapter 2] [1].

The exceptional behaviour of heterounsaturated monomers of carbene-like structure, such as carbon monoxide and isocyanides, should be remembered here. Carbon monoxide readily undergoes copolymerisation with various unsaturated hydrocarbon monomers via coordination with transition metals [2]. By contrast, isocyanides are homopolymerised via coordination with a transition metal of the same type to yield polymers with a carbon–carbon main chain, i.e. poly(iminoethylene)s [3].

The application of coordination catalysts for polymerisations of heterocyclic monomers has broadened the synthetic feasibility of these polymerisations as regards the possibility of obtaining high molecular weight homopolymers that might be characterised by a high degree of regio- and stereoregularity, and the possibility of obtaining copolymers with heteroatom-containing cyclic and acyclic monomers.

The coordination polymerisation and copolymerisation of heterocyclic monomers have been restricted in industry to a much smaller volume than the polymerisation and copolymerisation of hydrocarbon monomers; polyether elastomers from epichlorohydrin and ethylene oxide or propylene oxide, and allyl glycidyl ether as the vulcanisable monomeric unit, are produced on a larger scale [4–7].

As regards the coordination homopolymerisation of heterounsaturated monomers, it does not play such an important role as in the case of heterocyclic monomers (with the exception of carbon monoxide). This is because of the high polymerisability of heterounsaturated monomers in the presence of ionic initiators which is taken into account in some industrial processes (e.g. polyformaldehyde).

9.1 Monomers and Catalysts

A wide variety of heterocyclic monomers have been polymerised successfully by ring-opening polymerisation with coordination catalysts, yielding high molecular weight polymers. This includes monomers containing an endocyclic heteroatom, as well as both endocyclic and exocyclic heteroatoms, in the molecule. Heteroatom-containing cyclic monomers capable of coordination homopolymerisation and copolymerisation with other heterocyclic or heterounsaturated monomers comprise oxiranes, tiiranes, oxetanes, lactones, lactides, alkylene carbonates and thiocarbonates, bisalkylenespiroorthocarbonates, dicarboxylic acid anhydrides and α-aminoacid *N*-carboxy anhydrides, morpholinedione and oxyphosphonoyloxytrimethylene. Representative examples of the coordination polymerisation of heterocyclic monomers of both types, i.e. those with an endocyclic heteroatom (oxa- and thiacyclic monomers) and with both endocyclic and exocyclic heteroatoms (oxa-, thia-, aza- and phosphacyclic monomers), are shown in Tables 9.1 and 9.2 respectively [1].

The data presented in Tables 9.1 and 9.2 focus on homopolymerisations; copolymerisations of the mentioned heterocyclic monomers of both types proceed similarly, involving catalysts that promote their homopolymerisations.

Heterounsaturated monomers such as aldehydes, ketones, ketones, isocyanates and isocyanides, which have been reported to undergo a polymerisation in the presence of coordination catalysts, are listed in Table 9.3 [1,3].

Table 9.1 Homopolymerisation of heterocyclic monomers, containing an endocyclic heteroatom, in the presence of coordination catalysts[a]

Monomer	Polymer
Oxacyclic monomer:	
 R = H, hydrocarbyl	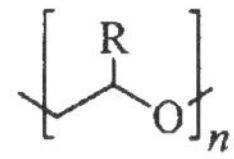
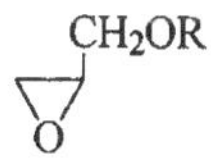 R = hydrocarbyl	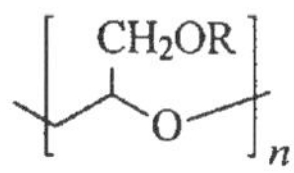
 R = CH_3 *cis* and *trans*; R–R = $(CH_2)_4$ *cis*	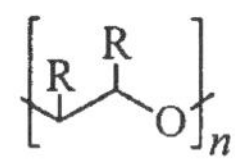
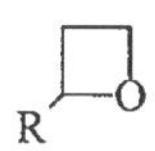 R = H, CH_3	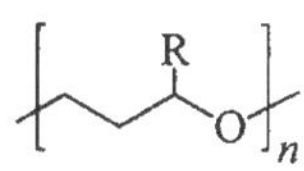
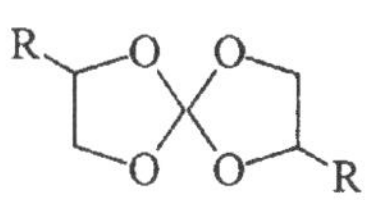 R = CH_3	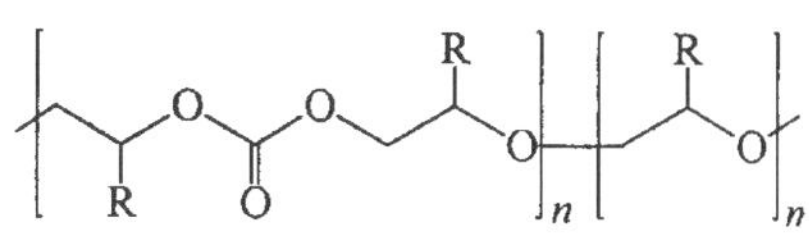
Thiacyclic monomer:	
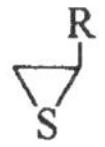 R = alkyl	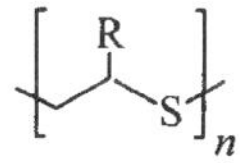
 R = CH_3 *cis*, R–R = $(CH_2)_4$ *cis*	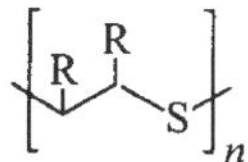

[a] According to Ref. 1.

Table 9.2 Homopolymerisation of heterocyclic monomers, containing both endocyclic and exocyclic heteroatoms, in the presence of coordination catalysts[a]

Monomer	Polymer
Oxacyclic monomer:	
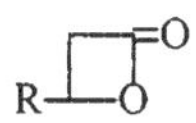 R = H, alkyl	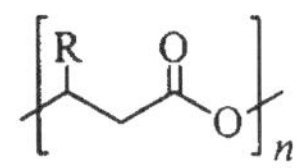
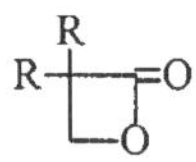 R = alkyl	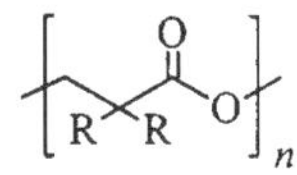
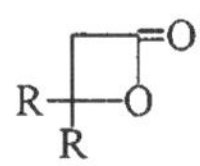 R = CH_3	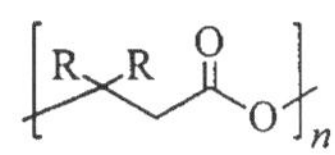
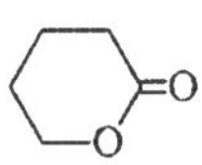	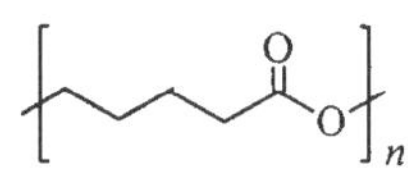
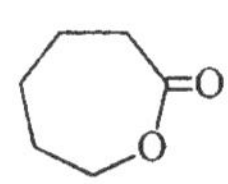	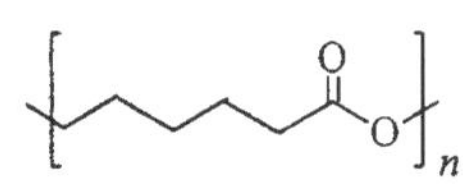
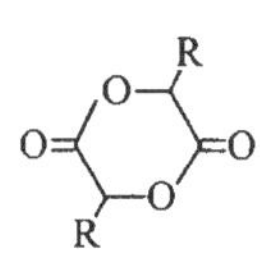 R = CH_3	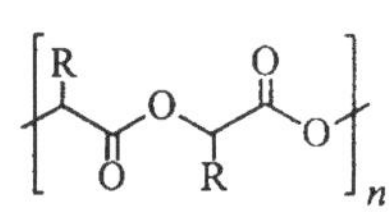
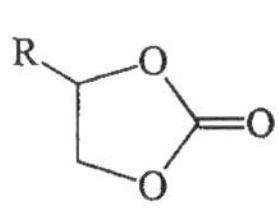 R = CH_3	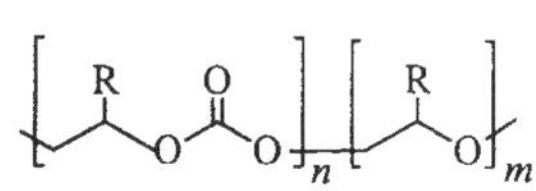
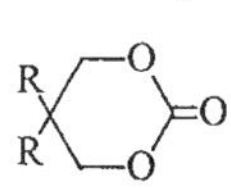 R = H, CH_3	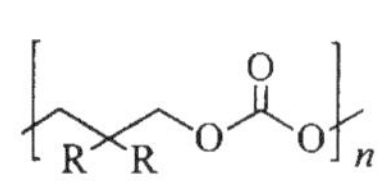

Table 9.2 (*continued*)

Monomer	Polymer
Thiacyclic monomer:	
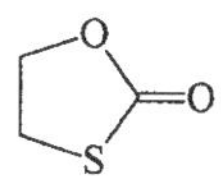	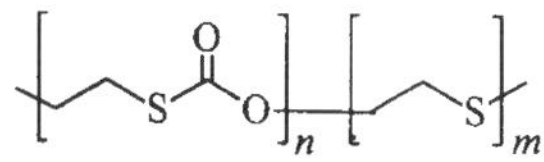
Azacyclic monomer:	
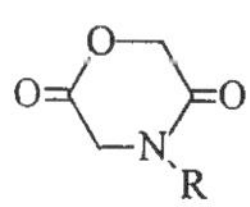 R = H, CH_3	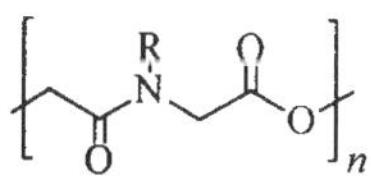
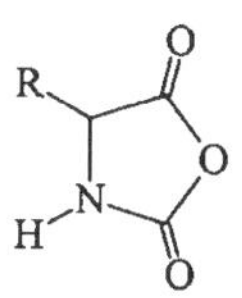 R = CH_3	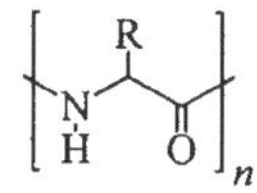
Phosphacyclic monomer:	
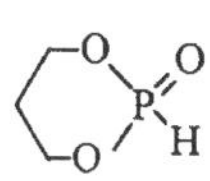	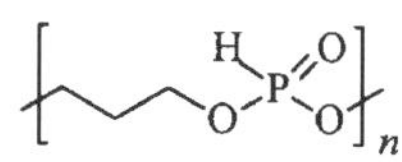

[a] According to Ref. 1.

Table 9.3 Homopolymerisation of heterounsaturated monomers in the presence of coordination catalysts[a]

Monomer	Polymer
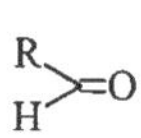 R = alkyl, CH_2Cl	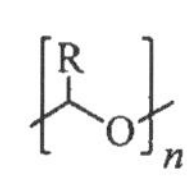
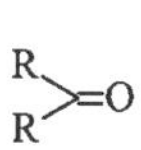 R = CH_3	

continued overleaf

Table 9.3 (*continued*)

Monomer	Polymer
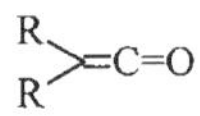 R = CH_3	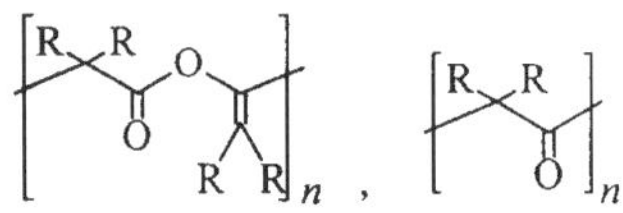
R−N=C=O R = n-C_4H_9	
R−N≡C R = t-C_4H_9 [b]	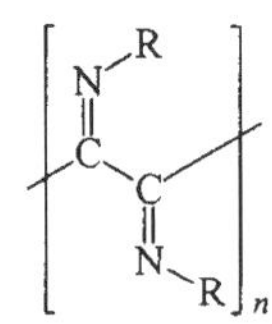

[a] According to Ref. 1.
[b] According to Ref. 3.

However, the most important goal that might be reached by the application of coordination catalysts for the polymerisation of heterounsaturated monomers is the possibility of the enchainment of heterounsaturated monomers, not susceptible to homopropagation, via their copolymerisation with heterocyclic monomers. This concerns primarily the coordination copolymerisation of carbon dioxide and oxacyclic monomers such as epoxides, leading to aliphatic polycarbonates [8–12]. Representative examples of the copolymerisations of heterocyclic monomers and hardly homopolymerisable heterocumulenes, in the presence of coordination catalysts, are listed in Table 9.4 [1].

The coordination catalysts usually applied for the polymerisation and copolymerisation of heterocyclic and heterounsaturated monomers involve a wide range of metal derivatives that are characterised by moderate nucleophilicity and relatively high Lewis acidity. Metal derivatives possessing free p, d or f orbitals of favourable energy are used as catalysts for epoxide polymerisation. In particular, compounds of group 2 and 3 metals, such as zinc, cadmium and aluminum, and transition metals, such as iron, as well as lanthanum and yttrium, are representative coordination catalysts. The appropriate Lewis acidity of the metal and the appropriate nucleophilicity of the metal substituent in these catalysts make the monomer coordination favourable prior to the nucleophilic attack. The nucleophilic attack of the covalently bound metal substituent on the monomer molecule coordinated at the metal atom at the catalyst active

Table 9.4 Copolymerisation of heterocyclic monomers with heterocumulenes in the presence of coordination catalysts[a]

	Comonomers	Copolymer
CO_2 *and:*		
	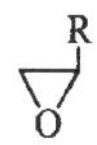 R = H, hydrocarbyl	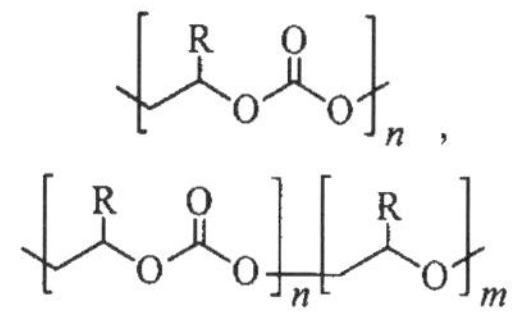
	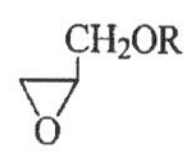 R = hydrocarbyl	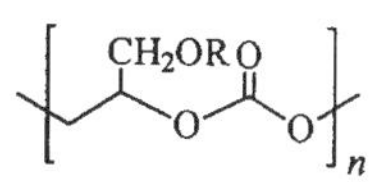
	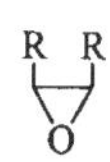 R = CH_3; R–R = $(CH_2)_4$	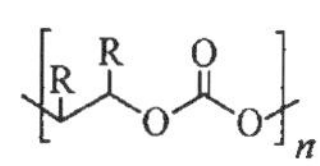
	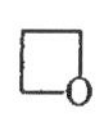	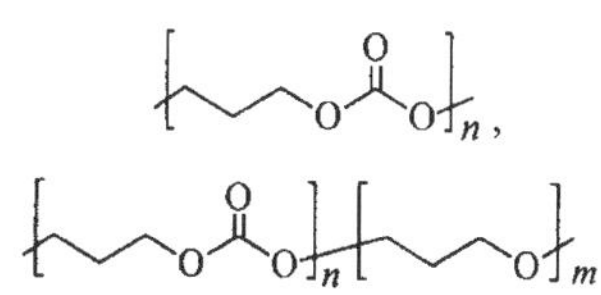
	 R = CH_3	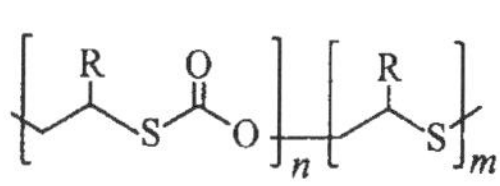
CS_2 *and:*		
	 R = CH_3	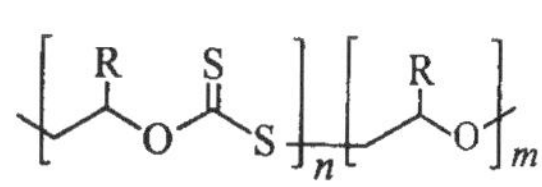
	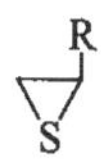 R = H, CH_3	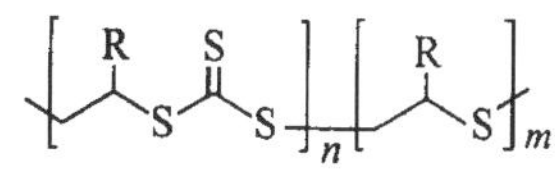

continued overleaf

Table 9.4 (*continued*)

Comonomers	Copolymer
SO_2 *and:*	
 R = CH_3	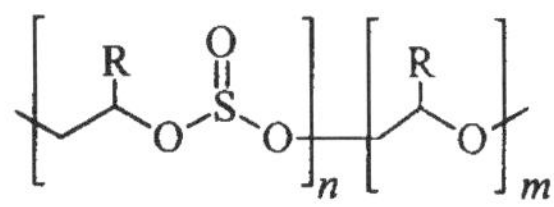
Ph–N=C=O *and:*	
	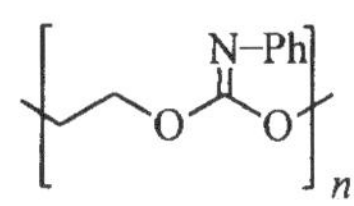

[a] According to Ref. 1.

site is the most important mechanistic feature of the coordination polymerisation of heterocyclic and heterounsaturated monomers.

Coordination of the monomer increases the capacity of the monomer for attack by the nucleophilic metal substituent and makes this nucleophilic attack simpler on account of repulsion of the negative charge from the monomer via the metal to its substituent [schemes (6) to (9) in Chapter 2].

It is worth mentioning in this connection that the anionic polymerisation of heterocyclic and heterounsaturated monomers requires the application of nucleophilic initiators, involving mostly alkali metal compounds, which are characterised by a high nucleophilicity of the monomer attacking agent and by low Lewis acidity of the positive counterion in the initiator. Thus, nucleophilic initiation of anionic polymerisation does not require any monomer coordination on to the metal, although interaction of the monomer with an electrophilic counterion is considered commonly to occur to some extent.

The difference between the coordination polymerisation of heterocyclic and heterounsaturated monomers and their anionic polymerisation lies in the covalent nature of the metal–heteroatom bond (which, however, is polarised) in the coordination catalyst, which activates the monomer by its coordination, enhancing the nucleophilicity of the metal substituent simultaneously, and in the ionic character of the metal–heteroatom bond in the nucleophilic initiators which do not need such monomer activation by coordination prior to the nucleophilic attack. Thus, treatment of the coordination polymerisation of heterocyclic and heterounsaturated monomers as 'pseudo-anionic' polymerisation is not correct.

Most coordination catalysts have been reported to be formed in binary or ternary component systems consisting of an alkylmetal compound and a protic compound. Catalysts formed in such systems contain associated multinuclear species with a metal (Mt)–heteroatom (X) active bond (→Mt–X→Mt–X→ or →Mt–X–Mt–X→; Mt=Al, Zn, Cd and X=O, S, N most frequently) or non-associated mononuclear species with an Mt–X active bond (Mt=Al, Zn and X=Cl, O, S most frequently). Metal alkyls, such as triethylaluminium, diethylzinc and diethylcadmium, without pretreatment with protic compounds, have also been reported as coordination polymerisation catalysts. In such a case, the metal–heteroatom bond active in the propagation step is formed by the reaction of the metal–carbon bond with the coordinating monomer. Some coordination catalysts, such as those with metal alkoxide or phenoxide moieties, can be prepared in other ways, without using metal alkyls. There are also catalysts consisting of a metal alkoxide or related compound and a Lewis acid [1].

One should add, in relation to this, that catalysts that are applicable for carbon monoxide copolymerisation and isocyanide homopolymerisation are of a different nature, since they exhibit the capability of coordinating the monomer via complexes involving π^0 orbitals of these monomers.

9.2 Polymerisation of Oxacyclic Monomers

Oxacyclic monomers constitute the most widely investigated class of heterocyclic monomers regarding both academic and industrial interest. In particular, the coordination polymerisation of cyclic ethers such as epoxides (oxiranes) and of cyclic esters such as lactones, lactides and cyclic carbonates has been considered.

9.2.1 Polymerisation of Cyclic Ethers

The coordination polymerisation of cyclic ethers has been limited to those containing a three- and four-membered ring in the molecule. However, most of the literature data reporting on the homopolymerisation and copolymerisation of cyclic ethers in the presence of coordination catalysts concern epoxide monomers (Tables 9.1 and 9.4).

9.2.1.1 Stereoisomerism of Epoxide Polymers

Epoxide polymers exhibit stereoisomerism originating from the chirality of tertiary carbon atoms present in the polymer main chain. The stereoisomers of epoxide polymers are therefore tactic polymers; their tacticity is connected with the structure of epoxide monomers undergoing polymerisation. Epoxide polymers with regular head-to-tail linkages, such as poly(propylene oxide) which is formed from chiral 1,2-epoxypropane (propylene oxide), can appear as isotactic polymers; it is to be emphasised that the tacticity of polymers of chiral alkylene oxides is due to the initial presence of the chiral site in the monomer molecule but does not arise from the formation of a chiral site as a consequence of bond formation in the polymerisation as in the case of prochiral α-olefins and aldehydes. It should also be noted that no syndiotactic polymers of chiral alkylene oxides have ever been obtained. Disubstituted chiral epoxides such as *trans*-2,3-epoxybutane (*trans*-2-butane oxide) can give *erythro*-diisotactic polymers. Achiral epoxides such as 1,2-epoxycyclohexane (cyclohexene oxide) and *cis*-2,3-epoxybutane (*cis*-2-butane oxide) can appear instead as *threo*-disyndiotactic polymers. The structure of epoxide polymers of various tacticity that have been obtained from mono- and disubstituted epoxides is shown in Figure 9.1.

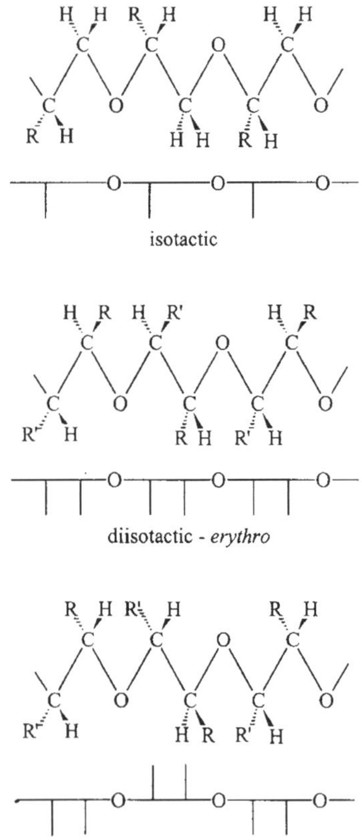

Figure 9.1 Structure of epoxide polymers of various tacticity obtained from mono- and disubstituted epoxides

Note that stereoregular polymers of disubstituted epoxides with a *threo*-diisotactic and *erythro*-disyndiotactic structure have not been synthesised.

Let us recall in this relation some definitions concerning the polymerisation of chiral epoxide monomers. Racemic monomers of such a type can undergo enantiosymmetric (stereoselective) polymerisation or enantioasymmetric (asymmetric stereoselective, stereoelective) polymerisation. One should consider that enantiosymmetric polymerisation involves the polymerisation of *R* and *S* enantiomers separately, which results in the formation of polymer chains consisting of the same enantiomeric units, *R* or *S*. Enantiosymmetric polymerisation therefore leads to a racemic mixture of macromolecules with chiral carbon atoms of identical configuration, *R* or *S*, in each polymer chain. By contrast, enantioasymmetric polymerisation, one of the enantiomers of the racemic monomer undergoes a faster polymerisation than its counterpart, which leads to a polymer containing chains with the chirality originating from the configuration of this enantiomer.

9.2.1.2 Catalysts for Polymerisation of Epoxides

The first report on the coordination polymerisation of epoxide, leading to a stereoregular (isotactic) polymer, concerned the polymerisation of propylene oxide in the presence of a ferric chloride–propylene oxide catalyst; the respective patent appeared in 1955 [13]. In this catalyst, which is referred to as the Pruitt–Baggett adduct of the general formula $Cl(C_3H_6O)_xFe(Cl)(OC_3H_6)_yCl$, two substituents of the alcoholate type formed by the addition of propylene oxide to Fe–Cl bonds and one chlorine atom at the iron atom are present [14]. A few years later, various types of catalyst effective for stereoselective polymerisation of propylene oxide were found and developed: aluminium isopropoxide–zinc chloride [15], dialkylzinc–water [16], dialkylzinc–alcohol [16], trialkylaluminium–water [17] and trialkylaluminium–water–acetylacetone [18] and trialkylaluminium–lanthanide triacetylacetonate–H_2O [19]. Other important catalysts for the stereoselective polymerisation of propylene oxide, such as bimetallic μ-oxoalkoxides of the $[(RO)_2AlO]_2Zn$ type, were obtained by condensation of zinc acetate with aluminium isopropoxide in a 1:2 molar ratio of reactants [20–22].

The first enantioasymmetric polymerisation of (*R*, *S*)-propylene oxide, reported in 1962, was carried out in the presence of the diethylzinc–(+)-borneol catalyst [23,24]. Other optically active catalysts for propylene oxide stereoelective polymerisation, e.g. diethylzinc–(*R*)-(−)-3,3-dimethyl-1,2-butanediol were described later on [25].

Special interest has been devoted to such systems as diethylzinc–methanol (7:8) [26–28] and diethylzinc–1-methoxy-2-propanol (2:3) [29], since well-defined catalysts were isolated in the form of single crystals in these systems, namely $[Zn(OMe)_2]\cdot[EtZnOMe]_6$ and $[Zn(OCH(Me)CH_2OMe)_2]_2\cdot[EtZnOCH$

(Me)$CH_2OMe]_2$ respectively. Molecules consist of two enantiomorphic distorted cubes which share a corner, the centre of symmetry consisting of the octahedral zinc atom. By contrast, molecules of $[Zn(OCH(Me)CH_2OMe)_2]_2$. $[EtZnOCH(Me)CH_2OMe]_2$ are characterised by the appearance of a centrosymmetric chair-form skeleton consisting of four zinc and four oxygen atoms. Each molecule has two octahedral zinc atoms, one of which is surrounded by three (*R*)-methoxyisopropoxy groups, and the other by three (*S*)-methoxyisopropoxy groups. Two of the three methoxyisopropoxy groups are *exo* and *endo* coordinated, involving an etheral oxygen atom, on to the octahedral zinc atom, and one methoxyisopropoxy group remains non-coordinated:

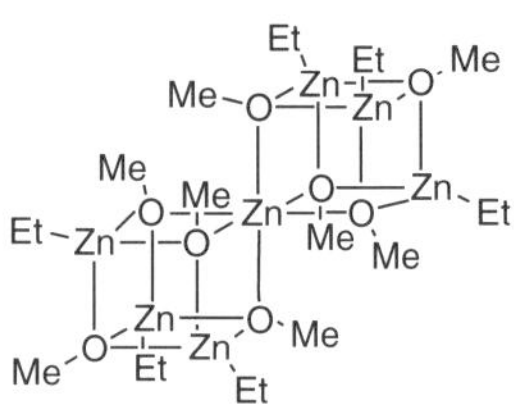

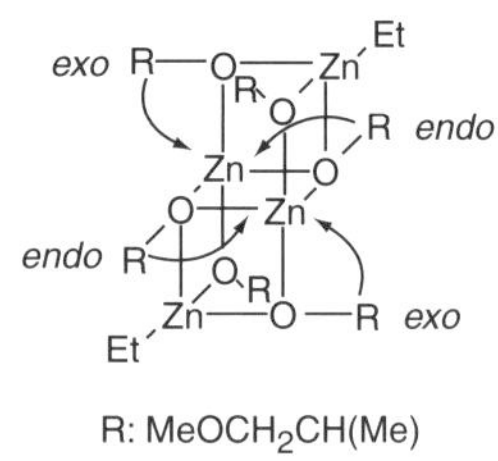

$[Zn(OMe)_2].[EtZnOMe]_6$ $[Zn(OCH(Me)CH_2OMe)_2]_2$. $[EtZnOCH(Me)CH_2OMe]_2$

The formation of epoxide polymers with a very high molecular weight by the discussed catalysts containing associated multinuclear species (→Zn–O→Zn–O→) indicates that only a small fraction of the metal species in the catalyst is effective for the polymerisation. The broad molecular weight distribution of polymers yielded by these catalysts corresponds to the existence of various active sites [30].

Catalysts obtained from metal alkyl and a bi- or trifunctional protic compound, e.g. in systems such as $AlEt_3$–H_2O [17], $ZnEt_2$–H_2O [16] and $ZnEt_2$–$Ar(OH)_3$ [31], which are characterised by the appearance of associated multinuclear species with condensed metal atoms (→Mt–O–Mt–O→), also form epoxide polymers with a very high molecular weight and broad molecular weight distribution; therefore, in this case also, only a small fraction of the metal species in the catalyst is effective for the polymerisation.

Catalysts for epoxide polymerisation of quite different characteristics comprise metalloporphyrins of aluminium and zinc, such as (5,10,15,20-tetraphenylporphinato)aluminium chloride [(tpp)AlCl], methoxide [(Mtpp)AlOMe] or 1-propanethiolate [(tpp)AlSPr] and (5,10,15,20-tetraphenyl-21-methylporphinato)zinc methoxide [(Mtpp)ZnOMe] [32–35]:

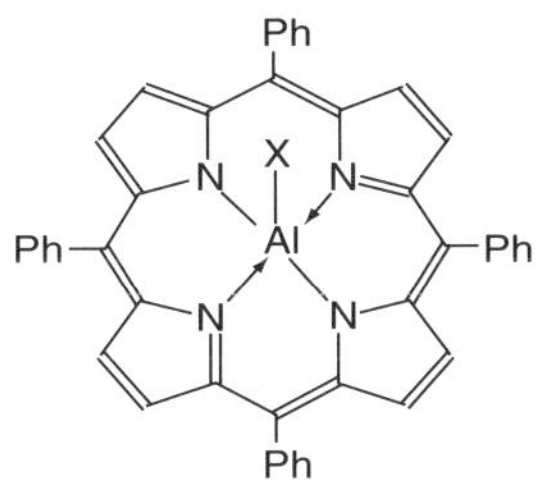

X: Cl, MeO

(tpp) ALX

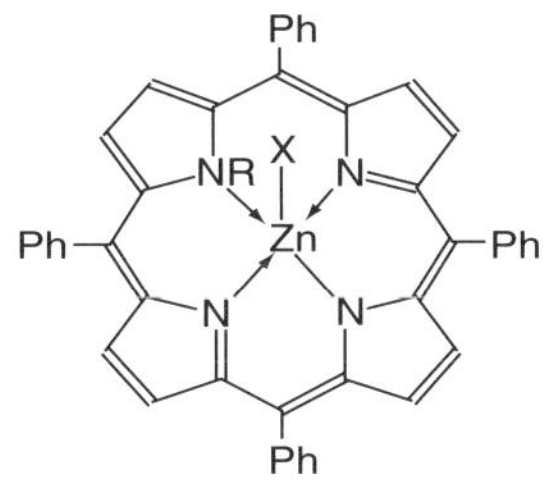

R: Me, X: MeO

(Mtpp) ZnX

Metalloporphyrin catalysts are characterised by their isolated metal atoms surrounded by a large, planar, rigid ring and, therefore, form monomeric mononuclear species (Mt–X). The polymerisation of epoxides with these well-defined catalysts appeared to exhibit a living character; the polyethers obtained are characterised by a small molecular weight and very narrow molecular weight distribution [30,36].

Other well-defined catalysts for epoxide polymerisations, containing an isolated metal atom, have been derived from the reaction of diethylaluminium chloride with a Schiff's base [37–40]. For instance, {2,2′-[(1*R*,2*R*)-1,2-cyclohexylenebis(nitrilomethylidene)]diphenolato} aluminium chloride [(sal)AlCl] appeared to produce low molecular weight poly(propylene oxide) characterised by a narrow distribution of molecular weights [40]:

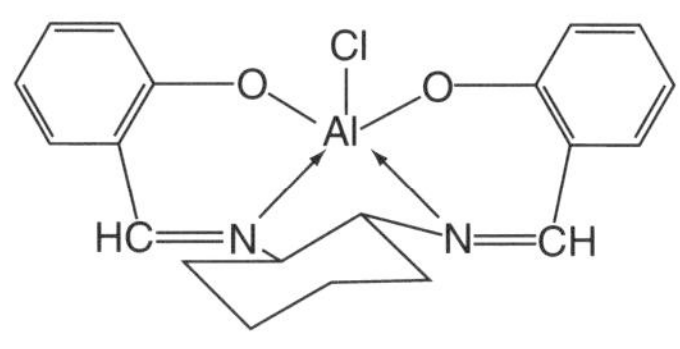

(sal)AlCl

The salcenatoaluminium chloride, (sal)AlCl, is chiral, which makes it different from the achiral (tpp)AlCl in terms of the stereochemical behaviour of epoxide polymerisation in the presence of these catalysts [30,39].

Well-defined catalysts with isolated mononuclear metal species have also been obtained from reactions of diethylaluminium compounds with calix[4]-arenediols. For instance, a catalyst such as (25,27-dimethoxy-*p*-*t*-butylcalix[4]-arene-26,28-diolato)aluminium chloride [(dmca)AlCl], characterised by a rigid structure of the AlO_4 moiety like a distorted trigonal bypyramid, forms a monomeric species in a benzene solution at freezing temperature. It has been successfully applied for propylene oxide and cyclohexene oxide polymerisations, leading to bifunctional low molecular weight polyethers of relatively narrow molecular weight distribution [41]. The achirality of the (dmca)AlCl catalyst should be noted:

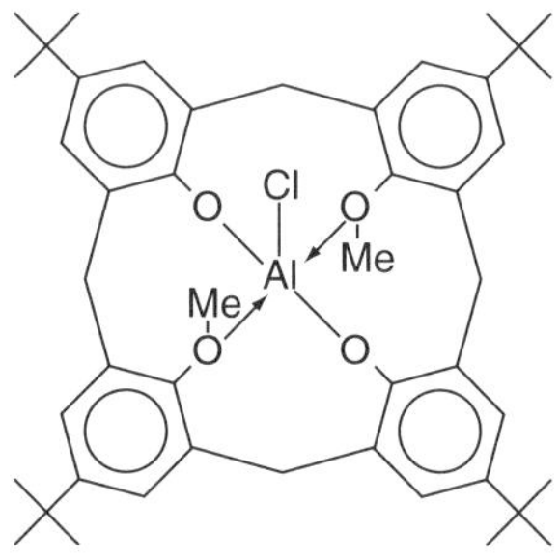

(dmca)AlCl

9.2.1.3 Mode of Epoxide Ring Opening

The polymerisation of racemic propylene oxide with coordination catalysts leads to a polymer that can be fractionated into crystalline and amorphous poly(propylene oxide)s:

The crystalline fraction was found to be formed via a cleavage of the C_β–O(CH_2–O) bond in the monomer molecule and proved many years ago to be an isotactic polymer (with regular head-to-tail linkages) (Figure 9.1) [42]. The structure of the amorphous fraction, on the other hand, varies depending on the kind of catalyst. Some amorphous poly(propylene oxide)s prepared with catalysts such as diethylzinc–methanol [43] or aluminium isopropoxide–zinc chloride [44] consist of regular head-to-tail linked units, but they are atactic (the mole fraction of isotactic diads is less than 0.6) [43]. Some other amorphous poly(propylene oxide)s obtained with catalysts derived from reactions in the diethylzinc–water [44,45], and triethylaluminium–water [46] systems, and with aluminium isopropoxide [44], have been found to be irregular, i.e. to contain head-to-head and tail-to-tail enchained monomer units.

One can note, in this connection, that styrene oxide undergoes polymerisation in the presence of aluminium isopropoxide which involves a selective cleavage of the C_α–O[CH(Ph)–O] bond [47]. On the other hand, polymerisation of styrene oxide with the diethylzinc–water catalyst was found to proceed via C_β–O(CH_2–O) bond scission [48].

It has been found that the polymerisation of propylene oxide with catalysts characterised by an isolated metal atom surrounded by a porphyrin [49,50], Schiff's base [40,51] or calix[4]arene [41] moiety also proceeded by C_β–O bond cleavage.

The polymerisation of epoxides with coordination catalysts has been established to proceed with the inversion of the configuration at the carbon atom of the epoxide ring where cleaved:

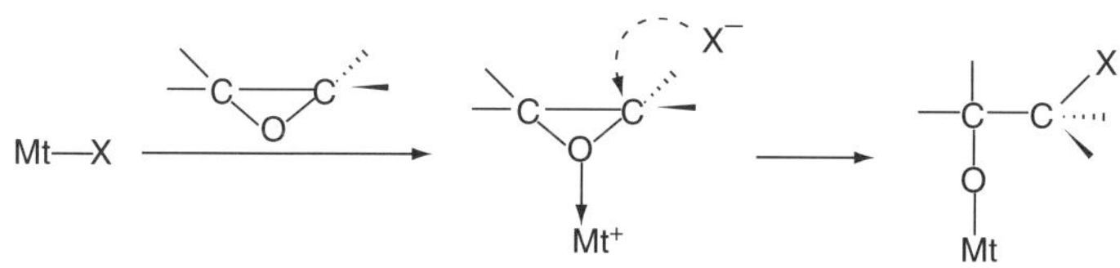

(1)

Such inversion of the configuration has been demonstrated in ring-opening polymerisations of various epoxides, carried out in the presence of catalysts containing multinuclear species, as well as in the presence of those containing mononuclear species with an isolated metal atom [41,52,53]. The crystalline fraction of poly[*trans*-(β-d)-propylene oxide] obtained by polymerisation with the diethylzinc–methanol catalyst was shown to have an *erythro*-diisotactic structure [43,54]. It was concluded, therefore, that epoxide ring opening took place by cleavage of the C_β–O bond with inversion of the configuration at the C_β atom. Configuration inversion at the carbon atom of the epoxide ring where cleaved has also been found for the polymerisation of appropriately deuterated ethylene oxide and propylene oxide with triethylaluminium–water (1:1) and diethylzinc–water (1:1) catalysts [55–59]. The inversion in the course of the epoxide ring cleavage coincides with those reported for the polymerisation of isomeric 2-butene oxide with catalysts such as triethylaluminium–water–acetylacetone (2:1:1) [18,60,61] and metalloporphyrins such as tetraphenylporphinatoaluminium chloride or methoxide and N-methyltetraphenylporphinatozinc methoxide [62] in which the *cis*- and *trans*-epoxide was converted into a *threo*- and *erythro*-enchained unit respectively. Moreover, it has been established that, in the polymerisation of cyclohexene oxide (1,2-epoxycyclohexane) (*cis*-epoxide) with catalysts such as diethylzinc–methanol (7:8), diethylzinc–1-methoxy-2-propanol (2:3) [63,64], diethylzinc–4-*t*-butylcatechol–phenol (2:1:1), diethylzinc–4-*t*-butylcatechol–1-phenoxy-2-propanol (2:1:1) [65] and dimethoxy-*p*-*t*-butylcalix[4]arene-diolatoaluminium chloride [41], the polymer formed was characterised by the appearance of *threo*-enchained monomer units. This testifies to epoxide ring opening with inversion of the configuration at the carbon atom of the cleaved C–O bond.

As regards the ring-opening reaction with retention of the configuration at the carbon atom of the epoxide ring where cleaved [scheme (2)], there are very few examples of such a reaction in the literature [66,67]:

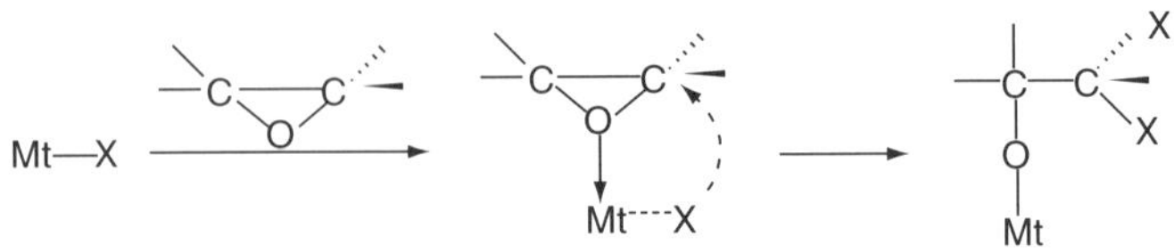 (2)

For instance, the reaction of *cis*- and *trans*-2-butene oxide with aluminium chloride yields mainly *erythro*- and *threo*-chlorohydrin (after hydrolysis) respectively [66].

9.2.1.4 Models of the Active Sites, Epoxide Polymerisation Mechanism and Stereochemistry

Although any coordination polymerisation involves a monomer coordination in each step, literature data that might concern complex formation between the monomer heteroatom and the metal atom at the active site are rather scant. This is due to the lability of such complexes of monomers with active sites; if they were stable enough, they would not undergo any further rearrangement, which might lead to polymer chain growth. Thus, models have been studied that have consisted of a metal complex, of no effectiveness as a catalyst, and a monomer, or of a catalyst and a non-polymerisable heterocycle.

The first isolated and characterised species that could be envisioned as intermediates in the initiation step for the coordination polymerisation of epoxides when using metal carboxylate catalysts were complexes formed between cadmium carboxylates, solubilised in organic solvents by the tris-3-phenylpyrazole hydroborate ligand, and epoxides such as propylene oxide and cyclohexene oxide [68]. Other epoxide complexes with various metal derivatives have also been reported in the literature [69–72].

The inversion of the configuration at the carbon atom of the epoxide ring where cleaved in the course of the ring opening during polymerisation indicates that the monomer complexed with the metal atom is attacked from the back side by the nucleophilic substituent X [scheme (1)]. If front side nucleophilic attack of this substituent occurred on the coordinated monomer, i.e. via the four-membered transition state as in scheme (2), no inversion but rather the retention of the configuration at the epoxide ring carbon atom where cleaved should be observed; however, this is not the case.

Taking into account the associated structures of catalysts for the coordination polymerisation of epoxides and considering that the coordinated epoxide is attacked by the nucleophile from the back side [scheme (1)], the catalyst must engage its two metal atoms in order to make the polymerisation possible. In fact, zinc-based catalysts containing associated multinuclear species (→Zn–O–Zn–O→), including those with condensed zinc atoms (→Zn–O–Zn–O→), are all characterised by the appearance of active sites with two zinc atoms bridged via a nucleophilic oxygen atom in which the O atom, bound covalently to the octahedral zinc atom, is coordinated to the adjacent tetrahedral Zn atom (Zn–O→Zn).

Models of active sites in catalysts formed in systems such as $ZnEt_2$–$MeOCH_2CH(Me)OH$ (2:3) [29], $ZnEt_2$–$C_6H_3(t-Bu)(OH)_2$–PhOH (2:1:1) [65] and $ZnEt_2$–$C_6H_3(OH)_3$ (2:1) [31,73] have been proposed schematically [1,74] as follows:

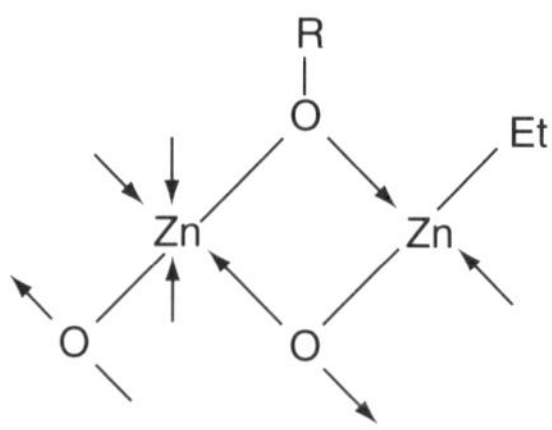

R = 1-methoxy-2-propanol moiety

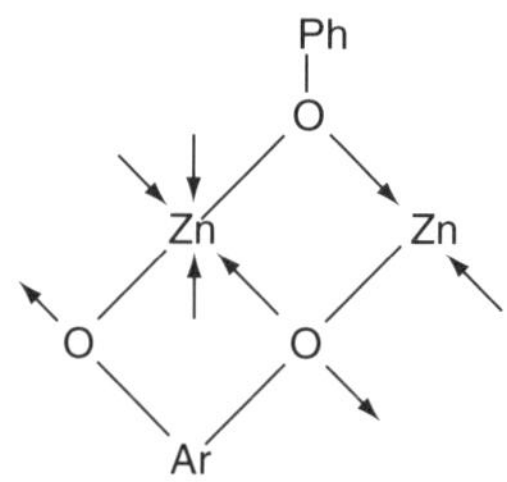

Ar = 4-*t*-butyl-l,2-dihydroxybenzene (4-*t*-butylcatechol)moiety

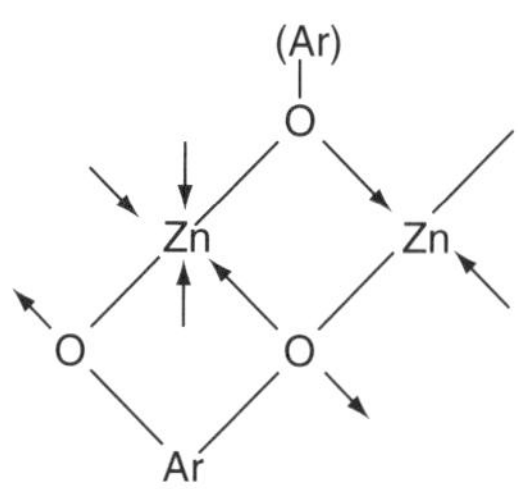

Ar = 1,2,3-trihydroxybenzene (pyrogallol)moiety

Characteristic of these models of active sites is the appearance of the OZn–O→Zn(O)Et species. Considering the structural properties of the discussed catalysts as well as the polymer chain microstructure and the structure of the end groups of poly(propylene oxide) obtained with the $PhOZnOC_6H_3$ (*t*-Bu)OZnEt catalyst, a concerted mechanism of epoxide ring-opening polymerisation with catalysts containing multinuclear species, including those with condensed zinc atoms, has been postulated [65,74]:

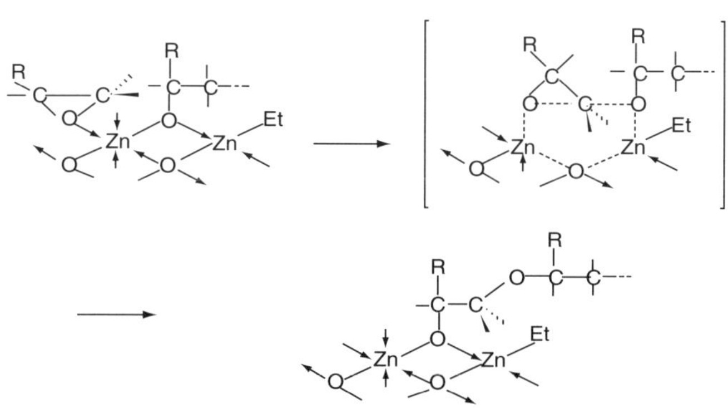

(3)

According to this mechanism, the attack of the epoxide on the catalyst active site involves its coordination with the more electrophilic zinc atom, which is that of the OZnO unit containing an octahedral Zn atom. Such a coordination results in enhancement of the electrophilicity of the attacked monomer carbon atom as well as the nucleophilicity of the attacking oxygen atom. The mutual activation of the epoxide monomer and zinc–oxygen species in the resulting complex of appropriate geometry and charge distribution accommodates the nucleophilic attack of the oxygen atom, carried out by the adjacent tetrahedral zinc atom, on the coordinating epoxide molecule from its back side. Such rearward nucleophilic attack on the coordinating epoxide molecule in the six-centered activated complex involving two metal atoms results in the inversion of the configuration at the carbon atom of the epoxide ring where cleaved [65].

The following mesomeric forms, which can be proposed for the complex of the epoxide with the catalyst active site, seem to present the nature of the active species in a more illustrative way [74]:

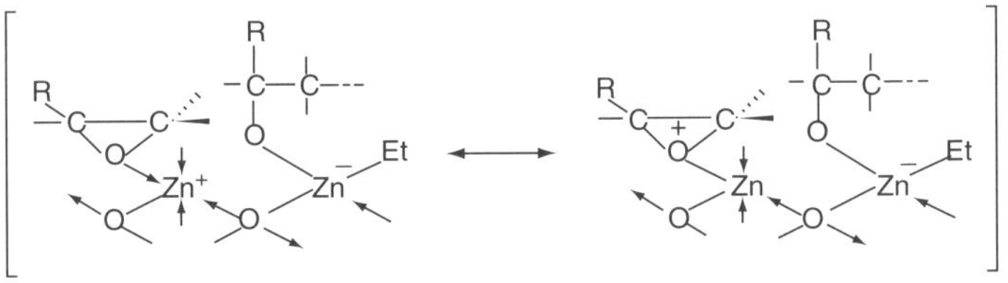

The mechanism of stereoregulation in the stereoselective polymerisation of propylene oxide with zinc dialkoxide and related zinc dialkoxide–ethylzinc alkoxide complexes has been satisfactorily explained by the enantiomorphic catalyst sites model prepared by Tsuruta *et al.* [52,75]. According to this model, the presence of chiral sites with a central octahedral zinc atom, bearing the polymer chain and coordinating the monomer, was assumed to be the origin of the stereoregulation mechanism.

The R^* sites accept (*R*)-propylene oxide in preference to the (*S*)-enantiomer, resulting in the formation of *RRR* isotactic sequences of polymer chains. On the other hand, the S^* sites accept (*S*)-propylene oxide in preference to the (*R*)-enantiomer, resulting in the formation of *SSS* isotactic sequences of polymer chains.

Propylene oxide polymerisation with catalysts derived from reactions in the diethylzinc–phenol or alcohol and/or polyhydric phenol system exhibited an enhancement in the average molecular weight of the poly(propylene oxide) and in the crystallinity (isotacticity) as the reaction progressed [65]. Triad tacticities of the poly(propylene oxide) formed testified to the polymerisation propagation step proceeding according to the mechanism of enantiomorphic catalyst site control. Similar results concerning triad tacticities have been obtained from studies of the polymerisation of propylene oxide with heterogeneous supported catalysts, such as diethylzinc, ethylzinc phenoxide or ethylzinc 1-phenoxy-2-

propoxide–γ-alumina [76]. It was concluded, therefore, that the zinc atom of the OZnO unit, bearing the growing polyether chain at the catalyst active site, could attain a saturated octahedral configuration because of its complexation with oxygen atoms of the chain of an appropriate length. Such a six-coordinate zinc atom forms an enantiomorphic site, R^* or S^*, with the adjacent zinc atom of the catalyst containing multinuclear species, capable of stereoselective propylene oxide reaction in the propagation step [65,76].

It is to be noted that enantiomorphic Zn-based catalysts as well as other catalysts for the polymerisation did not recognise the mode of orientation of cyclohexene oxide which is an achiral *cis*-disubstituted epoxide. The catalysts discussed above serve as bulky groups, terminating the growing polymer chains, which makes simpler, in combination with strong steric effects exerted from the last and penultimate monomer units, the syndiotactic enchainment of cyclohexene oxide (*threo*-disyndiotactic polymer) [64,65,77]. It is interesting that poly (cyclohexene oxide) chains are ended by unsaturated groups to some extent which, considering the H^+ transfer reaction, testifies to the occurrence of the cationic propagation reaction [64,65,77]. Cationic propagation may also take place in the copolymerisation of cyclohexene oxide and tetrahydrofuran with ethylzinc methoxide [77] and *trans*-2-butene oxide and tetrahydrofuran with the $AlEt_3$–H_2O–Acac catalyst [78]. The presence of consecutive sequences of tetrahydrofuran units in the copolymers obtained testifies to the cationic mechanism of copolymerisation including propagation steps via the formation of oxonium ions; this indicates that zinc- and aluminium-based catalysts containing multinuclear species can generate cationic propagation sites.

The behaviour of catalysts derived from reactions of diethylzinc with chiral glycols deserves special attention in the polymerisation of propylene oxide. It is worth noting that, for this polymerisation with the catalyst obtained in the diethylzinc–(*R*)-(−)-3,3-dimethyl-1,2-butanediol (1:1) system, the enantiomer with the same spatial configuration as the diol was chosen preferentially for the polymerisation and the tacticity of the poly(propylene oxide) yielded obeyed the enantiomorphic catalyst site model [25,79]. Spassky *et al.* [80] named this type of polymerisation, in which the chiral initiator preferentially polymerises the enantiomer of the same spatial configuration, homosteric enantioasymmetric polymerisation. Instead, the polymerisation involving a preferentially polymerised enantiomer of the opposite spatial configuration to that of the chiral diol used with diethylzinc for the catalyst preparation was named antisteric (heterosteric) enantioasymmetric polymerisation.

In connection with the discussed mechanism of epoxide polymerisation in the presence of associated multinuclear catalysts [65], it is worth noting the 'flip-flap' mechanism proposed by Vandenberg to operate in epoxide coordination polymerisation [18,60,61] as the earliest explanatory attempt involving the formation of an intermediate that could accommodate the essentially linear, three-centred transition state necessary for inversion of the configuration at the epoxide ring carbon atom where cleaved. The 'flip-flap' mechanism assumed the epoxide molecule to be coordinated to one aluminium atom prior to the rearward nucleophilic attack on it by the polymer chain bound to the adjacent aluminium atom. However, this mechanism is now out-of-date and cannot be accepted.

The mechanism of epoxide polymerisation with Zn-, Al- or other metal-based coordination catalysts containing multinuclear species can be presented schematically as in scheme (4a) (the structures and charge distribution are simplified in the scheme, without differentiation of + or − and δ^+ or δ^- charges) [1]:

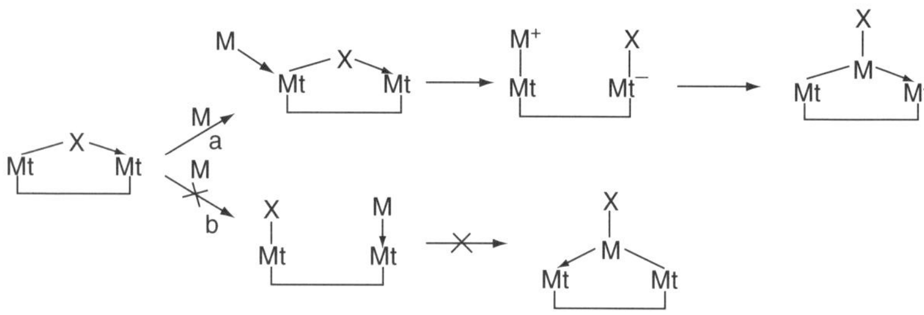

(4)

The reaction pathway according to scheme (4b) (the 'flip-flap' mechanism) does not take place.

It is worth adding a comment concerning the cationic propagation, apart from the propagation according to the coordination mechanism involving a nucleophilic attack on the coordinating monomer, in some polymerisation systems with cyclohexene oxide in the presence of coordination catalysts [64,65,77]. Cationic propagation may occur when the incoming monomer attacks the coordinated monomer at the active site of the multinuclear catalyst [scheme (5)] instead of the nucleophilic attack on the coordinated monomer that might be carried by the nucleophilic polymer chain end as in scheme (4a):

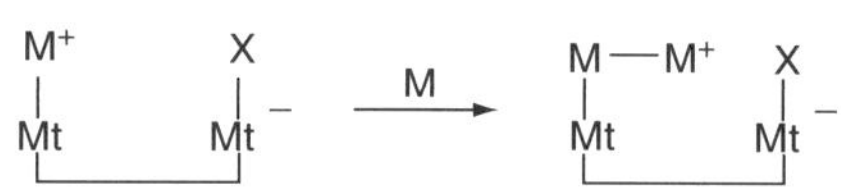

(5)

Aluminium-based catalysts with tetraphenylporphinato [32,35,38,81], Schiff's base [37–40] and calix[4]arene ligands [41] are characterised by the appearance of a non-associated, isolated pentacoordinate metal atom forming an active bond Mt–X with nucleophilic substituent X. As stated above, epoxide polymerisation with these catalysts involves the rearward attack of the nucleophilic substituent on the coordinating epoxide molecule [scheme (1)]. In order to explain this, a mechanism involving the simultaneous participation of two catalyst molecules in the initiation and propagation reaction has been proposed [40,41,62,82]. According to this mechanism, the rearward attack of the nucleophilic substituent on the coordinating monomer is carried out by the six-coordinate aluminium species. These species can appear as neutral epoxide complexes with catalyst molecules containing a chlorine atom or the growing polymer chain, and as species involving ions or ion pairs consisting of positively charged aluminium species with the coordinated epoxide molecule and negatively charged aluminium species bearing the attacking nucleophile.

Scheme (6) shows the epoxide polymerisation mechanism in a system with (25,27-dimethoxy-*p*-*t*-butylcalix[4]arene-26,28-diolato)aluminium chloride [41]:

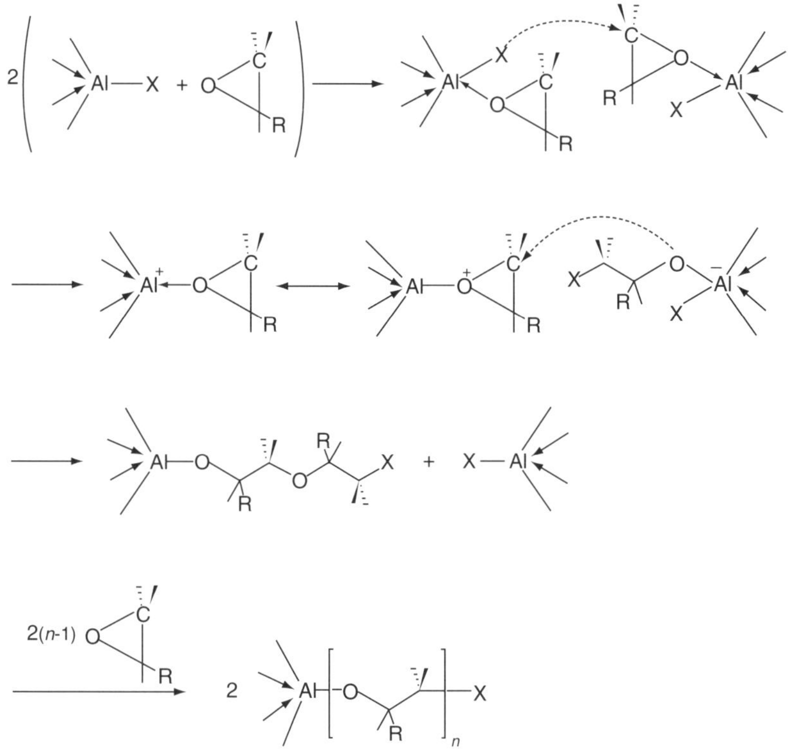

(6)

Instead, a general mechanistic pathway for epoxide polymerisation with catalysts containing mononuclear species (Mt–X) can be presented schematically as follows (simplified, without differentiation of ions or ion pairs) [1]:

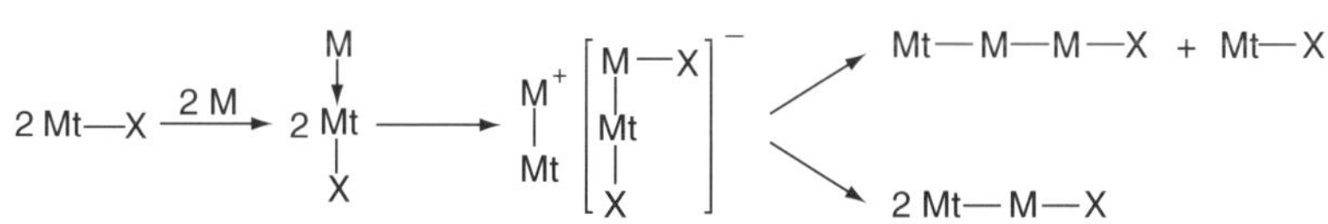

(7)

Cationic propagation involving mononuclear catalyst species is illustrated schematically as follows [1]:

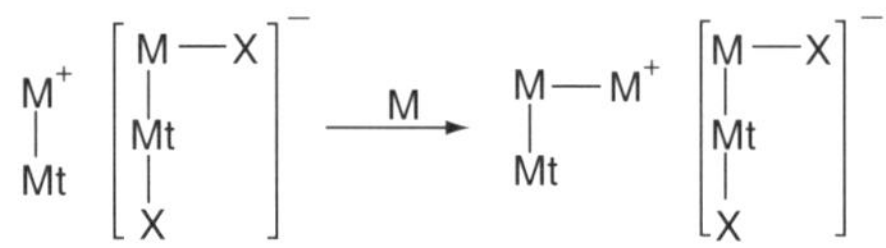

9.2.1.5 Polymerisation of Oxetanes

Oxetane, a four-membered cyclic ether, is highly susceptible to cationic polymerisation [83]. However, this monomer also undergoes coordination polymerisation in the presence of catalysts such as zinc dimethoxide [84], triethylaluminium–water–acetylacetone [85–87], aluminium isopropoxide–zinc chloride and diethylzinc–water [87,88], as well as tetraphenylporphinatoaluminium chloride–methylaluminium di(2,6-di-*t*-butyl-4-methylphenoxide) [89]. Studies of the microstructure of the polymer derived from the polymerisation of 2-methyloxetane with the triethylaluminium–water–acetylacetone (2:1:2) catalyst showed that the polyether obtained consisted of regioregular monomer unit sequences, fairly rich in isotactic triads [87]:

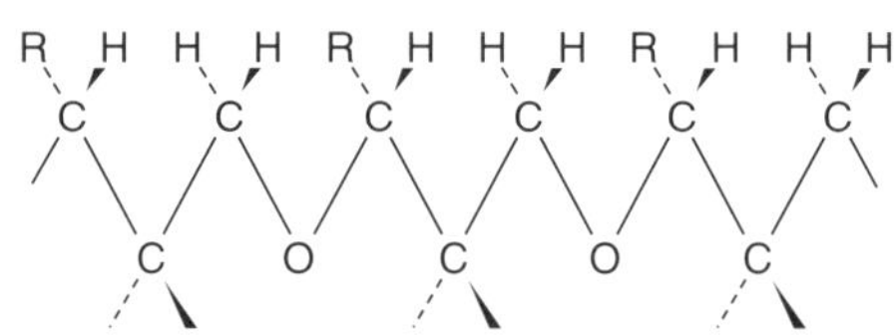

Such a polymer microstructure appeared to be different from that of the poly(2-methyloxetane) formed with the triethylaluminium–water (1:1) system, which was irregular [87]. This testifies to the coordination mechanism of 2-methyloxetane polymerisation in the former case, and to the cationic polymerisation mechanism in the latter case.

At the end of considerations of the polymerisation of cyclic ethers with coordination catalysts, one should emphasise the versatility of catalysts used for polymerisations, which is manifested in the formation of various active polymerisation sites capable of promoting propagation according to various mechanisms.

9.2.2 Polymerisation of Cyclic Esters

The coordination polymerisation of cyclic esters concerns mostly lactones, especially those containing a four-membered ring in the molecule. There is, however, an interest in the coordination polymerisation of such oxacyclic ester monomers as lactide and alkylene carbonate and, to a lesser extent, in the coordination polymerisation of thiacyclic, azacyclic and phosphacyclic esters (Table 9.2).

9.2.2.1 Stereoisomerism of Lactone Polymers

Lactone polymers can exhibit tacticity owing to the presence of chiral tertiary carbon atoms in the polymer main chain. Of particular interest are polymers of lactones with a four-membered ring (β-lactones), such as β-butyrolactone. For instance, poly[(*R*)-β-butyrolactone] {poly[(*R*)-3-hydroxybutyrate]} is one of the most common members of the group of poly(β-hydroxyalkanoate)s which are present in a wide range of microorganisms [90].

The structures of isotactic and syndiotactic stereoisomers of poly(β-butyrolactone) are presented in Figure 9.2.

9.2.2.2 Polymerisation of Lactones

The ring-opening polymerisation of lactones containing a four-membered ring (β-lactones) in the presence of coordination catalysts can proceed according to two mechanistic pathways that involve C(O)–O bond cleavage to form a metal alkoxide growing species [scheme (9)] or C_β–O bond cleavage to form a metal carboxylate growing species [scheme (10)]; the mode of the lactone ring opening depends on the kind of monomer and catalyst:

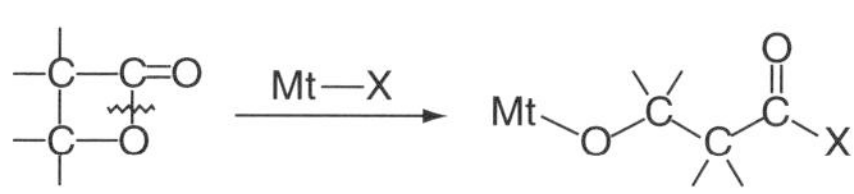

(9)

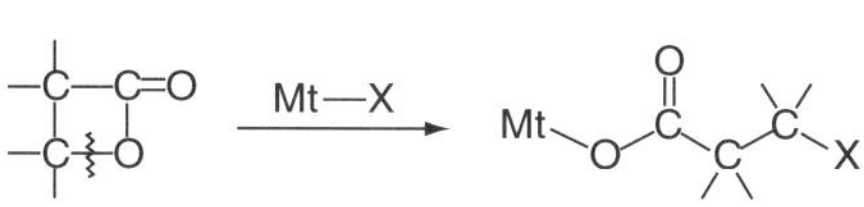

(10)

The coordination polymerisation of lactones with a six- and seven-membered ring (δ- and ε-lactones respectively) occurs via ring opening at the C(O)–O linkage to generate metal alkoxide chain terminals, following a reaction analogous to that presented by scheme (9).

Aluminium alkoxides (especially aluminium isopropoxide), dialkylaluminium alkoxides, yttrium alkoxides, zinc alkoxides, aluminoxanes, zincoxanes, bimetallic μ-oxoalkoxides, aluminium porphyrins and aluminium Schiff's base complexes are the most representative coordination catalysts, containing multinuclear or mononuclear species, for lactone polymerisations (Table 9.5).

Of the catalysts evaluated, those effective for the polymerisation of β-lactones with ring opening by C(O)–O bond cleavage [scheme (9)], as well as for the polymerisation of δ- and ε-lactones (also with C(O)–O bond cleavage), have

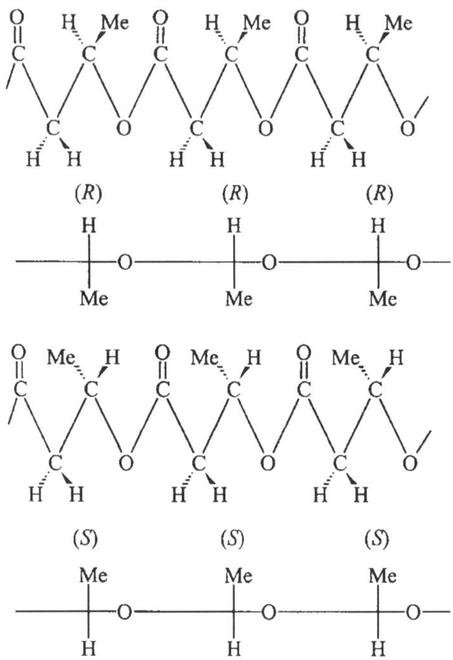

isotactic poly(β-butyrolactone)

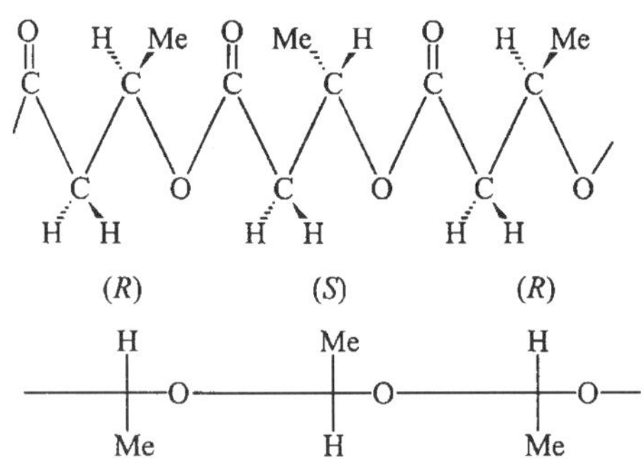

syndiotactic poly(β-butyrolactone)

Figure 9.2 Structures of isotactic and syndiotactic stereoisomers of poly(β-butyrolactone)

been aluminium and dialkylaluminum alkoxides [91–110], yttrium alkoxides [99,100], zinc alkoxides [111–114], aluminoxanes [115–120], zincoxanes [117,121], bimetallic μ-oxoalkoxides [122–124] and aluminium porphyrins of the (tpp)AlOR type [81,125–127]. On the other hand, aluminium porphyrins

Table 9.5 Representative examples of coordination catalysts for polymerisations of heterocyclic and heterounsaturated monomers[a]

Catalyst[b]	Monomer	Reference
Catalysts with multinuclear species:		
$FeCl_3$–PO	Epoxide	[267,290–296]
	Isocyanate	[268]
$AlEt_3/H_2O$	Epoxide	[17,18,46,61]
	Lactone	[117]
AlR_3/H_2O; R: Me, Et, *i*-Bu	Lactone	[115,116,119,120]
$AlEt_3/H_2O$ / AcacH	Epoxide	[15,18]
	Oxetane	[86,87]
	Oxetane/CO_2[c]	[245]
$ZnEt_2/H_2O$	Epoxide	[16,45,61,297–303]
	Epoxide / CO_2[c]	[137,199–201]
	Tiirane	[154,155]
$ZnEt_2/H_2O/CO_2$	Epoxide / CO_2[c]	[202]
Al(O-*i*-Pr)$_3$	Epoxide	[57]
	Lactone	[91,92,97,98,106–110]
	Lactide	[108,117]
Al(O-*i*-Pr)$_3$/$ZnCl_2$	Epoxide	[15,44]
$[(RO)_2AlOZnOAl(OR)_2]_x$; R: Bu	Epoxide	[20–22,69,197]
	Lactone	[122,124,125]
$ZnEt_2$ / MeOH (1:2)	Epoxide	[16,42,43,54,64,84,246]
	Oxetane	[84]
	Lactone	[113]
$ZnEt_2$ / MeOH (7:8)	Epoxide	[26–28,64]
$ZnEt_2/MeOCH_2CH_2OH$ (7:8)	Epoxide	[304]
$ZnEt_2/MeOCH_2$ CH(Me)OH (2:3)	Epoxide	[29,64,77]
$ZnEt_2$/(*R*)-(−)-HOCH(*t*-Bu) CH_2OH (1:1)	Epoxide	[79,80]
	Tiirane	[79,153,307–307]
	Lactone	[111,112]
$CdMe_2$/(*R*)-(−)-HOCH(*t*-Bu) CH_2OH (1:1)	Tiirane	[159]
	Lactone	[114]
$ZnEt_2/HOC_6H_4OH$ (1:1)	Epoxide / CO_2[c]	[203–205,308,309]
$ZnEt_2/HOC_6H_3$(*t*-Bu)OH (1:1)	Epoxide	[65]
	Epoxide / CO_2[c]	[207]
$ZnEt_2/HOC_6H_4$ C(O)OH (1:1)	Epoxide / CO_2[c]	[212]
$ZnEt_2/HOC_6H_4NH_2$ (1:1)	Epoxide / CO_2[c]	[208]
$ZnEt_2/HOC(O)C_6H_4$ C(O)OH (1:1)	Epoxide / CO_2[c]	[212]
$ZnEt_2/HSC_6H_4SH$ (1:1)	Epoxide / CO_2[c]	[213]
$ZnEt_2/H_2NC_6H_4NH_2$ (3:2)	Epoxide / CO_2[c]	[213]
$ZnEt_2/C_6H_3(OH)_3$ (2:1)	Epoxide	[31]
	Epoxide / CO_2[c]	[73,208–211]
$ZnEt_2/C_6H_3(OH)_3/H_2O$ (2:1:0.5)	Epoxide / CO_2[c]	[211]
$CdEt_2/C_6H_3(OH)_3$ (2:1)	Epoxide / CO_2[c]	[209]
$AlEt_3/C_6H_3(OH)_3$ (2:1)	Epoxide / CO_2[c]	[209]
	Tiirane/CO_2[c]	[247]

continued overleaf

Table 9.5 *(continued)*

Catalyst[b]	Monomer	Reference
$Y[OP(O)(OR)_2]_3/Al(i\text{-}Bu)_3/C_3H_5(OH)_3$; R: $CH_3(CH_2)_3CH(Et)CH_2$	Epoxide / CO_2[c]	[227]
$Y[OP(O)(OR)_2]_3/Al(i\text{-}Bu)_3$; R: $CH_3(CH_2)_3CH(Et)CH_2$	Epoxide / CO_2[c]	[228]
$Y(OCH_2CH_2OMe)_3$	Lactone	[69]
$Y_5(\mu-O)(O\text{-}i\text{-}Pr)_{13}$	Lactone/lactide[c]	[100]
$(EtZnOMe)_4$	Epoxide	[64]
$ZnEt_2/\gamma\text{-}Al_2O_3$	Epoxide	[76,310,311]
	Epoxide/CO_2[c]	[76,218,219]
$ZnEt_2/PhOH/\gamma\text{-}Al_2O_3$	Epoxide	[76]
	Epoxide / CO_2[c]	[76]
$ZnO/HOC(O)C_6H_4C(O)OH$	Epoxide / CO_2[c]	[221,222]
$ZnO/HOC(O)(CH_2)_xC(O)OH$; x=3,4	Epoxide / CO_2[c]	[223]
Cd (*S*)-cysteinate	Tiirane	[157,160,162–164]
Catalysts with mononuclear species:		
$Zn[OC_6H_3(Ph)_2]_2.2Et_2O$	Epoxide / CO_2[c]	[238]
$Zn[OC_6H_2(t\text{-}Bu)_3]_2.2Et_2O$	Epoxide / CO_2[c]	[238]
(tpp) AlX; X: Cl, SPr, OMe	Epoxide	[32,35,49,81]
	Epoxide/cyclic Carbonate[c]	[149]
	Cyclic carbonate	[149]
	Lactone	[81,128]
(tpp)AlCl	Epoxide	[40,49,242,312,313]
(tpp)AlOMe	Lactone	[126,127]
	Lactide	[141]
(tpp)AlOC(O)R	Lactone	[129]
(tpp)AlCl/MeOH	Epoxide	[196]
(tpp)AlCl/Q^+X^-	Epoxide / CO_2[c]	[241]
	Epoxide/cyclic Acid anhydride[c]	[188,189]
(tpp)AlCl/MeAl $(OAr)_2$	Epoxide	[314–317]
	Lactone	[140]
(Mtpp)ZnOMe	Epoxide	[62]
(Mtpp)ZSPr	Tiirane	[125,165]
(sal)AlCl	Epoxide	[36,37,39,40,50,318]
	Lactone	[130]
(sal)AlOMe	Lactide	[130]
(dmca)AlCl	Epoxide	[41,244]
	Epoxide / CO_2[c]	[41,244]

[a] According to Ref. 1.

[b] Me = CH_3; Et = $CH_3(CH)_2$; Pr = $(CH_3)(CH_2)_2$*i*-Pr = $(CH_3)_2CH$; Bu = $CH_3(CH_2)_3$; *i*-Bu = $(CH_3)_2CHCH_2$; *t*-Bu = $(CH_3)_3C$; Ph = C_6H_5; Ar = 2,6-di-*t*-butylphenyl, 2,6-di-*t*-butyl-4-methylphenyl; AcacH = acetylacetone; $salH_2$ = salcene (*N,N*′-disalicylidene-(1*R*,2*R*)-1,2-cyclohexanediyldiamine, or *N,N*′-bis(2-hydroxybenzylidene)-(1*R*, 2*R*)-1,2-cyclohexanediamine, or 2,2-(1*R*,2*R*)-1,2-cyclohehylenebis(nitrilomethylidyne)diphenol); $tppH_2$ = 5,10,15,10-tetraphenylporphyrin; Mttp = *N*-methyl-5,10,15,20-tetraphenylporphyrin; Q^+X^- = quaternary phosphonium halide; $dmcaH_2$ = 25,27-dimethoxy-26,28-dihydroxy-*p*-*t*-butylcalix[4]arene, or 25,27-dimethoxy-*p*-*t*-butylcalix[4]arene-26,28-diol.

[c] Copolymerisation.

such as (tpp)AlCl or (tpp)AlOC(O)R [125,128,129] and an aluminium Schiff's base complex such as (sal)AlCl [130] promote the polymerisation where the lactone ring is cleaved at the C_β–O bond, the growing species being an aluminium carboxylate, in the case of β-lactones [scheme (10)].

The polymerisation of lactones via C(O)–O bond cleavage has been explained in terms of the nucleophilic attack on the carbon atom of the carbonyl group in the coordinated monomer molecule. This attack can be carried out by the adjacent metal atom of the catalyst containing the multinuclear species [82]:

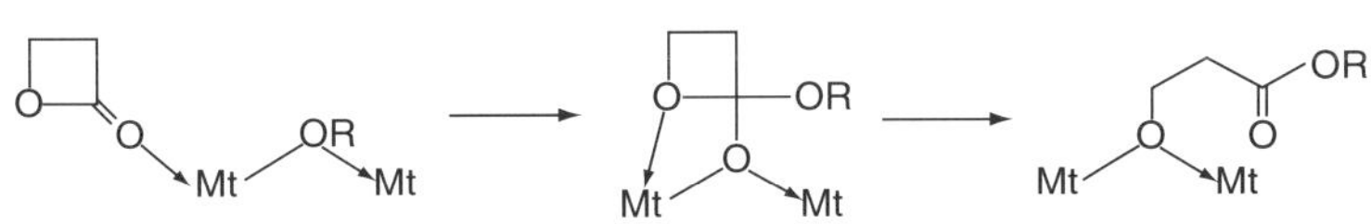

(11)

Recently, however, the polymerisation of lactones in the presence of aluminoxanes has been reported by Lenz *et al.* [119,120,131] to be the most effective with regard to the polymerisation of racemic (*R*, *S*)-β-substituted β-lactones to isotactic polymers of reasonably high molecular weight. The aluminoxanes most thoroughly investigated have been those containing either methyl or higher alkyl groups such as ethyl and isobutyl. The aluminoxane-catalysed polymerisations of β-substituted β-lactones have some shortcomings. Among these shortcomings are the long polymerisation times required, the broad molecular weight distributions of the polyesters obtained and the formation of a product mixture of isotactic and atactic polymers of various molecular weights [118]. However, optimisation of the polymerisation of racemic (*R*,*S*)-β-butyrolactone with alkylaluminoxanes has led to a very effective method for obtaining high molecular weight and highly isotactic poly(β-butyrolactone), especially when isobutylaluminoxane is used as the polymerisation catalyst [120]. Also, highly syndiotactic or atactic poly(β-butyrolactone)s could be formed, depending on the composition and method of preparation of the aluminoxane and the conditions of polymerisation [131].

More recently, a very efficient yttrium-based catalyst, yttrium 2-methoxyethoxide, has been applied successfully for β-butyrolactone polymerisation which proceeded easily at room temperature [99]. It is worth mentioning that rare-earth metal alkoxides (derived from yttrium and lanthanum) exhibit outstanding efficiency as catalysts for the polymerisation of cyclic esters such as ε-caprolactone [132] and lactide [133].

When catalysed by bimetallic μ-oxoalkoxides, the polymerisation of β-, δ- and ε-lactones involves the alkoxide group as the attacking nucleophile. Thus, the mean degree of association of the catalyst exerts a further decisive effect on the kinetics of the polymerisation [122,125]. Bimetallic μ-oxoalkoxides and aluminium alkoxides [106] have the great advantage of polymerising β-propiolactone and ε-caprolactone in a living manner.

The polymerisation of ε-caprolactone with aluminium isopropoxide obeys the same mechanism as that operating for bimetallic μ-oxoalkoxides [107]. This polymerisation with aluminium isopropoxide trimer and tetramer appears [96–98] to have been initiated by the trimer much more readily than by the tetramer, and the tetramer-to-trimer interconversion appears to proceed relatively slowly [98]. Therefore, when an aluminium isopropoxide trimer/tetramer mixture is used for the polymerisation of ε-caprolactone, the trimer is consumed completely whereas the tetramer remains unconverted, at least within the time required for the complete polymerisation. In the aluminium isopropoxide molecule reacted with the monomer, all three isopropoxide groups are engaged in the initiation step and transferred within the propagation step into the poly(ε-caprolactone) molecules as chain terminals [98,108]. Thus, three polymer chains grow from one aluminium atom. The ^{27}Al NMR spectra of the living poly(ε-caprolactone) suggest the coordination of the aluminium atoms with the acyl oxygen atoms of the growing polymer chain and indicate the coexistence of tetra-, penta- and hexacoordinate aluminium atoms during the polymerisation [98].

The enantioasymmetric polymerisation of racemic β-butyrolactone has been carried out with the diethylzinc–(*R*)-(−)-3,3-dimethyl-1,2-butanediol catalyst. The unreacted monomer became enriched in the *S* enantiomer and the enantioelection during the polymerisation was of the homosteric type [114]. The behaviour of the diethylzinc–(*R*)-(−)-3,3-dimethyl-1,2-butanediol catalyst in terms of its stereoselectivity contrasted with that of the diethylzinc–water or diethylzinc–methanol catalysts, the latter of which had no stereoselectivity. Also, racemic α-ethyl-α-methyl-β-propiolactone [(*R*)-(+)- and (*S*)-(−)-] was found [113] to give only an atactic polymer in the presence of the diethylzinc–methanol catalyst. The β,β-disubstituted β-lactones were also found [134] to yield only atactic polymers in the presence of the diethylzinc–water catalyst. This indicates that the enantiomorphic sites of the zinc coordination catalyst are not able to recognise the chirality of the lactone monomer used for the polymerisation. However, the polymerisation of racemic α-ethyl-α-methyl-β-propiolactone with the diethylzinc–(*R*)-(−)-3,3-dimethyl-1,2-butanediol catalyst was found [135] to yield an optically active polymer obeying stereoelection of the homosteric type. The enantioasymmetric polymerisation of racemic α-methyl-α-propyl-β-propiolactone using the diethylzinc–(*R*)-(−)-3,3-dimethyl-1,2-butanediol catalyst gave a similar result to that of α-ethyl-α-methyl-β-propiolactone [114]. An antisteric type of stereoelection in the enantioasymmetric polymerisation of racemic α-methyl-α-propyl-β-propiolactone was also reported [114] when the dimethylcadmium–(*R*)-(−)-3,3-dimethyl-1,2-butanediol catalyst was used for the polymerisation.

In the case of lactone polymerisation with the (tpp)AlOR catalyst, ring opening via C(O)–O bond cleavage involves the participation of two molecules of the catalyst [36,136] which can be presented schematically for the polymerisation of β-lactone as follows:

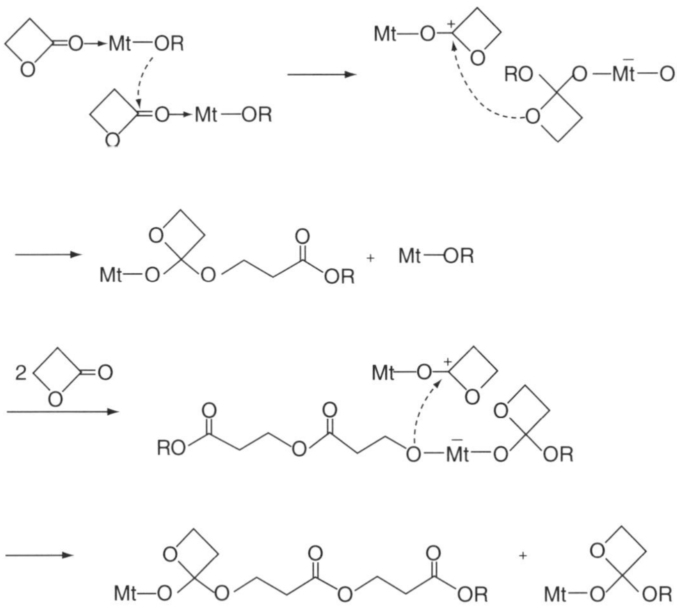

(12)

The polymerisation of β-butyrolactone in the presence of (tpp)AlCl, (tpp)AlO-C(O)R or (sal)AlCl catalysts, which involves C_β–O bond cleavage [125,129, 130,137], can be explained in terms of the nucleophilic attack being carried out with the participation of another catalyst molecule on the coordinated monomer:

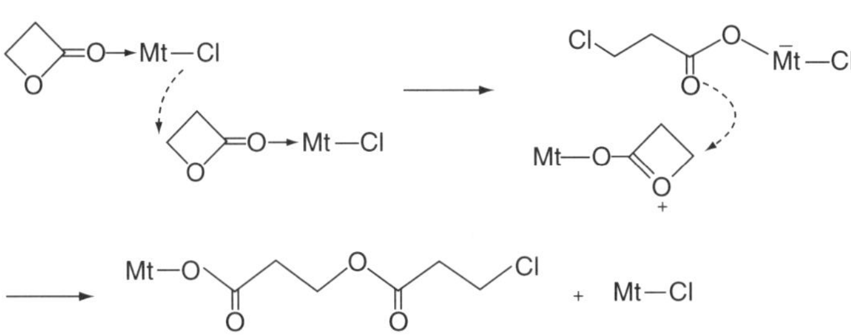

(13)

Of interest is the fact that a Lewis acid with bulky substituents, such as methylaluminium di(2,6-di-*t*-butyl-4-methylphenoxide), also has an accelerating effect on the polymerisation of β-lactone, the extent of acceleration being dependent on the mode of lactone ring cleavage. The polymerisation of β-butyrolactone in the presence of (tpp)AlOMe [scheme (9)] was slower than that in the presence of (tpp)AlCl [scheme (10)], but the accelerating effect of the bulky Lewis acid was more significant for the (tpp)AlOMe catalyst [125]. Thus, the acceleration effect is considered [125] to be due to the coordination of the carbonyl oxygen atom of the lactone at the aluminium atom of the Lewis acid which affects the reactivity directly when the carbonyl group is attacked by the growing species [polymerisation with the (tpp)AlOMe catalyst], while the attack at the C_β atom, remote from the carbonyl group, is less affected by the coordination [polymerisation with the (tpp)AlCl catalyst]. This results in different extents of polymerisation acceleration, depending on the mode of lactone ring cleavage.

In connection with the above, a significant acceleration effect of (tpp)AlCl on the polymerisation of δ-valerolactone with the (tpp)AlOMe catalyst [126], in spite of the fact that (tpp)AlCl itself is not capable of promoting this polymerisation, should be mentioned. The coordination of δ-valerolactone with (tpp)AlCl, followed by the nucleophilic attack of the alcoholate species formed from the (tpp)AlOMe catalyst, has been considered to take place during the polymerisation. Since the Lewis acidity of (tpp)AlCl is much higher than that of (tpp)AlOR, the polymerisation is accelerated dramatically by (tpp)AlCl [126].

Apart from the coordination polymerisation of lactones, the polymerisation of lactide (3,6-dimethyl-1,4-dioxacyclohexane-2,5-dione) with various coordination catalysts is of growing interest. This is connected with some advantageous properties of polyesters obtained from *rac*-D,L-lactide and *meso*-D,L-lactide as potentially useful materials for biomedical and pharmaceutical applications in terms of their low toxicity and favourable biodegradability. A number of catalysts containing multinuclear species, such as aluminium isopropoxide [108,138,139], triethylaluminium–neopentyl (1:1) [139], triethylaluminium–(+)-menthol (1:1) [139], methylaluminoxane [139] and bimetallic μ-oxoalkoxide [140], as well as catalysts containing mononuclear species, such as (tpp)AlOMe [141] and (sal)AlOMe [130], have been used for the polymerisation of lactide.

When using the aluminium isopropoxide catalyst to promote the polymerisation of lactide, all three isopropoxide groups have been found to initiate the polymerisation [137,139]. Ring cleavage in the polymerisation with aluminium isopropoxide and diethylaluminium isopropoxide occurred at the C(O)–O bond, leading to aluminium alcoholate propagating species [139]. Lactide polymerisation with other catalysts containing multinuclear species proceeded similarly [138–140]. Analysis of stereosequences of lactide polymers obtained with catalysts such as aluminium isopropoxide, diethylaluminium alcoholate and methylaluminoxane indicates that none of these catalysts favours the formation of isotactic blocks [139].

The polymerisation of lactide in the presence of catalysts containing mononuclear species, such as (tpp)AlOMe and (sal)AlOMe, was found to proceed by C(O)–O bond cleavage [130,141]. It is interesting that corresponding chloride-containing catalysts such as (tpp)AlCl and (sal)AlCl appeared to be unable to promote the polymerisation of lactide [130,141].

9.2.2.3 Polymerisation of Cyclic Carbonates

The ring-opening polymerisation of cyclic carbonates with coordination catalysts concerns monomers containing a five-membered ring or a six-membered ring in thc molccule.

Five-membered cyclic carbonates such as ethylene carbonate (1,3-dioxolan-2-one) and propylene carbonate (4-methyl-1,3-dioxolan-2-one) [142–146] hardly undergo a ring opening. Their polymerisation, however, has been reported to proceed in the presence of metal alkoxides, metal acetylacetonates and metal alkyls as catalysts. It is, however, characteristic that no polymerisation was found below 100 °C for ethylene carbonate [145], whereas in the case of propylene carbonate the polymerisation did not take place significantly even at 140 °C within a few days [143]. The polymerisation involved partial decarboxylation, and the loss of carbon dioxide during the polymerisation surpassed 50 mol.-% irrespective of the polymerisation conditions. Thus, the polymerisation of alkylene carbonates with a five-membered ring in the molecule fails to produce thermodynamically disfavoured poly(alkylene carbonate)s but leads to poly (alkylene ether–carbonate)s with a content of alkylene carbonate units not exceeding 50 mol.-% (Table 9.2) [142–145].

One of the mechanisms taken into consideration for the polymerisation of propylene carbonate was that this proceeded via orthocarbonate species. The likelihood of such a mechanism might be ascertained by the polymerisation of bispropylene spiroorthocarbonate with diethylzinc as the catalyst at 160 °C, which was found to yield poly(propylene ether–carbonate) [143].

It seems that the alkylene carbonate polymerisation proceeds via monomer decarboxylation in the first reaction step. The decarboxylation most probably involves metal carbonate species owing to ring opening of the 1,3-dioxolan-2-one via C_{β}–*O* bond cleavage [146,147]. A possible reaction scheme in the presence of zinc-based coordination catalysts is presented by scheme (14), in which, for the sake of clarity, participation of the adjacent zinc atom as the nucleophilic attack carrier is omitted:

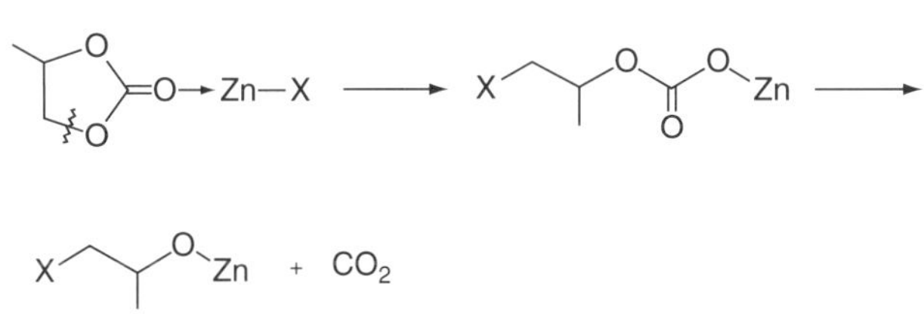

(14)

The polymerisation of propylene carbonate can be proposed to proceed via C(O)–O bond cleavage which is presented in a simplified way, omitting the participation of the adjacent zinc atom, by the following scheme:

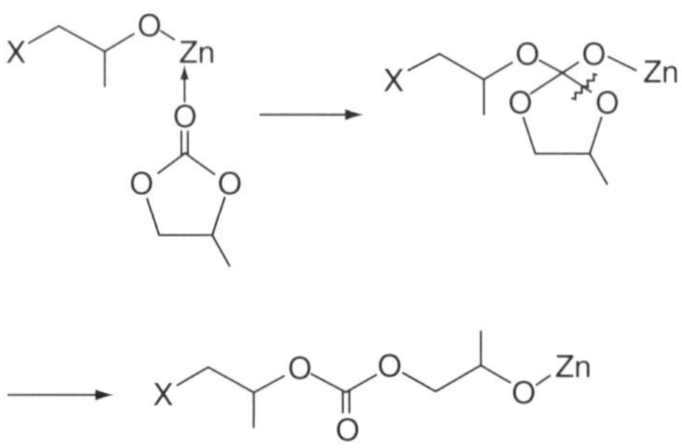

(15)

It seems that in alkylene carbonate polymerisation systems the reaction pathway involves cleavage of the C_β–O bond, followed by decarboxylation of the metal carbonate species formed [scheme (14)], to a higher extent than the C(O)–O bond cleavage leading to propagation via the metal alcoholate species [scheme (15)]. Thus, the resulting polymer is the alkylene poly(ether–carbonate) with prevailing ether linkages.

The polymerisation of the bispropylene spiroorthocarbonate with zinc-based coordination catalysts probably involves zinc alcoholate propagating species, which is shown schematically as follows:

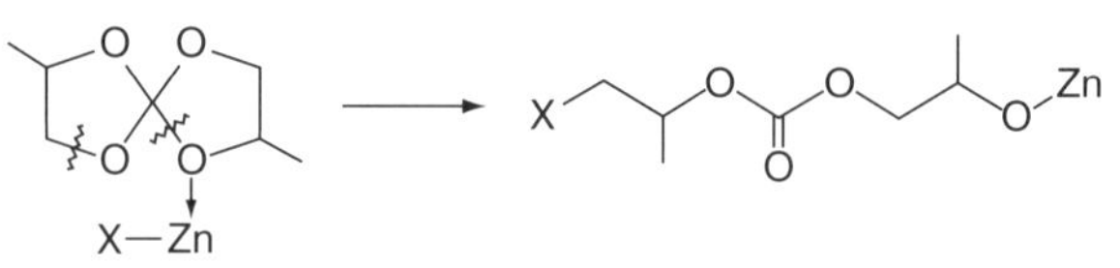

(16)

In this simplified scheme (the participation of the adjacent zinc atom as the nucleophilic attack carrier is omitted), the formation of poly(propylene ether–carbonate) characterised by a regioregular head-to-tail structure is shown. This is possible because of the nucleophilic attack on the C_β atom of the neighbouring ring to that coordinating with its more basic oxygen atom to the metal atom. Actually, the predominant head-to-tail structure of the poly(propylene ether–carbonate) obtained at 80 °C with the diethylzinc–pyrogallol–water (3:1:0.5) catalyst has been proved [146].

The coordination polymerisation of cyclic carbonates with a six-membered ring in the molecule, such as trimethylene carbonate (1,3-dioxan-2-one) and 2,2-dimethyltrimethylene carbonate (5,5-dimethyl-1,3-dioxan-2-one) [148–150], carried out in the presence of metal carboxylates {e.g. zinc stearate, tin-based catalysts such as the di(*n*-butyl)stannic diiodide–triphenylphosphine system [151] or porphinatoaluminium compounds such as (tpp)AlOR [149]} is not accompanied with decarboxylation and yields the respective polycarbonates (Table 9.2). The ring cleavage during the polymerisation of trimethylene carbonate and 2,2-dimethyltrimethylene carbonate in the presence of the above catalysts has been found [148,149,151] to occur at the C(O)–O bond, resulting in the formation of metal alcoholate propagating species. The living character of the polymerisation with the (tpp)AlOR catalyst is of interest [149].

9.3 Polymerisation of Thiacyclic Monomers

Although thiacyclic monomers do not play such an important role in coordination polymerisation as oxacyclic monomers, some polymerisation of tiiranes with coordination catalysts deserves attention, especially considering the polymerisation stereochemistry:

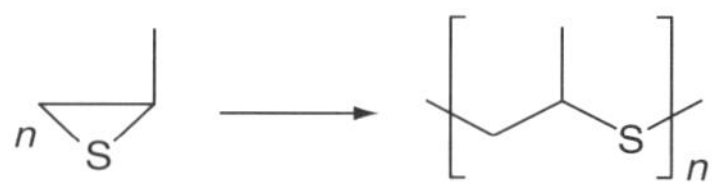

(17)

Note that some general considerations of ring-opening polymerisation originate from studies of the coordination polymerisation of tiiranes. Other thiacyclic monomers that contain an endocyclic sulphur atom and an exocyclic oxygen atom, such as monothiocarbonates, are instead analogous to five-membered cyclic carbonates.

9.3.1 Polymerisation of Tiiranes

The ruthenium(II) – ethylene episulphide complex [152] may serve as a model for the coordination of tiiranes at the catalyst active site.

Polymerisations of tiiranes have been carried out in the presence of various catalysts containing multinuclear species, such as those formed in diethylzinc or dimethylcadmium–water, alcohol, diol, thiol or carboxylic acid systems, as well as catalysts containing mononuclear species, such as porphinatozinc thiolate (Tables 9.1 and 9.5).

9.3.1.1 Tiirane Coordination Polymerisation with Catalysts Containing Multinuclear Species

Polymerisations of tiiranes in the presence of coordination catalysts containing multinuclear species have been extensively studied in terms of their stereoselective and stereoelective behaviour. For monosubstituted tiiranes, the polymerisation can proceed enantiosymmetrically and lead to a mixture of isotactic chains of opposite configurations. By using optically active catalysts, the polymerisation may occur enantioasymmetrically, with the enchainment of only one of the two enantiomers.

The enantiomorphic catalyst sites control mechanism was found to operate in the stereospecific polymerisation of tiiranes. Sigwalt *et al.* [79,153] found that the diethylzinc–(*R*)-(−)-3,3-dimethyl-1,2-butanediol (1:1) catalyst promotes the enantioasymmetric polymerisation of propylene sulphide and *t*-butylethylene sulphide. It appears also that *t*-butylethylene sulphide gives pure isotactic chains with most of the stereospecific catalysts, and that, when polymerising an enantiomerically enriched *t*-butylethylene sulphide in the presence of the diethylzinc–water catalyst, one can separate the polymer into a pure polyenantiomer partly soluble in chloroform and a racemic stereocomplex that is insoluble in chloroform [154,155].

Some cadmium compounds, including simple salts, were revealed to be excellent catalysts for the enantiosymmetric polymerisation of propylene sulphide [156,157]. For instance, the proportion of isotactic diads in the poly(propylene sulphide) sample obtained in polymerisation with the cadmium (*R*)-tartrate catalyst was more than 95%, higher than the 69% which was characteristic of a polymer sample prepared using the zinc (*R*)-tartrate catalyst [158]. The superior stereoselectivity of the cadmium (*R*)-tartrate catalyst is also borne out by the more effective separation into fractions having opposite optical rotations of the poly(propylene sulphide) yielded by cadmium tartrate, compared with that yielded by zinc (*R*)-tartrate. Note the quite different behaviour of these two catalysts in terms of their stereoelectivity in the polymerisation of propylene sulphide; only very slight optical activity was found for the poly(propylene sulphide) sample prepared using cadmium tartrate, whereas that associated with the polymer sample obtained with zinc tartrate was found to have a much higher value [158].

As regards the dimethylcadmium–(*R*)-(−)-3,3-diethyl-1,2-butanediol (1:1) catalyst, it was characterised by a lower stereoelectivity than the diethylzinc–(*R*)-(−)-3,3-dimethyl-1,2-butanediol (1:1) catalyst in enantioasymmetric polymerisations of propylene sulphide, *cis*-2-butene sulphide and cyclohexene sulphide, but the elected chirality was opposite to that found for polymers obtained with the zinc-based catalyst [159].

The above stereospecific tiirane polymerisations have generally been run in heterogeneous systems. Such conditions essentially make it impossible to determine the detailed structure of active species involved in these polymerisations. Thus, enantiosymmetric and enantioasymmetric polymerisations of propylene sulphide have also been studied in a homogeneous phase by using chiral cadmium thiolates of cysteine esters and chiral cadmium carboxylates of cysteine and methionine [157,160–164]. The most studied is living polymerisation using the cadmium derivative of the isopropyl ester of (*S*)-cysteine [160]:

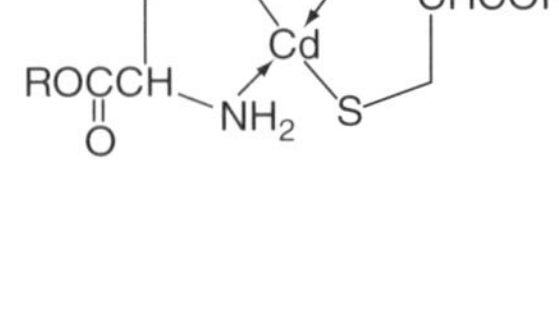

Poly(propylene sulphide)s prepared by the polymerisation of propylene sulphide with bis[isopropyl-(*S*)-cysteinato]cadmium as the catalyst are highly isotactic and optically active, although the optical rotation is not of large value. It must be emphasised that poly(propylene sulphide)s obtained with cadmium bis[(*S*)-cystcinatc] or cadmium bis[(*S*)-methioninate] (cadmium carboxylates) are also isotactic, but they are optically inactive. A lack of stereoelectivity of these carboxylate catalysts in propylene sulphide polymerisation should be noted in connection with the behaviour of cadmium (*R*)-tartrate. Further studies of propylene sulphide polymerisation with the bis[isopropyl-(*S*)-cysteinato]cadmium catalyst (cadmium dithiolate) revealed that stereospecificity (both stereoselection and stereoelection) appears only in systems in which the polymer molecular weight reaches 6000 and is dependent on the polymerisation temperature; with decreasing temperature, the polymer isotacticity increases and stereoelection of the polymerisation can be inverted. The polymers with a molecular weight lower than 6000 are amorphous and optically inactive. No stereoregularity or enantioasymmetry is observed in this range, but they both appear for polymers with molecular weights higher than 6000 [157,161]. Tacticities of polymers correspond to the high molecular weight fraction of poly (propylene sulphide) only, which was separated from the oligomers by precipitation in methanol. The tacticity coefficient σ (mole fraction of isotactic diads determined by ^{13}C NMR) increases very slowly with molecular weight. This can be explained in terms of the formation of an atactic segment with a number-average molecular weight M_n of 6 000 and a partially isotactic segment increasing in size with molecular weight. The variation in the tacticity, σ, of a polymer segment according to its location in the main chain is shown in Figure 9.3 [163,164].

It can be seen from Figure 9.3 that stereoregularity appears for molecular weights higher than 6000 and grows regularly, and there is a maximum ($\sigma = 1$) for a molecular weight of 35 000. For higher values of $\bar{M}_n$ (e.g. 50 000), there is

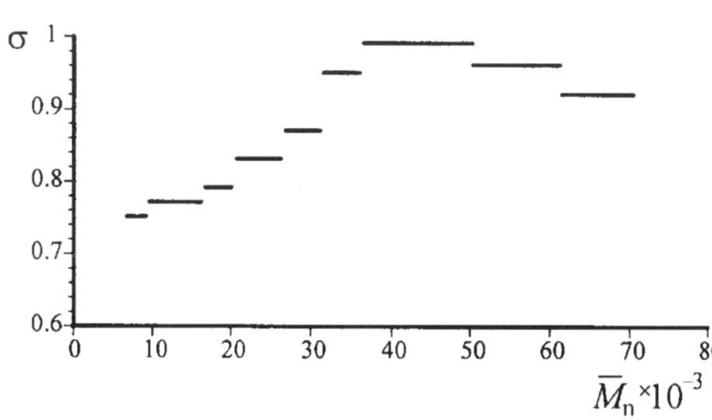

Figure 9.3 Variation in tacticity, σ, of a polymer segment according to its location in the main chain

a small decrease in stereoregularity. It is clear that the distribution of the two configurational units changes regularly with the length of the growing polymer chain. These data suggest the participation of the polymer chain in the structure of active sites in the propylene sulphide polymerisation system with the bis [isopropyl-(*S*)-cysteinato]cadmium catalyst, probably by complexation of at least one sulphur atom with the cadmium atom. This polymer chain might be the main polymer growing on one cadmium valency, or the oligomer growing very slowly on the second cadmium valency [163,164].

In order to explain the enantioasymmetry observed with this catalyst, the influence of the chiral cysteinato ligand is obvious and should occur through complexation of its nitrogen atom with the cadmium atom at the beginning of the polymerisation. The catalyst association and the tiirane monomer coordination are necessary to interpret the kinetic data obtained [164]. The nucleophilic attack on the C_β atom of the coordinating propylene sulphide molecule involves a neighbouring thiolate group. When the polymer chain reaches an appropriate length, a second coordination by a sulphur atom of this chain becomes possible, leading to both a stereospecific and an enantioasymmetric propagation [163,164]:

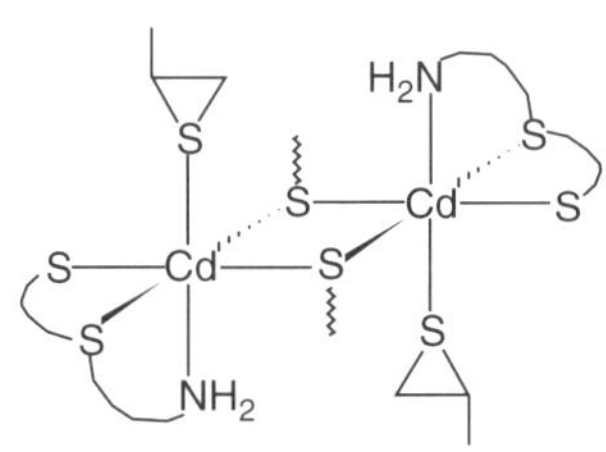

This stereospecific growth may then occur on both the oligomer and the high polymer. The active sites may be considered as diastereoisomers, since cadmium atoms of the associated catalyst are all coordinated to an (*S*)-cysteinate group but may also be complexed with either an (*R*)-polymer sequence or with an (*S*)-polymer sequence. Since they are not enantiomers, different rate constants for the propagation may be observed. It is worth noting the possibility of complexation of the cadmium atom by the sulphur atom of the next polymer chain instead of the amino group, as well as the complexation of the cadmium atom by two sulphur atoms of the high molecular weight polymer chain, in competition with the complexation involving the oligomers [163,164].

Considering the results obtained from enantioasymmetric polymerisations of tiiranes (and oxiranes) with a variety of chiral catalysts, Spassky *et al.* [25,80,159] formulated configurational rules determining the stereochemistry of ring-opening polymerisation with such catalysts. Substituents were classified into three categories with respect to their spatial configuration. For instance, molecules of (*R*)-(−)-3,3-dimethyl-1,2-butanediol and (*R*)-(+)-propylene sulphide have been classified into the same category of spatial configuration:

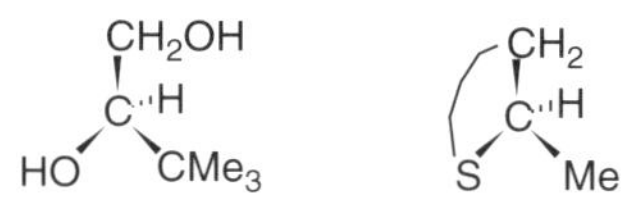

When a catalyst prepared in the reaction of diethylzinc with (*R*)-(−)-3,3-dimethyl-1,2-butanediol was used for the polymerisation of racemic propylene sulphide, the (*R*)-(+)-enantiomer, having the same spatial configuration as the diol, was preferentially chosen in the propagation (homosteric enantioasymmetric polymerisation). In contrast to this homosteric catalyst, a catalyst derived from the dimethylcadmium–(*R*)-(−)-3,3-dimethyl-1,2-butanediol reaction system polymerised preferentially the (*S*)-(−)-enantiomer, being of opposite spatial configuration to that of the diol used for the catalyst preparation (antisteric enantioasymmetric polymerisation).

It may be noted that all of the homosteric catalysts were chiral zinc systems, the composition of which was expressed as $[RZnOR^*]_x[R^*OZnOR^*]_y (x/y < 1)$. Antisteric catalysts were chiral zinc- or cadmium-based systems of a common compositional feature expressed as $[RZnOR^*]_x[R^*OZnOR^*]_y$ or $[RCdOR^*]_x[R^*OCdOR^*]_y (x/y > 2)$. The nature of homosteric and antisteric stereoelection has not yet been elucidated fully at the molecular level because the structure of the operating species and the polymerisation mechanism with these catalysts are not clearly established [52].

9.3.1.2 Tiirane Polymerisation with Catalysts Containing Mononuclear Species

An excellent catalyst for the living polymerisation of propylene sulphide appeared to be *N*-methyl-tetraphenylporphinatozinc propanethiolate [(Mtpp)ZnSPr]. Polymerisation with this catalyst proceeds at room temperature via (*N*-methyl-tetraphenylporhinato)zinc thiolate as the growing species [165]:

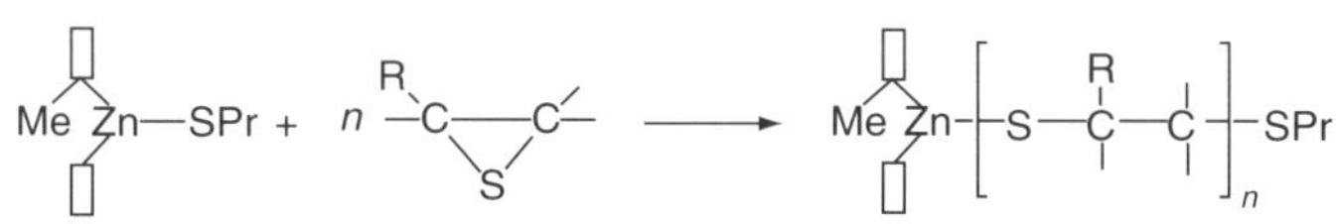

(18)

The polymerisabilities of tiiranes are usually much greater than those of oxiranes, and thus no reports on tiirane–oxirane copolymerisation under common conditions are available in the literature. However, under irradiation with visible light, these monomers can be copolymerised randomly when using the (mtpp)ZnSPr catalyst. The effect of irradiation was found [166] to be more significant on the cross-propagation steps than on the homopropagation steps. Such copolymerisation deserves particular attention, since it is the first example of photoinduced copolymerisability enhancement [30,36].

9.3.2 Polymerisation of Cyclic Thiocarbonates

The polymerisation of ethylene monothiocarbonate (2-oxo-1,3-oxathiolane) has been carried out in the presence of metal alkyls such as diethylzinc or diethylcadmium and metal alkoxides such as magnesium methoxide, aluminium *n*-butoxide or titanium *n*-butoxide. Polymerisation with these catalysts was accompanied with the elimination of carbon dioxide (analogously to the polymerisation of cyclic carbonates) and the polymers obtained appeared to be poly(ethylene sulphide–monothiocarbonate)s (Table 9.2). The content of ethylene monothiocarbonate (oxycarbonylthioethylene) units and ethylene sulphide (thioethylene) units in the polymers produced was dependent on the catalysts used, but all of the polymers yielded from the polymerisation of ethylene monothiocarbonate at 80 °C contained less than 50 % monothiocarbonate units, e.g. 49.8 % for diethylcadmium, 21.2 % for diethylzinc and 15.9 % for magnesium methoxide as catalysts. However, the polymerisation run at 40 °C with titanium *n*-butoxide as the catalyst yielded a polymer having a content of these units of ca 77 % [167].

Compared with the polymerisation of ethylene carbonate, the polymerisation of ethylene monothiocarbonate with coordination catalysts proceeds more easily and also may involve decarboxylation to a lesser extent. Decarboxylation during ethylene monothiocarbonate polymerisation seems to involve the metal monothiocarbonate species which are analogous to those formed in the system with ethylene carbonate [scheme (14)]. However, ethylene monothiocarbonate decarboxylation leads to metal thiolate species which may be presented schematically as follows:

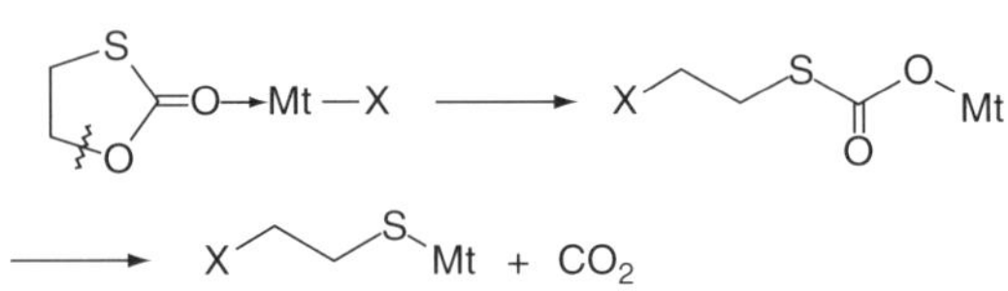

(19)

The metal thiolate species are more reactive towards the coordinating monothiocarbonate monomer during the polymerisation presented by scheme (20) than the corresponding metal alcoholate species operating in ethylene carbonate polymerisation [scheme (15)].

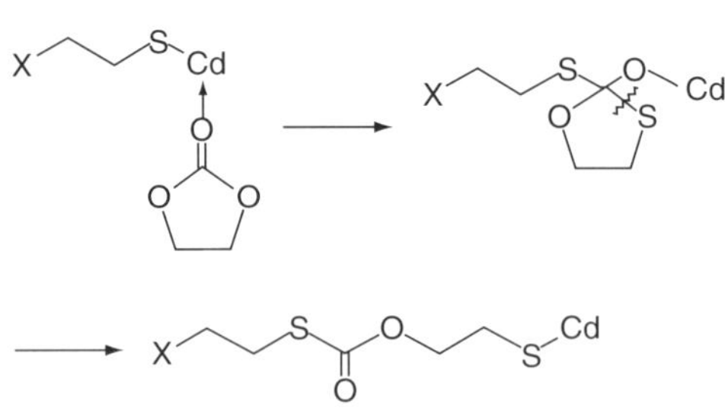

(20)

The higher relative efficiency of a diethylcadmium catalyst in the polymerisation of ethylene monothiocarbonate (at 80 °C), with respect to the high content of ethylene monothiocarbonate units in the polymer obtained, probably results from the softness of both cadmium and sulphur atoms fitting to each other, according to HSAB theory, to participate in covalent bonding. Thus, decarboxylation, which should involve the cadmium monothiocarbonate species with the metal–oxygen bond [scheme (19)], and not the metal–sulphur bond, occurs to a lesser extent. Moreover, propagation according to scheme (20), involving the formation of a cadmium–sulphur bond leading to a metal thiolate species of relatively high activity, is more likely to occur than that involving the formation of a less reactive metal alcoholate species.

9.4 Polymerisation of Azacyclic Monomers

Azacyclic monomers containing an endocyclic nitrogen atom which are most commonly subjected to polymerisation in the presence of coordination catalysts include α-aminoacid *N*-carboxyanhydrides. However, other monomers of this type, such as morpholine-2,5-dione, have also been subjected to coordination polymerisation.

9.4.1 Polymerisation of α-Aminoacid N-Carboxyanhydrides

The polymerisation of α-aminoacid *N*-carboxyanhydride, which is accompanied with the elimination of carbon dioxide, constitutes a convenient method for preparing high molecular weight polypeptide. This polymerisation, with the use of α-aminopropionic acid (alanine) *N*-carboxylic acid anhydride as a monomer, was also carried out in the presence of coordination catalysts such as group 2 and 3 metal alkyls [168–174] or their combinations with water, secondary amine or alcohol [168,173] and yielded polyalanine (Table 9.2).

Studies of the polymerisation of alanine *N*-carboxylic acid anhydride with diethylzinc or triisobutylaluminium as the catalyst have revealed the relatively fast deprotonation of the monomer NH group by the metal alkyl as the first reaction step. This reaction (easily detectable by evolution of the ethane or isobutane) produces *N*-metallated initiating species (substituted metal carbamates) which, however, are not monomeric but undergo association via metal–heteroatom coordination bonds [75,175].

The structure of the associated zinc carbamate species has been elucidated in ome simple systems [176,177] and may be presented schematically as follows:

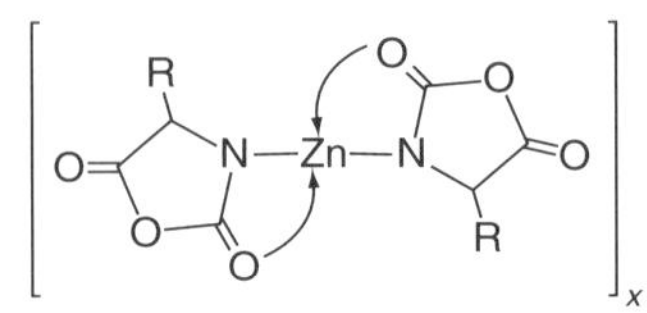

The polymerisation of alanine *N*-carboxylic acid anhydride involves the participation of metal–oxygen species which undergo decarboxylation yielding metal–nitrogen propagating species. This may be shown schematically as follows:

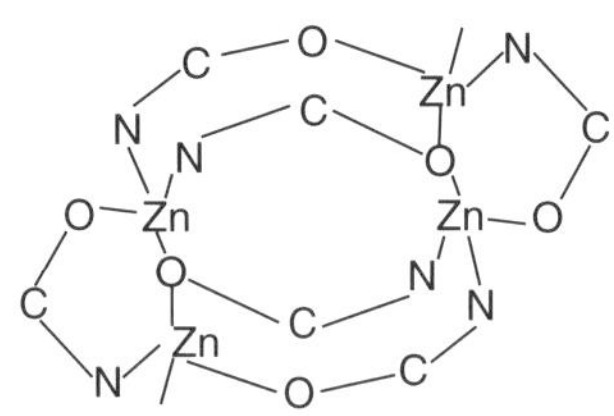

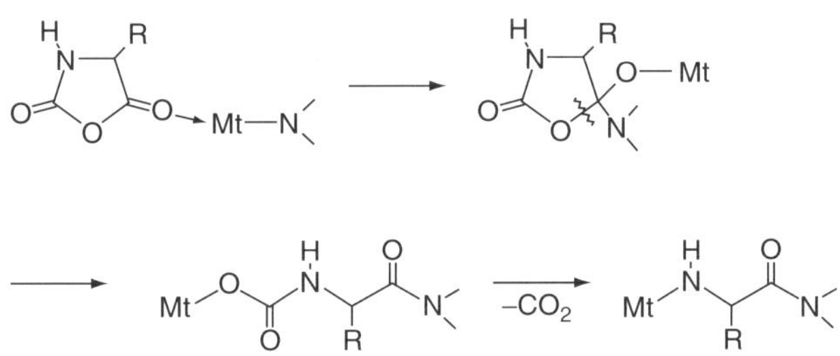

(21)

When D, L-alanine *N*-carboxylic acid anhydride was subjected to polymerisation with triethylaluminium as the catalyst, the polymer formed, polyalanine, was found [168,174] to consist of two fractions: a water-insoluble fraction which was rich in isotactic sequences and a water-soluble atactic fraction. This indicated that some stereoselective polymerisation had occurred [75].

In the copolymerisation of D- and L-enantiomers of alanine *N*-carboxylic acid anhydride in systems containing an excess of the L-enantiomer with respect to the D-enantiomer and with diisobutylaluminium *N*,*N*-diethylamide as the catalyst, the content of L-enantiomer units in the polymer formed was always higher than the content of this enantiomer in the monomer feed [173]. This fact shows the enantiomer present in excess in the starting monomer mixture to have been polymerised preferentially throughout the copolymerisation [75].

In the copolymerisation of D- and L-enantiomers of alanine *N*-carboxylic acid anhydride in the presence of triethylaluminium, the content of L-enantiomer units in the polymer formed in the early stage was found [173] to be lower than the content of this enantiomer in the monomer feed, resulting in an increased L-enantiomer content in the non-converted monomer. However, the L-enantiomer content in the non-converted monomer was found [173] to continue to decrease as the copolymerisation progressed, and in the final stage it attained lower values than the L-enantiomer unit content of the polymer phase of that stage. Correlations of the L- and D-enantiomer conversion with the total conversion in the polymerisation catalysed by triethylaluminium suggest that the amount of monomer polymerised before the crossover point is about five molecules per aluminium atom. It seems most reasonable to consider that equal amounts of the L- and D-enantiomers have reacted before the crossover point, and thereafter the L-enantiomer reacted preferentially over the D-enantiomer.

Such behaviour of the alanine *N*-carboxylic acid anhydride polymerisation systems studied makes it possible to exclude the enantiomorphic catalyst sites control mechanism from consideration. If the polymerisation with triethylaluminium were controlled by this mechanism, the L-enantiomer content in both monomer and polymer phases would be unchanged after the rapid initiation reaction. It should be noted in this connection that the metal atoms of metal carbamates do not attain the maximum coordination saturation owing to their association, which is shown schematically above for the case of zinc carbamates. Recalling that the enantiomorphism of zinc-based coordination catalysts (and also aluminium-based and related coordination catalysts) containing multinuclear species is connected with the octahedral configuration of the central metal atom bearing the covalently bound nucleophilic initiating group or polymer chain at the active site [52], one can easily notice the lack of any enantiomorphism of the zinc carbamate species that appear in the tetrahedral configuration [176,177].

Considering the spatial cooperation between the ultimate unit and the approaching monomer molecule through complexation with the metal atom, the growing chain end control mechanism has been postulated to be valid for the coordination polymerisation of alanine *N*-carboxyanhydride in the presence of metal alkyls [75].

9.4.2 Polymerisation of Morpholinedione

Morpholine-2,5-dione (1,4-oxazacyclohexane-2,5-dione) is characterised by the appearance in its molecule of both the δ-lactam and δ-lactone functions. The coordination polymerisation of this monomer [scheme (22)] has been reported for systems with diethylzinc [178] and tin bis(ethylhexanoate) [150] as catalysts:

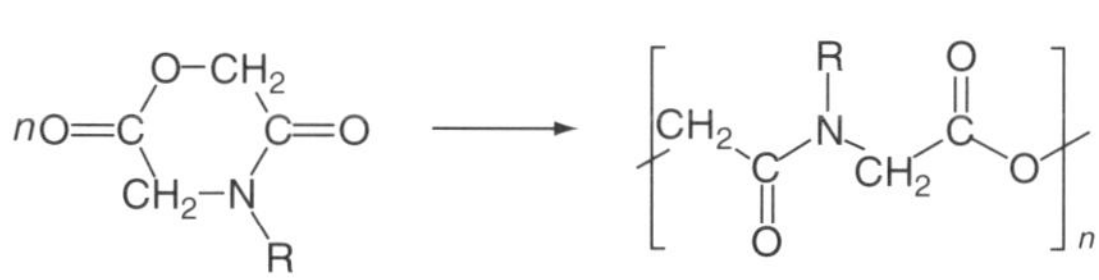

(22)

The polymer formed, poly(morpholine-2,5-dione), is a polydepsipeptide consisting of alternating sequences of α-amino acid and α-hydroxyacid (Table 9.2).

Morpholine-2,5-dione and its substituted analogue, 6-methylmorpholine-2,5-dione, appeared to undergo copolymerisation with lactide in the presence of the zinc oxide catalyst. The copolymers obtained were characterised by an increased content of α-hydroxyacid units. The copolymerisability of the unsubstituted monomer was comparable with that of lactide. However, the *N*-methylated monomer was much less susceptible than the unsubstituted one to copolymerisation with lactide [178].

It may be interesting to mention that other azacyclic monomers with various functions (including the lactam function), e.g. substituted ε-lactam ethers such as 2,7-dimethyltetrahydro-1,4-oxazepin-5-one [179] and substituted ε-lactam thioethers such as 2,7-dimethyltetrahydro-1,4-thiazepin-5-one [180], have been polymerised with diethylzinc, triethylaluminium or related ethylaluminium chlorides, yielding crystalline poly(amide–ether)s and poly(amide–thioether)s respectively. It was, found, however, that the ring-opening reaction of the lactam ether and the lactam thioether occurred easily at the C–O bond and the C–S bond of the ether group and the thioether group respectively, throughout the polymerisation, and the propagation was postulated to proceed according to a cationic mechanism.

9.5 Polymerisation of Phosphacyclic Monomers

An interesting example of the coordination polymerisation of a phosphacyclic monomer is represented by the polymerisation of oxyphosphonoyloxytrimethylene (2-hydro-2-oxo-1,3,2-dioxaphosphorinane) in the presence of triisobutylaluminium as the catalyst [181,182]. This polymerisation, carried out at a temperature of 25 °C, produced, in relatively high yield, poly(oxyphosphonoyloxytrimethylene) which appeared to be a high molecular weight polymer (Table 9.2).

There is rather scant information in the literature that might indicate the mechanism operating in the polymerisation of oxyphosphonoyloxytrimethylene with triisobutylaluminium. It seems most likely, however, that the propagation step involves ring opening at the P–O bond of the coordinating monomer, leading to the aluminium alcoholate species, as illustrated schematically below:

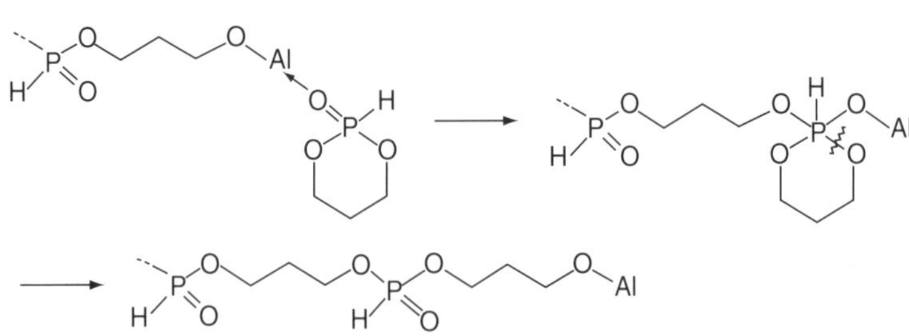

(23)

9.6 Copolymerisation of Heterocyclic Monomers

The multiplicity of species that may appear in polymerisation systems with coordination catalysts, together with side reactions often accompanying the propagation, make it difficult sometimes to evaluate the active sites accurately. However, varying the nature of the active site resulting from the cross-propagation, especially in alternating copolymerisation systems of comonomers differing from each other in their functional groups, should make it easier to gain insight into the mechanism of enchainment of individual monomers. The synthetic aspect of the copolymerisation must also reconsidered.

Since oxiranes are representative heterocyclic monomers containing an endocyclic heteroatom, and the most commonly polymerised of such monomers, they have been subjected to copolymerisations with heterocyclic monomers containing both an endocyclic and an exocyclic heteroatom. Coordination copolymerisations of heterocyclic monomers with different functions are focused on oxirane copolymerisation with cyclic dicarboxylic acid anhydride and cyclic carbonate. However, the statistical copolymerisation of heterocyclic monomers with an endocyclic heteroatom and monomers with both endocyclic and exocyclic heteroatoms have only a limited importance. Also, the block copolymerisation of oxirane with lactone or cyclic dicarboxylic acid anhydride is of interest both from the synthetic and from the mechanistic point of view. Block copolymerisation deserves special interest in terms of the exceptionally wide potential utility of block copolymers obtained from comonomers with various functions. It should be noted, however, that the variety of comonomers that might be subjected to a random, alternating and block polymerisation involving a nucleophilic attack on the coordinating monomer is rather small.

9.6.1 Copolymerisation of Oxiranes and Cyclic Acid Anhydrides

The copolymerisation of oxiranes (ethylene oxide, propylene oxide, epichlorohydrin) and cyclic acid anhydrides (phthalic anhydride, maleic anhydride) in the presence of coordination catalysts containing multinuclear species, such as triisobutylaluminium [183,184], diethylzinc [185–187], diethylzinc–methanol, acetic acid, acetylacetone or phenol [186,187], and in the presence of coordination catalysts containing mononuclear species, such as tetraphenylporphinatoaluminium chloride coupled with the quaternary onium salt (tpp)AlCl–$EtPh_3PBr$ [188,189], produces low molecular weight copolymers of an alternating distribution of comonomer units, i.e. polyesters or random copolymers containing an excess of oxirane units with respect to those of the anhydride [poly(ester–ether)s] (Table 9.2).

Although rather scant information concerning the structure of active sites in the above copolymerisation systems is available [183–189], cyclic acid anhydrides can be considered as coordinating to metal species via the carbonyl oxygen atom and reacting by nucleophilic attack of the metal substituent on the carbon atom of the coordinated carbonyl group [190,191]. Thus, the oxirane/cyclic acid anhydride copolymerisation pathway may be presented schematically as follows [82]:

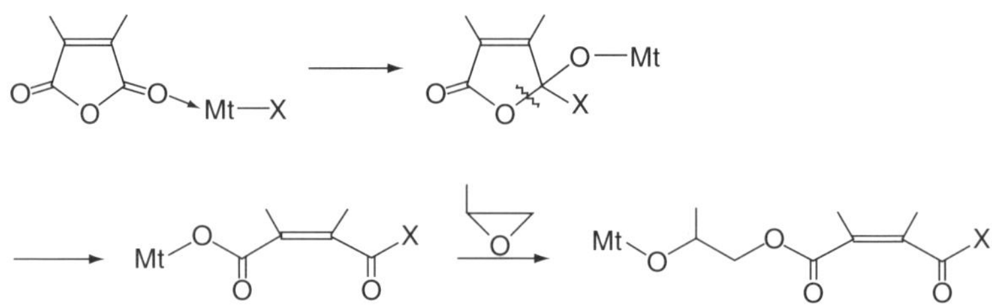

(24)

Concerning metal alkyl catalysts such as diethylzinc for the copolymerisation, it was established that they formed zinc carboxylate species owing to the reaction with maleic anhydride [192].

The tendency towards alternation in the oxirane/cyclic acid anhydride system with the zinc-based coordination catalyst is connected, according to literature data [193], with the stronger nucleophilic properties of the oxygen atom of the zinc alcoholate species and the weaker electrophilic properties of the zinc atom in such species compared with the respective properties of zinc carboxylate species. Studies of the copolymerisation of propylene oxide and maleic anhydride in the presence of catalysts such as diethylzinc–monoprotic compound (1:1) showed an increasing tendency towards alternation in systems with catalysts of decreasing electrophilicity of the zinc atom [186], which may corroborate the coordination mechanism proposed scheme (24).

Oxirane/cyclic acid anhydride alternating copolymers of controlled molecular weight with a narrow molecular weight distribution were found by Aida *et al.* [188,189] to be formed under mild conditions when copolymerising ethylene oxide and phthalic anhydride in the presence of the (tpp)AlCl–quaternary phosphonium salt catalyst system. The copolymerisation carried out with (tpp)AlCl alone proceeded very slowly, and the product was not poly(ethylene terephthalate) but contained ether linkages in considerable amount. The development of the living character and the tendency towards alternation of the copolymerisation when using the aluminium porphyrin catalyst, coupled with a quaternary salt, have been postulated [188,189] to be due to the formation of the six-coordinate aluminium porphyrin anion in which the reaction proceeds on both sides of the aluminium porphyrin plane. This has been presented schematically as follows [53]:

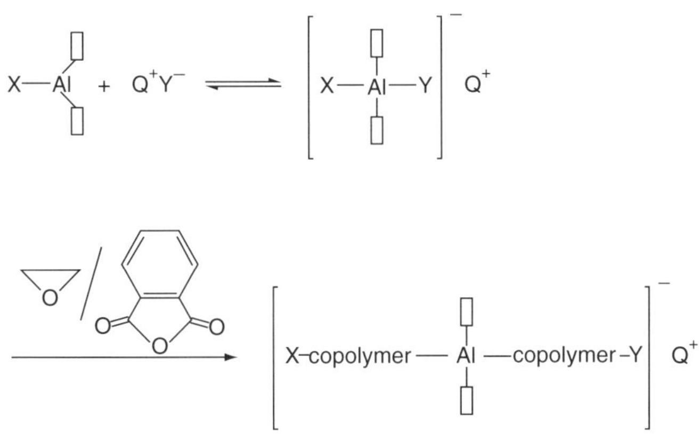

(25)

The polymerisation in ternary comonomer systems composed of ethylene oxide/phthalic anhydride/tetrahydrofuran with triisobutylaluminium [183,184] and propylene oxide/maleic anhydride/tetrahydrofuran with diethylzinc [187] yielded respective terpolymers of ca 1:1:1 composition and nearly perfect alternating structure. However, more detailed studies involving ^{1}H and ^{13}C NMR analyses of propylene oxide/maleic anhydride/tetrahydrofuran terpolymers and VPC analysis of glycols obtained from the hydrolysis of these terpolymers revealed the presence of oxypropylene–oxytetramethylene–oxypropylene and oxytetramethylene–oxypropylene–oxytetramethylene units as well as the predominant alternating terpolymer sequences [187]. The presence of ether linkages originating from tetrahydrofuran enchainment in the propylene oxide/maleic anhydride/tetrahydrofuran terpolymerisation indicates that zinc-based coordination catalysts can exert various effects on the coordinating monomer, involving the formation of a variety of polymerisation active sites that differ from one another. It seems that other coordination catalysts, including those containing a central aluminium atom, may behave similarly in terpolymerisation systems.

Taking the above into account, the oxirane/cyclic acid anhydride/tetrahydrofuran terpolymerisations should rather be considered to involve active sites of a cationic nature, such as those, shown in scheme (26), that are responsible for the formation of oxyalkylene sequences [82]:

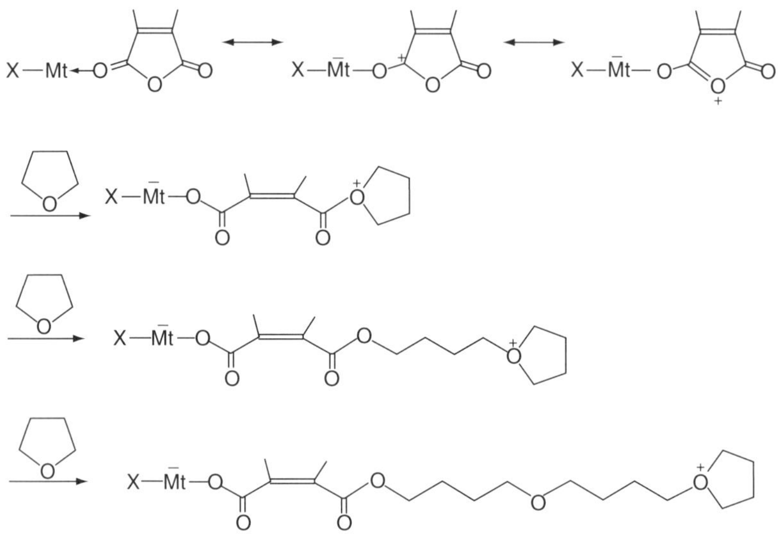

(26)

9.6.2 Copolymerisation of Oxiranes and Cyclic Carbonates

It has been found by Kuran and Listos [146] that hardly polymerisable cyclic carbonates with an unstrained, five-membered 1,3-dioxolan-2-one ring could easily undergo a copolymerisation with oxiranes in the presence of catalysts formed in the diethylzinc–phenol and/or polyhydric phenol systems. This copolymerisation yielded corresponding low molecular weight poly(ether–carbonate)s with prevailing contents of ether linkages (Table 9.2). A predominantly head-to-tail structure with almost equal proportions of isotactic and syndiotactic diads involving carbonate groups has been proved, by ^{13}C NMR spectroscopy, for copolymers of propylene oxide and propylene carbonate [146].

It seems that the initiation step of the copolymerisation most likely involves the oxirane reaction [according to scheme (3)]. Zinc alcoholate species formed in this reaction can easily propagate the copolymer chain, coordinating and enchaining both the oxirane [scheme (3)] and the cyclic carbonate [scheme (15)] comonomers. However, in the case of the cyclic carbonate, its enchainment may also proceed according to scheme (14), leading to decarboxylation. Thus, the obtained poly(ether–carbonate)s are characterised by a lower content of carbonate units with respect to the ether units [82,146].

Fairly good confirmation of the postulated mechanism of cyclic carbonate enchainment according to scheme (15) may be the obtaining of poly(oxypropylene-*co*-oxycarbonyloxypropylene) of predominant head-to-tail regioregularity from bispropylene spiroorthocarbonate with a zinc-based coordination catalyst [146]. It must be emphasised that the copolymerisation of propylene oxide and propylene carbonate, as well as bispropylene spiroorthocarbonate polymerisation with electrophilic initiators such as $Et_2O.BF_3$, were found to produce poly(oxypropylene-*co*-oxycarbonyloxypropylene)s of predominantly tail-to-tail regioregularity [143,194].

9.6.3 Block Copolymerisation of Oxiranes and Lactones or Cyclic Acid Anhydrides

Block copolymers characterised by different backbone structures of well-defined block lengths have been obtained from oxiranes and other heterocyclic monomers in the presence of catalysts that are effective at bringing about living polymerisations. Aida *et al.* [127,188,189,195,196] applied aluminium porphyrins and Teyssié *et al.* [125,197,198] applied bimetallic μ-oxoalkoxides for block copolymerisations in systems involving oxirane–lactone, oxirane–oxirane/cyclic acid anhydride, and oxirane/cyclic acid anhydride–lactone as block forming units and obtained respective polyether–polyester and polyester–polyester block copolymers. Such copolymers seem to be of exceptionally wide potential utility [53].

The growing species in living block copolymerisation systems may change when changing the comonomer to build the next block or they may retain their structure. For instance, the polymerisation of β-butyrolactone with an aluminium porphyrin catalyst such as (tpp)AlCl proceeds via aluminium carboxylate species [scheme (10)] which are converted to aluminium alcoholate species when the polymerisation of propylene epoxide is carried out from the living polyester as shown schematically below [195]:

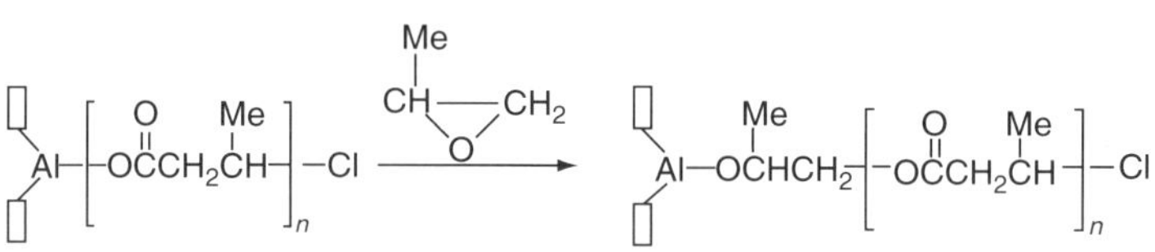

(27)

On the other hand, when propylene oxide polymerisation is carried out from the living polyester afforded by the polymerisation of β-butyrolactone with (tpp)AlOMe, the growing aluminium alcoholate species [scheme (9)] remain unchanged.

9.7 Copolymerisation of Heterocyclic and Heterounsaturated Monomers

Copolymerisations of heterocyclic and heterounsaturated monomers generally lead to random copolymers with a prevailing content of the heterocyclic monomer and to alternating or nearly alternating copolymers, depending on the kind of comonomer and catalyst (Tables 9.4 and 9.5). The copolymerisation of oxiranes and carbon dioxide is the most representative example of such copolymerisations. The chemistry of carbon dioxide has received much attention during recent decades in terms of polymer synthesis, since carbon dioxide is easily available and a practically unfailing resource.

9.7.1 Copolymerisation of Oxiranes and Carbon Dioxide

The copolymerisation of propylene oxide and carbon dioxide was carried out for the first time in the presence of the diethylzinc–water (1:1) catalyst by Inoue *et al.* [199,200] who found it to yield a high molecular weight copolymer of almost completely alternating monomer distribution, poly(alkylene carbonate) (Table 9.4). Since this discovery, more effective zinc-based coordination catalysts containing multinuclear species for oxirane/carbon dioxide copolymerisation have been found and developed (Table 9.5). According to later studies, binary and ternary catalyst systems, containing a diprotic (or multiprotic) compound as the partner component with diethylzinc, formed effective catalysts for propylene oxide/carbon dioxide alternating copolymerisation. Water [137,201,202], primary amine [201], dihydric phenol [203–207], trihydric phenol [208–211], aminophenol [208] or other aromatic derivatives such as hydroxycarboxylic acid [212], dicarboxylic acid [212], diamine [213], dithiol [213] and other related protic compounds, as well as polymeric supports with labile hydrogen atoms such as poly(*p*-hydroxystyrene) [214,215] poly(styrene-*co*-acrylic acid) [216,217] or γ-alumina [76,218,219], were used to form with diethylzinc effective catalysts for propylene oxide/carbon dioxide alternating copolymerisation. Also, other organometallic compounds such as diethylmagnesium [220] and diethylcadmium [209] were applied instead of diethylzinc to form, with protic compounds such as water or resorcinol and pyrogallol, effective catalysts for this copolymerisation. Inorganic zinc compounds such as zinc oxide and zinc hydroxide combined with dicarboxylic acid appeared to operate as effective catalysts, again yielding high molecular weight poly(propylene carbonate) from the propylene oxide/carbon dioxide copolymerisation [221–224].

Catalysts derived from reaction systems such as triethylaluminium–water and triethylaluminium–water–acetylacetone [225], triethylaluminium–triphenylphosphine [226], triethylaluminium–pyrogallol [209] and rare-earth metal phosphonate– triisobutylaluminium–glycerol [227] appeared to be effective in the copolymerisation of propylene oxide and carbon dioxide, yielding high molecular weight poly(propylene ether–carbonate)s (Table 9.4) but not the respective alternating copolymer which is poly(propylene carbonate).

Although propylene oxide has been the oxirane most widely studied in copolymerisation with carbon dioxide, there are a variety of other oxiranes capable of coordination copolymerisation with carbon dioxide (Table 9.4).

Recently, Shen *et al.* [228] reported on the copolymerisation of epichlorohydrin and carbon dioxide, yielding high molecular weight poly(ether–carbonate) in the presence of a catalyst formed in the rare-earth metal (e.g. yttrium) phosphonate–triisobutylaluminium (1:8) system. The alkylation of yttrium species by triisobutylaluminium in this catalyst system has been found to occur. On the basis of ^{31}P NMR studies of this system, the active species were assumed to involve oxygen-bridged yttrium and aluminium atoms. Taking into account that chain terminals of the copolymer formed were *i*-BuC(O)O groups, the catalyst active species were postulated to contain metal–carbon bonds and to react with carbon dioxide in the initiation step [228]. This agrees with conclusions from studies of reactions of methylaluminium compounds with propylene oxide which showed them to proceed with difficulty [72].

Among other oxirane monomers, cyclohexene oxide deserves attention, since its alternating copolymer with carbon dioxide gives a *threo*-1,2-cyclohexanediol upon alkaline hydrolysis, indicating inversion of the configuration at the carbon atom of the cycloxene oxide (*cis*-epoxide) ring where cleaved [207,229]. The stereochemistry of cyclohexene oxide/carbon dioxide alternating copolymerisation is shown schematically below [53]:

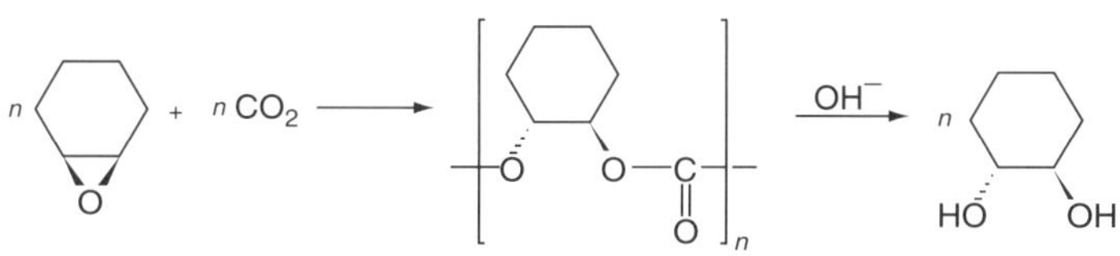

(28)

Another interesting monomer for copolymerisation with carbon dioxide is isomeric 2-butene oxide. In copolymerisation in a ternary comonomer system consisting of 2-butene oxide, 1-butene oxide and carbon dioxide with the diethylzinc–water catalyst, *cis*-2-butene oxide was incorporated in the copolymer, while *trans*-2-butene oxide hardly underwent an enchainment [230]. Thus, the smaller steric hindrance for the *cis*-isomer than for the *trans*-isomer throughout the coordination copolymerisation with carbon dioxide is to be taken into account.

Copolymerisation of propylene oxide as well as other oxiranes with carbon dioxide in the presence of zinc-based coordination catalysts is generally accompanied with the formation of a cyclic five-membered carbonate, propylene carbonate or another alkylene carbonate [147,206,207,210,212,230]. The alkylene carbonate, however, is not the precursor for poly(alkylene carbonate), since it hardly undergoes a polymerisation under the given conditions [142–146].

Studies of the regioselectivity of oxirane/carbon dioxide copolymerisation showed the polar effect exerted by the ring substituent, but not the bulkiness, to be the factor determining the direction of ring opening [231,232]. In the case of propylene oxide/carbon dioxide copolymerisation, C_β–O bond cleavage predominates, which was shown by applying optically active propylene oxide as the comonomer for copolymerisation with carbon dioxide [137,233] and by ^{1}H and ^{13}C NMR studies of the poly(propylene carbonate) formed [199,200,233, 234]. However, more detailed ^{13}C NMR investigations on propylene oxide/carbon dioxide copolymers derived from copolymerisation with various zinc-based coordination catalysts showed them to contain small amounts of ether linkages and inverted propylene oxide units [235].

A concerted mechanism of oxirane/carbon dioxide copolymerisation with catalysts formed in diethylzinc–dihydric phenol systems and related systems has been proposed [76,206,207]. According to this mechanism, the oxirane coordination and enchainment occur as shown by the following scheme:

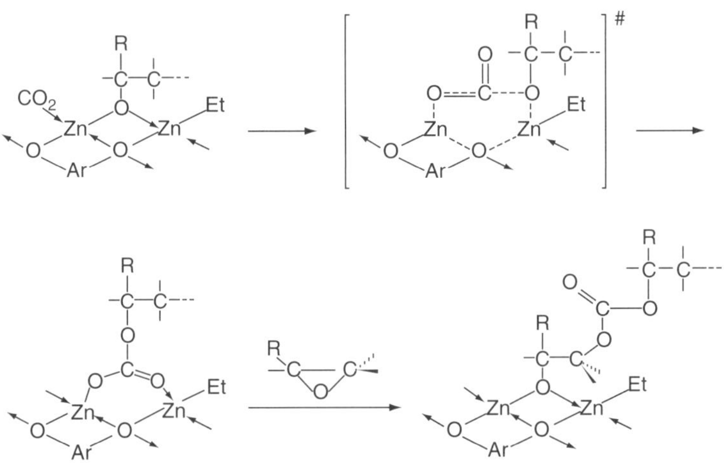

(29)

The mode of nucleophilic attack postulated in scheme (29) results in inversion of the configuration at the carbon atom of the oxirane ring where it is cleaved, which agrees with the experimental data. Both the nucleophilicity of the attacking oxygen atom and the electrophilicity of the zinc atom bound covalently with this oxygen atom fluctuate in a way conducive to the cross-propagation of comonomers in oxirane/carbon dioxide copolymerisation. The stronger nucleophilicity of the oxygen atom in the zinc alcoholate species and the weaker electrophilicity of the zinc atom in such species [scheme (3)] compared with the respective properties of zinc carbonate species [scheme (27)] result in an alternation throughout the copolymerisation.

The formation of a cyclic carbonate, e.g. propylene carbonate, accompanying the copolymerisation has been explained in terms of the backbiting reaction involving zinc alcoholate species [206,207]. This has been confirmed recently [147] by degrading poly(propylene carbonate) by using catalysts with zinc phenolate species. The degraded copolymer thus obtained was terminated in its chains with zinc alcoholate species and phenylcarbonate groups. The course of the degradation, involving zinc orthocarbonate species, may be presented schematically as follows (participation of the adjacent zinc atom in the active site is omitted for the sake of clarity):

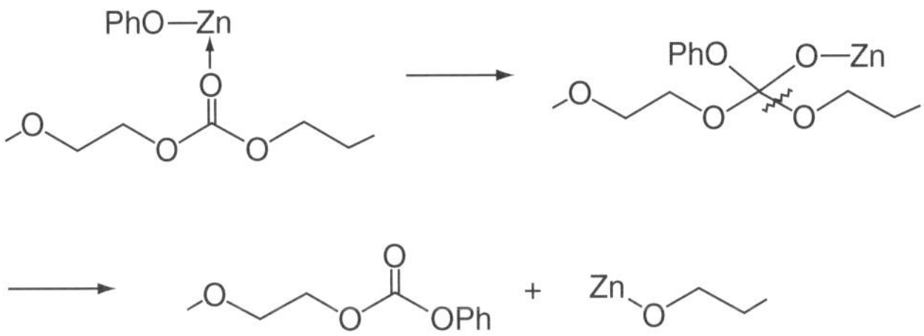

(30)

On the other hand, poly(propylene carbonate) degradation run in the presence of catalysts containing zinc alcoholate species resulted in the formation of degraded copolymer chains terminated with zinc carbonate species [scheme (31)] [147]:

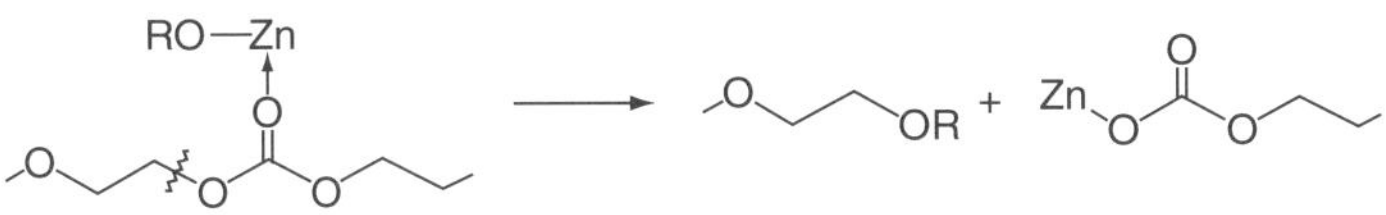

(31)

Zinc carbonate species appeared to undergo slow decarboxylation [scheme (14)] and then, as zinc alcoholate species, were involved in chain depolymerisation via backbiting reaction according to the following scheme: [147]:

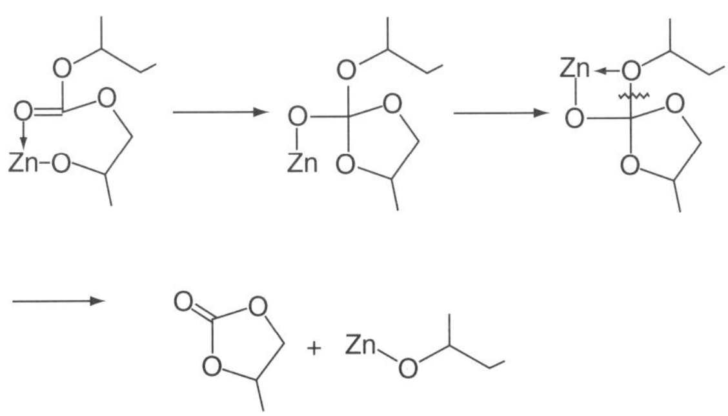

(32)

It may be mentioned that the use of ionic nucleophilic initiators, instead of zinc-based coordination catalysts, in order to promote propylene oxide/carbon dioxide copolymerisation, did not result in the formation of any copolymer but led to the cyclic carbonate, propylene carbonate [194,236,237]. Also, zinc-based coordination catalysts with non-condensed zinc atoms in their molecules (formed by the reaction of diethylzinc with a monoprotic compound such as phenol or alcohol) appeared to promote the cyclisation of propylene oxide and carbon dioxide and, apart from propylene carbonate, no copolymer was formed [236,237]. Considering this, it has been suggested [207] that, in systems with catalysts such as diethylzinc–phenol or alcohol, coordination of the carbonate moiety, followed by its subsequent reaction leading to propylene carbonate [scheme (32)], occurred much more readily than in the propylene oxide/carbon dioxide copolymerisation system with catalysts having condensed zinc atoms, such as diethylzinc–catechol [206] or pyrogallol [210]. However, as copolymerisation progresses [scheme (30)], the carbon dioxide concentration becomes lower and the cyclisation [scheme (31)] can occur to a significant extent in some cases [206]. Considering the course of the backbiting reaction [scheme (32)], when the growing polycarbonate chain coordinates at the active site, it seems that the more electrophilic central zinc atom of this site cannot attain more than four ligands around it and remains in the tetrahedral configuration. However, when the same catalyst operates with propylene oxide as the sole monomer in the homopolymerisation system, the backbiting reaction that might produce the propylene oxide cyclic dimer does not take place, but the coordination of the growing poly(propylene oxide) chain at the central zinc atom leads to the attainment of an octahedral configuration by this atom, which results in an enhancement of the catalyst stereoselectivity [65].

The difference between the reactivity of condensed and non-condensed zinc species can be attributed to steric as well as electronic factors. Non-condensed zinc species seem to be less sterically hindered for coordination of the carbonyl oxygen atom from the carbonate moiety, which is bulkier than the carbon dioxide molecule. On the other hand, condensed zinc species seem to be more sterically crowded for coordination of the carbonate moiety of the growing polymer chain, despite the fact that the carbonyl oxygen atom of the carbonate moiety is more nucleophilic than that in carbon dioxide [207]. Validation of the above considerations is provided by a recent report by Darensbourg *et al.* [238], which shows that a catalyst containing non-condensed metal species but with bulky substituents, such as zinc di(2,6-diphenylphenoxide) dietherate or zinc di(2,4,6-tri-*t*-butylphenoxide) dietherate, is capable of promoting oxirane/carbon dioxide copolymerisation. The copolymerisation of cyclohexene oxide and carbon dioxide in the presence of the above-mentioned zinc diphenoxides with bulky substituents yielded poly(cyclohexene ether–carbonate) which contained more than 90-mol.%-carbonate units, whereas the propylene oxide/carbon dioxide reaction with these catalysts yielded predominantly the cyclic carbonate, propylene carbonate. Hence, it seems that the copolymer formation may be explained in terms of greater steric hindrances for the backbiting reaction in the case of cyclohexene oxide/carbon dioxide copolymerisation in comparison with the propylene oxide/carbon dioxide reaction.

At the end of considerations dealing with oxirane/carbon dioxide copolymerisation, copolymerisation run under supercritical conditions in carbon dioxide as the reaction medium should be mentioned [239].

It may be of interest that no essential differences between zinc-based catalysts containing non-condensed and condensed species were observed when using these catalysts for promoting the copolymerisation of oxiranes with other than carbon dioxide comonomers. This concerns, for instance, oxirane/cyclic acid anhydride and epoxide cyclic carbonate systems [82] as well as the aziridine/carbon dioxide system [240]. However, in the latter system, with an aziridine such as 2-methylaziridine or 1-phenethylaziridine and a catalyst such as diethylzinc, ethylzinc phenoxide, diethylzinc–pyrogallol or diethylzinc-*o*-aminophenol, aziridine/carbon dioxide alternating copolymerisation leading to low molecular weight polyurethanes has been postulated to proceed according to a different mechanism to that operating in the epoxide/carbon dioxide copolymerisation [240].

The copolymerisation of oxirane and carbon dioxide has also been carried out in the presence of catalysts with mononuclear species such as porphinatoaluminium [241] and calixarenediolatoaluminium [41] catalysts. Of particular interest from a synthetic point of view is the first successful example for control of the molecular weight of the copolymer. Aida *et al.* [241] found that the copolymerisation of cyclohexene oxide and carbon dioxide with the metalloporphyrin–onium salt combination as a catalyst, (tpp) AlCl–$EtPh_3PBr$ (1:1), yield an alternating copolymer, poly(cyclohexene carbonate), of very narrow molecular weight distribution ($M_w/M_n = 1.06$). In the absence of the quaternary onium salt, the copolymerisation still produced a copolymer of narrow molecular weight distribution, but the copolymer was not an alternating one and the content of carbon dioxide units was low [242]. We are reminded that the molecular weight of oxirane homopolymers and copolymers yielded by zinc- or aluminium-based coordination catalysts with multinuclear species is generally high, but the distribution of molecular weight is very broad [30,36,52,53,136].

The mechanism of living copolymerisation in the presence of aluminium porphyrin coupled with quaternary onium salt was proposed to involve activation of the anion of the onium salt as the nucleophile by the metalloporphyrin; the structure of the active species was found to be six-coordinate aluminium prophyrin carrying one reactive axial ligand on both sides of its square planar AlN_4 skeleton [188]:

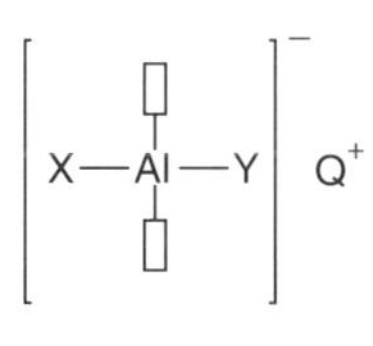

Considerations about the possible mechanism of copolymerisation with the (tpp) AlCl–$EtPh_3PBr$ catalyst should include the fact that quaternary onium salts themselves promote the cyclisation of epoxide and carbon dioxide [243].

Thus, the possibility of the nucleophilic attack of the anion from the quaternary onium salt on the oxirane coordinated at the aluminium atom should also be taken into account:

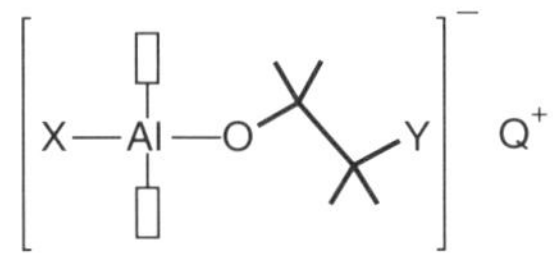

As regards oxirane/carbon dioxide copolymerisation with (dmca) AlCl, it yields low molecular weight propylene oxide/carbon dioxide copolymers with a prevailing content of ether linkages as well as cyclohexene oxide/carbon dioxide copolymers of predominantly carbonate linkages.

It is interesting that the reaction in the oxirane–carbon dioxide–tetrahydrofuran system with (dmca)AlCl produced the respective terpolymer, especially when cyclohexene oxide was used as the oxirane. The cyclohexene oxide–tetrahydrofuran–carbon dioxide terpolymer appeared to contain up to 29 mol.-% carbon dioxide and 20 mol.-% tetrahydrofuran units [244].

Taking the above into account, oxirane–carbon dioxide–tetrahydrofuran terpolymerisation should be considered to involve active sites that change in nature as the reaction progresses an account of cross-propagation, like those appearing and disappearing in schemes (6) and (7) [41,74,244].

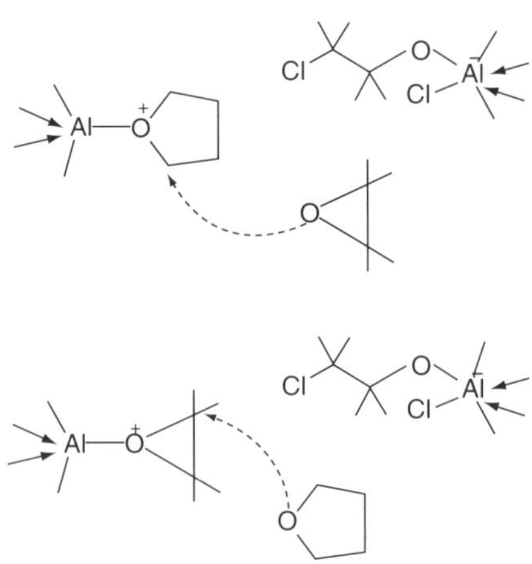

It is interesting, however, to compare the above mechanisms in terms of the different polymerisation behaviour observed for systems with catalysts containing multinuclear species and mononuclear species. In the case of polymerisations with catalysts possessing mononuclear species, two catalyst molecules must approach one another to form the active site for the propagation [scheme (7)]. Such a requirement is not needed for propagation involving the active sites of multinuclear catalysts, in which the more electrophilic metal atom (i.e. that coordinating the monomer) and the adjacent metal atom (i.e. the nucleophilic attack carrier) are already fixed to each other [scheme (4a)]. Thus, it seems that the difference in behaviour of polymerisation systems with these catalysts lies, perhaps, among other factors, in diffusion control of the polymerisation. As polymerisation progresses, apart from the monomer diffusion which concerns both cases, in systems with mononuclear catalysts the additional diffusion factor connected with the need for cooperation of two catalyst molecules bearing the growing polymer chains of lowering mobility may become important. The use of a cocatalyst, which serves only for coordination of the monomer and as the nucleophilic attack carrier, but cannot permanently anchor the growing polymer chain at its own metal atom, and thus still remains mobile, liberates the polymerisation from the diffusion control connected with the need for the approach of two metal atoms that could form an active site. In such a system, high-speed polymerisation can proceed.

The possibility of changing the polymerisation mechanism depending on the kind of monomer as well as the catalyst, especially in binary or ternary comonomer systems, is obvious. This may also concern change in the nature of the growing species throughout the propagation of one polymer chain in the presence of both multinuclear catalysts [scheme (33)] and mononuclear catalysts [scheme (34)]:

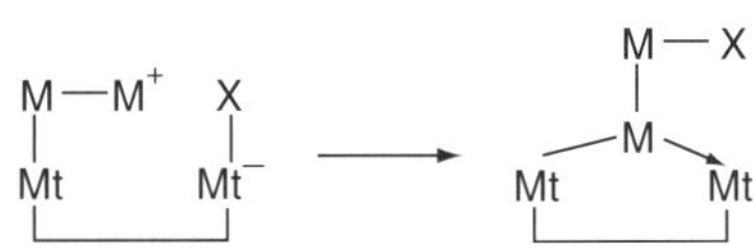

(33)

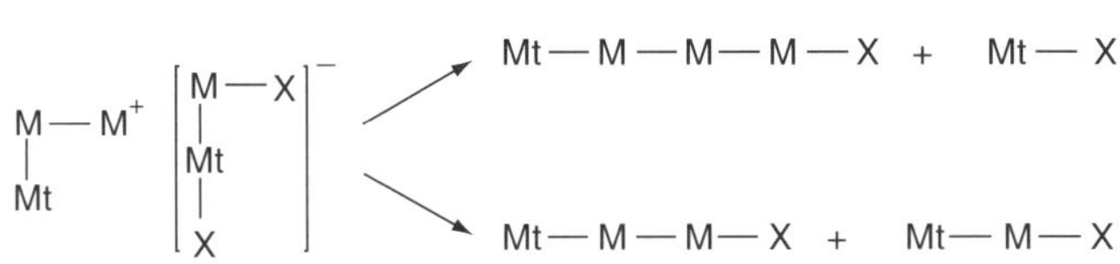

(34)

9.7.2 *Copolymerisation of Oxetanes and Carbon Dioxide*

Oxetane (oxacyclobutane) has been copolymerised successfully with carbon dioxide to give poly(trimethylene ether–carbonate) in the presence of the triethylaluminium–water–acetylacetone (2:1:1) catalyst (Table 9.4) [scheme (35)]. The carbon dioxide content in the copolymer produced was ca 20 mol.-%. Attempts to carry out oxetane/carbon dioxide copolymerisation with the diethylzinc–water (1:1) catalyst failed to give any copolymer [245].

However, organotin-based catalyst systems such as Bu_2SnI_2 or $Bu_3SnI–PBu_3$, PPh_3 or NEt_3 have been found to promote the alternating copolymerisation of oxetane and carbon dioxide to yield poly(trimethylene carbonate):

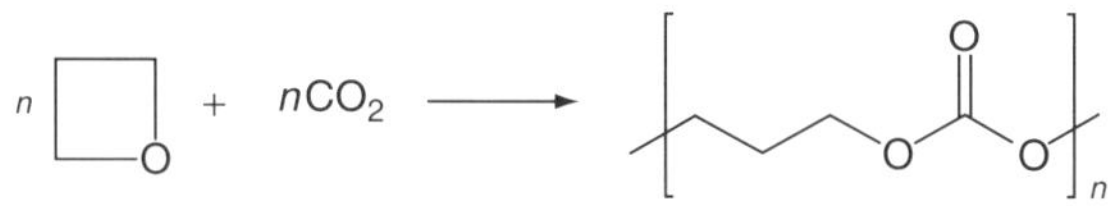

(35)

It may be of interest that, in the presence of the $Bu_2SnO(CH_2)_3I–PPh_3$ catalyst system, only the cyclic carbonate, trimethylene carbonate, was formed [151].

9.7.3 Copolymerisation of Tiiranes and Carbon Dioxide

Propylene sulphide was found [246] to undergo copolymerisation with carbon dioxide in the presence of the triethylaluminium–pyrogallol (2:1) catalyst, yielding copolymers, poly(propylene thioether–monothiocarbonate)s, containing propylene thiocarbonate units in an amount reaching 42 mol.-%, apart from the predominating propylene sulphide units (Table 9.4):

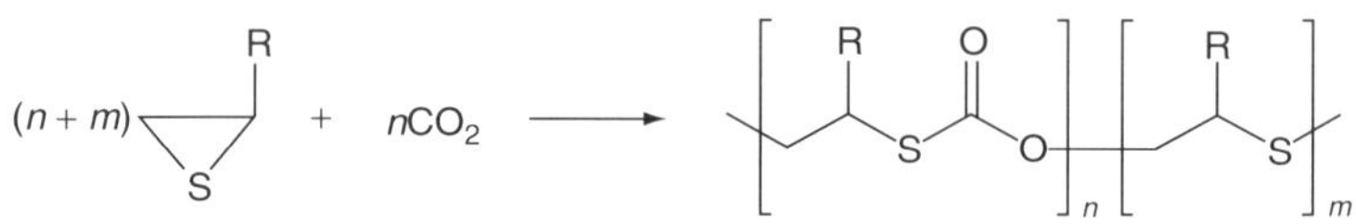

(36)

Attempts to use the diethylzinc–pyrogallol (2:1) catalyst to copolymerise propylene sulphide and carbon dioxide failed, since the content of propylene thiocarbonate units in the copolymers formed was small and did not exceed 10 mol.-%. It has also been observed that the presence of carbon dioxide in this copolymerisation system causes a lowering of the molecular weight and yield of the copolymer formed. Thus, it has been suggested that propylene sulphide homopolymerisation was favoured over cross-propagation with carbon dioxide in the presence of a zinc-based coordination catalyst because of higher HSAB symmetry of the system in the former case. The zinc atom in the Zn–S unit of the catalyst is a rather soft acid and will prefer reaction with a soft base such as propylene sulphide rather than with hard carbon dioxide. The presence of a hard acid centre in the triethylaluminium-based catalyst should result in an increase in the affinity of the catalyst towards carbon dioxide [247].

9.7.4 Copolymerisation of Oxiranes and Carbon Disulphide

The copolymerisation of propylene oxide and carbon disulphide was carried out with a catalyst consisting of diethylzinc and an electron donor, such as tertiary amine, tertiary phosphine or hexamethylphosphoric triamide, in equimolar amounts [248]. The low molecular weight copolymers obtained, poly(propylene ether–dithiocarbonate)s (Table 9.4), were characterised by a comonomer unit ratio of 0.5–0.7. Among the randomly distributed comonomer units, the copolymer also contained thiocarbonate carbonyl units which were formed during the copolymerisation by an exchange reaction between the propylene oxide oxygen atom and carbon disulphide sulphur atom. The copolymerisation was found to be accompanied with the formation of the cyclic propylene dithiocarbonate, 1,3-oxathioran-4-methyl-2-thione, but the latter proved not to be an intermediate in the copolymerisation:

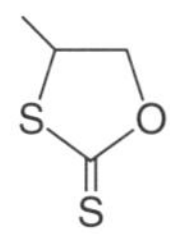

It is worth noting that catalysts formed in systems such as diethylzinc–water, alcohol and primary or secondary amine, exhibiting high activity for the homopolymerisation of propylene oxide, were not effective for the copolymerisation of propylene oxide and carbon disulphide [248].

9.7.5 Copolymerisation of Tiiranes and Carbon Disulphide

Ethylene sulphide and propylene sulphide have been reported by Soga *et al.* [249,250] to copolymerise with carbon disulphide in the presence of catalysts such as diethylzinc, diethylcadmium and mercury bis(*n*-butanethiolate), yielding poly(alkylene thioether–trithiocarbonate) copolymers (Table 9.4). The content of ethylene trithiocarbonate units in the ethylene sulphide/carbon disulphide copolymer obtained with the most efficient catalyst, mercury bis(*n*-butanethiolate), was in the range 50–70 mol.-% [249].

In order to clarify the mechanism, the reaction of carbon disulphide with mercury bis(*n*-butanethiolate) was studied. On the basis of results obtained, it was suggested that this reaction involved the formation of a coordination complex, followed by the formation of active species containing the Hg-SC(S) bond. Moreover, the cyclic trithiocarbonate, ethylene trithiocarbonate, found to be present in trace amounts in copolymerisation products, was excluded as a possible intermediate for the copolymer formation, since it did not undergo any polymerisation under the given conditions [249].

9.7.6 Copolymerisation of Oxiranes and Sulphur Dioxide

The copolymerisation of propylene oxide and sulphur dioxide in the presence of diethylzinc has been found [251] to produce poly(propylene ether–sulphite) (Table 9.4). However, the content of propylene sulphite units in the resulting copolymer appeared to be rather small. A triethylaluminium-based catalyst was more efficient and the oxirane/sulphur dioxide copolymer contained a considerable amount of propylene sulphite units. It was revealed that oxiranes very readily underwent an enchainment with aluminium alcoholate species as well as aluminium sulphinate species [252,253], whereas they hardly underwent the reaction with alkylaluminium (Al–C) species [72]. The aluminium sulphinate copolymer chain terminals were suggested to form dimeric species [252,253]:

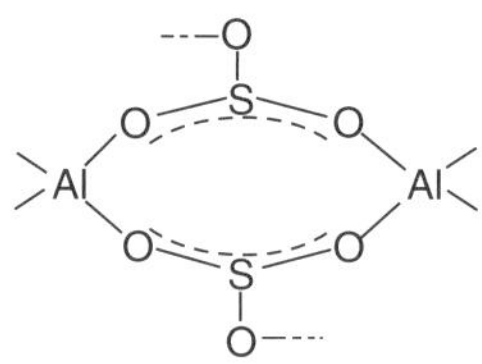

9.8 Polymerisation of Heterounsaturated Monomers

Heterounsaturated monomers that undergo coordination polymerisation or copolymerisation with other monomers can be divided into two classes: monomers with a carbene-like structure such as isocyanides and carbon monoxide which are coordinated by π complex formation with the transition metal atom at the catalyst active site, and monomers such as isocyanates, aldehydes, ketones and ketenes which are coordinated via 5-bond formation with the metal atom at the catalyst active site.

9.8.1 Polymerisation of Isocyanides

Polymerisation of isocyanides, R–N$\equiv$C, leads to polyisocyanides [poly(iminomethylene)s, poly(carboimidoyl)s] [254]:

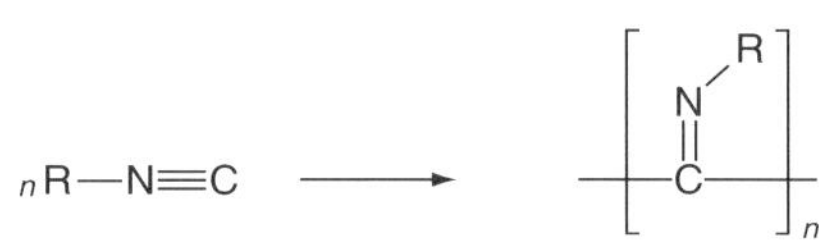

(37)

Polyisocyanides have a rigid helical structure with four monomeric units per helix turn [254,255]. The polymers of isocyanides do not display any stereoisomerism, since the carbon atoms in the polymer backbone are not chirotropic. However, it is worth noting that isocyanide polymerisation is stereoselective as regards the screw sense of the polymeric helix (atropisomerism). Therefore, right- and left-handed helices are formed from achiral isocyanides [256]. When the monomer is achiral, a racemic mixture of right- and left-handed helices is obtained, e.g. in the case of the polymerisation of *t*-butylisocyanide [257,258]. When the alkyl substituent R at the nitrogen atom in the monomer, R–N≡C, is bulky (like *t*-butyl and higher branched alkyls), polymer helices do not interconvert at room temperature [257] and represent a very interesting model for the investigation of the solution properties of conformationally rigid macromolecules. The polymerisation of chiral isocyanides, R*–N≡C, in an enantiomerically pure from yields preferentially right-handed or left-handed helices, which demonstrates the stereoselectivity of the polymerisation with respect to the screw sense [256].

The coordination polymerisation of isocyanides proceeds in the presence of various catalysts. Isocyanides can be rapidly polymerised by transition metal-based catalysts [255,259–261]. Some coordination complexes of Ni(II), such as NiX_2 (where X denotes a very weak nucleophilic anion), are of practically use for obtaining polymers of medium molecular weight. Polymers of lower molecular weight are formed using Ni(0) complexes with a ligand such as PPh_3, CO or RNC [254]. Generally, the average molecular weight of polyisocyanides lies in the range 4×10^4–1.5×10^5. The polymerisation of isocyanides involves an insertion of the monomer into a transition metal–carbon single bond [261–263].

9.8.2 Polymerisation of Isocyanates

Coordination polymerisations of alkyl isocyanates have not been widely studied, since these monomers could be polymerised via their C=N bond by using anionic initiators. However, such anionic polymerisations require low-temperature conditions [264]. It has been found recently [265] that alkyl isocyanates are capable of polymerisation in the presence of coordination catalysts at ambient temperature. By contrast, phenyl isocyanate appeared capable of coordination copolymerisation with oxirane [266].

9.8.2.1 Polymerisation of Alkyl Isocyanates

Polymerisation of *n*-butyl isocyanate was carried out with the Pruitt–Baggett adduct $Re(OR)_2Cl$ [267] at ambient temperature, yielding polyamide, a polymer of the nylon-1 type (Table 9.3) [268]. For more detailed mechanistic studies, *n*-butyl isocyanate was also subjected to polymerisation with a catalyst derived from the reaction of the Pruitt–Baggett adduct with *β*-naphthyl-*N*-(*n*-butyl) urethane (for the sake of clarity, the catalyst association and internal complexation of the Fe atom by etheral O atoms of the OR substituent are omitted in formulae):

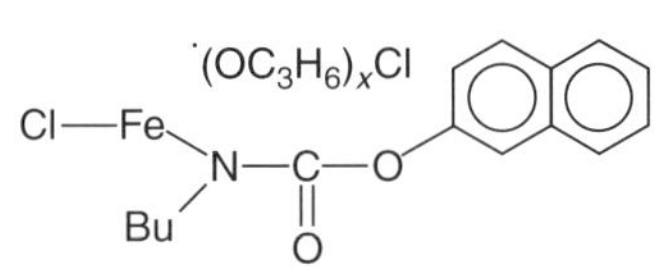

The presence of the naphthyl group of the naphthyl urethane-terminated chains in the polymer obtained with the catalyst containing a naphthyl urethane group was detected by means of UV spectroscopy. Other results obtained indicate the involvement of the monomer nitrogen atom in coordination at the catalyst iron atom. Thus, the coordinating monomer then undergoes an enchainment to yield a urethane function, which may be shown schematically as follows [268]:

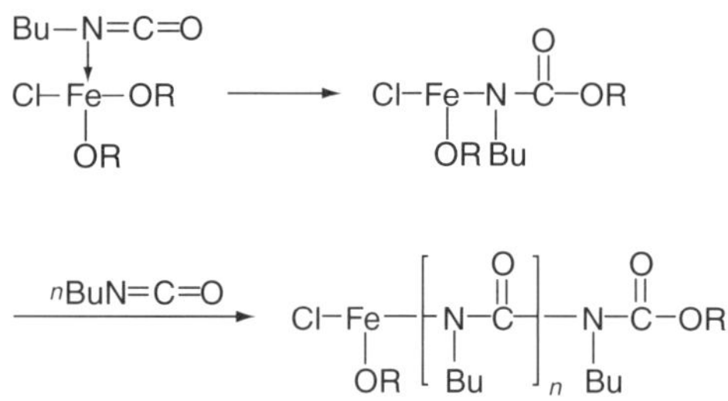

(38)

Polymerisation of n-butyl isocyanate was accompanied with cyclic trimer formation, which has been explained in terms of the backbiting reaction [268].

It should be noted that, in the presence of zinc-based coordination catalysts of the EtZnX type, the cyclotrimerisation of aliphatic as well as aromatic isocyanates occurred exclusively [269].

9.8.2.2 Copolymerisation of Phenyl Isocyanate and Oxirane

The copolymerisation of ethylene oxide and phenyl isocyanate has been found [266] to proceed in the presence of the triethylaluminium–water (2:1) catalyst, although phenyl isocyanate alone could not be polymerised by the same catalyst. The copolymer formed was characterised by an alternating comonomer distribution [scheme (39)] and contained acetalic units in its chains (Table 9.4):

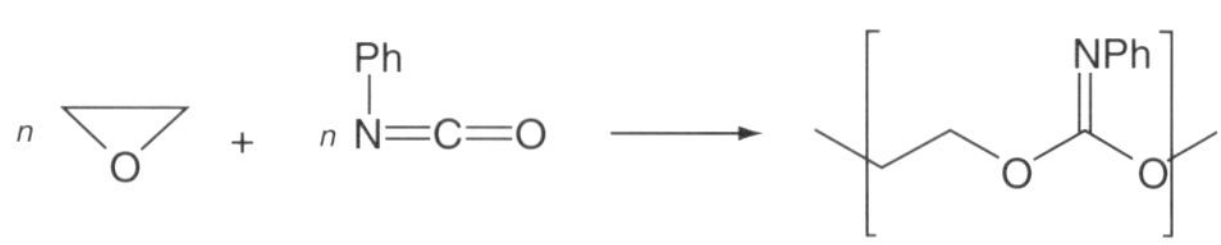

(39)

The structure of the acetalic copolymer indicates that the carbonyl group of phenyl isocyanate is involved in the copolymerisation. This is connected with the presence of the phenyl substituent at the isocyanate nitrogen atom. Thus, alternating coordinations of comonomers via oxygen atoms of the isocyanate carbonyl group and the oxirane, followed by coordinating comonomer enchainments, have been postulated to take place throughout the copolymerisation [266].

9.8.3 Polymerisation of Carbonyl Monomers

The coordination polymerisation of carbonyl monomers by their carbonyl group concerns mostly acetaldehyde, trichloroacetaldehyde, propionaldehyde and butyraldehyde. One basic problem with all polyaldehydes, and especially with polyketones, is not the polymerisation itself but the stabilization of resulting polymers (or copolymers) against thermal degradation.

The simplest carbonyl monomer, i.e. formaldehyde, can be polymerised to a high molecular weight polymer of acetalic structure by ionic initiators. Higher aldehydes, upon low-temperature polymerisation with a range of cationic or anionic initiators, yield isotactic crystalline polymers. It was Vogl [270–273] who correlated the polymerisation of formaldehyde, the higher aliphatic aldehydes and trihaloacetaldehydes within the framework of a comprehensive understanding of the field of polymer science. Extensive progress in the field of trihaloacetaldehyde polymerisation has led to the development of cryotachensic polymerisation, the concept of optical activity based on macromolecular asymmetry and ultimately the recognition of the importance of the study of stereoregular oligomers [273].

9.8.3.1 Polymerisation of Aldehydes

Furukawa *et al.* [274] and Natta *et al.* [275,276] succeeded independently in the preparation of crystalline polyacetaldehyde by using some organometallic compounds, such as diethylzinc or triethylaluminium, for the low-temperature polymerisation of acetaldehyde. Metal alkyls and metal alkoxides, e.g. aluminium isopropoxide, zinc ethoxide or ethyl orthotitanate, have also polymerised other aldehydes such as propionaldehyde and trichloroacetaldehyde to give crystalline polymers (Table 9.3) [270,275,277]. A highly crystalline isotactic polymer has been obtained from the polymerisation of *n*-butyraldehyde with triethylaluminium or titanium tetrachloride–triethylaluminium (1:3) catalysts. Combinations of metal alkyl, e.g. diethylzinc, with water [278] or amine [279] appeared to give very efficient catalysts for aldehyde polymerisations.

The formation of a complex between the carbonyl oxygen atom of the aldehyde monomer and the catalyst metal atom, followed by its rearrangement to the respective metal alkoxide of secondary alcohol, has been suggested to be the first polymerisation step [274,277,280,281]. In the case of alumina as a catalyst, the aldehyde coordination was confirmed by IR spectroscopy [282].

On the basis of studies of the reactions of formaldehyde, acetaldehyde and butyraldehyde with ethylzinc diphenylamide as a model catalyst of moderate effectiveness, Van der Kerk [279] revealed the mechanism of aldehyde coordination polymerisation by means of organozinc catalysts:

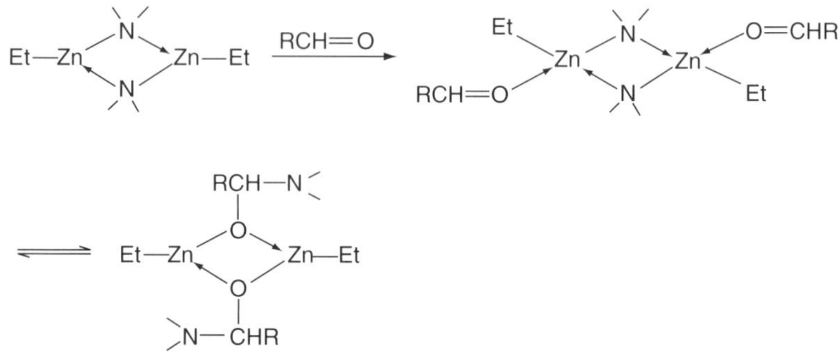

(40)

Both the dimeric species, containing Zn–N bonds with coordinated aldehyde and Zn–O bonds with enchained aldehyde, are in equilibrium with monomeric species as well. The coordinatively unsaturated dimeric species with an initiated polymer chain may be able to coordinate aldehyde, but above the ceiling temperature no enchainment of the coordinating carbonyl group occurs. Below the ceiling temperature, a fast propagation of the coordinated monomer becomes operative, which leads to highly stereoregular polyaldehyde (isotactic; no syndiotactic polyaldehyde has ever been obtained). The catalyst stereospecificity may be the result of stereocontrol of both the monomer coordination and the coordinated monomer incorporation into a growing polymer chain [279].

9.8.3.2 Polymerisation of Ketones

There are a few reports dealing with the coordination polymerisation of acetone in the presence of organometallic catalysts. Such a situation results from long-term instability of acetone polymers [283]. The polymerisation of acetone with metal alkyls such as triethylaluminium as catalysts has been reported [284,285] to produce polymers containing acetalic units (Table 9.3), which was confirmed by IR spectroscopy.

However, in order to obtain polymers that might be relatively stable, polymerisations were carried out in two-component comonomer systems such as acetone/aldehyde or acetone/epoxide. Acetone copolymers, obtained in the presence of organometallic catalysts in this way, exhibited high crystallinity [284].

9.8.4 Polymerisation of Ketenes

9.8.4.1 Polymerisation of Dialkylketenes

Dimethylketene was subjected to polymerisations catalysed with various organometallic compounds. It was found to form highly crystalline polymers, the structure of which depended on the kind of organometallic catalyst used.

Polyesters were yielded [scheme (41)] by polymerisations run with diethylzinc as the catalyst [281,286,287], but polyketones were obtained [scheme (42)] when dimethylketene was polymerised with triethylaluminium or diethylaluminium chloride [288] (Table 9.3):

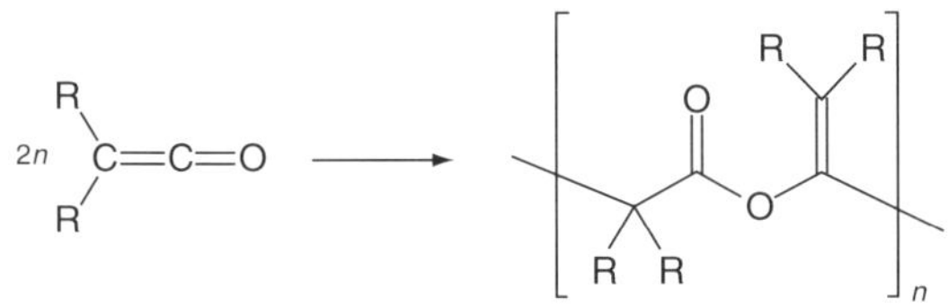

(41)

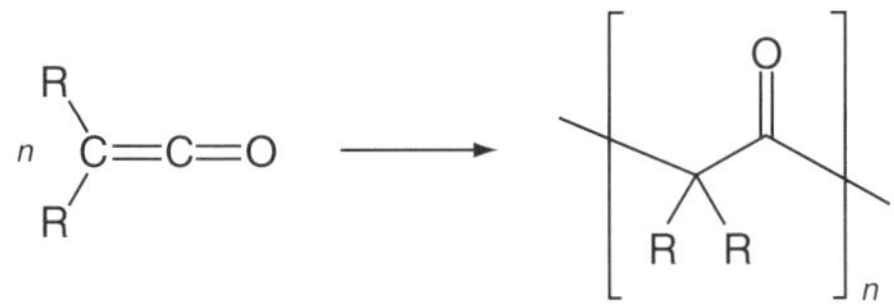

(42)

9.8.4.2 Copolymerisation of Ketenes and Aldehydes

The copolymerisation of ketene and acetaldehyde was carried out with ethylzinc diphenylamide as the catalyst. It was found to yield an alternating copolymer [scheme (43)], the respective polyester of highly isotactic structure (Table 9.3) [279]:

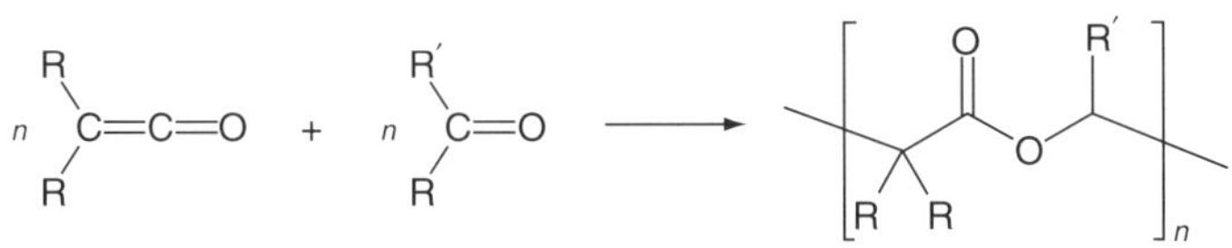

(43)

R = H, R' = Me

It may be interesting to note that such a polyester was also obtained in the homopolymerisation of β-butyrolactone under the same polymerisation conditions [279].

The copolymerisation of dimethylketene and acetaldehyde with diethylzinc as a catalyst produced a crystalline copolymer of alternating comonomer distribution (Table 9.3) [289]. Other aldehydes such as *n*-butyraldehyde, *i*-butyraldehyde or benzaldehyde were also copolymerised with dimethylketene to produce the respective polyesters [289–318].

At the end of considerations concerning the coordination polymerisation of heteroatom-containing cyclic and acyclic monomers, it is obvious that future development (beyond carbon monoxide copolymerisations) is to be anticipated, especially concerning new catalytic processes, including both new monomers for coordination polymerisation and new catalysts. Some new areas have been opened to study recently: for instance, the coordination polymerisation of cyclic siloxanes [319] which, by the nature of things, forms polymers with an inorganic backbone, or the coordination polymerisation of cyclic ethers, taking advantage of transition metal-based catalysts [320].

References

1. Kuran, W., *Prog. Polym. Sci.*, **23**, 919 (1998).
2. Sen, A., *Acc. Chem. Res.*, **26**, 303 (1993).
3. Pasquon, I., Porri, L. and Giannini, U., in *Encyclopedia of Polymer Science and Engineering*, Wiley-Interscience, John Wiley & Sons, New York, 1989, Vol. 15, pp. 708–709.
4. Vandenberg, E. J., in *Catalysis in Polymer Synthesis*, Washington, DC, 1992, ACS Symp. Ser. 496, pp. 10–14.
5. Maeda, A. and Inagami, N., *Rubber & Plastics News*, **42**, 1 (1987).
6. Meissner, B., Schätz M. and Brajko, V., in *Elastomers and Rubber Compounding Materials*, Elsevier, Amsterdam, 1989, Studies in Polymer Science, Vol. 1, pp. 274–278.
7. Owens, K. and Kyllingstad, V. R., in *Kirk-Othmer Encyclopedia of Chemical Technology*, Wiley-Interscience, John Wiley & Sons, New York, 1993, Vol. 8, pp. 1079–1093.
8. Inoue, S., *Chemtech*, **6**, 588 (1976).
9. Rokicki, A. and Kuran, W., *J. Macromolec. Sci. – Rev. Macromolec. Chem. C*, **21**, 135 (1981).
10. Kuran, W. and Listos, T., *Polish J. Chem.*, **68**, 1071 (1994).
11. Darensbourg, D. J. and Holtcamp, N. W., *Coord. Chem. Rev.*, **153**, 155 (1996).
12. Super, M. S. and Beckman, E. J., *Trends Polym. Sci.*, **5**, 236 (1997).
13. Pruitt, M. E. and Baggett, J. M., US Pat. 2 706 181 (to Dow Chemical Co.) (1955).
14. Borkovec, A. B., *J. Org. Chem.*, **23**, 828 (1958).
15. Osgan, M. and Price, C. C., *J. Polym. Sci.*, **34**, 153 (1959).
16. Furukawa, J., Tsuruta, T., Sakata, R., Saegusa, T. and Kawasaki, A., *Makromol. Chem.*, **32**, 90 (1959).
17. Colclough, R. O., Gee, G. and Jagger, A. H., *J. Polym. Sci.*, **48**, 273 (1960).
18. Vandenberg, E. J., *J. Polym. Sci.*, **47**, 486 (1960).
19. Wu, J. and Shen, Z., *J. Polym. Sci. Polym. Chem. Ed.*, **28**, 1995 (1990).
20. Osgan, M. and Teyssié, Ph., *J. Polym. Sci. B*, **5**, 789 (1967).
21. Kohler, N., Osgan, M. and Teyssié, Ph., *J. Polym. Sci. B*, **6**, 559 (1968).
22. Osgan, M. and Teyssié, Ph., *J. Polym. Sci. B*, **8**, 319 (1970).
23. Inoue, S., Tsuruta, T. and Furukawa, J., *Makromol. Chem.*, **53**, 215 (1962).
24. Tsuruta, T., Inoue, S., Yoshida, N. and Furukawa, J., *Makromol. Chem.*, **55**, 230 (1962).
25. Spassky, N., Le Borgne, A. and Sepulchre, M., *Pure Appl. Chem.*, **53**, 1735 (1981).
26. Ishimori, M., Higawara, T., Tsuruta, T., Kai, Y., Yasuoka, N. and Kasai, N., *Bull. Chem. Soc. Jpn*, **49**, 1165 (1976).
27. Ishimori, M., Higawara, T. and Tsuruta, T., *Makromol. Chem.*, **179**, 2337 (1978).
28. Higawara, T., Ishimori, M. and Tsuruta, T., *Makromol. Chem.*, **182**, 501 (1981).
29. Kageyama, H., Kai, Y., Suzuki, C., Yoshino, N. and Tsuruta, T., *Makromol. Chem., Rapid Commun.*, **5**, 89 (1984).

30. Inoue, S., in *Catalysis in Polymer Synthesis*, Washington, DC, 1992, ACS Symp. Ser. 496, pp. 194–204.
31. Kuran, W., Rokicki, A. and Pienkowski, J., *J. Polym. Sci., Polym. Chem. Ed.*, **17**, 1235 (1979).
32. Aida, T. and Inoue, S., *Macromolecules*, **14**, 1166 (1981).
33. Aida, T., Wada, K. and Inouc, S., *Macromolecules*, **20**, 237 (1987).
34. Yasuda, T., Aida, T. and Inoue, S., *Bull. Chem. Soc. Jpn.*, **59**, 3931 (1986).
35. Yasuda, T., Aida, T. and Inoue, S., *J. Macromol. Sci.–Chem. A*, **21**, 1035 (1984).
36. Inoue, S. and Aida, T., in *Ring-Opening Polymerization*, D. J. Brunelle (Ed.), Hanser Verlag, Munich, 1993, pp. 187–215.
37. Dzugan, S. G. and Goedkrn, V. L., *Inorg. Chem.*, **25**, 2858 (1986).
38. Jun, C. L., Le Borgne, A. and Spassky, N., *J. Polym. Sci., Polym. Symp. D*, **74**, 31 (1986).
39. Vincens, V., Le Borgne, A. and Spassky, N., in *Catalysis in Polymer Synthesis*, Washington, DC, 1992, ACS Symp. Ser. 496, pp. 205–214.
40. Vincens, V., Le Borgne, A. and Spassky, N., *Makromol. Chem., Rapid Commun.*, **10**, 623 (1989).
41. Kuran, W., Listos, T., Abramczyk, M. and Dawidek, A., *J. Macromol. Sci. – Pure Appl. Chem. A*, **35**, 427 (1998).
42. Kawasaki, A., Furukawa, J., Tsuruta, T., Saegusa, T., Kakogawa, G. and Sakata, R., *Polymer*, **1**, 315 (1960).
43. Hirano, T., Khanh, P. H. and Tsuruta, T., *Makromol. Chem.*, **153**, 331 (1972).
44. Jedlinski, Z., Dworak, A. and Bero, M., *Makromol. Chem.*, **180**, 949 (1979).
45. Tsuchiya, S. and Tsuruta, T., *Makromol. Chem.*, **110**, 123 (1967).
46. Oguni, N., Lee, K. and Tani, H., *Macromolecules*, **5**, 819 (1972).
47. Jedlinski, Z., Kasperczyk, J. and Dworak, A., *Eur. Polym. J.*, **19**, 899 (1983).
48. Rabagliati, F. M. and Contreras, J. M., *Eur. Polym. J.*, **23**, 63 (1987).
49. Aida, T., Mizuta, R. and Yoshida, Y., *Makromol. Chem.*, **110**, 123 (1967).
50. Le Borgne, M., Spassky, N., Jun, C. L. and Momtaz, A., *Makromol. Chem.*, **189**, 637 (1988).
51. Le Borgne, A., Moreau, M. and Vincens, V., *Macromol. Chem. Phys.*, **195**, 375 (1994).
52. Tsuruta, T. and Kawakami, Y., in *Comprehensive Polymer Science*, G. Allen and J. C. Bevington (Eds), Pergamon Press, Oxford, 1989, Vol. 3, pp. 489–500.
53. Inoue, S. and Aida, T., in *Comprehensive Polymer Science*, G. Allen and J. C. Bevington (Eds), Pergamon Press, Oxford, 1989, Vol. 3, pp. 553–569.
54. Khanh, P. H., Hirano, T. and Tsuruta, T., *J. Macromol. Sci. – Chem. A*, **5**, 1287 (1971).
55. Price, C. C. and Spector, R., *J. Am. Chem. Soc.*, **88**, 4171 (1966).
56. Tani, H., Oguni, N. and Watanabe, S., *J. Polym. Sci. B*, **6**, 577 (1968).
57. Price, C. C., Akkapeddi, M. K., De Bona, B. T. and Furie, B. C., *J. Am. Chem. Soc.*, **94**, 3964 (1972).
58. Oguni, N., Watanabe, S., Maki, M. and Tani, H., *Macromolecules*, **6**, 195 (1973).
59. Oguni, N., Maeda, S. and Tani, H., *Macromolecules*, **6**, 459 (1973).
60. Vandenberg, E. J., *Pure Appl. Chem.*, **48**, 295 (1976).
61. Vandenberg, E. J., *J. Polym. Sci. A-1*, **7**, 525 (1969).
62. Watanabe, Y., Yasuda, T., Aida, T. and Inoue, S., *Macromolecules*, **25**, 1396 (1992).
63. Tsuruta, T., *Makromol. Chem., Macromol. Symp.*, **6**, 23 (1986).
64. Hasebe, Y. and Tsuruta, T., *Makromol. Chem.*, **188**, 1403 (1987).
65. Kuran, W. and Listos, T., *Macromol. Chem. Phys.*, **195**, 401 (1994).
66. Inoue, M., Sugita, T., Kiso, Y. and Ichikawa, K., *Bull. Chem. Soc. Jpn.*, **49**, 1063 (1976).

67. Eisch, J. J., Liu, Z.-R. and Singh, M., *J. Org. Chem.*, **52**, 1618 (1992).
68. Darensbourg, D. J., Holtcamp, M. W., Khandelwal, B., Kalusmeyer, K. K. and Reibenspies, J. H., *J. Am. Chem. Soc.*, **117**, 538 (1995).
69. Condé, Ph., Hocks, L., Teyssié, Ph. and Vwrin, R., in *Catalysis in Polymer Synthesis*, Washington, DC, 1992, ACS Symp. Ser. 496, pp. 149–156.
70. Groves, J. T., Han, Y. and Van Engen, D., *J. Chem. Soc., Chem. Commun.*, **1990**, 436 (1990).
71. Dumas, Ph. and Guérin, Ph., *Can. J. Chem.*, **56**, 925 (1978).
72. Kuran, W., Pasynkiewicz, S. and Serzysko, J., *J. Organometall. Chem.*, **73**, 187 (1974).
73. Górecki, P. and Kuran, W., *J. Organometall. Chem.*, **312**, 1 (1986).
74. Kuran, W. and Listos, T., *Polish J. Chem.*, **68**, 643 (1994).
75. Tsuruta, T., *J. Polym. Sci., Polym Symp. D*, **7**, 179 (1972).
76. Listos, T., Kuran, W. and Siwiec, R., *J. Macromol. Sci. – Pure Appl. Chem. A*, **32**, 393 (1995).
77. Hasebe, Y., Izumitani, K., Torii, M. and Tsuruta, T., *Makromol. Chem.*, **191**, 107 (1990).
78. Vandenberg, E. J. and Mulis, J. C., *J. Polym. Sci., Polym. Chem. Ed.*, **29**, 1421 (1991).
79. Coulon, C., Spassky, N. and Sigwalt, P., *Polymer*, **17**, 821 (1976).
80. Deffieux, A., Sepulchre, M., Spassky, N. and Sigwalt P., *Makromol. Chem.*, **175**(2), 339 (1974).
81. Asano, S., Aida, T. and Inoue, S., *Macromolecules*, **18**, 2057 (1985).
82. Kuran, W. and Listos, T., *Polish J. Chem.*, **68**, 1071 (1994).
83. Penczek, S., Kubisa, P. and Matyjaszewski, K., *Adv. Polym. Sci.*, **37**, 1 (1980).
84. Ishimori, M., Hsiue, G. and Tsuruta, T., *Makromol. Chem.*, **128**, 52 (1969).
85. Vandenberg, E. J. and Robinson, A. E., in *Polyethers*, Washington, DC, 1975, ACS Symp. Ser. 6, pp. 101–119.
86. Vandenberg, E. J. and Robinson, A. E., *Polym. Prepr. Am. Chem. Soc., Div. Polym. Chem.*, **15**, 208 (1974).
87. Oguni, N. and Hyoda, J., *Macromolecules*, **13**, 1687 (1980).
88. Kops, J. and Spanggard, H., *Macromolecules*, **15**, 1200 (1982).
89. Inoue, S. IUPAC 35th International Symposium on *Macromolecules*, Akron, Ohio, USA, 1994, Abstracts, p. 60.
90. Okamoto, Y. and Nakano, T., *Chem. Rev.*, **94**, 349 (1994).
91. Jacobs, C., Dubois, Ph., Jérôme, R. and Teyssié, Ph., *Macromolecules*, **24**, 3027 (1991).
92. Ropson, N., Dubois, Ph., Jérôme, R. and Teyssié, Ph., *Macromolecules*, **27**, 5950 (1994).
93. Hofman, A., Slomkowski, S. and Penczek, S., *Makromol. Chem., Rapid Commun.*, **8**, 387 (1987).
94. Duda, A., Florjanczyk, Z., Hofman, A., Slomkowski, S. and Penczek, S., *Macromolecules*, **24**, 1640 (1990).
95. Duda, A. and Penczek, S., *Makromol. Chem., Macromol. Symp.*, **47**, 127 (1990).
96. Duda, A. and Penczek, S., *Macromol. Chem., Rapid Commun.*, **15**, 559 (1994).
97. Duda, A. and Penczek, S., *Macromol. Chem., Rapid Commun.*, **16**, 67 (1995).
98. Duda, A. and Penczek, S., *Macromolecules*, **28**, 5981 (1995).
99. Le Borgne, A., Pluta, C. and Spassky, N., *Macromol. Chem., Rapid Commun.*, **15**, 955 (1994).
100. Stevels, M. W., Anconé, M. J. K., Dijkstra, P. J. and Feijen, J., *Macromol. Chem. Phys.*, **196**, 1153 (1995).
101. Cherdron, H., Ohse, H. and Corte, F., *Makromol. Chem.*, **56**, 187 (1962).
102. Dubois, Ph., Jérôme, R. and Teyssié, Ph., *Polym. Bull.*, **22**, 475 (1989).

103. Dubois, Ph., Degée, Ph., Jérôme, R. and Teyssié, Ph., *Macromolecules*, **25**, 2614 (1992).
104. Duda, A., *Macromolecules*, **27**, 576 (1994).
105. Duda, A., *Macromolecules*, **28**, 1399 (1995).
106. Ouhadi, T., Stevens, C. and Teyssié, Ph., *Makromol. Chem., Suppl.*, **1**, 191 (1975).
107. Vion, J. M., Jérôme, R., Teyssié, Ph., Aubin, M. and Prud'homme, R. E., *Macromolecules*, **19**, 1828 (1986).
108. Kricheldorf, H. R., Berl, M. and Scharnagel, N., *Macromolecules*, **21**, 286 (1988).
109. Wurm, B., Keul, H. and Höcker, H., *Macromol. Chem. Phys.*, **195**, 1011 (1994).
110. Löfgren, A., Albertsson, A.-C., Dubois, Ph., Jérôme, R., Teyssié, Ph., Aubin, M. and Prud'homme, R. E., *Macromolecules*, **27**, 5556 (1994).
111. Le Borgne, A. and Spassky, N., *Polymer*, **30**, 2312
112. Grenier, D., Prud'homme, R. E., Le Borgne, A. and Spassky, N., *J. Polym. Sci., Polym. Chem. Edn. A*, **19**, 1781 (1981).
113. Spassky, N., Le Borgne, A. and Hull, W. E., *Macromolecules*, **16**, 608 (1983).
114. Le Borgne, A., Spassky, N. and Sigwalt, P., *Polym. Bull.*, **1**, 825 (1979).
115. Tani, H., Yamashita, S. and Teranishi, K., *Polym. J.*, **3**, 417 (1972).
116. Teranishi, K., Iida, M., Araki, T., Yamashita, S. and Tani, H., *Macromolecules*, **7**, 421 (1974).
117. Agostini, D. E., Lando, J. B. and Shelton, J. R., *J. Polym. Sci., Polym. Chem. Ed. A-1*, **9**, 2775 (1971).
118. Gross, R. A., Zhang, Y., Konrad, G. and Lenz, R. W., *Macromolecules*, **21**, 2657 (1988).
119. Benvenuti, M. and Lenz, R. W., *J. Polym. Sci., Polym. Chem. Edn A*, **29**, 793 (1991).
120. Pajersky, A. D. and Lenz, R. W., *Makromol. Chem. Macromol. Symp.*, **73**, 7 (1993).
121. Lavallée, C., Le Borgne, A., Spassky, N. and Prud'homme, R. E., *J. Polym. Sci. Polym., Chem. Ed. A*, **25**, 1315 (1987).
122. Ouhadi, T. and Heuschen, J. M., *J. Macromol. Sci. – Chem. A*, **9**, 927 (1975).
123. Ouhadi, T., Hamitou, A., Jérôme, R. and Teyssié, Ph., *Macromolecules*, **9**, 173 (1976).
124. Hamitou, A., Ouhadi, T., Jérôme, R. and Teyssié, Ph., *J. Polym. Sci., Polym. Chem. Ed*, **15**, 865 (1977).
125. Inoue, S. and Aida, A., *Makromol. Chem., Macromol. Symp.*, **73**, 27 (1993).
126. Shimasaki, K., Aida, T. and Inoue, S., *Macromolecules*, **20**, 3076 (1987).
127. Endo, M., Aida, T. and Inoue, S., *Macromolecules*, **20**, 2982 (1987).
128. Yasuda, T., Aida, T. and Inoue, S., *Macromolecules*, **16**, 1792 (1983).
129. Sugimoto, H., Aida, T. and Inoue, S., *Macromolecules*, **23**, 2869 (1990).
130. Le Borgne, A., Vincens, V., Jouglard, M. and Spassky, N., *Makromol. Chem., Macromol. Symp.*, **73**, 37 (1993).
131. Lenz, R. W., IUPAC International Symposium on *Ionic Polymerization*, Istanbul, Turkey, 1995, Abstracts, p. 18.
132. McLain, S. J. and Drysdale, N. E., *Polym. Prepr. Am. Chem. Soc., Div. Polym. Chem.*, **33**(1), 174 (1992).
133. McLain, S. J., Ford, T. M. and Drysdale, N. E., *Polym. Prepr. Am. Chem. Soc., Div. Polym. Chem.*, **33**(2), 463 (1992).
134. Hmamouchi, M., Lavallée, C., Prud'homme, R. E., Le Borgne, A. and Spassky, N., *Macromolecules*, **22**, 130 (1989).
135. Le Borgne, A., Grenier, D., Prud'homme, R. E. and Spassky, N., *Eur. Polym. J.*, **17**, 1103 (1981).
136. Inoue, S. and Aida, T., in *Encyclopedia of Polymer Science and Engineering*, Wiley-Interscience, John Wiley & Sons, New York, 1990, Supplement Vol., pp. 412–420.

137. Inoue, S., Koinuma, H. and Tsuruta, T., *Polym. J.*, **2**, 220 (1971).
138. Dubois, Ph., Jacobs, C., Jérôme, R. and Teyssié, Ph., *Macromolecules*, **24**, 2266 (1991).
139. Kricheldorf, H. R. and Boettcher, C., *Makromol. Chem.*, **194**, 1653 (1993).
140. Feng, X. D., Song, C. X. and Chen, W. Y., *J. Polym. Sci., Polym. Lett. Ed.*, **21**, 593 (1983).
141. Trofimoff, L. R., Aida, T. and Inoue, S., *Chem. Lett.*, **1987**, 991 (1987).
142. Soga, K., Hosoda, S., Tazuke, Y. and Ikeda, S., *J. Polym. Sci., Polym. Lett. Ed.*, **14**, 161 (1976).
143. Soga, K., Tazuke, Y., Hosoda, S. and Ikeda, S., *J. Polym. Sci., Polym. Lett. Ed.*, **15**, 219 (1977).
144. Vogdanis, L. and Heitz, W., *Makromol. Chem., Rapid Commun.*, **7**, 543 (1986).
145. Vogdanis, L., Martens, B., Uchtmann, H., Henzel, F. and Heitz, W., *Makromol. Chem.*, **191**, 465 (1990).
146. Kuran, W. and Listos. T., *Makromol. Chem.*, **193**, 945 (1992).
147. Kuran, W. and Listos, T., *Macromol. Chem. Phys.*, **195**, 1011 (1994).
148. Kricheldorf, H. R., Jenssen, J. and Kreiser-Sanders, I., *Makromol. Chem.*, **192**, 2391 (1991).
149. Hovestatd, W., Keul, H. and Höcker, H., *Polymer*, **33**, 1941 (1992).
150. Heldrr, J., Kohn, F. E., Sato, S., van den Berg, J. W. and Feijen, J., *Makromol. Chem., Rapid Commun.*, **6**, 9 (1975).
151. Baba, A., Meishou, H. and Matsuda, H., *Makromol. Chem., Rapid Commun.*, **5**, 665 (1974).
152. Amarascker, J., Rauchfuss, T. B. and Wilson, S. R., *J. Am. Chem. Soc.*, **110**, 2332 (1988).
153. Sigwalt, P., *Pure Appl. Chem.*, **48**, 257 (1976).
154. Dumas, Ph., Spassky, N. and Sigwalt, P., *Makromol. Chem.*, **156**, 55 (1972).
155. Spassky, N., Dumas, Ph., Sepulchre, M. and Sigwalt, P., *J. Polym. Sci., Polym. Symp. D*, **52**, 327 (1975).
156. Spassky, N. and Sigwalt, P., *Bull. Soc. Chim. France*, **1968**, 4617 (1968).
157. Dumas, Ph., Sigwalt, P. and Guérin, Ph., *Makromol. Chem.*, **182**, 2225 (1981).
158. Marchett, M., Chellini, E., Sepulchre, M. and Spassky, N., *Makromol. Chem.*, **180**, 1305 (1979).
159. Spassky, N., Le Borgne, A., Momtaz, A. and Sepulchre, M., *J. Polym. Sci., Polym. Lett. Ed.*, **18**, 3089 (1980).
160. Dumas, Ph., Guérin, Ph. and Sigwalt, P., *Nouv. J. Chim.*, **4**, 95 (1980).
161. Dumas, Ph., Sigwalt, P. and Guérin, Ph., *Makromol. Chem.*, **185**, 1317 (1984).
162. Guérin, Ph., Boileau, S. and Sigwalt, P., *Eur. Polym. J.*, **10**, 13 (1974).
163. Dumas, Ph., and Sigwalt, P., *Chirality*, **3**, 484 (1993).
164. Dumas, Ph., Sigwalt, P. and Guérin, Ph., *Makromol. Chem.*, **1193**, 1709 (1992).
165. Aida, T., Kawaguchi, K. and Inoue, S., *Macromolecules*, **23**, 3887 (1990).
166. Watanabe, Y., Aida, T. and Inoue, S., *Macromolecules*, **24**, 3970 (1991).
167. Soga, K., Imamura, H. and Ikeda, S., *Makromol. Chem.*, **176**, 807 (1975).
168. Matsuura, K., Inoue, S. and Tsuruta, T., *Makromol. Chem.*, **80**, 149 (1964).
169. Tsuruta, T., Matsuura, K. and Inoue, S., *Makromol. Chem.*, **83**, 289 (1965).
170. Makino, T., Inoue, S. and Tsuruta, T., *Makromol. Chem.*, **83**, 316 (1965).
171. Matsuura, K., Inoue, S. and Tsuruta, T., *Makromol. Chem.*, **103**, 140 (1967).
172. Makino, T., Inoue, S. and Tsuruta, T., *Makromol. Chem.*, **131**, 147 (1970).
173. Makino, T., Inoue, S. and Tsuruta, T., *Makromol. Chem.*, **133**, 137 (1971).
174. Tsuruta, T., Inoue, S. and Matsuura, K., *Makromol. Chem.*, **63**, 219 (1963).
175. Kricheldorf, H. R., in *Comprehensive Polymer Science*, G. Allen and J. C. Bevington (Eds), Pergamon Press, Oxford, 1989, Vol. 3, pp. 531–551.

176. van Santvoort, F. A. J. J., Krabbendam, H., Spek, A. L. and Boersma, J., *Inorg. Chem.*, **17**, 388 (1978).
177. Boersma, J., in *Comprehensive Organometallic Chemistry*, G. Wilkinson, G. A. Stone and E. W. Abel (Eds), Pergamon Press, Oxford, 1983, Vol. 2, pp. 823–862.
178. Yonezawa, N., Toda, F. and Hasegawa, M., *Makromol. Chem., Rapid Commun.*, **6**, 607 (1985).
179. Ogata, N., Asahara, T. and Tohoyama, S., *J. Polym. Sci. A-1*, **4**, 1359 (1966).
180. Ogata, N., Tanaka, K. and Takayama, S., *Makromol. Chem.*, **119**, 161 (1968).
181. Kaluzynski, K., Libiszowski, J. and Penczek, S., *Makromol. Chem.*, **178**, 2943 (1977).
182. Libiszowski, J., Kaluzynski, K. and Penczek, S., *J. Polym. Sci., Polym. Chem. Ed.*, **16**, 1275 (1978).
183. Hsieh, L. H., *Polym. Prepr. Chem. Am. Soc., Div. Polym. Chem.*, **13**, 157 (1972).
184. Hsieh, L. H., *J. Macromol. Sci. – Chem. A*, **7**, 1526 (1973).
185. Inoue, S., Kitamura, K. and Tsuruta, T., *Makromol. Chem.*, **126**, 250 (1969).
186. Kuran, W. and Nieslochowski, A., *Polym. Bull.*, **2**, 411 (1980).
187. Kuran, W. and Nieslochowski, A., *J. Macromol. Sci. – Chem. A*, **15**, 1567 (1981).
188. Aida, T. and Inoue, S., *J. Am. Chem. Soc.*, **107**, 1358 (1985).
189. Aida, T., Sanuki, S. and Inoue, S., *Macromolecules*, **18**, 1049 (1985).
190. Reinheckel, H. and Haage, K., *J. Organometall. Chem.*, **10**, 29 (1967).
191. Zweifel, H., Lolinger, J. and Volker, T., *Makromol. Chem.*, **153**, 125 (1972).
192. Kuran, W. and Nieslochowski, A., *Synth. React. Inorg. Met.-Org. Chem.*, **11**, 79 (1981).
193. Inoue, S., Kobayashi, M. and Tozuka, T., *J. Organometall. Chem.*, **81**, 17 (1974).
194. Kuran, W., Listos, T., Iwaniuk, R. and Rokicki, G., *Polimery*, **38**, 405 (1993).
195. Yasuda, T., Aida, T. and Inoue, S., *Macromolecules*, **17**, 2217 (1984).
196. Asano, S., Aida, T. and Inoue, S., *J. Chem. Soc., Chem. Commun.*, **1985**, 1148 (1985).
197. Teyssié, Ph., Bioul, J. P., Condé, P., Druet, J., Heuschen, J., Jérôme, R., Ouhadi, T. and Warin, R., in *Metal-Alcoholate Initiators*, Washington, DC, 1985, ACS Symp. Ser. 286, p. 97.
198. Hamitou, A., Jérôme, R., Hubert, A. and Teyssié, Ph., *Macromolecules*, **6**, 651 (1973).
199. Inoue, S., Koinuma, H. and Tsuruta, T., *Makromol. Chem.*, **130**, 210 (1969).
200. Inoue, S., Koinuma, H. and Tsuruta, T., *J. Polym. Sci. B.*, **7**, 287 (1969).
201. Inoue, S., Kobayashi, M., Koinuma, H. and Tsuruta, T., *Makromol. Chem.*, **155**, 61 (1972).
202. Rätsch, M. and Haubold, W., *Faserforsch. Textiltech., Z. Polymerforsch.*, **28**, 15 (1977).
203. Kobayashi, M., Inoue, S. and Tsuruta, T., *Macromolecules*, **4**, 658 (1971).
204. Kobayashi, M., Tang, Y.-L., Tsuruta, T. and Inoue, S., *Makromol. Chem.*, **169**, 69 (1973).
205. Kuran, W., Pasynkiewicz, S. and Skupinska, J., *Makromol. Chem.*, **177**, 1283 (1976).
206. Kuran, W., *Appl. Organometall.*, **5**, 191 (1991).
207. Kuran, W. and Listos, T., *Macromol. Chem. Phys.*, **194**, 977 (1994).
208. Kuran, W., Pasynkiewicz, S., Skupinska, J. and Rokicki, A., *Makromol. Chem.*, **177**, 11 (1976).
209. Rokicki, A. and Kuran, W., *Makromol. Chem.*, **180**, 2153 (1979).
210. Górecki, P. and Kuran, W., *J. Polym. Sci., Polym. Lett. Ed.*, **23**, 299 (1985).
211. Kuran, W., Listos, T. and Kulpa, A., Polish Pat. 165,616 (1994).
212. Kobayashi, M., Inoue, S. and Tsuruta, T., *J. Polym. Sci., Polym. Chem. Ed.*, **11**, 2383 (1973).

213. Kuran, W., Rokicki, A. and Wilinska, E., *Makromol. Chem.*, **180**, 361 (1979).
214. Nishimura, M., Kasai, M. and Tsuchida, E., *Makromol. Chem.*, **179**, 1913 (1978).
215. Tsuchida, E. and Kasai, M., *Makromol. Chem.*, **181**, 1612 (1980).
216. Chen, L.-B., Chen. H.-S. and Lin, J., *J. Macromol. Sci. – Chem. A*, **24**, 253 (1987).
217. Chen, L.-B., *Makromol. Chem., Macromol. Symp.*, **59**, 75 (1992).
218. Soga, K., Hyakkoku, K. and Ikeda, S., *Makromol. Chem.*, **179**, 2837 (1978).
219. Soga, K., Hyakkoku, K., Izumi, K. and Ikeda, S., *J. Polym. Sci., Polym. Chem. Ed.*, **16**, 2383 (1978).
220. Inoue, S., in *Progress in Polymer Science*, K. Imahori and T. Higashimura (Eds), Kodansha, Tokyo, 1975, Vol. 8, p. 1.
221. Inoue, S., Takada, T. and Tatsu, H., *Makromol. Chem., Rapid Commun.*, **1**, 775 (1980).
222. Hino, Y., Yoshida, Y. and Inoue, S., *Polym. J.*, **16**, 159 (1984).
223. Rokicki, A., US Pat. 4,943,677 (1990).
224. Soga, K., Imai, E. and Hattori, I., *Polym. J.*, **13**, 407 (1981).
225. Inoue, S., Koinuma, H. and Tsuruta, T., US Pat. 3,585,168 (1971).
226. Koinuma, H. and Hirai, H., *Makromol. Chem.*, **178**, 1283 (1977).
227. Chen, X., Shen, Z. and Zhang, Y., *Macromolecules*, **24**, 5305 (1991).
228. Shen, Z., Chen, X. and Zhang, Y., *Macromol. Chem. Phys.*, **195**, 2003 (1994).
229. Inoue, S., Koinuma, H., Yokoo, Y. and Tsuruta, T., *Makromol. Chem.*, **143**, 97 (1971).
230. Inoue, S., Matsumoto, K. and Yoshida, Y., *Makromol. Chem.*, **181**, 2287 (1980).
231. Hirano, T., Inoue, S. and Tsuruta, T. Y., *Makromol. Chem.*, **177**, 3237 (1976).
232. Hirano, T., Inoue, S. and Tsuruta, T. Y., *Makromol. Chem.*, **177**, 3245 (1976).
233. Inoue, S., Hirano, T. and Tsuruta, T., *Polym. J.*, **9**, 101 (1977).
234. Udipi, K. and Gillham, J. K., *J. Appl. Polym. Sci.*, **18**, 1575 (1974).
235. Lednor, P. W. and Rol, N. C., *J. Chem. Soc., Chem. Commun.*, **1985**, 598 (1985).
236. Koinuma, H., Naito, K. and Hirai, H., *Makromol. Chem., Rapid Commun.*, **1**, 493 (1980).
237. Rokicki, G., Kuran. W. and Pogorzelska-Marciniak, B., *Monatsh. Chem.*, **115**, 205 (1984).
238. Darensbourg, D. J. and Holtcamp, M. W., *Macromolecules*, **28**, 7577 (1995).
239. Super, M., Berluche, E., Costello, C. and Beckman, E., *Macromolecules*, **30**, 362 (1997).
240. Kuran, W., Rokicki, A. and Romanowska, D., *J. Polym. Sci., Polym. Chem. Ed.*, **17**, 2003 (1979).
241. Aida, T., Ishikawa, M. and Inoue, S., *Macromolecules*, **19**, 8 (1986).
242. Aida, T. and Inoue, S., *Macromolecules*, **15**, 628 (1982).
243. Nishikumbo, T., Kameyama, A., Yamashita, J., Tomoi, M. and Fukuda, W., *J. Polym. Sci., A, Polym. Chem.*, **31**, 939 (1993).
244. Abramczyk, M., Kuran, W. and Listos, T., in *The XVIIIth International Conference on Organometallic Chemistry*, Munich, Germany, 1998, Book of Abstracts, p. A39 (Vth Regional Seminar of PhD-Students on *Organometallic and Organophosphorus Chemistry*, Sec, Czech Republic, 1999, Abstracts, p. 35).
245. Koinuma, H. and Hirai, H., *Makromol. Chem.*, **178**, 241 (1977).
246. Tsuruta, T., Inoue, S., Ishimori, M. and Yoshida, N., *J. Polym. Sci. C*, **4**, 407 (1964).
247. Kuran, W., Rokicki, A. and Wielgopolan, W., *Makromol. Chem.*, **179**, 2545 (1978).
248. Adachi, N., Kida, Y. and Shikata, K., *J. Polym. Sci., Polym. Chem. Ed.*, **15**, 937 (1977).
249. Soga, K., Imamura, H., Sato, M. and Ikeda, S., *J. Polym. Sci., Polym. Chem. Ed.*, **14**, 677 (1976).

250. Soga, K., Sato, M., Imamura, H. and Ikeda, S., *J. Polym. Sci., Polym. Lett. Ed.*, **13**, 167 (1975).
251. Schaefer, J., Kern, R. J. and Katnic, R. J., *Macromolecules*, **1**, 107 (1968).
252. Ziegler, K., Krupp, F., Weyer, K. and Larbig, W., *Ann. Chem.*, **629**, 251 (1960).
253. Coates, G. E. and Mukherjee, R. N., *J. Chem. Soc.*, **1964**, 1295 (1964).
254. Millich, F., *J. Polym. Sci., Macromol. Rev.*, **15**, 207 (1980).
255. Drenth, W. and Nolte, R. J. M., *Acc. Chem. Res.*, **12**, 30 (1979).
256. van Beijnen, A. J. M., Nolte, R. J. M., Naaktgeboren, A. J., Zwikker, J. W., Drenth, W. and Hezemans, A. M. F., *Macromolecules*, **16**, 1679 (1983).
257. Nolte, R. J. M., van Beijnen, A. J. M. and Drenth, W., *J. Am. Chem. Soc.*, **96**, 5932 (1974).
258. van Beijnen, A. J. M., Nolte, R. J. M. and Drenth, W., *Recl. Trav. Chim. Pays-Bas*, **99**, 121 (1980).
259. Yamamoto, Y., Takizawa, T. and Hagihara, N., *Nippon Kagaku Zasshi*, **87**, 1355 (1966).
260. Nolte, R. J. M., Stephany, R. W. and Drenth, W., *Recl. Trav. Chim. Pays-Bas*, **92**, 83 (1973).
261. Millich, F. and Baker, G. K., *Macromolecules*, **2**, 122 (1969).
262. Millich, F., *Adv. Polym. Sci.*, **19**, 117 (1975).
263. Nolte, R. J. M. and Drenth, W., *Recl. Trav. Chim. Pays-Bas*, **92**, 788 (1973).
264. Burr, A. J. and Fetters, L. J., *Chem. Rev.*, **76**, 727 (1976).
265. Yilmaz, O., Usanmaz, A. and Alyürük, K., *J. Polym. Sci. C*, **28**, 341 (1990).
266. Furukawa, J., Yamashita, S., Maruhashi, M. and Harada, K., *Makromol. Chem.*, **85**, 80 (1965).
267. Colclough, R. O., Gee, G., Higginson, W. C. E., Jackson, J. B. and Litt, M., *J. Polym. Sci.*, **36**, 541 (1959).
268. Yilmaz, O., Usanmaz, A. and Alyürük, K., *Eur. Polym. J.*, **28**, 1351 (1992).
269. Noltes, J. G. and Boersma, J., *J. Organometall. Chem.*, **7**, 6 (1967).
270. Vogl, O., *J. Polym. Sci. A*, **2**, 4607 (1964).
271. Neeld, K. and Vogl, O., *Macromol. Rev.*, **16**, 1 (1981).
272. Vogl, O., in *Encyclopedia of Polymer Science and Engineering*, H. F. Mark, N. M. Bikales, C. G. Overberger and G. Menges (Eds), Wiley-Interscience, New York, 1987, Vol. 1, pp. 623–643.
273. Vogl, O., *J. Macromol. Sci. – Pure Appl. Chem A*, **29**, 1085 (1992).
274. Furukawa, J., Saegusa, T., Fujii, H., Kawasaki, A., Imai, H. and Fujii, Y., *Makromol. Chem.*, **37**, 149 (1960).
275. Natta, G., Corradini, P. and Bassi, I. W., *J. Polym. Sci.*, **51**, 505 (1965).
276. Natta, G., Mazzanti, G., Corradini, P. and Bassi, I. W., *Makromol. Chem.*, **37**, 156 (1960).
277. Furukawa, J., Saegusa, T. and Fujii, H., *Makromol. Chem.*, **44–46**, 398 (1961).
278. Ishimori, M., Nakasugi, O., Takeda, N. and Tsuruta, T., *Makromol. Chem.*, **115**, 103 (1968).
279. van der Kerk, G. J., *Pure Appl. Chem.*, **30**, 389 (1972).
280. Natta, G., Mazzanti, G., Corradini, P., Chini, P. and Bassi, I. W., *Atti Accad. Naz. Lincei, Rend. Classe Sci. Fis. Mat. Nat.*, **28**, 8 (1960).
281. Furukawa, J., Saegusa, T., Tsuruta, T., Fujii, H., Kawasaki, A. and Tatano, T., *J. Polym. Sci.*, **36**, 546 (1959).
282. Furukawa, J., Saegusa, T., Tsuruta, T., Fujii, H., Kawasaki, A. and Tatano, T., *Makromol. Chem.*, **33**, 32 (1959).
283. Colomb, H. O., Bailey Jr, F. E. and Lundberg, R. D., *J. Polym. Sci., Polym. Lett. Ed.*, **16**, 507 (1978).

284. Furukawa, J., Saegusa, T., Tsuruta, T., Ohta, S. and Wasai, G., *Makromol. Chem.*, **52**, 230 (1962).
285. Kawai, W., *Bull. Chem. Soc. Jpn*, **35**, 516 (1962).
286. Natta, G., Mazzanti, G., Pregaglia, G. F., Binaghia, M. and Peraldo, M., *J. Am. Chem. Soc.*, **82**, 4742 (1960).
287. Natta, G., Mazzanti, G., Pregaglia, G. F., Binaghia, M. and Peraldo, M., *Makromol. Chem.*, **44–46**, 537 (1961).
288. Yamashita, Y. and Nanumoto, S., *Makromol. Chem.*, **58**, 244 (1962).
289. Miller, R. G. J., Nield, E. and Turner-Jones, A., *Chem. Ind (Lond.)*, **1962**, 181 (1962).
290. Pruitt, M. E. and Baggett, J. M., US Pat. 2,706,181 (1955).
291. Pierre, St. and Price, C. C., *J. Am. Chem. Soc.*, **78**, 3432 (1956).
292. Price, C. C. and Osgan, M., *J. Am. Chem. Soc.*, **78**, 690 (1956).
293. Price, C. C. and Osgan, M., *J. Am. Chem. Soc.*, **78**, 4789 (1956).
294. Ishida, S. and Murashashi, A., *J. Polym. Sci.*, **40**, 571 (1959).
295. Ishida, S., *Bull. Chem. Soc. Jpn.*, **33**, 727 (1960).
296. Ishida, S., *Bull. Chem. Soc. Jpn.*, **33**, 731 (1960).
297. Sakata, R., Tsuruta, T., Saegusa, T. and Furukawa, J., *Makromol. Chem.*, **40**, 64 (1960).
298. Booth, C., Higginson, W. C. E. and Powell, E., *Polymer*, **5**, 479 (1964).
299. Bruce, J. M. and Hurst, S. J., *Polymer*, **7**, 1 (1966).
300. Kern, R. J., *Makromol. Chem.*, **81**, 261 (1965).
301. Booth, C. and Price, C. C., *Polymer*, **7**, 85 (1966).
302. Kasperczyk, J. and Jedlinski, Z., *Makromol. Chem.*, **187**, 2215 (1986).
303. Tsuruta, T., Inoue, S. and Koinuma, H., *Makromol. Chem.*, **112**, 58 (1968).
304. Hasegawa, H., Miki, K., Tanaka, N., Kasai, N., Ishimori, M., Heki, T. and Tsuruta, T., *Makromol. Chem., Rapid Commun.*, **3**, 947 (1982).
305. Dumas, Ph., Spassky, N. and Sigwalt, P., *J. Polym. Sci., Polym. Chem. Ed.*, **17**, 1583 (1979).
306. Dumas, Ph., Spassky, N. and Sigwalt, P., *J. Polym. Sci., Polym. Chem. Ed.*, **17**, 1595 (1979).
307. Dumas, Ph., Spassky, N. and Sigwalt, P., *J. Polym. Sci., Polym. Chem. Ed.*, **17**, 1605 (1979).
308. Kuran, W. and Mazanek, E., *J. Organometall. Chem.*, **384**, 13 (1990).
309. Kuran, W. and Mazanek, E., *Main Group Metal Chem.*, **12**, 241 (1989).
310. Furukawa, J., Saegusa, T., Tsuruta, T. and Kakogawa, G., *J. Polym. Sci.*, **36**, 541 (1959).
311. Furukawa, J., Saegusa, T., Tsuruta, T. and Kakogawa, G., *Makromol. Chem.*, **36**, 25 (1960).
312. Aida, T. and Inoue, S., *Makromol. Chem., Rapid Commun.*, **1**, 677 (1980).
313. Vincens, V., Le Borgne, A. and Spassky, N., *Makromol. Chem., Macromol. Symp.*, **47**, 285 (1991).
314. Kuroki, M., Watanabe, T., Aida, T. and Inoue, S., *J. Am. Chem. Soc.*, **113**, 5903 (1991).
315. Watanabe, T., Aida, T. and Inoue, S., *Macromolecules*, **23**, 2612 (1990).
316. Inoue, S. and Aida, T., *Macromol. Symp.*, **88**, 117 (1994).
317. Inoue, S. and Aida, T., *Chemtech*, **24**, 28 (1994).
318. Taton, D., Le Borgne, A., Sepulchre, M. and Spassky, N., *Macromol. Chem. Phys.*, **195**, 139 (1994).
319. Yoshinaga, K. and Iida, I. *Chem. Lett.*, **1991**, 1057 (1991).
320. Crivello, J. V. and Fan, M., *Makromol. Chem. Macromol. Symp.*, **54/55**, 179 (1992).

Further Reading

Kuran. W., 'Coordination Polymerization of Heterocyclic and Heterounsaturated Monomers', *Prog. Polym. Sci.*, **23**, 919 (1998).
Vandemberg, E. J., 'A Key to Advances in Applied Polymer Science', in *Catalysis in Polymer Synthesis*, Washington, DC, 1992, ACS Symp. Ser. 496, pp. 10–14.
Inoue, S., 'High Polymers from CO_2', *Chemtech*, **6**, 588 (1976).
Darensbourg, D. J. and Holtcamp, N. W., 'Catalysts for the Reactions of Epoxides and Carbon Dioxide', *Coord. Chem. Rev.*, **153**, 155 (1996).
Kuran, W., 'Poly(propylene oxide)', in *Polymeric Materials Encyclopedia*, CRC Press, Boca Raton, 1996, Vol. 9, pp. 6657–6662.
Inoue, S., 'Metalloporphyrin Catalysts for Control of Polymerization', in *Catalysis in Polymer Synthesis*, Washington, DC, 1992, ACS Symp. Ser. 496, pp. 194–204.
Inoue, S. and Aida, T., 'Anionic Ring-opening Polymerization: Copolymerization', in *Comprehensive Polymer Science*, G. Allen and J. C. Bevington (Eds), Pergamon Press, Oxford, 1989, Vol. 3, pp. 553–569.
Vincens, V., Le Borgne, A. and Spassky, N., 'Oligomerization of Oxiranes with Aluminum Complexes as Initiators', in *Catalysis in Polymer Synthesis*, Washington, DC, 1992, ACS Symp. Ser. 496, pp. 205–214.
Kuran, W., 'Poly(propylene carbonate)', in *Polymeric Materials Encyclopedia*, CRC Press, Boca Raton, 1996, Vol. 9, pp. 6623–6630.

Problems

1. Name heteroatom-containing cyclic and acyclic monomers capable of undergoing coordination polymerisation. Indicate racemic monomers among them.
2. Name polymers formed from heterocyclic monomers that display stereoisomerism.
3. Name polymers formed from heterounsaturated monomers that display stereoisomerism.
4. Name all possible stereoregular polymers of oxiranes and give their formulae (zigzag and adopted Fischer projections). Indicate which of these stereoregular polymers are obtainable and which have never been obtained.
5. Characterise the main features of catalysts for the coordination polymerisation of heterocyclic and heterounsaturated monomers.
6. Explain why inversion of the configuration at the carbon atom of the oxirane ring where cleaved occurs during polymerisation with coordination catalysts.
7. Explain why tetrahydrofuran can be concatenated into the polymer chain in copolymerisation systems with coordination catalysts. Give examples of copolymers and terpolymers, obtained by coordination polymerisation, that contain oxytetramethylene units in the main chain.
8. Explain why propylene carbonate undergoes relatively easily coordination copolymerisation with propylene oxide but hardly undergoes homopolymerisation at moderate temperature.
9. Explain why carbon dioxide can form with propylene oxide in the presence of coordination catalysts a linear polymer, poly(propylene carbonate), but with anionic initiators gives only a cyclic carbonate, propylene carbonate.
10. What is the difference between catalysts containing multinuclear metal species and catalysts containing mononuclear species? Which are the main features of the mechanism operating in the polymerisation of oxiranes with both these catalysts?

11. Indicate tactic or ditactic polymers that can be obtained by the coordination polymerisation of oxiranes.
12. Characterise the stereoselective (enantiosymmetric) polymerisation and stereoselective (enantioasymmetric) polymerisation of heterocyclic monomers. Give examples.

Index

Note: page numbers in *italic* type refer to diagrams; those in **bold** type refer to tables.

Index complied by Colin Will